W0037011

Lecture Notes in Mobility

Series Editor

Gereon Meyer🆔, *VDI/VDE Innovation + Technik GmbH, Berlin, Germany*

Editorial Board Members

Sven Beiker, *Stanford University, Palo Alto, USA*
Evangelos Bekiaris, *Hellenic Institute of Transport (HIT), Centre for Research and Technology Hellas, Thermi, Greece*
Henriette Cornet🆔, *University of San Francisco, San Francisco, CA, USA*
Marcio de Almeida D'Agosto, *COPPE-UFJR, Federal University of Rio de Janeiro, Rio de Janeiro, Brazil*
Nevio Di Giusto, *Fiat Research Centre, Orbassano, Italy*
Jean-Luc di Paola-Galloni, *Sustainable Development and External Affairs, Valeo Group, Paris, France*
Karsten Hofmann, *Continental Automotive GmbH, Regensburg, Germany*
Tatiana Kováčiková, *University of Žilina, Žilina, Slovakia*
Jochen Langheim, *STMicroelectronics, Montrouge, France*
Joeri Van Mierlo, *Mobility, Logistics and Automotive Technology Research Centre, Vrije Universiteit Brussel, Brussel, Belgium*
Tom Voege, *GRS Service GmbH, Brussels, Belgium*

The book series Lecture Notes in Mobility (LNMOB) reports on innovative, peer-reviewed research and developments in intelligent, connected and sustainable transportation systems of the future. It covers technological advances, research, developments and applications, as well as business models, management systems and policy implementation relating to: zero-emission, electric and energy-efficient vehicles; alternative and optimized powertrains; vehicle automation and cooperation; clean, user-centric and on-demand transport systems; shared mobility services and intermodal hubs; energy, data and communication infrastructure for transportation; and micromobility and soft urban modes, among other topics. The series gives a special emphasis to sustainable, seamless and inclusive transformation strategies and covers both traditional and any new transportation modes for passengers and goods. Cutting-edge findings from public research funding programs in Europe, America and Asia do represent an important source of content for this series. PhD thesis of exceptional value may also be considered for publication. Supervised by a scientific advisory board of world-leading scholars and professionals, the Lecture Notes in Mobility are intended to offer an authoritative and comprehensive source of information on the latest transportation technology and mobility trends to an audience of researchers, practitioners, policymakers, and advanced-level students, and a multidisciplinary platform fostering the exchange of ideas and collaboration between the different groups.

Ciaran McNally · Páraic Carroll ·
Beatriz Martinez-Pastor · Bidisha Ghosh ·
Marina Efthymiou · Nikolaos Valantasis-Kanellos
Editors

Transport Transitions: Advancing Sustainable and Inclusive Mobility

Proceedings of the 10th TRA
Conference, 2024, Dublin, Ireland
- Volume 2: Sustainable Transport
Development

Springer

Editors
Ciaran McNally
School of Civil Engineering
University College Dublin
Dublin, Ireland

Beatriz Martinez-Pastor
School of Civil Engineering
University College Dublin
Dublin, Ireland

Marina Efthymiou
DCU Business School
Dublin City University
Dublin, Ireland

Páraic Carroll
Faculty of Architecture, Building
and Planning
University of Melbourne
Parkville, VIC, Australia

Bidisha Ghosh
Department of Civil, Structural
and Environmental Engineering
Trinity College Dublin
Dublin, Ireland

Nikolaos Valantasis-Kanellos
School of Business Technology, Retail
and Supply Chain
Technological University Dublin
Dublin, Ireland

ISSN 2196-5544　　　　ISSN 2196-5552 (electronic)
Lecture Notes in Mobility
ISBN 978-3-031-85577-1　　ISBN 978-3-031-85578-8 (eBook)
https://doi.org/10.1007/978-3-031-85578-8

© The Editor(s) (if applicable) and The Author(s) 2025. This book is an open access publication.

Open Access This book is licensed under the terms of the Creative Commons Attribution 4.0 International License (http://creativecommons.org/licenses/by/4.0/), which permits use, sharing, adaptation, distribution and reproduction in any medium or format, as long as you give appropriate credit to the original author(s) and the source, provide a link to the Creative Commons license and indicate if changes were made.
The images or other third party material in this book are included in the book's Creative Commons license, unless indicated otherwise in a credit line to the material. If material is not included in the book's Creative Commons license and your intended use is not permitted by statutory regulation or exceeds the permitted use, you will need to obtain permission directly from the copyright holder.
The use of general descriptive names, registered names, trademarks, service marks, etc. in this publication does not imply, even in the absence of a specific statement, that such names are exempt from the relevant protective laws and regulations and therefore free for general use.
The publisher, the authors and the editors are safe to assume that the advice and information in this book are believed to be true and accurate at the date of publication. Neither the publisher nor the authors or the editors give a warranty, expressed or implied, with respect to the material contained herein or for any errors or omissions that may have been made. The publisher remains neutral with regard to jurisdictional claims in published maps and institutional affiliations.

This Springer imprint is published by the registered company Springer Nature Switzerland AG
The registered company address is: Gewerbestrasse 11, 6330 Cham, Switzerland

If disposing of this product, please recycle the paper.

Preface

We are pleased to publish the Conference Proceedings of the 10th Transport Research Arena (TRA 2024), held on April 15–18, 2024, in Dublin, Ireland. The conference brought together 4500 delegates from 57 countries who came together to discuss research findings, the latest innovations in policy, technology and practice, and the future directions of mobility and transport.

The conference tagline was *Transport Transitions: Advancing Sustainable and Inclusive Mobility*, and four primary conference themes were defined, namely

- Safe & Inclusive Transport.
- Sustainable Mobility of People and Goods.
- Efficient & Resilient Systems.
- Collaborative Digitalization.

TRA takes place every 2 years, and TRA2024 featured an array of plenary sessions, ministerial sessions, strategic sessions and special sessions which took place alongside the technical programme. A call for papers was issued in early 2023 which resulted in 1182 submissions. A double-blind peer review process was initiated, which ultimately resulted in 784 papers that were chosen for presentation at the conference (66% conversion rate). These papers were presented in a combination of oral or poster presentations over the course of the conference.

All accepted papers presented at TRA 2024 are published in a topical collection of the journal European Transport Research Review (ETRR) or within these proceedings. Both are published in a fully open-access format.

TRA is a multi-modal conference that draws on the support of key stakeholders. These include the European Commission, ACARE (Advisory Council for Aviation Research and Innovation in Europe), ALICE (Alliance for Logistics Innovation through Collaboration in Europe), CEDR (Conference of European Road Directorates), ECTP (European Construction Technology Platform), ERRAC (European Rail Research Advisory Council), ERTRAC (European Road Transport Research Advisory Council), ETRA (European Transport Research Alliance), and the Waterborne technology platform. Key Irish supporters of the event were Transport Infrastructure Ireland, Enterprise Ireland and the Irish Government's Department of Transport.

The editors would like to express their thanks to the presenters, authors, reviewers, session chairs, committee members and sponsors for helping deliver such a successful event. TRA 2026 will take place in Budapest, Hungary.

Páraic Carroll
Beatriz Martinez-Pastor
Bidisha Ghosh
Marina Efthymiou
Nikolaos Valantasis-Kanellos
Ciaran McNally

Contents

viii   Contents

Future Workforce and Skills

Urban, Regional and Rural Transport

How the Sensoriality of Experience Contributes to Behavior Change: The Case of Daily Mobility

Jeanne Lallement[1]([⊠]), Florence de Ferran[2], and Sarah Machat[1]

[1] NUDD, IUT La Rochelle Université, La Rochelle, France
Jeanne.lallement@univ-lr.fr
[2] NUDD, IAE La Rochelle Université, La Rochelle, France

Abstract. The transport sector accounts for a third of greenhouse gas emissions, and over half of these are generated by households. How can we effectively leverage a change towards less carbon-intensive mobility? Motivations for modal choice have been studied by various disciplines, and there is a consensus that the determinants are essentially rational and cognitive. However, research over the past thirty years has demonstrated the effectiveness of sensory triggers. The present study, in the field of sensory marketing, focuses on the complex experience of mobility change, notably the consumer's senses. We adopt a qualitative, exploratory approach. Two series of interviews were carried out with 18 people who had recently changed their means of mobility. Considering the user as a meaning-producing actor, the present study examines both individual and contextual motivations, and socio-cultural, experiential, and symbolic aspects of mobility. We analyze the corpus of verbatim from the perspective of the framework developed by Krishna, and highlight both the multiple sensory characteristics of the new mobility experience, and their consequences on attitudes and behaviors. Attitude, memory, and behavior are affected by both grounded cognition and grounded emotion. Our results have implications for policies that encourage people to test soft mobility options.

Keywords: Mobility · sensory marketing · Behavior change

1 Introduction

Fighting against climate change is one of the major challenges of the 21st century. The transport sector accounts for a third of greenhouse gas emissions, and more than half of these are generated by households. The challenge is therefore to encourage less carbon-intensive forms of transport. What are the most effective levers for driving this change?

Individuals' modal choice [5] appears to be a combination of environmental factors (the presence of infrastructures, its characteristics, quality and availability), situational factors (linked to the reasons for travel, frequency, duration and weather conditions), socio-demographic factors (in particular, family structure, age, employment, income, means of transport owned), and psychological factors (norms, values, attitudes, and intentions). The determinants spontaneously mentioned by people are generally inherent

© The Author(s) 2025
C. McNally et al. (Eds.): TRAconference 2024, LNMOB, pp. 3–10, 2025.
https://doi.org/10.1007/978-3-031-85578-8_1

to modes of transport and infrastructures (cost, time, distance, comfort, accessibility). The main motivations have been studied by various disciplines in the human sciences, and all agree that the explanatory factors are essentially rational. However, over the past thirty years, research into consumer behavior has incorporated elements relating to haptics, vision, taste, smell and hearing, and has demonstrated the effectiveness of sensory triggers. The present study, in the field of sensory marketing, focuses on the complexity of mobility change experiences, which involve "*all of consumers' senses and affect their perceptions, judgement and behaviour*" [8]. Affective reactions do not necessarily depend on cognition, and the present study explores the affects and senses associated with different mobility options.

We focus on periods of mobility change, and seek to understand both the role played by all of the senses, and how they can change behavior. Many models have focused on individual behavior change; typically, they recognize that it takes time for an individual to change his behavior, and that change is gradual [2, 3, 11]. In particular, the model developed by Bamberg [2] is interesting for two reasons. First, it has already been applied to reducing car use [2, 3, 12, 14], and, second, it is less cognitivist than other models, and incorporates emotions, particularly in the pre-decisional phase of mobility. Our questions are therefore as follows: what role do the senses play in a new mobility experience, and how do they influence the process of changing behavior?

2   Changing Mobility: An Essentially Cognitivist Approach that Neglects Sensory Dimensions

Mobility change is generally studied by focusing on the **effects of public policies** on individual behavioral change [7]. This body of work, grouped under the term *Travel Demand Management*, has studied the effects of coercive and incentive measures that aim to limit car use. The combination of punishment and awareness-raising has been shown to be effective [17], notably making alternatives to the car more attractive, and more carbon-intensive modes of transport less attractive. In this context, the decision to change mobility is presented as the result of a cost-benefit analysis by a rational user, who seeks, above all, to minimize travel time and cost.

The same rationalist perspective has been adopted in research examining **decision-making**. In this micro-economic approach, the user chooses the mode of travel that maximizes his or her utility [5]. Most determining variables relate to the cost and time of the various means of transport, and draw upon a series of spatial, socio-demographic and trip-characteristic explanatory indicators. Although many researchers have stressed the importance of subjective aspects of journeys [3], to date, only psychological factors have been taken into account, linked to experience, habits, or perceptions. The sensorial aspects of the mobility experience—touch, sight, hearing, taste, and smell—have received little or no attention.

Mobility change has been studied by a number of different models, of which Prochaska and DiClemente [11] is probably the most popular. Change is described as gradual, with each stage characterized by specific intentions and behaviors. Although highly operational, this model has been criticized for being descriptive and scientifically questionable [18]. Another model, which is more comprehensive and adapted to mobility, has been proposed by Bamberg [2, 3, 12, 14]. In the latter, four stages characterize behavior change. The transition from one stage to the next occurs through the formulation of specific intentions, themselves linked to numerous socio-psychological factors. However, beyond the positive and negative emotions involved in the pre-decisional stage, no emotional or affective elements are included in subsequent stages. The sensorial and bodily experience seems to be ignored in models describing mobility choices. Hence, the study of change through the prism of sensory marketing can provide us with a better understanding of the phenomenon.

Sensory marketing is an analytical framework that enriches the explanation of behavioral change. It is defined as *"marketing that engages the consumer's senses and affects their perception, judgement and behaviour"* [8]. The approach provides an analytical framework in which cognition is based on the senses. Subconscious triggers characterize consumer perceptions, which, in turn, explain many behaviors. Touch, rarely studied in marketing research, is described by Aristotle as the most important sense, and it offers a faithful image of the intrinsic nature of the object. A cat's soft coat indicates the gentleness of its character. 'Warm' and 'cold' personalities are often described by their haptic characteristics. The sense of smell affects learning, as studies in biology and anatomy have shown. Proust's *madeleine* is an illustration of how olfactory information is permanently encoded. Hearing is widely used in marketing communications. The sound of words, of a language, of a voice, or the sounds produced by handling a product have a meaning, and influence consumer reactions. Although human beings can only distinguish five pure tastes (sweet, salty, sour, bitter, and umami), the richness of our taste descriptions indicates the importance of the combination of the five senses in the perception of taste. All of these sensory experiences affect attitudes and behaviors. The grounded cognition school [4] suggests that these experiences engage bodily states that lead to behavior and thought processes. The bodily experience is a source of information [9] that the consumer will use in the same way as any other source of information. This raises the question of the consequences of changing mobility on an individual's body, which seems relevant given the lack of knowledge highlighted in the literature regarding the determinants of individual mobility change. This line of research favors a comprehensive vision of mobility, as it integrates sensory dimensions, and is based on individual experience, both real and imagined [8].

3   Methodology

Our aim is to understand the role of bodily experience in the context of a change in daily mobility. Mobility change is often associated with contextual or structural change [6]. In particular, and for a variety of reasons, restrictions due to the Covid pandemic were an opportunity for mobility change. In this study, we adopted an exploratory approach, and a qualitative methodology. Two series of interviews were conducted, first in April–May 2020, then in January–May 2021. Three themes were addressed: (1) daily home-to-work habits before and after the change in mobility; (2) the reasons and contexts for choosing this mobility (infrastructure, family structure, transport options owned and available); and (3) the individual's environmental sensitivity. This approach made it possible to identify the meaning that users themselves attributed to their mobility, as well as the reasoning underlying their mobility choices. Considering the user as a meaning-producing actor [15], the present study takes into account individual and contextual motivations, together with socio-cultural, experiential, and symbolic aspects of mobility [1]. In the following part, we will only cover the elements mentioned during the interviews in relation to our issues.

Eighteen people, aged 18–60 from different geographical areas in France (urban/rural, and multiple regions) were interviewed (cf. table 1 below). They were recruited on the basis of a call for applications distributed by various means (online and offline social networks, posters placed in several bicycle stores). The only inclusion criterion was to answer yes to the following question: "Have you recently changed how you move around on a daily basis?" Interviews were transcribed, forming a 274-pages corpus that was analyzed by the authors. Initial, *a priori*, coding referred to concepts from the literature on the antecedents of mobility choices, notably those distinguishing between intra- and inter-individual differences. A second round of *a posteriori* coding supported a more interpretative analysis of the interviews. The analytical method proposed by Spiggle (1994) was used [13].

Table 1. Sample description and change of mean of mobility

Respondents #, gender, age	Occupation	City	Change of mean of mobility	
			Before	After
#1, male, 58	Training consultant	Suburb of Paris	Scooter	Bike, Walk
#2, female, 50	Digital executive	Downtown Paris	Metro, Walk	Walk, Kick Scooter
#3, female, 49	Publishing executive	Downtown Paris	Commuter train	Electric bike
#4, female, 45	Secretary	Suburb of La Rochelle	Thermic train	Electric car
#5, female, 30	Communication manager	Suburb of La Rochelle	Bus	Electric bike
#6, male, 60	Professor in economics	Downtown Paris	Car	Metro, Commuter train, Walk
#7, female, 50	Professor in Marketing	Amiens	Car	Bike
#8, female, 41	Intermediary profession	Annecy	Car	Longboard Skate, Walk
#9, male, 50	Consultant	Marseille	Car	Less Car
#10, female, 38	Consultant	Paris	Metro, Car, Commuter train	Metro, Walk, Commuter train

(*continued*)

Table 1. (*continued*)

Respondents #, gender, age	Occupation	City	Change of mobility	
			Before	After
#11, female, 35	Computer scientist	Lyon	Bike, Commuter train	Bike
#12, male, 40	Consultant	Paris	Metro, walk	Metro, Walk, Self-service Bike
#13, male, 33	Computer engineer	Suburb of Paris	Car, Bike, Commuter train	Bus, Bike, Commuter train
#14, female, 31	Graphic designer	Toulouse	Car	Shared Car
#15, female, 45	Consultant	Paris	Commuter train	Commuter train, Metro, Walk
#16, male, 49	Human ressources manager	Suburb of Paris	Scooter	Scooter, Self-service Bike
#17, male, 31	Logistic employee	Suburb of Nantes	Car	Motorbike
#18, female, 34	Technician	Rennes	Car	Kick Scooter, Car

4   Changing Mobility, a Sensory Experience

An analysis of the verbatim based on the framework developed by Krishna [8] high-lights that the new mobility experience has multiple sensory characteristics, and its consequences on attitudes and behaviors. Twelve respondents are part of a trend towards a gentler mode of individual transport, favored, above all, by changes in material conditions. This new mobility is embodied in body movements and new sensations. This is the case not only for walking and skateboarding, but also when respondents talk about bicycles (electric or muscle bikes). The whole body is impacted, notably its relationship with the outside world. Drawing upon Krishna's model, Fig. 1 plots the significant elements of an experience involving four of the senses.

Touch is the sense that respondents most spontaneously mention as a determining factor when describing new mobility experiences. Soft mobility, supported by circumstances linked to Covid restrictions, materializes in a new sense of touch and, more specifically, movement involving different parts of the body (the heartbeat and walking). This new mobility has resulted in a different relationship with the outside world, linked to the air, wind, space, and crowds. In some cases, the desire to stop touching was also a trigger for change, to avoid contamination. **Sight** was also described as a characteristic element of the change in mobility. Respondents mentioned reading while travelling on the metro, meeting new people (when travelling by bike), or seeing new landscapes, as they were able to move in a more 'conscious' time. While **smell** was not used to directly describe the new mobility experience, the semantics associated with smell are symptomatic of its unconscious role. Users 'felt better', and 'smelled the wind'.

Finally, **hearing** contributes to the construction of a different experience, notably the importance of silence (during Covid), and a different musicality while cycling or when using public transport.

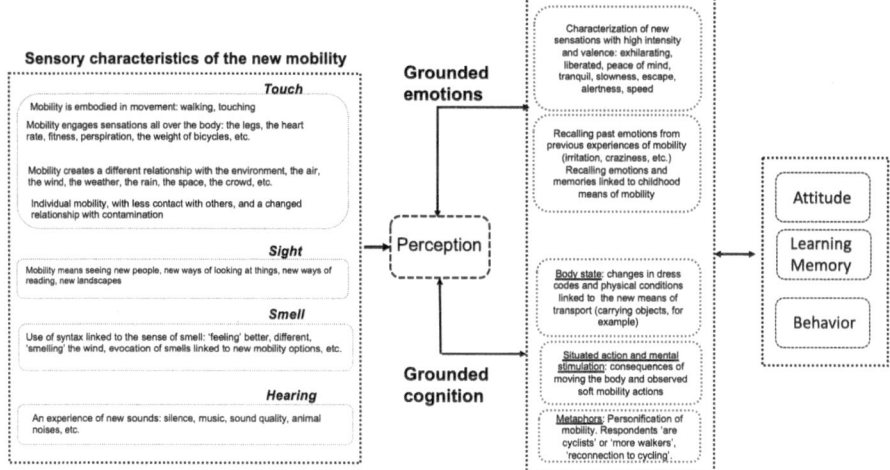

Fig. 1. Adaptation of the sensory marketing conceptual framework to new mobility experiences

All of the described sensations contribute to a perception that then takes two, complementary paths: Grounded emotions and Grounded cognition. **Grounded cognition** refers to the work of Barsalou [4], who indicates that our bodily state and the mental images (metaphors) linked to these actions generate cognitive activity which then affects our attitudes, behavior, and memory. One bodily state, in particular, seems important to note. With soft mobility, notably cycling and walking, dress codes have changed, enabling more active options, particularly for women (*"What makes it possible to switch to cycling is that habits, dress codes, have completely changed. And everyone wears sneakers, even for important client meetings... And for girls, it makes a huge difference, because cycling in heels!"*) The importance of the metaphors used underline the richness of the sensations experienced. The syntax used by respondents reflects the personification of the means of transport. Word combinations form an abstract system that describes the latter's very special status—the means of transport is often seen as a living being: *"I've abandoned the car"* or *"cars don't respect bicycles"*. This personification is taken to the extreme when the mean of transport becomes an expression of one's own personality (*"the neighbors are more car than cyclist"* or *"I'm less cyclist than walker"*).

Grounded emotions describe a whole range of emotions of opposing valence. Some relate to states of excitement (*exhilarating, escape, alertness, speed*), while others relate to quieter moments (*liberated, peace of mind*). This polarity is particularly evident in the affect associated with car journeys, which runs along a continuum from pleasure and a sense of freedom (*"I can go to my mates, I can go to the theater, I can carry the cases of wine. I can go to Leroy Merlin on a whim..."*) to disgust (*"the car is the black spot in our lives"*). Respondents may have changed their mobility option because they have had a bad experience with their current choice (*"The metro was a nightmare. There are always problems, using my car was too long... Too many traffic jams. It was stressful"*) or because they had a good experience with a new mode of transport when they were children (*"I used to walk when I was little and I really enjoyed it"*), or due to a recent experiment with friends and family (*"We all visited Paris together with my sister and our children on a scooter and it was great"*). A positive new mobility experience is important as it is a first step towards more frequent use. The processes reported by respondents describe an individual trade-off (mixing economic and human gains and costs) in which affect is exacerbated. Drawing on the theory of construct levels [16], anchored emotions that involve physical experiences have different construct levels, compared to experiences that are only envisaged and described. Specifically, lived experiences are more directly persuasive.

The integration of the sensory marketing framework complements the model of behavior change adapted to mobility. While the senses are presented as forming a 'whole', which encompasses anchored emotions and cognition, it would also be interesting to study the role of both their weight and their interactions in the repetition of behavior and, therefore, its anchoring and crystallization, in the face of other factors impacting habit change. From a managerial point of view, and in line with work on travel demand management, our study argues in favor of the use of measures that encourage people to test new soft mobility methods, such as long-term rental, shared electric cars, or self-service bicycles.

References

1. Arnould, E.J., Thompson, C.J.: Consumer culture theory (CCT): twenty years of research. J. Consumer Research **31**, 868–882 (2005)
2. Bamberg, S.: Applying the stage model of self-regulated behavior change in a car use reduction intervention. J. Environ. Psychol. **33**, 68–75 (2013)
3. Bamberg S., et al.: Development of a theory-driven, web-based behavioral change support system for environmentally friendly mobility behavior. Social marketing: global perspectives, strategies and effects on consumer behavior, pp.91–108 (2015)
4. Barsalou, L.W.: Grounded cognition. Annu. Rev. Psychol. **59**(1), 617–645 (2008)
5. De Witte, A., Hollevoet, J., Dobruszkes, F., Hubert, M., Macharis, C.: Linking modal choice to motility: a comprehensive review. Transportation Research Part A: Policy and Practice **49**, 329–341 (2013)
6. Fujii, S., Gärling, T., Kitamura, R.: Changes in drivers' perceptions and use of public transport during a freeway closure: effects of temporary structural change on cooperation in a real-life social dilemma. Environ. Dev. **33**, 796–808 (2001)
7. Garling T., et al.: A conceptual analysis of the impact of travel demand management on private car use. Trans. Policy **9**, 59–70 (2002)
8. Krishna, A.: An integrative review of sensory marketing: Engaging the senses to affect perception, judgment and behavior. J. Consum. Psychol. **22**(3), 332–351 (2012)
9. Krishna, A., Schwarz, N.: Sensory marketing, embodiment, and grounded cognition: a review and introduction. J. Consum. Psychol. **24**(2), 159–168 (2014)
10. Loukopoulos, P., Jakobsson, C., Gärling, T., Schneider, C.M., Fujii, S.: Car-user responses to travel demand management measures: goal setting and choice of adaptation alternatives. Transp. Res. Part D: Transp. Environ. **9**(4), 263–280 (2004)

11. Prochaska, J.O., Di Clemente, C.C.: Stages and processes of self-change of smoking: toward an integrative model of change. J. Consult. Clin. Psychol. **51**(3), 390–395 (1983)
12. Shiraz, A., Adan, M., Janssens, D., Wets, G.: A personalized mobility-based intervention to promote pro-environmental travel behavior. Sustain. Cities Soc. **62**, 1–12 (2020)
13. Spiggle, S.: Analysis and interpretation of qualitative data in consumer research. J. Consumer Res. **21**(3), 491–503 (1994)
14. Sunio, V., Schmöcker, J.-D., Kim, J.: Understanding the stages and pathways of travel behavior change induced by the technology-based intervention among university students. Trans. Res. Part F: Psychol. Behav. **59**, 98–114 (2018)
15. Thompson, C.: Interpreting consumers: a hermeneutical framework for deriving marketing insights from the texts of consumers' consumption stories. J. Mark. Res. **34**(4), 438–455 (1997)
16. Trope, Y., Liberman, N.: Construal-level theory of psychological distance. Psychol. Rev. **117**(2), 440–463 (2010)
17. Wang, Y., Geng, K., May, A.D., Zhou, H.: The impact of traffic demand management policy mix on commuter travel choices. Transp. Policy **117**, 74–87 (2022)
18. West, R.: Time for a change: putting the transtheoretical (stages of change) model to rest. Addiction **100**(8), 1036–103

Open Access This chapter is licensed under the terms of the Creative Commons Attribution 4.0 International License (http://creativecommons.org/licenses/by/4.0/), which permits use, sharing, adaptation, distribution and reproduction in any medium or format, as long as you give appropriate credit to the original author(s) and the source, provide a link to the Creative Commons license and indicate if changes were made.

The images or other third party material in this chapter are included in the chapter's Creative Commons license, unless indicated otherwise in a credit line to the material. If material is not included in the chapter's Creative Commons license and your intended use is not permitted by statutory regulation or exceeds the permitted use, you will need to obtain permission directly from the copyright holder.

Mind the Gap! A Mixed Methods Study on Inequalities in Accessibility: Evidence from Brussels and Strasbourg

Alexis Conesa[1](✉), Armand Pons[1], and Arthur Nihoul[2]

[1] Université de Strasbourg, CNRS, LIVE, 67000 Strasbourg, France
conesa@unistra.fr
[2] Université Catholique de Louvain, CREAT, 1348 Ottignies-Louvain-la-Neuve, Belgium

Abstract. With the growing need to limit car dependence, public transportation appears as a key means of ensuring the transition to greener cities. However, access inequalities in undergrounds, tramways and buses are still considerable and intra-urban accessibility for all is still far from being achieved. Not only the usually labelled "people with reduced mobility (PRM)" but everyone could endure some types of barriers in their daily transit routes. The extent and severity of the resulting disparities are still seldom studied, accessibility being generally considered as the same for everyone. Conversely, this paper adopts a people-centered perspective to analyze and model constraints faced by selected vulnerable users in their use of public transport. We used mixed-methods, combining different survey and modelling techniques to enrich the usual accessibility approaches. The study is grounded in two cities—Brussels and Strasbourg —, enabling the development of adaptive metrics in various geographical environments. It focuses mainly on the impact of physical disability, visual impairment, old age and poverty on transit users' behaviors. This approach leads to identifying failures and disincentives of each transport system. We conclude by discussing dissemination to transport operators and future research prospects.

Keywords: Urban Accessibility · Mobility Justice · Inclusive Transportation

1 Introduction

Municipalities worldwide have implemented various policies in order to improve urban accessibility. However, whereas most of these policies are meant to be inclusive, the complex and multifaceted nature of accessibility could lead them to exclude some individual behaviors and widen the accessibility gaps [1]. These gaps are even wider when specific, vulnerable groups are considered. Therefore, there is a clear need for relevant analytical frameworks that can assess these little-known gaps in accessibility to urban opportunities. Although a relevant theoretical framework has actually been designed and enhanced [2], accessibility studies are still struggling to approach the experience and perceptions of Public Transport (PT) users [3]. On the other hand, qualitative research

© The Author(s) 2025
C. McNally et al. (Eds.): TRAconference 2024, LNMOB, pp. 11–17, 2025.
https://doi.org/10.1007/978-3-031-85578-8_2

assessing PT accessibility often observe individual mobility constraints without display-ing any measure or indicator, preventing generalization or comparison. Social, material, and socio-cognitive constraints impairing accessibility for vulnerable groups remain sel-dom modelled, whereas public spaces, PT systems and buildings are still not designed to be fully inclusive. These inequalities foster resentment, worsened by the exclusion of vulnerable users from decision-making processes. Drawing on the capability approach [4], our research proposes a people-oriented, intra-urban PT accessibility study aim-ing at revealing both accessibility gaps and perceptions. The subsequent goal is to help decision-makers to build a comprehensive inclusive strategy for public spaces and PT networks.

2 Methodology

2.1 Theory and Background

After having developed accessibility alternatively as a system goal or as an evaluative metric, transportation research has recently suggested it could be a justice imperative [5], drawing on highlighted theoretical links between accessibility measures and justice concerns [6]. Since the 1990s, the concept of justice has accordingly been mobilized to demonstrate the importance of addressing the issues of socio-economic inequality mainly. Unlike traditional transport appraisals that neglect the vulnerable populations, accessibility to key activities seems to be able to "cater for more vulnerable groups" [7]. In this context, moving away from John Rawls' universalist position, accessibility is considered as a combined capability that "draws out the spatial dimension in moral concerns over equality of opportunities" [8]. According to this framework, although PT is one of the main tools to improve spatial justice, justice policies are hindered by the systemic nature of accessibility, that relies on morphology of the urban fabric, availability of transport modes, level of PT service, opening hours of the targeted activity venues, physical impairments, psychological/social/cognitive constraints, social environment, etc. Accessibility is then uneasy to define and measure, especially in different urban environments and for specific populations. Despite the growing interest in people with disabilities [9], extensive thoughts on inclusive urban design have not led to adequate accessibility metrics. Usual indicators do not specifically consider vulnerable users while composite indexes are hard to understand and compare. These issues have to be addressed with a relevant accessibility approach.

2.2 Combining Methods

We detail hereunder (Fig. 1) an approach for systematically assessing the accessibility of individuals facing physical, cognitive and social constraints when traveling – for this purpose we select four target groups (physically impaired, visually impaired, elderly and socio-economically underprivileged). A convergent parallel design is proposed to combine the assets of the existing frameworks. It consists in four phases:

1/ A participative co-construction phase identifying the main mobility constraints as well as feelings of injustice endured by the target groups (2021). This phase has involved local stakeholders as well with NGOs representatives for the four target groups;

2/ The modelling of mobility barriers endured by the target groups in order to measure several accessibility indices, using the OpenTripPlanner (OTP) router linked with GIS mapping tools (2022);

3/ The enrichment of the measurements with qualitative insights thanks to commented trips and focus group discussions (2023);

4/ The building of inclusive strategies in cooperation with the stakeholders, *i.e.* the PT operators and local authorities (from fall 2023 to spring 2024).

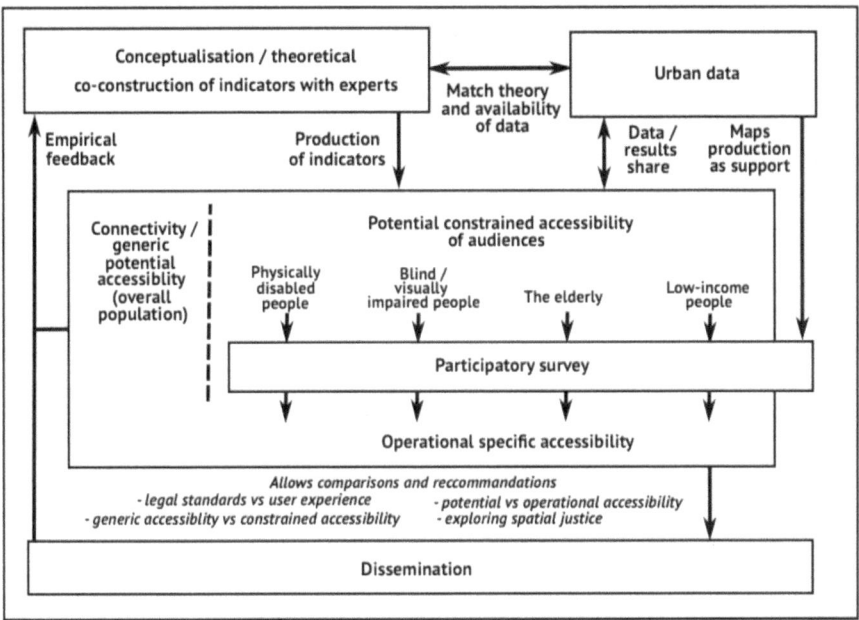

Fig. 1. A mixed-methodological framework for accessibility planning

3 Cases of Study

First, we confronted the literature findings with the experience of both decision-makers and target groups representatives, through a set of interviews. This co-construction resulted in a shared conceptual framework, as well as constraints and indicators (travel time, walking distance and number of transfers) subsequently used in our metrics—depending on data availability. The overall method was implemented in two cases of study: a) *Brussels-Capital Region*, with a population over 1.2 million inhabitants, which encompass roughly the core and the inner suburbs of the largest Belgian metropolitan region; b) *Eurométropole de Strasbourg* (*EMS* - Greater Strasbourg), around half a million inhabitants, which can be considered a second-rate European city despite its modest size. Unlike *Brussels-Capital Region*, the considered urban area includes less dense peri-urban areas. The differences in demographics and in terms of age and extent of the PT networks provided a decent overview of the issues encountered in most European cities. Strasbourg is one of the cities with a mainly surface network and tramway lines built to PRM standards. Brussels, by contrast, has a predominantly underground network that is older and more complex. This dual survey enabled us to investigate the effects of some constraints where they are most important. For instance, the need to avoid stairs and escalators were obviously of much greater importance in Brussels.

In the second phase, OTP router was fed with i) OpenStreetMap files, sometimes modified thanks to municipal databases regarding the accessibility of public spaces or PT networks, ii) General Transit Feed Specification (GTFS) files to model the PT service and schedules. Specific data (transit station lifts, tactile bands, steps, etc.) were added to the model to measure accessibility under different combinations of constraints such as *Route without platform-vehicle gap, Route avoiding the use of stairs, Route with tactile paths, Route avoiding complex stations, Bus aversion, etc.* These accessibilities were compared with a generic accessibility (i.e. for *Joe Public*).

Third, in order to comprehensively assess specific populations' accessibility, *in situ* experiments and focus groups were carried out. We collected users' feelings and tracked their specific socio-cognitive barriers, notably gender-sensitive ones. These qualitative results were used to calibrate and refine our first measurements. Based on those observations, we also have overlaid constraints in order to match the user experience as closely as possible. For instance, a profile combining the need of tactile paths, the aversion of complex stations, and the dependence to bus signage was built to render account of the high vulnerability to the absence of information and markings for some of the interviewees. On the other hand, we were struck by the autonomy of certain people. Many individuals have managed to overcome modelled obstacles which has led us to develop a finer gradient of constraints, more or less restrictive (*e.g.* size of gaps), to better represent the diversity of our audiences. Similarly, the relevance of a series of theoretical routes to a large number of locations was not appropriate in our case, as most constrained users had to prepare—if not learn—their route. Hence, measuring accessibility to a limited number of symbolic locations spread evenly across the urban area was more relevant.

By adopting this perspective, we considered that the users don't always choose their route according to the shortest path between their origin and their destination, but by arbitrating between paths that are more or less demanding. The users actually select the route that minimizes the walk distance of their trip, but they accept detours in the event of a difficult crossing, an impossible transfer, or to avoid congested streets and use wider, better-signposted roads as alternative routes. Our streamlined modal selection (briefly exemplified in Fig. 2) facilitates the efficient computation of route decisions, especially when contrasted with more intricate stochastic models: if all major PT services (colored lines) are accessible, generic users choose the fastest route (**1**) from their origin (pentagon) to a destination (star); alternatively, if some PT are not step-free, some of them will take an alternative path (**2**); but certain users would avoid a transfer in a complex station and may prefer a detour (**3**). It is to note that these additional efforts can be dissuasive. Thus, the optimized route is not necessarily the shortest path. Accompanied walks also impact the walking speed, lowered from 4 to 2 kph for certain profiles in order to match the actual trips as closely as possible. This means that the time cost of walking is particularly high in some of our calculations, as it is particularly burdensome for some of the target groups.

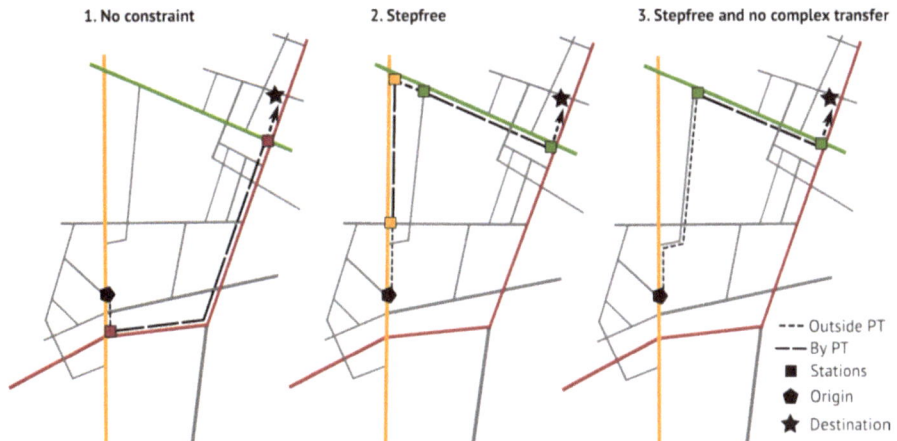

Fig. 2. User's route choices balance speed, safety and accessibility

All these measures were used to characterize and map potential accessibility on the scale of both cities, enabling to identify the failures and dissuasive effects of each transport system. Given that quantitative results were not sufficient to point out a sense of exclusion—because these concepts deal with moral concerns that vary according to specific population subgroups [10] —, we also used the qualitative dataset to complement our analysis and suggest a set of recommendations.

4   Discussion

Which groups suffer the most of constrained accessibility? What population is the most at risk of transport exclusion? Do specific neighborhoods should be targeted by PT policies? All these issues have been discussed in a decision-oriented approach, supported by an interactive atlas for operators, enabling them to quickly visualize and spatialize accessibility deficits, depending on the modelled constraints. Accordingly, in the two cities, co-constructed recommendations and scenarios have been discussed.

In Brussels, the physical access to the stations and the rolling stocks remains the most crucial issue. For instance, travelling in full autonomy considering a step-free route appears to be difficult, pointing out a population suffering from impactful accessibility gaps. The incomplete tactile paving and roadways obstacles were also discussed with NGOs and PT operators, allowing to define prospective scenarios of priority actions.

In Strasbourg, given the scarce and sometimes rural nature of the peripheral areas, accessibility gaps are mainly displayed due to geographical features. Hence, all the groups of users suffering from walking difficulties (slow speed or reluctance) seems particularly vulnerable to public transport poverty. In addition, thanks to an *EMS*-managed database, it has also been shown that a significant part of the streets was not wheelchair-accessible according to French legal standards. Last, the quality of audible beacons and more generally blind-friendly signposting was also highlighted thanks to the interviews. Co-construction has provided practical recommendations and also advocated the need for raising awareness and communicate about vulnerabilities.

In general, notwithstanding the essential differences between the two cities, some accessibility barriers were shared. Besides roadways and sidewalks accessibility, inclusive guiding, PT information, and more generally consideration for the vulnerable groups were mentioned and are probably universal issues. This experiment shows that the involvement of vulnerable groups in participatory planning can not only lead to a more inclusive PT planning but also alleviate the resentment of these groups.

Finally, one of the limitations of this approach is the lack of data to assess exhaustively and in real time the physical obstacles in public spaces to safe travel. Addressing this issue in future research could be gathering real-time feedback from citizens. In this context, the design of an intuitive user interface could make it possible to engage directly with the audiences concerned, in a participatory planning rationale.

References

1. Lucas, K.: Transport and social exclusion: where are we now?, Trans. Policy **20**(C), 105–113 (2012)
2. Martens, K.: Justice in transport as justice in accessibility: applying Walzer's 'Spheres of Justice' to the transport sector. Transportation **39**(6), 1035–1053 (2012)
3. Ryan, J., Pereira, R.H.M.: What are we missing when we measure accessibility? comparing calculated and self-reported accounts among older people. J. Transp. Geogr. **93**, 103086 (2021)
4. Vecchio, G., Martens, K.: Accessibility and the capabilities approach: a review of the literature and proposal for conceptual advancements. Trans. Rev. (2021)
5. Soja, E.W.: Seeking Spatial Justice. University of Minnesota Press, Minneapolis (2010)
6. Martens, K.: Transport Justice: Designing fair transportation systems. Routledge, New York (2016)
7. Levitas, R., Pantazis, C., Fahmy, E., Gordon, D., Lloyd, E., Patsios, D.: The multi-dimensional analysis of social exclusion. Department of Sociology and School for Social Policy. Report by Townsend Centre for the International Study of Poverty and Bristol Institute for Public Affairs. University of Bristol, Bristol (2007)
8. Pereira R., Schwanen T., Banister D.: Distributive justice and equity in transportation, Trans. Rev. (2016)
9. Imrie, R.: Disabling environments and the geography of access policies and practices. Disability Soc. **15**(1), 5–24 (2000)
10. van Wee, B., Geurs, K.T.: Discussing equity and social exclusion in accessibility evaluations. Eur. J. Transp. Infrastruct. Res.Infrastruct. Res. **11**(4), 350–367 (2011)

Open Access This chapter is licensed under the terms of the Creative Commons Attribution 4.0 International License (http://creativecommons.org/licenses/by/4.0/), which permits use, sharing, adaptation, distribution and reproduction in any medium or format, as long as you give appropriate credit to the original author(s) and the source, provide a link to the Creative Commons license and indicate if changes were made.

The images or other third party material in this chapter are included in the chapter's Creative Commons license, unless indicated otherwise in a credit line to the material. If material is not included in the chapter's Creative Commons license and your intended use is not permitted by statutory regulation or exceeds the permitted use, you will need to obtain permission directly from the copyright holder.

N/M20: Our Transport Solution

Tara O'Leary[1]([✉]) and Alex Thomas[2]

[1] NM20 Project Team, Sweco, Cork, Ireland
tara.oleary@sweco.ie
[2] NM20 Project Team, Sweco, Glasgow, UK
alex.thomas@sweco.co.uk

Abstract. The N/M20 Cork to Limerick Project is a key element of Project Ireland 2040, the Government's long-term strategy to make Ireland a better country for all. The NDP outlines that the project would improve connectivity between Ireland's second and third largest cities by enhancing the transport network.

A multi-disciplinary team have formed with sustainability and future-proofing fundamental to the planning and design process, responding to European goals to reduce transport related emissions. From Phase 1, road and rail were investigated, looking at use of the corridor and the opportunities available for behavioural change.

At Phase 2, attributes including accessibility and integration were addressed, analysing the potential socio-economic impact of the scheme on vulnerable groups, journey times and multi-modal travel. Emerging guidance documents, including NIFTI, were managed proactively, with their requirements incorporated into the appraisal process at the earliest opportunity. Phase 2's conclusion saw "**Our Transport Solution**" emerge, a multi-modal proposal incorporating road, rail, bus, active travel and environmental aspects to address the issues and opportunities identified in the corridor.

Phase 3 has seen the refinement of "Our Transport Solution", responding to further changes in Irish and European guidance and investigating opportunities for a mobility hub network within the study area.

Keywords: Multi-modal · multi-discipline · active travel · inclusion for all · mobility hubs

1 Background

1.1 Phase 1

Limerick City and County Council, in partnership with Cork County Council, Cork City Council, Transport Infrastructure Ireland and the Department of Transport are developing the N/M20 Cork to Limerick project. Limerick City and County Council, as lead authority, have appointed Barry Transportation and its project partners Sweco and WSP (BSW) as Technical Advisors to progress the planning and design for the scheme (Fig. 1).

The 2040 National Planning Framework and the National Development Plan 2018–2027 establish a minimum target population growth of 50% by 2040 for Cork and

© The Author(s) 2025
C. McNally et al. (Eds.): TRAconference 2024, LNMOB, pp. 18–23, 2025.
https://doi.org/10.1007/978-3-031-85578-8_3

Our Transport Solution

Fig. 1. N/M20 Cork to Limerick Project: Our Transport Solution

Limerick, with these cities being the second and third largest in Ireland respectively, by population. Although these cities are only approximately 100km apart, there are limited economic links and little evidence of agglomeration (economic interaction) between the two. This is due, in part, to the shortcomings in transport links. Good transport links are essential for economic sustainability and development. Improving connectivity by reducing journey times, improving journey time reliability, and improving safety will support in the enhancement of economic growth. An improved transport network between Cork and Limerick has been identified in the 2018–2027 National Development Plan as a major enabler for balanced regional development. The overarching project objective for the N/M20 Cork to Limerick project is "To enable national and regional planning policies, particularly those supporting the National Strategic Outcomes of the National Planning Framework to promote balanced regional development, through enhanced population and economic growth." This is to be achieved by improving connectivity between the cities of Cork and Limerick, and ultimately Galway, by:

- Reduced land transport journey times,
- Improved journey time reliability, and.
- Facilitating the safe and efficient movement of people, goods, and services on the transport network both now and in the future.

Since the identification and appraisal of the preferred route for the M20 Cork Limerick Motorway Scheme in 2010, there have been many changes to design standards, environmental legislation, and economic appraisal assessment criteria for locating and developing schemes. Therefore, the scheme had to 'start again' and be assessed in accordance with current legislation, design standards and appraisal requirements in line with the Government's Public Spending Code.

2   Multi-modal Approach

2.1   Phase 1

The work completed in Phase 1 (Concept and Feasibility), identified the preferred road-based scenario as being broadly within the existing N20 corridor running past two large intermediate settlements, these being Charleville and Mallow. This road-based scenario performed best overall in relation to the project objectives following appraisal of seven road-based scenarios. As part of the assessment of alternative options, two rail-based scenarios were identified in Phase 1, one involving improved service frequency with through services at Limerick Junction on existing lines, the other providing a new rail line directly between Charleville and Limerick Both rail-based scenarios were taken forward for further examination in Phase 2.

2.2   Phase 2

Phase 2 Options Development saw the continuation and broadening of the multi-modal approach to this transformational project. The Phase 2 appraisal process followed the guidance outlined in TII Project Management Guidelines, and saw three distinct stages for road-based options:

- **Stage 1: Preliminary Options Assessment** – Multi-criteria analysis (MCA) against Engineering, Economy and Environment criteria
- **Stage 2: Project Appraisal Matrix** – MCA and cost benefit analysis (CBA) against the six Common Appraisal Framework criteria
- **Stage 3: Selection of a Preferred Option** – Preparation of a Project Appraisal Balance Sheet (PABS) to summarise the impact of the preferred option.

The project team was judged as being very proactive and groundbreaking in its consideration of emerging policies such as the National Investment Framework for Transport in Ireland (NIFTI), with its requirements incorporated into the options appraisal process at the earliest opportunity to ensure sustainability and climate aspects were considered. During Phase 2, Stage 2, in the absence of formal guidance, the project team was required to develop a bespoke approach to appraisal, shaped around the NIFTI Investment Priorities, the NIFTI Modal Hierarchy, and the NIFTI Intervention Hierarchy.

The four NIFTI Investment Priorities are as follows (Fig. 2):

The NIFTI Modal Hierarchy sees the prioritisation of sustainable modes to support the significant shift from low-occupancy private vehicles and to help meet the National Strategic Outcomes (Fig. 3).

The NIFTI Intervention Hierarchy is a framework which ensures that investment is proportionate to the problem identified, comprising (Fig. 4):

All shortlisted road-based options were appraised with respect to the criteria outlined in the Investment Priorities and the two hierarchies, with a focus on how the shortlisted options could meet them. For example, the extent to which a new route would fall within or without the existing transport corridor and therefore be judged against the Maintain, Optimise, Improve and New elements of the Intervention Hierarchy.

Fig. 2. NIFTI Investment Priorities

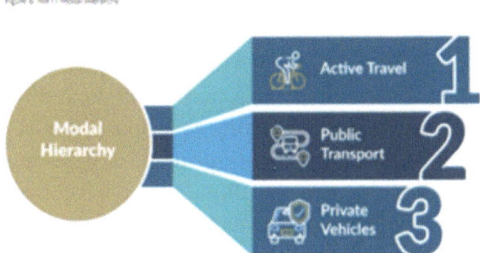

Fig. 3. NIFTI Modal Hierarchy

Fig. 4. NIFTI Intervention Hierarchy

The proactive and innovative approach to the emerging NIFTI ensured the prioritisation of sustainability in the project and the preferred option.

As well as ensuring sustainability has been inbuilt into the project, the Phase 2 work has also seen the complete inclusion of all modes of travel. This has included the appraisal of rail-based options using the full TII National Model for Ireland which incorporates a rail module as part of its variable demand capabilities. The options appraised included running increased frequency of services between Cork and Limerick, either on the existing network or on new routing options involving new rail lines between Charleville and

Limerick The results of the modelling indicated that, although rail demand increased in all options, the extra demand was largely driven by reassignment from bus mode, rather than from the private car. The conclusion was that increasing service frequency would be overall beneficial to travelers and should be taken forward, but a large expenditure of new lines would not represent extra value for money.

The potential of each road-based option to improve express bus service times was considered in the Phase 2 appraisal work. For example, high quality divided carriageway options would help achieve the goal of Bus Éireann of improving journey times between Cork and Galway from 165 min to 150 min. The benefits to more local bus services were also considered taking into account journey times between major settlements and key destinations, such as hospitals, employment and tertiary educational institutions, recorded for each option against the existing situation where bus services get caught up in recurrent congestion in urban part of the N20 routes. The findings of the analysis highlighted considerable journey time savings.

An Active Travel Strategy was developed for the study area at the start of Phase 2 Option Selection, taking account of national, regional, and local strategy documents, public consultation findings, and additional external data sources including Census data, Strava Metro and Garmin Connect. The research highlighted desire lines between key settlements, public transport hubs and existing & proposed active travel infrastructure, culminating in the development of a series of principles, opportunities, and objectives, all with multi-modal, sustainable, equitable mobility at the forefront. These would shape the development of the active travel elements for future stages of the project, ensuring integration and accessibility by focusing on the study area holistically, rather than just the corridors of the various route options.

The culmination of Phase 2 saw the emergence of "Our Transport Solution", a multi-modal proposal incorporating environment, road, rail, bus, and active travel aspects to address the issues and opportunities identified in the current transport corridor.

2.3 Phase 3

Phase 3 Design has seen the development of the design of the preferred option which has reflected the multi-modal, integrated offering of the scheme as a whole. This has included a "carrot-and-stick" methodology whereby interventions to increase and inte-grate sustainable travel into the scheme have been explored alongside potential demand management options.

"Carrots" include the provision of active travel facilities for the full length of the scheme and also the identification of a number of mobility hubs at key interchange loca-tions. A tiered approach to mobility hubs has been determined, focusing on sites which can act as Park & Ride / Park & Share sites, medium sized "interchange" hubs, and smaller, locally focused mobility hubs. These have emerged following a demand anal-ysis exercise utilising various outputs from Census data etc. "Sticks" could potentially encompass taxation and road user charging.

A major public consultation exercise was undertaken during Phase 2 and will be repeated and expanded in Phase 3. The plan is to undertake a range of feedback sur-veys from the public to gain more detailed knowledge on travel behaviors and what opportunities would interest the local communities.

3 Continual Challenges

The Project Team has met many challenges along the way for this scheme with numerous changes to European, national and regional guidance. There has been a major shift in how schemes are appraised, and the investment rationale needed for a scheme. A scheme is also so much more than just one or two elements, it is a large range of elements that need to be bought together and consideration given to exactly how they will interact and be delivered. Guidance has been changing so quickly this team are the first to use many guidance documents and continue to set a high standard on how these are applied across topics such as cross section, EV charging, Alternative funding, services areas etc.

The emergence of the Climate Action Plan 2023, and the ambitious targets that it outlines, including a 20% reduction in total vehicles kilometres by 2030, has reiterated the importance of incorporating sustainable transport into the project and providing travel choice to users so that the car does not have to be the de facto mode of transport.

The project team has therefore utilised emerging tools to inform the wider appraisal of 'Our Transport Solution', including the TII Carbon tool and the TII Roads Emissions Model. The latter takes modelled traffic volume and link speed outputs and calculates potential emission outputs and economic impact of a scheme.

Open Access This chapter is licensed under the terms of the Creative Commons Attribution 4.0 International License (http://creativecommons.org/licenses/by/4.0/), which permits use, sharing, adaptation, distribution and reproduction in any medium or format, as long as you give appropriate credit to the original author(s) and the source, provide a link to the Creative Commons license and indicate if changes were made.

The images or other third party material in this chapter are included in the chapter's Creative Commons license, unless indicated otherwise in a credit line to the material. If material is not included in the chapter's Creative Commons license and your intended use is not permitted by statutory regulation or exceeds the permitted use, you will need to obtain permission directly from the copyright holder.

Sustainable Urban Mobility: Implementation and Impact

William Hynes and Michal Gelbart[✉]

KPMG Ireland, Stokes Place, Dublin 2, Ireland
michal.reut-gelbart@kpmg.ie

Abstract. As society strives to tackle the challenges of global population growth, increasing urbanisation, continuous transportation emissions and climate change, the ongoing development of sustainable urban mobility is of critical importance. Putting sustainable mobility at the forefront of the discussions and debates not only can tackle these challenges but also provide an opportunity to enable communities to have a stronger voice, ensuring the development of more liveable places, improved active travel, i.e. walking, cycling, increasing the use of public transport, and overall improving citizens' health, well-being, and quality of life. Achieving these will have a direct consequence of reducing level of social isolation, allowing for more efficient movement, facilitating access to education and employment opportunities, and enabling economic growth. Moving towards the implementation of sustainable urban mobility, and understanding and assessing its impact, can successfully be achieved through effective planning and collaboration. According to the European Commission, this planning and collaboration can be undertaken through the application of Sustainable Urban Mobility Plans (SUMPs) and these are considered essential urban land use and transport planning tools for urban areas that are looking to move towards more sustainable and citizen-centric mobility. This paper will consider the importance of sustainable urban mobility regarding implementation and impact within an ever changing and evolving urban society and the importance of citizen collaboration.

Keywords: Sustainable Urban Mobility · Sustainable Urban Mobility Plans (SUMPs) · Active Mobility

1 Sustainable Urban Mobility (SUM)

1.1 Intro: What is SUM?

Against a backdrop of growing urbanisation, congestion, and near ubiquitous net zero commitments, policymakers and urban leaders are under pressure to modernise transport systems to drive efficiencies and to reduced greenhouse gas (GHG) emissions. Sustainable urban mobility (SUM) is the concept that captures this aspiration, referring to a desired state of more low impact transport modes and reduced dependence on high-impact ones.

© The Author(s) 2025
C. McNally et al. (Eds.): TRAconference 2024, LNMOB, pp. 24–30, 2025.
https://doi.org/10.1007/978-3-031-85578-8_4

SUM frameworks vary significantly from city to city, but share an emphasis on public transport, Mobility-as-a-Service (MaaS), Intelligent Transport Systems (ITS), active mobility modes such as walking and cycling, zero-emission logistics, and vehicle sharing, as well as multimodal optimisation via digital platforms and multimodal hubs. SUM is grounded in a range of foundational technologies including smartphones, application programming interfaces (APIs), internet of things (IoT), cloud computing and data analytics.

1.2 SUM: Why is It Important?

Given the huge complexity of realising SUM programmes at scale, their advocates can expect to face significant scrutiny over their costs and benefits.

Whilst up-front costs can be significant, SUM promises many benefits to stakeholders, particularly communities and citizens, across the urban landscape. These benefits carry profound long-term implications for spending on human health and wellbeing as well as productivity, which need to be factored into any comprehensive cost benefit analyses.

The transition to lower-emission transport modes and away from fossil fuel-powered vehicles promises to improve air quality in cities currently experiencing chronic pollution-related pulmonary health issues. Reduced noise pollution and congestion can deliver improved quality of life and diminished stress. Active mobility options, such as walking and cycling, are proven to improve overall health and fitness, reducing susceptibility to a range of salient health issues. The 'fifteen-minute' city/neighbourhood concept, in which urban environments are planned to give all residents access to needed services, facilities, and amenities within a 15-min radius of home, can also improve wellbeing, equity, and social inclusion. Social isolation and loneliness, major modern public health preoccupations, can be alleviated by more efficient and effective urban connectedness and permeability.

From an economic perspective, reduced urban transport congestion can deliver significant productivity gains in the many cities where journey times are currently lengthening, thus supporting economic growth as well as facilitating public access to commerce, services and amenities.

1.3 International Context

SUM is a worldwide goal, not merely a preoccupation of advanced economies. In most of the world, rapid urbanisation remains a salient trend, with implications for pollution, air quality, congestion, inequality, and public health. Mobility that functions effectively whilst also reducing emissions and congestion is a priority for all major cities, due to its acknowledged positive benefits across the many metrics mentioned above. In addition, arguments for SUM are strengthened by its clear alignment with a range of strategic plans at the global and regional levels, including:

- The UN's Sustainable Development Goals (which aspire to achieve inclusive, sustainable cities, as well as reduced emissions) [1];
- The European Green Deal (which mandates climate neutrality by 2050) [2];

- The European Platform on Sustainable Urban Mobility Plans (which promotes a transition to resource-efficient mobility in Europe) [3]; and
- The Paris Agreement (which commits to limit the global temperature increase in this century to 2 degrees Celsius) [4].

As a consequence, SUM strategies are being pursued in a wide range of geographies, including: Chicago, London, Paris, Shanghai, Singapore, Tokyo, and many more.

1.4 Irish Context

In Ireland, a range of indicators point to the potential for SUM. Ireland's urbanisation grew by 4% between 2019 and 2022, despite the pandemic [5]. Transport is currently responsible for some 20% of Ireland's GHG emissions, and Ireland's current modal split reflects a heavy dependence on car travel, with over 80% of land passenger transport accomplished by car in 2019 [6] (Table 1).

Table 1. Ireland's modal split [6].

Modal	%
Passenger cars	81.5%
Buses and coaches	14.8%
Railways	3.3%
Tram and Metro	0.4%

Spending on new road infrastructure remains high, with Transport Infrastructure Ireland's (TII) allocated monetary spending for new roads to 2030 standing at approximately €5.1bn, and is an active source of tension at governmental level as prioritisation is disputed between roads and public transport projects such as Dublin's Metrolink, Dart +, Cork's Community Rail investment programme, Limerick's Cycling Network and Park and Ride, Galway's City Cycle Network and redevelopment of Ceannt Station Quarter, and BusConnects Projects across Ireland's major cities [7].

Perhaps unsurprisingly in this context, the Irish Government's National Sustainable Mobility Plan (2022) has already set out a vision for growing active travel and public transportation options substantially by 2030, as well as alternatives to car travel by 2025 [8]. The current National Development Plan (NDP) (2021–2030) envisages significant capacity expansion for Dublin's Luas tram system, as well as major new transport infrastructure in the form of MetroLink [9]. At the regional and local level, Dublin's City Centre Transport Plan, Cork's Transport and Mobility Plan, Galway's Galway Transport Strategy, Limerick's Sustainable Mobility and Transport Strategy, and Waterford's City & County Development Plan all envisage enhancements aimed at facilitating active mobility options and decarbonisation.

It remains to be seen, of course, to what extent central and regional government and local authorities can deliver on these visions, given the challenges they face in the form of interagency collaboration, planning challenges, political resistance, and budgetary constraints.

1.5 SUM: Making It Happen

By its nature, SUM requires an ambitious strategy, bold leadership, and multi-agency cooperation over multi-year time horizons. How can it be realised?

Inevitably, different cities will have different priorities and challenges to address, and every SUM strategy must, ultimately, be developed for its unique circumstances. At the same time, sufficient experience has been accumulated around the world in recent years to sketch the parameters of best practice that are likely to apply. These include:

- Multi-stakeholder **collaboration**: while specific circumstances will differ by city, stakeholders will typically include citizens and communities, businesses, policymakers, municipal and national governments, transport authorities, and regulators. Effective development of SUM is impossible without the cooperation and consent of all these parties, from an early stage.
- For this reason, building SUM projects at scale strongly favours **public-private partnerships**, which have the ability to manage the inevitable risk, significant costs, and legal and regulatory complexity involved.
- Encouragement of **micro mobility** solutions (e.g. bike sharing, e-scooters) via appropriate regulation and engagement with service providers.
- Active reallocation of road space to **active mobility** pathways.
- Redirection of **major infrastructure spend** from roads to metropolitan transport.
- Discouragement of fossil fuel-powered private vehicles, via taxation, smart charging, and other incentivisation.
- Active planning for **advanced aerial mobility (AAM)**.

1.6 SUM Challenges

Inevitably, efforts to radically evolve legacy urban mobility systems, which usually demand widespread behavioral change, face a range of serious challenges. Relevant initiatives have faced significant political pushback on a variety of grounds, and risk being caricatured as 'anti-car', unfair, or unduly expensive, making them vulnerable to political sabotage or curtailment. As a result, early engagement at the citizen level, via participatory planning, consultation, and effective communication, are critical to build the foundation for durable SUM success.

In addition to political hostility, SUM programmes must overcome a host of practical difficulties, which are likely to include ageing infrastructure, serious regulatory and technological complexity, disparate data-sharing regimes, evolving cybersecurity risks, and a breakneck pace of technological change.

2 SUM: Best Practice Examples

Sustainable urban mobility is already a work in progress in numerous cities around the world. The following are a selection of notable exemplars:

2.1 New York City

As recently as 2009, Times Square was a byword for multi-lane traffic gridlock. In little more than a decade, it has been reclaimed for pedestrians, with vehicular traffic largely removed and over 100,000 square feet of pedestrian space created. As a result, Times Square is once again a celebrated thoroughfare and public square, with improved safety as well as public transport connectivity. The rejuvenation of Times Square is now a celebrated example of tactical urbanism.

2.2 Copenhagen

Copenhagen has set a goal for 75% of all urban travel to take place by foot, bike, or public transport by 2025. The city's metro has been expanded, bicycles have been more deeply integrated with public transport modes, and car use is discouraged via taxation and parking charges. As a result of sustained investment in cycling routes and safety, nearly half of all Copenhagen's commuting takes place by bike, city dwellers own five times more bikes than cars [10], and Copenhagen's travel time per 10km stands at 16 min 50 s, compared to Dublin's 28 min 30 s, almost double [11].

2.3 Oslo

Oslo has targeted significant emissions reduction via a comprehensive strategy, a major pillar of which is the creation of a car-free city centre and a 10-min city. The strategy also mandates more concentrated urban development, major public transport investments, and deliberate construction in the vicinity of public transport hubs [12]. Oslo is currently replacing its diesel-fueled buses with electric ones, has already electrified most of its ferries, and has also implemented bike sharing schemes and routes, smart traffic management systems, and other measures, as a result of which Oslo's travel time per 10km stands at 15 min 10 s, very similar to that of Copenhagen [11], and private cars account for only around 30% of city travel.

2.4 Curitiba, Brazil

Situated in one of the world's richest biodiversity hotspots, Brazil's Curitiba has adopted an urban development policy aimed at the preservation of its biodiversity, to which end it has instituted multiple new parks, ponds, and innovative Bus Rapid Transit (BRT) networks. Today, Curitiba has an enviable 64 square metres of green space per inhabitant, a bus network serving more than 60% of the city's commuters, and little more than 20% of urban travel is accomplished via personal car [13–16].

3 Implications: The Irish Context

In Ireland, where road transport accounts for well over 90% of transport emissions and private car use accounts for nearly 75% of all journeys, the government's sustainable mobility policy to 2030 was announced last year (2022), with clearly stated goals including:

- Major contribution to emissions reduction targets (51% by 2030);
- Delivery of 500,000 additional daily active travel and public transport journeys by 2030; and
- A 10% reduction in the number of kilometres driven by fossil fueled cars.

  Measures to achieve these goals include [17]:

- Demand management and behavioural change measures to reduce private car journeys.
- Safety and accessibility enhancements for active mobility routes and public transport.
- Decarbonising public transport vehicles.
- BusConnects and Connecting Ireland.
- Reallocation of road space to active mobility; and
- Placing new housing developments closer to public transport.

As ever, much depends on execution. The Irish strategy aims to enhance delivery through a range of measures including a mobility-focused leadership group and a National Sustainable Mobility Forum for engagement with stakeholders, of whom there are many, including: national, regional and local government bodies, mobility services providers, infrastructure operators, businesses, voters, and major public bodies.

Nonetheless, Ireland's SUM platform inevitably faces a range of challenges, some consistent with other markets and some particular to Ireland. Broadly across the globe, efforts to develop and deploy SUM will require as prerequisites: robust cybersecurity and data sharing arrangements, appropriate evolution of the regulatory framework, and integration with ageing and inadequate infrastructure. More particularly, Ireland's SUM platform must account for: population growth of potentially 1 million people in the next decades; lengthy permitting, contracting, and planning processes; questions over EV charging capacity and grid readiness; areas of strong pro-motorist public opinion; real political differences over the construction of new roads.

4 Conclusions

Sustainable urban mobility is not merely a nice to have, but an indispensable prerequisite for the world's cities in the light of net-zero commitments and demographic trends. The real-world evidence from today's pioneer cities is bearing out the legion benefits that SUM can provide to urban citizens, including: cleaner air; reduced congestion, isolation, and loneliness; enhanced equity and inclusion; greater economic growth. The next generation of mobility solutions promise to grant citizens around the world a higher quality of life, health, and well-being. It is an opportunity, and a challenge, that no city can afford to ignore.

References

1. https://sdgs.un.org/goals
2. https://commission.europa.eu/strategy-and-policy/priorities-2019-2024/european-green-deal_en

3. https://www.eltis.org/mobility-plans/european-platform
4. https://www.un.org/en/climatechange/paris-agreement
5. https://www.macrotrends.net/countries/IRL/ireland/urban-population
6. https://op.europa.eu/en/publication-detail/-/publication/14d7e768-1b50-11ec-b4fe-01aa75ed71a1
7. https://www.irishtimes.com/politics/2023/09/29/tensions-rise-over-roads-spending-as-fine-gael-and-fianna-fail-press-eamon-ryan/
8. https://www.gov.ie/en/publication/848df-national-sustainable-mobility-policy/
9. https://www.gov.ie/en/publication/774e2-national-development-plan-2021-2030/
10. https://radkompetenz.at/wp-content/uploads/2022/03/Copenhagen-Mobility-facts-and-figures-2021.pdf
11. TomTom Traffic Index (2023)
12. https://www.toi.no/getfile.php/1347125-1518715339/mmarkiv/Forside%202018/Elle%20de%20Vibe.pdf
13. https://thecityfix.com/blog/curitiba-is-evolving-but-remains-a-model-for-urban-sustainability/
14. https://www.unescap.org/sites/default/files/15.%20CS-Curitiba-Brazil-transport-and-zoning-policies.pdf
15. https://www.greencitytimes.com/curitiba/
16. https//sustainablemobility.iclei.org/ecomobility-alliance/curitiba-brazil/
17. https://www.gov.ie/en/press-release/dd807-irelands-new-sustainable-mobility-policy-is-a-priority-in-our-climate-and-energy-use-plans-for-the-future/

Open Access This chapter is licensed under the terms of the Creative Commons Attribution 4.0 International License (http://creativecommons.org/licenses/by/4.0/), which permits use, sharing, adaptation, distribution and reproduction in any medium or format, as long as you give appropriate credit to the original author(s) and the source, provide a link to the Creative Commons license and indicate if changes were made.

The images or other third party material in this chapter are included in the chapter's Creative Commons license, unless indicated otherwise in a credit line to the material. If material is not included in the chapter's Creative Commons license and your intended use is not permitted by statutory regulation or exceeds the permitted use, you will need to obtain permission directly from the copyright holder.

Empowering Cities: Unveiling the Innovative Urban Mobility Ecosystem to Support the Transition Through City-Led Innovations

Dimos Touloumidis[✉], Elpida Xenou, and Georgia Ayfantopoulou

Centre for Research and Technology Hellas (CERTH) - Hellenic Institute of Transport (HIT), Thessaloniki 57001, Greece
dtouloumidis@certh.gr

Abstract. Purpose: The primary objective of this paper is to develop a framework that will support municipal governments and policymakers to capture a city's capacity and maturity for adopting and implementing innovative mobility solutions for both passenger and freight.

Methods: This paper adopts a three-step which starts with the definition of the "city-led innovation" and "Innovation Readiness" for mobility planning based on literature and lessons learned from the European funded H2020 SPROUT project. Then, it identifies the elements and sub-elements of the urban mobility ecosystem through systematic literature review and validates them through workshops with SPROUT mobility experts. The third step focuses on assigning weights to these elements and sub-elements by applying an AHP to the opinion of specific mobility experts.

Results: The main finding of this study is the definition of the Innovative Urban Mobility ecosystem which consists of six elements along with eleven sub-elements that capture the city's readiness for innovation. The prioritization methodology highlighted that the most important element is the Innovative Governance & Growth which holds the importance at 40% followed by Innovative People & Stakeholders (20%) while the least important was Smart & Easily Accessible with 9.50%. A further investigation on sub-elements level revealed that the most important factors are Mobility Planning (11.66%), Public Investments (16.52%) and Cities Capacity (13.67%) while the least important sub-elements are Inter-departmental coordination (5.70%), Openness (4.61%) and Industry Diversity (6.32%) highlighting the important role of public authorities.

Conclusion: In conclusion, this article suggests that by using the proposed framework, cities can better prepare the conditions to harness innovation by implementing city-led, instead of industry-led, initiatives in mobility. Furthermore, it proposes that this framework can serve as a foundation for creating a self-assessment tool that allows cities to evaluate their strengths and weaknesses in relation to elements within the ecosystem.

© The Author(s) 2025
C. McNally et al. (Eds.): TRAconference 2024, LNMOB, pp. 31–45, 2025.
https://doi.org/10.1007/978-3-031-85578-8_5

Keywords: Innovative Urban Mobility Ecosystem · Innovative
Mobility Solutions · Smart Cities · City-Led Innovation · Digital
Transformation

1   Introduction and Background

Urban mobility poses a concern for cities worldwide as they struggle to offer
faster and more comfortable transportation options to their residents. As urbanization and mobility continue to grow both within countries and across borders,
inefficient and unsustainable transportation systems can have negative impact to
the life of the city. These effects include lower quality of life, decreased productivity, increased energy consumption, traffic congestion, pollution, and compromised safety [1]. According to [2] one of the challenges in transportation arises
when urban transport systems fail to meet the demands and expectations of
stakeholders and the capacity of the city while they also highlight that the complexity and potential disruptions, in urban transport systems escalate with city
size, requiring management strategies.

In the urban environment, the industrial sector often plays a crucial role in
driving advancements in transportation technology. While the aim is usually to
revolutionize or improve how people get around these innovations, cities can face
obstacles if they need to be adequately prepared for these developments. The gap
between industry's rapid use of innovations and city's ability to effectively integrate these advancements often leads to challenges and in some cases the failure
of new mobility services. A clear example of this can be seen with the introduction of scooters by companies like Lime in various European cities before the
COVID-19 pandemic. Although this new mobility mean was originally designed
as an environmentally friendly alternative to transportation options, these scooters faced multiple difficulties during both the installation and the implementation phase [3]. Analytically, cities were not prepared to provide infrastructure
(e.g. lanes) for scooters and also clear guidelines regarding operational zones and
speed limits together with safety challenges. These issues encompassed the problem of spaces being obstructed by chaotically parked scooters, resulting in both
noise and visual pollution. Additionally, the presence of scooters led to conflicts,
with transportation modes that worsen congestion problems and safety concerns
[4]. Another example is the operation of automated public transport vehicles,
which faced obstacles such as regulatory barriers and a series of high profile
accidents that eroded public trust slowing down the adoption of self driving
cars over human operated vehicles on roads [5,6]. Thus, the case underscores
the need for more coordinated and preemptive planning strategies on the part
of municipal authorities to ensure that industrial innovations in urban mobility
are effectively integrated, thereby fulfilling their intended role of augmenting the
quality of urban life.

To comprehend the framework and fundamental elements of an urban mobility system recent research has introduced the concept of an ecosystem approach
drawing inspiration from the natural sciences; specifically [7] discusses this approach and highlights how it mirrors the functioning of ecosystems in cities. Reference [8] proposes a framework that identifies technology, policy, community, and

place as development drivers working toward goals, such as sustainability, livability, productivity, and governance. Expanding on this idea [9] connects cities with resilience by emphasizing their capacity to absorb shocks through six dimensions that encompass social, economic, and digital aspects. IoT technologies play a role in this system as highlighted by [10,11] who discuss its applications in domains such, as smart mobility and energy management. Additionally [12] emphasizes the role of Information and Communication Technology (ICT) while [13] focuses on state of the art mobility technologies that enable these applications. According to [14], the brand strategy plays a role, in cities that require a clear vision, identity, and effective communication with stakeholders. [15] explores how branding also applies to tourism, where ICT can enhance the competitiveness of tourist destinations through information and active participation [16,17]. In addition to that, [18] emphasizes the importance of prioritizing the needs and well being of citizens by adopting design methodologies. The study of [19] highlights renewable energy powered electromobility as a component of smart city ecosystems. From a different aspect, [20] discusses the impact of shifts and climate change on domains like healthcare and water management presenting both challenges and opportunities. Looking into the future [21] looks at street designs integrated with technologies to improve mobility and livability. Additionally infrastructure serves as the framework for smart city ecosystems by integrating ICTs with systems, like energy, transportation, health and education [16,17]. Furthermore accessibility encompasses the inclusivity and usability of city solutions, across user demographics, including citizens, businesses and public authorities. This takes into account their requirements, preferences, and capabilities [22,23]. In addition, the environment plays a role in the functioning of smart city ecosystems. It both influences and is influenced by the sustainability and resilience of city solutions [24,25]. Significantly, [25] underscored the city's typology which refers on how smart city ecosystems are classified based on their characteristics such as focus, scope, maturity and performance. Lastly safety measures the level of protection and security that smart city ecosystems offer to their users and assets against cyber threats like accidents, disasters and attacks [26]. These studies led to the definition of different ecosystems, which include different elements such as smart governance, smart economy, smart people, smart environment, smart infrastructure, etc. It is worth mentioning the contribution made by [27] who emphasized the importance of mobility planning within an ecosystem approach that considers innovation. This study resulted in six elements that set the ground for the development of the Innovative Urban Mobility ecosystem of this study. Analytically, [27] created an environment with minimum carbon emissions and clean energy sources; incorporating smart growth principles that offer a mix of mobility options including shared modes; focusing on sociability and livability to enhance social capital and foster innovation among urban residents; ensuring easy access to all types of mobility services, through technology enabled solutions; promoting sustainable safety measures that reduce crash risk for all road users regardless of age or gender; investing in innovative infrastructure that prioritizes sustainable uses while accommodating various modes of transportation.

The extensive research on the smart cities and smart cities' ecosystem created the need for a framework that assesses the innovation level of an urban mobility ecosystem. The study of [28] developed a framework for measuring the strengths and weaknesses of a city's transportation system that can then be used by policymakers and city planners to develop sustainable and fair mobility solutions. Finally, the recent study by [29] developed a quantitative model to assess and predict the innovation readiness of cities within the context of urban mobility and logistics. This model, used qualitative data from cities across the EU, Eastern Asia, and the USA and identified planning and data collection processes as the most significant factors in fostering innovation in urban landscape.

The main objective of this study is to define the concept of city-led innovation and the city's Innovation Readiness in mobility planning and to explore the main elements of an Innovative Urban Mobility ecosystem and their respective significance, in determining a city's maturity for implementing innovative solutions. The research addresses the following gaps in existing literature. Firstly, it introduces the concept of city-led innovation which highlights the necessity of cities to adopt innovative solutions based on their capacity which considers the lessons learnt from real life implementation. Then, it proposes a framework to support a city's mobility planning process that considers all the different dimensions of a city's urban mobility system in the context of understanding a city's readiness to adopt innovations. Thirdly, it quantitatively determines the importance of these dimensions that affect a city's readiness, which fills the gap in limited evidence data on this matter. Lastly, while many previous studies focus primarily on technological aspects, this research emphasizes the significance of governance, societal attitudes, and cultural factors thus providing a holistic perspective. By doing so, it advances the understanding of how these factors contribute to a city's Innovation Readiness to implement innovative urban mobility solutions.

2   Methodology

This study follows a methodology of three major steps; at first, the lessons learned from the H2020 SPROUT program together with the results of an extended literature review were used to define the term of a "city-led innovation" and the city's Innovation Readiness in mobility planning. Then, the Preferred Reporting Items for Systematic Reviews and Meta-Analyses (PRISMA) methodology is being conducted to identify the elements of an Innovative Urban Mobility ecosystem that describe city's readiness. Finally, these elements were further analyzed into sub-elements, while in the last step, the sub-elements were being prioritized through an expert group to identify the importance of each sub-element.

2.1   Systematic Literature Review

A systematic review of the literature on the elements of an Innovative Urban Mobility ecosystem was conducted using the Preferred Reporting Items for Systematic Reviews and Meta-Analyses (PRISMA) method [30]. PRISMA is a widely accepted framework for reporting systematic reviews and meta-analyses, which aims to improve the transparency, completeness and accuracy of the research process and findings. The PRISMA checklist and flow diagram were followed to identify, screen, select, and synthesize relevant studies from various sources.

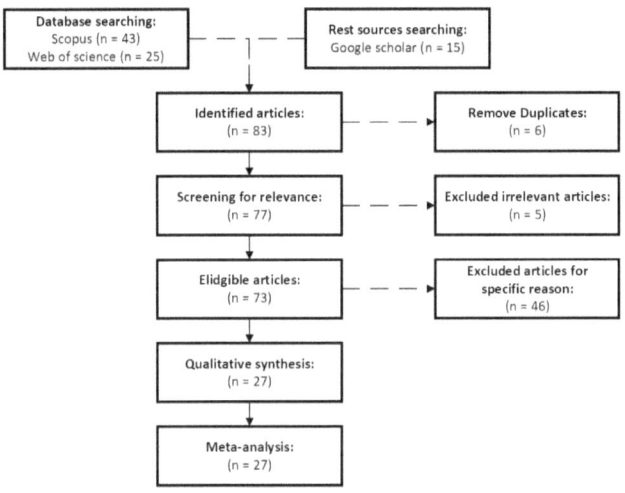

Fig. 1. The flowchart of the PRISMA methodology

Specifically, 68 articles were identified from the database of scientific journals and 15 articles were collected from other sources. Before the selection process, the entire list of articles was eliminated by removing 6 duplicates. Then, the title and the abstract of the articles was assessed and the irrelevant (5) articles were excluded from the analysis. The remaining articles were assessed to their full text, and the non-elidgible articles (not accessible, not relevant, not peer reviewed, not identifying smart city elements) were also excluded, and the final list was narrowed down to 27 articles. The final articles were used for both the next two steps, qualitative synthesis and meta-analysis. The initial search for articles was performed using the following key words: "Innovative Urban Mobility ecosystem" or "smart mobility ecosystem" or "smart city elements". It should be mentioned that since the smart city concept is relevant for Innovative Urban Mobility, the initial search also included the smart city elements. The flowchart of the PRISMA methodology followed can be seen in Fig. 1.

After the definition of the Innovative Urban Mobility ecosystem elements, these elements were further analyzed into sub-elements that were determined important to capture the Innovation Readiness of a city using the results of the systematic literature review and the opinion of SPROUT experts.

2.2   Prioritization Methodology (AHP)

The identified sub-elements were then prioritized following the Analytic Hierarchy Process (AHP) which is a decision-making framework developed by [31] that helps in order to make complex decisions by systematically evaluating and prioritizing multiple criteria and alternatives (Fig. 2). AHP operates on the principle that decision-making involves comparing and weighing various factors and provides a structured methodology to do so. The process begins by defining the decision problem and identifying the criteria and alternatives involved. Next, pairwise comparisons are made to assess the relative importance or preference of criteria and alternatives. These comparisons generate a matrix of values, which is then processed using mathematical algorithms to calculate priority scores and consistency ratios. Finally, the results are interpreted to determine the best choice or ranking of alternatives.

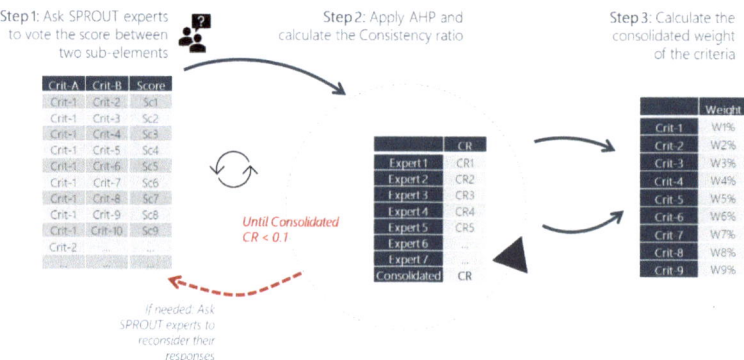

Fig. 2. The AHP methodology to capture the weight of the Innovative UM ecosystem sub-elements

The assignment of each sub-element with a specific weight was determined through the Analytic Hierarchy Process (AHP). More specifically, 12 SPROUT experts participated in this process and were asked to prioritize the sub-elements mentioned above: 6 experts from Centre for Research & Technology Hellas (CERTH), 2 experts from Wuppertal Institut für Klima (WI), 2 experts from POLIS Network, 1 from the Zaragoza Logistics Center (ZLC) and 1 from Vrije Universiteit Brussel (VUB). The SPROUT experts voted: i) which of the sub-elements is more important and ii) how much more important it is with a score between 0–9.

3   Results and Discussion

3.1   The Role of City-Led Innovation and Innovation Readiness in Shaping the Future of Urban Mobility

In the changing world of transportation both industry driven and city driven advancements present unique advantages and challenges that need to be balanced for sustainable progress. Industry-led innovations often bring in technologies and services at a pace shaping the urban landscape swiftly [32]. While these innovations hold potential they can also bring consequences that local authorities might not be fully prepared to handle [33]. As a result many groundbreaking mobility solutions are often limited to pilot stages without realizing their benefits, for society. To overcome this obstacle cities must equip themselves with the knowledge to create conditions for these innovations [34]. Additionally urban authorities should develop the expertise to formulate policies that not encourage the integration of solutions but also extend their application to maximize their impact on urban transportation systems. On the hand city-led innovation in mobility is expected to be a complex blend of various technologies and services in the future. This complexity will require harmonized approaches in planning and operations. Cities will also need to be increasingly adaptable, by tailoring their transportation infrastructure to cater to emerging needs and fluctuating demand patterns. This dynamism will push cities to address questions, about developing infrastructure and managing space especially in densely populated inner city areas. By combining the flexibility of innovation driven by industry with the planning that comes with city-led approaches the future of mobility can be both liveable and environmentally friendly. Private partnerships have the potential to be highly beneficial as they bring together the sectors technical expertise and innovative thinking with the long term planning and socially responsible approach of public agencies. In this context city-led innovation can be described as:

"City-led innovation aims at empowering leaders (city authorities and policymakers) to develop policies that steer the introduction of innovative solutions and gear innovation towards implementation solutions. This establishes the policy in a way that is more effective and legitimate and allows for more active participation of stakeholders."

The concept of city-led innovations indicates that cities should enhance their readiness for innovation to effectively implement measures. Specifically, the Innovation Readiness can help in identifying if the city is capable and ready in deploying or enabling the deployment of mobility innovations. Capturing the Innovation Readiness of the city by studying the city's organization structure, capability and capacity, current regulation practices, engagement practices, financial strength and investment opportunities etc. The dimension of the Innovation Readiness will be the base of the analysis to identify the Innovative Urban Mobility ecosystem elements and sub-elements.

3.2   The Innovative Urban Mobility Ecosystem

The systematic review of the literature resulted in six elements that were determined necessary to define an ecosystem of Innovative Urban Mobility. The first element is "Climate & City Typology" which describes the geophysical and environmental conditions of a city and the element "Smart & Easily accessible" contains the level of accessibility of New Mobility Services (NMS) & Transportation services. Afterwards, the element "Safe & Security" assesses the level of security and safety of the current mobility services and the element Smart & "Innovative Resources & infrastructure available" refers to the Availability of smart resources, and physical and digital infrastructure. Moreover, the element "Innovative People & Stakeholders" concerns the behavior of people, smart entrepreneurialism, etc. Finally, the element "Innovative Governance & Growth" describes the integrated planning and current political framework. Then, the 6 elements were further analyzed into 11 sub-elements to better understand. The description of the six (6) Innovative Urban Mobility ecosystem that were identified can be found below:

- **Climate & City Typology:** A city that embraces green and innovative mobility solutions by creating synergies among various sectors and domains, and has high standards of education and research in the field of mobility [9,35].
- **Smart & Easily accessible:** A city that offers accessible, affordable, and inclusive mobility services and governmental processes that enables citizens to have a voice and a choice in the decision-making [8,36].
- **Safe & Secure:** A city that ensures the safety and security of its users when they use innovative urban mobility solutions, and that reduces the number of accidents and deaths related to mobility. A city that is also resilient to shocks and disasters that may affect the mobility system [27,37].
- **Smart & Innovative Resources & infrastructure available:** A city that has a smart and integrated infrastructure that supports multimodal and innovative mobility solutions with high efficiency and low congestion, and that uses data-driven methods for mobility planning and management. A city that also has a skilled workforce in the mobility sector [10,36].
- **Innovative People & Stakeholders:** A city that has a culture of innovation and diversity in its mobility sector, and that engages stakeholders with different levels of smartness and behavior in the adoption of innovative mobility solutions [19,38].
- **Innovative Governance & Growth:** A city that has a clear strategy and investment plan for innovative mobility, and that uses cross-sectoral collaboration and innovative procurement practices to facilitate the deployment of innovative mobility solutions. A city that also fosters co-creation among stakeholders in the design and implementation of innovative mobility solutions [27,39].

These elements were further analysed into 11 sub-elements that capture the Innovation Readiness of a city:

Fig. 3. The Innovative Urban Mobility ecosystem elements and sub-elements

- **Inter-departmental coordination:** It describes the city's organizational structure and dedicated departments, essential for the effective implementation of innovative mobility solutions [27].
- **Mobility Planning:** Outlines the current regulatory framework related to passenger and freight transportation within the city, including the presence of Sustainable Urban Mobility Plans (SUMPs) and Sustainable Urban Logistic Plans (SULPs) [27].
- **Liaison:** Indicates the level of engagement between the public and private sectors, as well as corporate participation in city mobility initiatives and partnerships [40].
- **Public Investments:** Measures the extent to which the city has secured funding and resources to support the implementation of innovative mobility solutions [40].
- **Openness:** Evaluates city's connectivity on national and international levels, assessing synergies and collaborations among the city's institutions [21].
- **Science & Education:** Considers the educational attainment of the city's residents and the presence of research institutes and universities within the city [21].
- **Transparency & Accountability:** Reflects the transparency of government processes and the accessibility and availability of UM data [21].
- **Data Availability:** Pertains to the development of the city's data collection systems, including methods such as physical surveys and infrastructure for data observation [21].
- **Cities Capacity:** Explains the city's capacity to embrace innovation, including the availability of specialists, evidence-driven policymaking, and the presence of infrastructure to support innovative mobility solutions [41].
- **Culture:** Indicates city's orientation toward innovation, considering its past experience in implementing innovative business models and the prevailing user trends favoring environmentally friendly modes of transportation [41].

– **Industry Diversity:** Assesses the diversity and intelligence of the city's industries, taking into account the presence of major innovators, including startups and high-tech companies, established within the city [41].

It's important to note that the aspect of Safety & Security is not considered as an element, within the context of the Innovative Urban Mobility ecosystem. This is because it does not directly reflect the city's readiness to embrace mobility solutions but rather focuses on the satisfaction and trust of users. While safety and security are aspects of any mobility system, they primarily depend on the operational aspects of the specific solution rather than being influenced by general attributes or environment of the city.

3.3 Prioritization of Innovation Readiness Elements and Sub-Elements

The Analytic Hierarchy Process (AHP) is a multicriteria decision-making method that allows comparing alternatives based on their relative importance and performance. In this study, the AHP was applied to assign weights to the 11 sub-elements identified in the Innovative Urban Mobility ecosystem. The AHP Consistency Ratio (CR) of the consolidated matrix was calculated at 3.45%, which is below the acceptable threshold of 10% and therefore, the AHP methodology was accepted [42]. The following table (Table 1) contains the weight of the elements and sub-elements of the Innovative Urban Mobility ecosystem based on the results of the AHP.

Table 1. The weight of the Innovative Urban Mobility ecosystem sub-elements

Element	Weight	Sub-element	Weight
Innovative Governance & Growth	39.86%	Inter-departmental coordination	5.70%
		Mobility Planning	11.66%
		Liaise	5.98%
		Public Investments	16.52%
Climate & City Typology	10.89%	Openness	4.61%
		Science & Education	6.28%
Smart & Easily Accessible	9.50%	Transparency & Accountability	9.50%
Smart & Innovative Resources & Infra	25.91%	Data availability	12.24%
		Cities Capacity	13.67%
Innovative People & Stakeholders	13.84%	Culture	7.52%
		Industry Diversity	6.32%

Based on the outcomes of the AHP it is evident that the crucial aspect for a city to embrace urban mobility solutions is "Public Investments", which holds the highest weight of 16.52% which implies that financial resources are considered vital in guiding transition. Following closely are "Cities Capacity" and "Data Availability" with weights of 13.67% and 12.24%, respectively, highlighting the significance of being operationally ready and having data collection systems in place. Also, "Mobility Planning" and "Transparency & Accountability" also received attention with weights of 11.66% and 9.50% respectively underlining the importance of frameworks and effective governance. Other factors such as "Culture", "Science & Education" and "Industry Diversity" hold weights indicating their roles in nurturing an innovative ecosystem. Surprisingly traditional governance elements like "Inter Coordination", "Liaison", and "Interdepartmental coordination" were given lower weights suggesting that stakeholders believe these aspects will naturally fall into place once other critical elements are established. Lastly, "Openness" received the weight overall, suggesting that while external connections are beneficial, they are not deemed as crucial as factors such as investment capacity to foster Innovative Urban Mobility.

The combined outcomes of the Analytic Hierarchy Process provide a viewpoint, on what is considered crucial for a city to be prepared for implementing Innovative Urban Mobility. The top priority is Innovative Governance & Growth, with a weight of 39.86% emphasizing how policy making, coordination among departments and financial investment play a role in shaping the mobility landscape. This is also reflected in weights given to Mobility Planning (11.66%). Public Investments (16.52%) within this category. Next on the list is Smart & Innovative Resources and Infrastructure which carries a weight of 25.91% highlighting the importance of data driven infrastructure and resource allocation. Other elements such as Innovative People & Stakeholders (13.84%) and Climate & City Typology (10.89%) illustrate significant aspects like stakeholder engagement and the city's focus on education and environmental awareness respectively. Even though it has the weight at 9.50% the Smart & Easily Accessible element still indicates the importance of transparency and accessibility within the ecosystem. Overall these weightings demonstrate that governance and robust infrastructure are foundations, for innovative solutions while stakeholder involvement and a supportive city environment serve as important enablers.

4   Conclusion

This research study focuses on identifying, assessing and comprehending the factors that affect a city's readiness in innovative urban planning, to implement Innovative Urban Mobility solutions. By conducting a review of existing literature, it is first established the foundation by defining concepts like "city-led innovation" and "Innovation Readiness," explaining why these aspects are crucial for modern city planning. This study identified 6 elements within the ecosystem that range from infrastructure to governance and further dissected them into 11 sub-elements that capture a city's readiness for innovation. To evaluate these sub-elements the Analytic Hierarchy Process (AHP), which is a decision making method based on multiple criteria, was deployed. Through this process, empirical weights were assigned to each sub-element emphasizing the role of Public Investments, Cities Capacity, and Data Availability which were validated the AHP results using a consistency ratio. The study concluded with an integrated overview that highlights the components of an urban mobility ecosystem; strong governance structures, state-of-the-art infrastructure and resources, informed and receptive people, as well as a resolute commitment to inclusivity and sustainability. This robust analytical framework can serve as a resource, for policymakers, urban planners, and other stakeholders seeking to enhance their city's ability to embrace urban mobility solutions.

The findings from the Hierarchy Process (AHP) highlight the role of public investments in driving innovation in urban mobility. Additionally the study emphasizes the significance of governance and policy frameworks "Innovative Governance & Growth" which underscores the importance of governance across departments and well defined investment strategies. The availability of data and operational readiness are also identified as factors that underscore the importance of data-driven decision making in developing sustainable mobility solutions. Interestingly societal attitudes and stakeholder involvement play a essential role calling for targeted public awareness campaigns and educational initiatives. The study also emphasizes the need for integrated planning that combines industry-led innovations with policy considerations. While less weight is given to connections and openness, they still provide added value by suggesting that international collaborations can enhance efforts. Furthermore although not directly weighted safety and security emerged as factors influencing user adoption. With these weighted elements considered, cities can adopt an approach to become more ready, for innovation. It highlights the significance of having a organized system of governance, advanced resources an open minded public and a strong dedication, to inclusivity and sustainability. Hence, this research offers a guide for city officials, policymakers, and industry stakeholders who aim to promote an urban transportation system.

While this research has set the foundation for understanding the complexities of mobility future studies could explore various aspects that were not covered in this research. First, it would be beneficial to investigate how global trends like climate change or economic shifts influence the prioritization of the identified elements in the ecosystem. Moreover, comparative studies across settings could provide valuable insights into how cultural, political, and economic contexts shape the adoption and integration of innovative transportation solutions. Examining the effectiveness of partnerships in driving innovation in this field through case studies or long-term analyses could also offer refined understanding of successful collaboration models. Lastly, exploring the use of (machine learning) algorithms to predict a city's Innovation Readiness would provide decision makers with a powerful tool.

References

1. El-Sherif, D.M.: 4 - Urban mobility systems components. In: Vacca, J.R. (ed.) Solving Urban Infrastructure Problems Using Smart City Technologies, pp. 89-106. Elsevier (2021). https://doi.org/10.1016/B978-0-12-816816-5.00004-8

2. Bucchiarone, A., De Sanctis, M., Marconi, A., Pistore, M., Traverso, P.: Incremental composition for adaptive by-design service based systems. In: 2016 IEEE International Conference on Web Services (ICWS), pp. 236-243. IEEE Press (2016). https://doi.org/10.1109/ICWS.2016.38

3. CNN: E-scooters were supposed to fix travel in Rome. then they became a major problem. CNN (2022). https://edition.cnn.com/travel/article/rome-scooter-problems/index.html. Accessed 22 Mar 2023

4. Gössling, S.: Integrating e-scooters in urban transportation: problems, policies, and the prospect of system change. Transp. Res. D: Transp. Environ. **79**, 102230 (2020). https://doi.org/10.1016/j.trd.2020.102230

5. Guardian: Self-driving bus involved in crash less than two hours after Las Vegas launch. The Guardian (2017). https://www.theguardian.com/technology/2017/nov/09/self-driving-bus-crashes-two-hours-after-las-vegas-launch-truck-autonomous-vehicle. Accessed 06 Feb 2023]

6. Tirone, J.: Driverless bus hits pedestrian in Vienna. Automotive News Europe (2019). https://europe.autonews.com/automakers/driverless-bus-hits-pedestrian-vienna. Accessed 29 July 2023

7. Flügge, B. (ed.): Smart Mobility – Connecting Everyone. Springer, Wiesbaden (2017). https://doi.org/10.1007/978-3-658-15622-0

8. Yigitcanlar, T., et al.: Understanding 'smart cities': intertwining development drivers with desired outcomes in a multidimensional framework. Cities **81**, 145–160 (2018). https://doi.org/10.1016/j.cities.2018.04.003

9. Apostu, S.A., Vasile, V., Vasile, R., Rosak-Szyrocka, J.: Do smart cities represent the key to urban resilience? rethinking urban resilience. Int. J. Environ. Res. Public Health **19**(22), 15410 (2022). https://doi.org/10.3390/ijerph192215410

10. Bellini, P., Nesi, P., Pantaleo, G.: IoT-enabled smart cities: a review of concepts frameworks key technologies. Appl. Sci. **12**(3), 1607 (2022). https://doi.org/10.3390/app12031607

11. Attaran, H., Kheibari, N., Bahrepour, D.: Toward integrated smart city: a new model for implementation and design challenges. GeoJournal. **87**(4), 511-526 (2022). https://doi.org/10.1007/s10708-021-10560-w

12. Portmann, E., Finger, M., Engesser, H.: Smart Cities. Informatik-Spektrum **40**(1), 1–5 (2016). https://doi.org/10.1007/s00287-016-1000-7

13. Savithramma, R.M., Ashwini, B.P., Sumathi, R.: Smart mobility implementation in smart cities: a comprehensive review on state-of-art technologies. In: 2022 4th International Conference on Smart Systems and Inventive Technology (ICSSIT), pp. 10-17. IEEE Press, Tirunelveli (2022). https://doi.org/10.1109/ICSSIT53264.2022.9716288

14. Augustyn, A.: Smart Cities - brand cities of the future. In: The Business of Place: Critical, Practical and Pragmatic Perspectives, pp. 1-11. Manchester (2013)

15. Matos, A., Pinto, B., Barros, F., Martins, S., Martins, J., Au-Yong-Oliveira, M.: Smart cities and smart tourism: what future do they bring? In: Rocha, Á., Adeli, H., Reis, L.P., Costanzo, S. (eds.) WorldCIST'19 2019. AISC, vol. 932, pp. 358–370. Springer, Cham (2019). https://doi.org/10.1007/978-3-030-16187-3_35

16. Girardi, P., Temporelli, P.: Smartainability: a methodology for assessing the sustainability of the smart city. Environ. Dev. Sustain. **19**(4), 1577–1593 (2017). https://doi.org/10.1016/j.egypro.2017.03.243

17. Dawes, S.S., Vidiasova, L., Parkhimovich, O.: Smart cities: contradicting definitions and unclear measures. In: Proceedings of the 17th Annual International Conference on Digital Government Research, vol. 4128, pp. 591-592. Springer, Heidelberg (2016). https://doi.org/10.13140/2.1.1756.5120

18. Khamis, A.: Toward a people-centric smart city. In: Smart Mobility, pp. 1-9. Apress, Berkeley, CA (2021). https://doi.org/10.1007/978-1-4842-7101-8_1

19. Curiel-Ramirez, L.A., Ramirez-Mendoza, R.A., Bustamante-Bello, M.R., Morales-Menendez, R., Galvan, J.A., de J. Lozoya-Santos, J.: Smart Electromobility: interactive ecosystem of research, innovation, engineering, and entrepreneurship. Inter. J. Interactive Design Manufact. (IJIDeM) **14**(4), 1443–1459 (2020). https://doi.org/10.1007/s12008-020-00710-8

20. Eggers, W.D., Skowron, J.: Forces of change: Smart cities. Deloitte Insights (2018). https://www2.deloitte.com/xe/en/insights/focus/smart-city/overview.html. Accessed 12 May 2023

21. Rui, J., Othengrafen, F.: examining the role of innovative streets in enhancing urban mobility and livability for sustainable urban transition: a review. Sustainability **15**(7), 5709 (2023). https://doi.org/10.3390/su15075709

22. Meijer, A., Rodriguez, B., Pedro, M.: Governing the smart city: a review of the literature on smart urban governance. Int. Rev. Adm. Sci. **82**(2), 392–408 (2016). https://doi.org/10.1177/0020852314564308

23. Fernandez-Anez, V., Fernandez-Guell, J.M., Giffinger, R.: Smart City implementation and discourses: an integrated conceptual model. case of Vienna. Cities. **78**, 4–16 (2018). https://doi.org/10.1016/j.cities.2017.12.004

24. Bundgaard, L., Borrás, S.: City-wide scale-up of smart city pilot projects: governance conditions. Technol. Forecast. Soc. Change. **172** (2021). https://doi.org/10.1016/j.techfore.2021.121014

25. Mora, L., Deakin, M., Reid, A.: Strategic principles for smart city development: a multiple case study analysis of European best practices. Technol. Forecast. Soc. Change. **142**, 70–97 (2019). https://doi.org/10.1016/j.techfore.2018.0

26. Torfing, J., Sørensen, E.: Enhancing collaborative innovation in the public sector. Adm. Soc. **43**(8) (2011). https://doi.org/10.1177/0095399711418768

27. Karim, D.M.: Creating an innovative mobility ecosystem for urban planning areas. In: Meyer, G., Shaheen, S. (eds.) Disrupting Mobility. LNM, pp. 21–47. Springer, Cham (2017). https://doi.org/10.1007/978-3-319-51602-8_2

28. UITP: Urban Mobility Innovation Index (UMII): Leading transformations with innovation for inclusive, sustainable and resilient urban mobility (2021)

29. Ayfantopoulou, G., Touloumidis, D., Mallidis, I., Xenou, E.: A quantitative model of innovation readiness in urban mobility: a comparative study of smart cities in the EU, Eastern Asia, and USA Regions. smart cities. **6**(6) (2023).https://doi.org/10.3390/smartcities6060148

30. Rethlefsen, M.L., et al.: Prisma-S: an extension to the PRISMA statement for reporting literature searches in systematic reviews. Syst. Rev. **10**(1) (2021). https://doi.org/10.1186/s13643-020-01542-z

31. Saaty, T.L.: Decision making with the analytic hierarchy process. Int. J. Serv. Sci. **1**(1), 83 (2008). https://doi.org/10.1504/ijssci.2008.017590

32. Dia, H., Bagloee, S., Ghaderi, H.: Technology-led disruptions and innovations: the trends transforming urban mobility. In: Augusto, J.C. (ed.) Handbook of Smart Cities, pp. 1-36. Springer International Publishing, Cham (2020). https://doi.org/10.1007/978-3-030-15145-4_51-1

33. Mazzarino, M., Braidotti, L., Royo, B., de la Cruz, T.: Innovative technologies and systems for urban mobility: the case of padua. In: Nathanail, E.G., Gavanas, N., Adamos, G. (eds.) Smart Energy for Smart Transport, pp. 504-519. Springer Nature Switzerland, Cham (2023).https://doi.org/10.1007/978-3-031-23721-8_42

34. McKinsey: Why the Automotive Future Is Electric. McKinsey & Company (2021). www.mckinsey.com/industries/automotive-and-assembly/our-insights/why-the-automotive-future-is-electric. Accessed 14 Sep 2023
35. Nasir, S.: Sustainable and Efficient City Logistics. In: Bányai, T., Bányai, Á., Kaczmar, I. (eds.) Supply Chain - Recent Advances and New Perspectives in the Industry 4.0 Era. IntechOpen (2022). https://doi.org/10.5772/intechopen.104413
36. Khan, Z., Anjum, A., Soomro, K., Tahir, M.A.: Towards cloud based big data analytics for smart future cities. J. Cloud Comput. 4(1), 1–11 (2015). https://doi.org/10.1186/s13677-015-0026-8
37. Taniguchi, E., Fwa, T.F., Thompson, R.G.: Urban Transportation and Logistics: Health, Safety, and Security Concerns. 1st edn. Taylor & Francis, Boca Raton (2013). https://doi.org/10.1201/b16346
38. Taniguchi, E., Thompson, R.G., Yamada, T.: New opportunities and challenges for city logistics. Transp. Res. Procedia. 12, 5–13 (2016). https://doi.org/10.1016/j.trpro.2016.02.004
39. Xenou, E., Madas, M., Ayfandopoulou, G.: Developing a smart city logistics assessment framework (SCLAF): a conceptual tool for identifying the level of smartness of a city logistics system. Sustainability 14(10), 6039 (2022). https://doi.org/10.3390/su14106039
40. Bouton, S., Canales, D., Trimble, E.: Public-private collaborations for transforming Urban Mobility. McKinsey & Company (2017). https://www.mckinsey.com/capabilities/sustainability/our-insights/public-private-collaborations-for-transforming-urban-mobility. [Accessed 14 June 2023]
41. Zhang, J.X., Cheng, J.W., Philbin, S.P., Ballesteros-Perez, P., Skitmore, M., Wang, G.: Influencing factors of urban innovation and development: a grounded theory analysis. Environ. Dev. Sustain. 25(3), 2079–2104 (2023). https://doi.org/10.1007/s10668-022-02151-7
42. Goepel, K.D.: Comparison of judgment scales of the analytical hierarchy process - a new approach. Int. J. Inf. Technol. Decis. Mak. 18(2), 445–463 (2019). https://doi.org/10.1142/s0219622019500044

Open Access This chapter is licensed under the terms of the Creative Commons Attribution 4.0 International License (http://creativecommons.org/licenses/by/4.0/), which permits use, sharing, adaptation, distribution and reproduction in any medium or format, as long as you give appropriate credit to the original author(s) and the source, provide a link to the Creative Commons license and indicate if changes were made.

The images or other third party material in this chapter are included in the chapter's Creative Commons license, unless indicated otherwise in a credit line to the material. If material is not included in the chapter's Creative Commons license and your intended use is not permitted by statutory regulation or exceeds the permitted use, you will need to obtain permission directly from the copyright holder.

Effects of Street Space Redesign on Travel Demand in Berlin, Germany

Simon Nieland[(✉)] [iD], Daniel Krajzewicz[iD], Jan Weschke[iD], and Julia Schuppan[iD]

Institute of Transport Research, German Aerospace Center, Rudower Chaussee 7, 12489 Berlin, Germany

simon.nieland@dlr.de

Abstract. Many cities today are struggling with increasing urban car traffic and the associated negative effects such as emissions, air pollution, accidents and noise. To overcome these challenges, cities are investing in public transport services or reducing fares, building cycle paths or implementing measures to reduce car traffic. These include measures such as one-way streets, reduced speed limits or reduced parking spaces and increased parking fees. However, it is important for city administrations to know the effects of their measures in advance, both to anticipate their impact and to discuss them with the public and legitimize their actions. Therefore, this study describes the application of an agent-based travel demand model in combination with a microscopic traffic simulation model to simulate the effects of a baseline scenario and a redesign scenario of urban transport infrastructure on a small spatial scale of an inner-city neighborhood. From the results, we can infer the impact of different measures on mode choice and traffic volumes. The presented methodology show a high potential for planning, evaluation and decision making in the field of sustainable urban mobility.

Keywords: Traffic Management Measures · Simulation · Agent-based

1 Introduction

Cities around the world are facing enormous challenges in terms of traffic-related emissions and increasing motorized traffic. Many of them are trying to create a more accessible and sustainable environment by promoting mixed-use neighborhoods and shifting urban mobility to more sustainable modes of transport. Recent successful examples (e.g. Barcelona and Paris) have implemented traffic-reducing measures such as redesigning streetscapes to encourage active modes of transportation such as walking and cycling. In addition, the redesign of streetscapes is a key element in achieving more sustainable cities with less negative impact from traffic [1, 2]. The design and allocation of street space in cities has a significant impact on the traffic connections and mobility patterns of residents, which affects their quality of life and road safety (e.g. [3]). Therefore, city governments are keen to take effective policy measures to achieve a shift towards sustainable mobility and more liveable neighborhoods. In order to support city and local governments in their decision-making and to legitimize their street redesign measures,

© The Author(s) 2025

C. McNally et al. (Eds.): TRAconference 2024, LNMOB, pp. 46–51, 2025.
https://doi.org/10.1007/978-3-031-85578-8_6

it is crucial to quantify ex-ante the expected impacts on transport demand, traffic volumes, accessibility, land use, air quality and noise (e.g. [4]). Modeling and simulating the impact of policy measures provides administrations with data and visualizations to discuss the design and implementation plans with citizens and thus create acceptance. On this basis, we present a modeling approach that quantifies the impact of several traffic reduction measures and apply the model to a real-life example in the Kreuzberg district of Berlin, Germany. The presented results show the potential of the method for planning, evaluation and decision making in the field of sustainable urban mobility.

2   Method

This section describes the methodological framework of this work and illustrates the redesign measures implemented.

2.1   Modelling of Transport Demand and Traffic Volume

To model the transport demand and the traffic volume for each edge of the road network, a model of the traffic in the city of Berlin is generated using the agent-based traffic demand model TAPAS[1] [5] and the traffic simulation package SUMO[2] [6]. TAPAS is used to calculate the mobility behavior of a synthetic population of Berlin, including the choice of means of transport and the destination locations. The daily mobility pattern of a single simulated person is retrieved from about 50,000 mobility diaries from mobility surveys, randomly selecting one that matches the person's socio-demographic characteristics. For the selected mobility diary, the places where the activities listed in the diary can be performed and the means of transportation by which they can be reached are calculated, again using functions derived from mobility surveys. The result of a TAPAS run are daily trip chains for all modeled persons.

These trip chains, i.e. the trips included that are made by private car, are passed on to SUMO. SUMO simulates the movement of the corresponding vehicles through the road network with time steps of one second and taking into account the road network and its traffic rules. The result of the simulation of private car journeys in SUMO are both traffic measurements, including average speeds or traffic flow, as well as derived measurements, such as the amount of pollutants emitted.

2.2   Study Area and Redesign Measures

The study area and its location within the city of Berlin are shown in Fig. 1. Since urban redevelopment is usually carried out in small urban areas, the application of microscopic demand modeling at a low aggregation level is of great interest. In total, two scenarios were created. The first is the baseline scenario, which reflects the current state of transportation demand and traffic volumes in the study area. In a second scenario, specific measures are implemented in a neighborhood around Lausitzer Platz in the

---

[1] https://github.com/DLR-VF/TAPAS.

[2] https://github.com/eclipse/sumo.

Friedrichshain-Kreuzberg district of Berlin. This redesign scenario includes a number of infrastructural and regulatory changes, in particular speed reduction, reduction of parking space, increase of parking fees, introduction of protected bicycle lanes as well as modal filters and one-way streets (see Fig. 2).

Fig. 1. Left: City boundary of Berlin, Germany. Right: Study area "Lausitzer Platz".

Modal filters describe road closures that allow access by bicycle and on foot, but not by car. Cycle lanes have been introduced on some of the main roads around the district, while the speed limit on some sections of road has been reduced from 30 or 50 km per hour to walking speed.

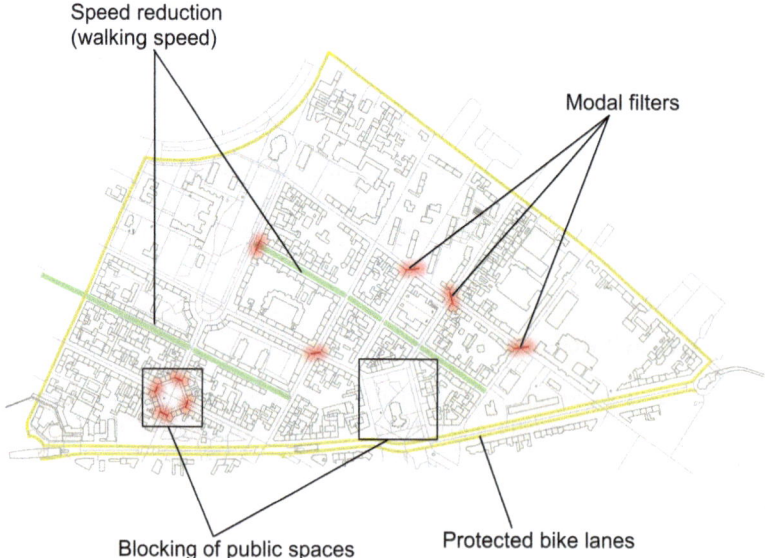

Fig. 2. Implemented measures.

The measures are intended to achieve three effects: firstly, to calm traffic in certain streets by introducing a speed reduction; secondly, to impede through traffic by means of traffic filters and one-way streets; and thirdly, to reduce the number of journeys by private motorized transport that end in the area by introducing parking space management and reducing the number of parking spaces. The selection of the measures to be implemented is the result of public participation procedures performed by the local administration. Although the study area only covers a very small part of Berlin, the simulation covers the entire city in order to also examine the effects on the surrounding areas.

3   Results

This section illustrates the results of the modeling approach with a focus on the differences betweet the baseline and the redesign scenario.

3.1   Effects on Transport Demand

The results of the redesign scenario show a significant reduction in motorized traffic (from 13.2% to 10.1%) for trips that start in the study area. In addition, a clear shift towards sustainable modes of walking and cycling can be observed, increasing from 28.9% to 30.4% and from 19.8% to 21.7% respectively (see Fig. 3). The total number of trips starting or ending within the area decreases by 4% overall, with an increase in the number of trips made by bicycle (+5.5%) and on foot (1.1%), while the number of trips made by car (-26.5%), as a passenger (-15.9%) and by public transport (-1.8%) decreases.

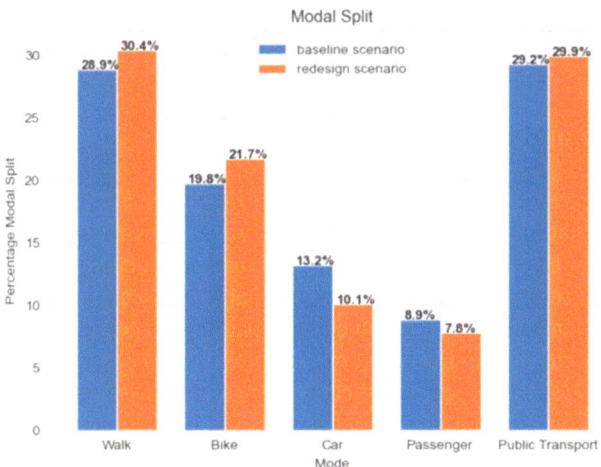

Fig. 3. Modal Split of baseline and redesign scenario.

3.2 Effects on Traffic Volume

The modal filters, speed reduction and parking space management lead to a reduction in car traffic within the study area of around 30% (see Fig. 4); the volume of car traffic in the streets on the edge of the study area increases sharply in some cases, but overall there is also a reduction of 10% here. The decrease in traffic within the area is not evenly distributed throughout the day. The strongest decline can be observed especially after the rush hour peak times (09:00–11:30 and 18:00–10:00).

Changes in car traffic
volumne between
baseline and redesign
scenario

▬▬	-1229 - -764
▬▬	-764 - -300
▬▬	-300 - 165
▬▬	165 - 630
▬▬	630 - 1094
▬▬	1094 - 1559

0   100   200 m

Fig. 4. Difference in traffic volume (cars per day) between the baseline and the redesign scenario.

4   Discussion and Conclusion

In summary, the results of the redesign scenario compared to the baseline scenario inevitably show that infrastructural and regulatory measures related to urban transport have an impact on travel demand and traffic volumes in inner-city neighborhoods and can be part of the solution to enable more sustainable urban environments and address urban challenges. In particular, the reduction of parking spaces and the introduction of parking management play an important role here, as these measures effectively prevent trips that end in the study area or shift them to more sustainable modes of transportation. The speed reduction does not appear to have a major impact on traffic demand, as car travel times change only slightly. Nevertheless, an increase in road safety due to this measure can be assumed [8].

The presented approach shows that by using travel demand and traffic simulation models, combinations of measures can be analyzed in terms of their impacts, informing stakeholders ex-ante about the expected outcomes of urban measures. This helps local authorities to make decisions and develop applicable solutions for practitioners and to discuss action plans with the public. However, the calibration and the complex and static input data for demand modeling limit the predictive power at the district level, but encourage the derivation of trends and tendencies. In the future, further scenarios will be added, derived in additional iterations with local authorities, including assessment of the impact of individual measures on transport demand and traffic volumes to generate ideal bundles of measures that optimize the desired effects.

References

1. City of Barcelona: Pla de Mobilitat Urbana 2024 (2023). https://www.barcelona.cat/mobilitat/ca/qui-som/pla-de-mobilitat-urbana-2024, Accessed: 24 March 2023
2. City of Paris: Paris ville du quart d'heure, ou le pari de la proximité (2022). https://www.paris.fr/dossiers/paris-ville-du-quart-d-heure-ou-le-pari-de-la-proximite-37, Accessed 24 March 2023
3. Holz-Rau, C., Scheiner, J.: Land-use and transport planning–A field of complex cause-impact relationships. Thoughts on transport growth, greenhouse gas emissions and the built environment. Transp. Policy **74**, 127–137 (2019)
4. Li, M., Zhao, J.: Gaining acceptance by informing the people? public knowledge, attitudes, and acceptance of transportation policies. J. Plan. Educ. Res. **39**(2), 166–183 (2019)
5. Heinrichs, M., Krajzewicz, D., Cyganski, R., von Schmidt, A.: Introduction of car sharing into existing car fleets in microscopic travel demand modelling. Pers. Ubiquit. Comput. **21**, 1055–1065 (2017)
6. Alvarez Lopez, P., et al.: Microscopic Traffic Simulation using SUMO. In: 2019 IEEE Intelligent Transportation Systems Conference (ITSC), The 21st IEEE International Conference on Intelligent Transportation Systems, pp. 2575–2582. IEEE. (2018). https://doi.org/10.1109/ITSC.2018.8569938
7. Krajzewicz, D., Heinrichs, D., Cyganski, R.: Intermodal contour accessibility measures computation using the'UrMo accessibility computer'. Inter. J. Adv. Syst. Measurem. **10**(3&4), 111–123 (2017)
8. Wang, C., Quddus, M.A., Ison, S.G.: The effect of traffic and road characteristics on road safety: a review and future research direction. Saf. Sci.. Sci. **57**, 264–275 (2013)

Open Access This chapter is licensed under the terms of the Creative Commons Attribution 4.0 International License (http://creativecommons.org/licenses/by/4.0/), which permits use, sharing, adaptation, distribution and reproduction in any medium or format, as long as you give appropriate credit to the original author(s) and the source, provide a link to the Creative Commons license and indicate if changes were made.

The images or other third party material in this chapter are included in the chapter's Creative Commons license, unless indicated otherwise in a credit line to the material. If material is not included in the chapter's Creative Commons license and your intended use is not permitted by statutory regulation or exceeds the permitted use, you will need to obtain permission directly from the copyright holder.

Netnography Study to Investigate End Users' Satisfaction with Public Transport in UPPER Project's Living Labs

Carol Soriano[✉], Amparo López-Vicente, Juan F. Giménez, Raquel Marzo, José Solaz, and Elisa Signes

Instituto de Biomecánica de Valencia, Universitat Politècnica de València, Camino de Vera s/n, Edificio 9C , Valencia, Spain

`carol.soriano@ibv.org`

Abstract. *UPPER* is a HorizonEurope innovation project aimed at spearheading a Public Transport revolution, to support European cities in its objective of climate-neutrality by 2030, in line with the goal of Cities Mission. The project is implementing more than 80 mobility measures in ten European cities, being eight of them included in the Cities Missions program. To achieve its main objectives, the project is performing a user research to map the citizens mobility needs and patterns. As part of the qualitative user research, this paper presents a *Netnography* study, performed to investigate the citizens satisfaction with different transport modes, in five cities that are part of the Living Labs' of the UPPER project: *València, Île-de-France, Rome, Oslo* and *Mannheim*. More than fifteen thousand comments and reviews have been collected from sources like *TripAdvisor*, *Google Reviews* and *Twitter*, to assess six different transport modes: *shared bike, bus, tram&subway, taxi, shared LEV* and *shared car*. The comments have been analysed employing natural language algorithms, that allowed us to extract the *sentiment polarity* (positive, negative, neutral), and the *emotions* (anger, joy, sadness) including the *hate level*. The results are presented aggregated per transport mode for all the cities, showing differences in the way citizens perceive and are satisfied with the communal transport modes, and the individual and shared transport modes. The results also present the differences in the topics per gender.

Keywords: citizen science · *Netnography* · natural processing language (NPL) · transport modes · Cities-Missions · sentiment polarity · emotions

1 Introduction

Public transport keeps cities moving, is essential to reduce congestion and air pollution (a severe health issue, according to WHO [1]), and the cleaner air resulting from a stronger use of public transport benefits society as whole, not just those who use it. On the other hand, the objectives of the EU Green Deal [2] and the EU Cities Mission to reduce climate emissions by 55% by 2030 and to become climate-neutral by 2050 cannot be met only with car decarbonization, and strengthening the role of public transport emerges as the backbone of mobility in cities.

© The Author(s) 2025

C. McNally et al. (Eds.): TRAconference 2024, LNMOB, pp. 52–58, 2025.

https://doi.org/10.1007/978-3-031-85578-8_7

Modal shifting to collective public transport combined with active modes in daily mobility requires a change in the mindset and culture of the users. To promote citizens' behavioural change in mobility, cities need to act [3], diagnosing their initial situation, identifying users' needs, raising awareness [4, 5] and promoting transport services that facilitate new transport attitudes [6].

UPPER [7] is an innovation project aimed at implementing mobility measures in ten cities from nine EU countries, to foster sustainable mobility by employing the PT. The project includes a definition of users' requirements and the mapping of citizens' mobility needs, that we have tackled by performing a user research task, focused on identifying key points and critical factors to increase the use of PT.

As a first intervention of the user research, we have performed a collection of users' insights in an open way, by reviewing online chats and social networks following the Netnography methodology [8]. In these channels, users rate different transport modes and make comments about their mobility experience, in the cities where they live or they visit as tourists. We collected data from five living labs of the UPPER project: València, Île-de-France, Rome, Oslo and Mannheim.

The paper presents the main results generated in the Netnography study, comparing the users' satisfaction with six different transport systems, including communal and individual transport modes, with the aim of detecting aspects that generate dissatisfaction to transform them into improvements.

2   Methodology

To perform the online observation, we have applied Netnography [8]. This is an online research method aimed in understanding social interaction in contemporary digital communications contexts.

The main aim of this *Netnography* intervention was to analyse citizen transport (in its different modes), through the reuse and analysis of online data, mainly comments and assessments (ratings). The methodology consisted of analysing 5 representative cities in EU that participate in the UPPER project as Living Labs (*València, Ile de France, Rome, Oslo* and *Mannheim*). The transport modes analysed were: *Bus, Subway/Tram, Taxi, Shared bike, Shared LEV* (motorbike and/or e-scooter), and Shared car. The methodological approach followed included these steps:

1. Search for online information sources about the different modes of public transport. *TripAdvisor, Google Reviews* and also other social networks like *Twitter* were the main data sources of the study. The data was collected from middle January 2023 to the end of February 2023.
2. Web scraping techniques were employed to extract data and discern gender and residency-related characteristics (tourists vs. local residents). Language extraction and gender detection tools such as *ScrapeHero* or *Gender API* were employed, in addition to the evaluation of ratings (Fig. 1).

3. Annual review count, to assess the usage evolution.
4. Textual data analysis employing *Natural Language Processing* (NLP), represented in:

TYPE OF TRANSPORT:	SAMPLE: CITIES:										
	VALENCIA (SPAIN)		ILE DE FRANCE (FRANCE)		ROME (ITALY)		OSLO (NORWAY)		MANNHEIM (GERMANY)		TOTAL:
	Nº Reviews	Nº Comments	Nº Reviews	Nº Comments	Nº Reviews	Nº Comments	Nº Reviews	Nº Comments	Nº Reviews	Nº Comments	Nº Reviews Nº Comments
a. SHARED BIKE	387	292	1.194	1.049	-	-	49	49	32	19	1.662 1.409
b. BUS	623	363	952	512	1.087	835	251	140	44	18	2.957 1.868
c. SUBWAY /TRAM	847	847	2.923	2.923	2.377	942	459	336	187	101	6.793 5.149
d. TAXI	1.506	910	2.341	1.647	2.126	829	1.251	662	2.095	1.036	9.319 5.084
e. SHARED LEV	309	174	620	410	699	622	85	75	105	105	1.818 1.386
f. SHARED CAR	93	64	237	191	133	127	608	371	109	105	1.180 858
TOTAL:	3.765	2.650	8.267	6.322	6.422	3.355	2.703	1.633	2.572	1.384	23.729 15.344

UPPER

Fig. 1. Netnography Study Sample Size

a. Sentiment-polarity analysis: classifying comments as *positive, negative, mixed*, or *neutral* (*Amazon Comprehend* algorithm).
b. *Emotion* and *hate/aggression* level analysis of comments using the *pysentimiento* algorithm.
c. frequency of occurrence (*WordStream Maker* program).
5. Semantic analysis, reading a predefined set or a representative sample of responses, typically around 100 (expert-level categorization).
6. Extraction of characteristic verbatim, capturing representative excerpts from the comments, once the relevant topics have been identified.
7. Comparative analysis by transport mode using aggregated data.

3   Results

3.1   Results on the Use and Satisfaction of Public Transportation

This study shows that public transportation use declined during the pandemic, with a noticeable recovery starting in 2022. A strong negative correlation (-0.88) exists between rising customer reviews, indicating increased use, and decreasing average ratings, reflecting declining satisfaction levels from 2015 to 2022, dropping from an average rating of 3.8 to 3.

In terms of satisfaction, an analysis of all transportation modes across five cities yields an average rating of 3.2 out of 5.

3.2   Satisfaction Levels by Type of Transportation.

The comments classification shows that Subway/Tram, Taxi, Shared LEV and Shared car have obtained more positive comments than negative comments, while for Bus and Shared bike this ratio changes.

On the other hand, the best rating (Subway/Tram) were not obtained by the transport mode with more positive comments, but that with the best ratio positive/negative comments (3 for Subway/Tram vs. 2.5 for Taxi). So according to this ratio, and considering that positive comments and negative comments are related to fulfilling users' expectations, we get another transport mode classification where Subway/Tram and Taxi are transport modes that cover reasonably user's expectations and Shared LEV, Shared car, Shared bike and Bus does not.

3.3 Improvements Extracted from Negative and Positive Comments

In order to have a deeper understanding about what is wrong with those transport modes that are part of the second group (Fig. 2.), we can explore the terms that users are employing when they make positive and negative comments (verbatim analysis). Indeed, Fig. 2 presents the semantic analysis of the comments collected in the five living labs for the Shared Bike.

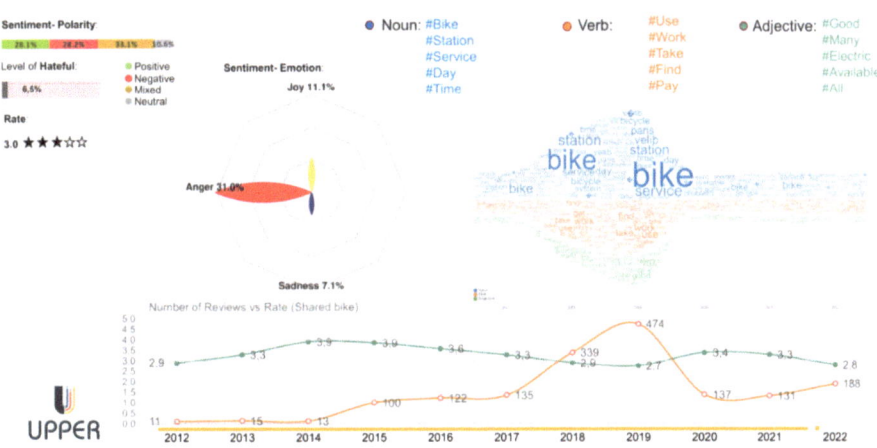

Fig. 2. Aggregated results for the Shared Bike's assessment in five UPPER's Living Lab.

The bubbles graph included in Fig. 3. Presents the terms related to positive comments and negative comments for the Shared bike. The terms *customer*, *terminal*, *broken* and *electric* used exclusively in negative comments, jointly with *bicycle*, *service*, *station* and *application* (employed in both, positive and negative comments), suggest that users consider bikes and docks are not properly maintained, e-bikes could be an interesting alternative, and the customer service should improve.

3.4 Results on Extreme Emotions (Hate) and Analysis of Gender Differences

To monitor levels of hate and aggression is critical to identify the triggering topics in user comments. Among the most frequently mentioned concerns in comments containing hate were *Ticket issues* (e.g. problems related to ticket purchase, including long queues

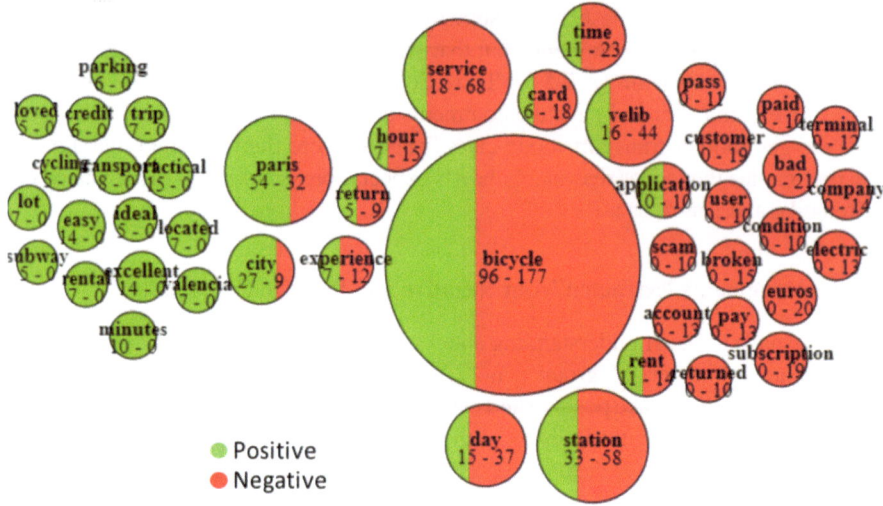

Fig. 3. Positive and Negative terms related to Shared Bike

or malfunctioning machines), *subway challenges* (e.g. concerns about security, aging of train carriages, or overcrowding), *bus problems* (e.g. complaints about old, poorly maintained, and unclean buses, or aggressive and unsafe driving), and *Station-related issues* (e.g. poorly designed and maintained stations, inaccessibility with long corridors, or difficulties for carrying luggage).

In the context of gender-specific analysis, several notable trends have emerged. Firstly, a gender bias becomes apparent in the usage of shared transportation modes (including bicycles, light electric vehicles, and cars), with approximately 67.7% of comments originating from male users. Secondly, women tend to favour buses, taxis, and subways more frequently, while demonstrating lower utilization of shared transportation options. Lastly, men exhibit a greater propensity for criticism towards public transportation, as evidenced by a higher incidence of negative feedback when compared to women.

4   Conclusions

This study provides valuable insights into the state of public transportation through non-participatory *Netnography*. We have harnessed a vast array of information, encompassing text, images, and georeferenced data, spontaneously shared by users on the internet. This approach allows us to evaluate various modes of transportation across different global cities. The information derived from *Netnography* is particularly significant, as it offers a rich blend of qualitative context and quantitative data, thanks to the sheer volume of data. Notably, *Netnography*'s insights complement those obtained from other objective and subjective studies, such as opinion surveys and sensor data. Furthermore, the extraction and analysis of this data can be systematized to create a real-time "dashboard" that provides insights into the current status and evolution of different modes of transportation.

This holistic approach enhances our ability to make informed decisions and improvements in the realm of public transportation. The comprehensive UPPER *Netnography* study is available on the project website [7].

In general, the results have shown that the European Public Transportation system is adequate and satisfactory, although there are areas for improvement depending on the type of transportation. Shared public transportation options such as bicycles, e-scooters, and cars need to enhance vehicle maintenance and upgrades, as well as provide a swift and satisfactory customer service experience. Regarding the predominant mass transit system, which includes subway and tram services, it is highly rated due to the absence of traffic congestion and shared infrastructure, unlike buses. Buses, on the other hand, require higher frequency, speed, and safer driving to meet passenger expectations. Both, buses and subway/tram services should also improve maintenance and cleanliness. Additionally, subway/tram services should prioritize safety measures against theft and other security concerns.

The authors want to thank all the *UPPER* consortium members. This project has received funding from the European Union's Horizon Europe research and innovation programme under grant agreement No [101095904].

References

1. World Health Organization - Regional Office for Europe, Data and statistics. https://www.euro. who.int/en/health-topics/environment-and-health/air-quality/data-and-statistics
2. European Commission, Communicaaion from the Commision: The European Green Deal. COM (2019) 640, Brussels (2019)
3. Promoting mobility behaviour change. Practical guidance for inspiring more walking, cycling and public transport and minimising car use (June 2019). https://ec.europa.eu/futurium/en/sys tem/files/ged/promoting_behaviour_change.pdf
4. INVOLVE, Nudge, think or shove? Shifting values and attitudes towards sustainability. A briefing for sustainable development practitioners (November 2010)
5. CIVITAS Initiative, Behavioural change & mobility management: Influencing and changing attitudes and travel behaviour through soft measures. https://civitas.eu/thematic-areas/behavi oural-change-mobility-management
6. Planning for attractive Public Transport, October 2022. https://civitas.eu/news/planning-for-attractive-public-transport-submit-feedback-now-on-the-draft-sump-topic-guide
7. https://www.upperprojecteu.eu/
8. Kozinets, R.V.: On Netnography: initial reflections on consumer research investigations of cyberculture. In: Alba, J.W.,   Wesley Hutchinson, J. (eds.) NA - Advances in Consumer Research, vol.25, pp. 366–371. Association for Consumer Research, Provo, UT (1998)

Open Access This chapter is licensed under the terms of the Creative Commons Attribution 4.0 International License (http://creativecommons.org/licenses/by/4.0/), which permits use, sharing, adaptation, distribution and reproduction in any medium or format, as long as you give appropriate credit to the original author(s) and the source, provide a link to the Creative Commons license and indicate if changes were made.

The images or other third party material in this chapter are included in the chapter's Creative Commons license, unless indicated otherwise in a credit line to the material. If material is not included in the chapter's Creative Commons license and your intended use is not permitted by statutory regulation or exceeds the permitted use, you will need to obtain permission directly from the copyright holder.

Choice of a Transport Mode for Trips to Large Environmentally Protected Areas in Czechia

Hana Brůhová Foltýnová$^{(\boxtimes)}$, Jan Brůha, and Jaromír Tonner

Faculty of Social and Economic Studies, Jan Evangelista Purkyně University in Ústí nad Labem, Ústí nad Labem, Czech Republic
hana.bruhova@ujep.cz

Abstract. In this paper, we study the determinants of the transport mode choice of the trips to and within protected areas among the Czech adult population and the potential to change from cars to more sustainable transport modes. To do so, we designed a series of choice experiments. We distinguish one-day trips by couples (without children), one-day trips by families with children, weekly visits by couples, and weekly visits by families with children. The scenarios consist of various combinations of monetary costs of trips, travel time, characteristics of the environment around the parking lot/public transport stop, and the parking fee. In addition, various socio-demographic characteristics of the respondents were collected.

Parking fees effectively reduce car usage even for travelling within protected areas. In the case of both couples and families, an increase in parking fees by 50% would decrease the share of cars by 6 p.p., and other modes (public transport, walking and cycling) would gain about 2 p.p. each. The travel time is also important: the reduction of the time spent in public transport by 20% would lead to an increase in public transport share by 3 p.p. and a decrease in car share by about 2 p.p. The relative strength of travel time and financial costs differ from travels to the protected areas. For families with children, the parameters of public transport (travel time, price) have negligible effects on the stated modal split.

Keywords: Nature protected areas · Transport mode choice · Sustainable mobility · Choice experiment

1 Introduction

Nature protected areas are faced with an increase in the number of visitors and the negative impacts associated with their transport, and these problems can be expected to further increase. The available studies indicate that for trips to these areas, the most common means of transport is clearly the car [1]. However, it is above all car usage which causes negative impacts on the environment and nature protection and the quality of the visitors´ experience [2]. The ramifications of car usage span a spectrum of issues, including heightened demand for parking spaces and instances of illegal parking, elevated risks on roads accommodating multiple users (e.g., cars, pedestrians, cyclists), and the disruption of the visual harmony of landscapes. Additionally, inappropriate parking

© The Author(s) 2025
C. McNally et al. (Eds.): TRAconference 2024, LNMOB, pp. 59–65, 2025.
https://doi.org/10.1007/978-3-031-85578-8_8

practices contribute to landscape degradation and harm to nature. Transport has an obvious negative effect on environmental protection and its components, e.g. by contributing to air and noise pollution, water and soil pollution, landscape fragmentation or pollution by possible fuel leakage or tire abrasion [3]. Achieving a delicate balance in safeguarding and enhancing resources while reconciling the diverse interests of stakeholders is imperative for fostering sustainable tourism development in nature-protected areas. This entails a concerted effort to mitigate the negative impacts of transportation while facilitating access to these areas in a manner that preserves their ecological integrity and enhances visitor experiences [4].

According to Czech data [5, 6], the car remains the predominant mode of transportation, with 73% of respondents opting for it during one-day trips and 76% during multi-day stays, whether for the entire weekend or a full week. In contrast, public transport (hereinafter, PT) is the primary choice for a mere 8% of respondents during one-day trips, 7% for weekend getaways, and 6% for week-long vacations. Some respondents utilize combinations of different transport modes alongside a car, such as car and PT (6%) or car and bike/e-bike (3%). Despite the availability of alternatives, it is evident that car usage significantly contributes to negative impacts on the environment and nature conservation efforts.

In this paper, we investigate the factors influencing the choice of transportation modes for trips to and within protected areas among the adult population of the Czech Republic. Our aim is to explore the potential for shifting from car-centric transportation to more sustainable modes. To accomplish this, we conducted a series of four choice experiments:

1. The first experiment centered on one-day trips undertaken by couples (without children).
2. The second experiment focused on one-day trips taken by families with children.
3. The third experiment examined weekly visits made by couples.
4. The fourth experiment delved into weekly visits by families with children.

Through these experiments, we sought to discern the determinants guiding transport mode preferences and assess the feasibility of transitioning towards more eco-friendly transportation options within protected areas.

The next chapter describes data and methodology used, Sect. 3. presents and discusses results of analyses and the final chapter concludes.

2 Data and Methodology

The data collection process involved administering an online questionnaire to a sample of the Czech adult population from May 19 to May 30, 2022. Participants were selected based on specific criteria: they must have visited a protected area within the year preceding the survey and had access to a car for their visits. This yielded a final sample size of 1024 respondents. Table 1 provides a summary of descriptive statistics for the sample, offering insights into key characteristics of the respondents.

The respondents had to state their preferred travel mode to and within a protected area. The set of possible options for travelling to the protected area for one-day trips includes

Table 1. Descriptive statistics of selected socio-demographic traits

	Frequency	Percentage
Gender		
Female	270	47%
Male	241	53%
Age		
18–29	76	15%
30–44	325	64%
45–59	88	17%
60 +	22	4%
Education		
Elementary; high school (without a diploma)	180	35%
High school graduate (with a diploma)	186	36%
University degree	145	28%
Income		
Below average	51	10%
Average	359	70%
Above average	84	16%
Do not want to answer	17	3%
Cars available in the household		
0	9	2%
1	278	54%
2	200	39%
3 +	24	5%

Source: Own computation on survey data

car, public transport and the combination of car and public transport. The set of possible options for travelling inside protected areas consists of car, public transport, cycling and walking. The scenarios consist of various combinations of monetary costs of trips, travel time, characteristics of the environment around the parking lot/public transport stop, and the parking fee. In addition, various socio-demographic characteristics of the respondents were collected. We estimate a discrete choice model that explains the respondents´ stated preferences regarding the mode choice. We ask what factors would be effective for reducing car transport.

3 Results and Discussion

In Table 2, we present the results derived from the multinomial logit model. Incorporating an income variable for above-average earning groups yields intuitive insights. Specifically, the cost parameter alone is negative, while the cost parameter multiplied by the categorical variable income is positive (and usually smaller in absolute value), suggesting that people with higher incomes are less sensitive to the expenditure costs.

Across generalized cost factors such as journey duration, financial costs, parking fees, and travel time, the expected negative signs align with theoretical expectations. Essentially, an increase in the cost associated with a particular transportation option, while holding other costs constant (such as those for cars, public transport, combinations of both, bicycling, or walking), corresponds to a decreased probability of selecting that option.

For week-long vacations, all cost components including travel time, financial costs, parking fees, and duration of reaching the destination prove statistically significant. However, for one-day trips for couples, financial costs emerge as statistically insignificant.

The analysis of the surroundings was conducted separately, revealing that Area 3 holds particular significance. This area entails the availability of a kiosk offering refreshments at the boarding point, along with access to separate pedestrian infrastructure, especially appealing for one-day trips by couples and families. The positive coefficient suggests that the likelihood of visiting a location within this neighborhood surpasses that of other neighborhoods. For families embarking on one-day trips, there is a clear aversion to environments characterized by walking along busy roads. Conversely, there is a strong preference for settings featuring amenities like a snack kiosk at the boarding point and access to dedicated pedestrian pathways. When it comes to week-long holidays, both couples and families exhibit a preference for surroundings equipped with separate pedestrian infrastructure.

Age was examined separately, revealing significant associations with travel preferences. Specifically, for day trips by couples, there's a discernible trend where older age correlates with a decreased preference for public transport. Similarly, in the context of week-long stays for couples, advancing age corresponds to a reduced inclination towards walking. For families, age also emerges as a significant factor influencing travel choices. In the case of day trips, older family members exhibit a decreased preference for a combination of car and public transport. Furthermore, when considering week-long vacations, older family members tend to show a heightened preference for walking. This highlights the nuanced impact of age on travel preferences, suggesting differing priorities and comfort levels across various demographic groups.

The analysis of the highest level of education unveils intriguing patterns in transportation preferences. Notably, couples and families with higher education opt for public transport for both one-day trips and week-long stays. Additionally, couples with advanced education exhibit a strong preference for cycling during weekly vacations, with walking trips as a secondary choice. Conversely, families with higher education solely favor walking over cycling for their week-long vacations, indicating a distinct preference within this demographic group. These findings underscore the influence of educational attainment on travel choices, suggesting a correlation between higher education and a propensity towards utilizing public transport, cycling, and walking for both short and extended trips.

Table 2. Results for the experiments using multinomial logit models.

Parameters	Couples: one-day trips	Families: one-day trips
(Intercept) × Car+PT	-0.711**	-1.089***
(Intercept) × PT	-0.755*	-1.352***
Time of the trip	-0.008*	-0.007
Costs of the trip	-0.003+	-0.007***
Costs above average income	0.004**	-0.001
Parking costs	-0.002***	-0.002***
Time: walking to the starting point	-0.014*	-0.012*
Utility: travel time by car	0.006+	0.005
Utility: travel time by PT	0.000	-0.004
Utility: surrounding – Car	0.012	-0.008
Utility: surrounding – PT	-0.020	-0.072**
Utility: surrounding – Car+PT	-0.081*	-0.118***
Statistics		
Num. of Observations	7665	7695
AIC	5065	4882
RMSE	0.613	0.594
R2	0.026	0.027

Parameters	Couples: weekly holidays	Families: weekly holidays
(Intercept) × PT	0.069	-0.253
(Intercept) × Bike	-0.627+	-0.981**
(Intercept) × Walk	0.986**	0.098
Time of the trip	-0.020***	-0.014***
Costs of the trip	-0.002***	-0.001***
Costs above average income	0.001	0.001*
Parking costs	-0.002***	-0.001***
Utility: surrounding – Car	-0.001	-0.271**
Utility: travel time by car	0.011*	0.001
Utility: travel time by PT	0.004	0.002
Utility: travel time by Bike	0.014*	0.009
Utility: surrounding – PT	-0.247**	-0.474***
Utility: surrounding – Bike	-0.189	-0.323**
Utility: surrounding – Walking	-0.365***	-0.418***
Statistics		
Num. of Observations	10220	10260
AIC	6708	6743
RMSE	0.719	0.719
R2	0.018	0.019

Source: Own calculations

Note: *** significant at 1%, ** significant at 5%, * significant at 10%, + significant at 15%.

4 Conclusion

Parking fees demonstrate a significant influence on car usage, even within large environmentally protected areas. For both couples and families, a 50% increase in parking fees would result in a notable decrease in the share of cars by 6 percentage points (p.p.), while other modes of transportation such as public transport, walking, and cycling would each gain approximately 2 p.p. in share.

In the case of couples, the impact of public transport pricing is minimal; even if public transport were made free, its share would only increase by less than 2 p.p. However, travel time proves to be a more influential factor. A 20% reduction in the time spent on public transport would lead to a 3 p.p. increase in its share and a decrease in the share of cars by about 2 p.p. This indicates a shift in the relative importance of these factors for travel within protected areas.

On the other hand, for families with children, the parameters related to public transport, such as travel time and price, have negligible effects on the stated modal split. This suggests that other factors may play a more significant role in shaping transportation choices for this demographic group.

Acknowledgement. This paper is enabled by support from TA CR Grant no. CK01000067 "Analysis of alternative solutions for mobility planning in environmentally sensitive areas".

References

1. Brůhová Foltýnová, H., Rybová, K., Timoftej, R., Vácha, O., Jordová, R.: Potential of sustainable mobility management in Czech large environmentally sensitive areas. In: Shibayama, T., Emberger G. (eds.), Beiträge zu einer ökologisch und sozial verträglichen Verkehrsplanung. 1/2022. Institut für Verkehrswissenschaften. Forschungsbereich für Verkehrsplanung und Verkehrstechnik, Technische Universität Wien, 37–46. https://www.fvv.tuwien.ac.at/forschung/publikationen/institutsschriftenreihe/wctr-sig-g2-2021 (2022)
2. González, R.M., Román, C., Ortúzar, J.D.: Preferences for sustainable mobility in natural areas: the case of teide national park. J. Transp. Geogr. **76**, 42–51 (2019). https://doi.org/10.1016/j.jtrangeo.2019.03.002
3. Kramer, M.G.: Our Built and Natural Environments: A Technical Review of the Interactions Among Land Use, Transportation, and Environmental Quality. U.S. Environmental Protection Agency (EPA), 2nd edition, (2013). https://www.epa.gov/sites/default/files/2014-03/documents/our-built-and-natural-environments.pdf
4. Buckley, R.: Neat trends: current issues in nature, eco- and adventure tourism. Int. J. Tour. Res. **2**, 437–444 (2000)
5. Timoftej, R., Brůhová Foltýnová, H.: Possibilities of sustainable mobility and tourism management in large environmentally sensitive areas in the Czech Republic. Czech Journal of Tourism **9**(1), 68–82 (2020)
6. Brůhová Foltýnová, H., Brůha, J., Tonner, J.: What determines the travel behaviour of visitors of large environmentally protected areas? A case study from Czechia. International Journal of Tourism Research. Under review

Open Access This chapter is licensed under the terms of the Creative Commons Attribution 4.0 International License (http://creativecommons.org/licenses/by/4.0/), which permits use, sharing, adaptation, distribution and reproduction in any medium or format, as long as you give appropriate credit to the original author(s) and the source, provide a link to the Creative Commons license and indicate if changes were made.

The images or other third party material in this chapter are included in the chapter's Creative Commons license, unless indicated otherwise in a credit line to the material. If material is not included in the chapter's Creative Commons license and your intended use is not permitted by statutory regulation or exceeds the permitted use, you will need to obtain permission directly from the copyright holder.

Preferences of Rural Travelers Towards Demand Responsive Transport

Dániel Tordai[1,2]($\boxtimes$) , József Pál Lieszkovszky[1] , and András Munkácsy[1]

[1] KTI Hungarian Institute for Transport Sciences and Logistics, Budapest, Hungary
tordai.daniel@kti.hu

[2] Department of Transport Technology and Economics, Budapest University of Technology and Economics, Budapest, Hungary

Abstract. Demand responsive transport (DRT) has received significant attention in recent years as a transport mode that can bridge the gap between personal motorized travel and public transport. It combines the best of individual cars and public transport: it can be more flexible, than traditional public transport, resembling the convenience of owning a car, but it does not come with the high cost of owning and maintaining a vehicle for the user. We conducted a stated preference (SP) survey about potential users' preferences towards demand responsive transport at a rural Eastern European town in Hungary, Kiskunhalas, Hungary, and modelled individuals' preferences towards DRT using a multinomial logit model. We had 6012 responses from a sample of 501 individuals, that was representative of the settlement with respect to age and gender. The results show that all else being equal, individuals find the DRT service the second most attractive mode of transport after cycling, meaning that a DRT service could have a potentially large uptake in the population. Our results could be used by decision makers and service providers for the design of a DRT service.

Keywords: rural transportation · demand responsive transport · choice modelling

1 Introduction

Demand responsive transport (DRT) has received significant attention in recent years as a transport mode that can bridge the gap between personal motorized travel and public transport. This type of transport service has no fixed schedule; instead, passengers have to signal their intention to travel to the service provider in advance, via either a phone call or a mobile app, otherwise the vehicle might not depart. DRT can have different configurations, some have fixed stops, others' departure and arrival points are flexible, some have predefined schedule, others operate without one.

DRT combines the best of individual cars and public transport: it can be more flexible than traditional public transport, resembling the convenience of owning a car, but it does not come with the high cost of owning and maintaining a vehicle for the user. Looking at it from a service provider point of view, providing services with a flexible

© The Author(s) 2025

C. McNally et al. (Eds.): TRAconference 2024, LNMOB, pp. 66–71, 2025.
https://doi.org/10.1007/978-3-031-85578-8_9

schedule instead of a fixed one can result in significant cost reduction, making public transportation service more viable at more sparsely populated areas. Providing high quality public transportation in rural regions have been a challenge financially, and DRT can provide a solution.

Understanding potential users' taste is key for the design of a successful service, by helping to set the parameters of the service according to the users' needs. Over the last decade, a number of studies have looked at preferences of potential users towards DRT services, i.e. Alonso-González et al. [1], Choudhury et al. [2], or Frei et al. [3]. However, all of these studies investigate users' attitude in dense urban or sub-urban environments, where high quality public transport is already present, or could be provided in a financially sustainable way, and to our knowledge, no study has been conducted in rural area, where this type of service could be the most beneficial, and where transport demand has different characteristics. Bronsvoort et al. [4] estimate users' preferences in rural areas of the Netherlands, but due to the nature of that country, that can be still characterized as a relatively densely populated area comparted to other rural regions in the world. König & Grippenkoven, [5] analyzed a household survey conducted in rural Germany, but used different modelling techniques, and did not estimate utility functions. Schasché et al. [6] have done a review on the factors influencing user acceptance of rural demand responsive transport.

2   Data and Methods

We are conducted a stated preference (SP) survey about potential users preferences towards demand responsive transport at Kiskunhalas, Hungary. This town has about 26 000 inhabitants, and is located at the southern part of Hungary, one of the most sparsely populated areas of the country. Major cities are more than 50 km away, only smaller settlements can be found in its close neighborhood. The town is completely flat and has good cycling infrastructure along the main roads of the settlement, so it is very suitable for active travel modes. Local, and interurban transportation is provided by buses. The settlement has two railway lines, one of which is closed for years due to a major reconstruction, while the other only has a minor role in public transportation provision.

We conducted a household survey with 501 inhabitants of the town, a representative sample of its citizens in terms of age and gender, asking them to make decisions in mode choice scenarios. First, we asked the respondents about their most typical travel: what is the purpose of this travel, where is their destination, what transportation mode do they use, and how long does this trip usually take. Based on the responses to these questions, we asked the respondents to make decisions in hypothetical mode-choice scenarios with regard to their most typical travel. The available modes depended on the answers of the individuals: in case the most typical travel happens within the settlement the possible alternatives were passenger car, local bus, bicycle, and walking and demand responsive bus, in case the most typical travel was inter-urban, local bus option was replaced by coach, and walking was not an option. The alternatives had different attributes, as can be seen in Table 1. Each attribute had three levels, each chosen to be realistic to the reported typical travel.

Table 1. Alternatives and their attributes in the SP survey

Alternative	Attributes
car	travel time, cost
bus, demand responsive bus	travel time, cost, frequency, egress and excess time
walking, cycling	travel time

The demand responsive bus services was described to the respondents in details: it would run on a flexible route with predefined stops, and would depart only in case a passenger indicated her intent to travel in advance. In the SP experiment, demand responsive bus service had higher frequency, shorter travel time, and mostly shorter egress and excess walking time, but higher cost than the traditional bus service, to be able to capture the trade of between the two different type of bus services.

To estimate the potential users' preferences, we used a multinomial logit model. These preferences are represented by utility functions that can be written in the following forms for each mode:

$$U_{car} = asc_{car} + \beta_{travel\ time} * travel\ time + \beta_{cost} * cost \tag{1}$$

$$U_{bus} = asc_{bus} + \beta_{travel\ time} * travel\ time + \beta_{egr} * egress + \beta_{cost} * cost + \beta_{frequency} * frequency \tag{2}$$

$$U_{drtbus} = asc_{drtbus} + \beta_{travel\ time} * travel\ time + \beta_{egr} * egress + \beta_{cost} * cost + \beta_{frequency} * frequency \tag{3}$$

$$U_{bicycle} = asc_{bicycle} + \beta_{travel\ time} * travel\ time \tag{4}$$

$$U_{walk} = asc_{walk} + \beta_{travel\ time} * travel\ time \tag{5}$$

where *asc* stands for alternative specific constant, *travel time* is in vehicle travel time, *cost* is financial cost, *egress* is the time it takes to walk to and from the bus stops, and *frequency* shows how often the bus serves the given route.

3   Results

Out of the 501 respondents, 471 stated that their most typical travel occurs within Kiskunhalas. Out of these 471 respondents, 174 had chosen the demand responsive at least once in the SP study. In the case of the intra urban travelers, it was 9 out of 30 how have chosen at least in one choice situation the on-demand bus service.

In terms of the modal split of the most typical travel, 222 respondents reported to use a car either as a driver or as a passenger. Notably, 226 people reported to use some form of active mobility for this most typical travel, such as cycling or scooter. Further 26 people walk, and 22 uses public transport. In terms of the goal of their travel, 293 respondents' most typical travel is work, while the second most populous category is shopping with 152 responses. School, healthcare, administration and the option 'other' were selected only in a few dozen cases.

The results of our choice model can be seen in Table 2:

Table 2. Results of the multinomial logit model

asc_{car}	2.32***
	(0.155)
asc_{bus}	2.06***
	(0.208)
$asc_{drt\ bus}$	2.84***
	(0.161)
asc_{bike}	3.68***
	(0.146)
β_{cost}	0.0019***
	(0.000281)
$\beta_{travel\ time}$	−0.0625***
	(0.00397)
$\beta_{frequency}$	−0.0157***
	(0.00147)
β_{egr}	−0.0347***
	(0.0132)

Standard errors in parentheses.

The results show that everything being equal, residents prefer cycling the most out of the available modes, and that is followed by the hypothetical demand responsive bus service. This is a very encouraging result for the potential new service, showing that it could potentially be very competitive with other modes. The least preferred mode is walking (the alternative specific constant of which mode has been normalized to zero). This can be explained by the fact that the town is rather dispersed with relatively large distances within the settlement, so that mode is usually not competitive with the others.

Travel time, frequency and egress time have the expected negative sign. An interesting result is that the coefficient of egress time is way smaller than the coefficient of in vehicle travel time, which shows that individuals face a larger disutility while on a vehicle than on the way to/from the stop.

The coefficient of cost is positive. The reason behind this phenomenon has to be investigated further, but as a consequence, no meaningful value of travel time can be estimated.

4 Conclusions

In this contribution we have estimated user's preferences of potential rural travelers towards a hypothetical on-demand bus service by using a stated preference survey.

As most typical travels by this city's residents are done locally, within the settlement the local decision makers can have a large impact using traffic management tools. Introducing more attractive public transportation services can reduce car dependency and reduce traffic in the town.

Based on our results, a number of policy recommendations could potentially be formulated. It can be seen from the results that a relatively large portion of individuals did not find demand responsive transport attractive at a pricing level that resembled the price of the normal bus service. This might be due to the unfamiliarity of this type of service to the local community, which might make individuals overestimate the mental cost of indicating the will to travel to the service provider in advance. In case a DRT service would be introduced, a well targeted advertisement campaign could play a key role in informing individuals about the nature of a service like this. However, the large alternative specific constant of the DRT service shows that those who would use such a service would also prefer it compared to the traditional bus service.

On the one hand, the modal share of public transit is relatively low in this settlement, as it can be expected in an urban environment with low population density and regular, but not very frequent bus service, which means that there might be room for the demand responsive transit to take up potential users from that group that travels by car currently. However, the design of the service has to pay attention to the fact that 50% of the travelers reported to use some sort of active travel, either cycling or walking, which is already very environmentally friendly and cost effective on a societal level. Making them switch to DRT, a motorized and subsidized on-demand bus service would result a decrease of social welfare and would go against potential environmental goals. Designing a service that is attractive to current public transportation users and car users, but not for people walking or cycling is a challenge, but our research can help decision makers set up the service properly. For example, the fact that travelers show higher affection towards shorter egress and excess times than in other cities can mean that a high number of stops should be defined as part of the demand responsive service, so the service can be attractive to a larger audience.

Further research should work out the financial details of the service based on the demand that can be estimated from the results of this study and show the potential financial benefit for the service provider of converting the traditional bus services into a demand responsive one, as this question is out of the scope of this paper. Another research direction could be to use RP data in a rural town where on-demand bus service is already introduced and estimate users' preferences based on real-life mode choice data.

References

1. Alonso-González, M.J., van Oort, N., Cats, O., Hoogendoorn-Lanser, S., Hoogendoorn, S.: Value of time and reliability for urban pooled on-demand services. Trans. Res. Part C: Emerging Technol. **115**, 102621 (2020). https://doi.org/10.1016/j.trc.2020.102621
2. Choudhury, C.F., Yang, L., de Abreu e Silva, J., Ben-Akiva, M.: Modelling preferences for smart modes and services: a case study in Lisbon. Trans. Research Part A: Policy Pract. **115**, 15–31 (2018). https://doi.org/10.1016/j.tra.2017.07.005
3. Frei, C., Hyland, M., Mahmassani, H.S.: Flexing service schedules: assessing the potential for demand-adaptive hybrid transit via a stated preference approach. Trans. Res. Part C: Emerging Technol. **76**, 71–89 (2017). https://doi.org/10.1016/j.trc.2016.12.017
4. Bronsvoort, K., Alonso-González, M., Van Oort, N., Molin, E., Hoogendoorn, S.: Preferences toward bus alternatives in rural areas of the netherlands: a stated choice experiment. Transp. Res. Rec. **2675**(12), 524–533 (2021). https://doi.org/10.1177/03611981211029919
5. König, A., Grippenkoven, J.: The actual demand behind demand-responsive transport: Assessing behavioral intention to use DRT systems in two rural areas in Germany. Case Stud. Trans. Policy **8**(3), 954–962 (2020). https://doi.org/10.1016/j.cstp.2020.04.011
6. Schasché, S.E., Sposato, R.G., Hampl, N.: The dilemma of demand-responsive transport services in rural areas: conflicting expectations and weak user acceptance. Transp. Policy **126**, 43–54 (2022). https://doi.org/10.1016/j.tranpol.2022.06.015

Open Access This chapter is licensed under the terms of the Creative Commons Attribution 4.0 International License (http://creativecommons.org/licenses/by/4.0/), which permits use, sharing, adaptation, distribution and reproduction in any medium or format, as long as you give appropriate credit to the original author(s) and the source, provide a link to the Creative Commons license and indicate if changes were made.

The images or other third party material in this chapter are included in the chapter's Creative Commons license, unless indicated otherwise in a credit line to the material. If material is not included in the chapter's Creative Commons license and your intended use is not permitted by statutory regulation or exceeds the permitted use, you will need to obtain permission directly from the copyright holder.

Empowering Sustainable Tourism Mobility Through Multimodal Transportation in Rural Areas

Mirjam Baumann[✉], Iljana Schubert, and Andrea Del Duce

ZHAW - Zurich University of Applied Science, Winterthur, Switzerland
mirjam.baumann@zhaw.ch

Abstract. The mobility sector is a major contributor to climate change in Switzerland which is mostly due to its high reliance on private passenger cars. While initiatives, policies and alternative mobility solutions aimed at promoting car reduction are growing in urban areas, rural regions, which face unique challenges due to higher car dependency and limited alternatives, are lagging behind. This paper investigates current mobility behaviour, needs, and attitudes towards new mobility services in rural regions focusing on the Swiss touristic region of Toggenburg. A survey conducted among tourists (N = 572) travelling to Toggenburg identifies several key determinants influencing car use reduction. In particular, it reveals that women, low-income households and resident of more urban areas tend to be more receptive to the idea of reducing their car use. Other factors contributing to this receptiveness include biospheric and egoistic values, as well as the availability and affordability of alternative transportation options, such as car sharing or on-demand services.

Keywords: rural region mobility concept · tourism transport · sustainable mobility · decarbonization · multimodal mobility

1 Introduction

The mobility sector significantly contributes to greenhouse gas emissions (GHG) across Europe, including Switzerland, where it accounts for about 30% of total emissions [1], with private cars being responsible for the largest share of GHG. Moreover, in the context of leisure mobility, more than two-thirds (69%) of trips are made by car in Switzerland [2]. While the electrification of passenger cars is expected to play a key role in decarbonising the mobility sector, there also needs to be a shift towards higher use of public and active transport, of carpooling and sharing of vehicles - hence multimodal mobility. Although cities are implementing policies to increase public transport use and multimodal options, rural regions, which have higher car dependency and less available alternatives, are facing bigger challenges [3]. One primary disadvantage of public transport in rural areas is the absence of railway stations at start or end-points of individual leisure journeys, called the "first or last mile" problem [4]. Solutions may

© The Author(s) 2025
C. McNally et al. (Eds.): TRAconference 2024, LNMOB, pp. 72–78, 2025.
https://doi.org/10.1007/978-3-031-85578-8_10

include (on-demand) shuttles or bike-sharing systems, e.g. providing tourists with convenient options for first and last mile problems in rural areas [4–6]. Choice experiment research into transportation preferences, focusing on Swiss rural tourism, has shown that car-dependency could be reduced if attractive alternatives, like electric bicycles or shuttles, are available [6]. Additionally, on-demand transport services or car sharing options at public transportation endpoints can reduce reliance on cars for personal travel, thus allowing tourists easier access to public transport [4, 7, 8].

While these examples highlight that multimodal approaches can and should play a significant role in rural areas, in practice rural mobility is still dominated by cars. Additionally, strong habits and beliefs of no viable alternative to cars play a relevant part in the resistance to change [3]. Further knowledge of mobility users' needs and attitudes towards alternative mobility approaches is required to understand barriers to adopting novel mobility solutions and how local contexts can influence behaviour change [7].

Thus, focusing on the rural region of Toggenburg in Switzerland, we answer the following research questions (RQs): RQ1 - What are rural tourists' mobility behaviours, attitudes and needs? RQ2 - Which core values and socio-demographic characteristics explain the openness towards car use reduction? RQ3 - What factors affect tourists' willingness to change towards alternative rural transportation services? Results can support mobility providers, tourism organisations and policy makers in developing car reduction paths and tailored sustainable mobility solutions for rural areas.

2   Methodology

2.1 Survey and Data Collection

We used an online survey to collect data from tourists visiting a rural area of Toggenburg (Oberes Toggenburg), known for its hiking and skiing possibilities, from 26.12.2022 to 19.02.2023. It must be noted that these hiking and skiing resorts cannot be reached directly by train but that a change from train to a bus is needed when travelling by public transport. We gathered information on tourists' mobility behaviour, details regarding their Toggenburg trip, such as mode of transport, openness to using alternative modes, luggage transport needs, duration of stay and purpose. Additionally, we collected information on openness to reduce car use (using the stage model of behaviour change [9]), core values (biospheric, altruistic, egoistic and hedonic), socio-demographic characteristics and willingness to reduce car use.

The target group consisted of individuals visiting recreational areas for hiking and skiing. The survey was advertised via flyers in local restaurants, cable cars, hotels, and the tourism office. Participation was incentivised through a raffle entry to win 1 of 12 free one-day ski passes for each completed survey. A total of 572 respondents participated.

2.2 Sample and Analysis

Gender and age distribution of respondents are balanced, with 49.9% male and 49.2% female participants and across various well-represented age groups (18–29: 12.2%; 30–44: 43.4%; 45–65: 38.2%; > 65: 6.1%). However, the sample shows a relatively high level of education and income, with 45.7% holding a university degree and 46.5% belonging to the higher income groups ($\geq$ 9,000 Swiss Francs/months).[1] Most tourists are from urban areas (40.5%), with 45.2% coming to the region for day trips and 24.7% spending 4–7 days. To answer our RQs, we perform descriptive analysis (RQ1) and estimate a multinomial logistic regression (RQ2) and a binary logistic regression (RQ3) using SPSS.

3 Results

With respect to RQ1, we find that most tourists (92.7%) report travelling to the hiking and skiing resorts comprised in the study by car. Only 7.3% of respondents use alternative modes such as public transport, bicycles, or motorcycles. Findings show a differentiation in transport modes depending on activity, although the car is always the main choice. Specifically, 11% of hikers and 33% of mountain bikers report using alternative modes of transport, and the minority of ski tourists (5.7%) arrive by means other than car.

When asked how important it is to travel in an environmentally-friendly way, 73.4% state that it is important or rather important. 12.7% say they don't know and 13.9% mention that it is less important or not important. With respect to the openness to change (i.e., to reduce their car use) 46.7% of respondents answer that they primarily use their car for most of their trips and are satisfied with their current level of car use, hence they see no reason to reduce it (category "none", Table 1 reference category). 33.7% of participants express a desire to reduce their car use but lack knowledge or have not yet started using alternative modes of transportation (category "yes-not active", Table 1). Finally, 19.7% either regularly use alternative transport, or do not own a car (category "yes-active", Table 1). When asked if they would be willing to drive less (Table 2), 25.7% said yes, and the remaining 74.3% said no. These results show a discrepancy between the expressed importance for travelling in an environmentally-friendly way and the openness or willingness to change mobility behaviour regarding car use.

To answer RQ2, we estimate the relationship between openness to change car use (none (reference), yes-not active, yes-active) with socio-demographics and core values using a multinomial logistic regression (Table 1). Results indicate that high biospheric values are positively related to the openness to reduce car use, whereas high egoistic values act as deterrents against reducing car usage. Furthermore, people living in suburban areas were less open to reduce their car use, compared to those in urban areas. For those already using alternative transport or who don't own a car (yes-active), a similar pattern emerges in terms of values and spatial typology. Tourists from rural or semi-rural areas are less likely to opt for alternative modes. Income is also a key factor here, as lower-income households are more prone to using alternatives.

---

[1] 1 Swiss Franc $\approx$ 1.1 Dollar.

Table 1. Multinomial logistic regression: stage model of car usage behaviour change

Category	Variable	B	SE	p-value
Yes- not active (Desire to reduce car usage but have not yet started[a])	Biospheric value	0.664	0.254	**0.009**
	Egoistic value	−0.490	0.151	**0.001**
	Spatial typology (rural)	−0.616	0.362	0.089
	Spatial typology (suburban)	−0.678	0.317	**0.033**
	Spatial typology (urban)	Reference category		
Yes- active (Use alternative transport modes or do not own a car [b])	Biospheric value	1.125	0.351	**0.001**
	Egoistic value	-0.360	0.180	**0.045**
	Income (<6'000)	1.177	0.466	**0.012**
	Income (6'000–8'999)	0.105	0.396	0.792
	Income (9'000 +)	Reference category		
	Spatial typology (rural)	−1.020	0.460	**0.026**
	Spatial typology (suburban)	−1.072	0.396	**0.007**
	Spatial typology (urban)	Reference category		

Model fit: Chi2(24, N = 327) = 79.624, Nagelkerke R2 = 0.246, p <.001

Reference category: No desire to reduce car usage (none): 1st predecisional. [a]Combined stages: 2nd predecisional, preactional, or actional stage, [b] Combined stages: postactional /captives; [9]. *Bold figures indicate significant values.*

To answer RQ3, we investigated factors explaining tourists' willingness to change their behaviour and reduce their car usage through a binary logistic regression analysis (Table 2).

Table 2. Binary logistic regression: willingness to reduce car use

Variable	B	SE	p-value
Gender (0 = male, 1 = female)	−0.963	0.275	**0.000**
Spatial typology			0.114
Spatial typology (rural)*	−0.792	0.382	**0.038**
Spatial typology (suburban)*	−0.485	0.333	0.145

(*continued*)

Table 2. (*continued*)

Variable	B	SE	p-value
Biospheric value	0.693	0.237	**0.004**
Car sharing: Car sharing stations in Nesslau (end point of train)	1.516	0.584	**0.009**
Car sharing: Cheaper fares e.g. for weekend trips	0.976	0.304	**0.001**
On-demand service: Attractive prices	0.516	0.273	0.059
On-demand service: Provision of information	1.531	0.485	**0.002**
On-demand service: booking option via an app	0.681	0.308	**0.027**
Model fit: Chi2(12, N = 371) = 65.663, Nagelkerke R^2 = 0.236, p <.001			

[*] Reference category: urban
What would you like to change about the way you travel? drive less.
Bold figures indicate significant values.

In this case, gender emerged as a significant factor, indicating that women were significantly more willing to reduce their car use. Although the overall spatial typology variable did not show any significance, residents of rural regions showed a significantly lower willingness to reduce car use compared to city dwellers. Biospheric values continued to play an important role in this context. Furthermore, several key factors were identified that could increase people's willingness to reduce car usage in the touristic region. Firstly, the availability of more car sharing options, particularly at the terminal train station of Nesslau, which lies about 10 to 15 km away from the major touristic attractions in the area, along with more affordable car sharing options, especially for weekend trips. An on-demand service with convenient booking options via an app and the provision of comprehensive information were also identified as key factors. Interestingly, improvements in public transport or carpooling, as well as bike-sharing options did not significantly influence people's willingness to reduce car use.

4 Discussion and Conclusion

Our results reveal that more than 90% of the tourists travelling to the hiking and skiing resorts in Toggenburg covered by the study rely heavily on private cars, which is significantly higher than the average use of private cars for leisure travel in Switzerland (69%) [2]. This suggests that touristic mobility in rural areas deserves special attention because of its potential for high car dependency, even though almost three quarters of the participants stated that environmentally-friendly mobility behaviour was important or rather important to them. It is plausible that many people want to use eco-friendly transportation modes but are unaware of the impacts from their private cars as opposed to available sustainable alternatives. To overcome this attitude–behaviour gap, a promising approach involves prioritizing information campaigns and sustainable communication [10].

With respect to values, participants with strong biospheric values, rooted in environmental concerns, were more likely to consider reducing their car use. Conversely, those with egoistic value orientations, which prioritize individual convenience and comfort, were less inclined to embrace such change. These results highlight the importance of aligning sustainability efforts with personal values to drive behaviour change effectively. Overall, many people have strong biospheric values [11] and intervention efforts should focus on bringing biospheric values to the foreground, making them more salient.

In the area of innovative mobility solutions, the integration of car sharing and on-demand services with public transport, accompanied by user-friendly booking processes and comprehensive information, has been shown to significantly increase the willingness to reduce car use. These findings align with previous research, reinforcing the potential of these alternatives to transform mobility behaviours [7, 8]. It's worth noting that the choice of transportation mode also depends on the purpose of the trip. When undertaking activities such as skiing that require a significant amount of luggage, it is essential that alternative options take into account the need for luggage transportation [12]. Overall, the results suggest a complex interplay between socio-demographic factors, personal values, specific multimodal criteria and information campaigns as influential elements in the shaping of attitudes towards car use reduction. Understanding these nuances can help policymakers and transport planners to design effective strategies to promote sustainable and environmentally friendly tourism transportation in the future.

Acknowledgements. The project was funded by the Swiss Federal Office of Energy. Responsibility for content and conclusions lies only with the authors. We are grateful to Marius Schmidt and Markus Erne (SOB), Daniel Wittenwiler (Energieagentur St. Gallen), as well as Raphael Hoerler and Uros Tomic (ZHAW) for feedback and support in our research work.

References

1. Bafu, K.: Das Wichtigste in Kürze. https://www.bafu.admin.ch/bafu/de/home/themen/thema-klima/klima--das-wichtigste-in-kuerze.html, Accessed 8 Sep 2023
2. BFS. Mobilitätsverhalten der Bevölkerung. Ergebnisse des Mikrozensus Mobilität und Verkehr 2021. https://www.bfs.admin.ch/asset/de/24165261, Accessed 8 Sep 2023
3. Baumgartner, A., Schubert, I., Sohre, A., Tomic, U., Moser, C., Burger, P.: Toward a reduction of car-based leisure travel: an analysis of determinants and potential measures. Int. J. Sustain. Transp. **17**(8), 911–930 (2023)
4. Bauchinger, L., Reichenberger, A., Goodwin-Hawkins, B., Kobal, J., Hrabar, M., Oedl-Wieser, T.: Developing sustainable and flexible rural–urban connectivity through complementary mobility services. Sustainability **13**(3) (2021)
5. Scappini, B., Zucca, V., Meloni, I., Piras, F.: The regional cycle network of Sardinia: Upgrading the accessibility of rural areas through a comprehensive island-wide cycle network. Eur. Transp. Res. Rev. **14**(1), 10 (2022)
6. Curtale, R., Sarman, I., Evler, J.: Traffic congestion in rural tourist areas and sustainable mobility services. the case of ticino (Switzerland) Valleys. Tourism Planning Develop., 1–25 (2021)

7. Poltimäe, H., Rehema, M., Raun, J., Poom, A.: In search of sustainable and inclusive mobility solutions for rural areas. Eur. Transp. Res. Rev. **14**(1), 13 (2022)
8. Shibayama, T., Lemmerer, H., Winder, M., Pfaffenbichler, P.: Cooperative car sharing in small cities and scarcely populated rural area—An experiment in Austria, vol. 198 (2013)
9. Bamberg, S.: Applying the stage model of self-regulated behavioral change in a car use reduction intervention. J. Environ. Psychol. **33**, 68–75 (2013)
10. Tölkes, C.: The role of sustainability communication in the attitude–behaviour gap of sustainable tourism. Tourism Hospitality Res. **20**(1), 117–128 (2020)
11. Bouman, T., van der Werff, E., Perlaviciute, G., Steg, L.: Environmental values and identities at the personal and group level. Curr. Opin. Behav. Sci. **42**, 47–53 (2021)
12. Bursa, B., Mailer, M.: Car-less on holiday? Sustainable tourist travel in Alpine regions. Tourism naturally conference (2018)

Open Access This chapter is licensed under the terms of the Creative Commons Attribution 4.0 International License (http://creativecommons.org/licenses/by/4.0/), which permits use, sharing, adaptation, distribution and reproduction in any medium or format, as long as you give appropriate credit to the original author(s) and the source, provide a link to the Creative Commons license and indicate if changes were made.

The images or other third party material in this chapter are included in the chapter's Creative Commons license, unless indicated otherwise in a credit line to the material. If material is not included in the chapter's Creative Commons license and your intended use is not permitted by statutory regulation or exceeds the permitted use, you will need to obtain permission directly from the copyright holder.

Investigate the Rural Mobility and Accessibility Challenges of Seniors

Tom Ryan[✉]

National Transport Authority, Dublin, EU, Ireland
tom.ryan@nationaltransport.ie

Abstract. This paper investigates the rural mobility and accessibility challenges of a specific target group - Seniors. The target group is those over 66 years of age entitled to use the Public Transport (PT) Free Travel Scheme in rural Ireland. The paper explores at a high level some of the projected rural PT challenges and requirements over the next twenty years, noting that statistical predictions show that there will be a significant population demographic shift.

Astonishingly, the Central Statistics Office (CSO) has forecasted that by the middle of this century, the Irish Senior's age profile will have increased by 98%. By 2051, over 1.6 million over 65's will be living in the Republic of Ireland [1].

Using the Political, Economic, Social, Technological, Environmental, and Legal factors (PESTEL) framework, the literature review explored existing research concerning mobility and accessibility challenges faced by Seniors.

Twenty-seven qualitative, in-depth interviews with stakeholders within the ecosystem were undertaken. The stakeholders included rural PT customers, Local-Link managers, National Transport Authority (NTA) senior management, a Minister of State, and a European parliament policymaker.

To address the breadth and range of the research, the interviewees were sub-divided into two stakeholder groupings, Tier 1 and Tier 2. Tier 1 interviewee feedback spotlights that the PT network system does not exist for rural patients to access hospital facilities. The findings indicated that Mobility as a Service (MaaS) is potentially revolutionary in the PT arena to help address transport poverty in rural locations.

Finally, this paper suggests several short-, medium- and long-term recommendations based on the research findings. These recommendations are a potential springboard to ensure that rural PT is suitable for future Irish generations.

Keywords: Accessibility · Active ageing · Car dependence · Isolation · Seniors' health issues · Behavioural change · Environmental challenges · Demand-responsive · Mobility as a Service (MaaS)

1 The Motivation Behind the Research

The motivation behind the research - a significant population shift will occur in Ireland over the next twenty years. The transport industry is evolving, and new technologies challenge how customers access and use PT in rural regions. The research investigated

© The Author(s) 2025
C. McNally et al. (Eds.): TRAconference 2024, LNMOB, pp. 79–84, 2025.
https://doi.org/10.1007/978-3-031-85578-8_11

how other regions utilise technology solutions to address and solve mobility issues for older people. The investigated areas included Demand Responsive Transport (DRT) solutions and trip-sharing services. A review of published literature was undertaken. The researcher explored what is known about the topic area and determined other researchers' approaches. A literature review determined whether this research was 'exploratory' and could help 'advance what is already known about the topic' [2].

2   Research Methodology

Strategy/research design - Qualitative research methods explored the barriers experienced by users of PT services in rural areas. Time has been spent analysing target audiences and reviewing population demographic shifts over the next twenty years. The methodology chosen for this research was based on using the funnelling process. The funnelling process helped 'frame' the research paper question [3]. Work experience and background reading have shown the researcher the importance of awareness of the needs and wants of end-users.

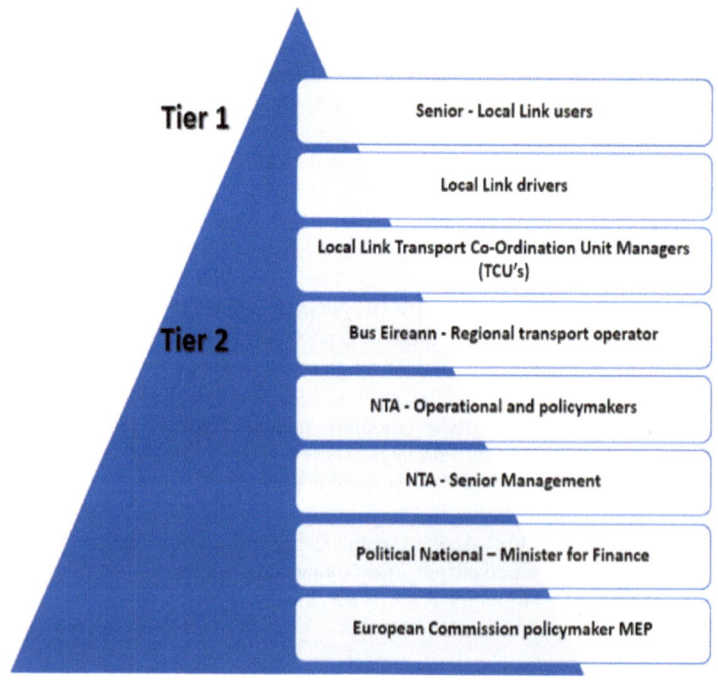

Fig. 1. Tier 1 and Tier 2 research interview stakeholders.

Tier 1: Stakeholder collective group one was defined as Tier 1 Senior rural Local Link customers/users, rural accessibility wheelchair users, rural Local Link drivers and rural Local Link regional transport managers. People who need and use the services, people who need but can access services, drivers of vehicles who provide rural transport services and finally, rural transport office-based service providers. The demographic makeup of this Tier 1 group was two females and eight males.

Tier 2: Stakeholder group two was defined as Tier 2. This research stakeholder group was focused on the creation and delivery of rural transport policy, international policy perspective, policymakers in the arena of the target group seniors and national and European rural transport policymakers. The demographic makeup of this Tier 2 group was two females and ten males.

3   Seniors in Rural Ireland

The picture is very different for Seniors in rural Ireland without access to a car or PT. The practical impact of poor or no access can result in not having the independence to get to the local shop, collect the pension, get to the doctor for a general check-up, or collect a prescription on time [4]. This research will review the impact of rural transport on Seniors who live in rural areas. Preliminary interviews indicated that accessibility to the community is critical. If Seniors are not mobile or do not have access to PT options, the long-term impact can be very detrimental, potentially resulting in isolation and loneliness. 'Mobility does not always generate movement, but it does generate happiness' [5]. In practical terms, poor accessibility can result in sections of the rural countryside not meeting or speaking to neighbours from one week to the next.

4   Research Findings/Results

The research findings have been segmented into multiple themes. Each interview was transcribed and reviewed. Common themes were distilled and coded for the Tier 1 and Tier 2 stakeholders. The following Venn diagram Fig. 2 highlights the research themes found during the thematical analysis phase. The findings explored first-hand interview references to support the primary research interview themes in greater detail. The research findings indicate that health policy creates urban centres of excellence, staffed to provide a best-in-class medical system. However, the Tier 1 interviewee feed-back spotlights that the Public Transport network system does not exist for rural patients to access these hospitals. There was no evidence from the Tier 2 research findings to show that health policymakers and transport planners are working to deliver a national transport solution to support patients getting access to hospital appointments.

5   Lack of Synergies - Connecting State-Run Services

Linking up different state-run systems could potentially bridge the current gaps, and a degree of retrofitting may be required to resolve these physical issues 'the services are not necessarily built in the locations with the thought of older people accessing them' [6]. O'Mahoney discussed the potential of linking up hospital appointments and transport offerings. The concept is that if you opted into a medical appointment, then transport to the hospital could be offered. This collaborative thinking could potentially

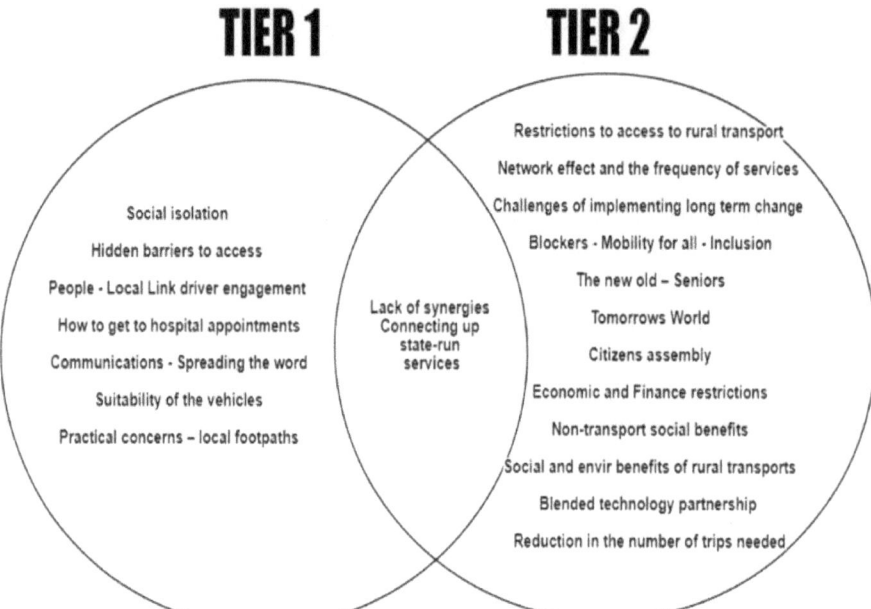

Fig. 2. Venn diagram of research findings

help bridge GDPR compliance. When accepting your hospital appointment, you could also be 'automatically offered a public transport trip' [7]. This offering could reduce/limit the current problems faced by rural outpatient customers.

6 The New Old - Who Are They and What They Want and Need – Seniors

Several stakeholders had interesting views on the new-old and those entering this demographic. Tier 2 stakeholders spoke about the new old - who they are and what they want and need. When discussing the growth in the Seniors demographic, O'Mahoney felt that 'our (Bus Éireann) newer customer database will be much more tech-savvy' [7]. He felt that 'the rural over 65's will be an evolving market.' 'The way things will be delivered will be different,' and the type of products delivered by rural transport services will need to change [7]. How Seniors use transport will possibly also change. 'Their use of public transport would be to get them from point A to point B, rather than being a social interaction' [7]. To deliver the products and services to Seniors, he felt that the system needed to 'adapt' due to 'changing patterns or changing travel needs' [7]. Due to external factors, it is possible that the current system needs to be re-planned, re-designed and changed for the better to meet the needs of rural Seniors 'life has changed completely' [8].

7 Recommendations

This paper suggests several short-, medium- and long-term time frame recommendations based on the research findings. These recommendations are a potential springboard to ensure that rural Public Transport is suitable for future Irish generations. All concepts in the recommendations section are distilled ideas that have been evidence driven. These ideas have been informed during the literature review, primary qualitative research, investigation of gaps in current literature, and reflection during the research process.

8 Conclusion

In conclusion, this research paper investigates Seniors' rural mobility and accessibility challenges using the Free Travel Scheme. The research objectives are to explore the rural Public Transport system and how it delivers services to Seniors living in rural Ireland. The research paper reviews and shows how the rural transport system might manage and address the demographic shift of Seniors living in Ireland over the next twenty-five years.

Parting glass, a quote from the wise…
'In a world where we can expect to see more and more people leading significantly longer lives, innovative and creative thought around the ageing process will become increasingly important' [9].

Disclaimer:
The opinions expressed in this abstract are those of the author – Tom Ryan, they do not purport to reflect the opinions or views of the National Transport Authority.

References

1. CSO. 1 Sep (2023). https://www.cso.ie/en/csolatestnews/pressreleases/2023pressreleases/pressstatementcensusofpopulation2022-summaryresults/. https://www.cso.ie, Retrieved Sept 4, 2021, from
2. Teherani, et al.: (2015: 669). https://doi.org/10.4300/JGME-D-15-00414.1. Retrieved 9 July 2021
3. Burns, N., Grove, S.: 1 Dec 2021 (2007). https://www.scirp.org/(S(351jmbntvnsjt1aadkposzje))/reference/ReferencesPapers.aspx?ReferenceID=574704. https://www.scirp.org
4. Maynooth University 1 Dec 2021 (2008). https://mural.maynoothuniversity.ie/1052/1/Flexibus_Booklet_MAy_2008.pdf. https://mural.maynoothuniversity.ie , Retrieved 1 Sept 2021
5. Walker, J. (2012). Human Transit. In Human Transit How Cleaner Thinking about public Transit Can enrich our communities and our lives (p. 19). Washington: Island Press. Retrieved Oct 2021
6. Graham. Ms Graham Chief Executive Officer NTA. (T. Ryan, Interviewer), 4 Oct (2021).
7. O'Mahoney. Mr O'Mahoney, Head of PSO Contracts Bus Éireann. (T. Ryan, Interviewer), 1 Oct (2021)
8. Creegan. Mr Hugh Creegan Director of Transport Planning Investment, Deputy Chief Executive NTA. (T. Ryan, Interviewer), 30 Sept (2021).
9. President Higgins-TILDA. (2016, Dec 2021). https://tilda.tcd.ie/publications/reviews-newsletters/pdf/Newsletter_2016.pdf. https://tilda.tcd.ie:

Open Access This chapter is licensed under the terms of the Creative Commons Attribution 4.0 International License (http://creativecommons.org/licenses/by/4.0/), which permits use, sharing, adaptation, distribution and reproduction in any medium or format, as long as you give appropriate credit to the original author(s) and the source, provide a link to the Creative Commons license and indicate if changes were made.

The images or other third party material in this chapter are included in the chapter's Creative Commons license, unless indicated otherwise in a credit line to the material. If material is not included in the chapter's Creative Commons license and your intended use is not permitted by statutory regulation or exceeds the permitted use, you will need to obtain permission directly from the copyright holder.

Ecological Sustainability and Passengers' Values in Mobility – A Case of Fell Lapland Region

Valtteri Ahonen[1]([✉]) [iD], Maria Hakkarainen[2] [iD], Erkki Nykänen[1],
and Pekka Leviäkangas[1] [iD]

[1] University of Oulu, P.O. Box 4300, 90014 Oulu, Finland
`valtteri.ahonen@oulu.fi`
[2] University of Lapland, P.O. Box 122, 96101 Rovaniemi, Finland

Abstract. This paper studies whether and how different demographical and societal determinants affect people's valuation of ecological sustainability in mobility. The data has been collected in Fell Lapland in the spring of 2022. The analysis is done by using multinomial logistic regression. The results show that age, certain occupations, and car ownership do affect the responses. *Teacher or researcher,* or *expert* as an occupation increases the value of ecological sustainability in responses and *car ownership* reduces it. The effect of the age determinant cannot be interpreted unambiguously. Other occupations, gender, tourism, being an inhabitant of the region, and obtaining a driver's license have also been tested as determinants and found to be not significant in valuing ecological sustainability in mobility.

Keywords: Ecological sustainability · mobility · multinominal logistic regression

1 Introduction

Determinants that influence the decision of mode of choice usually relate to convenience or the absence of mobility alternatives to private vehicle use [1]. In addition, price, speed, personal attitudes, status, and personal preferences affect travel behaviour [2]. Environmental consciousness has also been recognised as one of the determinants of choosing the transport mode [3].

Demographic and social determinants those relations to environmental consciousness have been studied are gender, age, annual house income, level of education, profession, employment status, political ideology, race, income, religion, and urban-rural differences (e.g. [4, 5]). The results in the studies are study-dependent, some of the studies have found other determinants relevant that others have not. There is a consensus on some of these determinants which have been reviewed by Gifford & Nilsson [4]. They state that factors such as female gender, higher education, better knowledge of environmental issues, and middle- or upper-middle social class tend to correlate to stronger environmental consciousness. Recent studies (e.g. [6]) regarding environmental consciousness in Finland have similar conclusions.

© The Author(s) 2025
C. McNally et al. (Eds.): TRAconference 2024, LNMOB, pp. 85–90, 2025.
https://doi.org/10.1007/978-3-031-85578-8_12

This paper investigates the determinants that affect environmental consciousness related to mobility in a survey made in Fell Lapland. The novelty of the paper relies on the sample that is under research. The sample is varied containing people from different backrounds sharing different connections to Fell Lapland. The aim is to study valuing of sustainability among the local and non-local inhabitants of Fell Lapland which has not been done yet in this scale. The study investigates the determinants in user profiles affecting the valuing of sustainability in mobility (RQ1), and how the determinants affect on valuing of sustainable mobility (RQ2).

Data and Methodology

The data consists of a survey collected in the Fell Lapland region during 2022 studying the mobility of local and non-local people (e.g. tourists, seasonal workers, second home residents) in the area (n = 1189). In the survey question that is analysed in this paper, the respondents were asked to put in order the mobility preferences: speed, easiness, cost, and ecological sustainability. The responses to ecological sustainability preferences were quantified to values 1–4, number 1 meaning that ecological sustainability is the most important value among 4 different values, number 2 meaning that ecological sustainability is the second most important value, and so forth.

The method used was multinomial logistic regression analysis. This method assumes that the independent variables determine the probability of the possible values of the dependent variable. Mathematically this is written as follows:

$$\log \left(p_k / p_1 \right) = b_{k0} + b_{k1} \cdot x_1 + b_{k2} \cdot x_2 + \ldots \tag{1}$$

The Eq. (1) states that the ratio of probabilities p_k and p_1 for the values k and 1 (the reference level) of the dependent variable are connected exponentially to a linear combination of independent variables x_i with regression coefficients b_{ij}.

Multinomial models were created using the R statistical software and its *multinom* function. Model selection was done stepwise. First, the variable that created the lowest AIC value was chosen in one variable model. The second variable was the variable that together with the first variable produced the lowest AIC value in a two-variable model. This iteration was continued until adding new variables did not improve the model. Improvement of the model was tested using the likelihood ratio test of the models so that with and without the new variable it produced a p-value under 5%.

2 Analysis

The variables of the final model were *car ownership, age, teacher or researcher* as an occupation, *expert* as an occupation, and the *rest*. The *rest* variable was included in the model to make it complete, and it contains three groups with only a few responses: *some other profession* (n = 29), *unemployed* (n = 12), and *empty response* (n = 12). Some of the variables were tested but left out of the final model because they did not have significant response to the model. These variables are occupations *student, retired, management, middle-management,* and *entrepreneur*. Also, *inhabitant* of the region, *tourist,* obtaining a *driver's license,* obtaining *driver's license and car,* and gender *men* were discarded.

All variables except *age* were dichotomous variables. Variable *age* was treated as a continuous variable even though it was an integer number in the range of 1–7 referring to seven successive age groups (12–17 years the first one and over 65 years the last one). In addition, the square of the variable *age* was included as a variable since it seemed possible that the effect of age is not necessarily linear.

Table 1 presents relative risk ratios (rrr), p-values, and confidence intervals (C lmin, C lmax) for different variables and outcome categories. Relative risk ratio refers to the probability of choosing the outcome category (ecological values 2–4) that is studied over the base outcome category (ecological value 1). If the rrr of a certain level is greater than 1 the ratio of probabilities of this level and base outcome category increases as the value of the variable increases. Cases with rrr less than 1 refer to the opposite situation. In these cases, the ratio of the probability of the base category (ecological value 1) and the probability of this other level increases. The p-value states the statistical significance of the result. Only p-values less than 0.05 were acknowledged (highlighted in Table 1).

The model predicted that the variables *teacher* or *researcher*, *expert*, and *rest* increase the response value of ecological sustainability whereas car ownership reduced it. The effect of age could not be described with one word due to the seven levels and non-linearity, but it did affect the responses. Figure 1 shows the differences between different determining variables. The *age* variable is presented in Fig. 2.

For comparison ordinal logistic regression model and logistic regression models for values 1 vs 2–4, values 1–2 vs 3–4, and values 1–3 vs 4 were created. Ordinal logistic regression was not chosen as the final model due to the structure of the survey where the answers were not just values of ecological sustainability but permutations of numbers 1–4 to 4 different values, one of which was ecological sustainability. These permutations did not necessarily have a natural ordering and therefore the applicability of ordinal logistic regression is not certain. The results of these comparison models did not significantly differ from the results of the multinomial model. The significant variables and their order of importance were mostly the same with only minor differences.

3   Findings and Discussion

The results show that *age*, *profession*, and *car ownership* are determinants that affect valuing sustainability in this survey. The *age* determinant did not behave linear and is for that reason challenging to interpret. Figure 2 shows that there is a minor implication for young (12–17 and 18–25 years) and old age groups (56–65 and over 65 years) valuing sustainability more than other age groups but this can not be proven based on the figure. However, this observation complies with the newest climate barometer [6] which studied the views of the Finns on climate issues. It shows that the youngest and, in some cases, oldest respondents had increased environmental consciousness and willingness to change their actions. The results of scientific publications are inconclusive one reason being that there are overlapping determinants such as generational events and political climate that might affect environmental consciousness more than age [4].

Determinants *teacher or researcher* and *expert* as an occupation increased response value of ecological sustainability. The climate barometer [6] shows that higher education increases environmental awareness and willingness to take action against climate change,

Table 1. Multinominal regression model results.

Ecological value	Variable	rrr	p	Clmin	Clmax
2	**car**	**1.6038**	**0.0461**	**1.0082**	**2.5513**
2	age^2	1.0138	0.754	0.9306	1.1044
2	teacher or researcher	0.6333	0.2277	0.3014	1.3304
2	expert	1.1506	0.5721	0.7073	1.8718
2	age	0.8369	0.6613	0.3774	1.856
2	rest	0.5527	0.2249	0.2121	1.4399
3	**car**	**2.5643**	**1e-04**	**1.5979**	**4.1149**
3	**age^2**	**0.8903**	**0.0096**	**0.8153**	**0.9722**
3	**teacher or researcher**	**0.3886**	**0.0122**	**0.1856**	**0.8138**
3	expert	0.7231	0.1768	0.4518	1.1574
3	**age**	**2.6409**	**0.0205**	**1.1616**	**6.0041**
3	rest	0.5968	0.2348	0.2547	1.3984
4	**car**	**1.8990**	**0.0014**	**1.2811**	**2.8148**
4	age^2	0.9404	0.1042	0.8732	1.0128
4	**reacher or researcher**	**0.3768**	**0.0026**	**0.1995**	**0.7117**
4	**expert**	**0.6049**	**0.0221**	**0.3934**	**0.9303**
4	age	1.4727	0.2697	0.7406	2.9287
4	**rest**	**0.3018**	**0.0051**	**0.1304**	**0.6982**

which is often the education required for the aforementioned occupations. Car ownership reduces the response value of ecological sustainability, which is a similar result to the climate barometer [6].

Contradictory to the climate barometer [6] the analysis of the survey responses did not show indications that gender would affect valuing economical sustainability in mobility. Even though gender has been studied and shown to affect environmental consciousness (e.g. [4]) in some case studies (e.g. [7]) it did not have significant predictive value.

Limitations do exist. Firstly, survey data has responses from inhabitants living in all parts of Finland, which means that the results describe a larger group than just inhabitants in the Fell Lapland region. However, *inhabitant* of the region is not a determining factor valuing ecological sustainability in this study which can be interpreted as inhabitants having a similar view of sustainability related to mobility than e.g. tourists in the region. Secondly, the question under investigation was not directed enough toward ecological sustainability in mobility (although it is easily deduced from the context of the survey). This might lead to responses describing common views against ecological sustainability,

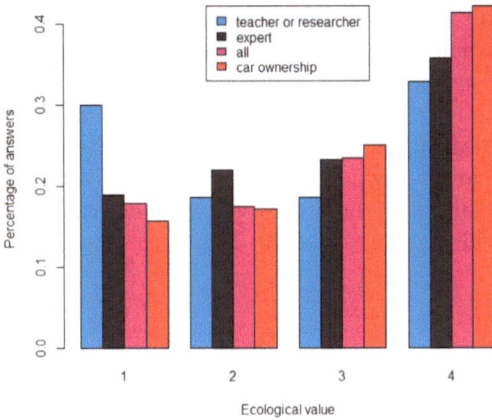

Fig. 1. Percentage of answers for valuing ecological sustainability by different determinants.

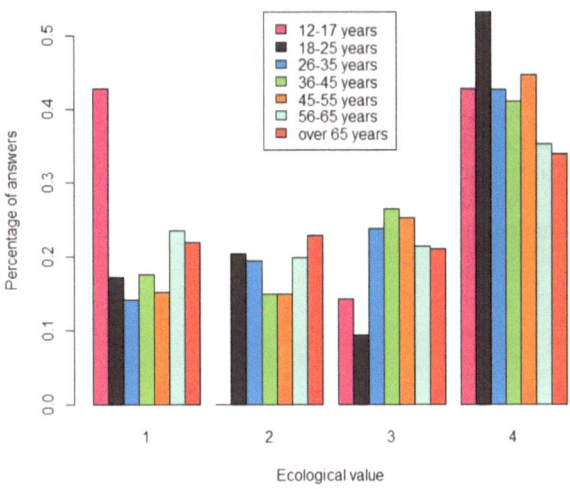

Fig. 2. Percentage of answers for valuing ecological sustainability by different age groups.

not directed exclusively to mobility. Finally, survey data has not been balanced which exposes results to bias.

4   Conclusion

This paper studies how different determinants affect the valuing of ecological sustainability in mobility. The sample of the survey is unbalanced and has other limitations that have been explained in Sect. 3. The results show that determinants affecting the valuing of ecological sustainability were *age,* occupation (teacher, researcher, or expert), and *car ownership* (RQ1). Being a *teacher or researcher* and being an *expert* increased responses that indicated a higher valuation of ecological sustainability and *car ownership* indicated a lower valuation (RQ2). *Age* did also affect, but due to non-linearity, it was not clear how. The results are similar to former studies made globally and in Finland.

Future research concerning this survey could include studying correlations between valuing ecological sustainability and using sustainable modes of travel. In addition, it should be studied whether there is a difference in valuing ecological sustainability between urban and rural residents. A similar study could be replicated in other regions of Finland or other countries to investigate the determinants affecting valuing of sustainability in mobility.

References

1. Poltimäe, H., Rehema, M., Raun, J., Poom, A.: In search of sustainable and inclusive mobility solutions for rural areas. Eur. Transp. Res. Rev. **14**(1), 13 (2022)
2. Van Acker, V. Goodwin, P., Witlox, F.: Key research themes on travel behaviour, lifestyle and sustainable urban mobility. Int. J. Sustain. Urban Transport. **10**(1), 25–32 (2016)
3. Shen, J., Sakata, Y., Hashimoto, Y.: Is individual environmental consciousness one of the determinants in transport mode choice? Appl. Econ. 40(10), 1229–1239 (2008)
4. Gifforf, R., Nilsson, A.: Personal and social factors that influence pro-environmental concern and behaviour: a review. Int. J. Psychol. **49**(3), 141–157 (2014)
5. McCright, A.: The effects of gender on climate change knowledge and concern in the American public. Popul. Environ. **32**(1), 66–87 (2016)
6. Ministry of the Environment. Climate Barometer 2023 (2023)
7. Xia, T., Zhang, Y., Braunack-Mayer, A., Crabb, S.: Public attitudes towards encouraging sustainable transportation: an Australian case study. Int. J. Sustain. Transport. **11**(8), 593–601 (2017)

Open Access This chapter is licensed under the terms of the Creative Commons Attribution 4.0 International License (http://creativecommons.org/licenses/by/4.0/), which permits use, sharing, adaptation, distribution and reproduction in any medium or format, as long as you give appropriate credit to the original author(s) and the source, provide a link to the Creative Commons license and indicate if changes were made.

The images or other third party material in this chapter are included in the chapter's Creative Commons license, unless indicated otherwise in a credit line to the material. If material is not included in the chapter's Creative Commons license and your intended use is not permitted by statutory regulation or exceeds the permitted use, you will need to obtain permission directly from the copyright holder.

Evaluation of Beyond MaaS Concept and Technical Solution: Case Enriched Travel Chains

Olli Pihlajamaa[1]([✉]), Toni Lusikka[2], Janne Lahti[2], Immo Heino[1], Jenni Vestinen[2], and Maria Hakkarainen[3]

[1] VTT Technical Research Centre of Finland Ltd., Tekniikantie 21, 02150 Espoo, Finland
olli.pihlajamaa@vtt.fi
[2] VTT Technical Research Centre of Finland Ltd., Kaitoväylä 1, 90570 Oulu, Finland
jenni.vestinen@vtt.fi
[3] University of Lapland, PL 16, 96301 Rovaniemi, Finland

Abstract. The Beyond MaaS concept aims at combining digital mobility services with the digitalized information about the contexts and reasons of mobility in an intelligent way. Trip planners and various degrees of MaaS approaches have made public transportation easier to use but the modal shift to more sustainable mobility is still lagging the goals. Integrating rich, machine-readable digital representation of destinations and services making people to travel and mobility service, it is possible to optimize human activities as a whole and create new schemes that attract people more often to use sustainable transport modes. This paper provides an example of the Beyond MaaS concept aimed at tourism in rural areas: Enriched Travel Chains. The concept aims at making multi-part public transportation travel chains more attractive by automatically suggesting the traveler PoIs and activities for the waiting times between the travel chain parts. Enriched Trip Planner (ETP) realizes the concept as a web application. The first experiences from the implementation reveals the challenges with scarcity and quality of tourist service and PoI data. On the other hand, first impressions from the tests of the ETP prototype provide promising directions for further development utilizing crowd-sourced or niche data for tourist mobility for special interests.

1 Introduction

Peripheral destinations, especially nature-related experiences, are key tourism assets for Finland, which Covid 2019 further increased in value. As more and more tourists seek these destinations, accessibility to sparsely populated areas (Lusikka et al. 2022) is a key factor for tourism business and regional development. In an era of environmental crisis, the demand for sustainable choices also affects the services provided by the tourism industry.

Sustainable tourism, especially in sparsely populated areas, can be increased by promoting the use of public transport as part of tourist mobility. This objective has highlighted the need for easily accessible digital information services on local transport. It is also clear to both locals and tourists that there are synergies in the overall development

© The Author(s) 2025

C. McNally et al. (Eds.): TRAconference 2024, LNMOB, pp. 91–96, 2025.
https://doi.org/10.1007/978-3-031-85578-8_13

of digital mobility services. Till these days, using public transport has not been always easy or even possible for tourists, as most services tend to focus on the needs of local users and disregard the needs of tourists and other travellers. The accessibility of remote areas in general, can be improved by new mobility services that integrate different types of transportation and exploit digitalization (Eckhardt et al. 2018). To include tourists as users of these mobility services requires understanding their tourism related destination service needs and combining these needs to the mobility service offering in the area.

To respond to the tourism mobility challenges and needs in rural areas, this paper dives into digital mobility services supporting *the Beyond MaaS concept*. The concept is based on the idea that mobility and transport are not absolute value themselves, but only means to fulfill other tasks that human-beings have. Thus, to serve better sustainable tourism, the digital mobility services at tourist destinations should be combined with the tourist-related contexts and reasons of mobility (Points-Of-Interests (PoIs), activities, services, social relationships etc.) (Pihlajamaa et al. 2022).

This paper describes an application of the Beyond MaaS concept developed for tourism mobility "Enriched Travel Chains" and its prototype implementation "Enriched Trip Planner". Also, the challenges and future potential of the concept is discussed.

2 Tourist Mobility Concept: Enriched Travel Chains

The challenges and the needs of the rural tourism destinations shortly introduced in Sect. 1 indicate that there is a need for information services that combine travel information services and points of interest in the tourist destination area. To meet this need, we developed different concept variations combining trip planner functionality and tourist information under the name "Enriched Travel Chains".

Travel chain enriched with destination services

Fig. 1. Enriching travel chain with optional visits during the waiting times of the trip.

One of the concept variations, *Travel chain enriched with destination services* (Fig. 1), is specifically aimed at multi-part travel chains that serve longer transitions within the tourist destination areas or between them. Especially in rural areas, the scarcity of the transport services leads often to relatively long waiting times between travel chain legs, which, in turn, may make tourists avoid such trips and use of public transportation. The concept introduced here aims at making such travel chains more attractive and, at the same time, allowing the tourist to find more experiences from the destination. In the concept, a digital trip planning service provides not only travel chains from origin to destination but also suggests services and PoIs to visit as meaningful options for "killing dead time" during the waiting times between the travel chain legs. Traveller may also prefer to have "loose schedule" when the service has more freedom to suggest intermediate locations to visit by choosing not so tightly scheduled travel chain.

3 Prototype Implementation of the Enriched Travel Planner (ETP)

The concept introduced above has been implemented as a Enriched Trip Planner (ETP) protype utilizing Finnish Digitransit[1] open-source journey planning solution based on Open Trip Planner (OTP)[2] route planning algorithms and APIs. In the following we shortly present the core functionality and system architecture of ETP.

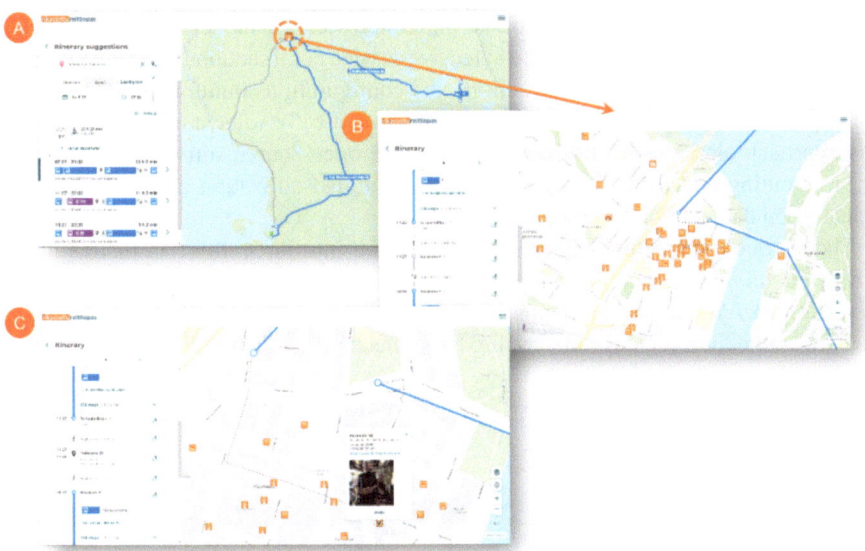

Fig. 2. Core functionality of the Enriched Trip Planner. (ETP by VTT [authors] which is based on Digitransit (https://digitransit.fi/en/developers/), Map data from OpenStreetMap (http://openstree tmap.org/copyright), and Service data from Visit Finland Datahub (https://datahub.visitfinland. com/)).

3.1 Functionality of the Enriched Trip Planner

The functionality of the Enriched Trip Planner is described here by an example, where Alice is a traveler and is arriving at Oulu airport. Her final destination is Ruka Ski Resort, located 250 km from Oulu airport. The search for connections to the destination produces several connections (Fig. 2, A). While looking closer the first connection she notices that there are over 6 h waiting time in Rovaniemi in the proposed travel chain.

---

[1] https://digitransit.fi/en/.

[2] https://www.opentripplanner.org/

Therefore, ETP has proposed different options for utilizing this time showing them as icons representing the possible PoIs and services (Fig. 2, B).

As Alice knows she needs to eat along the route, she decides to click restaurant icons to open more information about local restaurants. She is happy to find a restaurant serving typical Lappish food. She selects that restaurant as an intermediary stopping point in ETP and then it reroutes the travel chain to include the restaurant visit (Fig. 2, C). As the waiting time is long, ETP allows to add also more intermediate stopping points and routing through them if Alice wishes so.

4 Architecture and Implementation

The service architecture (Fig. 3) is based on a micro service model where stand-alone micro services jointly constitute more extensive service combinations. This enables easy and flexible introduction of new functions and configurations minimizing unnecessary replication of existing functionalities and focusing on truly value adding components. This approach also enables utilization of existing open-source software components, e.g., for routing and data management (OTP) and also utilization of open resources, such as open data and open or shared API's.

The service architecture can be roughly divided into three different main entities: *ETP Application UI*, which provides the web-based user interface handling and user interaction, *PoI Service* for handling the PoI data management, and *Integration Engine* for main application logic for integrating the route creation with PoI points.

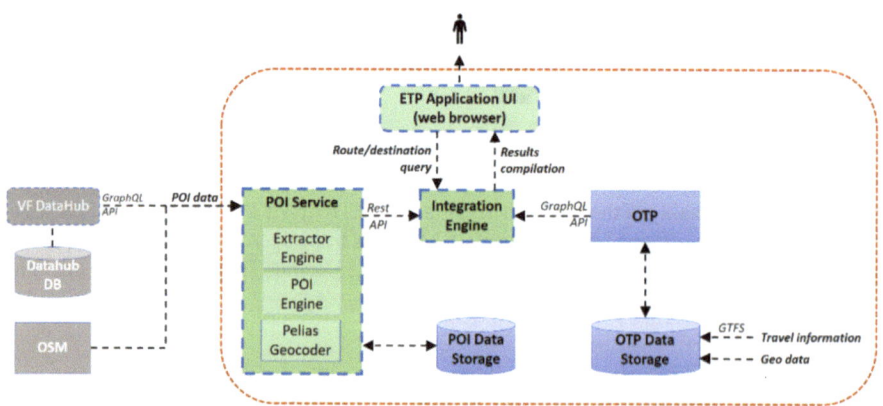

Fig. 3. High-level Service Architecture for the Enriched Trip Planner

ETP Application UI is based on Digitransit journey planning solution that is modified to integrate the PoI functionality for the trip planning UI. The main functionalities of the ETP UI include searching for multimodal travel chains and showing the potential PoI-destinations that could be selected for visiting during the transition points of the multi-part travel chain (Fig. 2).

Integration Engine (IE) contains the main application logic for selecting the potential PoI destinations that could be visited in conjunction with the overall trip plan. IE analyses the trip plan resulting from the route query passed to OTP and identifies potential transition points where traveler could visit PoI-destinations. The algorithm for selecting potential PoIs is based on calculation of new routes between transition points and PoIs, taking into account the PoI characteristics (typical duration of the visit and opening hours) while retaining the original route plan schedule.

PoI Service offers a REST API interface through which the applications can download information about attractions or services related to the stopping points on the route, such accommodations, restaurants, cafes etc. PoI Service uses open data services as information sources, including Visit Finland's (VF) DataHub and Open Street Map (OSM). VF Datahub is a database where Finnish companies producing tourism services can register and store information about their company and the products and services they offer.

Technically PoI service has been implemented as a set of web microservices that contains three subcomponents: Extractor engine, PoI Engine and Pelias geocoder. Extractor engine periodically clones information from VF DataHub downloading OSM information from Geofabrick's web site. Pelias service transforms OSM PoI information from WGS84 coordinates to street addresses.

5 Discussion

Implementing the ETP presented two types of challenges: data-related and routing-related issues. Firstly, availability of tourism-related data became an issue in early phases of implementation. Idea was to use new tourism DataHub offered by VF, but so far its data content is quite modest covering service descriptions only from 5% of Finnish tourism related companies. Furthermore, tourists need also the same basic services as local people from which adequate information is scattered or not available from any data source. In our tests, the scarcity of the available visiting options in the travel chain transits (especially in rural areas) caused frustration as the idea to offer meaningful doing for the long travel chain transits was often difficult to realize with real existing PoIs and services.

With the travel information providing digitalized routes, stops and timetables the situation is better as it is provided in de facto standard form (GTFS). However, in rural areas, the coverage of travel information is not perfect either. For tourists planning their often trips well in advance, one of the biggest problems is that digital travel information covers usually only near future (week or months). This highlights the different travel information needs for tourisms compared to trip planning of local people.

In routing, we relied on widely used OTP routing engine (Raptor algorithm) that in our tests produced sometimes surprising re-routing suggestions after adding intermediate points to the route when comparing to the original route. In the end user tests, these surprises caused puzzlement when in original route suggestion there was, for example, many hours of waiting time on some transit point of a multipart travel chain and adding one hour stop to the very same transition point resulted totally new (typically much longer) travel chain.

6   Conclusion

According to our experiences, available machine-readable PoI and service information is still too scarce, low quality and heterogeneous to combine with travel information for easy construction of well serving general ETP for tourism. However, there are very comprehensive, well-defined databases for certain areas of life in Finland (e.g., natural formations, ancient monuments and historical PoIs, crowd-sourced nature observations, cycling and hiking routes) that may very well serve tourists looking for such attractions with sustainable transportation means. Constructing first working applications for specialized (niche) tourism interests is probably better starting point for successful application than aiming at general tourist mobility application covering all tourist PoIs and services. For that we look forward to the ETP application that can be easily supplemented with different kinds of PoI layers for special interests.

For further development, we strongly suggest the development of digital PoI and service description standards to enable, for example Beyond MaaS type services that span service sectors and allow novel, intelligent ways to better support and optimize human activities and create new opportunities for out of the silo business ecosystems.

Acknowledgements. The authors gratefully acknowledge Business Finland for enabling and co-financing the FIT ME! (Foreign Individual Travelers' hospitality and Mobility Ecosystem) project, as well as project stakeholders for their contribution.

References

1. Eckhardt, J., Nykänen, L., Aapaoja, A., Niemi, P.: MaaS in rural areas - case Finland. Res. Transp. Bus. Manag. **27**, 75–83 (2018)
2. Lusikka, T., Eckhardt, J., Hakkarainen, M.: Moving beyond MaaS with ecosystemic way of work. Transport. Res. Procedia **72**, 1755–1762 (2022)
3. Pihlajamaa, O., Lusikka, T., Eckhardt, J.: Enriched travel chains with Beyond MaaS. In: 3rd International Conference on Mobility as a Service (ICoMaaS 2022), Tampere (2022)

Open Access This chapter is licensed under the terms of the Creative Commons Attribution 4.0 International License (http://creativecommons.org/licenses/by/4.0/), which permits use, sharing, adaptation, distribution and reproduction in any medium or format, as long as you give appropriate credit to the original author(s) and the source, provide a link to the Creative Commons license and indicate if changes were made.

The images or other third party material in this chapter are included in the chapter's Creative Commons license, unless indicated otherwise in a credit line to the material. If material is not included in the chapter's Creative Commons license and your intended use is not permitted by statutory regulation or exceeds the permitted use, you will need to obtain permission directly from the copyright holder.

Annual Mileage Development Over Passenger Car Age and Driving Power in Finland

Riku Viri$^{(\boxtimes)}$ ⓘ and Johanna Mäkinen ⓘ

Transport Research Centre Verne, Tampere University, Tampere, Finland
`riku.viri@tuni.fi`

Abstract. This study investigates how car age, driving power and area type affect the average annual mileage of passenger cars in Finland. For the analysis, data from the Finnish car fleet register and vehicle inspection records from 2012 to 2022 are used. The results of this study indicate that newer cars have a higher annual mileage compared to older ones. Regarding driving powers, diesel cars have the largest mileage in Finland, mainly due to the taxation system making diesel less expensive on high mileage. Battery electric vehicles on average have lower annual mileage compared to plug-in hybrid vehicles. For area types, cars have higher mileage in rural areas compared to the more urban areas. These findings are consistent with previous studies conducted in various countries and regions. The study contributes to the development of more accurate passenger car emission calculation models for Finland.

Keywords: Annual mileage · Car age · Driving power

1 Introduction

1.1 Background

Car fleet mileage is an important aspect when calculating the total emissions of the passenger car fleet. There have been different estimations through different models and surveys on how passenger car driving power and age affects the use of the car. In this study, a dataset of Finnish car fleet register is used to calculate the average annual mileage of cars. Based on the data, it is possible to investigate how different variables, such as car age, driving power or regional classification affect the mileage.

Through this type of approach, it is possible to investigate which variables induce specific effects on car fleet mileage. These results can then be used to further develop existing Finnish passenger car emission calculation models, such as the Finnish regional car fleet model SALAMA [1, 2]. In addition, the results will improve the authors' estimation of how a change in driving power should be projected in the model's mileage development. This would allow the model to produce more accurate estimate of the car fleet emissions development in Finland.

© The Author(s) 2025
C. McNally et al. (Eds.): TRAconference 2024, LNMOB, pp. 97–104, 2025.
https://doi.org/10.1007/978-3-031-85578-8_14

1.2   Annual Mileage

There have been different estimations through different models and surveys on how passenger car driving power and age affects the use of the car. Relationships between the age of passenger cars and their annual mileage have been often estimated by analyzing national vehicle registration data. For example, OECD [3] used Slovenia's vehicle registration data to investigate the relationship between the age of passenger cars and their annual mileage. The finding was that older cars accumulate lower annual mileage. Similar results were found in U.S. [4] and Norway [5], where the total annual mileage of passenger cars was found to decrease linearly as the vehicles age increases. In Finland, relationship between age and annual mileage have been studied in 1994. In this study, findings demonstrated a consistent decline in annual mileage with increasing car age [6], in harmony with research by OECD [3] and Lu [4]. Many Finnish national transport models have been relying on results from the study conducted in 1994. However, there is a need for updated insights into the relationship between car age and annual mileage. Therefore, the objective of this study is to provide current and relevant information on the relationship between age and annual mileage.

There has been less research on how driving power affects annual mileage. In Norway, on average, plug-in hybrids (including both petrol and diesel plug-in hybrids, PHEVs) have the highest annual mileage. Diesel, battery electric vehicles (BEV) and gas-powered vehicles have very similar annual mileage (around 12,000–13,000 km). Petrol vehicles have significantly lower annual mileage (7000–8000 km). [5] However, it should be noted that there have been shifts in annual mileage of alternative powered vehicles (BEV, PHEV and gas) during the last few years.

1.3   Data and Method

The dataset [7] used in the study contains the technical information of every passenger car in use in Finland on end of March 2022. In addition to the technical data of passenger cars, a separate data table containing the mileage information throughout the inspection of the cars is used to track the yearly mileage of all the cars in the dataset. Both datasets are not commonly available but obtained with special research permission. Mileage information is available from every vehicle inspection since the start of the year 2012. There are some limitations with this approach, as the first inspection in Finland is currently done when the car has been in use for four years and then biannually until the car age reaches 10 years, after which an annual inspection is needed. Thus, cars registered during the last four years have generally not yet been inspected and no mileage data is collected. This lowers the sample size of some cars, especially in the BEV and PHEV -category.

The mileage data is collected by hand as a part of the inspection process in Finland and thus, there can be some errors in the data. Therefore, a data cleaning process is done before the use of mileage data. In this data, there are own variables for storing information about mileage meter only having 5 digits, mileage meter storing values as miles instead of kilometers, and a non-existing mileage meter. In the first phase, any car having any of these values as true is removed from further processing. This is done, as a 5-digit meter generally rolls over throughout the lifetime of the car, thus creating wrong values after the first 100,000 km. The mile versions of the meters are ignored, and based on the dataset, they are not noted through all inspections throughout the car's lifetime, and thus, they are much more likely to cause unit conversion error at some point in the calculation. To start, the dataset has 2,750,181 cars, and through this cleaning process, 76,926 (~3%) cars are removed.

After this, the inspection date and mileage information are used to find all concurrent inspections and to calculate the days passed and mileage driven between each inspection for a car. For the first inspection, the days passed are calculated from the date of the registration and the milage is calculated from zero. Based on this information, the average annual mileage between inspections is then calculated as *kilometers driven / passed time in years*. At this point, some other possible errors are cleared. In case the car has multiple inspection records from the same day, the one with the highest mileage is used. If in any part of the car's history there is a decrease in mileage between two inspections, the car is removed from the calculation. Also, cars without any mileage information are removed from the dataset. In the end, there are 2,163,491 cars that have usable information for the analysis.

For the analysis with different variables, some data from the technical information of the cars is added to the mileage dataset, allowing the analysis of different driving powers (driving power the car is registered as) and area types (the area type the car user resides in). When analyzing the effect of the car age on the annual mileage, the age is calculated as years from the car registration and the annual average mileage record is linked to the full year closest to the car age at the inspection date. Since the inspections are annual only after 10-year-old cars, the years having no mileage data will use the closest following record. Thus, cars having information for 4th and 6th year, will have the average from 4th year record for years 1 to 4, and the average from 6th year record for years 5 and 6, and no information for year 7 and onwards.

2   Mileage Development

2.1   Milage by Age and Driving Powers

Based on the data, the average mileage per age of the car can be calculated, as is shown in Fig. 1.

When looking at the different driving powers combined, it can be seen that the average annual mileage on new cars sits just below 20,000 km, and there is a steady decrease that continues year by year. The annual average mileage for 5-year-old cars is about 17,500 km and for 10-year-old cars it is about 15,200 km. However, when looking at the data with the driving power information, it will show different trends. In Fig. 2, the average annual mileage by different driving powers per car age is shown. It should be noted that the average is only calculated when there is at least a sample size of 500 cars. Thus, not all driving powers have information after a certain year.

For new cars, diesel has the largest mileage (27,900 km), whereas petrol has the lowest (15,400 km). However, there are also some other trends that can be seen from Fig. 2. Petrol, having the lowest annual average mileage, seems to have a steady decline, similar to when looking at the data of all driving powers. Battery electric vehicles (BEV) seem to have a lower mileage on their first year, but then have an increased mileage from 4th and 5th year. Plug-in hybrid vehicles (PHEV) are used more than BEVs, which could

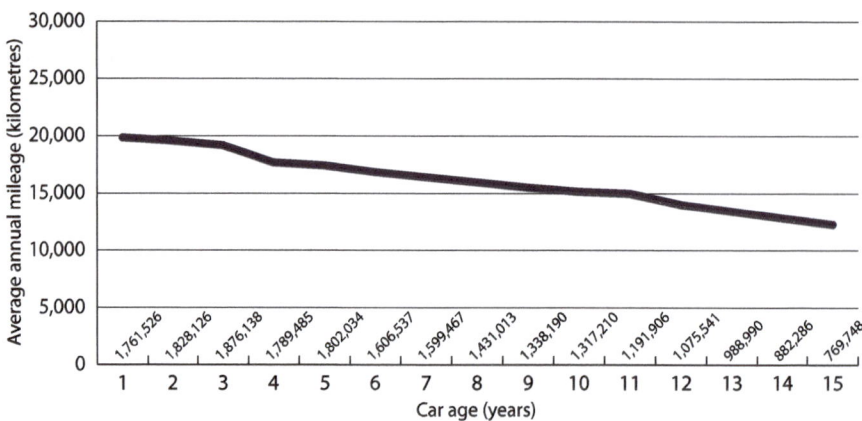

Fig. 1. Average mileage per car age, the number above year presents the sample size (n) of that year.

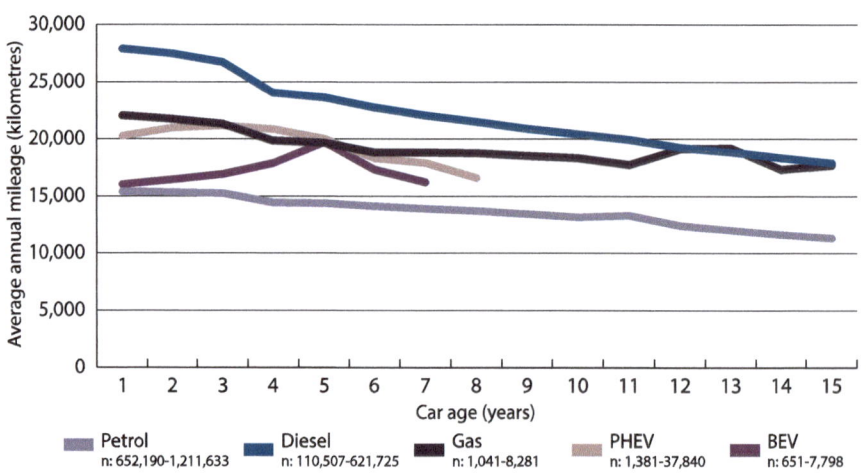

Fig. 2. Average mileage per car age by different driving powers (with a sample size of 500 cars or more).

be a result of users interested in EVs choosing a PHEV over a BEV due to higher mileage and range anxiety. Overall, gas-powered cars are used more than petrol, which is probably due to having lower use costs than petrol-powered cars.

In Finland, the driving-power taxation favors diesel for users with high mileage, as the tax added to pump prices of diesel is lower than in petrol, but the annual tax is higher. Thus, if the user drives enough to overcome the annual lump sum tax with the saved pump prices, diesel will be less expensive than petrol. This explains why the annual mileage of diesel is higher than in the other driving powers.

For both diesel and BEVs, a sudden change in mileage can be seen in Fig. 2 in 3rd to 4th year for diesels and between 4th to 6th years for BEVs. Without better ownership information, the reasons for these cannot be verified, but based on the data and used car market in Finland, it could be assumed that the reasons for these shifts come through an owner change. It could be assumed that at least some cars are moved from the first owner to the second owner somewhere around 3rd to 5th year. This would be supported by both the usual lengths of the warranties of the new cars as well as common lengths of different leasing contracts. Thus, an explanation could be that a new diesel car is purchased by a user having high mileage and the car is then sold after the warranty period. Then, the new user still has high mileage, but not as high as the original owner, who already moved to a newer vehicle.

For BEVs, these could apply the other way around. It should be noted that as most mileage data comes from older car stock than 2018, the data from BEVs are mostly from older, more expensive, models. Thus, in the case those BEVs are leased with a contract of 3 to 4 years, when they appear on the used cars market, they are then bought by the second owner having higher mileage, but who not necessarily had the ability to buy or lease them as new cars. Another potential explanation for increased mileage could be the introduction of more EV-charging infrastructure, enabling EV-users to cover longer distances as charging options became more accessible. However, it should be noted that these are assumptions, and more data is needed to verify these assumptions.

2.2 Mileage by Area Types

When looking at different areas, it should be noted that the data does not have full information about the past locations of cars. There is only information about the area of the current user, and when the car was obtained by the current user. This allows to limit the mileage calculation only for the inspection dates which affect the current user, but this approach cannot note, if the car is moved from area type to another with the current user. Due to this limitation, it was chosen that only data for inspections that have occurred 30 days prior the current user obtained the car or later are calculated. The 30-day period before the ownership change was chosen, as if the car was inspected by the previous owner and then immediately sold after, this mileage value could be used as the starting mileage for the new user. In all other cases, any current user would need to have at least two inspections done to have enough data to calculate the annual mileage. Due to this, only 957,915 cars have enough records available to be used for the area calculation.

For area types, the Finnish urban-rural spatial classification [8] was used to divide the cars in 7 different categories based on postal code information (described in [1]). After that, the average mileage was calculated for all current users residing in that area. The results, as well as the area definitions, can be seen in Table 1.

Inner- and outer urban areas have the lowest average mileage. In urban areas, the share of trips made by walking, cycling and public transport are higher than in rural regions. Moreover, the average trip length is shorter in densely built urban areas. In more rural and less densely built areas, there are fewer alternative transport modes available, as well as longer distances to services, thus causing higher annual mileage for cars. For R4, *local centres in rural areas*, the cause of lower-than-average mileage could be a result of local centres having services in close proximity, thus creating less mobility needs compared to other rural areas. However, it should also be noted that these areas are relatively small, the sample size is by far the lowest of all area types.

Table 1. Average annual mileage per area type in Finland.

Area type	Description [8]	Annual mileage (km)	Sample (n)
U1	Inner urban area	11,000	176,633
U2	Outer urban area	11,900	158,272
U3	Peri-urban area	12,400	252,957
R4	Local centres in rural areas	11,800	2,696
R5	Rural area close to urban area	13,400	116,286
R6	Rural heartland area	12,900	131,598
R7	Sparsely populated rural area	14,400	119,473
Total	**All area types**	**12,500**	**957,915**

3   Conclusions and Further Research

The aim of the study was to investigate how car age, driving power and area type affect the average annual mileage of passenger cars in Finland. In general, the newer the car, the higher the annual mileage is. This outcome had been previously demonstrated by 1994-data in Finland [6] and supported by additional studies in different countries [3, 4]. The findings of this study further substantiated these results. Figure 3 shows the fitted trend lines for both datasets; 2022-trendline is based on this research, while the 1994-trendline is fitted based on the formula in the prior study [6]. Both studies show a decreasing trend, but it can be seen that the annual average mileage was slightly higher in the 1994-dataset, and the rate at which milage decreases over age was also a bit steeper.

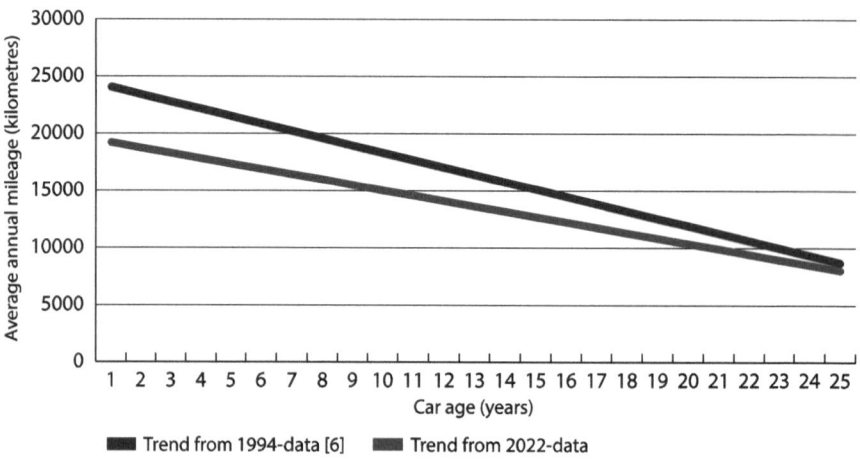

Fig. 3. Average mileage per car age in Finland, data from 1994 [6] and 2022.

While looking at different driving powers, it was seen that diesel had the highest annual mileage, while petrol had the lowest, which is caused by the Finnish taxation of driving powers. PHEVs were generally used more than BEVs, which would be in line with several EV-adoption studies stating that the range and range anxiety is the major barrier into adopting BEVs [e.g., 9, 10]. Therefore, with higher mileage, PHEV can be chosen instead of BEV to overcome the range anxiety. In addition, when compared to the Norwegian statistics, the results are similar, as PHEV has higher mileage than BEV and diesel has in general the largest mileage. However, the actual mileage number in Norway were much lower compared to Finnish one.

When considering various types of areas, a common trend emerges: the more rural the area, the greater the annual mileage tends to be. However, as no former research was found during this research, these results cannot be compared to other studies.

For further research, more accurate information about the ownership changes of the cars would be interesting, as it would allow to better understand the reasons behind the sudden changes in the annual mileage. Also, since the current data from 2022 have only a minor share of BEVs inspected, there is lack of new BEV mileage data. When repeating the same type of process after a couple of years, the data would represent much better the BEV-usage, and it would allow to better examine how BEVs affect the average annual mileage.

In general, this study shows that there are differences in annual mileage between driving powers, area types and the car age currently in Finland. This information is crucial for improving the accuracy of a passenger car emission calculation model. With this data, a more precise estimation of the relationship between driving powers and annual mileage can be developed for future years.

References

1. Viri, R., Mäkinen, J., Liimatainen, H.: Modelling car fleet renewal in Fin-land: a model and development speed-based scenarios. Transp. Policy **112**, 63–79 (2021)
2. Viri, R., Mäkinen, J.: The impact of modal shift on passenger car CO2 emis-sions in Tampere region. Case Stud. Transp. Policy **13**, 101066 (2023)
3. OECD/ITF. Tax Revenue Implications of Decarbonising Road Transport: Scenarios for Slovenia. OECD Publishing, Paris (2019)
4. Lu, S.: Vehicle Survivability and Travel Mileage Schedules. National Center for Statistics and Analysis (2006)
5. Statistics Norway. Road traffic volumes, by type of fuel, age, contents, vehicle type and year. Database. https://www.ssb.no/en/transport-og-reiseliv/landtransport. Accessed 27 Sept 2023
6. Mäkelä, K., Auvinen, H.: LIISA 2010 laskentajärjestelmä (In Finnish, translated: Road traffic exhaust gas emissions in Finland. LIISA 2010 calculation software). Technical Research Centre of Finland (VTT) (2011)
7. Finnish Transport and Communications Agency Traficom. Car register and inspection mileage data snapshots on 31st March 2022 provided for research use (2022)
8. Finnish Environment Institute. Urban-Rural classification 2018 [spatial dataset] (2020)
9. Vassileva, I., Campillo, J.: Adoption barriers for electric vehicles: experiences from early adopters in Sweden. Energy **120**, 632–641 (2017)
10. Berkeley, N., Bailey, D., Jones, A., Jarvis, D.: Assessing the transition towards Battery Electric Vehicles: a multi-level perspective on drivers of, and barriers to, take up. Transport. Res. Part A: Policy Pract. **106**, 320–332 (2017)

Open Access This chapter is licensed under the terms of the Creative Commons Attribution 4.0 International License (http://creativecommons.org/licenses/by/4.0/), which permits use, sharing, adaptation, distribution and reproduction in any medium or format, as long as you give appropriate credit to the original author(s) and the source, provide a link to the Creative Commons license and indicate if changes were made.

The images or other third party material in this chapter are included in the chapter's Creative Commons license, unless indicated otherwise in a credit line to the material. If material is not included in the chapter's Creative Commons license and your intended use is not permitted by statutory regulation or exceeds the permitted use, you will need to obtain permission directly from the copyright holder.

Parking in Macroscopic Transport Models: Modelling Parking Capacities in Traffic Assignment

Dawn Spruijtenburg[✉], Erwin Walraven, Reinier Sterkenburg,
and Marieke van der Tuin

TNO, Anna van Buerenplein 1, 2595 DA The Hague, The Netherlands
dawn.spruijtenburg@tno.nl

Abstract. Parking measures are typical for cities that aim to improve the liveability in terms of air quality, noise, congestion and space. The strategic transport models, used to determine the effects of (policy) measures, do not incorporate the effect of parking behaviour during the trip. The behaviour of parking the car near the destination, and walking the last bit, is not yet modelled in these models. In order to incorporate this behaviour in the static traffic assignment, a methodology is developed. The methodology is twofold. First, the parking capacities, i.e. the number of available parking spaces per destination zone are determined using spatial data. Second, these parking capacities are used in the traffic assignment, which uses parking and walking links. In this way, the extra search time for a parking space is modelled, as well as the diverting behaviour when (almost) all parking spaces are occupied. Simulations for a use case in Amsterdam show that the diverting behaviour is modelled and show possible effects of parking policies, for instance on the amount of car traffic in the city.

Keywords: transport modelling · static traffic assignment · parking capacities

1 Introduction

Different cities are facing challenges with liveability, and sometimes (re)designing city centers to become more liveable. This means better air quality, less noise, more space for moving around and recreation, and generally less (car) traffic. In some cases, the measures taken impact parking, for instance when removing parking spaces, or providing only few parking spaces in newly developed areas. Some measures taken are specifically aimed at parking, such as (higher) parking fees.

Before introducing such measures in the city, it is good to study the effects of these measures to make sure the overall effects are positive (e.g. less car traffic) and not negative (e.g. same amount of car visitors but generating more traffic due to searching for parking spots). For this, typically strategic transport models are used. These models compute the destination choice, mode choice, route choice and resulting network flows in a certain region. In some cases, parking is considered in the destination choice and

© The Author(s) 2025
C. McNally et al. (Eds.): TRAconference 2024, LNMOB, pp. 105–119, 2025.
https://doi.org/10.1007/978-3-031-85578-8_15

mode choice. The parking conditions (such as parking fee, walking time, (probability of) availability) are factors in the decisions made.

However, some parking behaviour only arises during the trip. For instance, when the location at which a person wanted to park has no available spaces. Then, this person might circle around to see and wait if a space becomes vacant, or drive to a next location at which parking might be available. This behaviour results in extra travel time and sometimes distance.

In this study, we propose a methodology that includes parking capacity in the static assignment of traffic, to get insight in the diverting behaviour and its effects. We do this by modeling a limited parking capacity per zone. The method builds upon existing knowledge and is incorporated in existing models.

Additionally, a methodology on determining parking capacities from geographic data is described. These parking capacities are used in the traffic assignment.

In the rest of this paper a literature review is presented (Sect. 2), followed by a description of the methodology (Sect. 3), use case (Sect. 4), results and discussion (Sect. 5), and a conclusion (Sect. 6).

2 Literature Review

It is important to consider parking and its resulting behaviour in strategic transport models, as some studies say that up to a third of traffic volume in city centers can be attributed to drivers that are in search of a vacant parking spot (Shoup 2006). A strategic transport model typically consists of four steps (Ortuzár et al. 2011): trip generation (determining the amount of trips that people will perform), trip distribution (responsible for making trips by linking origins and destinations), mode choice (determining whether a trip is performed by car, public transport or bike), and traffic assignment (determining the route through the network from origin to destination). The traffic assignment results in travel times (including congestion), which (iteratively) influences the mode choice (and sometimes also the destination choice).

The parking behaviour is mostly only considered in the mode choice, destination choice and sometimes time choice. Often, attributes of parking that are considered are: parking cost (parking fees), the (un)certainty of finding an available parking spot, estimated time searching for a parking spot, estimated walking time from the parking spot to the destination, and socio-economic characteristics of the driver (Chaniotakis et al. 2015).

However, after mode choice and destination choice, also in the network assignment the parking behaviour can be taken into account. This is the part in which a specific parking location (spot) is determined (or assumed, if not modelled explicitly). The (un)availability of parking spaces causes drivers to circle around, or to divert to a different parking location and walk to the destination. As a result, travel times (by car) to certain destinations with limited parking availability will increase, which will result in second-order effects such as choosing another destination (trip distribution) or choosing another mode (e.g. public transport) to reach the final destination.

This network aspect of parking is considered in several studies. For instance Lam et al. (2006) considers simultaneous departure time, route, parking location and parking duration choice, and implements this in a network equilibrium model. Boyles et al. (2015) considers parking searching in a traffic network assignment and considers uncertainty regarding the availability of vacant parking spaces. Both studies formulate a variational inequality problem, and solve it using a heuristic solution algorithm.

Leurent et al. (2014) use a static traffic assignment for reaching an equilibrium while taking into account route choice and parking location choice. It is assumed per parking location that a car can find a parking spot with a certain probability. The selection of parking locations and the routes taken to drive to those locations are outcomes of the model.

Pel et al. (2017) introduces the concept of *parking search routes*, where a driver visits a sequence of parking locations until finding a vacant parking spot. Here, drivers might go to locations that are close by, or locations that have a high certainty of available parking spots first, and then moving on. They use a queuing model and a stochastic user equilibrium to achieve a solution.

Lam et al. (1999) use a user equilibrium assignment with departure time choice and parking location choice. A walk network is used to walk from parking locations to destinations. In Lam et al. (2006) a travel time function (computing the travel time as a function of the number of cars and the capacity)) is used to model availability of parking spaces. As more people use a certain parking location, the time (cost) for using that parking location increases, and the parking location becomes less attractive to use.

Inspired by these approaches, and building upon modelling techniques developed previously and described by van der Tuin et al. (2021), we propose a method that incorporates parking behaviour such as diverting traffic and extra search time into the static traffic assignment. We use parking links, a walk network, and travel time functions to model parking and walking time, and the availability of parking spaces. The focus of this method is thus on the traffic network assignment, rather than mode or destination choice.

3   Methods

In this chapter, the methodology is described. First the implementation of the parking links with parking capacities is described, after which the method of determining the parking capacities is described.

3.1   Parking Links

The basis of our work is a static traffic assignment. A volume averaging algorithm is used, an heuristic to approach an equilibrium situation where no one can improve their travel time by choosing a different route (Ortuzár et al. 2011). An origin-destination matrix (OD matrix) containing all car trips is assigned to a network. In each iteration, the travel time and travel distance between each origin and destination zone are computed, and (a part of the) travelers change their route to take a shorter route between their origin and destination, if one exists.

To include parking in the static assignment, parking links are included in the network. Each zone (centroid) in the network gets one parking link with a set capacity, which represents the amount of parking spaces available. The parking search time is determined by a travel time (BPR-) function, as explained in more detail later. The more vehicles use a certain parking link (i.e. park in a certain zone), the higher the travel time (search time for a parking spot). At some point as the parking spaces fill up, the travel time becomes so high that it is quicker to park in a neighboring zone, and walk to the destination, instead of parking in the intended destination zone.

This mechanism is modelled by adding 4 different types of 'parking links' to the network: parking connector links, parking links, walk parking links and interzonal walk links. Figure 1 represents the situation where no parking links are added, and Fig. 2 shows an example where the 4 types of parking links have been added.

The parking links have limited capacity and cause the extra searching time. The walk parking links connect the parking links to the zones. The parking connector links connect the parking links to the car network, at the same locations as where the original connector links were connected. There can be several parking connector links for one zone, but each zone has only one parking link.

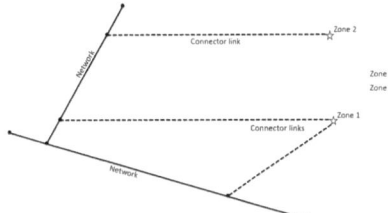

Fig. 1. Reference situation without parking links

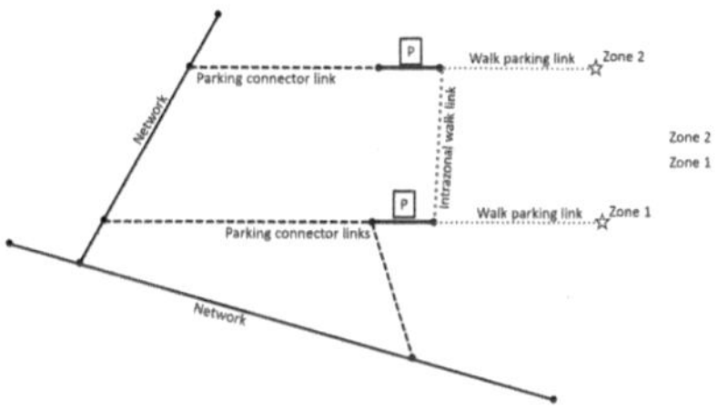

Fig. 2. Example of added parking links

The interzonal walking links are used for connecting the ends of parking links with those of other zones, and allow people to park in one zone and walk to another for their destination zone, when two zones are within a realistic walking distance. These interzonal walking links are added between each two zones that are within a set distance of each other. The walking time is determined by the distance between two zones (Euclidian distance) as well as the speed which is assumed for walking. This speed can be set a bit lower than the assumed walking speed, to correct for the fact that the real distance that has to be walked is probably more than Euclidian distance due to streets and corners.

Note that the parking links are only used at the destination-side of a trip, on the origin-side the shortest route through a parking link is used, and no capacity is considered. For walk parking links and intrazonal walk links, no capacity is used (infinite capacity).

For computing the travel time (i.e. search time) on a parking link, a (BPR) travel time function is used: $T_a(V_a) = T_0 \cdot (1 + \alpha \cdot (V_a/C_a)^\beta)$, where T_a is the travel time on link a, dependent on the traffic volume V_a, capacity C_a and free flow travel time T_0. The value for alpha is set at 300, and for beta at 4.1.

In the current implementation, only one time period is considered. This means that when the morning peak is modelled, all parking spots are assumed to be free. No car trips and resulting occupied parking spaces remaining from the time period before are used in the assignment.

In each iteration of the Volume Averaging assignment, the travel times are updated. It is determined how many parking spaces are occupied and the corresponding travel time (search time) on the parking link is computed. This usage is taken into account in the next iteration, until an equilibrium is achieved. After the final iteration, the search time and walking time (interzonal and to the destination) are added to the regular travel time, and saved in the level-of-service matrix. This level-of-service matrix is used in a next iteration of the connected mode choice model.

3.2 Parking Capacities

The method described above requires the capacity (maximum number of available parking spaces) per zone. Since such data is not readily available in the Netherlands, a methodology is presented to determine the amount of parking spaces. The method uses GIS (Geographic Information System) techniques, with data from several sources: Dutch cadaster (*Kadaster*), road data from the executive agency of the Dutch Ministry of Infrastructure and Water Management (*Rijkswaterstaat*), and data from the Dutch national parking registry (*Nationaal Parkeerregister*). Three different types of parking are distinguished: parking on own terrain (for homes), parking on parking lots and parking garages, and street side parking.

Parking on Own Terrain

For parking on own terrain, currently there is limited data available for who can park a car on their own terrain. Therefore a method is developed to make an estimate based on the amount of free space on a property, by checking whether there is sufficient unbuilt area available on the street-side of each home to park one or multiple cars.

Several data sets are used in this process. The registry of addresses and buildings (*Basisregistratie Adressen en Gebouwen*), for address points and street names, the registry of cadaster (*Basisregistratie Kadaster*) for properties and buildings, the national road registry (*Nationaal Wegenbestand*) for roads with street names, and the basic registry of topography (*Basisregistratie Grootschalige Topografie*). There is no data on 'land use' of private property, i.e. no data to distinguish between the garden and driveway.

For each home property, a raster of cells of 1 by 1 m is created. Each cell is then classified: built or unbuilt area, and inside or outside line of sight. With line of sight, it is checked whether the perpendicular line of that specific cell to the road is interrupted with built area and/or property boundaries. Then, cell clusters that are too narrow to drive a car past or to park a car are filtered out. A minimum length of 4.5 m and width of 2 m are used. The remaining cells determine the area suitable for parking on own terrain, see Fig. 3.

This is translated to a number of parking spots. It is assumed that part of the space is used for other purposes such as garden and/or walkway. The minimum required space for a parking spot would be approximately 11 square meter (2 m by 5.5 m). An area larger than 20 square meter is assumed to hold one parking spot, an area larger than 60 square meter is assumed to hold two parking spots. This conversion from square meter to number of parking spots has been verified by checking a small sample of home properties in Rotterdam, using Google Street View, and was deemed sufficiently accurate.

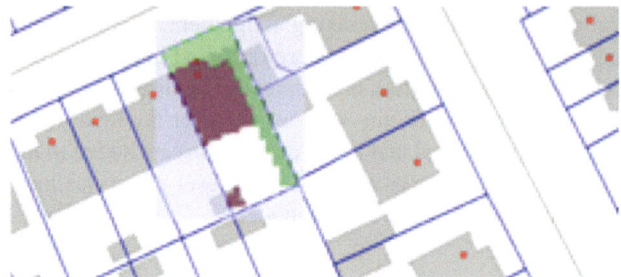

Fig. 3. Result of determining parking on own terrain cells (green) around a house (brown), using the line of sight

Parking on Parking Lots and Parking Garages

For parking in parking lots and garages, the method is more simple. For parking terrains, the basic registry topography (*Basisregistratie Grootschalige Topografie*) contains a field road part (*wegdeel*) with one of its classifications being 'space for parking'. The polygons with this classification are selected and the number of parking spots in each polygon is determined by dividing the area size by 11 square meter (the size of an average parking spot), and rounding down. For parking garages, the data set open parking data (*Open Parkeerdata*, by RDW) contains information on all public parking garages within Dutch municipalities. The set does not contain information on private parking garages (such as garages specific for shops or for apartment buildings). The number of parking spots in each garage is taken directly from this data set.

Street Side Parking

In many streets, either one or both sides of the street can be used to park alongside the street, this is noted by parking on the street side. For this type of parking, no direct data is available, so a method based on geographic data is developed. The national road registry (*Nationaal Wegenbestand*) is used in combination with OpenStreetMaps. Both data sets are combined, as OpenStreetMaps has a good classification of streets, and *Nationaal Wegenbestand* splits streets up into segments, to take into account crossings.

It is assumed that several street types are suitable for street side parking, and others are not. The suitable types are: access, bridleway, living_street, residential, road, tertiary, tertiary_link, unclassified. Highways and provincial roads are not suitable for parking and are left out of the selection of roads. It is assumed that 6 m of uninterrupted space is required for a parking spot. In order to determine these places, points are generated on the suitable streets each 3 m on both street sides. Then, points are eliminated. Points within 2 m from a crossing (beginning or end of a road segment) are removed, and points that are within 3 m of a parking on own terrain parking spot are also removed, to prevent blocking of driveways. Then, each set of three consecutive points (equal to 6 m) is counted as one parking spot. See Fig. 4 for an example of results.

When verifying this method with satellite pictures, it was found that the number of street side parking spaces is overestimated. In some areas, especially urban areas, street side parking is not possible due to narrow streets or parking restrictions. Therefore, the number of street side parking spaces is reduced by assuming only a percentage of the

Fig. 4. Parking alongside the street locations, blue dots represent the determined parking spaces, the red dots are filtered out. Aerial photo (luchtfoto) from PDOK, edited in QGIS

originally computed street side parking spaces to be available, dependent on the degree of urbanisation of the zone in which the street is located. In areas where the average number of inhabitants per square kilometer is higher than 85, only 25% of the street side parking spaces is available. For 50–85 inhabitants per square kilometer, a percentage of 50% is used, and for 25–50 inhabitants per square kilometer, 75% of street side parking spaces is assumed. These percentages have been verified using a sample with Google Maps, and seem realistic.

4   Use Case and Modelling Framework

The described assignment with parking links, using the parking capacities, is tested in a use case of the city of Amsterdam and surrounding areas. The model of Amsterdam is based on the *Verkeersmodel Amsterdam* (VMA), and contains 3035 zones and 77,668 links. The studied year is 2020, and the time period used is the morning peak period (7.00 – 9.00) (Gemeente Amsterdam, 2022).

The model is implemented in Urban Strategy (Lohman et al. 2023), a Digital Twin tool developed by TNO consisting of several models, such as an assignment module for car, freight and bike (Traffic+), for public transport (Public Transport), for mode choice (New Mobility Modeller), and modules for air quality, emissions and noise (Air and Noise). In this use case the module used is Traffic+. Although not demonstrated here, it is possible to use the methodology in a chain of modules, for instance running the assignment of cars in combination with (adjusted) mode choice.

In Fig. 5 a standard view in Urban Strategy is displayed. The traffic intensity of the morning peak of the reference scenario (described below) is displayed. The colour and bandwidth of the roads shows the amount of traffic on that road.

Parking capacities are determined by the methodology as described in Sect. 3.2. The zones on the edge of the network get unlimited parking capacity (999,999) so that all

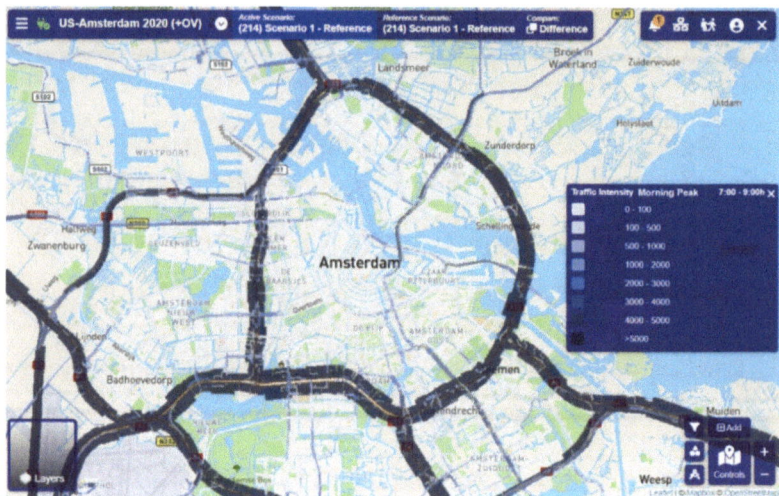

Fig. 5. Traffic intensities in Urban Strategy of morning peak period without parking links activated (scenario 1)

traffic can exit the network in the desired place. In total, this results in around 1.8 million parking spaces for the Amsterdam model (excluding the edge zones), of which around 9% are located in the city. The model allows for the application of *controls* with which multiple (policy) measures can be explored. In this case, the parking capacity (number of parking places) in specific zones can be adjusted using controls.

Table 1. Overview of scenarios

	Scenario	Use parking links	Capacity/controls
1	Reference	No	N.A
2	Parking capacity	Yes	Standard – no controls
3	Reduced parking capacity zone	Yes	Capacity in a zone reduced from 2042 to 50 parking spaces
4	Reduced parking capacity Amsterdam city center	Yes	Capacity in city center reduced by 75%

We present 4 scenarios for the city of Amsterdam, as shown in Table 1. The first scenario is a reference scenario, where the assignment is performed without using the parking links and capacities. In the second scenario, the parking links are used with the parking capacities as determined in Sect. 3.2, such that the effects of the parking links can be studied. In the third and fourth scenarios, controls are applied to see the effect on the parking behaviour. In the third scenario, a reduction in parking capacity in a zone near the Vondelpark in Amsterdam of 1992 places (from 2042 to 50 parking spaces) is realised. See Fig. 6 for the location, marked with the parking label (P). In the fourth scenario the parking space in the whole city center (within the ring road S100) is reduced by 75%. See Fig. 7 for the zones for which parking capacity is reduced, marked with pink dots. The total parking capacity of this area is reduced from 177,210 parking spots to 44,352 parking spaces.

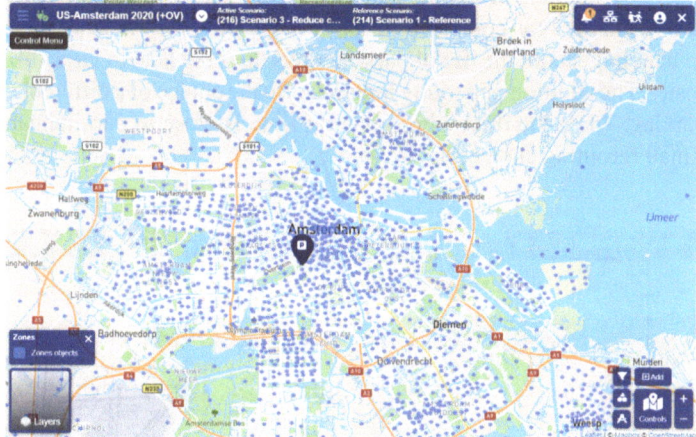

Fig. 6. Location of zone (parking marker) with reduced parking capacity in scenario 3

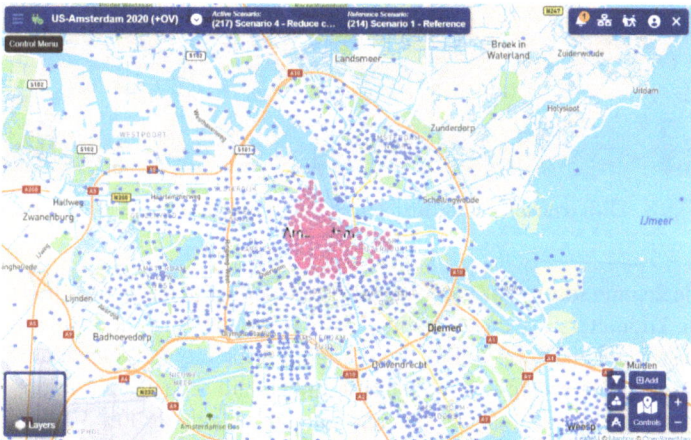

Fig. 7. Zones for which the parking capacity is reduced in scenario 4 marked with pink dots

5 Results and Discussion

In this chapter, the results of the 4 scenarios are presented. Intensity plots or difference plots are shown, to show the re-routing traffic due to limited capacity. Additionally, some indicators like vehicle loss hours (the amount of extra time spent in traffic as compared to a situation with no delay), vehicle kilometers driven, and occupation of parking spaces are shown.

In Fig. 8 the difference in car traffic intensity between scenario 2 and scenario 1 is shown. Figure 9 shows the same figure, but with a closer zoom to the city center of Amsterdam. The thickness of the bandwidth indicates the amount of cars on each link. The grey indicates the share of traffic that both scenarios have in common, red indicates an increase in scenario 2 (as compared to scenario 1), and green indicates a decrease. It can be seen that, in general, the use of parking links does not lead to large differences in traffic intensity. The addition of parking links leads to the choice of a different destination zone for parking, when the original destination zone is close to parking capacity, and thus only small rerouting choices are made.

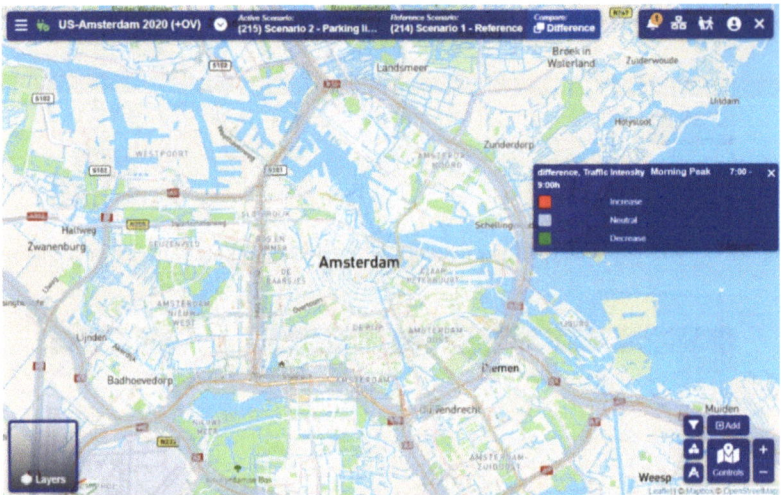

Fig. 8. Difference in car traffic intensity between scenario 2 and scenario 1

In Table 2 some results for the 4 scenarios are shown in terms of vehicle loss hours and vehicle kilometers driven. A distinction is made between the results for the total model (Amsterdam and surrounding areas), and just the city of Amsterdam.

When comparing scenario 1 (no parking links) with scenario 2 (parking links with standard capacity), it can be seen that the amount of vehicle loss hours and vehicle kilometers driven decreases. This can be explained by the fact that the introduction of parking links sometimes leads to shorter paths. This happens in the case where two zones are in close proximity to one another, but the car network between the two zones contains a large detour, for instance when one has to cross a bridge or park to get to the other zone. With the birds-flight distance that is assumed, parking in the one zone, and walking to the other zone is then a shorter route than driving to the original destination.

This issue needs some attention, and can be mitigated by defining zone-pairs between which cannot be walked. For instance, zone-pairs where for instance a canal or highway is located between both, and walking distance is actually larger than birds-flight distance

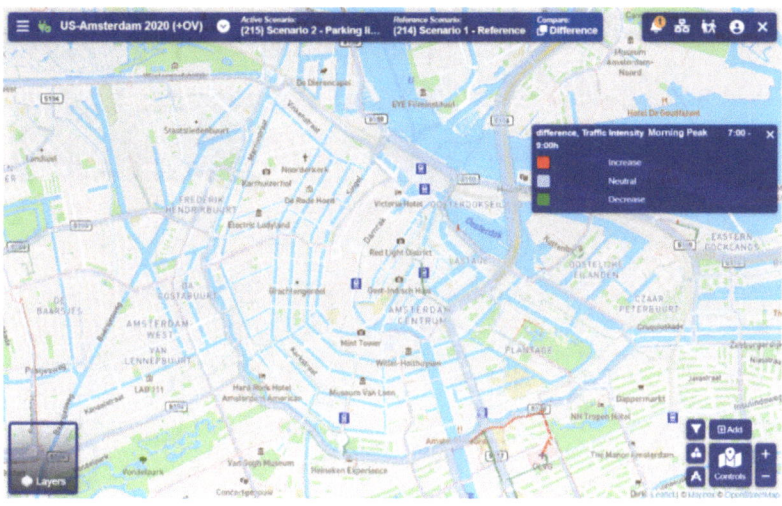

Fig. 9. Difference in car traffic intensity between scenario 2 and scenario 1 (closer view)

because people have to walk around. By carefully defining these zone-pairs, it is expected that the reduction in vehicle loss hours and vehicle kilometers driven becomes much less.

Table 2. Results of the 4 scenarios in terms of vehicle loss hours and vehicle kilometers driven

	Vehicle loss hours (total)	Vehicle kilometers driven (total) (*1,000)	Vehicle loss hours (Amsterdam)	Vehicle kilometers driven (Amsterdam) (*1,000)
Scenario 1	22076.1	5024.7	1243.2	1493.7
Scenario 2	21928.9	4999.2	1173.9	1475.4
Scenario 3	21929.0	4999.2	1173.9	1475.4
Scenario 4	21930.4	4999.0	1176.5	1475.0

Figure 10 shows the difference in car traffic intensity between scenario 3, where the parking capacity of a zone is reduced, and scenario 2. It can be seen that a small portion of traffic is rerouted because of this. In scenario 2, 209 cars parked in this zone, and in scenario 3 this is reduced to only 13 cars. The rest of the cars now park in a neighboring zone, and walk to their destination. The most used zones for this are just north of the zone with adjusted capacity. One can also see in the figure that the traffic towards the adjusted zone decreases slightly (green) and elsewhere in the network a small increase in traffic (red) can be observed. Of course, the effect is small, but this scenario serves as a showcase for the effect of the (reduced) capacity of the parking links. Also in Table 2 it can be seen that the effect of adjusting the parking capacity in one zone (scenario 3 compared to scenario 2) is almost nil.

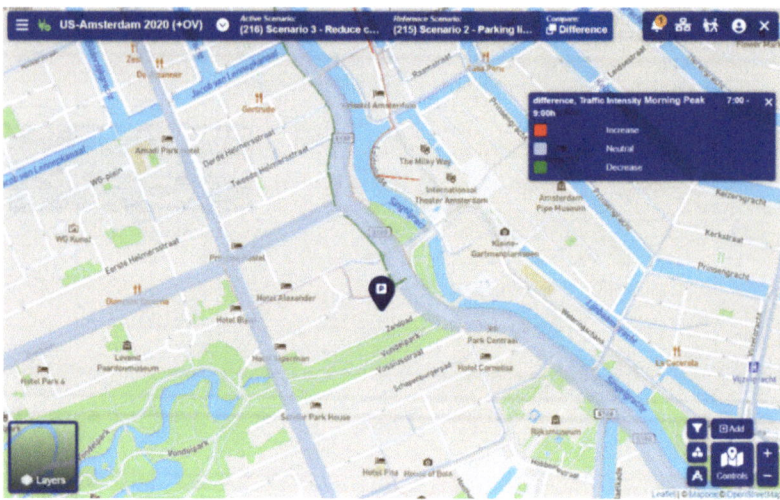

Fig. 10. Difference in car traffic intensity between scenario 3 and scenario 2

In Table 3 some results for the 4 scenarios in terms of occupied parking spaces (distinguished between the whole model, and the zones for which the capacity is reduced in scenario 4), and the number of cars that parked in a different zone than the destination, and walked to the destination are shown. For scenario 1 no results are shown, since the parking links are not used in this scenario. The number of occupied parking spaces is the same for the remaining three scenarios, which is logical, since in all scenarios all cars need to park somewhere. When comparing scenario 3 with scenario 2, an increase in cars parking in zones other than their destination can be seen. This can be explained by the reduced capacity for the zone. Since the zone for which the capacity is reduced is exactly on the edge (just outside) of the area in the city center (ring road S100), also an increase in the number of occupied parking spaces in the city center can be observed, since this is (part of) the traffic that parks in a different zone.

Table 3. Results of the 4 scenarios in terms of occupied parking spaces and rerouting traffic

	Number of occupied parking spaces	Number of occupied parking spaces within city center	Number of cars parked in zones other than destination zone
Scenario 1	N.A	N.A	N.A
Scenario 2	456313	8073	71958
Scenario 3	456313	8212	72015
Scenario 4	456313	6609	74352

When looking at parking capacity reduction for a larger area, as in scenario 4, larger effects can be observed, as seen in Fig. 11. Since the parking capacity for the whole city center is modelled, often driving to a different nearby zone is not a solution, since these zones also have reduced capacity. Some adjusted route choices can be seen, but the effect is mostly observed in the vehicle loss hours and vehicle kilometers driven (Table 2). The vehicle kilometers driven still decrease, because people take shorter (in terms of kilometers) routes, by parking and walking the last bit, resulting in less car vehicle kilometers driven. The vehicle loss hours increase, because of congestion that occurs on the parking links when zones (nearly) reach their parking capacity.

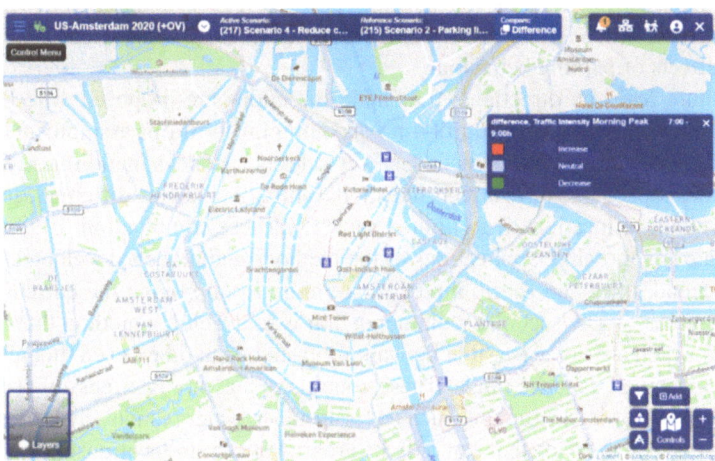

Fig. 11. Difference in car traffic intensity between scenario 4 and scenario 2

The increase in vehicle loss hours is, even with a 75% reduction of parking capacity in the city center, quite limited. This indicates that the amount of parking spaces estimated is still quite sufficient. Two explanations can be given for this fact. The first is that we model total parking capacities to be available, but only the traffic for a morning peak period. In reality, some spots are already taken at the start of the morning peak, which are not taken into account. An improvement of the method would be to keep track of parking capacities over the different time periods that can be modelled. Then, the parking capacity is used by the traffic that arrives at a parking location for the whole day rather than just a part of it. The second reason may be that the parking capacities are overestimated. Part of the parking spots are the spots estimated to be along the streets. Although the number of these should be limited in Amsterdam because of high urbanisation rates, in reality these parking spots are very rarely available in the city center of Amsterdam. An adjustment or calibration of the number of parking spots in urbanised city centers may be required.

Additionally, when comparing scenario 4 with scenario 2, it can be seen that much less cars are parked within the city center (Table 3). This is to be expected due to the reduced parking capacity. At the same time, the number of cars parked in a different zone than their destination increases.

For running the traffic assignment of 10 iterations volume averaging with parking links, the model takes 92 s on a standard desktop PC.

6 Conclusion and Recommendations

In this paper, a method is described to determine parking capacities and to include these parking capacities in the static traffic assignment. The method to determine parking capacities is novel, since data is not readily available due to many different types and owners of parking spaces. The usage of parking capacities in traffic assignment is demonstrated by application to a use case of Amsterdam, and it is shown to display realistic behaviour and fast computation times. This methodology allows cities to explore the effects of parking measures on for instance route choice, destination choice and mode choice in detail, which is very important if cities wish to be car-free without the possible negative effects of parking behaviour.

For future research directions, four main directions are proposed. The first, is to study the behaviour and model equilibrium of combining the traffic assignment including parking links with other travel choices such as mode choice or destination choice. Parking measures are expected to have effects like modal shift, or choosing different destinations when travelling to locations with limited capacity. These combinations of choices can demonstrate new possible applications of the methods and models developed.

The second, third and fourth direction are more in improving the current application and use cases. The second recommendation is to develop a method that describes zone-pairs which are not accessible by walking from one another, even though they are within the specified distance range. These zone-pairs may not be accessible because of physical barriers or infrastructure, such as for instance water (canals, rivers, lakes, etc.) or highways or other roads that can only be crossed at certain locations. Defining these zone-pairs will lead to more realistic results, because currently in some instances the model uses these zone-pairs to find (much) shorter routes which are not realistic because they cross water or infrastructure.

The third recommendation is to further verify and detail the method for determining the parking capacities. Although the method described is a good first step in defining data sets that are not readily available, improvement is possible. Especially in the city centers, very often parking along the streets is not possible or not allowed, and the currently estimated parking spaces are an overestimation of reality. Introducing more detail here will lead to more realistic parking capacities, which in turn lead to a better assignment of traffic to the network.

The fourth recommendation is related to the destination of trips. It is now assumed in the model that each traveler has a fixed destination, and they know where to drive to. Even when parking in a different zone, they still walk to this destination. In reality, travelers sometimes drive to a certain area and park where they see a free spot. This behaviour of circling around is not yet explicitly modelled, but would be a good next development.

Acknowledgements. This work is partially funded by the project XCARCITY, funded by the Dutch Research Council (NWO) under the Perspectief program.

References

Boyles, S.D., Tang, S., Unnikrishnan, A.: Parking search equilibrium on a network. Transport. Res. Part B: Methodol. **81**, 390–409 (2015)

Chaniotakis, E., Pel, A.J.: Drivers' parking location choice under uncertain parking availability and search times: a stated preference experiment. Transport. Res. Part A: Policy Pract. **82**, 228–239 (2015)

Gemeente Amsterdam, Verkeer en Openbare Ruimte. Uitgangspunten Verkeersmodel Amsterdam 4.0 (2022). https://data.amsterdam.nl/datasets/zTHEXVzwMqQI9w/verkeersprognoses-verkeersmodel-amsterdam/

Lam, W.H., Li, Z.C., Huang, H.J., Wong, S.C.: Modeling time-dependent travel choice problems in road networks with multiple user classes and multiple parking facilities. Transport. Res. Part B: Methodol. **40**(5), 368–395 (2006)

Lam, W.H., Tam, M.L., Yang, H., Wong, S.C.: Balance of demand and supply of parking spaces. In: 14th International Symposium on Transportation and Traffic Theory Transportation Research Institute (1999)

Leurent, F., Boujnah, H.: A user equilibrium, traffic assignment model of network route and parking lot choice, with search circuits and cruising flows. Transport. Res. Part C: Emerg. Technol. **47**, 28–46 (2014)

Lohman, W., Cornelissen, H., Borst, J., Klerkx, R., Araghi, Y., Walraven, E.: Building digital twins of cities using the Inter Model Broker framework. Futur. Gener. Comput. Syst. **148**, 501–513 (2023). https://doi.org/10.1016/j.future.2023.06.024

de Dios Ortúzar, J., Willumsen, L.G.: Modelling Transport. John wiley & sons, Hoboken (2011)

Pel, A.J., Chaniotakis, E.: Stochastic user equilibrium traffic assignment with equilibrated parking search routes. Transport. Res. Part B: Methodol. **101**, 123–139 (2017)

Shoup, D.C.: Cruising for parking. Transp. Policy **13**(6), 479–486 (2006)

Van der Tuin, M.S., de Romph, E., Pieters, M.: Grip op parkeercapaciteiten–modelaanpak en data-analyse. In: Bijdrage aan het Colloquium Vervoersplanologisch Speurwerk, Utrecht, 25 en 26 november 2021, p. 1 (2021)

Open Access This chapter is licensed under the terms of the Creative Commons Attribution 4.0 International License (http://creativecommons.org/licenses/by/4.0/), which permits use, sharing, adaptation, distribution and reproduction in any medium or format, as long as you give appropriate credit to the original author(s) and the source, provide a link to the Creative Commons license and indicate if changes were made.

The images or other third party material in this chapter are included in the chapter's Creative Commons license, unless indicated otherwise in a credit line to the material. If material is not included in the chapter's Creative Commons license and your intended use is not permitted by statutory regulation or exceeds the permitted use, you will need to obtain permission directly from the copyright holder.

Driving Style Characterisation and its Impact on Vehicle Energy Efficiency

Jaime Suarez[1]([✉]) [iD], Andres L. Marin[2] [iD], Dimitrios Komnos[1],
Alessandro Tansini[1] [iD], and Georgios Fontaras[1] [iD]

[1] European Commission, Joint Research Centre (JRC), Ispra, Italy
{Jaime.Suarez-Corujo,Georgios.Fontaras}@ec.europa.eu
[2] Universitat Politècnica de València, VRAIN, 46022 Valencia, Spain

Abstract. The driving style is a crucial variable for real-world vehicle energy performance. This study uses a state-of-the-art methodology to separately assess its impact on the energy consumption. A benchmark driving cycle is modified according to specific driving styles, and the corresponding energy consumption is theoretically simulated. The driving style characterisation includes acceleration, braking and gear-shifting patterns for each driver through ad-hoc metrics. The methodology establishes a comparative analysis regarding the energy impact of a pool of drivers. Furthermore, the variability of a driver's consumption on different vehicle technologies is assessed by considering an internal combustion engine- (ICE) vehicle and a plug-in hybrid electric vehicle (PHEV). The results reveal that each driver exhibits characteristic and heterogeneous acceleration and braking styles, with the gear-shifting style playing a significant role in the vehicle's fuel consumption. From a policy perspective, this work contributes in the understanding of the discrepancies in energy consumption between officially reported values and the real-world performance.

Keywords: Sustainable mobility · Transport emissions · Energy consumption · CO_2 emissions · driver characterisation · driving style

1 Introduction

In the route to zero emissions transport, increasing interest is growing in reducing the CO_2 emissions over real-world traffic conditions [1]. Vehicle energy efficiency varies due to the combination of vehicle technology, route, ambient conditions and driver-dependent factors. This variation is at the root the of the fuel consumption gap found between real-world data and the official values of the fleet [2]. Among the driver-dependent factors, the driving style is known to have a deep impact on the energy/fuel consumption. The driving style modulates key aspects of the vehicle operation, such as longitudinal accelerations and gear-shifting patterns. However, assessing the impact out of experimental observations is complex and frequently unapproachable due to the strong entanglement among the different driving factors.

Human drivers show a distinct heterogenic driving behaviour in real-world as a result of variable conditions [3], and the CO_2 footprint shows high inter-and intra-driver

© The Author(s) 2025
C. McNally et al. (Eds.): TRAconference 2024, LNMOB, pp. 120–126, 2025.
https://doi.org/10.1007/978-3-031-85578-8_16

variability [4]. The first one is the result of each driver having a characteristic behaviour, notably on the acceleration pattern. The intra-driver variability represents the influence of other factors in the behaviour of the same driver. An interesting factor to analyse is the behaviour of the same driver on vehicles of different technologies.

The scope of this study is to quantify the influence of the driving style on a comparative basis among different drivers and vehicle technologies. Using the methodology introduced in [5], the comparative analysis of the driver's energy footprint is benchmarked to a reference driving cycle (trip). The speed profile of the trip is conveniently modified according to the driving behaviours captured from the 15 different drivers that followed a real-world driving campaign, taking into account their respective acceleration and braking styles.

The first part of the work consists of characterising the stochastic driving style of each driver by probabilistic functions on the Independent Driving Style (IDS) metric [3]. In a second step, the same metric is used to generate the synthetic trajectories adapted to each driving style. Afterwards, the energy consumption and CO_2 emissions are calculated with a vehicle energy performance simulator for two vehicles representative of different technologies: a conventional internal combustion engine (ICE) vehicle and a plug-in hybrid electric vehicle (PHEV). As a novelty in this work, an integral characterisation of the driving style is adopted, considering both positive and negative longitudinal accelerations and different gear-shifting patterns.

Section 2 provides a short description of the methodology (for further details the reader is referred to [5]), including some details about the experimental driving campaign. The results of the comparative study according to the simulations are given in Sect. 3, while Sect. 4 includes a discussion on the conclusions that can be obtained.

2 Methodology

2.1 Experimental Driving Campaign

The empirical data used in this work have been obtained during the experimental driving campaign followed for two years [6, 7]. During a period of three weeks, each of the 15 drivers in the pool drove more than 1,000 km on each of the two vehicles detailed in Table 1 in the Annex. Both vehicles belong to the same manufacturer and have similar dimensions. The drivers covered a total mileage of 19,588 km on the ICE and 24,945 km on the PHEV. Most of the trips took place in the North of Italy, where the driving conditions are predominantly characterised by free-flow accelerations.

2.2 Characterisation of the Driving Style

The characterisation of the driving style is performed according to two main metrics introduced in previous works. The IDS characterises the dynamicity of the driver during the accelerations [3] and, as a novelty in this work, braking events. By normalising the instantaneous accelerations to the maximum potential acceleration at each speed, the metric characterises the driving style independently of the characteristics of the vehicle. Given the intra-driver heterogeneity, each human driver needs to be characterised by a certain probability distribution, rather than a single value of the IDS metric. The probability functions (see Fig. 1) have been generated out of the data collected during the experimental driving campaign.

The gear-shifting behaviour can also be reproduced using the Gear-Shifting Style (GSS) parameter introduced in [8]. This metric represents the shifting points when the driver shifts gear, tailoring the power curve of the vehicle that ultimately determines the vehicle speed. A higher value implies a behaviour where the vehicle reaches higher engine speeds before gear-shifting. For the PHEV vehicle, the transmission is automatic and the GSS is predefined by the vehicle and taken as GSS = 0.3 in this paper.

In this paper we explore the whole variability range of combinations of IDS and GSS that can be found in human driving heterogeneity.

2.3   Energy Performance Based on Modified Speed Profiles

The impact of the driving style on the energy consumption is benchmarked to the WLTC reference cycle followed in the Worldwide Harmonised Light Vehicle Test Procedure (WLTP), used in the certification of fuel consumption and CO_2 and pollutant emissions. The reference speed profile is modified according to new acceleration/braking and gear-shifting strategies, keeping constant the distance range of each acceleration event but changing the speed and duration. The methodology is as follows:

- Selection of the benchmark speed profile (time versus speed), obtained from the WLTC cycle, and identification of the acceleration events.
- Modification of the acceleration/braking events according to the required IDS and GSS values, using synthetic trajectories generated by the MFC model [8].
- For each ad-hoc modified speed profile, the vehicle energy performance is simulated using the CO_2MPAS simulation tool [9] for the ICE vehicle and the MFC model [10] for the PHEV vehicle. To better compare the energy performance of both vehicles, the CO_2 emissions from the ICE are translated into energy consumption [11].
- Analysis of the variability in the CO_2 emissions regarding different driving IDS and GSS values (shown in the Results section).

In the case where the target speed cannot be reached at the end of the acceleration segment, usually when the average IDS of the modified acceleration is lower than the WLTC one, the acceleration continues until the point where the speed matches again the WLTC speed profile.

3   Results

3.1   Acceleration Profile of the Drivers

The data shows distinct IDS distributions for each driver in acceleration and braking events for both vehicles. Figure 1 shows the acceleration distributions for all the drivers. The distributions show the heterogeneity in the acceleration behaviour of each driver, reflected in the broadness of the distributions. Differences are found both among drivers (inter-driver variability) and between the two driven vehicles (intra-driver variability). This variability encourages an analysis of its effect on CO_2 emissions, using a combination of gear-shifting, braking and acceleration behaviours on the WLTC cycle.

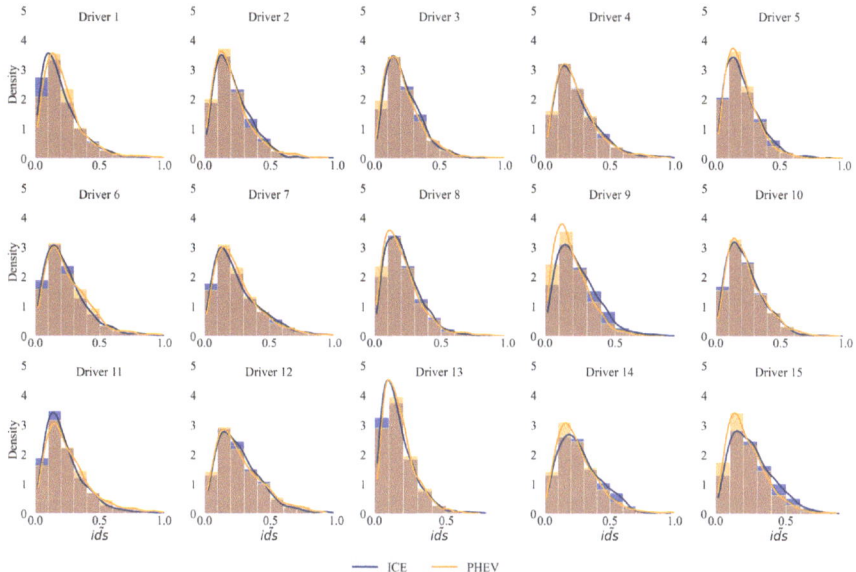

Fig. 1. IDS probability distribution for each driver in both the ICE (blue) and PHEV (orange) vehicles.

3.2   Energy Consumption and CO_2 Emissions

The vehicle's energy performance is simulated (in terms of energy consumption and/or CO_2 emissions) for each combination of IDS and GSS metrics explored by the drivers. In the case of the ICE vehicle, shown in Fig. 2 (top), the energy consumption is related to the acceleration (IDS) and gear-shifting behaviour, showing two clear trends. On the one hand, the energy consumption increases with the dynamicity in the acceleration style (IDS function), reaching a variability of 25% between the lowest and highest IDS values for low gear-shifting styles, while the variability is reduced to 10% for the most aggressive gear-shifting (GSS = 1). This pattern reflects the higher speeds that a higher IDS value involves, and the higher fuel consumption associated to a high-revolutions engine speed regime determined by the GSS. We also identify a certain saturation effect at high-IDS values, where most of the potential acceleration of the vehicle is demanded.

Figure 2 (bottom) shows the impact of the driving style on the PHEV, which represents the typical operation of the vehicle when performing a WLTC cycle (ca. 23 km) with more than 50% of the initial battery State of Charge (SOC). In this case, the gear-shifting style has been set to 0.3 reflecting the automatic transmission of this PHEV. In hybrid vehicles, the energy consumption is optimised by recovering part of the energy lost during the braking events (regenerative braking). The figure clearly shows that there is an evident link between the driver's style during braking and the impact on energy consumption and CO_2 emissions, with more aggressive braking patterns (higher IDS values) leading to higher energy consumption. The difference between the most conservative braking style and the most dynamic one considered is 34 Wh/km, which represents a 20%–25% range of the energy consumed.

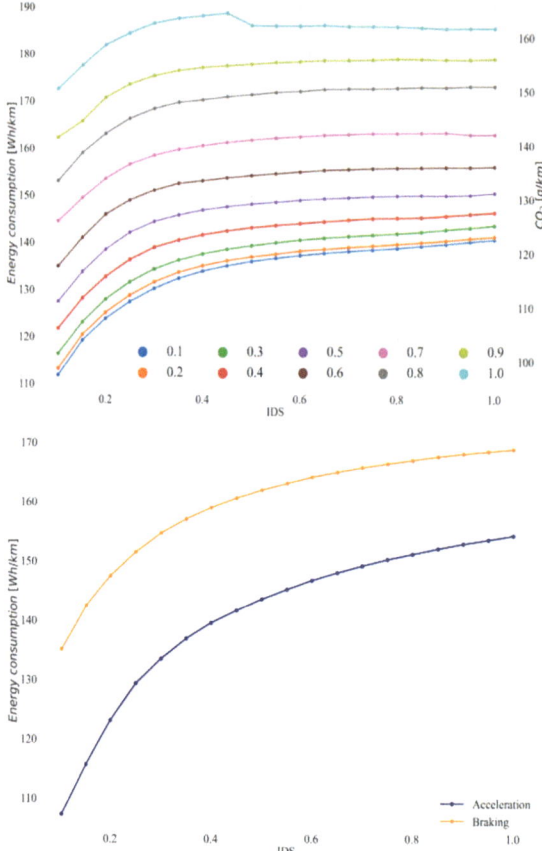

Fig. 2. Top: Energy consumption and CO_2 emissions as a function of IDS for the ICE vehicle for different gear-shifting styles (GSS). Bottom: Same as top, for the PHEV vehicle running on electric mode for different acceleration/braking strategies. A constant GSS value of 0.3 is assumed.

The impact of different acceleration styles on the variability of the energy consumption is higher than for braking, resulting in a 50 Wh/km range between the most moderate and the most dynamic acceleration style (yielding a variability range of 32%–45%). The curve of the braking falls above the acceleration one due to the fact that the WLTC assumes an acceleration style that is more dynamic compared to the braking style, which is quite conservative in WLTC.

Regarding the comparison between the energy consumption of the PHEV and the ICE vehicle using GSS = 0.3, the range for ICE extends from 117 Wh/km to 143 Wh/km from the lowest to highest IDS, while the PHEV covers a broader range [108 Wh/km–154 Wh/km]. The energy consumption of the PHEV in electric mode shows a rather monotonous and increasing trend regarding the IDS value. The efficiency of the electric engine is therefore less speed-dependent than for ICE operation. Regarding the total energy consumed, both vehicles have comparable values. The explanation is that the

ICE vehicle is a diesel car, while the petrol-PHEV is heavier because of the battery and the electric motor. Furthermore, we have calculated the energy consumption at the wheels for the ICE vehicle with a typical combustion efficiency of ca. 30%, notably lower than the energy conversion for electric engines.

4   Conclusions

The present work quantifies the impact of the driving style on the vehicle energy consumption by means of computational simulations. The driving style was characterised by the acceleration dynamicity (IDS) and gear-shifting pattern (GSS). The effect on different vehicle technologies is analysed, considering an internal combustion engine vehicle and a plug-in hybrid vehicle. The analysis of the energy consumption benchmarked to the WLTC cycle shows a high impact of the IDS and GSS parameters, highlighting the key role of the driver behaviour for reducing the vehicle's CO_2 emissions. The proposed approach of the driving style has demonstrated to be suitable for the describing the inter-driver and intra-driver characteristic variabilities.

Our simulations show that both the acceleration and the braking aggressiveness between initial and final fixed speeds have a direct impact on the consumption. In both vehicles, the energy impact increases with acceleration dynamicity. The variation in energy consumption is higher for the PHEV running in electric mode (up to 45%) than for internal combustion (20%-30%), where a saturation effect is found at dynamicity. The gear-shifting style plays also a role on the energy consumption, but the effect is lower for highly-dynamic trends. The role of the braking style is as important as the acceleration style in this type of hybrid vehicles, and it can lead to variations up to 25% in the energy consumption, with moderate braking improving the energy efficiency.

Appendix

Table 1. Characteristics of the ICE and PHEV vehicles in the driving campaign.

C-Segment vehicle	ICE	PHEV
Powertrain	Diesel	Gasoline/Electric
Transmission	Manual (5 gears)	Automatic
Engine size	1598 cc	1395 cc
Engine power	85 kW	110 kW
Electric range/motor power	–	70 km/80 kW
Vehicle Mass	1488 kg	1698 kg
Fuel Consumption	5.1 l/100 km	0.9 l/100 km

References

1. Mamarikas, S., Doulgeris, S., Samaras, Z., Ntziachristos, L.: Traffic impacts on energy consumption of electric and conventional vehicles. Transp. Res. Part Transp. Environ. **105**, 103231 (2022). https://doi.org/10.1016/j.trd.2022.103231
2. Pavlovic, J., et al.: Understanding the origins and variability of the fuel consumption gap. Environ. Sci. Eur. **32**, 53 (2020). https://doi.org/10.1186/s12302-020-00338-1
3. Makridis, M.A., Anesiadou, A., Mattas, K., Fontaras, G., Ciuffo, B.: Characterising driver heterogeneity within stochastic traffic simulation. Transport. B: Transp. Dyn. **11**(1), 725–743 (2023). https://doi.org/10.1080/21680566.2022.2125458
4. Fiori, C., Ahn, K., Rakha, H.A.: Microscopic series plug-in hybrid electric vehicle energy consumption model: Model development and validation. Transp. Res. Part Transp. Environ. **63**, 175–185 (2018). https://doi.org/10.1016/j.trd.2018.04.022
5. Suarez, J., Makridis, M., Anesiadou, A., Komnos, D., Ciuffo, B., Fontaras, G.: Benchmarking the driver acceleration impact on vehicle energy consumption and CO2 emissions. Transp. Res. Part Transp. Environ. **107**, 103282 (2022). https://doi.org/10.1016/j.trd.2022.103282
6. Suarez, J., Laverde, A., Tansini, A., Ktistakis, M.A., Komnos, D., Fontaras, G.: Observations on the Driving of Plug-In Hybrid Cars in Real-World Conditions. https://www.springerprofessional.de/en/observations-on-the-driving-of-plug-in-hybrid-cars-in-real-world/24611318
7. Ktistakis, M., Tansini, A., L. Marin, A., Suarez, J., Komnos, D., Fontaras, G.: Understanding the fuel consumption of plug-in hybrid electric vehicles: a real-world case study. In: THIESEL 2022 Conference on Thermo- and Fluid Dynamics and Clean propulsion Powerplants (2022)
8. Makridis, M., Fontaras, G., Ciuffo, B., Mattas, K.: MFC free-flow model: introducing vehicle dynamics in microsimulation. Transp. Res. Rec. J. Transp. Res. Board. **2673**, 762–777 (2019). https://doi.org/10.1177/0361198119838515
9. Fontaras, G., et al.: The development and validation of a vehicle simulator for the introduction of Worldwide Harmonized test protocol in the European light duty vehicle CO2 certification process. Appl. Energy **226**, 784–796 (2018). https://doi.org/10.1016/j.apenergy.2018.06.009
10. He, Y., Makridis, M., Mattas, K., Fontaras, G., Ciuffo, B., Xu, H.: Introducing electrified vehicle dynamics in traffic simulation. Transp. Res. Rec. J. Transp. Res. Board. **2674**, 036119812093184 (2020). https://doi.org/10.1177/0361198120931842
11. Noce, T., de Morais Hanriot, S., Sales, L.C.M., Sodré, J.R., de Novaes, M.B.: Energy conversion factor for gasoline engines in real-world driving emission cycle. Automot. Innov. **3**, 169–180 (2020). https://doi.org/10.1007/s42154-020-00098-x

Open Access This chapter is licensed under the terms of the Creative Commons Attribution 4.0 International License (http://creativecommons.org/licenses/by/4.0/), which permits use, sharing, adaptation, distribution and reproduction in any medium or format, as long as you give appropriate credit to the original author(s) and the source, provide a link to the Creative Commons license and indicate if changes were made.

The images or other third party material in this chapter are included in the chapter's Creative Commons license, unless indicated otherwise in a credit line to the material. If material is not included in the chapter's Creative Commons license and your intended use is not permitted by statutory regulation or exceeds the permitted use, you will need to obtain permission directly from the copyright holder.

Using a Motorcycle Probe Vehicle to Provide Infrastructure Information for Powered Two Wheelers

Andreas Hula$^{(\boxtimes)}$ ⬤, Christian Klösch⬤, Matthias Hahn⬤,
Bernhard Preiser-Kapeller⬤, Roland Spielhofer⬤, and Peter Saleh⬤

Austrian Institute of Technology, 1210 Vienna, Austria
`andreas.hula@ait.ac.at`

Abstract. Powered two wheelers are a consistently popular mode of transportation and riding them is a widely practiced recreational activity. However, with regards to safety powered two wheelers are clearly vulnerable road users, for whom accidents of all types yield more severe outcomes on average than for larger vehicles. One way to ensure the safety of powered two-wheeler riders is to identify challenging infrastructure properties, such as transversal evenness qualities or potholes, through the use of probe vehicles. We present the analysis of infrastructure properties through data collected with a motorcycle probe vehicle, previously employed to study human driving dynamics and assess the risk thereof. We present the first steps towards a new standardized evaluation of motorcycle driving dynamics data for this purpose and show that our outcomes can be achieved based on in-vehicle driving dynamics data. This holds the potential to enable the provision of safety relevant data from everyday driven vehicles, which represents the needs of powered two wheelers as much as those of passenger cars or larger vehicle-types and could serve as a template for similar analyses employing smaller probe vehicles like bicycles or (e-)scooters.

Keywords: Powered Two Wheelers · Vulnerable Road Users · Safety · Infrastructure Conditions · Connected Vehicles

1   Introduction

Motorcycling dynamics are noticeably affected by road condition and infrastructure conditions (see [1]). To capture the specific effects of road infrastructure properties on motorcyclists and their driving dynamics, a motorcycle equipped with several sensor systems, geolocalisation and access to in vehicle systems (the vehicle CAN-Bus, see [2]), can be employed. Such a system provides time series of rotational speeds (Yaw-, Roll- and Pitch-rates) and accelerations (X-, Y- and Z-accelerations), (see [3]), that can be investigated through the means of statistical modelbuilding/machine learning (see [4, 5]), to extract information on the road infrastructure reflected in the driving dynamics or in turn, try and predict dynamics data for a given rider based on infrastructure properties.

The interplay obtained this way could be used to find dangerous spots for particular rider types or predict other safety relevant information before having a rider even reach a future critical spot.

© The Author(s) 2025
C. McNally et al. (Eds.): TRAconference 2024, LNMOB, pp. 127–133, 2025.
https://doi.org/10.1007/978-3-031-85578-8_17

2 Materials & Methods

2.1 The Motorcycle Probe Vehicle (MoProVe)

The motorcycle probe vehicle (MoProVe, [6]) is a KTM (see [7]) 1290 Super Adventure motorcycle equipped with 2 front wheel cameras, access to the vehicle CAN-bus, an additional IMU and data logging device (see [8]) and multiple gelocalisation options. The motorcycle probe vehicle was ridden by an experienced motorcyclist, who gave informed consent to take part in this experiment and had PRIOR knowledge of the test track. Data obtained via the MoProVe was for vehicle accelerations (A_x, A_y, A_z) and rotational speeds (yaw-rate Y, roll-rate R and pitch-rate P) all obtained from the CAN-BUS.

2.2 Road Condition Data

Road condition data was acquired utilizing the RoadSTAR (see [9]) measurement systems. This system provides established procedures for measuring friction μ (see [10]), transverse evenness q (see [11]), longitudinal evenness s (see [12]) und finally for assessing road surface damage (total dem and cracks cr in percent on each meter) from camera data (see [13]). Additionally, road with w and curvature k are obtained during the measurement procedure. This data was essential for investigating the relation between road parameters and the driving dynamics of the MoProVe.

2.3 Test Track

The test track was located between the two Austrian federal states of Lower Austria and Burgenland, around the area known as Leithagebirge. Data was used from 26.3 driven kilometers and this track included an initial highly curved section through hilly and forested territory followed by a rural road, passing through 2 towns and finally a highway, forming almost a full circle back to the initial point of the test track. The test track was split into a training data set consisting of the first 8 km (the hilly part) and a test data set of the remaining 18.3 km consisting of the rural roads and highway. This was a deliberate choice, to investigate how well a model trained on the relatively most dynamic part might perform on other types of roads. Four repetitions were done for the MoProVe data collection.

3 Results

3.1 Analysis

Riding data was aligned to per meter values obtained from the RoadSTAR time series, based on GPS coordinate proximity. MoProVe GPS Data was inspected for poor alignment with the reference RoadSTAR trajectory and the respective segments replaced with a speed-based interpolation of on track position. Overall, only a small single digit percentage of signals had to be corrected, with this primarily affecting the initial hilly and forested section through the Leithagebirge.

Furthermore, for each meter on the track we calculated the max and the min value of the driving dynamics, as well as the averages, each per meter over all rides.

The resulting time series (values for each meter) for RoadSTAR data were then

$$D\_RS := (w, \mu, k, q, s, dem, cr) \tag{1}$$

and analogously, for the MoProVe the averages per meter over 4 rides

$$D\_MoPro\_av := (y, r, p, a\_x, a\_y, a\_z) \tag{2}$$

and the maximum and minimum values per meter extracted at these rides.

$$D\_MoPro\_ext := (y_{max}, r_{max}, p_{max}, a\_x_{max}, a\_y_{max}, a\_z_{max}, \\ y_{min}, r_{min}, p_{min}, a\_x_{min}, a\_y_{min}, a\_z_{min}) \tag{3}$$

These timeseries were subsequently used in a first regression analysis based supervised learning approach, as shown in the next subsections. Model were implemented in R [14] using the basic "stats" functionalities.

3.2   Prediction of Riding Dynamics Data from Infrastructure Data

The outcomes of

$$Y \sim c_w w + c_\mu \mu + c_k k + c_q q + c_s s + c_{dem} dem + c_{cr} cr + \in \tag{4}$$

for Y in D_MoPro_av encoded the results of regression of average riding dynamics parameters against select RoadSTAR parameters (with coefficients c_w, c_μ, c_k, c_q, c_s, c_{dem}, cr and a normally distributed error term ϵ). We show the results of variance explained for all target variables on the training set and the R^2 between prediction and true value on the test set in Table 1 below.

Table 1. Training Results (R^2) and Pearson-correlations between prediction and Y with test data for all target variables Y for predicting a rider's dynamics from road parameters.

Target Variable Y	Adjusted R2	Test Set R2	Training Correlation	Test Correlation
y	0.19	0.28	0.43	0.53
r	0.02	−0.02	0.14	0.12
p	0.11	−0.4	0.33	0.04
a_x	0.27	−0.01	0.52	0.15
a_y	0.14	0.24	0.37	0.55
a_z	0.03	−0.01	0.18	0.05

In Fig. 1 we depict the case of a_y prediction on the training and test data side by side.

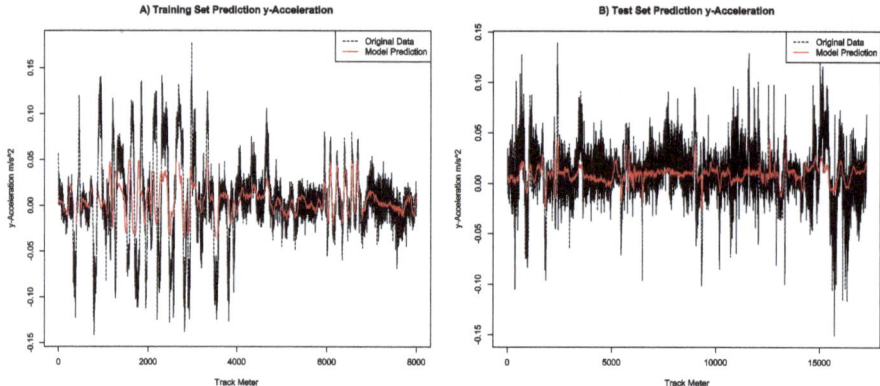

Fig. 1. A) Prediction (red, solid) of y-acc by road parameters on training data (black, dotted). B) Prediction (red, solid) of y-acc by road parameters on test data (black, dotted).

3.3 Prediction of Infrastructure Data from Riding Dynamics Data

Analogously to 3.2 the regression of

$$Y \sim D\_MoPro\_ext + D\_MoPro\_ext^2 + \in \tag{5}$$

encoded the results of predicting, with linear and squared terms, the road parameters (Y in D_RS) from the extremes of the riding dynamics data, on the grounds that these should be more sensitive to the same road segment meter being passed on changing motorcycle trajectories, compared to just averages.

We show the results of variance explained for all target variables on the training set and the correlation between prediction and true value on the test set in table 2 below.

Table 2. Training Results (R^2) and Pearson-correlations between prediction and Y with test data for all target variables Y for predicting road parameters from a rider's dynamics.

Target Variable Y	Adjusted R2	Test Set R2	Training Correlation	Test Correlation
w	0.16	−0.26	0.4	0.01
μ	0.29	−0.49	0.54	0.08
k	0.7	−0.38	0.84	0.34
q	0.48	0.2	0.69	0.45
s	0.44	−0.54	0.66	0.16
dem	0.11	−0.34	0.33	−0.01
cr	0.18	−3.54	0.43	0.01

In Fig. 2 we depict the case of transversal evenness prediction on training and test data side by side.

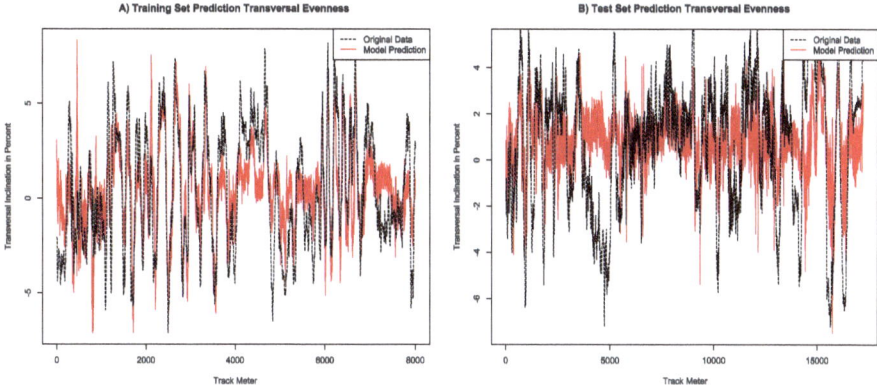

Fig. 2. A) Prediction (red, solid) of transversal evenness by riding dynamics on training data (black, dotted). B) Prediction (red, solid) of transversal evenness by riding dynamics on training test data (black, dotted).

4 Discussion

This work presented an investigation of a correlational analysis between road infrastructuredata, collected with high quality equipment and processed by long term experts, with riding dynamics data collected with a motorcycle probe vehicle, devised for the express purpose of investigating safety and reliability of infrastructure for motorcyclists. The analysis revealed strong correlations on both a training set and then afterwards on the test data, supporting the potential to investigate motorcycle relevant infrastructure conditions through probe vehicles, as well as predicting likely driving dynamics from known infrastructure parameters.

The current investigation suggests that y-dynamics (y-acceleration and yaw-rate) for the dynamics data and transversal evenness for the infrastructure data fit well, with predictions that generalize to new data. These parameters are quite safety relevant for motorcyclists, encouraging further research in this direction.

In particular, the prediction of driving dynamics (namely yaw-rate and y-acceleration) based on infrastructure quality parameters, might be very useful, following further research, in that these driving dynamics could be utilized in turn in a risk model (for instance [15]), to predict accident risk, without even necessarily running additional experiments/performing test rides, provided the infrastructure data has been obtained.

Conversely, predicting infrastructure features from motorcycle driving data shows some promise, as transverse evenness is predicted very well, for a model fit to a particular rider. However, there are limitations, as the vehicle covers only a few driven trajectories, compared to the full lane coverage of data collection by dedicated measurement vehicles.

The weaknesses of the prediction of infrastructure from driving dynamics could perhaps be counteracted through the use of fleet data i.e., a sufficiently large number of motorcycles with accessible CAN-Bus data, as well as geolocalisation. Future research efforts will aim to investigate this and quantify the potential safety benefits and infrastructure information.

The predictions face the challenge of having to align the high variance IMU data with relatively slow changing, highly structured road condition data. High correlations on the test set suggest a good on-average relation. Further research will need to address this, with candidates being primarily the use of crowd data to stabilize estimates and potentially the use of stronger filters on the sensor data and nonlinear predictions.

References

1. Yannis, G., Vlahogianni, E., Golias, J., Saleh, P.: Road infrastructure and safety of powered two wheelers. In: Proceedings of the 12th World Conference on Transport Research, Lisbon. (2010)
2. International Organization for Standardization: ISO 11898–1:2015 Road vehicles—Controller area network (CAN)—Part 1: Data link layer and physical signalling (2015). https://www.iso.org/standard/63648.html
3. Cossalter, V., Lot, R., Matteo M.: Motorcycle dynamics. In: Modelling, Simulation and Control of Two-Wheeled Vehicles, pp. 1–42 (2014)
4. Bishop, C.: Pattern Recognition and Machine Learning. Springer, Heidelberg (2016). http://www.springer.com/lncs
5. Duda, R.O., Hart, P.E., Stork, D.G.: Pattern Classification. Wiley, New York (2016)
6. Ecker, H., Saleh, P.: MoProVe - A probe vehicle for traffic accident research. In: EVU Proceedings 25th Annual Congress, EDIS Editing Centre of University of Zilina, pp. 29–41 (2016). ISBN: 978-80-554-1260-3. http://www.springer.com/lncs. Accessed 21 Nov 2016
7. KTM Sportmotorcycle GmbH. https://www.ktm.com/en-gb/models/travel/ktm-1290-super-adventures2023.html. Accessed 10 Oct 2023
8. VBOX automotive: VBOX 3i Dual Antenna—100 Hz Vehicle Dynamics Measurement. https://www.vboxautomotive.co.uk/index.php/en/products/data-loggers/vb3i. Accessed 6 Oct 2020
9. RoadSTAR reference page. https://www.ait.ac.at/en/labs/roadstar. Accessed 06 Oct 2023
10. Friction measurement reference. https://standards.iteh.ai/catalog/standards/cen/6e951634-9318-487a-8c5a-0000cf9f5cf5/cen-ts-15901-1-2009. Accessed 06 Oct 2023
11. Transverse evenness reference. https://standards.iteh.ai/catalog/standards/sist/3fdc4740-f2f9-462a-841c-8ff1099dd968/sist-en-13036-8-2009. Accessed 06 Oct 2023
12. Longitudinal profile reference. https://standards.iteh.ai/catalog/standards/cen/1898bd20-3a3a-4c29-a5f1-67f07554e059/en-13036-6-2008. Accessed 06 Oct 2023
13. Surface damage reference: RVS 11.06.69 Digitale Hochgeschwindigkeitsbilderfassung der Fahrbahnoberfläche mit dem System RoadSTAR – Grundtext (in German), Austrian Research Association for Roads, Railways and Transport (2009). http://www.fsv.at/shop/produktdetail.aspx?IDProdukt=8512b248-26e3-419e-8a80-7068c3c5d102
14. R: A language and environment for statistical computing. https://www.R-project.org
15. Hula, A., Fürnsinn, F., Schwieger, K., Saleh, P., Neumann, M., Ecker, H.: Deriving a joint risk estimate from dynamic data collected at motorcycle rides. Accid. Anal. Prevent. **159**, 106297 (2021). ISSN 0001-4575, https://doi.org/10.1016/j.aap.2021.106297

Open Access This chapter is licensed under the terms of the Creative Commons Attribution 4.0 International License (http://creativecommons.org/licenses/by/4.0/), which permits use, sharing, adaptation, distribution and reproduction in any medium or format, as long as you give appropriate credit to the original author(s) and the source, provide a link to the Creative Commons license and indicate if changes were made.

The images or other third party material in this chapter are included in the chapter's Creative Commons license, unless indicated otherwise in a credit line to the material. If material is not included in the chapter's Creative Commons license and your intended use is not permitted by statutory regulation or exceeds the permitted use, you will need to obtain permission directly from the copyright holder.

From Regular Cyclist to Cargo Bike User?
A Step Closer to Enhancing Cargo Bike Culture

Juliana Betancur Arenas[(✉)] [iD], Philippe Lebeau[iD], and Cathy Macharis[iD]

Mobilise Mobility and Logistics Research Group, Vrije Universiteit Brussel, Elsene, Belgium
juliana.betancur.arenas@vub.be

Abstract. An innovative, clean, and practical solution to overcome the lack of capacity of regular bikes compared to personal vehicles is the e-cargo bike. While most of the research on cargo bikes has been made on their usage for logistics purposes, only a few studies have been deepening into their personal mobility use. This paper aims to analyze and reveal the potential of regular cyclists to become cargo bike users by surveying a well-established cycling community that uses the services of the regional parking agency. Results show that among non-cargo bike users, 45% of them are potential users interested in using this innovative vehicle. The sample indicates that age, education, environmental awareness, and family situation related to children are factors that have a positive influence on cyclists to become cargo bike users. While the top barriers mentioned were associated to the lack of safe bike storage and cargo bike prices, motivators to use them are primarily related to benefits for the planet and money and timesaving.

Keywords: Cargo bike adoption · Cargo bike potential · Cycling mobility · Mobility behavior · Mode choice

1 Introduction

Amid our current climate crisis, cycling has been gaining more and more attention as one of the major solutions to the urban challenges created by road transport proliferation and pollutant emissions [2]. While significant achievements have been made regarding cycling infrastructure, promotion, and investment, barriers remain to achieve the desired sustainable mobility shift [7]. In this regard, e-cargo bikes appear as an innovative solution to overcome some of the regular bike's physical barriers, such as their lack of load capacity compared to vans or cars [3].

Cargo bikes can be defined as adapted cycles to carry loads or passengers. They have existed in Western countries since the early 1900s and have been mostly related to small commercial deliveries from merchants to factories until their passenger adaptation and development during the 1980s. Cargo bikes have seen, however, a revolution with the adaptation of the electric model, enhancing their load capacity, adaptability, and range [6]. By overcoming these barriers, the potential was identified first in cycle logistics for commercial transport [3].

© The Author(s) 2025
C. McNally et al. (Eds.): TRAconference 2024, LNMOB, pp. 134–139, 2025.
https://doi.org/10.1007/978-3-031-85578-8_18

Less is known, however, about the potential of cargo bikes for private transport. While several studies focus on understanding the user characteristics of regular bikes and e-bikes, there are fewer insights on cargo bike users [1, 3, 5]. More particularly, the potential of e-cargo bikes for private and shared schemes is very much unknown as they are still considered a niche mobility and transport alternative [4]. Therefore, cargo bikes need to be included in the conversation about sustainable mobility transition, as recognized by Carracedo and Mostofi [3].

In this paper, we contribute to the literature on sustainable transport behavioral change. We want to explore the potential of taking cargo bikes from niche innovation to mainstream by deepening the knowledge of the behavioral insights of the "easier-to-attract" cargo bike users and making the step from "early adopters" to the "early majority." More precisely, we aim to contribute to existing knowledge by investigating the behavioral insights of three groups of cyclists: the cargo bike users, the potential cargo bike users, and the undecided or uninterested cargo bike users.

2 Pedaling Perspectives: Surveying the Cyclist Community

One of the most influential factors in cargo bike adoption is having a strong previous cycling culture and knowledge [4]. Therefore, we focused our approach on the cyclist community by surveying cyclists registered at the regional parking agency of the Brussels Capital Region. We aim to understand regular cargo bike users' characteristics and behavior while instigating regular cyclists' deterrents to adopt cargo bike usage. We want to test to what extent regular bike users are prone to using a cargo bike and the primary motivators and deterrents they can encounter.

This paper uses a quantitative methodology based on purposive and online convenience sampling. Results are based on a descriptive online survey spread through an email list of about 9,850 members of Cycloparking Brussels. We received 1,061 responses, 828 of which were ruled as valid (N = 828), providing a margin of error of ± 3 at the 95% confidence interval.

We used a survey designed on Qualtrics XM to collect the respondents' answers. It took approximately 10 min to complete, and it explored three main factors: (1) the participant's everyday mobility behavior, including transport mode and parking accessibility; (2) exposure to cargo bikes, barriers, and motivators; and (3) individual mobility characteristics, including willingness to use cargo bikes. Data collection and processing comply with the legal principles imposed by the new European General Data Protection Regulation (GDPR or GDA), which has been in force since 25 May 2018. All responses were anonymous.

We used descriptive statistics for the analysis to understand and compare cargo bike users' and non-users' characteristics and behavior, including their motivators and deterrents. It is worth noting the limitations of this study, including the sample size and representativeness, since the targeted participants are already members of a cycling community. The aim was to analyze the established and easier-to-attract populations first. The data should be compared against other established cycling communities' results rather than a general audience.

3   Unraveling the Potential: Regular Cyclists' Transition to Cargo Bike Users

3.1   Profiling the Cargo Bike User

To unravel cargo bike users' potential in the cycling community, we first decided to expose the typical cargo bike profile by comparing it to cyclists who are non-cargo bike users. In total, 25% of the participants used a cargo bike. Among that group, 91% of them own one or more types of cargo bikes, and 9% use a sharing, renting or lending scheme. Considering the different cargo bike typologies in our survey, longtail bikes were the most common model with 60% of cargo bike users, followed by 35% of the participants using the "bakfiets" model, 16% using bike trailers and 2% classified as others. We could see a connection between the lack of space and a lighter or slimmer cargo bike typology chosen, since 51% of cargo bike owners said they had no garage to store their vehicles.

Regarding vehicle ownership compared to cargo bike users, the sample of 621 respondents who do not use cargo bikes had higher general car ownership (+5%) and lower bike ownership (-23% when comparing the participants who own more than three bikes). Additionally, this group stated that 67% do not have access to a vehicle storage place, exposing a bigger constraint to own a cargo bike due to the lack of parking space.

One crucial aspect of cargo bike adoption, apart from understanding their usage patterns, are the factors that influence individuals' decisions to incorporate them into their daily lives. According to the sample's sociodemographic information, cargo and non-cargo bike users showed gender parity in respondents. Both groups also share a marked trend of higher education profiles, with more than 80% of the sample achieving a university degree with a high representation of master's level. Regarding their employment status, full-time positions are the primary tendency in both groups. However, the more robust trends differentiating cargo bike users from non-cargo bike users are their age and their living situation. Most cargo bike users are between 35 and 44 years old (60%), while non-cargo bike users have a more comprehensive age range. More importantly, almost all cargo bike users live with children (83%). In contrast, regular cyclists' living situations have a greater range of diverse situations, including living in a couple without children (28%) and living on their own (17%), while still, the most common profile is living with partners and children (37%). This last group represents the highest potential to become cargo bike users.

Daily mobility patterns of cargo bike users prove a substantial proportion of bike usage for almost all their trips, especially compared to non-cargo bike users. In frequency, this finding confirmed high regular usage, with 67% stating that they used their cargo bike more than three days a week. For regular cyclists, the bike is still the primary mode of transport per daily trip (55% of the users started their daily mobility using a bike). Nevertheless, the non-cargo bike users show a more multimodal behavior with a higher use of public transport and walking to carry out their multiple journeys of the day. Which relates to some extent to their higher car ownership.

In concordance with former studies, children are the primary purpose for cargo bike users from a personal mobility perspective. 82% of the cargo bike users in this sample stated their primary use was kid's transport. They have a tendency to cover shorter distances and carry out chained multipurpose trips. The most common trip chaining observed in the daily mobility patterns was taking the children to school, going to their workplace, picking them up, or doing groceries before returning home. In contrast, for regular cyclists' non-cargo bike users, most everyday mobility trips start by going to work (73%). As mentioned earlier, these patterns are influenced by participants' more diverse living situations.

Cargo bike users' primary motivator is saving time compared to cars (chosen by 85% of the respondents), followed by the benefit for the planet (73%) and cost savings (60%). This result means that the vehicle's direct benefits and the cyclists' values play a role in their cargo bike culture. This outcome resembles the findings from Becker and Rudolf [1], who confirmed that many cargo bike users could be identified as environmentally conscious.

3.2   Willingness and Deterrents to Becoming a Cargo Bike User

While we first analyzed the sample with binomial regard, we used the replies to the Likert scale question "I would like to start using a cargo bike" to better understand the non-cargo bike users and their deterrents to adopt cargo bikes. From this approach, we could classify the non-cargo bike users into (1) potential cargo bike users (answering strongly agree and agree), (2) undecided or indifferent (answering indifferent), and (3) not interested (answering disagree and strongly disagree). Not considering 6% of non-cargo bike users who did not reply, 45% of respondents are interested in using cargo bikes, named for this study as *potential users*. On the other hand, 33% of participants are undecided, and 22% of regular cyclists are not interested in using a cargo bike.

Comparing the potential user's profile with the cargo bike user's profile shows, to a greater extent, some correlations in several factors and characteristics. Regarding sociodemographic, 52% of the potential users live with children, 28% are couples without children, and 11% live alone. In terms of age, there is more variety than cargo bike users, but the most represented group is still between 35 and 44 years old. Regarding gender, we see slightly greater interest from cyclists who identified as men (55%) than from women.

We found commonalities with the cargo bike users with a lower rate of car ownership and lesser usage of it. In their daily mobility, the potential profile showed more children-related trips than the undecided profile but fewer than the cargo bike users. Their first trip is usually related to work (72%), with groceries (32%) coming as second. Transporting kids comes in third place, with 19% of the participants choosing it as their trip purpose. Potential users were more optimistic about the cargo bike's capacity to supply most of their trips. They also stated that owning a cargo bike will have a more substantial effect on avoiding owning a car and said to have more external encouragement from their entourage to make the shift.

Despite the common traits, one of the main differences between the potential and the actual users is the safe storage availability. 73% of the potential users stated the lack of safe parking as their major deterrent, confirming the findings of other studies in the field [8]. While 40% of these users have three or more bikes, 69% do not have access to a storage place in their households. Directly related and further cited, the high price of the cargo bike and the fear of having it stolen became the most mentioned deterrents to having a cargo bike. Finally, the need for cycling infrastructure was also mentioned as an obstacle to making the final choice.

3.3 Not All Regular Cyclists Are Prone to Becoming Cargo Bike Users

Insights on the group of indecisive and uninterested cyclists towards cargo bike usage expose various living situations with less parental representation. However, the parents' category is still the most significant share in both groups. Another major difference relies on age terms, where there is more variety represented. In the group of uninterested cargo bike cyclists, there is a more significant proportion of people between 45 and 54 years old (34%). In both groups, almost the same number of men and women. Regarding jobs and education, we find a constant number of highly qualified profiles yet a higher representation of independents and retired participants than in the other groups.

When surveying the hypothetical use of cargo bikes, a significant difference among the groups was seen in cargo bike supply capacity, car ownership replacement, and external family and friends' usage encouragement. For the uninterested profiles, cargo bike capacity did not fulfill their daily needs. They have a higher usage of cars for longer distances and do not consider that owning a cargo bike would allow them to avoid their car usage.

Among the uninterested regular cyclists, 67% of them said they did not see the advantage nor need to own a cargo bike, becoming their major barrier. The lack of safe storage and the cargo bike's high price came next. The situation is the opposite for in-decisive users. Fewer users claimed to not see the need to own a cargo bike, with 48%. The most significant deterrent for them is the lack of storage at their home (61%). However, this last profile showed the highest interest in using a sharing cargo bike scheme (63%) compared to the possibility of owning the vehicle.

4 Conclusion

The use of cargo bikes as a sustainable mode of personal mobility has gained significant attention in recent years as cities grapple with congestion, pollution, and the need for eco-friendly transportation solutions. Its adoption process is not evident for all individuals and brings challenges. While not every cyclist will be interested in shifting to a cargo bike, this study exposes how age, living situation, parking, and car ownership influence cyclists to be more prone to use e-cargo bikes. Even though gender, education, and employment situation did not show a clear tendency, the sample exposed interest for both men and women with a high education level and full-time employment situation. In total, 25% of cyclists in this sample are cargo bike users. Among non-cargo bike users, 45% are interested in becoming cargo bike users. Motivators to use them are primarily related to benefits for the planet, money, and timesaving. To further leverage the potential of cargo bikes for our cities, the most pressing barriers should be overcome, such as the lack of safe cargo bike storage and the high cargo bike prices. There is a great interest and potential for cyclists to become cargo bike users, and sharing cargo bike systems could promote its usage among indecisive regular cyclists. Moreover, future research should focus on the different cargo bike typologies benefits and constraints, to have a more accurate offer and a more diversified mobility need satisfaction.

Acknowledgment. This research is a part of the project UIA05-156, Clean AIR - GO cargo BIKE, (CAIRGO BIKE), co-financed by the European Regional Development Fund through the Urban Innovative Actions initiative. Data collection was carried out with the support of our project partner parking brussels.

References

1. Becker, S., Rudolf, C.: Exploring the potential of free cargo-bikesharing for sustainable mobility. GAIA – Ecol. Perspect. Sci. Soc. **27**, 156–164 (2018). https://doi.org/10.14512/gaia.27.1.11
2. Brand, C., et al.: The climate change mitigation impacts of active travel: evidence from a longitudinal panel study in seven European cities. Glob. Environ. Chang. **67**, 102224 (2021). https://doi.org/10.1016/j.gloenvcha.2021.102224
3. Carracedo, D., Mostofi, H.: Electric cargo bikes in urban areas: a new mobility option for private transportation. Transport. Res. Interdisc. Perspect. **16**, 100705 (2022). https://doi.org/10.1016/j.trip.2022.100705
4. Dorner, F., Berger, M.: Peer-to-Peer cargo bike sharing: findings from LARA share project. In: Proceedings of 8th Transport Research Arena TRA 2020, Helsinki, Finland, 27–30 April 2020 (2020)
5. Hess, A.-K., Schubert, I.: Functional perceptions, barriers, and demographics concerning e-cargo bike sharing in Switzerland. Transp. Res. Part D: Transp. Environ. **71**, 153–168 (2019). https://doi.org/10.1016/j.trd.2018.12.013
6. Narayanan, S., Antoniou, C.: Electric cargo cycles—a comprehensive review. Transp. Policy **116**, 278–303 (2022). https://doi.org/10.1016/j.tranpol.2021.12.011
7. Riggs, W.: Cargo bikes as a growth area for bicycle vs. auto trips: exploring the potential for mode substitution behavior. Transport. Res. Part F: Traffic Psychol. Behav. **43**, 48–55 (2016). https://doi.org/10.1016/j.trf.2016.09.017
8. Thomas, A.: Electric bicycles, and cargo bikes—tools for parents to keep on biking in auto-centric communities? findings from a US metropolitan area. Int. J. Sustain. Transp. **16**(7), 637–646 (2022). https://doi.org/10.1080/15568318.2021.1914787

Open Access This chapter is licensed under the terms of the Creative Commons Attribution 4.0 International License (http://creativecommons.org/licenses/by/4.0/), which permits use, sharing, adaptation, distribution and reproduction in any medium or format, as long as you give appropriate credit to the original author(s) and the source, provide a link to the Creative Commons license and indicate if changes were made.

The images or other third party material in this chapter are included in the chapter's Creative Commons license, unless indicated otherwise in a credit line to the material. If material is not included in the chapter's Creative Commons license and your intended use is not permitted by statutory regulation or exceeds the permitted use, you will need to obtain permission directly from the copyright holder.

The Effect of Affluence and Gender on Active School Travel

Ross Higgins[(✉)] (iD)

University of Limerick, Limerick V94 T9PX, Ireland
Ross.Higgins@ul.ie

Abstract. Ireland's Climate Action Plan 2023 sets out aims to increase active travel trips by 50% and to shift 30% of escort-to-education trips to sustainable modes of travel by 2030. Promoting cycling to school requires a full understanding of the barriers to cycling to school for different groups. Previous work by the author identified gender differences in perceptions of cycling among secondary school students. In addition to gender, this study focuses on the affluence effect.

A survey was completed by 306 students across six schools in Limerick City and Suburbs. The Family Affluence Scale 2 (FAS 2) was used to categorise students into affluence groups and students responded to Likert Scale questions.

The results showed that as affluence increases, the gender gap in active travel (walking and cycling) rates increases. Among those in the highest affluence group, boys were approximately 2½ times more likely to walk or cycle to school than girls. Tukey-Kramer ad hoc tests comparing mean differences between responses in different affluence groups revealed that there is a particular affinity among the most disadvantaged boys and the most affluent girls towards travelling by car to school compared with other means of travel. These findings have implications on messaging to promote cycling to school to various affluence groups.

Keywords: cycling to school · active travel · affluence · SES · gender · walking

1 Introduction

Reducing greenhouse gas (GHG) emissions is considered necessary to limit global warming to 2 °C. Ireland's nationally determined contributions are contained within the Climate Action Plan 2023 (CAP23). It sets out aims to increase active travel (AT) trips by 50% and to shift 30% of escort-to-education trips to sustainable modes of travel by 2030 [1]. Promoting AT to school requires an understanding of the barriers to walking and cycling for different groups. Previous work [2] identified particular gender differences in perceptions of cycling among secondary school students (aged 12–18 years). Those findings indicated that girls were more affected than boys by traffic-, personal- and peer-related factors and this work examines the affluence effect on cycling.

Gender differences in cycling rates are observed across many developed countries. For example, in Ireland, according to 2022 census data, only 6.2% of boys cycle to secondary school but the corresponding rate for girls is only 0.8% [3]. However, boys and girls cycle at almost even rates in the Netherlands and Denmark [4].

© The Author(s) 2025
C. McNally et al. (Eds.): TRAconference 2024, LNMOB, pp. 140–146, 2025.
https://doi.org/10.1007/978-3-031-85578-8_19

These findings have implications for the design of interventions to promote cycling to school targeted by affluence to yield higher rates of cycling across the population.

2 Literature Review

This paper follows the work of the author's previous work [2] on gender differences in cycling which found that secondary school girls were more affected than boys in relation to many factors associated with cycling. That work found that perceptions of traffic risks, uniforms, heavy bags, the efforts associated with cycling, effect on personal appearance and peer-influences affected girls significantly more than boys. Subsequent work [5] found that the main components of the Theory of Planned Behaviour (TPB) was an appropriate model to explain cycling to school behaviours. None of the previous findings, however, were categorised by family affluence.

There has been a general trend in developed countries over the last few decades of increasing car ownership rates, leading to more car-travel, in turn, leading to less active travel and active commuting. While the modal shift among populations in fast-developing China suggest people are increasingly travelling by car [6], societies in some well-developed European countries have seen cycling rates increase [4].

Within individual countries research has found some associations and, in some instances, no associations between affluence and cycling commuting rates. Research from New Zealand indicates that while overall rates have dropped, the rates of active commuting among the most affluent have increased [7] and an analysis of census data from England and Wales has revealed that overall cycling rates have increased with the largest increases found in the most affluent areas [8] such that an association between affluence and cycle commuting rates no longer exist. However, a London-based case study found that cycle rates were higher (2.2% v 1.5%) among those whose household incomes were greater than £35k compared with those earning less than £15k [9].

Much less research on cycling by affluence level is available. Relevant findings relate to the combination of walking and cycling, termed active school travel (AST). Most findings indicate that higher affluence rates are associated with lower rates of AST, as in North America [10] and in New Zealand [11]. These were also the findings of longitudinal research across Scotland, Wales and the Czech Republic but not Norway where there was no relationship between family affluence and AST [12]. In Ireland, results from the Children's Sports Participation and Physical Activity (CSSPA) study, found that higher physical activities rates, were associated with higher affluence [13].

To our knowledge, no research has been completed to determine the factors, associated with cycling, that are influenced by affluence. This is the premise of this paper.

3 Methodology

In 2019, 306 students in Limerick City and Suburbs responded to a survey on school travel patterns and on perceptions associated with cycling. A question was asked to screen students that lived more than 8 km from school out of the survey as this was considered the distance threshold above which cycling becomes impractical [14].

The questionnaire consisted of a number of travel pattern related questions and thirty-six statements to which students could respond Strongly Agree (SA) to Strongly Disagree (SD) relating to perceptions of cycling to school.

To profile students by affluence level, the Health Behaviour in School-Aged Children (HBSC) Family Affluence Scale II (FAS 2) was used. This is a revision of the original FAS and is a widely used set of questions relating to household details. The scores from each question are added to yield a total which allows students to be profiled as Affluence Group 1 (AG1), 2 or 3 [15].

4 Results

Table 1 shows numbers and associated percentages of students that travelled to school actively and those that travelled using alternative modes and by gender. To determine if there is a statistical relationship between affluence group and reported active travel to school, chi-squared tests were carried out, one for males only and one for females only. For male students, the relationship was not significant, X^2 (2, N = 160) = 1.325, p = .516 but for girls, it was significant X^2 (2, N = 181) = 8.641, p = .013.

As affluence group increases, the gap between the active travel rates by gender increases. Among those in Affluence Group 3 (AF3), boys are approximately 2 ½ times more likely to actively commute compared to girls.

Table 1. Modal split for school travel during good-weather days by affluence (n = 341)

Non-active or active		Male			Female		
		Affluence Group			Affluence Group		
		1	2	3	1	2	3
Non-active	Count	15	50	20	16	61	41
	%	62.5%	50.0%	55.6%	57.1%	59.2%	82.0%
Active	Count	9	50	16	12	42	9
	%	37.5%	50.0%	44.4%	42.9%	40.8%	18.0%

Tukey Kramer tests were carried out to determine which affluence groups differed across each factor. Table 2 shows three rows of results for a number of factors, where statistically significant differences were found. The first row describes the mean difference between responses of AG1 and AG2, and so on. The significance levels are also shown and commentary on the statistically significant differences, at $\alpha = 0.05$, is provided below. Coding, of 1 for strongly agree to 5 for strongly disagree, was used.

For boys, the results indicated that there were significant differences between Affluence Group 1 (AG1) and AG2 in relation to the slopes being too steep with boys from the lower affluence groups reporting that slopes affect them more. Boys from AG3 reported being less affected by weather, thought that cycling was good for the environment and indicated that they did have access to a bicycle, significantly more than boys from AG1.

Boys from AG1 thought that they were not physically fit enough to cycle and that driving a car or getting a lift would be "cooler" compared with boys from the two other affluence groups. They also reported not feeling confident handling a bicycle compared to boys from AG3. Boys from AG2 indicated that they would get too hot and sweaty if they cycled compared to boys from AG3.

Among the girls, there were also factors which were significantly different by affluent group. Girls from AG1 were more likely than girls from AG3 to report that the distance was too short to consider cycling and more likely to indicate that they were not confident handling a bike but less likely to indicate that their bags were too heavy. Girls from AG1 indicated that their uniforms did not affect them as much as the other two groups and the differences were statistically significant.

Girls from AG3 reported that cycling would ruin their hair significantly more than girls of AG2. Girls from AG3 were also more likely to indicate, more than all other surveyed female students, that driving or getting a lift is "cooler" than cycling.

Table 2. Results of Tukey-Kramer post hoc tests comparing means of affluence groups

Factor	Affluence Grp		Male		Female	
	I	J	Mean Diff. (I-J)	Sig.	Mean Diff. (I-J)	Sig.
I don't own or have access to a bicycle	1	2	−0.416	.280	−0.326	.520
		3	−0.929	**.010**	−0.748	.064
	2	3	−0.513	.078	−0.422	.193
I am not physically fit enough to cycle	1	2	−0.752	**.007**	−0.317	.449
		3	−1.056	**.001**	−0.440	.283
	2	3	−0.304	.320	−0.123	.830
I do not feel confident handling a bike	1	2	−0.490	.114	−0.357	.347
		3	−0.788	**.017**	−0.725	**.031**
	2	3	−0.298	.339	−0.368	.183
I would get too hot and sweaty if I cycled	1	2	0.050	.984	−0.092	.939
		3	−0.550	.231	0.023	.997
	2	3	−0.600	**.046**	0.116	.862
My uniform does not lend itself to riding a bicycle	1	2	−0.084	.957	0.682	**.026**
		3	0.371	.535	0.736	**.031**
	2	3	0.456	.187	0.055	.964
Cycling would ruin my hair especially if I wore a helmet	1	2	−0.058	.979	−0.177	.790
		3	0.129	.925	0.366	.439
	2	3	0.186	.744	0.543	**.037**

(continued)

Table 2. (*continued*)

Factor	Affluence Grp		Male		Female	
	I	J	Mean Diff. (I-J)	Sig.	Mean Diff. (I-J)	Sig.
Driving a car or getting a lift in a car is cooler	1	2	−0.746	**.018**	0.137	.863
		3	−0.881	**.016**	0.708	**.045**
	2	3	−0.135	.832	0.571	**.024**

5 Discussion and Conclusion

Most previous relevant research indicated that lower affluence rates were associated with higher rates of active travel [10–12] and this was also the finding among the female respondents in this study; however, not among males. Most other research did not present relevant findings by gender; however, our findings indicate significant differences by gender: as affluence group increases, the gap between the active travel rates by gender increases. Among those in AG3, boys are approximately 2 ½ times more likely to actively commute compared to girls.

Tukey-Kramer ad hoc tests comparing mean differences between responses from students in different affluence groups revealed some interesting findings. For boys, the most disadvantaged (AG1 boys) have less access to bicycles than boys from other affluence groups and given that higher bicycle ownership rates are correlated with higher cycling commuting rates, at least among adults [16], the basic enabler of having access to a bicycle seems to disproportionately affect AG1 boys. This may indirectly explain other related findings of AG1 boys including not being physically fit enough to cycle than boys from the other affluent groups. They were also more affected by steep slopes than boys from AG2. This aligns with results from longitudinal studies on children's physical activity rates in Ireland which indicated that higher physical activity rates are associated with higher family affluence [13]. Their lower self-reported bicycle handling skills are also not at the same level as boys from AG3 which may be related to their lack of practice due to less bicycle availability. The idea of the car being a symbol of power, as described elsewhere [17], seems to have a higher value among disadvantages boys (AG1 boys), who think driving a car is "cooler" than cycling, compared to boys of higher affluent family.

As with AG1 boys, AG1 girls indicated that their bike handling skills were not as good as girls from other affluent groups. This is probably related to their lower bicycle access rates, compared to AG3 girls, which, although not significant, yielded an interesting p-value of .06. It is well-established that girls are more affected by their uniforms than boys in relation to cycling [2], however, it was unknown whether this also differed across affluent groups. Interestingly, this research found that girls from the most disadvantaged families (AG1) were not as affected as girls from the other affluent groups. In contrast, the motivation to align with an identity of femininity, with high values for personal appearance, seems to be stronger among the most affluent girls (AG3) who indicated that cycling would ruin their hair significantly more than AG2 girls. Aligning with the perceptions of AG1 boys, AG3 girls were also more likely, than AG1 and AG2 girls, to indicate that driving or getting a lift was "cooler" than cycling.

Previous work [2], indicated how promotions should be tailored by gender. This work concludes that the level of affluence also needs to be considered. Access to bicycles for boys and girls should be improved by specific interventions. Perhaps, the cycle -to-work scheme which allows taxpayers to purchase discounted bicycles [18], should be expanded to other groups of the community including the unemployed. Advertising promoting the benefits of cycling should include messages that the car is not a status-defining object, and that cycling may not negatively affect personal appearance. This would serve to counteract the car-based advertising that has been detrimental to the promotion of alternative forms of travel [19].

References

1. Department of the Environment Climate and Communications. Climate Action Plan 2023 (2023). https://www.gov.ie/en/publication/7bd8c-climate-action-plan-2023/. Accessed 3 Mar 2023
2. Higgins, R., Ahern, A.: Students' and parents' perceptions of barriers to cycling to school - an analysis by gender. Sustainability **13**(23), 13213 (2021)
3. Central Statistics Office. Census 2022 - Summary Results - FY079 - Means of Travel to Work, School or College (2023). Accessed 28 June 2023
4. Pucher, J., Buehler, R.: Making cycling irresistible: lessons from the Netherlands, Denmark and Germany. Transp. Rev. **28**(4), 495–528 (2008)
5. Higgins, R., Ahern, A.: Exploring why girls don't cycle to school Student and Parent Guardian focus group findings on barriers to cycling (2023)
6. Spinney, J.: Cycling the city: non-place and the sensory construction of meaning in a mobile practice, pp. 25–46 (2007)
7. McKim, L.: The economic geography of active commuting: regional insights from Wellington, New Zealand. Reg. Stud. Reg. Sci. **1**(1), 88–95 (2014)
8. Goodman, A.: Walking, cycling and driving to work in the English and Welsh 2011 census: trends, socio-economic patterning and relevance to travel behaviour in general. PLoS ONE **8**(8), e71790 (2013)
9. Steinbach, R., Green, J., Datta, J., Edwards, P.: Cycling and the city: a case study of how gendered, ethnic and class identities can shape healthy transport choices. Soc. Sci. Med. **72**(7), 1123–1130 (2011)
10. Rothman, L., Macpherson, A.K., Ross, T., Buliung, R.N.: The decline in active school transportation (AST): a systematic review of the factors related to AST and changes in school transport over time in North America. Prev. Med. **111**, 314–322 (2018)
11. Ikeda, E., et al.: Built environment associates of active school travel in New Zealand children and youth: a systematic meta-analysis using individual participant data. J. Transp. Health **9**, 117–131 (2018)
12. Haug, E., et al.: 12-year trends in active school transport across four european countries—findings from the health behaviour in school-aged children (HBSC) study. Int. J. Environ. Res. Public Health **18**(4), 2118 (2021)
13. McFlynn, P., et al.: Children's Sport Participation and Physical Activity Study 2022 (2023)
14. Van Dyck, D., De Bourdeaudhuij, I., Cardon, G., Deforche, B.: Criterion distances and correlates of active transportation to school in Belgian older adolescents. Int. J. Behav. Nutr. Phys. Act. **7**(1), 87 (2010)

15. Currie, C.E., Elton, R.A., Todd, J., Platt, S.: Indicators of socioeconomic status for adolescents: the WHO health behaviour in school-aged children Survey. Health Educ. Res. **12**(3), 385–397 (1997)
16. Owen, N., et al.: Bicycle use for transport in an Australian and a Belgian City: associations with built-environment attributes. J. Urban Health **87**(2), 189–198 (2010)
17. Gatersleben, B.: The car as a material possession: Exploring the link between materialism and car ownership and use, in Auto motives, pp. 137–148. Emerald Group Publishing Limited (2011)
18. Revenue. Cycle to Work Scheme (2022). Accessed 18 Jan 2023. https://www.revenue.ie/en/jobs-and-pensions/taxation-of-employer-benefits/cycle-to-work-scheme.aspx
19. Baslington, H.: Travel socialization: a social theory of travel mode behavior. Int. J. Sustain. Transp. **2**(2), 91–114 (2008)

Open Access This chapter is licensed under the terms of the Creative Commons Attribution 4.0 International License (http://creativecommons.org/licenses/by/4.0/), which permits use, sharing, adaptation, distribution and reproduction in any medium or format, as long as you give appropriate credit to the original author(s) and the source, provide a link to the Creative Commons license and indicate if changes were made.

The images or other third party material in this chapter are included in the chapter's Creative Commons license, unless indicated otherwise in a credit line to the material. If material is not included in the chapter's Creative Commons license and your intended use is not permitted by statutory regulation or exceeds the permitted use, you will need to obtain permission directly from the copyright holder.

Walkable and Cyclable Side Streets in Frankfurt – Appraising (Provisional) Traffic Calming Measures With a Multimethod Analysis

Dennis Knese[(⊠)] [iD]

Frankfurt University of Applied Sciences, Frankfurt Am Main, Germany
knese@fb1.fra-uas.de

Abstract. To encourage walking and cycling, Frankfurt is redesigning several side streets with a variety of provisional measures, including cycling priority, modal filters, multifunctional lanes, green elements, etc. If these experimental measures are effective and accepted, they will be made permanent. The effects are therefore evaluated using quantitative and qualitative research methods. The paper centres on the analyses of Oeder Weg, a bustling business and shopping street. So far, the surveys have revealed notable differences between various groups, such as the assessment of perceived traffic safety and the approval of individual measures. Many residents and almost all cyclists are in favour of the redesigned road space because they feel safer and the street is more attractive to use. Frequent car users complain of poorer accessibility and lack of parking facilities. Also, local businesses react heterogeneously. While restaurant owners are pleased about more outdoor space and an increased quality of stay, many retailers complain about the worsened conditions for customers who come by car. A large proportion of the customers surveyed does not share this assessment though, as many come by bicycle or public transport. The traffic observations have shown that objective traffic safety has improved at conflict points in the road space, e.g. through new markings in the dooring zone or curb extensions in intersection areas. Overall, the modal shares have shifted considerably. Motorised individual transport has decreased significantly in Oeder Weg, but has increased in some adjacent streets, which has led to an adjustment of the traffic guidance measures.

Keywords: Cycling · Walking · Transformation

1 Introduction

The promotion of walking and cycling is a central building block for the mobility transition in urban areas. Cities like Frankfurt, centre of the metropolitan area Rhine-Main, have to cope with substantial transport challenges. The city's popularity as a business hub and a centre of diverse functions has led to a significant influx of commuters from neighbouring regions. The high number of commuters has placed immense strain on Frankfurt's existing transportation infrastructure. The city's public transportation systems, including buses, trams, underground and suburban trains, have been stretched to

© The Author(s) 2025
C. McNally et al. (Eds.): TRAconference 2024, LNMOB, pp. 147–153, 2025.
https://doi.org/10.1007/978-3-031-85578-8_20

their limits, resulting in overcrowding during peak hours. Additionally, the road network has experienced congestion and bottlenecks, exacerbating air pollution and affecting overall traffic flow. Those who suffer are also the residents of streets that are used by motorised traffic as thoroughfares into the city centre.

To this end, Frankfurt's city parliament has passed the so-called "bicycle city resolution", which includes the design of eleven "bicycle-friendly side streets" over the next few years. Especially in residential areas, bicycle-friendly side streets should ensure more safety. It aims to promote walking and cycling by introducing a bicycle street (cars are allowed, but cyclists have priority), bicycle parking facilities on former car parking spaces, reduced speed limits, modal filters to restrict motorised traffic, red road markings at junctions, and other provisional measures. Further, the quality of life and quality of stay for residents and visitors is to be increased through more space for outdoor catering, recreational areas and green elements.

The associated projects mean significant interventions in the current layout of the road space. As this is not without controversy and has raised fears and concerns among residents, retailers, and other stakeholders (Bertolini 2020), the Frankfurt University of Applied Sciences is analysing the effects of the redesign. Decisive questions are whether the share of pedestrians and cyclists can be increased and motorised transport reduced by the planned measures, and how the quality of stay as well as the situation for businesses in the study areas will develop. The accompanying research will provide recommendations whether the measures should be structurally made permanent or if adjustments are necessary. So far, the provisional renovation measures have been implemented in three streets, namely Oeder Weg, Grüneburgweg and Kettenhofweg.

2   Methodical Approach

2.1   Applied Methods

A variety of quantitative and qualitative methods is used to analyse the impact and acceptance of the measures. To evaluate the perceived situation before and after the redesign processes in the three streets, quantitative pre- and post-surveys are being conducted among different user groups (online plus on-site). The surveys aim to determine behavioural patterns as well as the perception and acceptance of the reallocated public space. Since the study began after the implementation in Oeder Weg in 2021, no primary data could be collected before the street transformation. Instead, two post-surveys have been conducted, including questions about the before-situation. 925 fully completed questionnaires in the first survey and 1,944 in the second survey provided a comprehensive picture of the acceptance.

Qualitative expert interviews are conducted to explore the views of specific stakeholders in more detail. In this case, the focus is on the businesses located on each street, including retailers, restaurants, cafes, doctors' surgeries, pharmacies and other technical and commercial service providers. Oeder Weg in particular is an important shopping street, with a large number of shops, restaurants and other services. Their views and the impact on their business play an important role in the implementation and evaluation of the measures. 60 shop owners on Oeder Weg have been interviewed.

Traffic counts and traffic observations are used to study the changes in moving and stationary traffic. For example, previous points of conflict and new elements in the road space are observed with dedicated cameras to identify possible changes in road safety and road user behaviour. The objective road safety is also monitored using official road accident statistics. In addition, a special smartphone application developed in another research project called 'start2park' is used to measure parking search times.

2.2 Survey Sample

People aged between 12 and 88 took part in the Oeder Weg surveys. The largest share of respondents was between the ages of 50 and 59 (27% in both surveys), 30 and 39 (26% in the first survey/21% in the second survey), 40 to 49 (22%/24%). This is followed by the age groups 20 to 29 (11%/9%), and 60 to 69 (10%/13%). People over 70 and under 20 years represent the smallest groups. The gender distribution was 43% resp. 45% women and 54% resp. 52% men.

Since it was assumed that the perception of the redesign could vary greatly between user groups, the respondents were asked at the beginning of the survey to assign themselves to a user group and a mode of transport, with which they usually use Oeder Weg. Three quarters of respondents in the first survey, two thirds in the second survey identified themselves as residents, either of the Oeder Weg itself or of a surrounding street. Other groups included customers, visitors and through-traffic. The most common means of transport used by respondents on Oeder Weg is the bicycle (44%/46%), followed by walking (36%/31%), and cars (12%/13%).

3 Research Results

3.1 Changes in Mobility Patterns and Traffic

In the initial and subsequent surveys, 38% and 44% of the respondents reported that the street transformation has influenced their modal choices on Oeder Weg. 69% and 55% respectively opts to cycle and walk more frequently. Meanwhile, half has indicated a less frequent use of cars, see Fig. 1.

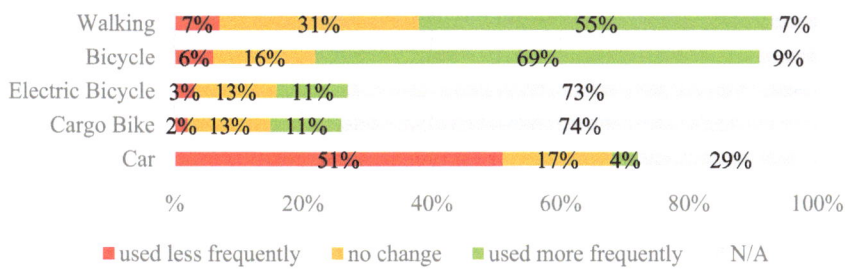

Fig. 1. Change in frequency of use of different transport modes at Oeder Weg (n = 849).

These results were confirmed by the traffic count data. The average daily traffic (ADT) within Oeder Weg was 7,683 vehicles and 3,763 bicycles in 2019; and that in 2023 was 4,444 vehicles and 4,706 bicycles. However, some of the traffic has shifted, as more vehicles have been counted on some nearby roads (City of Frankfurt 2023). In order to protect these streets from an increase in traffic, additional modal filters (bollards) were subsequently installed. Overall, however, the number of vehicles in the entire neighbourhood has decreased. This has been observed in many traffic calming projects around the world (Bauer et al. 2023), and it aligns with the concept of "traffic evaporation", wherein a drop in vehicle count, rather than traffic congestion, is observed as a long-term behavioural response to a reduced road capacity (Goodwin et al. 1998).

Two measures that are responsible for the reduction are the modal filter in the north (see Fig. 2) and the barrier to access from the direction of the city centre in the south of Oeder Weg. Camera-based traffic observations have shown that about 4% of drivers ignore the modal filter and drive through it.

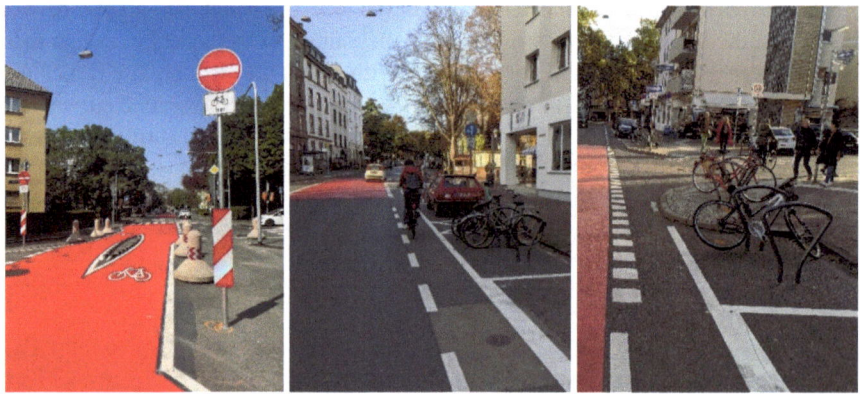

Fig. 2. Modal filter to restrict motorized through-traffic (left), marking to secure dooring zone (middle) and curb extension with bicycle parking (right) at Oeder Weg.

3.2 Evaluation of Road Safety and Quality of Stay

Respondents were asked to rate the before and after situation on Oeder Weg on a Likert scale from '1' (very good) to '5' (poor), based on various criteria. The average rating for the quality of stay (from '3' to '2'), the allocation and clarity of the road space as well as the noise levels (each from '4' to '2') has shifted in a positive direction, so that it can be assumed that most respondents perceive an improvement compared to the initial situation. A significant improvement is also perceived in terms of traffic safety, traffic flow and comfort when using the street, while accessibility and travel time are perceived as having changed little.

The survey participants were also asked to evaluate the impacts of the individual measures on road safety in the study area. Referring to Fig. 2, all the measures were considered to have demonstrated a positive contribution in this regard. The effectiveness of the safety measures, including curb extensions (68%), dooring zones (65%) and red markings (57%), was particularly acknowledged. Additionally, elements that influence the modal share and traffic flow, such as vehicle access barriers (65%), bicycle street (62%), and modal filters (59%), were also identified as helpful in promoting overall road safety in the area.

The assessment of the parking situation shows that the satisfaction for bicycles, cargo bikes and e-scooters has changed from '4' ('sufficient') to '2' ('good'), due to the increased number of parking facilities for micro-mobility. The median of the ratings for parking search time and parking spaces for motorised vehicles is '4' for both the before and after situation. However, the middle 50% of the rating has shifted to the range of '4' to '5', a slight deterioration compared to the previous situation.

The individual measures were evaluated very differently. Red markings at junctions is the one that was most noticed by the road users. Only 2% of the respondents had not noticed this measure. This contrasts with the introduction of delivery zones for commercial traffic, which was not noticed by 36% of respondents. While the installation of multifunctional areas in the side space was well received by the respondents, especially with regard to the quality of stay, more than a third of the respondents criticised the worsened accessibility of the destination due to the installation of the modal filter.

Irrespective of the individual measures, the effects on the quality of stay and road safety consistently received the most favourable ratings compared to other evaluation criteria. Overall, up to 80% of respondents indicated an improvement in the quality of stay and up to 68% an improvement in road safety because of the measures. Car users were the only group that indicated a negative change across all criteria. Overall, respondents under 30 years of age gave the most positive assessment of the after-situation compared to the other age groups. This divided picture is also evident in the open question about suggestions for improvement, where there are voices in favour of continuing or extending the traffic calming measures, but also wishes to reduce or dismantle them.

3.3 Impact on Businesses

The quantitative survey of business people and customers/visitors shows a different, sometimes contradictory perception of the changes. While several businesses complain about the negative impact on their turnover and number of customers, most customers/visitors stated that they visit the Oeder Weg more often, spend more money per visit and stay longer in the area since the changes. The in-depth interviews paint a more nuanced picture of the impact. Here, too, almost a third of the interviewees spoke of declining customer and sales figures. However, not all of them blame the reallocation at Oeder Weg. Other possible causes include covid effects, inflation and the general economic situation, or seasonal fluctuations.

Opinions also are mixed within sectors. Overall, the operators of restaurants, bars and cafes are significantly more satisfied with the measures than the retailers. Organisations in the health and social care sector as well as in technical and commercial services are much more divided. The reasons for this are manifold. Restaurant and cafe owners, for example, speak of additional walk-in customers due to more space for outdoor catering, seating and a "new flair" in the street. Some retailers criticise the loss of car parking spaces and poorer accessibility by car. Some shops have adapted their business model to sell only smaller items that can be transported by bike or on foot. Others are increasingly relying on delivery by bicycle instead of by car.

Those affected, such as shop owners and residents, wish to be more involved in the planning and implementation processes. Alongside critical comments from those who would prefer to see the area become more car-friendly again, there were also some dissenting voices. For example, the Oeder Weg could be made a pedestrian zone with designated customer parking only.

4   Conclusion and Outlook

The redesign of Frankfurt's side streets and the accompanying research are underway. The city has introduced various provisional measures that can be removed if they are not accepted. This allows new things to be tried out, people to be gently introduced to a changed streetscape and, ideally, sceptics to be won over in the long term.

So far, it has become apparent that the measures are being judged very differently. Discussions about the redesign are also very polarised on the streets and in the media. The example of the Oeder Weg shows first successes, especially with regard to a modal shift, even if not all provisional measures are fully accepted. Overall, cyclists rate almost all the measures very positively, while car drivers largely reject them. Pedestrians are particularly supportive of measures to improve road safety. Shopkeepers in particular would like to have more of a say and, in some cases, would like to see car restrictions lifted. The next surveys will provide further answers, especially with regard to possible changes in acceptance, after people have become familiar with the new design and have recognised new benefits, which often occurs in transport planning projects after a period of familiarisation (Zografos et al. 2020). With the results of the accompanying research, important lessons can be applied to the design of other walkable and cyclable streets in Frankfurt and beyond. Detailed results of the study can be found at www.relut.de (Knese et al. 2024).

References

City of Frankfurt. Rad- und Fußverkehrsfreundliche Umgestaltung Oeder Weg. Ergebnisse der begleitenden Verkehrszählungen (2023). https://www.radfahren-ffm.de/media/oeder_weg_obr-praesentation_2023-01-26.pdf. Accessed 24 Aug 2023

Bauer, U., Bettge, S., Stein, T.: Verkehrsberuhigung: Entlastung statt Kollaps! Maßnahmen und ihre Wirkungen in deutschen und europäischen Städten (Difu Policy Papers Nr. 2). Deutsches Institut für Urbanistik (2023)

Bertolini, L.: From "streets for traffic" to "streets for people": can street experiments transform urban mobility? Transp. Rev. 40(6), 734–753 (2020)

Goodwin, P., Hass-Klau, C., Cairns, S.: Evidence on the effects on road capacity reduction on traffic levels. Traffic Eng. Control 39(6), 348–354 (1998)

Knese, D., Künbet, S., Busse, J., Dehler, K., Wong, C.: Abschlussbericht Oeder Weg. Wissenschaftliche Begleitung der "fahrradfreundlichen Nebenstraßen" in Frankfurt am Main. Frankfurt University of Applied Sciences (2024). https://doi.org/10.48718/8q61-qn48

Zografos, C., Klause, K.A., Conolly, J.J.T., Anguelovski, I.: The everyday politics of urban transformational adaptation: struggles for authority and the Barcelone superblock project. Cities Int. J. Urban Policy Plan. 99 (2020)

Open Access This chapter is licensed under the terms of the Creative Commons Attribution 4.0 International License (http://creativecommons.org/licenses/by/4.0/), which permits use, sharing, adaptation, distribution and reproduction in any medium or format, as long as you give appropriate credit to the original author(s) and the source, provide a link to the Creative Commons license and indicate if changes were made.

The images or other third party material in this chapter are included in the chapter's Creative Commons license, unless indicated otherwise in a credit line to the material. If material is not included in the chapter's Creative Commons license and your intended use is not permitted by statutory regulation or exceeds the permitted use, you will need to obtain permission directly from the copyright holder.

Quantifying Cycling Infrastructure Investment Needs Across Europe Using OpenStreetMap Data

Aleksander Buczyński[(⊠)] and Andrea Chavez-Pacheco

European Cyclists' Federation, Brussels, Belgium
`a.buczynski@ecf.com`

Abstract. Using OpenStreetMap data, we approximated the scale of investments needed to make cycling a safe mobility option for the residents of EU major cities. We extracted data about existing cycle tracks, cycle lanes, and shared cycle and pedestrian tracks, within the administrative borders of 423 Trans-European Transport Network (TEN-T) urban nodes (as proposed by the European Commission in 2021). As of 2022, the infrastructure amounted to 61,862 km. Afterwards, we used a simplified version of the Sustainable Safety approach to evaluate for each of the cities whether the amount of infrastructure is sufficient. The public road network was divided into two categories: main roads (where segregated cycling infrastructure is necessary) and local roads (where, because of low volume and speed of motorised traffic, cyclists can safely share the carriageway with cars). By comparing the extent of the (selected) cycling infrastructure with the main road network we arrived at the investment needs, amounting to 97,000 km across the analysed cities. On average, current cycle infrastructure covers approximately 40% of the needs, but the percentage varies significantly between different countries – from less than 10% in Malta, Greece, Portugal and Cyprus to more than 80% in Netherlands and Finland.

Keywords: cycling infrastructure · sustainable safety · EU policy · TEN-T · urban nodes

1 Introduction

1.1 Policy Background

One of the challenges in enabling active mobility is assessing the scale of investment needs, considering universal, equitable access, and following that – defining investment priorities. Political declarations on the EU level aim to double the amount of safe cycle infrastructure [1], but there is no official data on the current amount on the European level, and early estimates were wrong by orders of magnitude. For example, the "Sustainable and Smart Mobility Strategy" [2], adopted by the European Commission in December 2020, set a target of only "5000 km in safe bike lanes" for the next decade. To address the knowledge gap, we used OpenStreetMap [3] data to approximate the scale of investments needed to make cycling a safe mobility option for the residents of EU major cities.

© The Author(s) 2025
C. McNally et al. (Eds.): TRAconference 2024, LNMOB, pp. 154–159, 2025.
https://doi.org/10.1007/978-3-031-85578-8_21

1.2 OpenStreetMap as a Source of Data

OpenStreetMap (OSM) – also known as "the Wikipedia of maps" – was founded in 2004 in the United Kingdom with the mission to create a free, editable, world-wide geographic dataset. Currently, many different users add or edit OSM data, including local or regional authorities, or commercial companies. The drawbacks of OSM include limited quality control, and varying level of accuracy in mapping different areas, with urban areas having usually more detailed descriptions.

Previous research has found that the accuracy of OSM data can be higher than official government or municipal sources [4, 5]. OSM data has already demonstrated its usefulness, for example, in generating predictions for better cycle network growth [6].

2 Methodology

2.1 Geographic Scope

We extracted cycle infrastructure within the administrative borders of 423 Trans-European Transport Network (TEN-T) urban nodes as proposed by the European Commission in 2021 [7]. The selected scope capitalises on the expected better quality of OSM data in urban areas, and the same time is of special relevance because of the ongoing discussion about the revision of the TEN-T regulation, stressing the need to promote cycling in TEN-T urban nodes. Wherever the name listed in Annex to the legislative proposal explicitly indicated that the extent of the urban nodes is wider than the city itself – for example, "Tricity" in Poland, or "Lyon Metropolitan Area (including Villeurbanne)" in France – additional municipalities were included in the evaluation. This increased the total number of administrative areas considered to 449.

The same data was also extracted for 104 major cities in 22 other countries. The additional data is available in the online dashboard, but, given the policy context, it was not used for further processing.

2.2 Data Extraction

In the first stage, we extracted data about existing cycling infrastructure: cycle tracks (physically segregated from motorised traffic), cycle lanes (designated with paint only), and shared cycle and pedestrian tracks.

The data about bus and cycle lanes, shared lanes and cycle streets was also collected, but after consideration not included in the second stage. Bus and cycle lanes and shared lanes are not suitable for all users, while the uptake, application and understanding of cycle streets varies significantly between different European countries.

In the process of extraction we kept track of whether a given segment of infrastructure is uni- or bidirectional. Wherever a total length of a given infrastructure type or the whole network is calculated, the length of unidirectional segments is divided by two (so, for example, two unidirectional cycle tracks on both sides on the street are equivalent in further analysis to one bidirectional cycle track).

2.3 Cycle Network Completeness Metrics

In the next step we tested different metrics to evaluate whether the amount of infrastructure is sufficient for a given city. Comparing the length of infrastructure with the population or the area of the city does not scale well with the different population densities and city topographies across Europe. On the other hand, more sophisticated methodologies, such as Bicycle Network Analysis [8], aiming to quantify what percentage of different short-distance trips is safely manageable by bicycle, are vulnerable to inaccuracies in data and a minor change in assumptions can significantly affect the result.

We settled on a simplified version of the Sustainable Safety approach [9]. The public road network was divided into two categories: main roads (trunk and distributor roads from the Sustainable Safety approach combined), where segregated cycling infrastructure is necessary for safe cycling, and local (access) roads, where, because of expected low volume and speed of motorised traffic, cyclists can comfortably share the carriageway with cars. Assuming the aim to enable everyone to be able to cycle safely in their daily trips, we arrived at the investment needs by comparing the extent of the (selected) cycling infrastructure with the main road network.

A similar approach was proposed as indicator 10 "Opportunity for active mobility" in the set of Sustainable Urban Mobility Indicators (SUMI) [10]. The difference is that the SUMI indicator tries to capture in a single number the share of the total road network adapted for both walking and cycling, while the approach presented in this article is more focused, limited to the main road network and cycling.

3 Results

3.1 Existing Cycle Infrastructure

Across the analysed 424 urban nodes, we identified, as of 2022:

- 38,913 km of cycle tracks,
- 5,204 km of cycle lanes,
- 17,745 km of shared cycle and pedestrian tracks.

summing up to 61,862 km of segregated cycle infrastructure. The main road network, where we expected motorised traffic volumes and/or speeds to be too high for cyclists to safely share the carriageway in mixed traffic, amounted to 156,141 km.

All the collected data, including also non-EU cities, is available in an interactive online dashboard – Quantifying Europe's Cycling Infrastructure using OSM [11].

3.2 Cycle Infrastructure Needs

After subtracting 2,625 km of "superfluous" cycle infrastructure (not addressing or exceeding the needs dictated by the main road network) from the identified existing cycle infrastructure, and comparing the remainder with the main road network, the estimated infrastructure needs amounted to 96,906 km (Table 1). On average, current cycle infrastructure covered approximately 40% of the needs, but the percentage varied

Table 1. Calculation of the cycle infrastructure needs in TEN-T urban nodes.

Main road networks	156,144 km
Existing cycle infrastructure	61,862 km
"Superfluous" cycle infrastructure	2,625 km
Cycle infrastructure needs	**96,906 km**

significantly between different countries – from less than 10% in Malta, Greece, Portugal and Cyprus to more than 80% in Netherlands and Finland.

In terms of absolute numbers, the biggest investments were estimated to be needed in Spain and Italy (nearly 15,000 km each), followed by Germany, France, Greece and Portugal. Figure 1 presents the identified cycle infrastructure needs across EU Member States in kilometres, while Fig. 2 – in percentage of the target network.

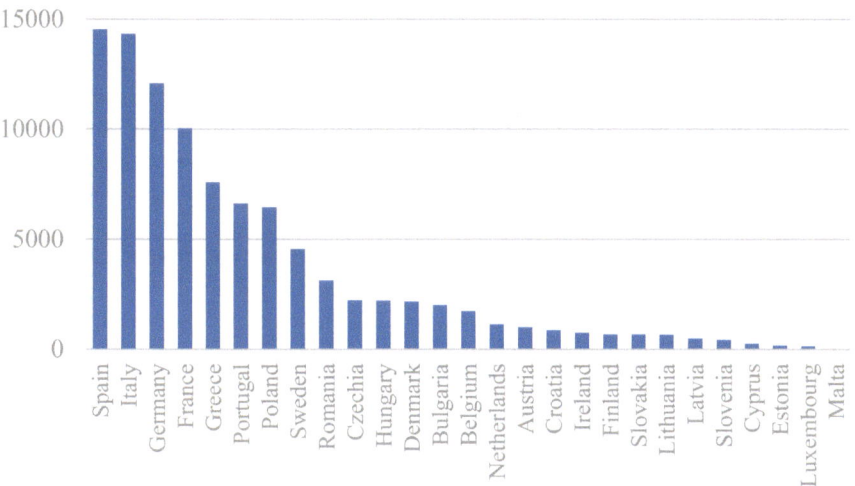

Fig. 1. Cycle infrastructure needs in kilometres in TEN-T urban nodes across EU Member States.

3.3 Limitations of the Approach

It should be noted that the calculated numbers are only an approximation. On one side, the assessment does not evaluate whether the current infrastructure type is adequate to the given location. It also does not take into account the quality of the existing infrastructure. Many existing cycle tracks and cycle lanes need to be improved in order to be safe to use, some common cycle and pedestrian tracks need to be reconstructed into separate tracks for cyclists and pedestrians, and so on. On the other side, on certain roads, even though in their current shape they have been classified as "main" roads, reduction and/or slowing down of car traffic might be a better approach than providing segregated infrastructure for cyclists.

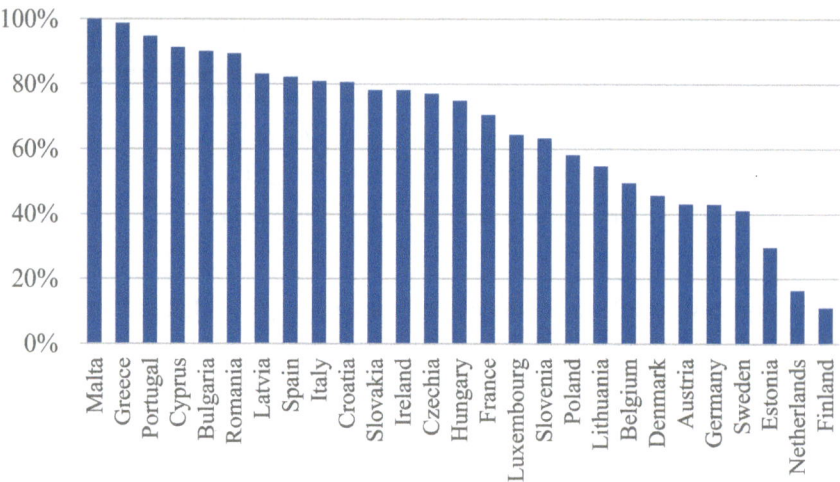

Fig. 2. Cycle infrastructure needs expressed as percentage of the main street network across EU Member States.

Moreover, the character of area included in the administrative borders varies across different countries, and also across different municipalities within specific countries. While the European Parliament proposed extending the definition of urban nodes to cover the whole functional urban areas, at the time of writing the final shape of the definition is still a subject of interinstitutional negotiations.

Finally, the approach provides only an estimate for linear infrastructure. A complete evaluation of a safe cycle network should also consider intersections between cycle routes and other traffic. In particular, major roads might need to be crossed with costly grade-separated solutions, bridges or tunnels.

4   Conclusions and Next Steps

While the calculated number of 97,000 km is only an approximate, it gives an orientation about the scale of investments needed to make cycling a safe mobility option for the residents of major EU cities. To the best of our knowledge it is the first data-driven attempt to quantify the cycle infrastructure needs on the European level.

The number is nearly 20 times higher than the 5,000 km target mentioned in the EU Sustainable and Smart Mobility strategy. The investment effort needs to be significantly upscaled in order to tap into the potential of cycling as a healthy and sustainable mode of transport.

Currently, we are working on expanding the methodology to rural areas, taking into account the need to adapt road classification (which categories of roads are suitable for cycling in mixed traffic), and different forms of infrastructure. For example, roads shared with agricultural vehicles might also under certain conditions be suitable for cycling, and they already constitute an important part of intercity cycle networks in several countries.

References

1. Frans Timmermans keynote speech at the Cycling Industries Europe Summit 2023. https://ec.europa.eu/commission/presscorner/detail/en/SPEECH_23_1561. Accessed 31 Aug 2023
2. Sustainable and Smart Mobility Strategy – putting European transport on track for the future. https://eur-lex.europa.eu/legal-content/EN/TXT/?uri=CELEX%3A52020DC0789. Accessed 31 Aug 2023
3. OpenStreetMap homepage. https://www.openstreetmap.org/. Accessed 31 Aug 2023
4. Ferster, C., Fischer, J., Manaugh, K., Nelson, T., Winters, M.: Using OpenStreetMap to inventory bicycle infrastructure: a comparison with open data from cities. Int. J. Sustain. Transp. **14**(1), 64–73 (2020)
5. Vierø, A. R., Vybornova, A., Szell, M.: BikeDNA: a tool for bicycle infrastructure data and network assessment. Environ. Plan. B: Urban Anal. City Sci. (2023)
6. Szell, M., Mimar, S., Perlman, T., Ghoshal, G., Sinatra, R.: Growing urban bicycle networks. Sci. Rep. **12**, 6765 (2022)
7. Proposal for a Regulation on Union guidelines for the development of the trans-European transport network. https://eur-lex.europa.eu/legalcontent/EN/ALL/?uri=COM%3A2021%3A812%3AFIN. Accessed 31 Aug 2023
8. European Parliament resolution of 16 February 2023 on developing an EU cycling strategy. https://www.europarl.europa.eu/doceo/document/TA-9-2023-0058_EN.html. Accessed 31 Aug 2023
9. Bicycle Network Analysis homepage. https://bna.peopleforbikes.org/. Accessed 31 Aug 2023
10. Sustainable Safety 3rd edition – The advanced vision for 2018–2030. https://swov.nl/nl/publicatie/sustainable-safety-3rd-edition-advanced-vision-2018-2030. Accessed 31 Aug 2023
11. Implementation of Sustainable Urban Mobility Indicators e-course. https://www.mobilityacademy.eu/enrol/index.php?id=109. Accessed 31 Aug 2023
12. Quantifying Europe's Cycling Infrastructure using OpenStreetMap v1.0. https://lookerstudio.google.com/u/0/reporting/81d2904d-7db5-4ed5-98e0-85af75b46577/page/p_qsvwe0yluc. Accessed 31 Aug 2023

Open Access This chapter is licensed under the terms of the Creative Commons Attribution 4.0 International License (http://creativecommons.org/licenses/by/4.0/), which permits use, sharing, adaptation, distribution and reproduction in any medium or format, as long as you give appropriate credit to the original author(s) and the source, provide a link to the Creative Commons license and indicate if changes were made.

The images or other third party material in this chapter are included in the chapter's Creative Commons license, unless indicated otherwise in a credit line to the material. If material is not included in the chapter's Creative Commons license and your intended use is not permitted by statutory regulation or exceeds the permitted use, you will need to obtain permission directly from the copyright holder.

Nudging Urban Cycling Through Gamification and Rewarding Schemes

Maria Konstantinidou[1]([⊠]) [iD], Josep Maria Salanova Grau[1] [iD], Annarita Leserri[2],
Fariya Sharmeen[3] [iD], and Matt Davoudizavareh[3] [iD]

[1] Centre for Research and Technology Hellas/CERTH-Hellenic Institute of Transport/HIT, 6Th
Km Charilaou-Thermi Road, 57001 Thessaloniki, Greece
`mariakon@certh.gr`
[2] PinBike, KM 32,700, SP231, 70033 Corato, Italy
[3] KTH Royal Institute of Technology, Brinellvägen 8, 114 28 Stockholm, Sweden

Abstract. Although behavioural nudging has been often used as a method to promote sustainable choices, its application is quite limited in the field of active mobility. Gamification and rewarding schemes appeared recently in mobility sector to nudge urban cycling. The scheme presented in this paper was based on a TRL9 antifraud system to certify, monitor, and reward urban bike rides in in 3 different cities during 2022. About 1,500 users were engaged in Braga, Istanbul, and Tallinn, saving more than 133 tons of CO2 in 6 months, and spending more than 85,000 euros in local shops. The paper presents the quantitative (KPIs based) and qualitative (users questionnaires based) assessment as well as the additional data analysis results that were carried out during the scheme testing. The correlation between trip, weather and socio-demographic data provided valuable knowledge that can be used to target vulnerable groups and to enhance the cycling incentives for bad weather days. The knowledge extracted can be integrated into the climate neutral strategies of the transition cities/regions for designing and implementing more efficient future actions and investments to successfully promote greener and more active mobility, ensuring high levels of cycling throughout the year.

Keywords: active mobility · gamification · planning strategy

1 Introduction

Although behavioral nudging has been often used as a method to promote sustainable choices, its application is quite limited in the field of active mobility. Gamification and rewarding schemes appeared recently in the mobility sector as an opportunity for behavioral nudging the potential day-to-day cyclist.

But what motivates communities to make a change? Is it an issue like poor air quality's negative impact on the health of our citizens? Or maybe cities are concerned about equity and whether all families, regardless of income or social status have access to work and school opportunities? Maybe it is the high cost of car ownership? Whatever the motivation, how can we influence behavior to change to a more desired transportation option?

© The Author(s) 2025
C. McNally et al. (Eds.): TRAconference 2024, LNMOB, pp. 160–172, 2025.
https://doi.org/10.1007/978-3-031-85578-8_22

The collective climate change goals now provide an opportunity to change the public's travel behavior because "a norm or practice that is understood to be emerging, or to be increasingly supported, can operate as a powerful nudge, even if it is not yet supported by the majority" (Thaler & Sunstein, 2021). Local governments are searching for ways to reduce their Greenhouse Gas emissions. And now they are supported by the motivation to attain their country-wide climate change goals. So, the decision makers are now in a place to support cleaner travel modes. And citizens are also motivated as climate change is now an eminent enough threat that people in all levels of society are discussing the consequences and possible solutions. People, based on their identity as Europeans, or citizens of a particular country, feel committed to acting against climate change. Or perhaps, on the issue of climate change, they have collectively moved past pluralistic ignorance, where individuals did not know that their neighbors were also interested in making a change (Thaler & Sunstein, 2021).

An additional motivation could be an increase in obesity rates and decrease in activity among citizens worldwide. Citizen inactivity is a greater health risk than the increased short-term air pollution exposure a cyclist might experience (de Hartog et al., 2010; de Nazelle et al., 2011; Giles-Corti et al., 2016; Rojas-Rueda et al., 2011). Inactivity and its consequences are now global concerns. Even in The Netherlands, a place known for high rates of bicycling, has a 62% sedentary rate for its population (de Hartog et al., 2010). The negative effects of this widespread lack of activity results in high rates of cardiovascular disease, as well as other negative health outcomes like diabetes, obesity, cancer, osteoporosis and depression (de Hartog et al., 2010). As a group, these are known as Non-Communicable Diseases (NCDs).

Of course, a bicyclist's exposure to air pollution carries a health risk. But the health problems occurring from inactivity and car culture have grown so enormous that we must find ways to encourage and promote active transportation (Giles-Corti et al., 2016). An evaluation of the launch of a bike-share program in Barcelona concluded that the health benefits of cycling using the sharing scheme were important compared with the risks from inhalation of air pollutants and road traffic incidents (Rojas-Rueda et al., 2011). Those that use cycling as a utilitarian mode (to travel from point A to B), often undervalue the health aspects of cycling.

The benefits of active modes are well established for several decades. Direct and indirect health benefits are indisputable facts of active travel modes such as cycling (Nazelle A. et. al. 2011, Green J. et al. 2013). Inducing 'active travel' via different strategies is beneficial for both mobility within our cities and our quality of life (Wanless D. et al. 2004, Mueller N. et al. 2015). Given the benefits, numerous governments across the world are gearing cities toward a bikeable built environment. An important strategy to increase active travel is the construction of new bicycle-friendly infrastructures (Pucher J. et al. 2011, Litman T. 2015, Chengxi L. et al. 2021). A study that was conducted in Paris and Lyon in France from 2014 to 2020 (Xiao S. et al. 2022) observed an increase of at least 14.7% and 8.2% in mean daily cycling counts in some identified locations in Paris and in Lyon respectively after 6 months from the development of new bicycle-friendly infrastructures. Another study in Glasgow Scotland observed that in the short-term introducing new cycling infrastructure, especially inside the city area will effectively induce cycling mode share (Hong J. et al. 2019). However, other studies have highlighted that constructing new cycle lanes may be necessary for cities to get more people on board when it comes to active travel mode share, but it is not sufficient to see significant changes in cycling levels (Tortosa et al. 2021, Rachel A. et al. 2013).

This lack of consistency across different studies could be a result of relying on measures that are prone to be biased. Therefore, more in-depth analyses are needed to get insights into contributing factors regarding cycling-induced policies, technologies and infrastructure. This will pave the way for decision-making authorities and policymakers when planning a sustainable and bikeable urban area.

One useful source of data to shed light on cycling behavior and route preferences is GPS-based bicycle data (Menghini J. et al. 2010, Hood J. et al. 2011, Hudson J. et al. 2012, Broach J. et al. 2012, Casello J. et al. 2014, Gustavo R. et al. 2015, Kristiann C. et al. 2016, Nikola M. et al. 2019, Chengming L. et al. 2019). Many GPS-based studies are conducted with a small number of candidates wearing a GPS device or via smartphones for a certain period of time. Therefore, they were not capable of recording for longer periods or with a large sample of users (Shen L. et al. 2014).

Many European countries have tried campaigns to promote active mobility. But the cycling mode share remains very low. How people choose to travel to work, school and even the grocery store is a behavior and a habit that is difficult to change, even as there are new modes available to choose from. The rewarding scheme presented in this paper aims to achieve a modal shift to bicycling, nudging participants to ride, by giving them tangible awards (both monetary and non-monetary) and widely communicating the message "you ride - you earn". The added value of the tested scheme is the amount of data being collected on a very large geographical and user scale (almost 1,500 users).

2 Methods

2.1 Scheme Description

The aim of the rewarding scheme was to motivate citizens to shift from motorized vehicles to cycling, through the implementation of gamification/rewarding techniques in 3 cities (Braga, Istanbul, Tallinn) that are already making great attempts to promote active mobility. The scheme was supported by an already high-TRL technological solution provided, consisting of a patented hardware and software, which is capable of accurately monitoring bicycle trips, and they engage:

1. Local authorities are the regulators of the scheme, set rules and prove rewards. Additionally, they collect bike trajectories data and extract useful information to create knowledge for the usage of active means of transportation.
2. A critical mass of users is the main actor of the scheme. They receive a bike kit, install the dedicated app and earn points and rewards as they ride.
3. Local shops are supporters of the scheme, and they provide discounts based on the collected points. Thus, an additional benefit for the local communities is the promotion of purchases from local shops instead of large shopping malls often located out of the city perimeter, and reachable only by private cars.

Although all the three pilots started in June 2022, the start dates were slightly different depending on the preparation activities of each city. Consequently, each pilot ended after the completion of 4 months between end of September and mid-October (as month is considered a 30-day duration and not a calendar month). During the whole duration each participant in all three cities was able to check the number of his/her rewards in the Pin Bike mobile application. For Braga and Tallinn, every 10 € accrued, the app automatically was creating a 10 € voucher that could be spent to purchase products/services offered by the local shop. The participants were able to see in the app the local shops that participated in the scheme. As soon as the vouchers were used by the participants, cities were responsible for the monthly reimbursements to the local shops.

In Istanbul, although vouchers were created automatically by the application for each 10 € earned, the amounts were transferred as a credit on the Istanbul Card of each participant. IstanbulCard is a city card that people can use for shopping, art activities, transportation, and other social activities.

After the 4-month pilot period in Tallinn, Braga and Istanbul, all three cities expressed their interest in continuing the pilots further until December 2022 since it was quite critical for them to incentivize urban cycling during the winter months.

2.2 Data Collection and Pre-processing

All three cities developed registration forms that were available to the citizens through the municipalities' websites during the pilots' preparation phase for stating their interest in participating in the scheme. Except of the users' profile data collected through these registration forms and the quantitative data collected during the pilots' implementation thanks to the GPS kits placed on the bicycles, a questionnaire was delivered to participants in Braga, Tallinn, and Istanbul after the end of the 4-month pilot period in order to collect their qualitative feedback about different elements of the system and the mobile application during the whole 4-month period.

For the qualitative evaluation of the pilots' extension period, an additional questionnaire was delivered to the users in December. A single questionnaire was drafted for the three pilot sites for allowing cross-area comparisons. Although some of the questions included in this last questionnaire were same with the ones of the previous questionnaires enabling the comparisons between summer and winter period, additional questions related to weather and sustainability factors were also included.

Moreover, weather data was collected and stored from an API (https://openweath ermap.org/). This includes weather category (clear, rain, cloudy, thunderstorm, etc.), wind speed and direction, amount of rain and other weather phenomena, at the time of which a trip was happening.

After data collection the raw data was cleaned and pre-processed. The outliers and errors collected by GPS kits were cleaned in a separate process before the analysis. Then the trip data was merged with weather data using start time of each trip with the corresponding weather data. Next, user data provided from registration forms was merged with new trip dataset, using user identifier (a code which is generated to keep user anonymized). The new merged data frame provides valuable knowledge on different dimensions, therefore for every trip trace we have the corresponding weather and sociodemographic. This process was repeated for all the three cities of Braga, Istanbul and Tallinn.

3 Results and Discussion

3.1 Users Profile

In Tallinn, the gender was quite balanced as 46% of the participants were women and 53% were men. In Istanbul, 85% of the participants were men and 15% were women. The participants in Braga were 31% women and 69% men. So, in Braga and Istanbul, the percentages were quite far from the balanced target values that was set by the municipalities before the start of the pilots. Age distribution, personal status and educational level were also among the asked characteristics.

People that were stated their interest in participating in the scheme were also asked about how often they used their bicycle before the start of the pilots. 34% of the registered citizens in Tallinn use their personal bicycle every day, including weekends, 27% of them every working day and 34% of them about two days a week. In Braga, the majority of the participants (36%) stated that they cycle 2–3 times a week. 29% stated that they cycle every day, including weekends, 19% only on the working days and 16% once a week. In Istanbul, the majority (46%) answered that they use their bicycle 2–3 times a week and 31% use it every day, including weekends (see Fig. 1).

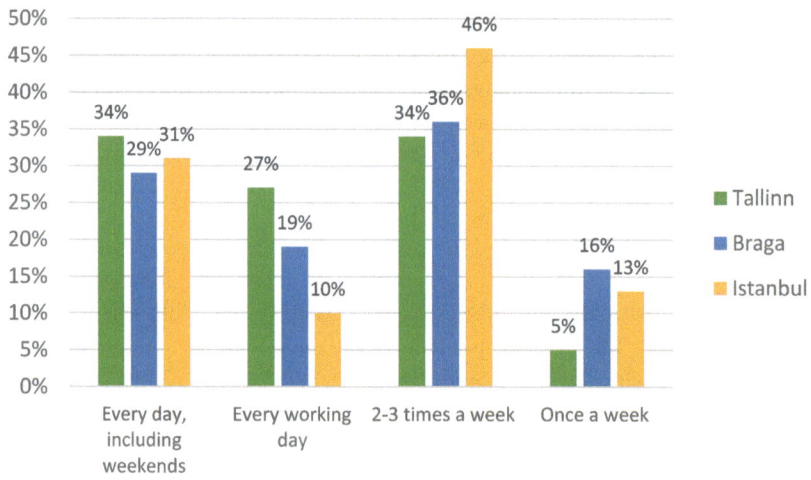

Fig. 1. Frequency of bicycle use before pilot's start

The question included in the registration form before the start of the pilot related to the frequency of the bicycle use enabled the examination of the contribution of the rewarding scheme in increasing the cycling levels in Tallinn, Braga, and Istanbul.

3.2 Quantitative Analysis

During the period of 6-month period important Key Performance Indicators (see Table 1) along with heatmaps of cycling trajectories were available to cities through the municipality dashboard to continuously monitor the cycling conditions in their cities including the routes used by citizens, the most used roads, starting and ending points, average and maximum speeds and preferred routes versus suggested routes.

Table 1. Key Performance Indicators in Tallinn, Braga and Istanbul

Key Performance Indicators	Tallinn	Braga	Istanbul
Active users	422	400	387
Cycling sessions	>38, 000	>31,320	>26, 200
Total distance travelled (kilometers)	237, 760	227, 143	370, 788
Average session distance (kilometers)	6.24	7.24	14.13
CO_2 saved (tons)	About 38	About 36	About 60 tons
Local shops participated	19	35	Istanbulcards shops
Total amount earnt (€)	About 26, 000	About 34,000	About 25, 400

As the start date of each pilot was slightly different, a month is considered as a duration of 30 days and not as a calendar month (except of the last month), to have comparable results in the pilot KPIs. The number of sessions in Braga and Istanbul follow a similar slightly decreasing trend by the 3rd month. The high numbers of the 1st month in all cities could be justified by the enthusiasm of the citizens to participate in a new mobility scheme. Some problems related to the allocation of the vouchers to the users during the 2nd month as well as some difficulties in the use of the mobile app and of the vouchers in the local shops could explain the slight decrease of the sessions. The 3rd month run during August in which is traditionally the summer vacation month in Braga and Istanbul, and thus the sessions appeared also decreased. In September, the sessions in Braga reached almost the number of the 1st month and then during October to 11th of December, they follow a decreasing trend due to weather conditions. In Istanbul, additional users joined the project in the 5th month, so this could be an explanation for the sessions increase.

A similar decrease between the 1st and the 2nd month was also noted in Tallinn for the same reasons mentioned above. However, the users in Tallinn cycled more during the 3rd month (August). A high decrease is also noted in September in Tallinn due to weather conditions. During the extension period of the pilots from October to 11th of December, the sessions in Tallinn were continuously decreasing due to weather. All the above observations are presented in Fig. 2.

The average sessions distance in Tallinn and Braga was similar (6.24 km in Tallinn and 7.24 km in Braga) while in Istanbul this KPI was almost double 14.13 km. This is due to the fact that Istanbul is a megacity with long distances and also in Istanbul, cyclists were allowed to put bicycles on ferries as we have noticed by looking at the trip trajectories on the map. Additionally, historical weather data during the pilot was compared to weather during the trip occurrence to see if and how the weather affects the cycling trips. A slight effect of rain on trips was noticed.

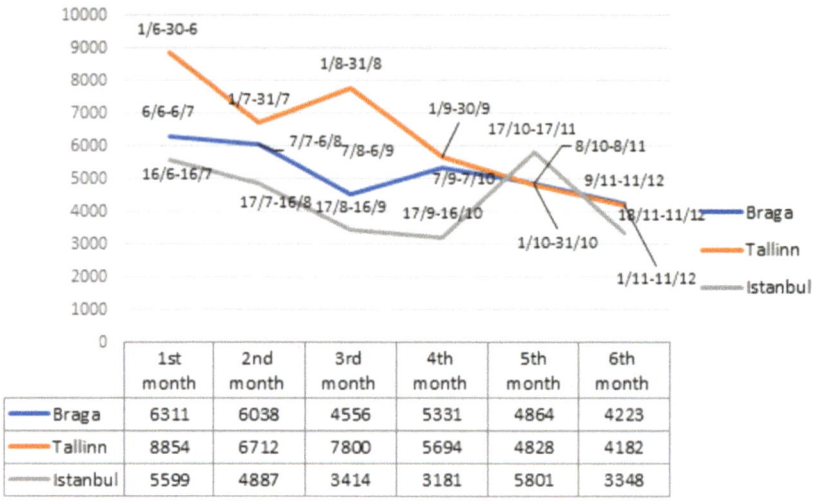

Fig. 2. Comparison of monthly trend of registered sessions among pilot cities

	1st month	2nd month	3rd month	4th month	5th month	6th month
Braga	6311	6038	4556	5331	4864	4223
Tallinn	8854	6712	7800	5694	4828	4182
Istanbul	5599	4887	3414	3181	5801	3348

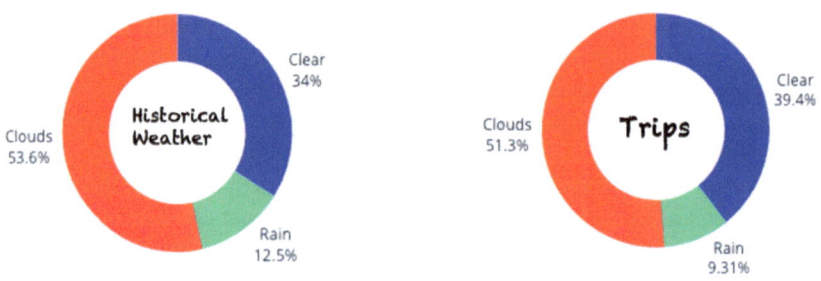

Fig. 3. Weather effect on Braga's trips

In Fig. 3 for instance it is observed that in Braga only 9% of trips took place in rainy conditions even though 12% of times the weather was rainy during the pilot's period. The same slight effect was observed in Istanbul but not in Tallinn.

3.3 Qualitative Analysis

The questionnaire to users in Braga and Tallinn was sent via notification in the mobile application. The answers collected were 166 in Braga, 194 in Tallinn and 253 in Istanbul.

Regarding the performance evaluation of the features of the system, the kit received the higher percentage of positive evaluation (4 or 5 value in Likert scale) in all three cities (84% in Braga, 85% in Tallinn and 78% in Istanbul). The second and the third higher percentage was noted for the rewarding system and the vouchers/rewards to be spent in local shops notifications in all cities. The mobile application was evaluated as a high-performance feature by more than 60% both in Braga and Istanbul. However, in Tallinn just 35% of the respondents evaluated mobile app with 4 or 5 values in Likert scale (see Fig. 4).

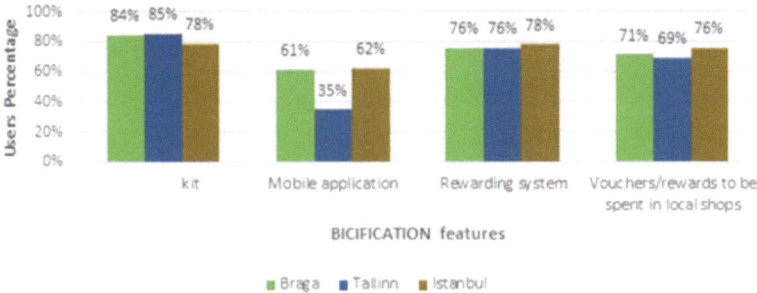

Fig. 4. Comparison of the performance evaluation results of system's features (4–5 values)

The positive evaluation (4 or 5 values in Likert scale) of the usefulness of the mobile app's features followed similar percentages in Braga and Istanbul, while in Tallinn the percentages were quite lower in all features except of the "sessions registration" and "rewards". The higher deviation was noted in the "CO2 savings" and the "help center" feature. In Braga and Istanbul, 64% and 69% of the participants evaluated positively the help center while in Tallinn the percentage was 38% (see Fig. 5).

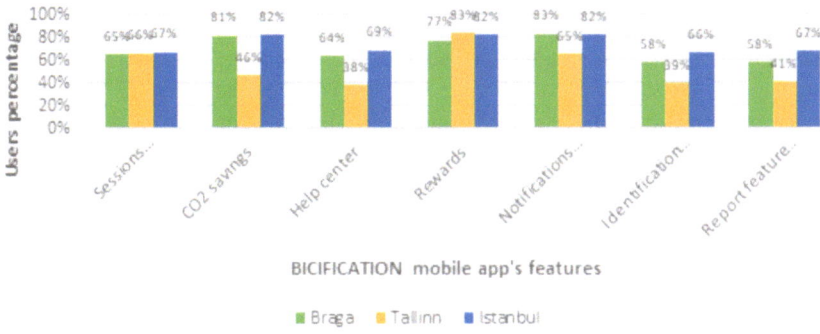

Fig. 5. Comparison of the usefulness evaluation results of system's features (4–5 values)

Most of the users in Braga (88%) evaluated their whole experience with the higher values (4 and 5 in Likert scale); a percentage like the one after the 4-month period. In the first questionnaire in Tallinn, more than half of the participants evaluated the experience with 4 and only 19% with 5. At the end of the pilot in December, the percentage of those that evaluated the whole experience with 3 decreased from 23% to 10% while 48% evaluated it with 5. In Istanbul, the percentage of those that evaluated with 5 increased from 7% (after the end of the 4month period) to 48%. Additionally, the number of participants that evaluated the overall experience negatively (values 1 or 2 in Likert scale) has almost eliminated (from over 25% after the 4-month period to 2% in December) (see Fig. 6).

In Tallinn, the percentages of those who use their bicycle "every day, included wee-ends" and "every working day" from October to the end of the pilot, were balanced to 21–22%. These percentages decreased compared to the ones of the first 4-month period

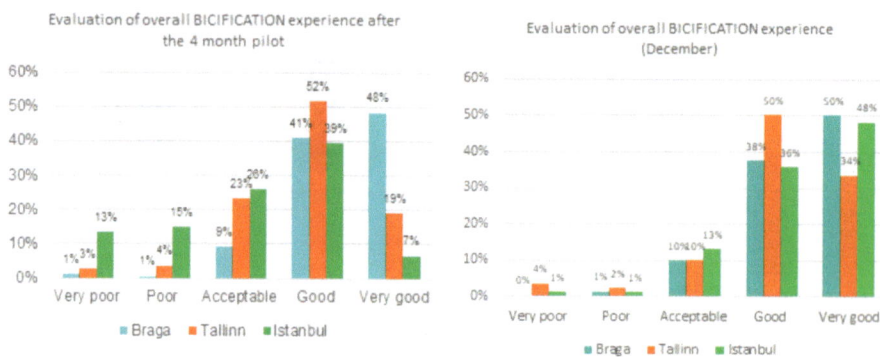

Fig. 6. Comparison of the evaluation results of the overall experience among pilot cities

due to weather conditions. In Braga, the percentage of the users that use their bicycle "every day included weekends" remained high both in the first 4-month period and the second period (30% in October-November compared to 36% in the 4-month period). The percentage of those that cycle "every working day" decreased from 27% to 18% while the percentage of those that cycle "once a week" increased from 3% to 18% due to weather conditions. Most users in Istanbul (32%) use their bicycle "once a week" during October-November while the percentage of those that use it "2–3 times/week" decreased from 41% in the 4-month period to 29%.

An important result derived from the analyses is presented in Fig. 7, in which we observed an increase of bicycle use frequency in the category "every working day" between what has been filled by the users in the registration form versus the use during the pilot.

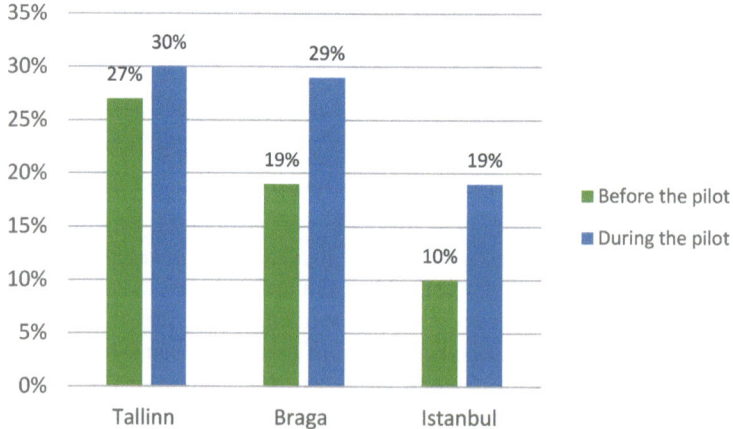

Fig. 7. Bicycle use "every working day" before and during the pilot period in Tallinn, Braga and Istanbul

Finally, in all three cities more than 90% of the users stated that it is quite (4) or very likely (5) to continue cycling after the end of the project and the stop of rewards provision (see Fig. 8).

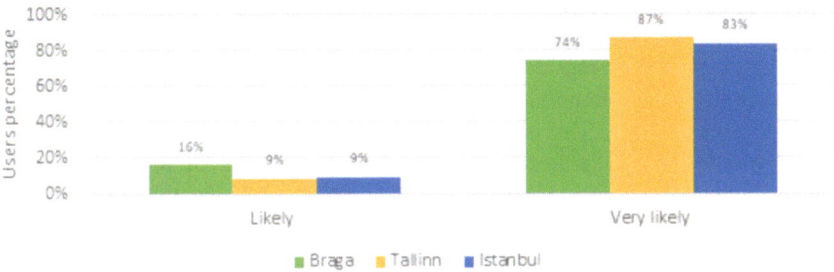

Fig. 8. Comparison of the likelihood of continuing cycling after the end of the project

This statement confirms and enhances the success of the scheme, as one of the main objectives was not only to support green and active mobility through the rewarding scheme but succeed a behavioral change aiming at the achievement of a long-term modal shift towards active mobility.

3.4   Limitations and Future Study

Regarding the weather correlation with trips, the analyses performed represent rough estimation and statistical analyses to detect significant correlations are suggested to be performed for getting better insights. Further analyses related to specific weather conditions such as the wind speed and direction, the temperature and other weather phenomena are needed to see to what extend different weather conditions can affect cyclists' behavior and frequency.

For future study, the trip trajectories will be matched with the built environment using QGIS and Open Street map, to further analyze the behavior of cyclists. This will provide valuable insights on built environment effects on cycling behavior. For instance, the presence of greenery, stores, bicycle infrastructure and their effects on trip frequency.

4   Conclusions

The scheme presented in this paper was based on a TRL9 antifraud system to certify, monitor and reward urban bike rides in Braga, Istanbul and Tallinn during 2022. Local authorities rewarded urban cyclists with monetary vouchers to be spent in local shops. Cities benefited from valuable cycling data collected for both evaluating already existing mobility measures and infrastructures and investing in data-driven policies guiding the transition towards climate neutrality. The knowledge extracted through the data analyses of the rewarding scheme can be integrated into the climate neutral strategies of the cities for designing and implementing more efficient future actions and investments to successfully promote greener and more active mobility, ensuring high levels of cycling throughout the year.

Additionally, the results of the analyses can be used from cities as a valuable material for designing educational campaigns on the health benefits of bicycling, highlighting the experiences of those who are already riding. Although these campaigns can motivate, with a light-touch behavioral nudge, those who are interested in trying to ride, they should be designed by a "choice architect", someone with the responsibility for organizing the context in which citizens make their decisions. The behavioral nudges to encourage higher ridership levels would aim to alter behavior in a predictable way without forbidding any options, or significantly changing a citizen's economic incentives (Thaler & Sunstein, 2021). This is exactly how the campaign of the scheme presented in this paper was designed. Certainly, the economic incentives per kilometer do not sound like much money. However, a monetary reward of this type fits in with the definition that a behavioral nudge does not significantly alter one's economic incentives, but it does give immediate feedback on making the desired choice.

Acknowledgements. The research is performed within the framework of BICIFICATION- Supporting modal shift and bicycle use through gamification and rewarding, co funded by EIT UM, TRANSFORMER- Designing long-term systemic transformation frameworks for regions, Accelerating the shift towards climate neutrality that has received funding from the European Climate, Infrastructure and Environment Executive Agency (CINEA) under grant agreement No 101069934 and «Safecity» (Project code: KMP6 0081411) under the framework of the Action «Investment Plans of Innovation» of the Operational Program «Central Macedonia 2014–2020», that is co-funded by the European Regional Development Fund and Greece.

References

Andersen, L.B., Schnohr, P., Schroll, M., Hein, H.O.: All-cause mortality associated with physical activity during leisure time, work, sports, and cycling to work. Arch. Intern. Med. **160**(11), 1621 (2000). https://doi.org/10.1001/archinte.160.11.1621

Broach, J., Dill, J., Gliebe, J.: Where do cyclists ride? A route choice model developed with revealed preference GPS data. Transp. Res. Part A: Pol. Pract. **46**(10), 1730–1740 (2012)

Casello, M.J., Usyukov, V.: Modeling cyclists' route choice based on GPS data. Transp. Res. Rec. **2430**(1), 155–161 (2014)

Chengxi, L., Tapani, A., Kristoffersson, I., Rydergren, C., Jonsson, D.: Appraisal of cycling infrastructure investments using a transport model with focus on cycling. Case Stud. Transp. Pol. **9**(1), 125–136 (2021). ISSN 2213-624X, https://doi.org/10.1016/j.cstp.2020.11.003

de Hartog, J.J., Boogaard, H., Nijland, H., Hoek, G.: Do the Health Benefits of Cycling Outweigh the Risks? Environ. Health Perspect. **118**(8), 1109–1116 (2010). https://doi.org/10.1289/ehp.0901747

de Nazelle, A., et al.: Improving health through policies that promote active travel: a review of evidence to support integrated health impact assessment. Environ. Int. **37**(4), 766–777 (2011). https://doi.org/10.1016/j.envint.2011.02.003

Giles-Corti, B., et al.: City planning and population health: a global challenge. Lancet **388**(10062), 2912–2924 (2016). https://doi.org/10.1016/S0140-6736(16)30066-6

Grabow, M.L., Spak, S.N., Holloway, T., Stone, B., Mednick, A.C., Patz, J.A.: Air quality and exercise-related health benefits from reduced car travel in the Midwestern United States. Environ. Health Perspect. **120**(1), 68–76 (2012). https://doi.org/10.1289/ehp.1103440

Heesch, K., Langdon, M.: The usefulness of GPS bicycle tracking data for evaluating the impact of infrastructure change on cycling behaviour. Health Promotion J. Australia **27** (2016). https://doi.org/10.1071/HE16032

Hong, J., McArthur, D.P., Stewart, L.J.: Can providing safe cycling infrastructure encourage people to cycle more when it rains? The use of crowdsourced cycling data (Strava). Transp. Res. Part A: Policy Pract. **133**, 109–121 (2020). ISSN 0965-8564, https://doi.org/10.1016/j.tra.2020.01.008

Hood, J., Sall, E., Charlton, B.: A GPS-based bicycle route choice model for San Francisco, California. Transp. Lett. **3**(1), 63–75 (2011)

Hudson, J.G., et al.: Using smartphones to collect bicycle travel data in Texas. No. UTCM 11-35-69. Texas Transportation Institute. University Transportation Center for Mobility (2012)

Marković, N., Sekuła, P., Vander Laan, Z., Andrienko, G., Andrienko, N.: Applications of trajectory data from the perspective of a road transportation agency: literature review and Maryland case study. IEEE Trans. Intell. Transp. Syst. **20**(5), 1858–1869 (2019). https://doi.org/10.1109/TITS.2018.2843298

Menghini G., Carrasco, N., Schüssler, N., Axhausen, K.W.: Route choice of cyclists in Zurich. Transp. Res. Part A Pol. Pract. **44**(9), 754–765 (2010). ISSN 0965-8564, https://doi.org/10.1016/j.tra.2010.07.008

Mueller N., et al.: Health impact assessment of active transportation: a systematic review. Prevent. Med. **76**, 103–114 (2015). ISSN 0091-7435, https://doi.org/10.1016/j.ypmed.2015.04.010

Li, C., Dai, Z., Peng, W., Shen, J.: Green travel mode: trajectory data cleansing method for shared electric bicycles. Sustainability **11**, 1429 (2019). https://doi.org/10.3390/su11051429

Litman, T.: Transportation and public health. Annu. Rev. Public Health **34**(1), 217–233 (2013). https://doi.org/10.1146/annurev-publhealth-031912-114502

Litman, T.: Evaluating public transit benefits and costs. Victoria, BC, Canada: Victoria Transport Policy Institute (2015)

Pucher, J., Buehler, R., Seinen, M.: Bicycling renaissance in North America? An update and re-appraisal of cycling trends and policies. Transp. Res. Part A: Pol. Pract. **45**(6), 451–475 (2011). ISSN 0965-8564, https://doi.org/10.1016/j.tra.2011.03.001

Rachel, A.: Who are Londoners on Bikes and what do they want? Negotiating identity and issue definition in a 'pop-up' cycle campaign. J. Transp. Geography **30**, 194–201 (2013). ISSN 0966-6923, https://doi.org/10.1016/j.jtrangeo.2013.01.005

Rojas-Rueda, D., de Nazelle, A., Tainio, M., Nieuwenhuijsen, M.J.: The health risks and benefits of cycling in urban environments compared with car use: health impact assessment study. Bmj-British Med. J. **343**, d4521 (2011). https://doi.org/10.1136/bmj.d4521

Romanillos, G., Austwick, M., Dick, E., de Kruijf, J.: Big Data and Cycling. Transport Reviews. **36**, 1–20 (2015). https://doi.org/10.1080/01441647.2015.1084067

Shen, L., Stopher, P.R.: Review of GPS travel survey and GPS data-processing methods. Transp. Rev. **34**(3), 316–334 (2014)

Thaler, R.H., Sunstein, C.R.: Nudge: The Final Edition. Yale University Press (2021)

Tortosa, E.V., et al.: Infrastructure is not enough: interactions between the environment, socioeconomic disadvantage, and cycling participation in England. J. Transp. Land Use, **14**(1), 693–714 (2021). JSTOR. https://www.jstor.org/stable/48646205. Accessed 25 June 2022

Wanless, D.: Securing good health for the whole population. 417–432 (2004)

Xiao, C.S., Sharp, S.J., van Sluijs, E.M.F., et al.: Impacts of new cycle infrastructure on cycling levels in two French cities: an interrupted time series analysis. Int. J. Behav. Nutr. Phys. Act. **19**(77), 2022 (2022). https://doi.org/10.1186/s12966-022-01313-0. Accessed 5 Sept 2022

Open Access This chapter is licensed under the terms of the Creative Commons Attribution 4.0 International License (http://creativecommons.org/licenses/by/4.0/), which permits use, sharing, adaptation, distribution and reproduction in any medium or format, as long as you give appropriate credit to the original author(s) and the source, provide a link to the Creative Commons license and indicate if changes were made.

The images or other third party material in this chapter are included in the chapter's Creative Commons license, unless indicated otherwise in a credit line to the material. If material is not included in the chapter's Creative Commons license and your intended use is not permitted by statutory regulation or exceeds the permitted use, you will need to obtain permission directly from the copyright holder.

Modelling Sustainable Transport – An Open Data Approach to Model Mode Shift Towards Net Zero

Stefan Huber🆔 and Iwan Porojkow$^{(\boxtimes)}$ 🆔

Faculty of Transport and Traffic Science "Friedrich List", Institute of Transport Planning and Road Traffic, Chair of Transport Ecology, Dresden University of Technology, 01062 Dresden, Germany
iwan.porojkow@tu-dresden.de

Abstract. The contribution presents a transport demand model based on open data and freely available information. The model is based on the classic sequential four-step model structure but runs on microscopic scale using disaggregate data. It includes the modes 'walk', 'cycle', 'car' and 'public transport'. Various open data sources and information are used to estimate and run the model, such as OpenStreetMap data and freely available census and postal data, freely accessible information (e.g., mobility indicators) from publications of national and urban household surveys on traffic behavior or freely available information on transport costs. The result of the study is a disaggregate microscopic transport model for passenger transport. The proof of concept reveals a high initial model accuracy and its responsiveness to different measures.

Keywords: Transport Demand Model · Open Data · Sustainable Transport

1 Introduction

The transportation sector is a significant contributor to greenhouse gas emissions as transportation is responsible for a substantial share of CO_2 emissions (see e.g. 20% in Germany [1]). Furthermore, the environmental impact extends beyond global CO_2 emissions, as traffic leads to extensive local noise pollution and poor air quality.

To address these challenges diverse strategies and related measures have emerged to avoid trips, shift them to more sustainable modes [2] or to enhance technological advancements [3]. However, most cities and municipalities struggle making notable progress because they need reliable data on the effects of corresponding measures. Cities often lack assessment tools to predict the effects 'ex ante' for justification.

Transport demand models are valuable tools for assessing the potential impacts of measures. However, most existing models do not consider sustainable modes of transport (e.g., cycling and walking) adequately, which impedes accurate impact assessments of potential measures. The promotion and choice of sustainable modes of transport also heavily depends on other modes competing with sustainable transport modes. Thus, multimodal demand models are needed for accurate assessment.

© The Author(s) 2025
C. McNally et al. (Eds.): TRAconference 2024, LNMOB, pp. 173–180, 2025.
https://doi.org/10.1007/978-3-031-85578-8_23

Literature review reveals few studies focusing on sustainable modes in demand modeling. [4] used a nested logit approach to model bicycle traffic in Stockholm. The utility is influenced by cycling infrastructure, land use, slope, and sociodemographic attributes. The authors use open data (census, household surveys) and proprietary data (traffic metrics, metadata) to predicted aggregated cycling demand. Difficulties emerge at the calibration stage due to insufficient count data. Sensitivity analysis is limited on abstract generalized cost changes, for which we consider additional scenarios in detail in our study. The study by [5] investigates requirements for modelling sustainable modes in macroscopic models. Environmental, on route and psychological factors are considered to determine utility for cycling. However, Data accessibility and availability is limited, as proprietary data is used. This study highlights the importance of proper zone sizes in order to model short trips. The "Propensity to Cycling Tool" (PCT) is a nationwide open-source bicycle model for the UK [6]. It enables to compare different scenarios where indicators act as variables. However, origin destination (OD) data is fixed and there is no interaction with other modes. This hampers impact assessment for measures that address all modes, which often interact. The multi-modal transport model for Hanover (Germany) presented by utilizes open data (OSM, census data and household surveys), resulting in a sophisticated transport supply model, calibrated with mobility-related attributes from the German household surveys [7]. However, the model reveals relatively low accuracy as activity plans are simplified. [8] present a microscopic modeling approach for Berlin, which also uses open data and the results represents the traffic as posted in household surveys. However, route choice is poorly considered as cyclists and pedestrians are 'teleported' to their destination.

This contribution presents a transferable open data approach, leveraging freely accessible data to model traffic demand and assess transport measures in any German city or municipality. We present the model's concept along with the utilized data and methods in the subsequent section. A proof of concept will illustrate accuracy and responsiveness of the model. The discussion and outlook section will close the contribution illustrating further steps.

2   Data and Modelling Method

The developed diSaggregate TranspORt Model (STORM) runs on microscopic scale including all inhabitants and their specific behavior. Centerpiece of the four-step modelling approach is the mode choice model. It is a key component for modelling mode shift reactions. Model structure and used data is described in the following sections.

In a **first step**, synthetic population is generated using open census data [9], which contains information about the population per traffic zone (100x100m) and their demographics. Census data provides information on income distribution [10]. Driver license availability and car ownership distribution per age, gender, and socio-economic status come from MiT data [11]. Census data is then enriched using the iterative proportional fitting (IPF) [12]. Thus, a distinct income class, driving license and car ownership is assigned to each person within the traffic zones. In a last stage, data from ImmoScout [13] is consulted, which reveals detailed information on buildings (location no. Apartments). Population is iteratively assigned to their respective buildings using a weighted household number marginally constrained Monte Carlo simulation method [14]. Activities of people are then determined by generating weekday schedules for each person. Household surveys [15] provide shares on activity patterns within the population. A fixed and constrained Monte Carlo approach subsequently generates activity patterns for each person by distributing activity patterns depending on the observed activity probability and omitting unsuitable patterns. The result is a digital twin of the population within the area of investigation.

In the **second step**, activities are distributed to activity locations. Therefore, activity locations are generated using point of interest (POI) data from OSM. Therefore, every POI receives one or more activity types by matching their attributes using a compiled key table. After assigning activities for each POI location, all feasible locations l are listed and the choice probability p for each individual activity a is determined via

$$p_{l,a}(d) = \frac{\frac{d_{max}}{d_{o,o}}}{\sum_{l=1}^{n} \frac{d_{max}}{d_{o,l}}} \tag{1}$$

with $d_{o,d}$ as the distance between the actual location o of a person and the activity location d, and d_{max} as the maximum possible distance within the investigation area. Location choice is further influenced by location specific marginal constraints due to capacity limits by computing gross floor areas from OSM data and employee density [16]. Location choice modelling is repeated if an activity chain consists of multiple activities. Trip chains are finally split into single trips between origins and destinations.

In the **third step**, mode choice is determined using a modified logit model based on [17]. The adapted logit model can distinctively reproduce mode choice probabilities across different distances – especially for short distance trips, which is relevant in microscopic models with high resolution. The model considers the modes 'walk', 'cycle', 'car', and 'public transport'. It currently takes into account travel time, travel cost, comfort, and mode affinity. Travel time for bike, walk and car is determined using shortest path search within the OSM network. For PT, GTFS travel times are considered [18]. Travel costs are composed of fixed and variable components for all modes (e.g., fuel, parking, tickets, etc.). Comfort is implemented as discomfort cost but currently fixed to zero (placeholder). Generalized costs are then determined from the aforementioned costs and the value-of-time from [19]. Mode affinity is represented by a constant that results from model calibration. It embodies e.g., constraints or preference for using a certain mode, which cannot be explained by the aforementioned factors (travel time, travel cost, comfort). The final cost function to determine generalized cost gc per trip purpose tp and mode m from origin o to destination d is

$$gc_{tp,m,o,d} = ftc_{m,o,d} + vtc_{m,o,d} * d_{m,o,d} + \left(ftt_{m,o,d} + \left[vtt_{m,o,d} * d_{m,o,d}\right]\right) * VoT_{tp,m} + dc_m \tag{2}$$

with fixed travel costs ftc, variable travel costs vtc, fixed travel time ftt, and variable travel time vtt, and the value of time VoT. Discomfort costs dc are currently fixed to 0. The utility u per tp and m from o to d is determined using the following utility function:

$$u_{tp,m,o,d} = \alpha_{tp,m} * e^{\left(-gc_{tp,m,o,d} * \beta_{tp,m}\right)} + \gamma_{tp,m} \tag{3}$$

α, β, and γ are model coefficients resulting from model calibration. The probability p of a mode is determined using the probability function

$$p_{tp,m,o,d}\left(u_{tp,m,o,d}\right) = \frac{u_{tp,m,o,d}}{\sum u_{tp,m,o,d}} \tag{4}$$

Modal share per trip distance is used as target value. We applied the simplex algorithm as a suitable linear optimization algorithm to determine the optimum parameters for a perfect fit of the model to the observed trip distance distribution [20].

In the **fourth step**, traffic assignment is performed on an enriched OSM network for all modes except PT as the routes are set by providers and retrieved from GTFS data. We assume people to choose the shortest route for PT access/egress and car route choice. Route choice of pedestrians is modelled using a distinct pedestrian route choice model that considers e.g., travel time, sidewalk width, land use, and speed of motorized traffic [21]. Route choice for cycling is modelled using a distinct route choice model, considering distance, cycling infrastructure, surface quality, slope, etc. [22].

3   Proof of Concept

The following sections demonstrate the usability and plausibility of the model for the case study city of Dresden (Germany). We furthermore present first results of traffic measure impact assessment focussing on the promotion of active mobility.

3.1   Basic Model Results

The generation step fully reproduces trip statistics. Overall, about 32,880 zones are used to cover the entire city area and 9,220 zones contain population. It consists of 569,173 people, which is distributed over 301,124 apartments. Driving license availability is around 88.5%, and car ownership rate is 47.6% (household car availability = 67%). Activity pattern distribution comes from household surveys [16] and is matched following the aforementioned method in chapter 2. Most common activity patterns are home-shopping-home (19.8%), home-leisure-home (19.2%), home-work-home (15.5%), home-education-home (8.7%). The remaining patterns sum up to 36.8%.

The approach formulated in chapter 2 identified 8,769 POIs where activities occur, further differentiating into leisure, education, shops, commercial, etc. Decomposing activity chains into single trips results in 2,016,142 trips (official statistics reveal 2.018.160 trips). The mode choice model reveals a high overall accuracy. Figure 1 shows that the model reveals a high overall prediction accuracy and fits well with statistics.

Each mode is predicted with an accuracy higher than 99%: car (99,5%), PT (99,4%), bike (99,7%), and walk (99,5%). We additionally apply a calibration variable δ to fit the predicted model values (p_model) to the observed ones (p_stat) per mode m via

$$\delta_m(p_m) = \frac{p\_model_m}{p\_stat_m} \tag{5}$$

Thus, predicted modal split values perfectly align to the observed values at the end of the calibration step. Traffic assignment is applied using the methods mentioned in chapter 2, resulting in a mileage illustrated in Fig. 2. Overall accuracy is 98.6%.

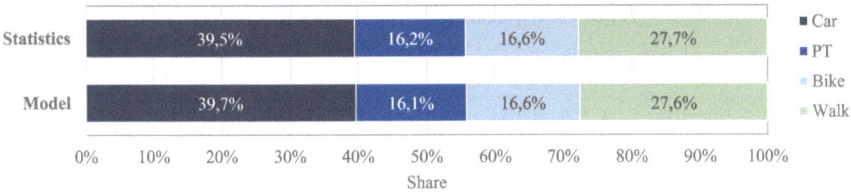

Fig. 1. Comparison of modal split values.

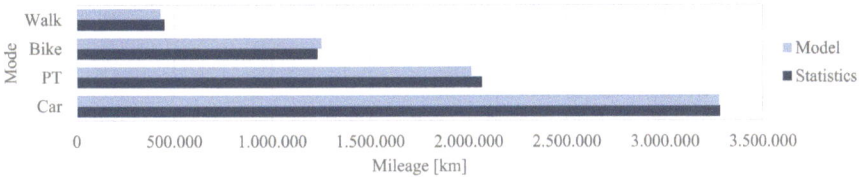

Fig. 2. Comparison of mileage per mode.

3.2 Scenario Test Case

We suggest a scenario test case to examine the usability and responsiveness of the model for traffic measure assessment. Demand for trip generation and distribution is assumed fixed. In order to evaluate mode shift reactions within the mode choice model we determine the following three measures:

a) Increase parking costs to 4€ and raise carbon pricing from 30 €/t to 275 €/t. impacting fuel cost in Germany, posing an externality price offset [23].
b) Subsidized PT at a subscription price of 1 € and a trip-based fare of 0,01 €.
c) Improve cycling infrastructure (network consistency, surface conditions)

As a result, fixed and variable costs for car usage increase while PT related costs decrease. Walking conditions remain the same. Travel time components for cycling decrease due to the improvement of infrastructure conditions.

Modelling results reveal a significant modal shift (see Fig. 3). Car use is considerably reduced (-234,969 trips). In contrast, the utilization of the other modes increases significantly (see Table 1).

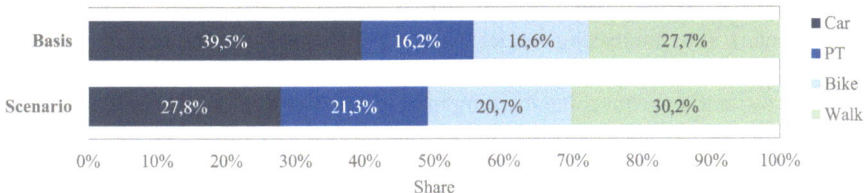

Fig. 3. Contrasting modal split values of the scenario to the basic modal split.

Sustainable modes gain a large share in the modal split. The share of PT increases by 5.1% to 21.3%. The share of cycling increases by 4.1% to 20.7%. Walking reveals a lower

Table 1. Trip number per mode and scenario

Mode	Basis	Test scenario	Difference	In-/decrease
Car	797.173	562.204	-234.969	-29,5%
PT	326.942	429.541	102.599	+ 31,4%
Bike	335.015	417.511	82.497	+ 24,6%
Walk	559.030	610.676	51.645	+ 9,2%

but still significant increase of 2.5% to 30.2%. Trip assignment reveals a corresponding de-/increase of mileage for the different modes.

4   Discussion and Outlook

The proof of concept reveals that the proposed model can model passenger traffic for the case study area and reproduce mobility behavior obtained by mobility surveys and documented official statistics. It shows a high modeling accuracy and can furthermore consider different measures to simulate the results for impact assessment. As the approach can be transferred to other cases (e.g., any city in Germany) this novel approach enables researchers and practitioners to apply the model and thus assist planners and decision makers with fact-based arguments.

However, additional work and research are inevitable regarding further model validation, quality assurance (ex-post analysis), and transferability to different study areas where corresponding data is available. Such a transfer will reveal practicability and necessity to, for example, adapt mobility indicators during trip generation. Similar applies to the cost functions we used. The generalized costs currently depend on fixed and variable components of travel times and travel costs per mode. Travel cost components are currently derived from general price statistics but could be adapted or explicitly modelled with local prices for the case study area. Modeling access time in more detail could improve travel time cost estimation, leading to more accurate model results. The integration of comfort criteria remains a task of major importance.

Beyond that, other models and algorithms, such as machine learning methods, could be used for modelling the different steps. These methods can get useful for datasets where linear regression methods are limited (e.g. overdispersed data, outliers, feature selection, etc.), thus increasing model accuracy. We used the simplex algorithm for mode choice model calibration. However, more sophisticated methods like maximum likelihood estimation or machine learning methods could be applied as well.

References

1. UBA - Umweltbundesamt: Klimaschutz im Verkehr. Rolle des Verkehrs bei den Treibhausgasemissionen in Deutschland. https://www.umweltbundesamt.de/themen/verkehr/klimaschutz-im-verkehr#rolle (2023). Accessed 20 August 2023

2. Nelldal, B.-L., Andersson, E.: Mode shift as a measure to reduce greenhouse gas emissions. Procedia Soc. Behav. Sci. (2012). https://doi.org/10.1016/j.sbspro.2012.06.1285

3. Siskos, P., Capros, P., de Vita, A.: CO2 and energy efficiency car standards in the EU in the context of a decarbonisation strategy: a model-based policy assessment. Energy Pol. (2015). https://doi.org/10.1016/j.enpol.2015.04.024

4. Liu, C., Tapani, A., Kristoffersson, I., Rydergren, C., Jonsson, D.: Development of a large-scale transport model with focus on cycling. Transp. Res. Part A: Pol. Pract. (2020). https://doi.org/10.1016/j.tra.2020.02.010

5. Gasser, P., et al.: Modélisation macroscopique de la circulation cycliste et piétonne - bases. Makroskopische Modellierung des Fuss- und Veloverkehrs- grundlagen. Bundesamt für Strassen (2017)

6. Lovelace, R., Parkin, J., Cohen, T.: Open access transport models: a leverage point in sustainable transport planning. Transp. Pol. (2020). https://doi.org/10.1016/j.tranpol.2020.06.015

7. Bienzeisler, L., Lelke, T., Wage, O., Thiel, F., Friedrich, B.: Development of an Agent-based transport model for the city of Hanover using empirical mobility data and data fusion. Transp. Res. Procedia (2020). https://doi.org/10.1016/j.trpro.2020.03.073

8. Ziemke, D., Kaddoura, I., Nagel, K.: The MATSim open berlin scenario: a multimodal agent-based transport simulation scenario based on synthetic demand modeling and open data. Procedia Comput. Sci. (2019). https://doi.org/10.1016/j.procs.2019.04.120

9. Statistische Ämter des Bundes und der Länder: Ergebnisse des Zensus 2011 zum Download - erweitert. https://www.zensus2011.de/DE/Home/Aktuelles/DemografischeGrunddaten.html (2018)

10. Statistisches Bundesamt: Die Datenbank des Statistischen Bundesamtes. Statistik: 12211 - Mikrozensus. https://www-genesis.destatis.de/genesis/online

11. infas, Deutsches Zentrum für Luft- und Raumfahrt e. V. (DLR), Institut für Verkehrsforschung, IVT Research: Mobilität in Tabellen (MiT 2017). https://mobilitaet-in-tabellen.dlr.de/ (2017)

12. Lomax, N., Norman, P.: Estimating Population Attribute Values in a Table: "Get Me Started in" Iterative Proportional Fitting. The Professional Geographer (2016). https://doi.org/10.1080/00330124.2015.1099449

13. ImmoScout24: Mietwohnung suchen. https://www.immobilienscout24.de/Suche/shape/wohnung-mieten

14. Moeckel, R., Spiekermann, K., Wegner, M.: Creating a synthetic population. In: 8th International Conference on Computers in Urban Planning and Urban Management (2003)

15. Gerike, R., Hubrich, S., Ließke, F., Wittig, S., Wittwer, R.: Sonderauswertung zum Forschungsprojekt „Mobilität in Städten – SrV 2018". SrV-Stadtgruppe: Mittelzentren, Topografie: hügelig. Professur für Integrierte Verkehrsplanung und Straßenverkehrstechnik (IVST) (2018)

16. Porojkow, I.: Abbildung der Fahrradnutzung auf Basis von Open Data. Diplomarbeit, Technische Universität Dresden, Professur für Verkehrsökologie (2021)

17. McFadden, D.: Conditional logit analysis of qualitative choice behavior. In: Zarembka, P. (ed.) Frontiers in Econometrics. Economic Theory and Mathematical Economics, pp. 105–142. Academic Press, New York (1974)

18. Brosi, P.: GTFS für Deutschland. https://gtfs.de/ (2023). Accessed 20 Aug 2023

19. Axhausen, K., et al.: Ermittlung von Bewertungsansätzen für Reisezeiten und Zuver-lässigkeit auf der Basis eines Modells für modale Verlagerungen im nicht-gewerblichen und gewerblichen Personenverkehr für die Bundesverkehrswegeplanung. Schlussbericht: FE-Projekt-Nr. 96.996/2011

20. Dantzig, G.B.: Linear Programming and Extensions. Princeton University Press, Princeton (1991)

21. Fritzsch, L.: Routenwahl im Fußverkehr. Welche Routen bevorzugen Fußgänger:innen und welche Faktoren beeinflussen die Wahl einer Route wie stark? Diplomarbeit, Technische Universität Dresden, Professur für Verkehrsökologie (2022)
22. Huber, S., Lissner, S., Lindemann, P., Muthmann, K., Schnabel, A., Friedl, J.: Modelling bicycle route choice in German cities using open data, MNL and the bikeSim web-app. In: 2021 7th International Conference on Models and Technologies for Intelligent Transportation Systems (MT-ITS). 2021 7th International Conference on Models and Technologies for Intelligent Transportation Systems (MT-ITS), Heraklion, Greece, 16.06.2021 - 17.06.2021, pp. 1–6. IEEE (2021). https://doi.org/10.1109/MT-ITS49943.2021.9529273
23. Projektionsbericht 2021 für Deutschland (2021). https://www.umweltbundesamt.de/sites/default/files/medien/372/dokumente/projektionsbericht_2021_uba_website.pdf

Open Access This chapter is licensed under the terms of the Creative Commons Attribution 4.0 International License (http://creativecommons.org/licenses/by/4.0/), which permits use, sharing, adaptation, distribution and reproduction in any medium or format, as long as you give appropriate credit to the original author(s) and the source, provide a link to the Creative Commons license and indicate if changes were made.

The images or other third party material in this chapter are included in the chapter's Creative Commons license, unless indicated otherwise in a credit line to the material. If material is not included in the chapter's Creative Commons license and your intended use is not permitted by statutory regulation or exceeds the permitted use, you will need to obtain permission directly from the copyright holder.

Evaluating Passenger Transport Emissions and Modal Shift with Elasticity-Based Regional Scenario Tool

Johanna Mäkinen$^{(\boxtimes)}$ (ID), Riku Viri (ID), and Jussi Sjögren (ID)

Transport Research Centre Verne, Tampere University, Tampere, Finland
johanna.m.makinen@tuni.fi

Abstract. Transport is one of the largest emitting sectors of greenhouse gas emissions and to reach climate targets, transport emissions must be reduced. In Finland, 97 municipalities have set ambitious target to reduce greenhouse gas emissions by 80% by 2030 from 2007 levels. Regions and municipalities need tools to evaluate whether their planned measures align with emission reduction targets. This study presents a novel regional scenario tool, specifically designed for estimating the influence of mobility measures on modal shift and, consequently, regional transport emissions. Moreover, the tool takes into account the evolution of regional car fleets, and their impact on regional passenger transport emissions. The effects of various mobility measures are estimated with demand elasticities and diversion factors. The developed scenario tool is demonstrated with two scenarios, which include changes in travel time and travel cost of both private cars and public transport. These alterations were found to result in shifts in modal share and further transport emissions. The results demonstrate that demand elasticities are stronger for travel time compared to travel cost. The tool created in this study is designed to assist local authorities with climate work and decision-making process, thus contributing to the achievement of climate targets.

Keywords: Sustainable transport · Emissions · Regional mobility plans · Scenario model

1 Introduction

Transport is one of the largest emitting sectors of greenhouse gas emissions and to reach climate targets, transport emissions must be reduced. Finland has committed to reduce domestic transport emissions by at least 50% by 2030. Many Finnish municipalities have set even more ambitious targets. For example, 97 municipalities have committed to reducing greenhouse gas emissions by 80% by 2030 from 2007 levels (Carbon Neutral Finland 2023). In order to reach emission targets, municipalities and regions are preparing and updating climate plans, which sets plans to reach emission targets. To support this regional climate work, municipalities need tools to evaluate the effect of transport measures on emissions. In addition, due to limited resources allocated to climate work at the regional and municipal levels, it's crucial to provide information on the most impactful climate actions.

© The Author(s) 2025
C. McNally et al. (Eds.): TRAconference 2024, LNMOB, pp. 181–195, 2025.
https://doi.org/10.1007/978-3-031-85578-8_24

In Finland, there has been lack of tools to analyse the impact of sustainable mobility measures on regional transport emissions. The objective of this study is to create a regional scenario tool specifically designed for estimating the influence of mobility measures on modal shift and, consequently, regional transport emissions. Moreover, the tool takes into account the evolution of regional car fleets, and their impact on regional passenger transport emissions. The tool calculates the total CO_2 emissions for domestic passenger transport (excluding air transport) within each region by 2035. Consequently, the tool can be used to create diverse scenarios on how car fleet development and mobility measures affect CO_2 emissions at the regional level. Such insights are important for regions and municipalities, as they must assess whether their planned measures align with emission reduction targets. To comprehensively evaluate the effects of various mobility measures, demand elasticities and diversion factors found from existing literature are used. The tool is designed to assist local authorities with climate work and decision-making process, and it will be made available for every Finnish subregion. The tool is designed to be regularly revised and updated.

Research behind demand elasticity primarily arises from the research around problems and assumptions of traffic and network equilibrium and traffic assignment (e.g., Beckmann et al. 1956; Beckmann 1967; Gartner 1980; Dafermos & Nagurney 1984a; 1984b; Yang 1997). Demand elasticities provide approximations of the probable impact of measures, polices, and other changes on travel demand of a certain mode of transport (de Jong & Gunn 2001). To simplify, in the context of scenario tool developed in this study, demand elasticities explain how a change in a factor such as cost of travel or travel time affect the demand of a certain transport mode (Litman 2023). This is calculated as a linear function where the percentage change in travel demand of a transport mode caused by a percentage change of a cost of travel or travel time is relative to the original travel demand, and cost of travel or travel time. An example of this could be a situation where the elasticity between the cost of travel for driving a car and the total distance driven by car is −0.3, and the increase in cost of travel for driving a car is 10%, which would lead to a decrease in the total distance driven by a car by 3% ($-0.3 \times 0.1 = -0.03$).

In order to maintain the state of equilibrium in the developed tool, it is necessary to reallocate the increased or decreased travel demand of a certain mode of transport to a new mode of transport or as disappearing/appearing traffic. To accomplish this, diversion factors are used. The role of diversion factors is to determine the impact of change within the demand of one mode of transport on the demand for other modes of transport and for new or disappearing travel demand (Dunkerley et al. 2018). Diversion factors presented as percentage values between 0 and 1. For the equilibrium in the tool, it is mandatory that all of the alternative diversion factors in each scenario sum up to 1, and by that it can be ensured that all of the changed travel demand is allocated. As an example, Flügel et al. (2018) has used diversion factors similarly in their model.

This study aims to address the following research questions:

1. What types of elasticities and diversion factors are suitable for a regional transport emission tool within the Finnish context?
2. How do various cost- and time-related factors influence regional emissions development when elasticity and diversion factors are incorporated into the model?

The rest of this paper is organized as follows. Section 2 describes the data and methodology used. Section 3 includes the results and discussion. The last chapter concludes the paper.

2 Methods

In this study, a scenario tool to estimate the effect of mobility measures on modal shift and further to regional transport emissions is developed. The following sections describes how existing Finnish car fleet model (Viri et al. 2021) is used to model the development of the Finnish car fleet and how data obtained from national travel survey is used to depict current travel behaviour within various regions. Furthermore, this section describes the use of demand elasticities and diversion factors, which are used to assess the impact of various mobility policy measures.

2.1 Car Fleet Model

The development and emissions of passenger cars are estimated with existing Finnish car fleet model SALAMA. A detailed explanation on how the model operates can be found in Viri et al. (2021) and the CO_2 emission calculation part is further described in Viri & Mäkinen (2023).

In general, the car fleet model uses current car fleet composition in Finland (as of end of March 2022) as the input and calculates the development of the total car fleet year by year until 2040. The car fleet development in the model is based on a set of user modifiable parameters. These include settings to change the probabilities for the users to switch to different driving power on a year-by-year basis in different area types and the possibility to change the turnover speed of the car fleet in general.

For this study, three different scenarios were generated, where these probability parameters were modified to describe various levels of interest in the adoption of electric vehicles. The same three scenarios were used as in Viri & Mäkinen (2023): Low EV ambition, EV-trend 2021, and High EV ambition. EV-trend 2021 scenario follows the current 2021 sales pattern and extends it to a forecast for upcoming years. Low and high EV scenarios follow the EV-trend 2021 -scenario, but with either increased or decreased probability of choosing an EV, resulting in 10%-point higher or lower probability to choose an EV in 2030. From the year 2030 onwards, probabilities for other driving powers than BEV start to gradually decrease toward zero, to fulfil the 2035 EU car sales targets. More specified information about these scenarios, and their effect on the emissions in Tampere region, Finland, are more specifically defined in Viri & Mäkinen (2023).

Based on the car fleet generated by the model for three different scenarios, the average CO_2 emissions of car fleet are calculated for each subregion and scenario per every year. As described in more detail in Viri & Mäkinen (2023), for each region, the average is calculated by the mileage weighted average of CO_2 values of the car fleet, with an added factor of 5% due to real-world usage against WLTP-calculation. All battery electric vehicles are calculated as 0 g CO_2/km and plug-in hybrid vehicles are calculated based on their registered CO_2 values. As per national calculation standards, the share of biofuels is not calculated as CO_2 emissions, and thus, only the fossil fuel share of CO_2 is noted in the calculation. The shares of biofuels are based on the newest proposal plans of biofuel uptake in Finland (Finnish Government 2023a).

In order to calculate annual emissions of buses, national transport emission projection and its' WAM-scenario by the VTT Technical Research Centre of Finland (VTT Technical Research Centre of Finland 2021) was used to estimate development of the bus fleet. In WAM-scenario, there are around 3 000 fully electric buses in 2035, accounting 20% of the entire bus fleet (VTT Technical Research Centre of Finland 2021). Electrification and improvement of energy efficiency in diesel vehicles decreases energy intensity of bus fleet. In this study, the average energy intensity of urban buses is assumed to be reduced from the 2022 level of 48 g CO_2/pkm to 37 g CO_2/pkm in 2035. For long distance buses, the average energy intensity decreases from 36 g CO_2/pkm to 31 g CO_2/pkm. Emission intensity of regional commuter trains is calculated to be 0 g CO_2/pkm, as all the commuter trains run on electricity. For long distance trains, emission intensity is assumed to be 1,5 g CO_2/pkm, as most of the train services in Finland are powered by electricity. Only railcars and some night trains in Northern Finland run on non-electrified routes, where diesel fleet is used. (VR Group 2023).

2.2 Travel Habits in Regions

The current travel habits are analysed based on the Finnish national travel survey. The national travel survey aims to monitor the mobility habits of Finns, and the survey is conducted using a travel diary method. The target group of the survey includes all Finns who were at least six years old excluding the residents of Åland. The national survey data from 2021 contains information on about 57,000 trips, including all travel modes. (Traficom 2023).

The national travel survey data was expanded to correspond the population of each Finnish subregion to enable regional analysis. Total number of trips, modal split and total distance travelled for each transport mode were calculated for every subregion based on the national travel survey data. The national travel survey data includes information on the starting and ending point of each trip. Moreover, data also contains the home region of the respondent. To be able to perform regional analysis, all trips were given information regarding in which region the trip is performed, in order to avoid calculating any trips to multiple regions. All trips were categorized based on starting and ending point of trip as follows:

- Starting and ending point of the trip located in the same region (intra-regional trips): Trip calculated to region, where the trip was made.
- Starting and ending point of the trip located in different regions (inter-regional trips): Trip calculated to the home region of the respondent.

Moreover, intra-regional trips were divided into following regional categories based on population: metropolitan region, large city regions (population > 100,000), medium-sized city regions (population of 40,000–100,000) and other regions.

Following the calculations, the modal split and total distance travelled per travel mode were obtained for each region. After that, the annual average CO_2 values for travel modes (see Sect. 2.1.) where combined with the total distance travelled. This enabled the calculation of annual CO_2 emission projections up to the year 2035.

2.3 Demand Elasticities and Diversion Factors

One of the targets of the scenario model is to determine the impacts of different transport measures on travel behaviour and further on transport emissions. The impact of measures on modal shift and total distance travelled by different modes are estimated with demand elasticities and diversion factors.

Factors of demand elasticity used in the scenario tool were gathered from an Emme based transport forecasting tool called 'Sampers 4'. Sampers 4 is the re-estimated national transport forecasting system developed by the Swedish transport administration, which is continuum of the Swedish transport forecasting systems that begin its development after the first version was developed in 1998 (Trafikverket, 2023). The reason for using elasticity factors presented in Sampers 4 for the tool is the relatively comparable transport systems of Finland and Sweden. Sampers 4 is a regional model, meaning that elasticity factors were calculated for each of the modelled regions, specifically Palt, Samm, Sydost, and Väst, in the year 2022 (Trafikverket, 2022).

As the Sampers model contains different elasticity factors for different regions, the most suitable regions were chosen to describe urban and rural areas in our scenario tool. For rural areas, the elasticity factors were collected from the elasticity factors of Palt. For urban areas, elasticity factors of County of Stockholm (Stockholm län) excluded from the rest of Samm were used. These areas were chosen as they were perceived as the most comparable regions with the separation of rural and urban areas in the tool. From elasticities presented for regions of Palt and County of Stockholm, elasticity factors of total distance travelled for commuting, business travel, and other travel in comparison to changes in cost of driving, cost of public transport, travel time with car, travel time on board with public transport, and headway of public transport were extracted. Elasticity factors of the cost and travel time of driving and public transport were used directly for rural areas, and for urban areas cost of travel and travel time of car were used directly, but the travel time of public transport was adjusted by summing 90% of the elasticity of travel time onboard with 10% of the elasticity regarding public transport headway. From this information, a list of demand elasticities for driving a car and public transport (rail and bus) were combined for the categories of commute, business travel, and other travel for the changes in travel time, and for the changes in cost of travel for both rural and urban areas.

Diversion factors were mainly extracted from a study conducted by Dunkerley et al. (2018), from which the diversion factors were derived from a substantial and diverse database. Due to the extensive database, it was decided to use averages of the average diversion factors for each of the affected mode of transport (car, bus, and rail) for different choice sets that were interpreted as viable for the usage of the scenario tool. From these values, a table of base values was concluded for the diversion factors to be used for the tool. These base values were applied to consider changes from car, bus, and rail to car, bus, rail, walking, cycling, and disappearing traffic (no travel) in urban and rural areas with long and short trips for commuting, business trips, and other travel such as leisure. The values were also divided into two sets, rail-access and no rail-access, as for areas without either urban or long-distance rail transport, there is no need to have a conversion towards/from rail, as it is not generally available as a travel mode.

Individual diversion factors for certain travel type variable (location, travel type, and mode of transport) combinations that could not be extracted from tables of Dunkerley et al. (2018) were created by developing coefficients for each travel type variable. These coefficients were generated by exploring relations between different travel type variables and interpolating the unknown coefficients based on known variables. Variables for rural travel type were generated from a study of Vicario (1999), while variables for long-distance travel were obtained from a study of Dargay (2010). With these coefficients, it was possible to create a holistic set of diversion factors to cover every possible combination of travel type variables within the scenario tool.

All the elasticity and diversion factors used in this study are presented in Appendix A and Appendix B.

3   Results and Discussion

3.1   Scenario Tool

The developed scenario tool calculates the total CO_2 emissions of domestic passenger transport for each subregion by 2035. The tool enables the user to choose a pre-calculated car fleet development scenario (three options, see Sect. 2.1.), modify the total mileage per travel mode and per trip purpose, which is listed as commuting, business or other (including leisure). In addition to these, the user can modify the average cost and travel time of different travel modes. All these modifications can be done on a year-by-year basis and all calculation is done cumulatively. Thus, every modification will be considered also in the values of future years automatically. As the data is currently from 2021, population development projections for the coming years are factored into the model. As a result, the total mileage per driving power in the subregions will adjust based on the forecasted population changes for the area (according to Statistics Finland 2021).

When the user modifies travel times or costs for a specific travel mode, the tool uses defined demand elasticities to initially calculate the impact on the altered travel mode. After that, it uses defined diversion factors to calculate the change caused to other travel modes, as some of the increased traffic for the modified mode will be diverted from other transport modes, while some will represent entirely new, appearing demand. Or, on the other hand, if the change results in a decrease in demand for the modified travel mode, the diversion factors will be used to calculate to which travel modes the decreased travel will move to and for which part of the travel will be disappearing traffic. For travel costs, the user can directly input a percentage change value for each travel mode and for short- and long-distance travel. Regarding travel times, user can input a minute change for travel time, and the original travel times in the calculation dataset will be used to calculate trip-by-trip change to be used for the elasticity calculation. By using this approach, time savings will automatically have a lesser impact for longer distances. However, based on this calculation, the change value is limited to 30% annually to prevent excessively large changes or the possibility of non-existent or negative travel times. All elasticities and diversions are also calculated cumulatively, so any changed demand will affect the calculation of the following years.

3.2   Scenarios

The developed scenario tool is demonstrated through presentation of two scenarios. These scenarios involve adjustments to travel time and costs for public transport and private cars, leading to shifts in the total distance travelled. Consequently, these changes have a direct impact on the total CO_2 emissions.

The first scenario, *"Car favour"*, is based on the government programme and budget proposal of Finnish Government for 2024–2027. According to the budget proposal, excise duty on transport fuels will be reduced as of the beginning of 2024 to compensate for the average fuel price increase (Ministry of Finance 2023). According to draft of government proposal, tax reduction is assumed to lower the price of diesel by 3% and price of petrol by 2% (Finnish Government 2023b). Moreover, the VAT rate of public transport will be increased from the current tax rate of 10% to 14%. Also, climate-based state aid grants, which have been targeted to large and medium-sized cities to purchase cleaner fleets and fuels and promote permanent increase in the modal share of public transport, will be cancelled (Finnish Government 2023c).

Consequently, adjustments made to travel time and costs in the first scenario align with measures proposed in government programme. In this context, the cost of private cars is decreased due to tax reduction of transport fuels. The cost of private cars is assumed to decrease 2% over the upcoming government term, starting from the next year (2024–2027). Conversely, the cost of public transport increases due to a rise in the VAT rate for public transport services and termination of climate-based grants aimed at enhancing public transport. The estimate is that the fare of urban buses will increase by 20%. For instance, the public transport authority of Jyväskylä region has estimated that the measures proposed in government programme will lead to a 20% price increase in public transport fares (Finnish Public Transport Association 2023). Moreover, the cost of long-distance bus and rail transport are expected to rise by 5%. The average travel time of private cars is assumed to decrease by one minute. Beyond 2028, it is assumed that the travel cost of private cars will increase by 1.5%. All the modifications made in this scenario are shown in Table 1.

Table 1. The modifications made in the *Car favour* -scenario.

Car favour	Travel mode	Travel cost	Travel time
2024–2027	Car	Long: −2% Short: −2%	−1 min/per trip
	Bus	Short: +20% Long: +5%	–
	Rail	Long: +5%	–
2028 →	Car	Long: +1.5% Short: +1.5%	+1 min/per trip
	Bus	–	–
	Rail	–	–

The second scenario, *"PT boost"*, demonstrates a situation where the quality of public transport services is significantly improved. This vision aligns with the goals set by numerous medium and large-sized city regions aiming to boost the usage of public transport, a goal often pursued through public transport service level improvements. In this scenario, the average travel time of bus is assumed to decrease by two minutes in 2025, followed by an additional two-minute decrease in 2027. Conversely, private car travel times are expected to increase by one minute in 2025 and an additional one minute in 2027. Achieving a significant reduction in the average travel time for public transport over a relatively short period hinges on prioritizing buses through initiatives such as dedicated bus lanes and signal priorities, as well as the implementation of Bus Rapid Transit systems. These efforts are anticipated to result in a slower average travel time for private cars. Furthermore, it is assumed that public transport improvement includes measures to reduce the average fares of public transport, while simultaneously anticipating a moderate increase in private car expenses. Table 2 shows changes in scenario *'PT Boost'*.

Table 2. The modifications made in the *PT Boost* -scenario.

PT Boost	Travel mode	Travel cost	Travel time
2025–2026	Car	Long: +2% Short: +2%	+1 min/per trip
	Bus	Short: −15% Long: −5%	−2 min/per trip
	Rail	–	–
2027 →	Car	–	+1 min/per trip
	Bus	–	−2 min/per trip
	Rail	–	–

Based on these two scenarios, two sets of scenarios are calculated for three Finnish subregions. Oulu subregion is located in Northern Ostrobothnia, and it consists of 7 municipalities. Oulu subregion has a population of 257,000 (Statistics Finland 2021). Jyväskylä subregion, located in Central Finland, is also formed by 7 municipalities, and has population of 188,000 (Statistics Finland 2021). The last subregion, Lahti, has population of 205,000 (Statistics Finland 2021) and it includes 10 municipalities.

Scenarios calculated for these subregions are presented in Figs. 1 and 2. Figure 1 presents the changes to emissions in *Car favour* -scenario, whereas Fig. 2 presents the similar changes in *PT Boost* -scenario.

When looking at the emissions development of the Car favour -scenario, it can be seen that there is a small increase in emissions from 2023 to 2024, as reduced travel time for cars will shift more trips towards passenger cars. Consequently, passenger car mileage increases. Throughout the time of less expensive car use, the scenario has higher total emissions, but when those changes are mostly reverted in 2028, there will be a switch

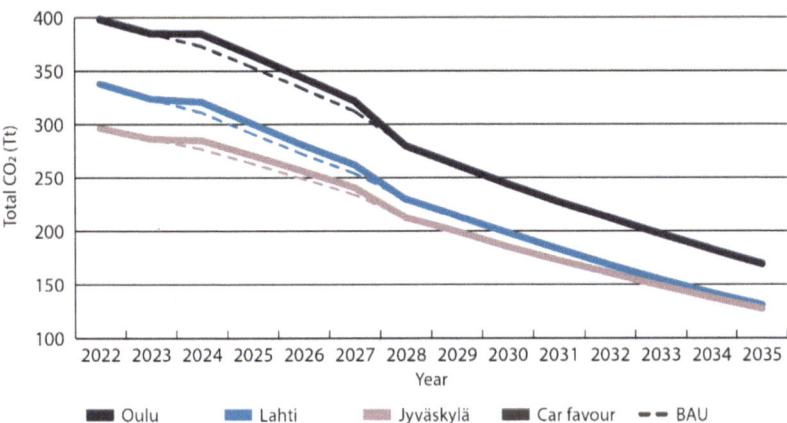

Fig. 1. Emission development in *Car favour* -scenario in 3 different Finnish subregions compared to BAU-scenario (dotted line).

back from passenger cars to other modes, and the development will then follow closely to the original BAU -scenario.

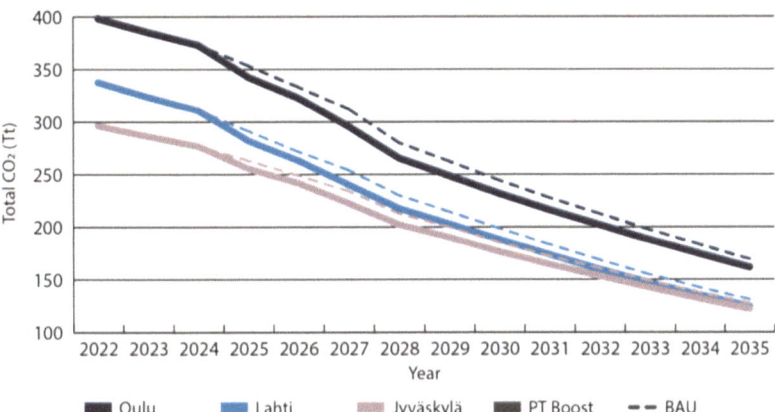

Fig. 2. Emission development in *PT Boost -scenario* in 3 different Finnish subregions compared to BAU-scenario (dotted line).

In the *PT Boost* -scenario, the decrease of emissions is stronger from the year 2025 as public transport is favoured through travel time and price. This shifts users from passenger cars towards public transport and hence, there is lower car mileage throughout the whole timeframe of the calculation. However, it should be noted that when going close to the end of the calculation, the difference between the scenarios decreases. This is due to electrification of the car fleet and higher biofuel usage. Therefore, even when there is still higher passenger car use, the cars in use have lower energy intensity and the used fuel has higher share of biocomponents, leading to lower average CO_2 values.

3.3 Discussion

The accuracy of the scenario tool could be improved by updating the data of mobility habits with the most recent national travel survey data from 2022 and 2023. In this study, data from 2021 was used, and due to the COVID-19 pandemic and the associated restrictions, people's mobility habits were different from previous years. For example, in the 2021 dataset, the total number of trips, trip lengths, and the modal share of public transport decreased compared to previous years (Traficom 2023). It should be noted that used modal share and total distance travelled by each mode influence the extent of the impact caused by modifications in travel time and cost. With variations in modal share and total distance travelled per mode, the scenario results could be different, and therefore, the results of the scenarios presented in this study should be viewed as example calculations rather than predictions.

For this study, the data from 2022 and 2023 was not yet available to be used, but since the model is built to note the structure of these datasets, the calculation model can be updated just by replacing the specific dataset, thus allowing the update of results as soon as the data is available.

With the aim of increasing the model's accuracy, it would be optimal to develop elasticity factors based on Finnish data. However, in these circumstances, it was not possible to develop new elasticity factors. Consequently, existing elasticity and diversion factors were found from literature, even if there have been some discussions over the transferability in time and region of elasticity factors (e.g., Gunn 2001; Karasmaa 2003; Wardman 2022). Nevertheless, the elasticity values used in scenario tool were found to be in line with other recommended values found in the literature (e.g., de Jong & Gunn 2001; Goodwin et al 2004; Paulley et al. 2006; Litman 2023).

Developing region-based elasticities for Finnish context would be a natural future research topic to improve usability of the scenario tool even further. In addition to region-based elasticities, future research and further development of scenario tool could include wider examination of micromobility, and shared mobility solutions as a part of the array of alternative modes of transport, especially if the data regarding shared micromobility would be available.

4   Conclusion

The aim of this study was to create a specialised regional scenario tool, specifically designed for assessing the impact of mobility measures on modal shift and, consequently, their influence on regional transport emissions. To comprehensively evaluate the effects of various mobility measures, demand elasticities and diversion factors from existing literature were incorporated. Additionally, the tool takes into account the evolution of regional car fleet, e.g., the number of electric vehicles in each region.

The research questions addressed in this paper were:

1. What types of elasticities and diversion factors are suitable for a regional transport emission tool within the Finnish context?
2. How do various cost- and time-related factors influence regional emissions development when elasticity and diversion factors are incorporated into the model?

The first research question was answered by exploring the existing literature regarding use of elasticities and diversion factors in transport context (Sect. 2.3). Based on prior research, elasticities can vary widely depending on specific contextual factors. As a result, data analysis is often recommended to determine elasticities for a particular region. However, this was not possible due to data availability and time limitations. Consequently, elasticity factors presented in a Swedish study (regional model Sampers 4) was used. The similarities in land use and transport systems between Finland and Sweden facilitated the transfer of these elasticities to the Finnish context. Moreover, Sampers 4 provides elasticity factors for different regions, which allows to apply different elasticity factors for urban and rural areas.

Regarding the second research question, this study introduces a novel scenario tool combining regional car fleet development with demand elasticities and diversion factors. The tool has been specifically designed to consider regional characteristics, as it takes into account regional car fleet development, regional mobility habits, and elasticity factors, which are considered separately for both urban and rural areas. The created scenario tool was demonstrated with two scenarios, which include changes in travel time and cost of both private cars and public transport. These alterations were found to result in shifts in modal share and further transport emissions. These scenarios were tested to three different subregions in Finland. The results demonstrate that demand elasticities are stronger for travel time compared to travel cost. Consequently, changes in the travel time of a particular transport mode are more likely to lead to significant shifts in the demand for that mode compared to changes in travel cost. In addition, it is important to highlight that the extent of demand change will be significantly influenced by the relative market share of the altered transport mode. In the tested subregions, private cars currently hold a dominant position in modal share. Consequently, measures affecting private cars have a more substantial influence because they impact a significant number of car-dependent journeys.

The method of this study holds the potential for adaptation to other countries with comparable regional car fleet data and travel survey data. To the authors' knowledge, there are not similar models in Finland, which enable the estimation of the impact of car fleet development and mobility measures on regional-level CO_2 emissions. Existing models primarily provide results at the national level, and results of these models are only partially applicable due to regional differences, such as variations in urban structure and mobility habits. Additionally, these models do not consider elasticities to assess the influence of mobility measures on modal shift. The tool created in this study is designed to assist local authorities with climate work and decision-making process, thus contributing to the achievement of climate targets.

Funding. This work was supported by the LIFE Programme of the European Union (LIFE17 IPC/FI/000002 LIFE-IP CANEMURE-FINLAND). The work reflects only the authors' views, and the EASME/Commission is not responsible for any use that may be made of the information it contains.

Appendix A: Used Elasticity Factors.

Area type	Trip purpose	Travel mode	Rail possibility	Price elasticity	Travel time elasticity
Urban	Commuting	Car	Rail+No rail	−0.3	−0.91
Urban	Business	Car	Rail+No rail	−0.03	−0.82
Urban	Other	Car	Rail+No rail	−0.39	−1.09
Urban	Commuting	Bus	Rail+No rail	−0.25	−0.604
Urban	Business	Bus	Rail+No rail	−0.11	−0.443
Urban	Other	Bus	Rail+No rail	−0.2	−0.634
Urban	Commuting	Rail	No rail	0	0
Urban	Business	Rail	No rail	0	0
Urban	Other	Rail	No rail	0	0
Urban	Commuting	Rail	Rail	−0.25	−0.604
Urban	Business	Rail	Rail	−0.11	−0.443
Urban	Other	Rail	Rail	−0.2	−0.634
Rural	Commuting	Car	Rail+No rail	−0.32	−0.73
Rural	Business	Car	Rail+No rail	−0.01	−0.91
Rural	Other	Car	Rail+No rail	−0.3	−0.77
Rural	Commuting	Bus	Rail+No rail	−0.32	−1.05
Rural	Business	Bus	Rail + No rail	−0.14	−0.67
Rural	Other	Bus	Rail+No rail	−0.3	−0.73
Rural	Commuting	Rail	No rail	0	0
Rural	Business	Rail	No rail	0	0
Rural	Other	Rail	No rail	0	0
Long d	Commuting	Car	Rail+No rail	−0.32	−0.73
Long d	Business	Car	Rail+No rail	−0.01	−0.91
Long d	Other	Car	Rail+No rail	−0.3	−0.77
Long d	Commuting	Bus	Rail+No rail	−0.32	−1.05
Long d	Business	Bus	Rail+No rail	−0.14	−0.67
Long d	Other	Bus	Rail+No rail	−0.3	−0.73
Long d	Commuting	Rail	No rail	0	0
Long d	Business	Rail	No rail	0	0
Long d	Other	Rail	No rail	0	0
Long d	Commuting	Rail	Rail	−0.32	−1.05
Long d	Business	Rail	Rail	−0.14	−0.67
Long d	Other	Rail	Rail	−0.3	−0.73

Long d. = Long-distance.

Appendix B: Used Diversion Factors.

Area type	Trip purpose	Rail possibility	Changed travel mode	From/to car	From/to bus	From/to rail	From/to walking	From/to cycling	Appearing/disappearing traffic
Urban	C	Rail	Car	0	0.275	0.511	0.101	0.060	0.052
Urban	B	Rail	Car	0	0.268	0.519	0.064	0.061	0.088
Urban	O	Rail	Car	0	0.267	0.323	0.075	0	0.336
Urban	C	No rail	Car	0	0.697	0	0.101	0.060	0.141
Urban	B	No rail	Car	0	0.645	0	0.064	0.061	0.229
Urban	O	No rail	Car	0	0.482	0	0.075	0	0.443
Urban	C	Rail	Bus	0.383	0	0.313	0.131	0.127	0.045
Urban	B	Rail	Bus	0.267	0	0.306	0.170	0.125	0.133
Urban	O	Rail	Bus	0.175	0	0.187	0.227	0.037	0.373
Urban	C	No rail	Bus	0.641	0	0	0.131	0.127	0.101
Urban	B	No rail	Bus	0.526	0	0	0.170	0.125	0.179
Urban	O	No rail	Bus	0.553	0	0	0	0	0.447
Urban	C	Rail	Rail	0.449	0.455	0	0	0	0.096
Urban	B	Rail	Rail	0.605	0.287	0	0	0	0.108
Urban	O	Rail	Rail	0.356	0.430	0	0	0	0.214
Rural	C	No rail	Car	0	0.827	0	0	0	0.173
Rural	B	No rail	Car	0	0.654	0	0	0	0.346
Rural	O	No rail	Car	0	0.599	0	0	0	0.401
Rural	C	No rail	Bus	0.609	0	0	0	0	0.391
Rural	B	No rail	Bus	0.838	0	0	0	0	0.162
Rural	O	No rail	Bus	0.551	0	0	0	0	0.449
Long d	C	Rail	Car	0	0.294	0.671	0	0	0.035
Long d	B	Rail	Car	0	0.104	0.663	0	0	0.233
Long d	O	Rail	Car	0	0.468	0.260	0	0	0.271
Long d	C	No rail	Car	0	0.828	0	0	0	0.172
Long d	B	No rail	Car	0	0.598	0	0	0	0.402
Long d	O	No rail	Car	0	0.582	0	0	0	0.418
Long d	C	Rail	Bus	0.142	0	0.443	0	0	0.415
Long d	B	Rail	Bus	0.527	0	0.363	0	0	0.110
Long d	O	Rail	Bus	0.546	0	0.275	0	0	0.179
Long d	C	No rail	Bus	0.481	0	0	0	0	0.519
Long d	B	No rail	Bus	0.847	0	0	0	0	0.153
Long d	O	No rail	Bus	0.690	0	0	0	0	0.310
Long d	C	Rail	Rail	0.398	0.479	0	0	0	0.123
Long d	B	Rail	Rail	0.655	0.257	0	0	0	0.088
Long d	O	Rail	Rail	0.384	0.268	0	0	0	0.348

Trip purpose: C = Commuting, B = Business, O = Other.
Area type: Long d. = Long-distance.

References

Beckmann, M., McGuire, C. B., Winsten, C.B.: Studies in the Economics of Transportation (1956)

Beckmann, M.J.: On the theory of traffic flow in networks. Traffic Quart. **21**(1) (1967)

Carbon Neutral Finland. Hinku municipalities (2023). https://www.hiilineutraalisuomi.fi/en-US/Hinku/Hinku_municipalities. Accessed 9 Oct 2023

Dafermos, S., Nagurney, A.: Sensitivity analysis for the asymmetric network equilibrium problem. Math. Program. **28**(2), 174–184 (1984)

Dafermos, S., Nagurney, A.: On some traffic equilibrium theory paradoxes. Transp. Res. Part B: Methodol. **18**(2), 101–110 (1984)

Dargay, J.: A forecasting model for long distance travel in Great Britain. European Transport Conference. Association for European Transport (AET) (2010)

de Jong, G., Gunn, H.: Recent evidence on car cost and time elasticities of travel demand in Europe. J. Transp. Econ. Pol. (JTEP) **35**(2), 137–160 (2001)

Dunkerley, F., Wardman, M., Rohr, C., Fearnley, N.: Bus fare and journey time elasticities and diversion factors for all modes. RAND Corporation (2018)

Finnish Government. Government proposal: Liikennepolttoaineiden jakeluvelvoite 13,5 % myös ensi vuonna (In Finnish, translated: Lower distribution obligation for transport fuels to continue in 2023) (2023)

Finnish Government. Draft of government proposal: Hallituksen esitys eduskunnalle laiksi nestemäisten polttoaineiden valmisteverosta annetun lain liitteen muuttamisesta (In Finnish, translated: Government's proposal to the parliament with a proposal for a law amending the annex to the law on excise duty on liquid fuels) (2023)

Finnish Government. A strong and committed Finland: Programme of Prime Minister Petteri Orpo's Government 20.6.2023. Publications of the Finnish Government 2023:60 (2023)

Finnish Public Transport Association. Hallitus ei välitä joukkoliikenteen käyttäjistä, vaan uhkaa nostaa joukkoliikenteen lipunhintaa verotuspäätöksillään (2023). News article, https://www.sttinfo.fi/tiedote/70016710/hallitus-ei-valita-joukkoliikenteen-kayttajista-vaan-uhkaa-nostaa-joukkoliikenteen-lipunhintaa-verotuspaatoksillaan?publisherId=69819173&lang=fi. Accessed 15 Nov 2023

Flügel, S., Fearnley, N., Toner, J.: What factors affect cross-modal substitution? – evidence from the Oslo area. Int. J. Transp. Dev. Integr. **2**(1), 11–29 (2017)

Gartner, N.H.: Optimal traffic assignment with elastic demands: a review part I. Analysis framework. Transp. Sci. **14**(2), 174–191 (1980)

Goodwin, P., Dargay, J., Hanly, M.: Elasticities of road traffic and fuel consumption with respect to price and income: a review. Transp. Rev. **24**(3), 275–292 (2004)

Gunn, H.: Spatial and temporal transferability of relationships between travel demand, trip cost and travel time. Transp. Res. Part E: Logist. Transp. Rev. **37**(2–3), 163–189 (2001)

Karasmaa, N.: The Transferability of Travel Demand Models: An analysis of transfer methods, data quality and model estimation. Helsinki University of Technology (2003)

Litman, T.: Understanding Transport Demands and Elasticities. Victoria Transport Policy Institute, Victoria (2023)

Ministry of Finance: Budget proposal: The budget proposal of Minister of Finance Riikka Purra implements decisions that strengthen general government finances. Press release (2023)

Paulley, N., et al.: The demand for public transport: the effects of fares, quality of service, income and car ownership. Transp. Policy **13**(4), 295–306 (2006)

Statistics Finland. Population projection 2021. Statistics Finland's PxWeb databases, statistics available (2021). http://pxnet2.stat.fi/PXWeb/pxweb/en/StatFin/

Traficom 2023. National Travel Survey 2021. In Finnish, abstract in English. Traficom Research Reports 1/2023. 118p

Trafikverket. 2022. Elasticiteter i Sampers 4

Trafikverket. Användarhandledning och riggningsbeskrivning - Sampers/Samkalk 4. Ärendenummer: 2020/55749 (2023)

Vicario, A.J.B.: Diversion factors and cross elasticities. Unpublished MA Dissertation, Institute for Transport Studies, University of Leeds (1999)

Viri, R., Mäkinen, J., Liimatainen, H.: Modelling car fleet renewal in Finland: a model and development speed-based scenarios. Transp. Policy **112**, 63–79 (2021)

Viri, R., Mäkinen, J.: The impact of modal shift on passenger car CO2 emissions in Tampere region. Case Stud. Transp. Pol. **13** (2023)

VR Group. Vastuullisuusraportti 2022 (In Finnish, translated: Sustainability report) (2023)

VTT Technical Research Centre of Finland. 2021. The baseline projection of transport emissions, WEM-scenario. Dataset

Wardman, M.: Meta-analysis of British time-related demand elasticity evidence: An update. Transportation Research Part A: Policy and Practice (2022)

Yang, H.: Sensitivity analysis for the elastic-demand network equilibrium problem with applications. Transp. Res. Part B: Methodol. **31**(1), 55–70 (1997)

Open Access This chapter is licensed under the terms of the Creative Commons Attribution 4.0 International License (http://creativecommons.org/licenses/by/4.0/), which permits use, sharing, adaptation, distribution and reproduction in any medium or format, as long as you give appropriate credit to the original author(s) and the source, provide a link to the Creative Commons license and indicate if changes were made.

The images or other third party material in this chapter are included in the chapter's Creative Commons license, unless indicated otherwise in a credit line to the material. If material is not included in the chapter's Creative Commons license and your intended use is not permitted by statutory regulation or exceeds the permitted use, you will need to obtain permission directly from the copyright holder.

Urban Ropeways – An Environmentally Friendly Alternative for Commuters?

Karl Hofer[✉] [iD], Lisa Kollenz, and Martin Fellendorf [iD]

Graz University of Technology – Institute of Highway Engineering and Transport Planning,
Graz, Austria
karl.hofer@tugraz.at

Abstract. Moderate sized European cities often face the problem of daily traffic congestions on entrance roads due to the high number of commuters. This affects regional buses and lead to delays. An urban ropeway in combination with park and ride lots at the city border could be an attractive supplement in entrance corridors without high performing public transit (PT) services and can lead to a modal shift to PT. Ropeways offer a high service rate that leads to no noticeable waiting times and no delays due to the usage of an independent track above ground level. In addition, the offered high capacities promise lower crowding levels than in regional buses. We conducted a stated choice experiment that includes the transport modes car, regional bus and ropeway in Graz, a moderate sized city in Austria. The results of the estimated ML-model with the attributes walking, waiting and in-vehicle-time, interchanges, crowding, reliability and travel costs as well as the personal attitude about an urban ropeway showed that mode-specific properties of the ropeway and personal attitude have a significant influence on choice behavior. In addition, interchanges in a ropeway trip chain have a less negative influence than in chains with a regional bus.

Keywords: Urban Ropeway · Commuting · Choice Modeling · Mixed Logit

1 Introduction

In numerous countries of Europe, the number of commuters increased in the last years. In 2021, the commuter rates in Austria and Germany was 53% to 59% of employees [1, 2]. Especially moderately sized cities of 250,000 to 500,000 inhabitants face the problem of daily traffic congestions due to the high number of commuters. Unfortunately, the quality of the city border crossing public transit (PT) does not reach the same high-quality level as in big cities with more than one million inhabitants. An urban ropeway in combination with park and ride lots at the city borders could be a cost-effective, attractive supplement in entrance corridors without a high performing PT-service (e.g. light rail). Ropeways have proved to be a suitable transit extension in several Latin American cities and offer a high service rate, no delays and an improved passenger comfort compared to usual regional buses.

The existing modeling literature for urban ropeways mainly focus on residents and trips inside cities and do not observe commuters and their choice behavior concerning

© The Author(s) 2025
C. McNally et al. (Eds.): TRAconference 2024, LNMOB, pp. 196–202, 2025.
https://doi.org/10.1007/978-3-031-85578-8_25

this new transport mode. Therefore, this paper seeks to estimate a mode choice model for commuters that includes a ropeway as a separate transport mode in Graz, Austria. Graz is a moderately sized city with 300.000 inhabitants and 90.000 commuters a day that cross the city border. Only 18% of these trips are PT-trips and therefore congestions on main entrance roads as well as air pollution in the city are a problem. Beside cars, these congestions also affect regional buses and lead to delays and unstable schedules.

2 Related Work

The available literature in the research field of urban ropeways raised in the last years. It reaches from general analyses of the system and system comparisons with traditional PT, extending to planning, construction and operation and also including case and feasibility studies. An example for a detailed literature overview can be found in Flesser and Friedrich [3]. Circulating ropeways are the appropriate system for a powerful and reliable service in an urban area. The advantages of these systems compared to reversible ropeways are the continuous service usually resulting in no noticeable waiting times at the station, higher capacities and intermediate stops as well as shorter distances between stops. The most commonly used systems are monocable detachable gondolas (MDG) and tricable detachable gondolas (TDG). Based on these cabin capacities, 4,000 to 6,000 passengers per hour and direction (pphpd) are the maximum capacities for these two systems and service intervals are in a range of 15 (MDG) to 90 s (TDG) [4].

This section gives a very short overview on modeling literature that can be divided in literature about existing ropeways, mainly in Latin America and modeling approaches for non-existing ropeways in Europe that focus on the mode choice of potential users. The Metrocable in Medellín and the TransMiCable in Bogotá were investigated the most in the last years. Studies in Medellín deal with the changed mobility behavior of users [5], perception of waiting time [6] and user behavior of passengers. In Bogotá, researchers analyze different aspects before and after the introduction of the TransMiCable. These aspects are the expectations and perceptions of users [7] or the influence of social capital on the willingness of using a ropeway.

Tiessler et al. [8] investigated the mode choice for a non-existing ropeway for inner city trips in Munich. Results showed that participants prefer connections that would also include a ropeway instead of a bus even when the bus is faster. They used the choice probability between the options car, PT and ropeway based on travel time to generate input data for the travel demand model of Munich. The influence of the mode-specific properties of an urban ropeway on mode choice for inner city trips was analyzed for people with and without car availability by Hofer et al. [9] and Hofer and Fellendorf [10]. Their estimated mixed logit models showed that crowding, reliability and the personal attitude have a statistically significant influence on the choice behavior of people with car availability. Their cycling behavior and the weather conditions additionally influences people without a car availability.

3   Methodology and Case Study

Our 4-step-methodology started with the preparation of the survey that included an analysis of the catchment area around Graz and the definition of point of interests (POI) for leisure, shopping and other trips in the city. We defined 10 corridors in the suburbs of Graz depending on topographic conditions and entrance roads. The survey was conducted in spring/summer 2023 with 121 persons who live in the suburbs and work or do activities in Graz. In step 2 we recruited participants at PT-stops in Graz, the city center of Graz and with promotion on social media. The survey included a general part with sociodemographic questions as well as questions concerning experiences with ropeways and the opinion about an urban ropeway in Graz. Depending on the recorded data, a personalized choice set of trip alternatives was generated in the second part of the survey. The final step included model estimations of a mode choice model and an analysis of results. The four-step-methodology is shown in Fig. 1.

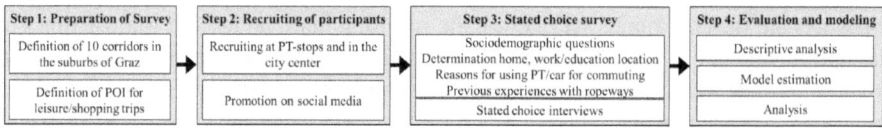

Fig. 1. Methodological approach of the study. (PT = public transit, POI = point of interests)

We initially analyzed the PT services of the corridors to the city center and the POIs. Afterwards, we calculated walking, waiting and travel times for all relations for regional buses and car. We defined the four crowding levels "low", "middle", "high" and "overcrowded". The levels use the approaches of the German HCM [11] and the Association of German Transport Companies [12]. With available passenger volumes data, it was possible to assign the PT-trips of the corridors to the crowding levels. The reliability for car and regional buses were calculated with a routing algorithm based on the quickest routes. We determined time of-day delays through variation of departure time (Fig. 2).

The general part of the survey included a socio-demographic section and questions about the motives of the usual commuting behavior of participants. However, the main part deals with questions about experiences with ropeways so far and their attitude about an introduction of an urban ropeway in Graz. A ropeway is an unusual urban public transit mode with no driving personnel in the vehicle, no possibility to get off at every time of the journey and a riding height of 30 to 50 m. This can lead to insecurity and fears of potential users during the usage. Therefore, we analyzed the attitude of participants concerning these topics by using Likert scales. We also investigated the attitude about the impact of an urban ropeway on the cityscape and acceptance of a ropeway crossing participant's houses/flats. The reported home municipality, work/education address and car availability delivered the basis for the personalized stated choice part of the survey. Beside the variations of travel, waiting and walking times and also crowding levels, we generated a variation of the reliability attribute through a deduction/addition of maximum 20%.

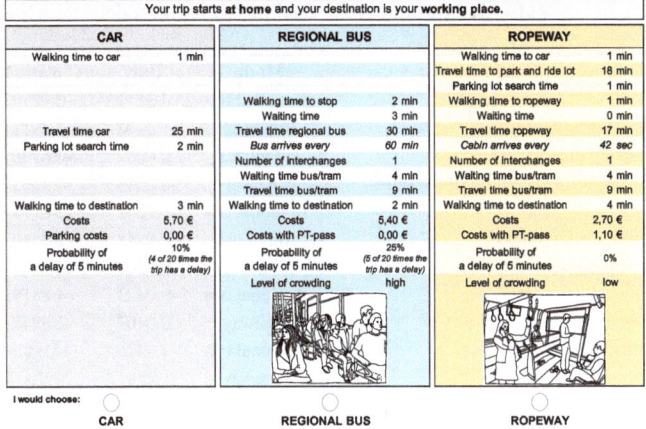

Fig. 2. One choice set of survey

The generation of ropeway trips considers the mode-specific properties of the transport mode. The fact that no traffic congestion is encountered means an urban ropeway has a very high reliability level and the probability of a delay for the ropeway in the questionnaires is thus always 0%. The planned TDG for Graz offers a continuous service with a service interval of 42 s and a capacity of 3,000 pphpd. This will usually lead to no noticeable waiting time and low occupancy rates for each cabin. For this reason, we assumed only three crowding levels, "low", "middle" and "high". Due to technically supervised weight limitations, overcrowded conditions will not occur in a cabin. However, crowded conditions can occur in the stations during peak hours at park and ride facilities and can lead to waiting times. Therefore, we added several levels of waiting time in addition to the usual not noticeable waiting time of 0 min.

4   Results

We received 118 interviews with nine choice sets for unemployed participants and twelve choice sets for employed participants and students. 51.3% of the participants were female and 48.7% male. 80.0% of them have a job and 13.9% study in Graz. A high proportion of participants already used a ropeway (97.4%) and the nearly 90% of them feel safe during using it. According to the "strongly agrees" from the Likert scaled questions we see that 8.7% have fear of heights and feel claustrophobic during ropeway usage. Strong wind conditions during ropeway usage lead to significant more fears than heavy rain. The not existing cabin personnel is a problem for only 4.7% of participants.

Approximately 59% of commuters in our survey have a good opinion about an urban ropeway in Graz. The main reasons for this positive opinion are the possible relief of high-volume roads, the environmental friendliness and the possible tourist attraction potential. The question about possible usage of this new system showed that 10.4% of participants would refuse to use an urban ropeway. Only 18.9% strongly agree that an urban ropeway has a negative impact on the cityscape of the city center. In comparison, 8.2% in the outskirts strongly agree. However, 57% strongly agrees that they would feel disturbed

Table 1.   Results of the final mixed logit model

Attribut	Mode	Par	(t-ratio)
Mode-specific constant	Regional bus	-0.3111	-0.9277
Mode-specific constant	Ropeway	-1.5178	-4.3913
Walking time (mins)	all modes	-0.0089	-0.6796
In-vehicle-time car to P+R-lot (mins)	Ropeway	-0.0580	-4.3662
Waiting time (mins)	Regional bus	-0.1212	-2.8170
Waiting time (mins)	Ropeway	-0.0988	-2.0510
In-vehicle-time (mins)	Car	-0.0829	-27.7655
In-vehicle-time (mins)	Regional bus	-0.0423	-4.8819
In-vehicle-time (mins)	Ropeway	-0.0407	-2.8847
Crowding (levels)	Regional bus	-2.0425	-13.6889
Crowding (levels)	Ropeway	-1.2882	-27.7007
Reliability/delays (percentage)	All modes	-0.3864	-5.5142
Interchanges (numbers)	Regional bus	-0.7631	-5.0199
Interchanges (numbers)	Ropeway	-0.3164	-2.1132
In-vehicle-time bus/tram (mins)	all modes	-0.0730	-5.4397
Costs (€)	Car	-1.8203	-56.5764
Fare (€)	All PT	-1.0746	-64.7586
Current car user (0,1)	Regional bus	-0.5582	-2.3748
Current car user (0,1)	Ropeway	0.7210	4.7761
Female (0,1)	Regional bus	-0.3716	-2.8688
Female (0,1)	Ropeway	-0.7579	-5.9611
Opinion ropeway (0,1)	Regional bus	-0.1743	-1.3788
Opinion ropeway (0,1)	Ropeway	1.3476	10.1273
Standard deviation parameter distribution of random parameters			
Crowding (levels)	Regional bus	44.2141	37.3183
Crowding (levels)	Ropeway	13.1341	108.5518
Costs (€)	Car	20.9374	128.6139
Fare (€)	All PT	7.0980	189.9295
Model fits / specification tests			
LL (beta)	-874.5		
LL (0)	-1,123.7		
Likelihood ratio test	498.63		
McFadden R²	0.2219		
Number of observations	1,047		

if a ropeway would cross their house/flat and 47% think that a ropeway line should only route on public property. In Table 1 the parameters and specification (likelihood ratio) tests for the final model are presented (n = 1,047 observations). We estimated several models (MNL, NL) and a mixed logit model reaches the highest McFadden R^2 with 0.2219. We used four random parameters (crowding and costs) with 500 Halton draws that use a log-normal distribution. Results show that interchanges in a trip chain with a ropeway has a lower negative influence than in trip chains with a regional bus. Crowding and waiting time for a ropeway have a statistically significant influence on mode choice, as well as the dummy coded attributes current car user, female gender and opinion about an urban ropeway in Graz (0 = bad, 1 = good). Fear of heights or claustrophobia did not show a statistically significant influence due to a too small proportion in our sample.

5 Conclusion and Future Work

In this paper we estimated a mode choice model for commuters in a moderately sized city with the inclusion of an urban ropeway. Ropeways with their mode-specific properties (high capacities, low crowding levels and no noticeable waiting times) could be an attractive PT-supplement in entrance corridors with delayed and crowded regional buses. In a survey we investigated commuters experiences with ropeways so far and their opinion about an urban ropeway in Graz. The stated choice part of the survey delivered data for the estimation of a mixed logit model. Participants tend to accept an interchange in a trip chain including a ropeway more than in chains with a regional bus. Mode-specific properties showed a statistically significant influence on mode choice.

Future work should compare corridors with a regional bus service with corridors offering a high performing PT-service (e.g. light rail). In addition, the sample size was sufficient, however a larger sample size for future experiments would be desirable.

References

1. Deutschlandatlas. https://www.deutschlandatlas.bund.de/DE/Karten/Wie-wir-uns-bewegen/100-Pendlerdistanzen-Pendlerverflechtungen.html. Accessed 06 Jan 2024
2. ÖAMTC. https://www.oeamtc.at/presse/studie-was-bewegt-oesterreichs-pendler-zum-ums teigen-43528943. Accessed 06 Jan 2024
3. Flesser, M., Friedrich, B.: Are we taking off? A critical review of urban aerial cable cars as an integrated part of sustainable transport. Sustainability **14**(13560) (2022)
4. Dale, S., Chu N., Imhäuser T., Cable Car Confidential – The essential guide to cable cars, urban gondolas & cable propelled transit. 1st ed. Creative Urban Project Inc., Toronto, Ontario, Canada (2013)
5. Goodship, P.: The impact of an urban cable-car transport system on the spatial configuration of an informal settlement. In: Proceedings of the 10th International Space Syntax Symposium, London, UK, vol. 131, pp. 1–17 (2015)
6. Jurado, D., Mesa-Arango, R.: Effect of queues, time perception, and user behavior on the selection of urban aerial ropeway transit: metrocable medellín. In: Presented at 98th annual meeting of the transportation research board, Washington, D.C., USA (2019)
7. Guzman, L.A., Cantillo-Garcia, V.A., Arellana, J., Sarmiento, O.L.: User expectations and perceptions towards new public transport infrastructure: evaluating a cable car in Bogotá. Transportation **50**(3), 751–771 (2022)
8. Tiessler, M., Engelhardt, R., Bogenberger, K., Hessel, C., Serwas-Klamouri, M.: Integration of an urban ropeway into Munich's transit system demand modeling. Transp. Res. Rec. J. Transp. Res. Board **2673**(10), 47–57 (2019)
9. Hofer, K., Haberl, M., Fellendorf, M.: Modeling of an urban ropeway integrated in a crowded transit system. Transp. Res. Rec. J. Transp. Res. Board (2023). https://doi.org/10.1177/036 11981231175889
10. Hofer, K., Fellendorf, M.: Cycling or Ropeway – two choices of environmentally friendly urban modes. In: Presented at WCTR 2023, Montréal, Canada (2023)

11. Road and Transportation Research Association (FGSV), German Highway Capacity Manual, FGSV-Verlag, Cologne, Germany (2015)
12. Association of German Transport Companies (VDV), Verkehrserschließung und Verkehrsangebot im ÖPNV, VDV-Schrift Nr. 4, Cologne, Germany (2001)

Open Access This chapter is licensed under the terms of the Creative Commons Attribution 4.0 International License (http://creativecommons.org/licenses/by/4.0/), which permits use, sharing, adaptation, distribution and reproduction in any medium or format, as long as you give appropriate credit to the original author(s) and the source, provide a link to the Creative Commons license and indicate if changes were made.

The images or other third party material in this chapter are included in the chapter's Creative Commons license, unless indicated otherwise in a credit line to the material. If material is not included in the chapter's Creative Commons license and your intended use is not permitted by statutory regulation or exceeds the permitted use, you will need to obtain permission directly from the copyright holder.

What Makes for a Successful E-Scooter Scheme? Best Practice from the West of England

Grace Packard[1]([✉]), Alex Bertram[2], George Lunt[2], and Oliver Coltman[3]

[1] AECOM, Edinburgh, UK
grace.packard@aecom.com
[2] AECOM, Bristol, UK
{alex.bertram,george.lunt}@aecom.com
[3] West of England Combined Authority, Bristol, UK
oliver.coltman@westofengland-ca.gov.uk

Abstract. In the mission to reduce transport carbon emissions, cities and towns are working to discover, develop and integrate more sustainable ways to travel. Micromobility brings great opportunity to aid this movement. This paper summarises the evolution of shared e-scooter schemes, using the UK's West of England e-scooter trial as a case study alongside associated research. It sets out three key properties required to support a successful scheme; usability, operational sustainability and policy delivery. The paper concludes that all three properties must work in balance, and this is achieved through stakeholder communication and collaboration, underpinned by effective scheme monitoring. It emphasises that no two schemes are the same and even the most successful schemes are required to continually adapt.

Keywords: micromobility · e-scooters · decarbonisation · innovation · best-practice

1 The Rise of E-Scooters

1.1 The Adoption of Micromobility

In the mission to reduce transport carbon emissions, cities and towns across the globe are working to discover, develop and integrate more sustainable ways to travel. Micromobility, a term that refers to a wide range of small lightweight vehicles including bikes, e-bikes and e-sooters, is gaining popularity as an opportunity to aid this movement. Economic and social benefits of micromobility can also be realised as a potentially more affordable and efficient mode of transport. Recent innovation in battery and Internet of Things technology has facilitated the growth of shared electric micromobility, with e-scooters providing great excitement and debate.

In the last five years, shared e-scooter schemes have been implemented in a significant number of places across Europe. Highlighting a true marriage of transport and technology, these schemes have already facilitated millions of sustainable trips, albeit

© The Author(s) 2025
C. McNally et al. (Eds.): TRAconference 2024, LNMOB, pp. 203–209, 2025.
https://doi.org/10.1007/978-3-031-85578-8_26

with implementation and management difficulties in some schemes due to local political factors and operational challenges, casting a shadow over further innovation. Where some schemes have initially flourished, a lack of funding has resulted in closures, removing a transport mode many had come to rely on. Towns and cities are working hard to understand how shared e-scooter schemes can benefit their areas. While some authorities and operators have up to five years' operating experience, knowledge transfer of the practical steps required to deliver a successful scheme has been limited.

1.2 This Paper

This paper fills the gap in practical knowledge transfer through the example of the West of England e-scooter trial, which has the highest demand in the UK [1] and is arguably one of the most successful schemes in Europe. Drawing detailed insights from the trial and associated research, this paper sets out the three aspects of a scheme deemed key to supporting a successful service: usability, operational sustainability and policy delivery.

These learnings from three years of scheme operations in the West of England region are highly transferable to other schemes, both in the UK and overseas. However, as with all transport schemes, it is important to recognize that one size does not fit all. The paper argues that local knowledge, user research and collaboration strengthen these aspects and importantly, these can only be developed at a local level.

2 E-scooter Schemes

2.1 E-scooter Scheme Overseas

In Europe, shared micromobility schemes have continued to grow and expand across more locations. By April 2023, it was estimated that there were over 600 shared micromobility schemes (both e-scooters and e-bikes) across the continent. Over the last year, nearly 200 new shared micromobility services were launched, contributing to a record 550 million trips being made, creating some €3.1 billion in revenue from end-users [2].

However, the expansion of micromobility services in Europe and further afield has not been without challenges. The pandemic had a significant effect on urban footfall in many cities. Different regulatory frameworks across countries have slowed scheme progression. Companies operating some schemes have faced a backlash from public authorities due to concerns about pedestrian safety and poor user behaviour, with political backlash causing some services to be shut down. The City of Paris recently held a referendum allowing citizens to vote on whether to allow the continuation of the shared e-scooter scheme. While 89% of voters chose to end the scheme, the voting turnout was below 8% [3] and privately owned e-scooters remain legal in France.

2.2 UK E-scooter Trials

Following the earlier launch of e-scooter schemes overseas, the UK Government fast-tracked the introduction of such services during the pandemic in 2020 to address the urgent need for sustainable, individual modes of transport. To navigate a difficult regulatory landscape, e-scooter trials were launched allowing areas across the UK to host schemes whilst more permanent legislation was developed. In July 2022, e-scooter trials were given the option to extend until the end of May 2024, which many but not all scheme sponsors agreed to. The UK Government has not currently provided any further indication as to when permanent micromobility legislation will be enacted.

Trial e-scooters have become a convenient mode of transport for many, with 24,365 e-scooters facilitating 34 million rides by January 2023 [4]. Many scheme sponsors quickly identified the significant level of management required to operate such a novel but well used scheme, resulting in a variety of management approaches across the UK.

2.3   The West of England E-scooter Trial

The micromobility industry and wide stakeholders continue to learn how these schemes can be managed. While each scheme operates differently, there remains a significant opportunity for knowledge transfer. As one of the largest and well used schemes in Europe, the West of England scheme acts as a great case study for others to learn from.

The shared e-scooter trial in the West of England was launched in October 2020 and continues to operate alongside 30 other trial schemes across the UK. The West of England Combined Authority manage the scheme, in agreement with three local authorities: Bath and North East Somerset, Bristol City and South Gloucestershire Councils. As of August 2023, the scheme attracted approximately 12 million trips, covering 30 million kilometres, indicating that by a stretch it is the most popular trial in the UK. Alongside the c. 3,500 hop-on-hop-off shared e-scooters which must be parked in geofenced zones termed 'Mandatory Parking Zones', a separate long-term rental scheme involves operators providing c. 500 users with their own e-scooter to take home. To understand the impacts of the scheme, the West of England Combined Authority commissioned the University of the West of England (UWE) to conduct a comprehensive evaluation of the impacts of the scheme [5]. The findings have helped to inform the development of the scheme which, alongside a new operator, introduced e-bikes and e-cargo bikes in October 2023.

3   Three-Legged Stool of Success

As with most transport schemes, multiple elements need to work together to deliver a holistically successful scheme. The analogy of a 'three-legged stool' of successful e-scooter schemes has therefore been developed (Fig. 1.).

In its most simple form, three core properties support e-scooter schemes: Usability, Operational Sustainability and Policy Delivery. Each aspect is required to be in balance with the other to support a scheme and this requires continuous collaboration between relevant stakeholders as the joints holding everything together, primarily, the highway authority and operator. Should one property be lacking, the scheme will struggle to be maintained and risks collapse. Importantly, non-users sit in the shadow of the scheme and will be minimally impacted if the stool is supported. However, should the stool collapse, non-users will be significantly impacted.

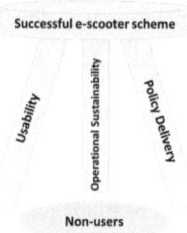

Fig. 1. The 'three-legged' stool of e-scooter scheme success (authors)

3.1 Usability

As introduced, the rapid development of e-scooter schemes across the UK was sparked by the pandemic, when alternative public transport modes were unable to operate to typical levels yet the need to travel remained for many. The growth in scheme usership over the three years following the pandemic, particularly in the West of England, indicates the continuing need for this mode of transport. Research undertaken by UWE [5] and AECOM [6] set out to understand possible reasons why.

Key findings from the UWE research [5] indicated that a significant attractor for users was the enhanced accessibility. 37% of scheme users do not own a car, while 39% in Bristol said that e-scooters enabled travel to places previously not possible. Users were motivated to use the vehicles due to their speed, convenience, costs and flexibility, with some also due to a perceived lack of alternatives such as poor public transport service or inability to walk to a destination. Destinations included work or education (four in ten trips), social/leisure purposes (four in ten trips) and personal errands (two in ten trips). AECOM [6] also found that most users were satisfied across their e-scooter experience, from signing-up, the journey itself and operator support.

These factors provide a true understanding of why the scheme continues to grow in popularity. This research, alongside collaboration with the operator, strong stakeholder relationships and communication with non-users has informed the scheme design, from where parking is located, through to where it can operate. Appropriate scheme design can in turn encourage usership, further enhanced by a smooth customer journey, from sign up, through to the gamification of safety education. Customer needs should be understood throughout the whole lifecycle of a scheme, from initial scheme design through to many years post implementation. Ultimately, economic and societal shifts may influence travel patterns, so, the scheme should be flexible to suit these needs.

3.2 Operational Sustainability

To provide a usable service, the operator and transport authority needs to ensure the scheme can balance commercial viability while remaining operationally sound. E-scooter schemes require significant human resource to operate through the distribution of e-scooters, maintenance and charging of vehicles in addition to management, research and development costs. When combined with capital costs including the purchase of vehicles and parking infrastructure, operators face significant challenges in being able to turn a profit for continued operations. While alternate transport authority vehicle ownership models can mitigate some financial pressure for the operator, ongoing maintenance and vehicle losses through theft and vandalism can mean even this model is unviable. True operational sustainability needs to be found for both parties.

In the West of England, strong scheme usership, significant research, consultations and financial modelling enabled the Combined Authority to develop a contractual model that aims to ensure commercial sustainability for operators, while also benefitting the Combined Authority and Unitary Authorities. While this bespoke revenue sharing model incentivises all parties to work together to enable the schemes long-term success, it cannot simply be replicated as is across other operating areas. To be successful, collaboration between all parties is crucial, with the contractual model seeking to reimburse the necessary public resource investments. However, the ultimate goal should remain: ensuring commercial and operational sustainability to enable all immediate stakeholders to benefit from the scheme's success. Indeed, the higher the revenue, the greater the investment can be used to develop sustainable transport infrastructure such as cycle lanes, something that the UWE research strongly indicated was needed if e-scooter and cyclist numbers were to thrive [5]. In turn, these infrastructure improvements further encourage usership and so a positive, sustainable transport feedback loop is developed, enabling local authorities to move ever closer to meeting policy goals.

3.3 Policy Delivery

The UK's Local Government Association [7] advises that 'plans and decisions are nothing without delivery, and the only way to be really thoughtful about delivery is through monitoring it properly'. Across the UK and beyond, transport schemes are implemented to connect people, support the economy and reduce carbon emissions. E-scooter schemes have proved their ability to contribute to these policy goals, but the extent of their contribution is only enhanced through collaboration with transport authority stakeholders. Tangible actions need to be delivered, such as expansion to areas of economic deprivation combined with appropriate discounts to encourage users in these areas. Any actions need to be carefully considered against anticipated and/or desired outcomes, such as an increase in usership in target areas or modal shift away from private cars, which can only be confirmed through scheme monitoring. To gain buy-in from all collaborators, the schemes must demonstrate they can delivery against policy goals. Only then can authorities and society get behind these sustainable transport schemes.

The provision of robust data sharing agreements alongside frequent reporting and continuous communication sits at the heart of a transparent, mutually beneficial scheme for both the operator and authorities in the West of England. Local policy goals include reducing transport emissions and improving access to work and skills provision, by developing an integrated transport system [8]. Since 2020, the e-scooter scheme has certainly made its mark in contributing to these goals. For example, the UWE research [5] indicated that between 24%–37% of e-scooter trips (dependent on area) are replacing car, taxi and ride-hailing. Further to this, carbon emission calculations suggested between 6–238 tonnes net reduction of CO_2 equivalent for Bristol in 2021. While the estimate includes a large variation, it is a reduction nonetheless.

3.4 Non-users

A sturdy three-legged stool of successful e-scooter schemes does not compromise the needs of non-users. Should one leg falter, non-users feel the impact first. In particular, pavement users such as the elderly, wheelchairs users, and blind and partially sighted pedestrians are significantly impacted by poor operations, including overcrowded parking areas, or the vehicle misuse by users. Furthermore, should non-user concerns not be addressed by all relevant stakeholders, scheme opposition may mobilise and grow.

The UWE research [5] indicated that 15% of on-street survey respondents felt discriminated against, many of which were disabled. Follow-along surveys with diverse non-users also highlighted a sense of risk and loss of pedestrian space. Actions to alleviate these concerns have been initiated by the authorities and operator, including the implementation of physical parking racks. Frequent stakeholder engagement meetings are held with members of the disabled community and the operator continues to improve parking accuracy through enhanced technology and user education.

4 Conclusion

Whilst still in their early stages, micromobility schemes offer great opportunities but need to be viewed holistically, as part of the wider transport network. The West of England scheme acts as a fine example of how an innovative transport scheme can quickly evolve thanks to the operators, local authorities and users. As with the three-legged stool, a careful balance needs to be found and this can only be achieved through continued communication and collaboration between stakeholders, underpinned by scheme monitoring. Communication, collaboration, and in-depth research has allowed the West of England scheme to continually improve based on the needs of users and non-users. While the stool provides a useful analogy, no two schemes are the same and even within each scheme, the needs of the operator, local authority and users will change over time.

References

1. Department for Transport: National evaluation of e-scooter trials, Findings report (2022)
2. Fluctuo: European Shared Micromobility Index (2022)
3. Politico. https://www.politico.eu/article/paris-bans-e-scooters-in-landmark-referendums/. Accessed 15 Sept 2023
4. CoMoUK, Overview and benefits. http://www.como.org.uk/shared-e-scooters/overview-and-benefits. Accessed 15 Sept 2023
5. Chatterjee, K., Parkin, J., Bozovic, T., Flower, J.: West of England E-scooter Trial Evaluation Final Report. Report to West of England Combined Authority. (2023)
6. AECOM: West of England User Experience Research. in publication (2023)
7. Local Government Association, Delivery. https://www.local.gov.uk/pas/topics/delivery. Accessed 19 Sept 2023
8. West of England Combined Authority. Joint Local Transport Plan (2020)

Open Access This chapter is licensed under the terms of the Creative Commons Attribution 4.0 International License (http://creativecommons.org/licenses/by/4.0/), which permits use, sharing, adaptation, distribution and reproduction in any medium or format, as long as you give appropriate credit to the original author(s) and the source, provide a link to the Creative Commons license and indicate if changes were made.

The images or other third party material in this chapter are included in the chapter's Creative Commons license, unless indicated otherwise in a credit line to the material. If material is not included in the chapter's Creative Commons license and your intended use is not permitted by statutory regulation or exceeds the permitted use, you will need to obtain permission directly from the copyright holder.

Design and Assessment of Shared and Electric Mobility Hubs for an Integrated Transport System

Pierluigi Coppola[1] ⓘ, Francesca Costa[2] ⓘ, Alessandro Luè[2(✉)] ⓘ,
Valerio Mazzeschi[2] ⓘ, Mara Tanelli[3] ⓘ, Vesna Janković Milić[4] ⓘ,
Ivana Marjanović[4] ⓘ, and Jelena Stanković[4] ⓘ

[1] Department of Mechanical Engineering, Politecnico di Milano, via G. La Masa 1,
20156 Milan, Italy
[2] Poliedra – Politecnico di Milano, via G. Colombo 40, 20133 Milan, Italy
{francesca.costa,alessandro.lue}@polimi.it
[3] Department of Electronics, Information and Bioengineering, Politecnico di Milano,
via G. Ponzio 34, 20133 Milan, MI, Italy
[4] Faculty of Economics, University of Niš, Trg Kralja Aleksandra Ujedinitelja 11,
18000 Niš, Serbia

Abstract. Electric shared mobility hubs are convenient access points, offering charging infrastructure, parking spaces, and user-friendly platforms/apps for seamless booking and return of vehicles. They encourage the usage of more sustainable alternatives to private cars. The paper describes three research projects, where we designed or implemented mobility hubs. The I-SharE LIFE project aims to demonstrate the practical and economic opportunities of various service models in small to mid-sized urban areas, to facilitate the development of potential hubs as alternative to private car and the behavioral change at neighborhood level for e-mobility. The second project is represented by the *Smart Sustainable District* research activity, carried out at Politecnico di Milano to draft practical guidelines as a strategic and methodological support for urban transformations, to aid public decision-makers and private operators. UR-DATA - a Horizon Europe project (Twinning for Excellence in Smart and Resilient Urban Development: Advanced Data Analytics Approach) shows how actions concerning the analysis of smart and sustainable city performances, in the city of Niš (Serbia), can contribute, at local district level, to the development and planning of a potential shared mobility.

Keywords: Shared mobility · mobility hub · smart district · sustainability · behavioural change

1 Introduction

Sharing mobility hubs serve as pivotal convergence points for effectively interconnecting transportation services [11]. These carefully designed strategic locations facilitate easy access and utilization of various shared mobility options, from car rentals to electric

© The Author(s) 2025
C. McNally et al. (Eds.): TRAconference 2024, LNMOB, pp. 210–215, 2025.
https://doi.org/10.1007/978-3-031-85578-8_27

bicycles. Their impact is substantial, reducing pollution and urban congestion while enhancing travel efficiency and promoting sustainable lifestyles. Moreover, these hubs serve as hubs for social and commercial interactions, fostering community cohesion and encouraging modal shifts. In summary, sharing mobility hubs play a fundamental role in promoting smart and responsible urbanization.

Furthermore, according to UN forecasts, 70% of the world's population will live in cities by 2050. In the latest years, old and new models of thinking about urban space have been advanced, such as the '15-min city' theorized by Carlos Moreno [1], Smart Sustainable Districts, Transit Oriented Development.

Therefore, dialogue between public and private stakeholders and citizens is becoming more and more crucial in order to improve and better design urban spaces to accommodate mobility that is increasingly integrated and sustainable, electric and shared.

2   HUBS to Promote Sustainable Mobility in the Balkan Area

The primary objective of this project [9] is to substantially enhance the research capabilities and scientific prowess of the Faculty of Economics at the University of Niš. This will be achieved through strategic collaborations with two research-intensive institutions, namely Poliedra – Politecnico di Milano and Oslo Metropolitan University. Together, these entities will engage in a comprehensive series of collaborative endeavors, focusing on the in-depth analysis of intelligent and sustainable urban performance. Notably, the re-search is centered on cities and municipalities within Serbia.

The thematic research initiatives are poised to bridge extant gaps in urban data analytics within the Western Balkans, with a specific focus on Serbia. These initiatives include urban mobility, spatial accessibility, and disparity, as well as urban dynamics in order to define the best mobility hubs. The Serbia's mobility ecosystem has inadequate infrastructure, outdated public transport networks, and burgeoning urbanization. In urban centers, congestion is exacerbated by a surge in private vehicle ownership, while public transit struggles to provide accessible and reliable alternatives. Additionally, inadequate pedestrian and cycling infrastructure further exacerbates the reliance on motorized transport, contributing to environmental degradation and reduced quality of life. The interplay of these factors underscores the pressing need for targeted interventions and policies that prioritize the enhancement of mobility solutions, aligning with the broader goals of sustainable urban development and economic growth.

The necessary data for these mobility analyses comes from a designated telecommunication operator in Serbia. The analyses include data-driven time-geographic assessments, investigations into daily mobility patterns, and their implications for societal inclusion and disparities. Additionally, the research delves into the intricacies of mobility inequalities and socio-spatial patterns, integrating transport and land use considerations, as well as the development of demand-responsive transportation services. Moreover, the insights gleaned from urban mobility data are instrumental in a series of studies aimed at estimating economic growth and resilience, predicated on flow patterns within Serbian urban areas.

The preliminary analysis was carried out on the example of the city of Niš, as a pilot city of the mobility study. In addition to data from telecommunication operator the analysis involved Global Positioning System (GPS) records pertaining to public transit, encompassing traffic metrics and passenger volume on specific routes; GPS data originating from taxi service providers; spatiotemporal records sourced from the Traffic Police division of the Niš Police Administration, encompassing data on traffic incidents and infractions; analogous spatiotemporal data pertaining to parking violations, sourced from the Municipal Police; detailed information on hourly occupancy rates within public garages and citywide parking facilities, furnished by the Public Utility "Parking Servis Niš". Spatial accessibility and inequality constitute critical dimensions that profoundly influence the lives of individuals and the planning of urban sustainable areas and sustainable mobility hubs. While accessibility has historically been an integral facet of urban planning ideals, it is only in recent decades that spatial accessibility has emerged as a distinct research discipline. Considering the advent of Smart Cities and the proliferation of geocoded data, fresh avenues for scrutinizing micro-level mobility patterns, diurnal patterns of segregation and integration, as well as sustainability and congruence between urban travel and accessibility have come to the fore. Within the framework of this theme, UR-DATA project leveraged and will leverage both newly generated geocoded data and existing data from Serbia to probe the spatial distribution of accessibility and its attendant inequalities and trace their evolution over time. The findings will be intertwined with research endeavors focused on local and regional economic development, as well as economic resilience.

3   Mobility Hubs to Promote Innovative Solutions of e-car Sharing in Medium and Small Size Urban Centers

I-SharE LIFE (www.i-sharelife.eu/) [2] has had the ambition to evolve the model of electric car-sharing, developed in large metropolitan cities, and to export it to the provinces and inland areas with low population density [3]. The innovation of I-SharE LIFE has had the sharing of electric cars among different users in different modes of transport and at different time intervals, in strategic points that become mobility hubs for the cities [4]. The project has tested the service's models in the 4 demonstration sites in Lombardy (Bergamo, Como, Bollate, and Busto Arsizio) and 1 in Croatia (Osijek), where the Local Public Transport integrated itself with other sustainable mobility services. For this purpose, five models of electric car sharing services and respective hub have been designed and tested to verify their effectiveness on the market and their environmental and economic sustainability in small and medium city contexts and in specific areas of use.

The e-car sharing model and the hub tested in the City of Bergamo were designed as a model for sharing electric vehicles between the City's own employees and citizens where the parking of the Municipality was the main hub in which also the charging station are installed. Other mobility hubs identified as strategic points for city users, that are not necessary for the Municipality's employees. Primarily, the City of Bergamo participated in the test by sharing the electric cars made available by the project, as a corporate fleet, with citizens. Sharing has well-defined schedules: employees of the Municipality use the electric cars during working hours, while in the evening/night and throughout the weekend the same cars can be used by citizens or city users of Bergamo [5].

During the testing phase in Bergamo (3 weeks in July 2019) were involved, as Beta Users: 4 Employee of Municipality and 6 Occasional Users; 4 e-care we used and have been travelled 2000 km in total. After testing in the city of Bergamo, the following data were obtained, with respect to the number of users and electric cars used. From July 2019 to March 2020: n° e-cars used inside Life Project: 5; total mileage done in this period: circa 9000 km; n° of users 127; n° of trips: 365. From September 2019 to February 2020 extra LIFE project: n° e-cars used from different Users extra Life Project:9; total mileage done in this period: circa 17000; n° of trips: 364. The exact number of users using the service could not be found.

To calculate the saving of air pollution emission we considered the distance that would have been travelled by private ICE cars in a hypothetical scenario where I-SharE LIFE project was not implemented. A shift from private vehicle use to car sharing results in lowered vehicle kilometers travelled (VKT). Car sharing users do shorter trips when using shared cars instead of their own car because of emphasizing variable driving costs, such as per hour and/or mileage. The choice of a car trips instead of other mobility options (e.g., TPL, bike) are also less frequent for the same reason. To include these facts in the estimate we considered a factor that transforms the kilometers travelled by I- SharE LIFE electric shared cars into kilometers not driven by ICE (Internal Combustion Engine) cars. According to the literature review [12], and considering the different service models, In Italy we chose a low value (1.3, corresponding to a 25% decrease) because models such as corporate (that are not studied yet in scientific literature) do not appear to produce a significant VKT reduction.

This model was the one most appreciated by users and was also the most popular model in the Lombardy Region. This model of a shared electric car sharing service and city hub has brought environmental benefits, in particular:

- the behavioral change of drivers towards more sustainable means. In particular, the choice of car sharing has caused many users to change their mode of travel and shift from exclusive use of private cars to electric cars combined with walking or biking from home to the mobility hub;
- the incentives for local public transport;
- the fleet of electric vehicles is more often renewed and more modern;
- the increase in active mobility, especially for people living and working in the city;
- the reduction of urban space abuses due to the parking of private cars.

4 Mobility Hubs to Implement Smart Sustainable Districts

The Smart Sustainable District (SSD) project started in 2021 and involved more than one hundred researchers at Politecnico di Milano [6]. The main outcome is a White paper that describes principles, solutions, tools for the realization of the SSD model in urban areas. Indeed, it can represent a strategic and methodological support for urban transformations, to support public decision-makers and private operators.

Different themes and best practices have been described inside the book relating the sustainable mobility and mobility hubs [7]. One of the main themes is the integration between service and infrastructural networks, focused on the integration of local public transport services at different scales by promoting intermodality between sustainable forms of mobility and the reduction of private motorization [8, 10].

The introduction of policies that limit the utilize of cars for local travel in of shared vehicles located in specific modal interchange nodes of local public transport (hub mobility) could encourage the growth of mobility hub for urban mobility and the active mobility at urban scale. To do this, it is necessary to work on two complementary levels. The first concerns the introduction of integrated (public-private) planning, while the other concerns the design of swap nodes. Mobility policies to support public and private operators could be: last mile logistics, i.e. the development and implementation of innovative urban logistics solutions. Lately there has been a change in consumer purchasing habits leading to the creation of new problems (e.g. small parcels) that require innovative solutions, like for example special structures that optimize parcel storage (parcel locker, micro-hub) or refers to the use of smaller, low-emission means of transport such as cargo-bike, electric vehicles and drones. In both cases, micro-hubs or mobility hubs seem to be the solution towards a change of mentality towards increasingly sustainable mobility.

5   Conclusions

From the three previous projects it can be seen that innovative shared mobility systems favor the modal shift towards low carbon mobility, active mobility and which therefore aims at the development of mobility hubs with the focus on a cultural change. This develops and can only be improved through a strong dialogue between public and private entities. In order to create sustainable mobility hubs, it will be necessary to intervene on integrated and easily available pricing schemes through specific digital access points. The basic objectives will therefore be to create valid integrated mobility services as alternatives to the private car, exploit new technologies to improve these services, provide mobility services that are environmentally friendly, accessible to all, socially inclusive and, finally, develop MaaS (Mobility as a Service) technologies and services that can facilitate the integration of offers.

Acknowledgements. The project I-SharE LIFE is co-financed by Life+, the EU's instrument supporting environmental projects. The activities described in this paper received funding from the European Horizon Europe Framework Programme (HORIZON) under grant agreement N. 101059994 (UR-DATA). The sole responsibility for the content of this publication lies with the authors. It does not necessarily reflect the opinion of the European Union. Neither the EASME nor the European Commission are responsible for any use that may be made of the information contained therein.

References

1. Moreno, C., Allam, Z., Chabaud, D., Gall, C., Pratlong, F.: Introducing the "15-minute city": Sustainability, resilience and place identity in future post-pandemic cities. Smart Cities 4(1), 93–111 (2021)
2. I-SharE LIFE project. https://www.i-sharelife.eu/. Accessed 23 Sept 2023

3. Wappelhorst, S., Sauer, M., Hinkeldein, D., Bocherding, A., Glaß, T.: Potential of electric carsharing in urban and rural areas. Transp. Res. Procedia **4**, 374–386 (2014)
4. Arena, M., Azzone, G., Colorni, A., Conte, A., Luè, A., Nocerino, R.: Service design in electric vehicle sharing: evidence from Italy. IET Intel. Transp. Syst. **9**(2), 145–155 (2015)
5. Bakker, S., Trip, J.J.: Policy options to support the adoption of electric vehicles in the urban environment. Transp. Res. Part D: Transp. Environ. **25**, 18–23 (2013)
6. SSD. https://www.poliedra.polimi.it/wp-content/uploads/SSD-SmartSustainableDistricts-LibroBianco_Poliedra-Polimi-1.pdf. Accessed 27 Sept 2023
7. Colorni, A., et al.: Smart Sustainable Districts – Un modello per progetti di rigenerazione urbana (Ricerca promossa dal Politecnico di Milano e coordinata da Consorzio Poliedra), pp. 73–89 (2022)
8. Colorni, A., et al.: The Smart Sustainable Districts framework to improve mobility in cities: opportunities, tools and barriers. The 51st European Transport Conference, Milan, 6–8 September 2023 (2023)
9. UR-DATA. https://urdataproject.com/. Accessed 02 Oct 2023
10. Gandini, P., Marchionni, G., Studer, L., Maja, R.: Sustainable and Aware Mobility Explained to Children (2019)
11. Aono, S.: UBC Sustainable Scholar: Identifying Best Practices for Mobility Hubs, pp. 3–6 (2019)
12. Shahee, S., Cohen, A., Farrar, E.: Carsharing's impact and future. Advances in Transport Policy and Planning, Volume 4, United States (2019)

Open Access This chapter is licensed under the terms of the Creative Commons Attribution 4.0 International License (http://creativecommons.org/licenses/by/4.0/), which permits use, sharing, adaptation, distribution and reproduction in any medium or format, as long as you give appropriate credit to the original author(s) and the source, provide a link to the Creative Commons license and indicate if changes were made.

The images or other third party material in this chapter are included in the chapter's Creative Commons license, unless indicated otherwise in a credit line to the material. If material is not included in the chapter's Creative Commons license and your intended use is not permitted by statutory regulation or exceeds the permitted use, you will need to obtain permission directly from the copyright holder.

Are Shared e-Bikes Disruptive of Established Shared e-Scooter Services? A Case Study of Braga, Portugal

Gabriel Dias[1]([✉]) ⓘ, Paulo Ribeiro[1] ⓘ, and Elisabete Arsenio[2] ⓘ

[1] Centre for Territory Environment and Construction, School of Engineering, University of Minho, 4710-057 Braga, Portugal
id8651@alunos.uminho.pt
[2] Department of Transport, LNEC, I.P., 1700-066 Lisbon, Portugal

Abstract. Free-floating shared micromobility services have been present in cities all around the world, however little is still known about the interaction between shared e-bikes and e-scooters. In the last few years shared e-scooters have experienced rapid growth worldwide, which, in some cities, jeopardizes the usage of shared e-bike services. Thus, this research work aims to explore if free-floating shared e-bikes can disrupt the usage of established e-scooter services. A case study in the city of Braga, north of Portugal, is developed from September of 2022 until May of 2023 in order to allow the comparison and contrast of the trips made by each micromobility mode, travel time, main origin and destinations of trips, as well as trip characteristics (e.g., vehicle rotation, the total number of trips per micromobility mode, total distance traveled). Results show that shared e-bikes and e-scooters are only used within city boundaries, and most of the trips originated in the parish where population density is higher. In Braga, riders prefer e-scooters when using a shared micro vehicle, since more than 98% of the trips made in the period studied were made by this mode. Also, shared e-scooters traveled more than 260,000 km in these nine months, while only 2,400 km were traveled in e-bikes. In short, Braga has experienced a rapid establishment of shared e-scooters instead of shared e-bikes, it can be due to the fact that trips on e-scooters are seen to be fun, pleasant, and quicker by riders.

Keywords: Shared micromobility · shared e-bike · shared e-scooter · micromobility trip data

1 Introduction

The Sustainable and Smart Mobility Strategy set out by the European Commission [1] has its backbone in the support of low- and zero-emission vehicles, which includes the usage of shared micromobility in cities. Hereof, the shared use of bicycles and e-scooters in cities has been associated with some positive impacts, such as cost savings and convenience, reduced vehicle miles traveled, reduced personal vehicle ownership, and reduced greenhouse gas (GHG) emissions [2].

© The Author(s) 2025
C. McNally et al. (Eds.): TRAconference 2024, LNMOB, pp. 216–221, 2025.
https://doi.org/10.1007/978-3-031-85578-8_28

Shared bicycle systems have been used since 1965, with the release of the White Bikes program [3], advancing through time until the development of technology that allowed peer-to-peer and dockless models. On the other hand, shared e-scooter services were introduced in cities only in 2017, with its expansion to Europe in the summer of 2018 [4]. Thus, it is crucial to acknowledge what happens in cities when both shared micromobility services are offered to dwellers.

A case study performed in Hungary shows that the population prefers to ride shared bicycles instead of shared e-scooters and free-floating options are preferred when compared to station-based shared services [5]. Younes and Baiocchi [6] analyzed the determinants of free-floating micromobility usage in four cities in the USA, and identified that built environment variables were strong predictors of shared micromobility trips, in addition, it was found that intersection density is positively associated with e-scooter trips. Despite the evidence of shared micromobility usage in cities, Lu et al. [7] suggest that shared e-bikes and e-scooters have different roles in urban mobility, since residents could have a preference for one mode over another based on the specific contexts, and travel purposes.

The current data available is confined to the cases where shared bicycles were implemented before or at the same time as shared e-scooters in cities. However, little is still known about how these micromobility modes interact when shared e-scooters are added to the urban space before any kind of shared bicycles are present. Thus, the present research work brings the novelty of understanding how both micromobility modes interact among themselves, and how usage patterns differ when shared e-bikes are introduced in the built environment after shared e-scooters are already established as a mode of transport.

For this, a case study in the city of Braga is proposed. This city is located in the north of Portugal and had the first shared e-scooter service implemented in the Summer of 2019, while the shared e-bike service was only presented to its population in the Fall of 2022. The methodology comprises the analysis and comparison of both microvehicles' trip data from September 2022 until May 2023 to answer the following research questions:

- How shared e-scooters and e-bike trip data are similar or different regarding the metrics of usage (e.g., number of trips, the distance of trips, duration of trips)?
- What happens to shared e-scooter usage after shared e-bikes are introduced in the city?

2 Methodology

The data used in this study were collected from the available API of two micromobility operators in the city of Braga, Portugal between September 01, 2022, and May 31, 2023. The starting date was selected following the implementation of the shared e-bike service in the city, and the last date comprises the availability of consistent data from the two shared micromobility modes while being operated simultaneously in the city. In addition, the main origins of trips in Braga were analyzed to better identify the hotspots for shared e-bikes and e-scooters in the city.

In order to identify the fleet size of shared e-scooters and e-bikes in the city, an instant KPI is obtained from the data received from the companies providing the service. This data is stored and organized by hour, day, and month. Every trip has its origin and destination point, as well as the label of the micromobility mode used (i.e. if it is a trip made by e-bike or e-scooter). This information is clustered with a spatio-temporal clustering (DBSCAN – Density-Based Spatial Clustering of Application with Noise) technique in order to discover the most used areas of the city.

The trip characteristics, such as the distance and duration, as well as the day of the week they occur have their descriptive statistics produced to allow a better understanding of the ridership and the differences in the usage of both e-bikes and e-scooters.

3   Results

Within the period of 273 days from September 2022 to May 2023, 185,355 trips were reported by both shared e-scooters and shared e-bikes in the city. The highest number of trips were made by e-scooters (182,850), while only 2,505 trips were made by e-bikes. The number of trips shows an increase in e-scooter and e-bike rides during the months of higher solar exposure and less rainfall, which are September, October, April, and May. However, during the colder and rainy months of November until mid-March, these modes of transport face a decline in the number of rides (Fig. 1).

Even if the pattern of usage throughout the timeframe studied seems similar for e-bikes and e-scooters, the number of trips made by each mode is different. E-scooters represent more than 97% of shared micromobility trips during all the months studied. These patterns can represent that e-scooters have more acceptance among users of shared micromobility in Braga, or that residents are more used to this mode, once it has been available to the population since 2019.

Another difference in e-scooter and e-bike usage in Braga is related to the characteristics of the trips regarding the distance traveled by each mode and the duration of these trips. Table 1 shows that the average distance traveled by e-scooters is higher than the ones made by e-bikes. However, e-bikes take longer to reach the destination.

Fig. 1. Number of trips per month by shared mode in Braga

Table 1. Descriptive statistics of shared micromobility usage in Braga

	Shared e-bike ($N = 2{,}505$)		Shared e-scooter ($N = 182{,}850$)	
	Trip distance (m)	Trip duration (min.)	Trip distance (m)	Trip duration (min.)
Mean	972.5	9.8	1508.1	8.6
Std. Dev	502.9	4.1	94.4	1.0
Min	118.5	1.7	1302.9	6.7
Max	3269.83	25.7	1822.1	11.6

Even though shared e-scooters are used more often than e-bikes, the trips made by e-bikes usually reach further distances in longer journeys. Figure 2 shows the percentage of trips made per day of the week for shared e-scooters and shared e-bikes. These trips are made mostly on the weekends, being the most used days Friday, Saturday, and Sunday, with a slight peak also on Tuesday. The e-scooter usage presents an almost even utilization rate during the week, reaching the peak of trip numbers on Friday, Saturday, and Sunday as well (Fig. 2), and these results are consistent with the temporal distribution of trips found by Lu et al. [7] in Austin, Texas, USA.

When comparing the main hotspots for shared micromobility in the city, it is possible to identify that e-bikes and e-scooters are used fairly in the same regions of the city, with special attention to the darker colors in Fig. 3, which represent the most populated parishes (smallest administrative municipal divisions) that comprise the biggest number of services for the population, such as the shopping mall, school districts, the University of Minho campus and city facilities.

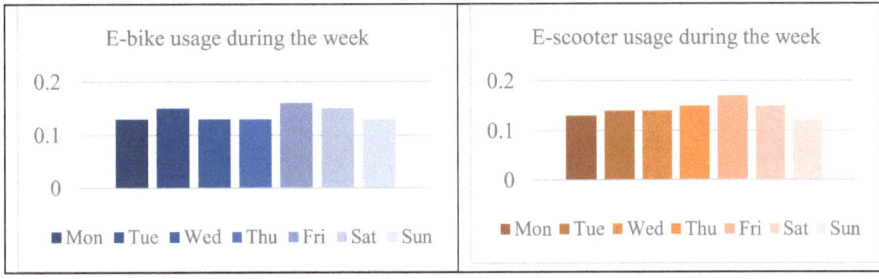

Fig. 2. Percentage of micromobility trips during the week in Braga

4 Discussion and Conclusions

Shared bicycles and e-bike services came to life earlier than shared e-scooters. Although, in Braga, the reverse implementation occurred, since shared e-scooters were introduced in the city in 2019, while e-bikes were only available three years later. The early rise of the former contributed to the familiarization of this mode to the general population,

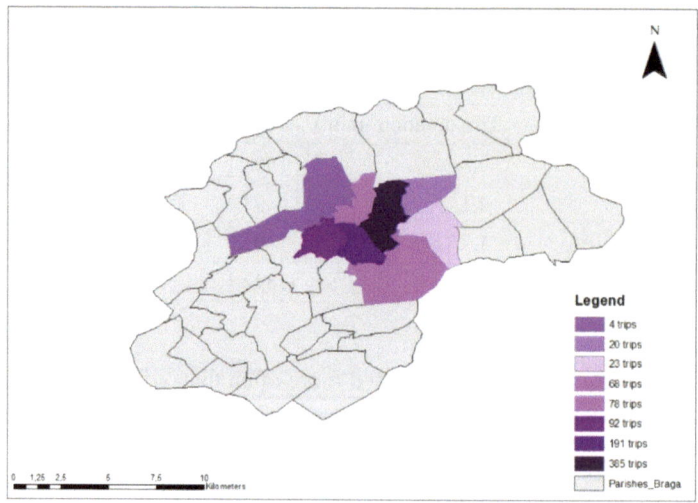

Fig. 3. Shared micromobility main area of usage in Braga

who consider e-scooters as a fun and pleasant mode of transport used when riders need to reach their destinations quickly [8].

The trip data shows an excessive disadvantage of e-bikes when considering the number of trips in the period studied, which can be due to the early familiarization of e-scooters. Another reason that can cause an advantage of e-scooter usage over e-bikes is the price of the trip. While e-scooters cost only 0.23EUR per minute traveled, e-bikes have a cost of 1EUR to unlock plus 0.18EUR per minute traveled.

The lack of dedicated infrastructure can be another crucial factor in determining the lack of competitiveness between e-bikes and e-scooters. This is because in Braga, in the area where most of the micromobility trips are made, cycle lanes are almost nonexistent, and the streets are narrow and paved with cobblestones. In this situation, e-scooters can have priority in usage, since they are easier to be ridden on the sidewalk, even though it is forbidden by law in Portugal [9].

In Braga, both modes are present and used in the same area of the city, which can represent that they are accessed in the same locations, which contrasts with results found by Chicco & Diana [10]. When comparing the usage throughout the week, patterns remain the same, with higher ridership rates on Fridays and weekends.

In short, shared e-scooters and e-bikes in Braga can coexist and provide a mobility option for short-distance trips. However, the attempt to introduce a mode of transport without the proper planning, or dedicated infrastructure can be crucial to the success of this mode. In Braga, in order to balance the usage of both micromobility modes, more efforts need to be made to minimize price differences among them and to provide equal opportunities for people to ride safely (i.e., provision of dedicated infrastructure), so dwellers can be familiarized with e-scooters as well as e-bikes and use them according to their mobility needs.

Acknowledgment. This research was funded by Fundação para a Ciência e a Tecnologia (FCT, I.P.), grant number 2020.05041.BD.

References

1. European Commission. Sustainable and smart mobility strategy: putting European transport on track for the future (2020)
2. Shaheen, S., Chan, N.: Mobility and the sharing economy: potential to facilitate the first-and last-mile public transit connections. Built. Environ. **42**, 573–588 (2016). https://doi.org/10.2148/benv.42.4.573
3. Shaheen, S., Guzman, S., Zhang, H.: Bikesharing in Europe, the Americas, and Asia. Transp. Res. Rec. 159–167 (2010). https://doi.org/10.3141/2143-20
4. Dias, G., Arsenio, E., Ribeiro, P.: The role of shared e-scooter systems in urban sustainability and resilience during the covid-19 mobility restrictions. Sustain **13** (2021). https://doi.org/10.3390/su13137084
5. Jaber, A., Hamadneh, J., Csonka, B.: The preferences of shared micro-mobility users in Urban Areas. IEEE Access (2023). https://doi.org/10.1109/ACCESS.2023.3297083
6. Younes, H., Baiocchi, G.: Analyzing the spatial determinants of dockless E-Scooter & E-Bike trips across four U.S. Cities. Int. J. Sustain. Transp. **17**, 870–882 (2022). https://doi.org/10.1080/15568318.2022.2119623
7. Lu, M., Traut, E.J., Ilgin Guler, S., Hu, X.: Analysis of spatial interactions among shared e-scooters, shared bikes, and public transit. J. Intell. Transp. Syst. Technol. Plann. Oper. **0**, 1–17 (2023). https://doi.org/10.1080/15472450.2023.2174803
8. Dias, G., Ribeiro, P., Arsenio, E.: Shared e-scooters and the promotion of equity across urban public spaces-a case study in Braga, Portugal. Appl. Sci. **13** (2023). https://doi.org/10.3390/app13063653
9. AMT: Linhas de orientação sobre a regulação da micromobilidade partilhada. Lisbon (2022)
10. Chicco, A., Diana, M.: Understanding micro-mobility usage patterns: a preliminary comparison between dockless bike sharing and e-scooters in the city of Turin (Italy). Transp. Res. Procedia **62**, 459–466 (2022). https://doi.org/10.1016/j.trpro.2022.02.057

Open Access This chapter is licensed under the terms of the Creative Commons Attribution 4.0 International License (http://creativecommons.org/licenses/by/4.0/), which permits use, sharing, adaptation, distribution and reproduction in any medium or format, as long as you give appropriate credit to the original author(s) and the source, provide a link to the Creative Commons license and indicate if changes were made.

The images or other third party material in this chapter are included in the chapter's Creative Commons license, unless indicated otherwise in a credit line to the material. If material is not included in the chapter's Creative Commons license and your intended use is not permitted by statutory regulation or exceeds the permitted use, you will need to obtain permission directly from the copyright holder.

Integrating Carpooling
in Mobility-as-a-Service Platforms:
Learnings from Co-design Activities

Francesca Cellina[1]([⊠]), Marco Derboni[2], Vincenzo Giuffrida[2], Mirko Baruffini[3],
Paolo Mastrobuono[3], Jan Trautmann[4], Uros Tomic[5], Raphael Hoerler[5],
and Camille Vedel[6]

[1] University of Applied Sciences and Arts of Southern Switzerland (SUPSI), Manno,
Switzerland
francesca.cellina@supsi.ch
[2] Dalle Molle Institute for Artificial Intelligence (IDSIA USI-SUPSI), Lugano,
Switzerland
[3] BePooler SA, Lugano, Switzerland
m.baruffini@bepooler.ch
[4] Lugano Living Lab, City of Lugano, Switzerland
[5] Zurich University of Applied Sciences (ZHAW), Winterthur, Switzerland
[6] Citec Ingénieurs Conseils SA, Geneva, Switzerland

Abstract. App-based Mobility as a Service (MaaS) platforms combining public transport, car- and micro-mobility-shared services with dynamic carpooling are emerging as viable alternatives to solo car use for sub-urban contexts. Insights from real-life implementation are however still limited. Which practical challenges affect MaaS platforms leveraging carpooling? We tackle this question from the perspective of potential users of the Swiss-based MixMyRide platform, engaging them in co-design workshops. We find four elements of practical interest, resonating with limitations already identified for carpooling. First, carpooling increases the number of inter-changes potentially affected by delays. This requires real-time traffic information data, re-scheduling tools, and features for quick interaction between users. Second, as social control is low, features to create trust between strangers are needed, which calls for trade-offs between strict identity checks and quick registration. Third, carpooling pick-up/drop-off may endanger safety if bus stops are used. This requires in-advance agreements, negatively affecting the MaaS' dynamism. Fourth, car-pooling offer is not granted. To accept possible discomfort, decreased flexibility, and effort to enter ride offers, drivers need incentives, such as sharing of travel expenses, virtual or tangible rewards by public institutions, and feedback on saved emissions.

Keywords: Mobility-as-a-Service · Carpooling · Co-Design

© The Author(s) 2025
C. McNally et al. (Eds.): TRAconference 2024, LNMOB, pp. 222–228, 2025.
https://doi.org/10.1007/978-3-031-85578-8_29

1   Introduction

App-based Mobility as a Service (MaaS) schemes that integrate public transport, car and micro-mobility (bicycle, e-scooter) shared services in the same trip enabled unprecedented opportunities to enhance inter-modal mobility and create viable alternatives to solo car use and even car ownership [1–5]. Low-density suburban contexts, typically under-covered by public transport and shared-mobility options, are not exempt from MaaS potential benefits, if real-time dynamic carpooling [6] is also integrated. As suggested by stated-preference surveys and simulations, car drivers already on the road could in fact help address critical first and last mile connection problems in sub-urban areas [7–10]. However, evidence from practical implementation is limited [11] and the feasibility of integrating carpooling into MaaS platforms has not been investigated yet: which practical conditions challenge the effectiveness of MaaS platforms leveraging carpooling?

We analyse the case of MixMyRide, an app-based inter-modal MaaS system that combines public transport, shared micro-mobility, and quasi real-time dynamic carpooling services, developed within a Swiss-based research project. To address our research question, we leverage a set of co-design workshops performed in late 2022, during the app features' development. Insights from the workshops allow us to uncover practical, general value challenges affecting MaaS platforms integrating carpooling, and to provide recommendations on how to tackle them in the design of future MaaS platforms.

2   Material and Methods

MixMyRide uses artificial intelligence algorithms fed by schedule public transport, carpooling, and shared micro-mobility data, to match mobility demand and supply. A typical trip suggested by MixMyRide is made-up of a carpooling leg from home to a railway station ("first mile"), then the main leg by train, and finally a leg by foot, shared bicycle, e-scooter, or carpooling again ("last mile"). The MixMyRide features were developed through a co-design living lab approach: a process to create and validate innovation within collaborative, real-world environments, by involving a broad range of stakeholders. Starting from a shared understanding of the users' needs and constraints around daily mobility practices, the co-design sessions aimed at identifying ideal MaaS solutions. Analysing the elements emerging from co-design discussions thus we can identify critical factors that can affect the real-life MaaS platform effectiveness, accounting for a multiplicity of user perspectives, competences, expectations, and needs.

Co-design activities were performed in late 2022 in the region of Lugano (Italian-speaking part of Switzerland), through three workshop sessions. In the first and third session (June, 7 and September, 19), two-hour in-depth discussions were performed, respectively with ten selected local experts of the mobility domain (transport operators, cantonal and municipal institutions, environmental NGOs), and with six voluntary citizens recruited via social media campaigns. The second session (October, 21) was instead organised within a Swiss-based

"National Open Innovation Camp" and engaged twenty-five entrepreneurs, start-uppers and researchers in the energy, mobility, and digitalisation domains from all over Switzerland in five 20-min small-group co-design workshops.

In the first two sessions, wireframe mock-up prototypes of the app, previously developed by our research team, were presented and discussed, in order to collect feedback on the goals, meaning, and main design choices around the app features. The discussion was organised around the three guiding items "I Like, I Wish, What If", typically used in Design Thinking processes. The third session consisted in fine-tuning the app's user experience and route planning algorithms. The insights we present here are drawn from the elements collected in the first two sessions, via word-per-word detailed notes (smooth verbatim transcripts) of the conversations occurring during workshop discussions. These were merged and classified in categories, according to a qualitative content analysis approach, followed by an in-depth literature review on the emerging topics.

3   Results and Discussion

Four key elements of general value and practical interest for future inter-modal MaaS systems integrating carpooling services emerged, at the crossroads between carpooling, transport infrastructure development, and big data.

3.1   Higher Risk of Being Affected by Traffic Perturbation and Delays

The strength of including carpooling rides in MaaS platforms lies in higher trip matching possibilities [10]. However, increasing the transport mode inter-changes in the same trip makes trip solutions more vulnerable and prone to the risk of trip chain disruption due to public transport delays and road traffic congestion: optimal trip solutions identified by MaaS route planning algorithms usually consider tight inter-change connections, thus could become unfeasible, even for small delays in planned travel times. To limit such problems, larger interchange times could be considered by the MaaS route planning algorithms. This would however reduce the chances that the MaaS trip solutions are welcomed as alternatives to solo car driving. At least, MaaS algorithms should rely on time-dependent travel time estimates, that account for average travel times previously registered in the same day of the week and time of the day as the prospected trip. This data is for instance automatically offered by Location services provided by Google for a fee. To our knowledge, however, such a piece of information is rarely available for public transport at the needed granularity level. In some cases, real-time public transport data is available —though this is not directly useful when MaaS platforms are used to plan trips in advance. Rather, long historical series of real-time public transport data would be needed, to compute average actual travel times.

In any case, possible disruptions that go beyond average actual travel times may always occur. To ensure that travellers can always complete their trips, MaaS systems could routinely use real-time data to check travel times when a planned trip is about to start (e.g. 30 min before it) and if needed send alternative trip solutions via push notifications. If disruption affects planned trips that include carpooling legs, in-app messaging systems (chats) become essential, to ensure quick interaction between carpooling companions. As carpooling drivers are not as strictly constrained to respecting time schedules as public transport, some carpooling rides might even be maintained under disruption, if both driver and rider have sufficient flexibility to wait for the companion to reach the planned interchange point. An in-app chat can favour swift re-arrangements, provided that carpoolers can access it very close to the scheduled trip.

3.2 Lack of Trust Between MaaS Users

In dynamic carpooling, social control is lower than in closed-community carpooling, where carpoolers are connected by other relationships besides sharing a ride –relationships that existed before the ride and will keep existing after the ride, such as for instance those between colleagues working for the same company, that occur in corporate carpooling schemes. As many people –especially female riders travelling after dark– would not feel comfortable in sharing car rides with unknown persons, the problem arises about how to build trust between strangers, by only relying on the piece of information offered by MaaS platforms.

An internal reviewing system, that, as soon as a carpooling trip has ended, invites users to leave a review of the travel companion (e.g. a star-based evaluation accompanied by a brief text, solicited via push notification), could serve the purpose. This procedure would not be sufficient, however, to fully ensure trust, as it would not cover new users with no or very few reviews. Co-design participants in fact indicated that they would feel safer (and thus be willing to share rides with unknown persons) if, at app registration, full personal details, including email address, phone number and personal identity card, are checked by the MaaS platform. Such a check of the digital identity could consist in "Know Your Customer" KYC procedures typically used to open bank accounts.

These procedures have however two direct drawbacks: operators are needed to certify prospect users' identity, and drop-out by prospect MaaS users could occur before their registration has been verified. To reduce drop-out risks, while ensuring in-depth checks of app users' identity, users might be registered in the MaaS platform via a small set of personal data, and full KYC procedures might be requested when they perform their first carpooling trip (as rider or driver).

3.3 Traffic Safety Problems

To optimise travel times, inter-changes between transport modes should ideally occur at public transport stops, where shared micro-mobility stations are usually also located. Carpooling inter-changes at bus or tram stops can be critical, as usually no temporary car parking slots are available. In many countries, road regulations explicitly prohibit cars from stopping at bus or tram stops, to avoid interference with public transport and risks for passengers' safety.

Current MaaS route planning algorithms lack the capability to identify safe inter-change points suitable to temporary parking: despite progress in artificial intelligence, carpoolers' personal knowledge of the inter-change area (or their capability to interact with an online map representing it) is still needed. To ensure safe pick-up and drop-off, carpooling companions thus need to invest time in identifying proper places and agree on them, via chat-based interactions. Though this is quite an easy task (especially for drivers that offer carpooling rides along commuting trips, that they frequently travel and know quite well), the need for such arrangements adds inconvenience and requires in-advance planning, negatively affecting MaaS platforms' dinamycity and flexibility.

3.4   Need for Incentives to Ensure Carpooling Ride Offer

The key benefits stemming from the integration of public transport, micro-mobility and carpooling depend on the availability of offered carpooling rides, that increase the amount of available connections between origins and destinations. Though simulations have shown that, compared to carpooling alone, less carpooling rides are sufficient for suitable trip matches if carpooling is combined with public transport [10], a critical mass of carpooling offers is still necessary.

Indeed, this is far from granted. Why should drivers accept discomfort in sharing personal space with unknown persons, minor deviations from their usual routes, potential losses in flexibility and independence, and make the effort to enter their trips in advance into a MaaS platform? Previous research has shown that even carpooling systems that are characterised by relatively stable arrangements and teams, where individuals have pre-existing relationships (or their relationships get easily shaped by daily ride repetitions) and usually alternate themselves in driving their own car (thus reducing the use of their own car for commuting and the consequent costs, emissions, and stress), are affected by a number of barriers and would benefit from specific incentives [6]. In dynamic carpooling systems integrated in MaaS platforms, barriers are likely to be perceived as even more relevant, as rides are potentially shared with always different companions. Thus, the social relation and reciprocal car driving benefits are missing.

To encourage prospect carpooling drivers to offer rides, MixMyRide workshops suggested several possibilities. First, to at least share travel expenses (the rider pays an agreed upon fee to the driver, based on a flat rate per travel kilometer). Second, to also to offer reward vouchers or tax discounts covered by public institutions (either based on the number of offered rides or on their kilometers). Third, to leverage virtual, non-tangible rewards, by means of gamification approaches (e.g. drivers are rewarded with virtual badges when they reach a given threshold or are better than other MaaS users in a ranking based on the number of offered rides, on the saved global CO_2 emissions, or on the number of other users they transported). The former strategy (sharing travel expenses) was already proven effective in carpooling interventions [12]. However, when used in MaaS systems, costs charged to carpooling riders would be additional to the costs they already face for public transport and/or shared micro-mobility trip legs. In regions where such costs are computed based on travel time, rather than specific travelled routes, riders might perceive the overall inter-modal trip cost as too high. Namely, there is the risk that the incentive for carpooling drivers turns into a disincentive for carpooling riders. Experimental research (e.g. randomised controlled trials) is thus needed to estimate the effects of such incentives.

4 Conclusions

Drawing from the MixMyRide co-design process performed in Southern Switzerland in late 2022, we have identified practical challenges potentially hindering the effectiveness of Mobility as a Service (MaaS) platforms that integrate public transport, (shared) micro-mobility and carpooling services, and suggested ways to address them. Though the stakeholder groups that joined our co-design sessions were broadly diverse, evidence stemming from real-life implementation is needed. Our learnings may in fact be influenced by the specific infrastructural and cultural conditions in which MixMyRide will operate, by the self-selection of participants, by their overall small number, and by the fact we were dealing with a prospect, hypothetical MaaS tool, rather than analysing users while interacting with it under their daily needs and constraints. Analysis of the real-life MixMyRide trial currently ongoing in the Swiss regions of Geneva, Winterthur-Zurich, and Lugano (March 2023 - February 2024) will provide useful materials to confirm, confute, and enrich the lessons we have learnt so far.

References

1. Sochor, J., Strömberg, H., Karlsson, I.M.: Implementing mobility as a service: challenges in integrating user, commercial, and societal perspectives. Transp. Res. Rec. **2536**(1), 1–9 (2015)
2. Matyas, M., Kamargianni, M.: The potential of mobility as a service bundles as a mobility management tool. Transportation **46**(5), 1951–1968 (2019)
3. Schikofsky, J., Dannewald, T., Kowald, M.: Exploring motivational mechanisms behind the intention to adopt mobility as a service (MaaS): insights from Germany. Transp. Res. Part A: Pol. Pract. **131**, 296–312 (2020)
4. Ho, C.Q., Mulley, C., Hensher, D.A.: Public preferences for mobility as a service: insights from stated preference surveys. Transp. Res. Part A: Pol. Pract. **131**, 70–90 (2020)
5. Hoerler, R., Stünzi, A., Patt, A., Del Duce, A.: What are the factors and needs promoting mobility-as-a-service? Findings from the Swiss Household Energy Demand Survey (SHEDS). Eur. Transp. Res. Rev. **12**(1), 1–16 (2020). https://doi.org/10.1186/s12544-020-00412-y
6. Créno, L.: User experience of dynamic carpooling: How to encourage drivers and passengers? In: Energy Consumption and Autonomous Driving: Proceedings of the 3rd CESA Congress, Paris, 2014, pp. 71–81. Springer (2016)
7. Wright, S., et al.: Public acceptance of SocialCar, a new mobility platform integrating public transport and car- pooling services: insights from a survey in five European cities. In: Proceedings of the Transport Research Arena TRA18 (2018)

8. Yan, X., Levine, J., Zhao, X.: Integrating ridesourcing services with public transit: an evaluation of traveler responses combining revealed and stated preference data. Transp. Rese. Part C: Emerg. Technol. **105**, 683–696 (2019)

9. Thao, V.T., Imhof, S., von Arx, W.: Integration of ridesharing with public transport in rural Switzerland: Practice and outcomes. Transp. Res. Interdiscipl. Perspect. **10**, 100,340 (2021)

10. Wright, S., Nelson, J.D., Cottrill, C.D.: MaaS for the suburban market: incorporating carpooling in the mix. Transp. Res. Part A: Pol. Pract. **131**, 206–218 (2020)

11. Shen, Q., Wang, Y., Gifford, C.: Exploring partnership between transit agency and shared mobility company: an incentive program for app-based carpooling. Transportation **48**(5), 2585–2603 (2021)

12. Bulteau, J., Feuillet, T., Dantan, S., Abbes, S.: Encouraging carpooling for commuting in the Paris area (France): which incentives and for whom? Transportation 1–20 (2021)

Open Access This chapter is licensed under the terms of the Creative Commons Attribution 4.0 International License (http://creativecommons.org/licenses/by/4.0/), which permits use, sharing, adaptation, distribution and reproduction in any medium or format, as long as you give appropriate credit to the original author(s) and the source, provide a link to the Creative Commons license and indicate if changes were made.

The images or other third party material in this chapter are included in the chapter's Creative Commons license, unless indicated otherwise in a credit line to the material. If material is not included in the chapter's Creative Commons license and your intended use is not permitted by statutory regulation or exceeds the permitted use, you will need to obtain permission directly from the copyright holder.

The Influence of Micro-Mobility Services on Peoples' Willingness to Join a MaaS Scheme

Panagiota Mavrogenidou[1], Amalia Polydoropoulou[1(✉)], and Athena Tsirimpa[1,2]

[1] Department of Shipping, Trade and Transport, University of the Aegean, Korai 2a, 82100 Chios, Greece
polydor@aegean.gr

[2] Department of Maritime Transport and Logistics, The American College of Greece, Gravias Street 6, Athens, Greece

Abstract. Mobility as a Service (MaaS) is a widely researched topic with increasing interest in recent years. While numerous factors have been examined in relation to users' willingness to embrace a MaaS scheme, a notable gap in the literature is being observed on how individuals' opinion for micro-mobility affects their tendency to adopt such systems. This study employs a hybrid logit model, that examines the interplay between these two emerging transportation services and investigates how micro-mobility perceptions shape individuals' choices. To accomplish this, a latent variable for micro-mobility positiveness is constructed to quantify the potential impact of micro-mobility on users' MaaS preferences. The study focuses on the Greek city "Thessaloniki", and a sample size of 392 individuals was used. The model provides important insights on the key determinants influencing users' choices regarding MaaS utilization and provides evidence that micro-mobility can be considered a positively influential factor to the wider adoption of such scheme. This research also reveals aspects that should be further analyzed to a greater extent. Specifically, understanding why workers, bus passengers, and people that frequent commute as car passengers are more willing to use a MaaS system, can offer valuable insights to researchers.

Keywords: micro-mobility · Mobility as a Service · Choice Model

1 Introduction

Mobility as a Service (MaaS) is a "user-centric, intelligent mobility distribution model", that provides users with a set of different mobility modes in a "single digital platform" that aggregates all mobility providers into one "sole mobility provider" [1]. Through a single interface, MaaS systems unify many different mobility services, including trip planning, service booking, service payments, and ticketing [2, 3]. A successful MaaS system requires the partnership and synergy of multiple and different actors [4], while the most important actors that a MaaS scheme consists of are the transport operators, the data providers, the technology and platform providers, the actors that are related to the ICT infrastructure, the insurance companies, regulatory organizations, as well as universities

© The Author(s) 2025
C. McNally et al. (Eds.): TRAconference 2024, LNMOB, pp. 229–235, 2025.
https://doi.org/10.1007/978-3-031-85578-8_30

and research institutions [1]. Additionally, a stable cooperation and competition between public and private actors is required [5]. MaaS is defined as a one-stop-shop, that provides users all the needed information, and the ability to purchase a single transport service or a combination of them.

To enhance sustainable transport services that benefits all users, it's important to create a mobility service that will serve all citizens without excluding individuals or groups of people (e.g., disabled people) [6]. Stakeholders' collaboration is crucial for achieving the anticipated sustainable outcomes in urban areas [7]. The main aim of these systems, is to create and promote sustainable transportation alternatives, that will reduce the levels of single occupied vehicles [2]. Mobility as a Service focuses on travelers' needs and promises a sustainable solution that will help reduce the number of traffic accidents and the congestion levels [8].

The Six categories that been identified as the most the influential factors on MaaS adoption are the socio-demographic characteristics, travel behavior, users' personality and attitudes, built environment, perceived usefulness and social influence [9]. Furthermore, studies have indicated that non-car users are less likely to adopt MaaS services, whereas infrequent or casual car users are potential MaaS adopters [10]. In the same context, frequent bus users have a higher probability to purchase a unimodal MaaS plan with a single transportation option of public bus. Furthermore, gender, age, income, and family status are the most influential factors on individuals' willingness to purchase a MaaS plan [7]. Specifically, males, high-income earners and families with children indicate lower probability to join MaaS scheme, and low-income earners, people that do not own a car, and younger people without children present an increased probability. Commuters favor MaaS plans with unlimited public transport, while adding sharing services (bike-sharing, e-scooter sharing, taxi rides, and car sharing) boosts adoption levels [10, 11].

To the researchers' knowledge there is limited research related to the relation of MaaS systems and individuals' opinion on micro-mobility. Micro-mobility refers to small, low-speed vehicles that are used for personal commuting, especially for the last and first mile of a trip [8]. Through a systematic literature review analysis and the keywords co-occurrence visualization, researchers found a link between MaaS and autonomous vehicles (AV), drones and micro-mobility services [9]. Micro-mobility usage levels and usage intention are strongly affected by users' socio-demographic characteristics, build environment (cycling infrastructure and topography), and weather parameters (air quality and temperature). To the researchers' knowledge there is no other study that examines the importance of micro-mobility services in a MaaS scheme based on individuals' characteristics and attitudes. This research aims to fill this gap by investigating individuals' willingness to use a MaaS system and by exploring how their level of positivity towards micro-mobility may influence their choice, contributing to a better understanding MaaS potentials on urban transportation.

2   Methodology

This paper uses survey data to analyze residents' willingness to use a MaaS system, and the contribution of micro-mobility services to users' final choice. This research is a case study for the Greek city Thessaloniki, which is considered having high usage levels of bicycles and e-scooters. To demonstrate the characteristics that influence individuals' final decisions and to determine their significance in individual choices, we developed a logit model with one latent variable that countifies micro-mobility's influence on citizens' choices. The survey was distributed online in the first half of 2021, and it was shared on websites, social media, emailed to organizations and businesses, and shared with the Aristotle University of Thessaloniki students and teachers. Further analysis of the distribution process and the questionnaire outline can be found online [12]. A valid sample of 392 individuals was collected and used to develop the proposed model.

2.1 Modelling Framework

According to a logit model, a user's probability to choose a mode of transport depends on other explanatory variables and factors, such as the individual's personal characteristics and attitudes. To analyze deeper into individuals decision-making process on MaaS acceptance, a logit model with a latent variable has been developed. The aim is to provide a comprehensive understanding of users' tendency to adopt MaaS system. The latent variable seeks to capture individuals' attitudes and positiveness regarding micro-mobility and serves as a tool to explore the relationship between users' openness to adopting a MaaS system and their opinion on micro-mobility. For the creation and estimation of the model, the open-source software "PandasBiogeme" (Bierlaire, 2020) was used. A hybrid choice model with one latent variable for the quantification of users' micro-mobility positiveness was developed.

For the proposed model, a latent variable was constructed using 10 indicators. Participants display a negative attitude towards commuting with bicycles (M = 1.69, STD = 1.31), or e-scooters (M = 1.21, STD = 0.71), while they seem to have safety concerns regarding these modes (Mbicycle = 2.42, STDbicycle = 1.17 / Me-scooter = 1.9, STDe-scooter = 1.04). Additionally, a more moderate attitude is observed towards the level of comfort provide by bicycles (M = 3.11, STD = 1.33) and scooters (M = 2.57, STD = 1.33). Furthermore, participants indicate a slightly more positive opinion regarding the reliability of both bicycles (M = 3.93, STD = 1.50) and e-scooters (M = 3.07, STD = 1.45). Finally, individuals consider that both bicycles (M = 5.62, STD = 0.91) and e-scooters (M = 4.61, STD = 1.32) have important environmental benefits for the society. Figure 1 presents the detail construction of the model and its latent variable, while below are given the equations used for the creation of the latent variable. For the latent variable model, the structural equation is taking into consideration participants age (under 35 = 41.1%, 35 and older = 58.9%), frequent pedestrians (frequent pedestrians = 19.9%, non-frequent pedestrians = 80.1%), and ownership of a bicycle (owns a bicycle = 33.7%, does not own a bicycle = 66.3%).

For the final estimation of the proposed choice model, three more variables are included. Specifically, the variable 'workers' identifies individuals that are currently employed (workers = 76%, unemployed = 24%), the 'FrequentPTusers' variable describes whether the responders are frequent public transport users (M = 3.03, STD = 1.71 - 6-point likert), and the variable 'FrequentCarPassengers' assesses whether individuals commute often as car passengers (M = 3.78, STD = 1.44 - 6-point likert).

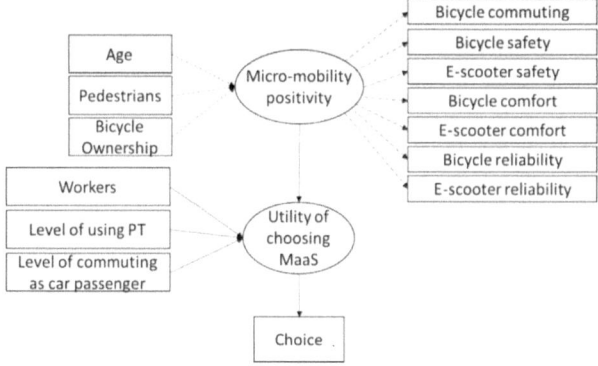

Fig. 1. The proposed model

3   Results

Table 1 presents the Latent Variable estimation results, while table 2 provides the estimation results of the choice model. For the proposed choice model, the initial log likelihood and the final log likelihood were calculated -271.7137 and -211.9284 respectively, while the R2 was estimated to 0.222. The interpretation of the result was carried out for a coefficient interval equal to 95%. Table 1 presents the results of the logit model that was developed.

Table 1. Latent Variable Estimation Results

Coefficients	Value	Rob. t-test	Rob. p-value
Structural Model			
Age (> = 34 years old)	−0.0426	−1.57	0.117
Bicycle Ownership	0.206	1.98	0.0479
Usually Commuting as Pedestrians	0.0372	1.25	0.21
intercept	−0.362	−5.57	2.55e−08
Measurement Model			
Bicycle commuting	11.4	1.94	0.0523
Bicycle comfort	2.24	1.98	0.0472
Bicycle reliability	1.87	2.25	0.0246
Bicycle safety	1.59	1.89	0.0585
E-scooter comfort	1.8	2.25	0.0244
E-scooter safety	1.7	2.14	0.0326

Based on the estimation results of the choice model residents of Thessaloniki have a moderate attitude towards a MaaS system. The estimated negative Alternative Specific

Table 2. Choice Model Estimation Results

Coefficients	Value	Rob. t-test	Rob. p-value
ASC_Yes	−0.0451	−0.134	0.894
Workers	0.594	2.11	0.0347
Frequent PT users	1.02	3.54	0.000404
Frequent Car Passengers	0.926	3.83	0.000128
Latent_ Micro-Positivity	1.27	1.36	0.173
sigma_s	−0.0281	−0.217	0.828
Summary Statistics			
Initial Log-Likelihoo	−271.7137		
Final Log-Likelihood	−211.9284		

Coefficient for the choosing a MaaS ("asc_Yes") indicate that for the examined sample, the residents are less likely to choose to use a MaaS system. However, no statistical significance is observed, and the negative trend that is observed needs further investigation. In addition, the analysis revealed a statistically significant relationship between the 'Workers' and individuals' choice to use a MaaS system, indicating that people that work are more likely to choose to adopt such a scheme. Furthermore, frequent public transport passengers, and people that frequently commute as car passengers, appear to have increased probabilities to choose a MaaS system. These relationships also demonstrate important statistical significances, underlining the significance of independence on MaaS adoption. Users' may feel that purchasing a MaaS package could increase their independence and their convenience.

Finally, the micro-mobility variable that was created shows that a positive attitude towards micro-mobility is associated with a positive attitude towards MaaS and increased likelihood of using the system. The constructed latent variable stands out with a coefficient value of 1.27, which is the highest among all the other variables in our utility function. On the other hand, the latent variable does not indicate a statistically significant, revealing the need of further investigation. The results indicate that the integration of micro-mobility modes in the provided MaaS options have the potential to increase the popularity of on demand transport. Further research may be able to specify the micromobility attitudes that influence MaaS adoption and explore potential variations based on different micro-mobility modes.

4   Conclusion

This paper examines individuals' willingness to use a MaaS system, and how micromobility may influence their final choice. A logit model with a latent variable for micromobility was developed, based on a sample of 392 individuals. The model showed that workers, bus passengers, and people that frequent commute as car passengers are more willing to use a MaaS system. The reasoning behind these results should be further analyzed with a more detailed survey on these specific groups, in order to determine the underlying factors and motivations that influence their preferences and choices related to MaaS. Finally, the latent variable that was created to determine the micro-mobility's influence on individuals' tendency to choose a MaaS system, showed that people that are positively preoccupied towards micro-mobility are more open to the idea of choosing a MaaS system. Future analysis should be made regarding micro-mobility's influence on peoples' acceptance of a MaaS system, to examine the level of influence different types of micro-mobility modes might have on the acceptance of a MaaS system.

References

1. Kamargianni, M., Matyas, M.: The Business Ecosystem of Mobility-as-a-Service (2017)
2. Aman, J.J.C., Smith-Colin, J.: Application of crowdsourced data to infer user satisfaction with Mobility as a Service (MaaS). Transp. Res. Interdisc. Perspect. **15**, 100672 (2022)
3. Esztergár-Kiss, D., Kerényi, T., Mátrai, T., Aba, A.: Exploring the MaaS market with systematic analysis. Eur. Transp. Res. Rev. **12**(1), 67 (2020)
4. Polydoropoulou, A., Pagoni, I., Tsirimpa, A.: Ready for Mobility as a Service? Insights from stakeholders and end-users. Travel Behav. Soc. **21**, 295–306 (2020)
5. Polydoropoulou, A., Pagoni, I., Tsirimpa, A., Roumboutsos, A., Kamargianni, M., Tsouros, I.: Prototype business models for Mobility-as-a-Service. Transp. Res. Part A: Pol. Pract. **131**, 149–162 (2020)
6. Dadashzadeh, N., Woods, L., Ouelhadj, D., Thomopoulos, N., Kamargianni, M., Antoniou, C.: Mobility as a Service Inclusion Index (MaaSINI): evaluation of inclusivity in MaaS systems and policy recommendations. Transp. Policy **127**, 191–202 (2022)
7. Arias-Molinares, D., García-Palomares, J.C.: The Ws of MaaS: understanding mobility as a service fromaliterature review. IATSS Res. **44**(3), 253–263 (2020)
8. Guyader, H., Friman, M., Olsson, L.E.: Shared mobility: evolving practices for sustainability. Sustainability. **13**(21), 12148 (2021)
9. Zhang, Y., Kamargianni, M.: A review on the factors influencing the adoption of new mobility technologies and services: autonomous vehicle, drone, micromobility and mobility as a service. Transp. Rev. **43**(3), 407–429 (2023)
10. Tsouros, I., Tsirimpa, A., Pagoni, I., Polydoropoulou, A.: MaaS users: Who they are and how much they are willing-to-pay. Transp. Res. Part A: Pol. Pract. **148**, 470–480 (2021)
11. Chen, C.F., He, M.L.: Exploring heterogeneous preferences for mobility-as-a-service bundles: a latent-class choice model approach. Res. Transp. Bus. Manag. **49**, 101014 (2023)
12. Mavrogenidou, P., Papagiannakis, A.: Διερεύνηση της αποδοχής της "Κινητικότητας ως υπηρεσία". Η περίπτωση της Θεσσαλονίκης (2022). Accessed 12 Oct 2023 Oct 12]; Available from: https://ikee.lib.auth.gr/record/340147

Open Access This chapter is licensed under the terms of the Creative Commons Attribution 4.0 International License (http://creativecommons.org/licenses/by/4.0/), which permits use, sharing, adaptation, distribution and reproduction in any medium or format, as long as you give appropriate credit to the original author(s) and the source, provide a link to the Creative Commons license and indicate if changes were made.

The images or other third party material in this chapter are included in the chapter's Creative Commons license, unless indicated otherwise in a credit line to the material. If material is not included in the chapter's Creative Commons license and your intended use is not permitted by statutory regulation or exceeds the permitted use, you will need to obtain permission directly from the copyright holder.

On-Demand Transport Services as a Supplement to Public Transport in Suburban Areas

A Comparison of Traditional and Modern Demand Responsive Transport Services in a Household Survey in the Suburban Area of Hamburg

Tyll Diebold$^{(\boxtimes)}$ (iD) and Carsten Gertz (iD)

Hamburg University of Technology, Am Schwarzenberg-Campus 1, 21073 Hamburg, Germany
tyll.diebold@tuhh.de

Abstract. App-based ridepooling offers the chance to provide a high-quality public transportation service in areas that have been inadequately served so far. One of the main strengths of ridepooling services is their ability to offer direct access to the public transport system and thus provide an attractive service on the first and last mile. While traditional on-demand services have so far been quite inflexible, due to having a fixed timetable, needing to be ordered in advance of the trip via a call centre, app-based ridepooling promises a true on-demand service for the user.

This study is based on a household survey carried out in June/July 2022 in suburban areas of Hamburg, Germany's second-largest city. We investigated one area with a modern app-based on-demand service, as well as one area with traditional dial-a-ride transport with telephone booking options. Similarities as well as differences in the perception of these different services were investigated.

The results show that the modern on-demand services are significantly better known and more frequently used than the traditional dial-a-ride transport services. At the same time, the same features of demand responsive transport (DRT) service are shown to be important to people, regardless of the type of DRT service offered. In particular, flexible booking options should be highlighted, which modern demand responsive transport is better able to fulfil.

Keywords: Mobility-on-Demand · Public Transport · Demand-responsive transport · household survey · Ridepooling · Ridesourcing · On-demand public transport · shared mobility · user analysis · ride-sharing

1 Introduction

The first research projects on demand-responsive transport (DRT) in a public transportation context started in Germany at the end of the 1970s. In 1977, the "Rufbus Friedrichshafen" was launched, and in 1978 a second research project, the "computer-controlled taxi-bus system", started in Wunstorf near Hannover [1]. In the 1990s and 2000s, more and more dial-a-ride transport (DART) systems were established in Germany as a supplement to, and in some cases as a replacement for, classic, scheduled public transport

© The Author(s) 2025
C. McNally et al. (Eds.): TRAconference 2024, LNMOB, pp. 236–251, 2025.
https://doi.org/10.1007/978-3-031-85578-8_31

services [2]. In 2018, the first modern, app-based DRT service integrated into the public transport system, the Wittlich-Shuttle and the ioki-Hamburg service, were launched [3, 4]. Since then, more than 80 additional modern DRT services have gone into operation in Germany as a result of funding programmes [5].

Traditional DRT services, such as DART services and classic Taxi services, and the modern DRT services, such as on-demand public transport (ODPT) services, on-demand ridepooling (ODR) services and Transportation Network Companies (TNCs), do not represent uniform product offerings. First of all, the services differ in terms of their funding. Public DRT services are financed by taxpayers as part of the public transport system, while commercial DRT services are financed by fare revenue and are intended to make a profit as well as taxi services. Furthermore, there are five central characteristics that can be used to fundamentally distinguish the various offers in the area of DRT services, which are shown in Table 1.

1. The most obvious distinction between the DART and the modern DRT services is the *method of booking*. In the case of the DART services, the booking is made via a telephone order, although in individual cases a booking via the internet may also be possible [2]. In contrast, the booking of modern DRT services is generally internet-based via an app, although some modern DRT services still allow ordering by phone [6].
2. Another way in which traditional and modern DRT services differ is the *mode of operation*. The DART services are usually timetable-based and use fixed routes, like a bus. In contrast, modern DRT services are timetable-free, area-based operations such as taxi services.
3. Traditional and modern DRT services also differ in terms of *ordering times*. The DART services in Germany usually have to be ordered 30 - 60 min before the scheduled departure time, sometimes even longer in advance [2]. The modern DRT services, on the other hand, offer the trips on demand without a specific pre-order time, depending on vehicle availability, but without the possibility of stopping the vehicle at the roadside, as is usual with taxis [6].
4. There are also differences between the DRT services in the way the services are operated *(Operation of the service)*. While the taxi services are of course operated by a taxi company, the DART services in Germany are usually also operated by a taxi company [2]. ODPT services, on the other hand, are mostly operated by local transportation companies [6, 7]. Similarly, ODR services operate its own fleet of vehicles with drivers [8]. TNCs, on the other hand, contract private drivers with their private vehicles [9].
5. The fifth characteristic that can be identified is the difference in the *fare system* between the services offered as public DRT and those offered as commercial DRT. The commercial DRT services generally have their own fare system, just as taxi services have their own fare system, whereas the public DRT services are usually integrated into the public transport fare system (sometimes with separate surcharges).

In this way, various subcategories can be defined under the term DRT, which are shown in Table 1. For the purpose of completeness, the taxi is also listed as a subcategory under the category of traditional DRT in Table 1.

Table 1. Definition of subcategories of demand responsive transport

	Demand responsive transport (DRT)				
	Traditional DRT		Modern DRT		
	Taxi	Public DRT	Commercial DRT		
		Dial-a-ride transport (DART)	On-Demand public transport (ODPT)	On-Demand ridepooling (ODR)	Transportation Network Companies (TNCs)
Method of booking	by telephone / by app / by stopping at the roadside	by telephone	by app	by app	by app
Mode of operation	timetable-free and area-based	timetable-based and line-oriented	timetable-free and area-based	timetable-free and area-based	timetable-free and area-based
Ordering times	no minimum pre-order time	30 min and more before scheduled time	no minimum pre-order time	no minimum pre-order time	no minimum pre-order time
Operation of the service	Taxi company	Taxi company	Transportation company	Transportation company	Private driver with private car
Fare system	Taxi fare system	Integrated into the public transport fare system or a separate fare system	Integrated into the public transport fare system or a separate fare system	Own fare system	Own fare system
Example		Anruf-Sammel-Taxi Stelle	hvv hop, elbMOBIL	MOIA	Uber, Lyft

There are some relevant publications on traditional DRT services, such as Mehlert [10] and Zietz et al. [2]. These publications mainly focus on operational, administrative, licensing and financial aspects of DART services. Mehlert [10] examines 13 different DART services on the basis of the service area, development, operation, key figures as well as problems and success or failure factors. Operational key figures such as the number of vehicles, the annual passenger volume, the annual vehicle performance and the occupancy rate of the vehicles are studied as well as key figures on economic efficiency such as costs, revenues, the cost recovery ratio and the deficit per passenger. Only in the case of the AnrufBus Leer, examined by Mehlert [10], statements are also made about the structure of users and the purposes of use: In 1998, 57% of the users were female, two thirds were younger than 20 years and 59% were younger than 14 years. The main purpose of travel was education (59%), while traditional commuting was under-represented (7%). The remaining third of the travel purposes was distributed between shopping (23%) and leisure (11%). In addition, it is noted that the share of passengers who combine journeys of the AnrufBus with journeys of the public transport system is only 5%. Furthermore, Mehlert [10] reports that 60% of the journeys take place within the own municipality, another 21% are journeys between the municipalities and the remaining 19% are journeys to nearby central cities [10, pp. 66–67].

In the more recent "Inventory analysis of flexible services and community buses in the Hamburg metropolitan region" by Zietz et al. [2], operational, tariff, licensing and infrastructural aspects are also the focus of the investigation. Only the absolute, annual passenger numbers, the timetable kilometres and the occupied kilometres of the services are examined. The number of passengers per year was available for just 27 of the 57 DRT services examined in the Hamburg metropolitan region. For the other two key figures, timetable kilometres and occupied kilometres, figures were only available for 7 and 11 services respectively [2]. This means that there is a lack of data about the users.

With regard to modern DRT services, research so far, both in German-speaking countries and in the international context, has focused mainly on commercial DRT services that are not integrated into the local public transport system [11–13]. Both Kostorz et al. [12] and Knie et al. [13] present initial findings on the user structures, but also on the journey purposes and the journey start times of the on-demand ridepooling services studied. In addition, Kostorz et al. [12] compare the structure, patterns and attitudes of users and non-users. This provides a deeper understanding of the user groups and non-user groups of a modern DRT service.

Findings on the users of ODPT services integrated into the local public transport system in urban areas are presented by Diebold et al. [14]. In addition, Diebold et al. [14] examine the effects of the introduction of a surcharge on top of a normal public transport ticket for the use of the ODPT service. Significant shifts in the age structure were observed, as well as significant effects on the choice of transport mode of the users due to the introduction of the surcharge. The reasons for the use of the service by the users are also presented, but no statements are made about non-users.

In consequence, there is a general research gap regarding the user groups and their usage patterns of DRT services, as well as regarding the comparison of traditional and modern DRT services. In addition, there is a lack of knowledge on DRT services in suburban and rural areas.

2   Methods

In the Hamburg metropolitan region, new DRT services were established at the end of 2020 in the county of Harburg, both west and east of the county seat Winsen/Luhe. To the west, in the municipality of Stelle, a DART service was introduced (see Fig. 1) and to the east, in the municipalities of Drage, Marschacht and Tespe, a modern ODPT service was introduced (see Fig. 1).

In both service areas, there are only bus routes in the north along the river to Winsen/Luhe. With the DART service, an inner-local public transport service was created for the first time and a connection to the railway station in Stelle was made possible. Similarly, the ODPT service also created an inner-local public transport service and also provided a connection to the railway station in Winsen(Luhe) via a satellite stop outside the main service area.

All four municipalities (Stelle, Drage, Marschacht and Tespe) can be assigned to the same spatial type "urban region - medium-sized city, urban area" according to the

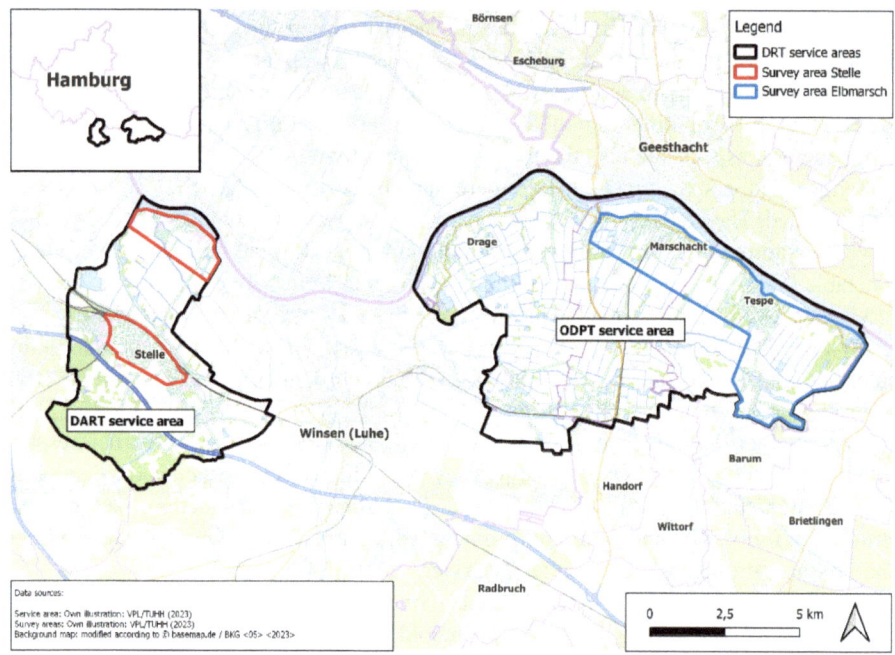

Fig. 1. Overview map of the DRT service areas and survey areas

spatial typology RegoiStaR 7 from the Federal Institute for Research on Building, Urban Affairs and Spatial Development [15]. In addition, it should be mentioned that a DART service existed in the ODPT service area until 2018, which was discontinued because no taxi company was found to continue operation. Table 2 shows the five characteristics presented in the introduction for the two new services:

The aim of this study is to obtain conclusions on users and non-users on the basis of the entire population by means of a representative household survey.

Accordingly, we aimed for a complete survey in the study areas. However, a complete survey of the entire service areas would have exceeded our capacities, which is why we decided to conduct a complete survey in representative sub-areas (see Fig. 1, survey area Stelle and Elbmarsch). In each of these sub-areas, complete surveys were carried out and every household in the survey area was asked to participate in the household survey by means of a postcard. These survey invitation postcards included an individual access code so that double participation could be identified and excluded from the survey. This also prevented the survey link from being distributed to relevant interest groups or outside the study area. It was requested that just one household member aged 14 or older participates in the survey.

Participation in the survey was possible both online and by post, with a written questionnaire. For this purpose, the distributed postcard had to be returned to us with a postal address. In addition, two filter questions had to be answered on the postcard. On this basis, we sent out the written questionnaire. Postage was paid for both the postcard

Table 2. Characterisation of the examined DRT services

	DART service [16]	ODPT service [6]
Method of booking	by telephone	by app / by telephone
Mode of operation	Timetable-based and line-oriented: Two corridor lines running from Monday to Saturday with five departures per day between 8 am and 7 pm	Timetable-free and area-based: Operation hours: Monday - Thursday: 5 am - 11 pm Friday: 5 am - 2 am Saturday: 8 am - 2 am Sunday: 8 am - 11 pm
Ordering times	60 min before scheduled time	no minimum pre-order time
Operation of the service	Taxi operator	Transportation company
Fare	Separate fare system: €3 for a journey within the municipality €5 for a journey to the county town Children under 14 years pay €1 less per journey	Integrated into the public transport fare system: Valid public transport ticket plus €1 per journey

and the return of the completed written questionnaire, so that no additional barriers were created. No incentive was offered for participating in the survey.

In the Elbmarsch study area, a total of about 2700 households were asked to participate in the survey. After data cleaning, 365 questionnaires could be used for the analysis in this study area. Of these, 78% of the questionnaires were answered online and 22% as a written questionnaire. The overall response rate was 13.5%. With a confidence level of 95%, the margin of error for this study area is 4,77%.

In the Stelle study area, a total of about 2900 households were asked to participate in the survey. After data cleaning, 393 questionnaires could be used for the analysis in this study area. Of these, 86% of the questionnaires were answered online and 14% as a written questionnaire. The overall response rate was 12.6%. With a confidence level of 95%, the margin of error for this study area is 4,60%.

Table 3 shows the main socio-demographic structural data of the two samples. Official comparative data is available for gender and age groups [17]. Overall, it can be seen that both samples are comparable with each other and also match the official data, which is why the values lie within the expected margin of error of 5%. No weighting of individual structural characteristics was carried out, as this study is intended to investigate the general influence of key structural characteristics on the use of DRT services.

Table 3. Socio-demographic structure of the samples Legend: Official numbers by [17] Sozio-economic status according to [18, pp. 17–18]

Gender	Elbmarsch/ODPT				Stelle/DART			
	Sample		official numbers		Sample		official numbers	
	absolute	relative	absolute	relative	absolute	relative	absolute	relative
male	163	45%	4469	51%	211	54%	5821	51%
female	196	54%	4368	49%	176	45%	5702	49%
Age	absolute	relative	absolute	relative	absolute	relative	absolute	relative
14–19	22	6%	393	5%	4	1%	460	5%
20–29	20	6%	713	9%	24	7%	1147	12%
30–39	32	9%	1056	14%	52	14%	1372	14%
40–49	44	13%	1065	14%	43	12%	1364	14%
50–59	69	20%	1666	22%	85	23%	2012	20%
60–69	69	20%	1392	18%	68	19%	1598	16%
70–79	55	16%	832	11%	51	14%	978	10%
80 or older	31	9%	571	7%	39	11%	944	10%
Child(ren) under 14 years in the household	absolute	relative			absolute	relative		
yes	56	15%			85	22%		
no	309	85%			308	78%		
Household size	absolute	relative			absolute	relative		
one person	70	20%			62	16%		
two person	164	46%			177	47%		
three persons or more	123	34%			140	37%		
Main activity	absolute	relative			absolute	relative		
working full time	130	36%			148	39%		
working part-time	47	13%			66	17%		
retired, pensioner	138	39%			141	37%		
housewife/ househusband	12	3%			10	3%		
pupil	19	5%			3	1%		
student	1	0%			8	2%		
unemployed	2	1%			5	1%		

(continued)

Table 3. (*continued*)

Gender	Elbmarsch/ODPT				Stelle/DART			
	Sample		official numbers		Sample		official numbers	
	absolute	relative	absolute	relative	absolute	relative	absolute	relative
trainiee	5	1%			0	0%		
other	4	1%			2	1%		
Socio-economic status	absolute	relative			absolute	relative		
low	33	13%			26	9%		
medium	76	29%			72	25%		
high	123	47%			153	53%		
very high	30	11%			40	14%		
Mobility restriction	absolute	relative			absolute	relative		
yes	76	22%			70	19%		
no	275	78%			300	81%		

3 Results

The results of this study are presented in two sections. First, the results from the two study areas are compared, followed by findings on users and non-users.

The first, most obvious question that arises when comparing the two new DRT services is the perception and use of these services. Here, very significant differences are immediately apparent (X^2: $p < 0.0001$, n = 758), as can be seen in Fig. 2. The ODPT service is known to almost all residents (95% know the service), while the DART service is known to only two thirds of the residents. A more detailed analysis shows that 43% (Elbmarsch/ODPT) and 44% (Stelle/DART) of the residents are aware of the DRT service stops. Therefore, there is no difference in the perception of the services. In contrast, the ODPT service's own vehicles (compare Fig. 3) were already perceived by 91% of the respondents. As the vehicles used for the DART service are conventional taxi vehicles, a comparable survey was not possible here. 53% of the respondents had heard about the ODPT service in other ways, such as from the newspaper or through family and friends. Lastly, 41% of the respondents heard about the DART service by other ways.

This significant difference in perception is also reflected in the use patterns of the two services. The ODPT service was already used by 30% of the residents, while the DART service was only used by 4% of the residents. This also shows that the ODPT service is not only generally used more frequently, but also more regularly by the residents. Thus, 63% of the ODPT users use it at least once a month respectively 23% at least weekly or more often, while the DART service is only used by 40% of the users at least once a month or more often.

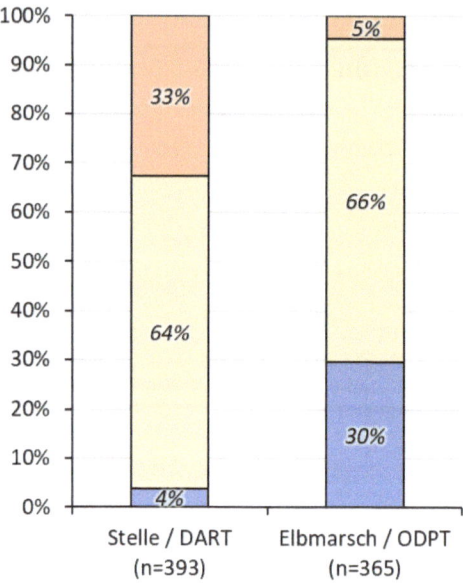

Fig. 2. Perception and usage of DRT services

Fig. 3. Vehicle for the ODPT service with brand name; Source: Tyll Diebold

This central result is also confirmed by a comparison with the use of the former DART service and the use of the current ODPT service in the same area. Only 23% of the ODPT users had also had experience of using the discontinued DART service. Therefore, 77% of the ODPT users were newly won for the ODPT service. Of all respondents who already had experience of using the former DART service, 80% also use the new ODPT service.

In the evaluation of six given characteristics of a DRT service, people rate them essentially the same regardless of the area, as shown in Fig. 4. For the rating, people were asked to rank the six characteristics from most important to least important. The most important characteristic of a DRT service is the "possibility of spontaneous booking". The characteristics " journeys if possible from or to the doorstep" and "operation also in the evening / night hours" are rated as the second and third most important characteristics in both study areas. The remaining three features are ranked in the same order of importance in both study areas, from more to less important as follows: " Link to public transport tickets", "Possibility of booking by telephone" and "Possibility of long advance booking".

In the context of the most important rated characteristic of DRT services, the possibility of spontaneous booking, it makes sense to consider the perception of bus headways with regard to one's own flexibility. For this purpose, respondents were asked whether they perceive a bus headway as a restriction to their personal flexibility or not. Regardless of the study area, this assessment changed for headways of between 15 and 30 min. Up to a 10-min interval, almost all respondents do not feel that their flexibility is restricted. At 20-min intervals, about 40% of the persons feel that their flexibility is restricted. At 30-min intervals, around 60% of persons feel that their flexibility is restricted. From a 40-min interval onwards, 80 to 90% of persons feel restricted in their own flexibility, compare Fig. 5. From this, it can be concluded for the possibility of a spontaneous booking that a journey within the next 10 min is accepted by almost everyone as a spontaneous booking. A journey within 15 to 30 min is still accepted, at least for the most

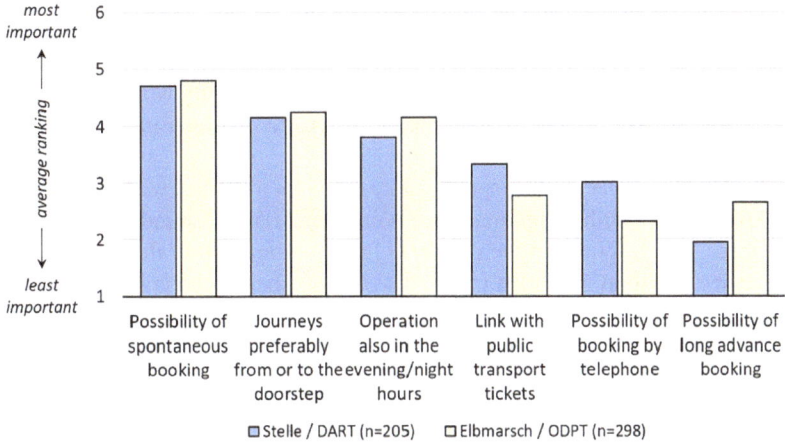

Fig. 4. Ranking of the characteristics of a DRT service

participants. On the other hand, a ride option in more than 30 min is no longer accepted as a flexible booking and can therefore not be regarded as a spontaneous booking option. Considering that the DART service studied requires a minimum pre-ordering time of 60 min before the scheduled departure time, this is likely one of the reasons for the limited usage of the DART service.

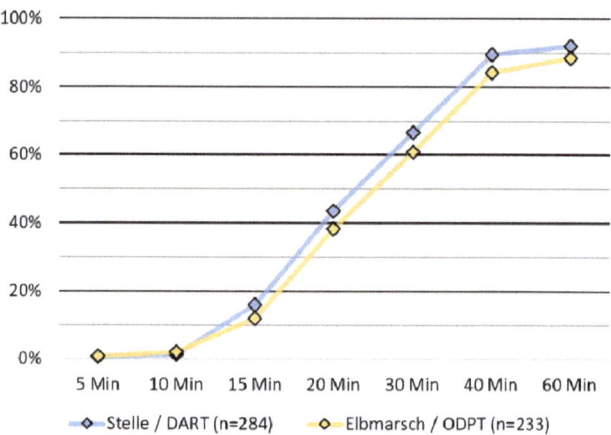

Fig. 5. Perception of a bus headway as a restriction of personal flexibility Legend: The Fig. 5 shows the percentage of persons who perceive that the bus headway restricts their personal flexibility.

Another aspect is the willingness to pay for the DRT services. A direct comparison is not possible due to the different fare systems, see also Table 2. For example, there is a separate fare for the DART service. A journey within the municipality costs €3.00 and a journey to the county seat Winsen/Luhe costs €5.00. These fares have to be paid by each passenger, regardless of whether the passenger has a public transport ticket or not. With the ODPT service, a valid public transport ticket is mandatory for the journey, and an additional surcharge of €1.00 per journey must be paid for the use of the ODPT. In the following, we look at the acceptance of the fare respectively the surcharge to the public transport ticket for a one-way trip within the municipality with the DRT service. To investigate this acceptance, we used the Price Sensitivity Meter method according to van Westendorp [19].

For the DART service, the acceptable price range for a journey within the municipality is €1.00 - €3.50, with an optimal price point of €2.05 and an indifferent price point of €2.10, see Fig. 6. With a price of €3.00 for a journey within the municipality, the DART service is within this price range, but only just below the upper price limit. For the ODPT service, the acceptable price range for the surcharge on the public transport ticket is from €0.55 to €3.00, with an optimal price point of €1.10 and an indifferent price point of €1.50, see Fig. 7. With a surcharge of €1.00 to the public transport ticket, the surcharge is within the price range and still below the optimal price point and the indifferent price point.

Overall, despite the different pricing systems, it can be said that the level of the surcharge for the ODPT service finds much greater acceptance than the price for the DART service. For the price of the DART service, which is just below the identified upper price limit, it can be concluded that the price dampens demand. For the level of the ODPT surcharge, there is again room for an upward price adjustment without expecting a decline in demand.

Another indicator of satisfaction with the DRT is the question about the continuation of the DRT service. Here again, there is a significant difference between the two DRT services. While there is a very clear support for the continuation of the ODPT service with 79%, the support for the DART service is significantly lower with 31% (X^2: p < 0.0001, n = 601), compare Fig. 8.

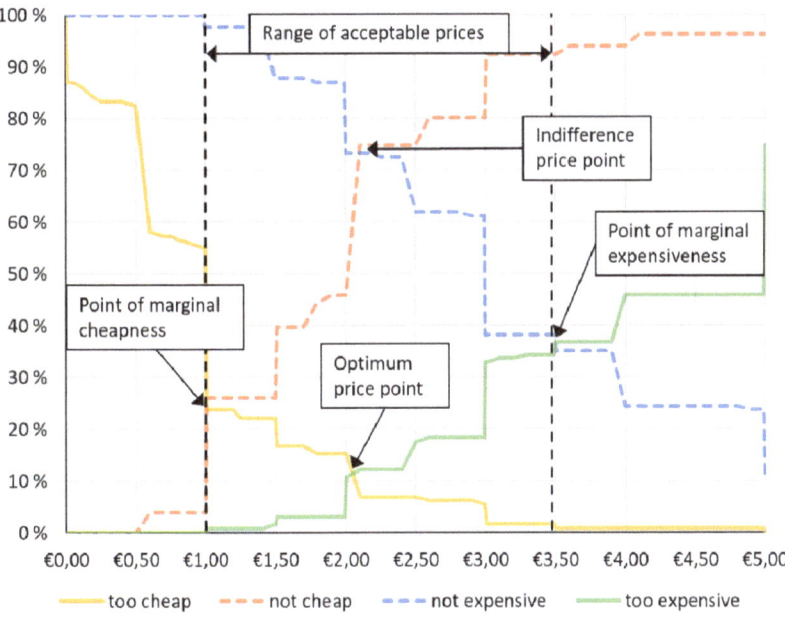

Fig. 6. Price Sensitivity Meter for the DART Service (n = 131)

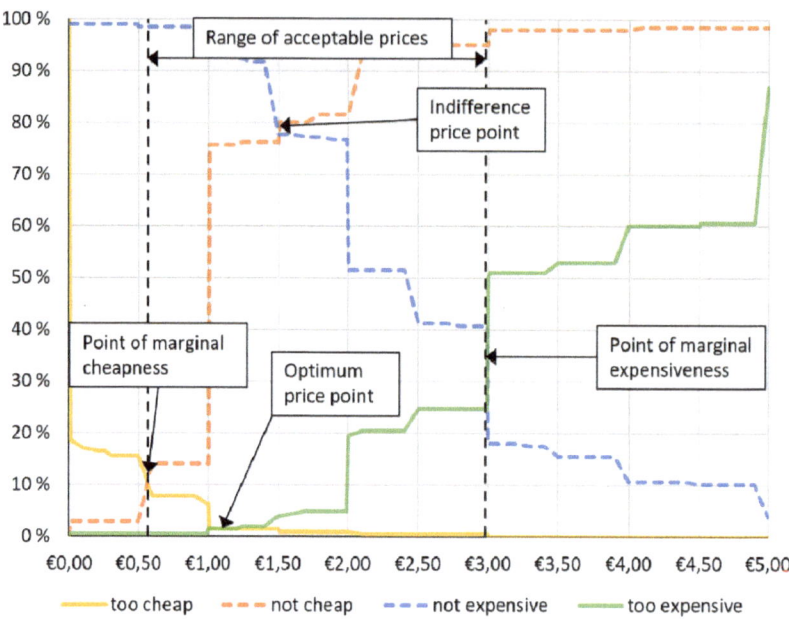

Fig. 7. Price Sensitivity Meter for the ODPT Service (n = 206)

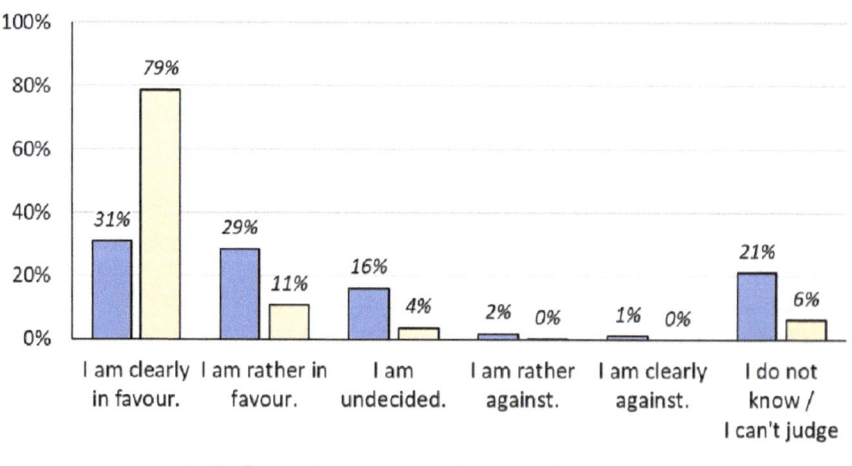

Fig. 8. Consent to the continuation of the DRT service

4 Discussion

The presented results of the representative household survey were able to show basic findings on DRT services within the local population. Previous studies such as Diebold et al. [14] or Knie et al. [13] or Kostorz et al. [12] focused on surveys of service users. Only the study by Kostorz et al. [12] additionally recruited non-users for the study via social media and an online panel. In contrast, this study made it possible to classify the results at the level of the local population on the basis of a representative household survey. In addition, this study investigated DRT services in suburban and rural areas, whereas the studies by Kostorz et al., Knie et al. and Diebold et al. [12–14] investigated DRT services in urban areas.

This study was able to gain insights into the perception of DRT services within the local population in a suburban area and to compare two different DRT services in their diffusion and use. Furthermore, findings on the evaluation of important characteristics of DRT services as well as on the price acceptance for journeys with DRT services could be obtained. In addition, there are the results on the perceived flexibility of a given bus interval and, as a result, the time within which people perceive a spontaneous booking to be attractive. These findings are probably the reason for the much greater popularity and use of the modern ODPT service compared to the traditional DART service. These findings should be fleshed out in further studies and reviewed with regard to transferability to other spatial structures, as well as generally validated.

Furthermore, there are also differences in frequency of use of modern DRT services. While Kostorz et al. [12] are reporting that only 9% of users use the ODR service at least weekly or more frequently, this study has shown that the ODPT service is used by 23% of the users at least weekly or more often. In this context, further research on the overlaps and differences of the user groups between the different modern DRT services and in comparison to other traditional DRT services would be useful. In addition, further studies should be carried out to investigate the effects of modern DRT services on the choice of transport mode, also depending on the trip purpose.

5 Conclusions

The results of the study have shown that a modern ODPT service achieves a significantly greater perception and greater usage within the population than a traditional DART service. In contrast, no difference can be found in the assessment of important features depending on the existing DRT service. On the contrary, people rate the importance of the characteristics of DRT services equally in both study areas. Furthermore, the acceptance of the prices for a journey with the DRT services could be determined. This indicates that the price of the surcharge for the ODPT service in addition to the public transport ticket corresponds very well to people's price expectations. In contrast, the price for a trip with the DART service is above people's price expectations and only just below the upper price limit determined in this paper. These findings are ultimately reflected in the very clear support for the continuation of the ODPT service, while support for the continuation of the DART service is noticeably more restrained.

Abbreviations

DART – Dial-a-ride transport
DRT   – Demand responsive transport
ODPT – On-demand public transport
ODR   – On-demand ridepooling
TNCs   – Transportation Network Companies

References

1. Mehlert, C., Schiefelbusch, M.: Der Rufbus Friedrichshafen: Lernen aus 40 Jahren flexiblem Nahverkehr (2018). https://www.nvbw.de/fileadmin/user_upload/aufgaben/planung_foerder ung/rufbusse/7_Festschrift_Rufbus_Friedrichshafen.pdf
2. Zietz, A., Mehlert, C., Brinkrolf, S., Michelmann, H., Märtens, W.: Bestands-analyse von felxiblen Angeboten und Bürgerbussen in der Metropolregion Hamburg (2016). https://metropolregion.hamburg.de/contentblob/5566666/ecf8fe326de5789933 d7009250030ae3/data/bestandsanalyse-flexible-bedienformen.pdf
3. ioki. Wittlich Shuttle: Erster digitaler Rufbus im ÖPNV - ioki. https://ioki.com/project/wit tlich-shuttle/. Accessed 26 Sep 2023
4. ioki. ioki Hamburg: ÖPNV integrierter On-Demand-Service. https://ioki.com/project/ioki-hamburg/. Accessed 26 Sept 2023
5. VDV. Hochlauf der On-Demand-Verkehre im ÖPNV. https://www.vdv.de/ondemandumfr age22.aspx. Accessed 26 Sept 2023
6. KVG Stade. KVG elbMOBIL. https://www.kvg-bus.de/fahrplaene/landkreis-harburg/elb mobil/. Accessed 26 Sept 2023
7. VHH. hvv hop - On-Demand Shuttle. https://vhh-mobility.de/hop/. Accessed 26 Sept 2023
8. MOIA. Dein Ridesharing Service für die Mobilität der Zukunft. https://www.moia.io/de-DE. Accessed 26 Sept 2023
9. State of California, Transportation Network Companies. https://www.cpuc.ca.gov/reg ulatory-services/licensing/transportation-licensing-and-analysis-branch/transportation-net work-companies. Accessed 26 Sept 2023
10. Mehlert, C.: Die Einführung des AnrufBus im ÖPNV: Praxiserfahrungen und Hand-lungsempfehlungen. Zugl.: Berlin, Techn. Univ., Diss. Bielefeld: Erich Schmidt (2001)
11. Henao, A., Marshall, W.E.: The impact of ride-hailing on vehicle miles traveled (in en). Transportation **46**(6), 2173–2194 (2019). https://doi.org/10.1007/s11116-018-9923-2
12. Kostorz, N., Fraedrich, E., Kagerbauer, M.: Usage and user characteristics—insights from MOIA, Europe's largest ridepooling service. Sustainability **13**(2), 958 (2021). https://doi.org/10.3390/su13020958
13. Knie, A., Ruhrort, L., Gödde, J., Pfaff, T.: Ride-Pooling-Dienste und ihre Bedeutung für den Verkehr. Nachfragemuster und Nutzungsmotive am Beispiel von "CleverShuttle" - eine Untersuchung auf Grundlage von Buchungsdaten und Kundenbefragungen in vier deutschen Städten," Berlin: Wissenschaftszentrum Berlin für Sozialforschung (WZB), WZB Discussion Paper SP III 2020–601 (2020). https://www.econstor.eu/handle/10419/220020
14. Diebold, T., Czarnetzki, F., Gertz, C.: On-Demand-Angebote als Bestandteil des ÖPNV: Nutzungsmuster und Auswirkungen auf die Verkehrsmittelentscheidung in einem Hamburger Stadtrandgebiet," (in de) (2021). https://doi.org/10.15480/882.3870
15. BMDV. BMDV - Regionalstatistische Raumtypologie (RegioStaR). https://bmdv.bund.de/ SharedDocs/DE/Artikel/G/regionalstatistische-raumtypologie.html. Accessed 26 Sept 2023
16. AST-Stelle. Anrufsammeltaxi der Gemeinde Stelle | AST-Stelle. https://www.ast-stelle.de/. Accessed 26 Sept 2023
17. Landesamt für Statistik Niedersachsen, LSN-Online. https://www1.nls.niedersachsen.de/sta tistik/html/default.asp. Accessed 26 Sept 2023
18. Nobis, C., Köhler, K.: Mobilität in Deutschland – MiD Nutzerhandbuch. BMVI, infas, DLR, IVT, infas 360. Bonn, Berlin (2018). https://www.mobilitaet-in-deutschland.de/archive/pdf/ MiD2017_Nutzerhandbuch.pdf
19. van Westendorp, P.: NSS-Price Sensitivity Meter (PSM): a new approach to study consumer perception of price. In: Proceedings of the 29th ESOMAR Congress (1976). Accessed 7 Sept 2023. https://archive.researchworld.com/a-new-approach-to-study-consumer-perception-of-price/

Open Access This chapter is licensed under the terms of the Creative Commons Attribution 4.0 International License (http://creativecommons.org/licenses/by/4.0/), which permits use, sharing, adaptation, distribution and reproduction in any medium or format, as long as you give appropriate credit to the original author(s) and the source, provide a link to the Creative Commons license and indicate if changes were made.

The images or other third party material in this chapter are included in the chapter's Creative Commons license, unless indicated otherwise in a credit line to the material. If material is not included in the chapter's Creative Commons license and your intended use is not permitted by statutory regulation or exceeds the permitted use, you will need to obtain permission directly from the copyright holder.

Accelerating Cross-Stakeholder Collaboration with Modern Traffic Management Systems and Strategies

Stavros Papadimitriou[1], Ioanna Pagoni[1], Adel Almohammad[2], Vasilis Gaitanidis[1], Nikolaos Efstathopoulos[1], Panos Georgakis[2], and Antulio Richetta[1] (✉)

[1] ARCADIS | IBI Group, 17 Andrea Papandreou Street, 15124 Maroussi, Athens, Greece
antulio.richetta@arcadis.com

[2] School of Architecture and Built Environment, University of Wolverhampton, Grimstone Street, Wolverhampton WV10 0JP, UK

Abstract. This paper introduces the Autonomous Network and Traffic Management Engine (ANTME). ANTME is an innovative network and traffic management solution – aimed at empowering corridor-level operators to create and implement collaborative response strategies for handling major planned and unplanned events. ANTME provides stakeholders with increased common situational awareness of corridor conditions and disruptions, seamless communication across entities for better response times, and efficient event response and management, based on actionable intelligence for complex decision-making. ANTME is built upon a scalable architecture that seamlessly integrates state-of-the-art technological components, striking a balance between various levels of automation and human interaction. The paper also presents the initial pilot round of the ANTME platform validation in Athens.

Keywords: Traffic Management Systems · Cross-stakeholder Collaboration · Operational Strategy

1 Introduction

In today's rapidly expanding societies, the demand for modern traffic management solutions is of outmost importance due to the complex challenges arising from increasing traffic congestion, unexpected climate change-related events, significant safety issues, disruptive technologies and evolving business models. To address those challenges, cities utilize a range of Traffic Management Systems (TMS), specifically crafted to mitigate traffic congestion and the related problems it brings [1, 2]. Technologies within TMS typically consist of a suite of application and management tools that incorporate communication, sensing, and processing technologies. These tools offer functionalities such as incident detection and response for various types of incidents [1].

Traditionally, TMS have primarily applied the deployment of centralized approaches [3], where a single central controller manages the entire traffic network. Nevertheless,

© The Author(s) 2025
C. McNally et al. (Eds.): TRAconference 2024, LNMOB, pp. 252–259, 2025.
https://doi.org/10.1007/978-3-031-85578-8_32

they often come with several constraints and disadvantages when addressing traffic in large-sized cities and complex networks [4]. Therefore, a shift towards the development of strategies with a decentralized logic has been observed [5]. Also, a significant portion of existing systems has explored a hybrid approach, where both centralized and decentralized controls co-exist in a hierarchical and interconnected manner to benefit from the advantages of each paradigm [6, 7].

In addition, recent surveys indicate that approximately 38% of network and traffic management stakeholders in Europe operate in silos, with only 31% collaborating with others often without formal agreements or an operational strategy [8]. This results in slow decision-making, duplicated efforts, inefficient allocation of resources, lack of accountability and responsibility, and ultimately, sub-optimal response plans and increased cost of operations. Despite current progress, limited research and market products focus on fostering multi-stakeholder collaboration in TMS, underscoring the need for technologically advanced solutions that prioritize stakeholder collaboration at the heart of development and operational efforts.

2 Autonomous Network and Traffic Management Engine

In response to the above, the Autonomous Network and Traffic Management Engine (ANTME) introduces an innovative multi-modal traffic management solution offering a blend of centralized and decentralized control elements. ANTME is designed to facilitate collaboration and interoperability among control centers dedicated to the management of diverse transportation services. The solution enhances decision-making during event management in multimodal transport corridors by providing actionable operational intelligence, delivering key advantages like enhanced common situational awareness of corridor conditions and disruptions; seamless communication across involved entities, leading to improved response times; and efficient incident response and management through collaborative, active, and proactive response generation plans.

2.1 Operational Concept

ANTME's operating environment and design principles focus on addressing the operational requirements of multi-modal transportation corridors, by harmoniously combining centralized automation with decentralized data collection and decision-making. The goal is to develop a collaborative response plan that optimally mitigates the impact of planned and unplanned events taking place in the corridor. ANTME's stakeholder map extends across various levels of the mobility ecosystem including infrastructure operators, transport operators, policymakers, commuters, logistics and technology providers.

The operational concept of ANTME (see Fig. 1) is based on the continuous monitoring of network conditions. The process starts by continuously detecting and gathering information about corridor events from various sources, both automated and non-automated. These events include planned or unplanned incidents like multi-casualty road accidents, metro station closures due to high threat levels, road spillages from trucks, and major sports events. Each of these can significantly impact other modes of transport within the corridor, requiring a proactive or reactive response from multiple stakeholders.

Once detected events are confirmed for validity, the response generation process begins. This process interacts with the simulation framework to identify an optimal response plan. The generated plan is then communicated to corridor operators, who decide whether to accept or reject it. If the operators accept the plan and its designated actions, they initiate their respective response mechanisms and execute them. However, if any operator rejects any actions in the plan, ANTME uses the collected feedback to customize or identify a new plan in a subsequent iteration.

Upon reaching an agreement, the response plan propagation phase begins by disseminating the response plan to the actors. Simultaneously, network conditions are monitored to assess plan effectiveness. If conditions improve, ANTME remains in a 'Do-nothing' mode until the incident ends. If the plan does not yield the expected improvements, ANTME triggers the re-generation of a new response plan within the event's lifetime. Throughout the event's lifecycle, ANTME's automations and actions specified for involved entities continuously interact under a common operational strategy [9].

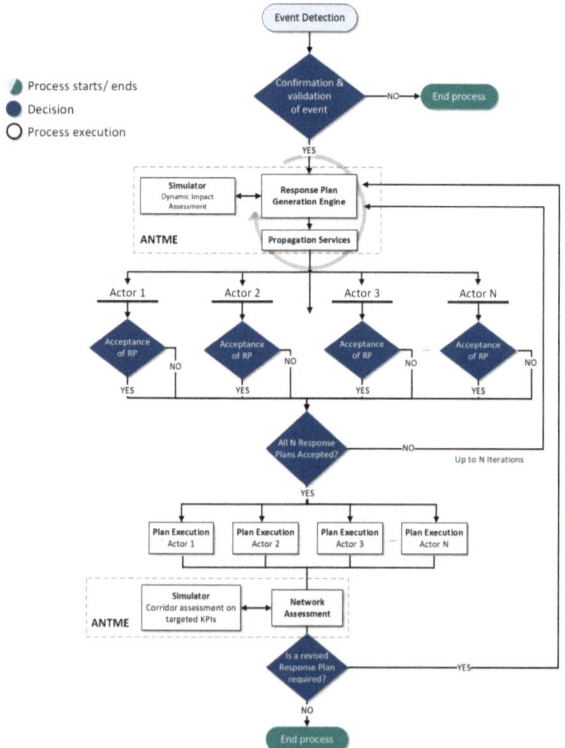

Fig. 1. Operational Concept Diagram.

2.2 Architecture Considerations

The degree of distribution of a system's architecture indicates its resilience, efficiency, and scalability [10, 11]. As previously discussed, many contemporary TMS adopt a hybrid architecture, which, by its nature, tends to favor a significant degree of centralization. Furthermore, this centralization is emphasized by the existence of a single command and control central system and the restricted independence granted to field controllers concerning operational modifications, alterations to response plans, and coordination, especially when their connection to the central system is disrupted for any reason. ANTME adheres to an architecture characterized by a substantial level of decentralization, a characteristic that becomes particularly evident due to its reliance on processing capacity situated within each corridor operator's domain. Moreover, on the operational front, the implementation of coordinated response plans, developed by ANTME, mandates unanimous consent from all corridor operators. This decentralized strategy reduces system overhead, boosts operational efficiency, and provides the overall system with increased scalability and resilience.

2.3 Functions

ANTME integrates various functions (see Fig. 2) to support multimodal traffic management services, from event detection to the monitoring of the executed response plans. The system facilitates continuous communication and interactions among stakeholders through its dedicated Graphical User Interface (GUI). ANTME's workflow manager controls the entire process, ensuring uninterrupted platform operation.

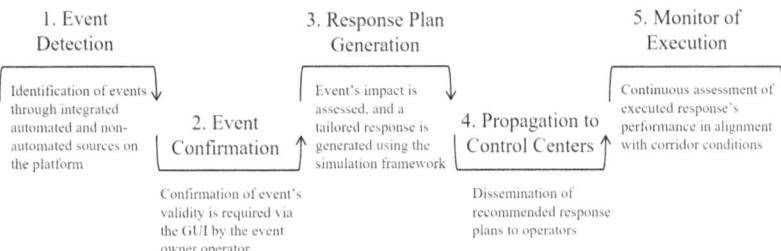

Fig. 2. ANTME Functions.

Event Detection. Event management begins with event detection (1) from various sources, both automatic (e.g. detection technologies, weather sensors) and non-automatic (e.g. reports from commuters, communication with external agencies), (2) ANTME's own detection means (automatic and/or non-automatic) or (3) reports from the mobile application used by corridor commuters. ANTME platform also enables users to expeditiously insert an event. This dynamic approach accelerates information's dissemination of corridor disruptions to ANTME and relevant stakeholders.

Event Confirmation. Once an event is added to the platform, the ANTME operator responsible for managing the lifecycle of the event (i.e., the event owner) verifies its accuracy. If additional information is required, a verification loop begins until the event is confirmed or rejected. Once confirmed, the workflow shifts to monitoring for acknowledgments from other actors via the GUI. This phase enhances ANTME operators' situational awareness of corridor disruptions and keeps them alert for potential related events, response actions, or communication needs.

Response Plan Generation. ANTME's Response Plans Generator component (see Fig. 3) begins its processes upon event confirmation via the GUI, receiving event details including incident timestamp, location, type, and severity. It starts with an identification process to check for missing data; if data is insufficient, it initiates a continuous request loop. If the necessary data cannot be obtained, ANTME terminates the event management process. Based on the event attributes, ANTME uses automation to generate a list of response plans based on a predefined library of templates. A scoring module assesses the impact of these plans on network performance via simulation. The generated and scored plans are sent to a recommendation services component, which ranks them and offers real-time recommendations for stakeholders, considering traffic measures and other factors such as environmental impact and passenger satisfaction.

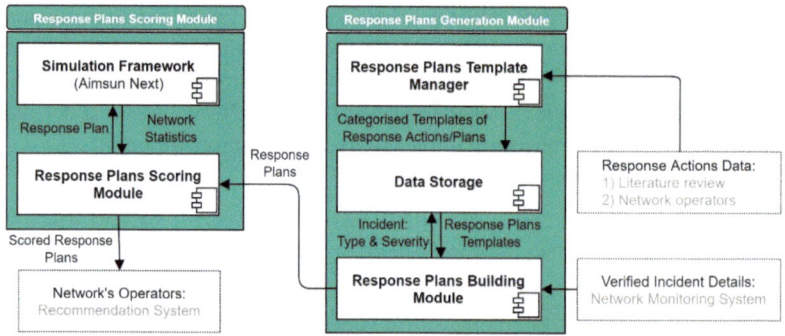

Fig. 3. ANTME's Response Plans Generator.

Propagation. Following the above, the recommended response plans are disseminated to operators' control centers via the GUI. Operators are prompted to select one of the highest-ranked plans. If all operators accept the plan, it proceeds to the execution phase. However, in the event of rejections, the response plan generator component along with an arbitration model is activated, initiating another cycle of response plan generation.

Monitor of Execution. Once the plan is accepted by operators, ANTME facilitates the execution management phase, where the corridor's state is assessed, and the effectiveness of the response plan is evaluated using a predefined set of Key Performance Indicators (KPIs). This process enables the platform to capture corridor-wide metrics, including average corridor delays, speed, journey time reliability, volume over capacity, and event clearance rate. ANTME generates a response plan evaluation record linked to the event, encompassing 3 evaluation states: (1) KPIs estimated by the response plan,

(2) estimated KPIs for normal corridor conditions, and (3) current corridor KPIs. Subsequently, ANTME requests a ranking from operators to assess the effectiveness of the implemented response plan and the event management process is concluded.

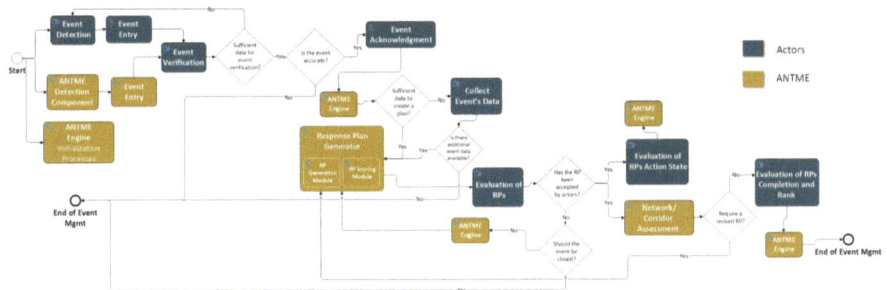

Fig. 4. Workflow Main Interactions between ANTME and Actors.

ANTME's Workflow Manager (WM) controls the entire process as a central hub, seamlessly integrating technology components to create a secure, transparent, collaborative, and scalable system from event detection to archival. The WM enables the system to remain informed of the process state and executes actions based on incoming requests and environmental changes. Figure 4 simplifies the workflow phases, highlighting key interactions between ANTME's automations and the actions for involved actors [12].

3   Pilot Validation – Athens, Greece

The main challenge in Athens' multimodal network is its siloed operations, lacking technical and organizational interfaces. This leads to minimal data sharing, poor communication, limited visibility of corridor disruptions, and weak response planning. As a result, travelers receive fragmented information from various centers and systems. ANTME addressed these issues through a live pilot on a segment of the Attiki Odos Motorway in Athens, chosen for its diverse transport modes (cars, buses, metro/rail). Three stakeholders—Attiki Odos, OASA, and Elliniko Metro—participated, testing two unplanned incidents: (1) a metro station closure due to a threat, and (2) a major motorway incident. Operators used ANTME in simulated role-play to test its functionalities guided by its workflow management process. In both cases, ANTME provided a response plan to restore network performance [13].

4   Conclusions

This paper introduces ANTME, a pioneering network and traffic management solution designed to facilitate collaboration among corridor-level operators and address major events by effectively combining automated central functions with decentralized data collection and decision-making. ANTME offers enhanced situational awareness, seamless

communication, and efficient event response, emphasizing cross-stakeholder cooperation throughout the entire event management lifecycle. The pilot validation in Athens
demonstrated its effectiveness in providing response plans for real-world event scenarios, addressing the growing demand for advanced solutions to prioritize collaboration
and coordination among stakeholders in the evolving field of traffic management.

Acknowledgements. This research is part of the FRONTIER project (Horizon 2020 grant No.
955317). It represents the authors' views and does not reflect the position of the European Union,
which is not responsible for any use of the information herein.

References

1. Djahel, S., Doolan, R., Muntean, G.-M., Murphy, J.: A communications-oriented perspective
   on traffic management systems for smart cities: challenges and innovative approaches. IEEE
   Commun. Surv. Tutor. **17**(1), 125–151 (2015)
2. De Souza, A.M., et al.: Improvement of traffic condition through an alerting and re-routing
   system. Comput. Netw. **110**, 118–132 (2016)
3. Papageorgiou, M., Diakaki, C., Dinopoulou, V., Kotsialos, A., Wang, Y.: Review of road
   traffic control strategies. Proc. IEEE **91**(12), 2043–2067 (2003)
4. Tettamanti, T., Varga, I.: Distributed traffic control system based on model predictive control.
   Periodica Polytechnica Civ. Eng. **54**(1), 3–9 (2010)
5. Manolis, D., Pappa, T., Diakaki, C., Papamichail, I., Papageorgiou, M.: Centralized versus
   decentralized signal control of large-scale urban road networks in real time: a simulation
   study. IET Intel. Transport Syst. **12**(8), 891–900 (2018)
6. Kouvelas, A., Aboudolas, K., Papageorgiou, M., Kosmatopoulos, E.: A hybrid strategy for
   real-time traffic signal control of urban road networks. IEEE Trans. Intell. Transp. Syst. **12**,
   884–894 (2011)
7. Faldu, P., Doshi, N., Patel, R.: Real time adaptive traffic control system: a hybrid approach. In:
   2019 IEEE 4th International Conference on Computer and Communication Systems (ICCCS),
   pp. 697–701. IEEE (2019)
8. FRONTIER EU Project. Deliverable 2.2: FRONTIER Requirements Specification -Initial
   version. Submitted to the European Commission on 31 January 2022 (2022)
9. FRONTIER EU Project: Deliverable 2.4: FRONTIER Use Cases and Conceptual Architecture. Submitted to the European Commission on 30 April 2021 (2021)
10. Gomides, T.S., Robson, E., de Souza, A.M., Souza, F.S., Villas, L.A., Guidoni, D.L.: An
    adaptive and distributed traffic management system using vehicular ad-hoc networks. Comput.
    Commun. **159**, 317–330 (2020)
11. Wuthishuwong, C., Traechtler, A.: Distributed control system architecture for balancing
    and stabilizing traffic in the network of multiple autonomous intersections using feedback
    consensus and route assignment method. Compl. Intell. Syst. **6**(1), 165–187 (2020)
12. FRONTIER EU Project. Deliverable 7.3: Integrated ANTME Platform – Initial Version.
    Submitted to the European Commission on 31 January 2023 (2023)
13. FRONTIER EU Project. Deliverable 8.3: Multimodal Mobility for Passengers and Freight
    and Cross-stakeholders Collaboration: 1st Iteration Results and Evaluation. Submitted to the
    European Commission on 31 July 2023 (2023)

Open Access This chapter is licensed under the terms of the Creative Commons Attribution 4.0 International License (http://creativecommons.org/licenses/by/4.0/), which permits use, sharing, adaptation, distribution and reproduction in any medium or format, as long as you give appropriate credit to the original author(s) and the source, provide a link to the Creative Commons license and indicate if changes were made.

The images or other third party material in this chapter are included in the chapter's Creative Commons license, unless indicated otherwise in a credit line to the material. If material is not included in the chapter's Creative Commons license and your intended use is not permitted by statutory regulation or exceeds the permitted use, you will need to obtain permission directly from the copyright holder.

Network-Wide Estimation of Average Daily Bicycle Traffic Based on Crowdsourced GPS Data and Permanent Counters

Emely Richter[✉] ⓘ, Joscha Raudszus ⓘ, Sven Lißner ⓘ, and Iwan Porojkow ⓘ

Chair of Transportation Ecology, Technische Universität Dresden, 01069 Dresden, Germany
{emely.richter,joscha.raudszus,sven.lissner,
iwan.porojkow}@tu-dresden.de

Abstract. This paper attempts to predict average daily bicycle volumes on a nationwide level using crowdsourced GPS data from the CITYCYCLING campaign in Germany. The data source was 514 permanent counting sites across the country as well as the campaign-generated GPS bicycle volumes of about 300,000 participants and 7.5 million tracks from a smartphone app. For model building, Gradient Boosting Regression and Support Vector Regression were selected. The results show a medium to high model fit for the prediction of bicycle volumes at sites with permanent counters. To illustrate this, the models are applied to the road network of a district of the city of Dresden, Germany.

Keywords: AADB · Crowdsourced Data · GPS Data

1 Introduction

In order to facilitate an environmentally friendly transport network, it is essential to provide safe and attractive cycling infrastructure [1, 2]. For planning and evaluating cycling measures and traffic safety assessment, knowledge of bicycle traffic volumes and/or cycling demand is required [3, 4]. Primary sources for bicycle volume data can be categorized into two groups. The first group consists of permanent automatic, temporary mobile, and manual bike counters. These are mainly located in strategic locations of high importance for the cycling network within cities. Although counters can provide counting data continuously, they lack spatial coverage. The second group includes GPS tracks of individual cycling trips, mainly recorded as part of smartphone-based cycling campaigns. Depending on the underlying campaign, the user group and the amount of tracked cycling trips in general, it represents a subset of the cycling population within a short time frame, but provides high spatial coverage. [5–8]

It is the intuitive next step to combine the advantages of both data sources to achieve a data set containing cycling traffic volumes representing the whole cycling population at a spatially high, i.e., network-wide, coverage. This holds especially true when considering an increasing spatial coverage of counters thus also including network sections outside the primary cycling network.

© The Author(s) 2025
C. McNally et al. (Eds.): TRAconference 2024, LNMOB, pp. 260–266, 2025.
https://doi.org/10.1007/978-3-031-85578-8_33

2   State of Literature

Previous studies, e.g., [5, 7], or [9], illustrate that it is generally possible to explain counter data with GPS tracks of a subpopulation of cyclists. However, as Bhowmick D, Saberi M, Stevenson M et al. (2023) report in their review of methods for estimating link-level cycling volumes, there are several limitations to existing studies [10]. Data sources were insufficiently reported or lacked model validation efforts, reducing the scientific value and transferability of the reported outcomes significantly [10]. Furthermore, most of the studies were conducted in the United States (representing 55 cities, e.g. [11, 12]), whereas other countries were represented by far fewer studies (representing 16 cities, e.g. [9]) [10].

The majority of up-to-date studies use linear or negative binomial regression models [9], Poisson regression [13]. These studies use, for example, the functional system of road segments, number of lanes or households with a certain income as predictors (e.g. [5]). The selection of predictor variables is questionable as it ignores the primary variables of the underlying data-gathering campaigns. As most of the existing studies focus on a single city, the number of participants in GPS data collection plays no role [5, 7–9]. This fact reduces the transferability to other study areas or other periods drastically.

3   Method

In this paper, we want to explore the possibilities of estimating network-wide average annual daily bicycle volumes (ADB) within several municipalities in Germany. The study is based on crowdsourced GPS data obtained by the nationwide CITYCYCLING (CC) campaign.

3.1   Data Types and Sources

In this study, two primary data sources were used. Data from 361 permanent automatic bicycle counters was obtained from the eco-counter API and complemented with other city-specific datasets from Germany. The counter data contains daily and weekly counts with a granularity of up to 15 minutes, latitude and longitude and in some cases whether it counts mono- or bi-directional. The gathered time span ranged from January 2019 to December 2022. Counting site locations are shown in Fig. 1. The federal state of Hessia represents a focal point as it established 275 permanent counting sites in 2021 and 2022.

Cycling volumes sourced by the nationwide CC campaign were used as the crowdsourced GPS-data. In year 2022, more than 2,400 cities partook within three weeks between the 1st of May and the 30th of September. Around 300,000 cyclists tracked about 7.5 million rides with their smartphones in 2022.

The modeling approach presented in this paper utilizes pre-processed campaign data from 2022, which was provided through the RiDE platform, and corresponding data of 514 permanent counters. Following previous studies, the campaign data is highly representative in terms of gender and age group representation [14].

Secondary data sources contain link attributes obtained from OSM. For application in this paper, the presence of bridges was considered. On the other hand, municipality-specific information was used, including the number and gender distribution of campaign participants, recorded trips within the municipality, their size classification after RegioStaR17 and population size and gender distribution. In the first step, we enriched the secondary data to the links close to the counters. The second step involves the entire network where we estimate the ADB. Both datasets, primary and secondary, were spatially joined using Python and QGIS.

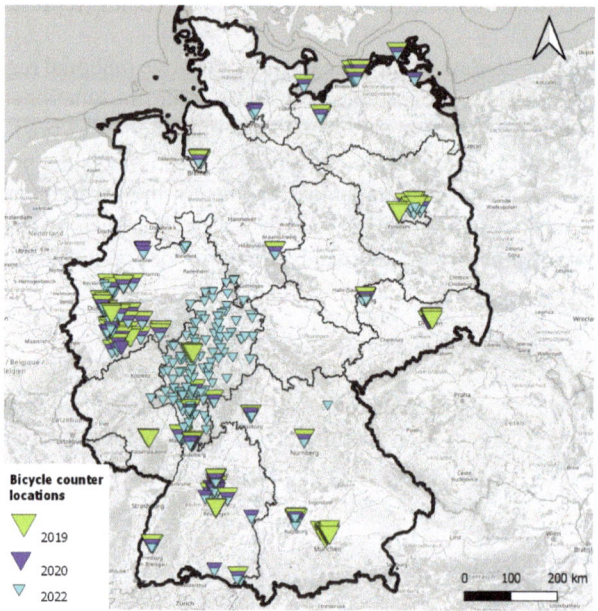

Fig. 1. Distribution of counting sites over different years. Map data from OpenStreetMap [15].

3.2   Model building and Analysis

Different approaches were used to model the ADB. In addition to the classical regression models (Linear/Poisson), Support Vector Regression (SVR), Multilayer Perceptron (MLP), and Gradient Boosting Regression (GBR) were carried out. SVR was chosen because it has excellent generalization capability and a high prediction accuracy. On the other hand, GBR can handle numerical and categorical data naturally and is robust to outliers. This is an essential feature as counting stations are highly susceptible to outliers with high cycling volumes if they are located on a bridge, for example. MLP can explain complex non-linear correlations.

To optimize model performance, the data was preprocessed. All numerical input variables were scaled using the Box-Cox transformation for linear regression, Poisson regression, SVR, and MLP. The categorical variables were encoded. A transformation of the target variable, i.e. the ADB, using a logarithmic and an exponential function ensured the output of non-negative estimates.

For model validation, a k-fold cross-validation was used, splitting the data set into k=10 subsets. This results in an almost unbiased error. R2, negative mean absolute percentage error (NMAPE), and computing time (CT) were analyzed to select the best model. Out-of-sample errors were used to prevent overfitting of the models.

4 Results

The following Table 1 presents all tested models' cross-validation errors and CT. Among them, SVR and GBR perform best for R2 and NMAPE, with the CT of GBR being comparatively high.

Table 1. Average cross-validation errors and associated computing times.

	Linear	Poisson	GBR	SVR	MLP
R2	0.451	0.563	0.715	0.726	0.672
NMAPE	−0.668	−0.909	−0.61	−0.581	−0.722
CT (sec)	0.39	2.37	115.02	5.14	764.27

GBR and SVR were applied to predict the ADB in a selected part within the city of Dresden. Only permanent counters from outside the city area were used to train the models. The estimated values were rounded to the nearest hundred. Fig. 2 shows the ADB predictions by the SVR (left) and the GBR model (right).

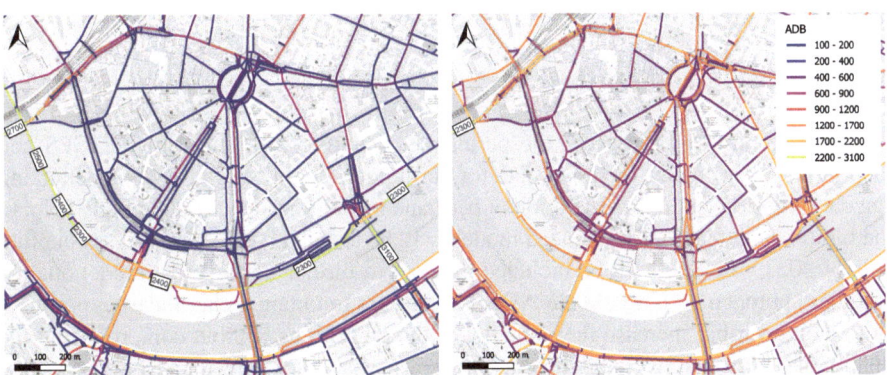

Fig. 2. Part of Dresden-Neustadt with model estimated ADB (SVR left, GBR right). Road sections with 2,200 or more estimated cyclists are labelled. Map data from OpenStreetMap [15].

The range of estimated values is larger for SVR than for GBR. No section is estimated by GBR with values below 400. For heavily frequented sections, the SVR model results are higher than with GBR. This is particularly evident for the major road sections in the primary network such as bridges, but also for the paths along the Elbe River, which hosts many cyclists. On the other hand, the values estimated by the SVR model on the secondary road network are lower than by the GBR model.

In the SVR model, low values are estimated more frequently, while higher values are becoming increasingly rare. The GBR estimates most of the road sections between 500 and 700 and around 1,300 cyclists. Significantly fewer cyclists are estimated between these maxima.

5 Discussion

Both models perform well, given that we predicted the ADB in Dresden using only permanent counting sites from outside the city. This highlights the potential to predict ADB in areas without permanent bicycle counters. The step that was most prone to potential errors was the assignment of GPS traffic volumes to the correct counting sites. Map matching plays an essential role in this step. Often, the OSM network is very detailed, visualizing bicycle infrastructure, car traffic lanes, and sidewalks as separate network links. The map-matching algorithm is susceptible to errors in correctly aligning GPS tracks to the utilized infrastructure, because of a lack of spatial accuracy. This complicates the correct alignment of permanent counting sites to GPS traffic volumes.

Another critical point concerns the exact positions of the permanent counting sites. The received coordinates often needed to be more accurate. Besides these facts, the information, if the counters count only one or both directions, could not be obtained automatically. This made a manual correction necessary.

Similar problems occurred for automatically assigning counting sites to the OSM network – especially at large junctions or bridges. The automatic assignment failed and needed to be corrected manually. Given these facts, the work with many permanent counting sites throughout Germany was challenging and prone to errors.

6 Conclusion and Outlook

Given more time to preprocess the permanent counting data and a higher GPS and map-matching accuracy, the input data for our models could be improved. A better and standardized description and location of permanent counting sites would greatly help. In the future, more and more advanced models will be tested, and GPS as well as counting data of 2023 and 2024 will be included as input to further improve the model quality. Another approach is to divide the ADB into classes to obtain a classification problem. This should further increase the accuracy of the algorithms without compromising the quality of the results. All in all, it could be shown, that combining data from permanent counters, GPS-tracks of the CC campaign, as well as some secondary data is a promising approach to the estimation of network-wide ADB.

Acknowledgement. The authors gratefully acknowledge the funding from the Federal Ministry of Transportation within the project 'MoveOn' (project number: VB2111A).

References

1. Buehler, R., Pucher, J.: Sustainable transport in Freiburg: Lessons from Germany's environmental capital. Int. J. Sustain. Transp. **5**(1), 43–70 (2011). https://doi.org/10.1080/155683110 03650531

2. Pucher, J., Dill, J., Handy, S.: Infrastructure, programs, and policies to increase bicycling: an international review. Prev. Med. (Baltim) **50**,no. SUPPL, S106–S125 (2010). https://doi.org/ 10.1016/j.ypmed.2009.07.028.

3. Fields, B., Cradock, A.L., Barrett, J.L., Hull, T., Melly, S.J.: Active transportation pilot program evaluation: A longitudinal assessment of bicycle facility density changes on use in Minneapolis. Transp. Res. Interdiscip. Perspect. **14**, June 2022. https://doi.org/10.1016/j.trip. 2022.100604

4. Umweltbundesamt, Evaluation zählt. Ein Anwendungshandbuch (2015). www.uba.de/evalua tion-verkehrsplanung

5. Dadashova, B., Griffin, G.P., Das, S., Turner, S., Sherman, B.: Estimation of Average Annual Daily Bicycle Counts using Crowdsourced Strava Data, September 2020. https://doi.org/10. 1177/0361198120946016.

6. Proulx, F.R., Pozdnukhov, A.: Bicycle Traffic Volume Estimation using Geo-graphically Weighted Data Fusion. J. Transp. Geogr., 1–14 (2017)

7. Hochmair, H.H., Bardin, E., Ahmouda, A.: Estimating bicycle trip volume for Miami-Dade county from Strava tracking data. J. Transp. Geogr. **75**(January), 58–69 (2019). https://doi. org/10.1016/j.jtrangeo.2019.01.013

8. Lin, Z., (David) Fan, W.: Modeling bicycle volume using crowdsourced data from Strava smartphone application. Int. J. Transp. Sci. Technol. (2020). https://doi.org/10.1016/j.ijtst. 2020.03.003.

9. Livingston, M., McArthur, D., Hong, J.: Predicting cycling volumes using crowdsourced activity data. Environ. Plann. B, 1–17 (2020). https://doi.org/10.1177/2399808320925822

10. Bhowmick, D., et al.: A systematic scoping review of methods for estimating link-level bicycling volumes. Transp. Rev. **43**(4), 622–651 (2023) https://doi.org/10.1080/01441647. 2022.2147240.

11. Boss, D., Nelson, T., Winters, M., Ferster, C.J.: Using crowdsourced data to monitor change in spatial patterns of bicycle ridership. J. Transp. Health **9**(February), 226–233 (2018). https:// doi.org/10.1016/j.jth.2018.02.008

12. Nelson, T., et al.: Generalized model for mapping bicycle ridership with crowdsourced data. Transp. Res. Part C **125**, 102981 (2020). https://doi.org/10.1016/j.trc.2021.102981.

13. Fagnant, D.J., Kockelman, K.: A direct-demand model for bicycle counts: the impacts of level of service and other factors. Environ. Plann. B Plann. Des. **43**(1), 93–107 (2016). https://doi. org/10.1177/0265813515602568

14. Lißner, S., Huber, S.: Influence of the cycling campaign CITY CYCLING on cycling behaviour in Germany. In: European Transport Conference (ETC) - Mila-no (2022). https:// doi.org/10.13140/RG.2.2.34516.48005.

15. OpenStreetMap. www.openstreetmap.org/copyright. Accessed 04 Apr 2024

Open Access This chapter is licensed under the terms of the Creative Commons Attribution 4.0 International License (http://creativecommons.org/licenses/by/4.0/), which permits use, sharing, adaptation, distribution and reproduction in any medium or format, as long as you give appropriate credit to the original author(s) and the source, provide a link to the Creative Commons license and indicate if changes were made.

The images or other third party material in this chapter are included in the chapter's Creative Commons license, unless indicated otherwise in a credit line to the material. If material is not included in the chapter's Creative Commons license and your intended use is not permitted by statutory regulation or exceeds the permitted use, you will need to obtain permission directly from the copyright holder.

GeTUP: An Internet of Things-Based Navigation System Architecture for High-Level Route Planning

Muhammad Tabish Bilal[1]([✉]) [iD] and Davide Giglio[2] [iD]

[1] Department of Mechanical, Energy, Management and Transport Engineering, Universita Degli Studi di Genova, 16145 Genova, Italy
muhammad.tabish.bilal@edu.unige.it, tabish.bilal@uet.edu.pk
[2] DIME and CEILI Department, Universita Degli Studi di Genova, 16145 Genova, Italy

Abstract. Modern drivers equipped with GPS-enabled devices digest traffic information and work as traffic information providers. Driving experiences and fuel consumption are also shared in novel systems to help users' route choices. Navigators in special terrain cities like Genova, Italy achieve sub-optimal results when they reroute the traffic greedily towards light-congested roads inducing new traffic jams. Furthermore, the absence of user preferences and advanced profiling curtail the competency of navigators to provide need-based responses to their users. This article aims to present a deliverable for Green MaaS for an adaptive urban planning project Liguria region in Italy to fill the gaps present in current navigators suggesting more optimal routing solutions considering the user requirements. The GeTUP application provides a specific pertinent and germane service equipped with an interface that is competent to acclimate to the characteristics of its users. The logical navigation algorithm is based on automated self-learning internet-of-things-based architecture that works on user choices to generate high-level route information. On receiving the request, the server executes the search algorithm based on the provided origin, destination and percentage of desired high-level information required based on the user profiling. The results are based on both the real-time traffic rhythm and near-future traffic distribution estimates based on the rerouting decisions. GeTUP application is implemented as a pilot MaaS solution for the city of Genova. The real-time data obtained upon evaluation revealed that the application contributes actively to coordinating the traffic based on user decisions and preferences between different alternatives and alleviates congestion situations.

Keywords: Route navigation · User preference · real-time routing · internet-of-things

1 Introduction

A skewed boom in the development and usage of mobile communication devices gave rise to the immense popularity of navigation applications leaving the map of the world at the palm of your hands. Google Trends estimated that 38% of smartphone users in

© The Author(s) 2025
C. McNally et al. (Eds.): TRAconference 2024, LNMOB, pp. 267–275, 2025.
https://doi.org/10.1007/978-3-031-85578-8_34

Italy got an advantage from famous navigation applications to reach their destinations in 2022. Whereas, this intense use of such applications sometimes creates bottleneck situations on road networks due to the nature of their backend navigation algorithm. These greedily converge the drivers to the fastest route to their destination based on previous traffic conditions causing congestion. Despite various efforts, the development of a cooperative path searching algorithm is still a need of an hour which should consider the situation of the network based on various usage rates of the application system. This article presents a first-of-its-kind navigation application which executes not only the data of traffic conditions, time-stamped location and route choice decisions but also the user preferences in information provision to give high-level path information between an OD. There have been many attempts in devising algorithms providing the solution to the problem of delivering high-level routing information; generally divided into three major categories. The first group of algorithms optimizes the user equilibrium giving the fastest route from an origin to a destination by computing the perceived journey times [1–3]. The second category delivers the objective of high-level information by aiming at optimal social benefits; not targeting a particular user but travel time cost for the entire user group of the service e.g. commuters using DRT service in a town [4–7]. Then comes the third class which uses some heuristics functions instead of perceived journey times to choose from the n-shortest paths based on the previously assigned traffic [8].

2   Internet of Things Enabled Navigator Aggregato

The tailor-made application of GeTUP relies on a powerful Navigator Aggregato (NA) algorithm. GetUP is meant to be installed on an Android cellular phone for the initial phase which has a GPS capability together with a communication module for connection with the AlgoWatt server. This module can be a DSRC. The server acts as the heart of the mechanism of dealing with user requests. It computes the route for an entered OD by the user based on a cooperative path computation algorithm enclosed in NA. The GeTUP application behaves as a multi-information system for the user by consuming the path computation results from the server. It reveals the high-level information on the links involved in the path as well as updates the route confirmation which consequently helps renew the network situation in real time to deal with subsequent OD navigation requests.

2.1   System Architecture

As illustrated in Fig. 1, the GeTUP application consists of four executable entities: the GeTUP application in mobile devices at the user end, the AlgoWatt server containing the NA algorithm hosting the application requests from users, the EXIS traffic analyser and the real-time data update API server.

The interaction between the application and server through an API is responsible for the smooth processing of the requests as well as the load balancing on the network links. Another interaction between the EXIS traffic analyser and the Algowatt server is done through a second API which estimates the segment travel times based on the demand being changed. This allows the algorithm to propose new routes in case of re-routing decision requests from the user. The NA computes the route candidate(s) links based on real-time data of predicted travel time on links and anticipated traffic load.

The information that the NA computation relies on is stored in two databases. Firstly on the internal embedded road network graph for the city of Genoa and secondly on the latest ETA of each road link; it is updated by the EXIS traffic analyser.

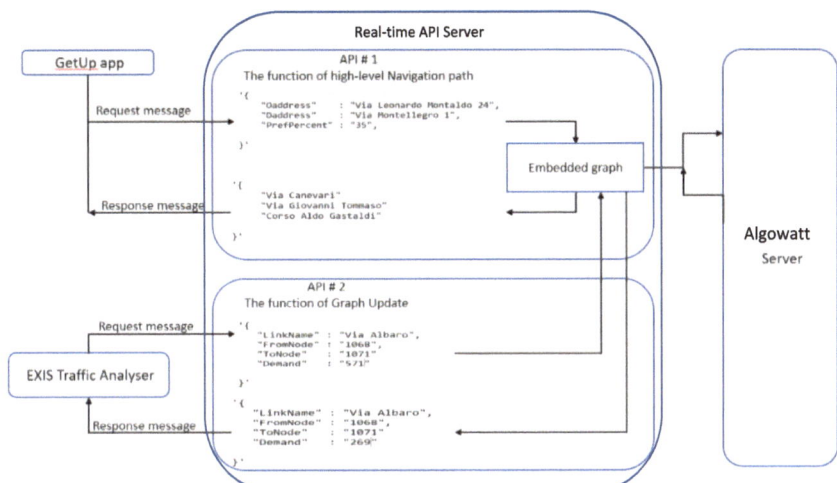

Fig. 1. GeTUP Service architecture

Navigation Process. As soon as the user places a routing call, the GeTUP application forwards it to the API server that calls the NA stored at the Algowatt server providing it with the OD coordinates/address provided by the user based on the embedded network graph. Apart from the origin and destination user also sets the percentage of preferred information he wants. This request is taken as input in the form of a JSON file and passed to the navigator algorithm. The algorithm checks the correctness of the information and passes the values to the city's already embedded transportation network graph. Snapping to the nearest possible starting point as of the origin and nearest possible ending point as of the destination via a map matching process that will be explained later. These points are passed to the second part of the algorithm which evaluates the more than one shortest paths that can be traversed depending upon the real-time situation on the network. Once calculated, the shortest path comprised of all the network links is passed to the third part of the algorithm which skims out the most important links based on the importance index of each link involved in the path as well as the user preference as mentioned earlier. This way a high-level search outcome is passed to the server which in return shows the output to the user. The routing results are returned to the GeTUP application via the API server. A confirmation of the selected route is meanwhile sent back to the database in the API server to update the situation at the links for use by the EXIS traffic analyzer.

Flow and travel time aggregation. During the map-matching process, the addresses are converted into WGS coordinates to match with the nodal points on the embedded graph map. Various points are matched with the data to reveal various possible routes based on the nearest start-end coordinates to an OD. GeTUP map matching is based on importance factor heuristics, which considers both the distance to nearby links, the snapped nodes and the context of each request. For example, it can be a possibility that a certain origin or destination is closer to one link but instead chooses another based on the importance index criteria ($ImpI_p$) as in (1). In GeTUP, the context information used to compute the transition probability between available points is the length of the candidate links, flow on them and real-time traffic conditions. The NA refines the selection of certain paths among various possibilities considering the above-mentioned factors in terms of a weight given to each link besides the length of that link.

$$ImpI_p = \sum_{l \in P} \omega_l * \pounds_l \tag{1}$$

where ω_l is the weight importance weight of the link l which belongs to the path p from the path set P for a particular OD.

Fig. 2. Polyline path from origin to destination developed in PTV Visum

Routing Algorithm. Bellman-Ford routing algorithm variation is implemented in the NA for route selection. The algorithm follows a dynamic programming approach calculating the shortest path in a bottom-up approach. The algorithm was designed and tested in a high-end sophisticated programming language (MATLAB). The program in the server acts upon the requests entered by the user considering the start and end point and calculating the shortest paths with at least one edge in the path. Then it steps up and goes for the path that is shortest and consists of at most two edges. The algorithm goes on to the point where it calculates the shortest paths at most "n" edges in the nth iteration. The maximum number of times that the algorithm loops is one less than the number of nodes. This eliminates the possibility of negative weight on edges.

The algorithm initiates with an assumption of zero distance as the shortest one from the start node. All distances are to be evaluated from the parent node (i.e. the origin/start point of the user). The theory of relaxation concept is used for devising the weight of each link. The algorithm stands on the fact that every link is traversed so weights for each link are calculated from the parent node to the destination node in the network routing request as shown in Fig. 2. a matrix M is crafted containing the link scores S_p as in (2) based on the importance index in (1) and estimated travel time ETT_p giving us the best possible shortest path with equilibrium cost depending on the weight of links involved in the embedded graph.

$$S_p = \frac{ETT_p}{\sum_{l \in p} ImpI_l} \tag{2}$$

To understand the granularity of the actual information and the technological efficiency of the aggregator algorithm, one of the tests is exemplified here. The server containing the NA algorithm gets a request with the origin (Corso Italia 44), destination (Stadio Luigi Ferraris) and percentage (20%) of the required information for a private mode. The traditional routing apps give us a polyline path from the origin to the destination for the same request time as shown in Fig. 2 with the possibility of detailed directions. However, for the same input attributes the NA computes the shortest path according to the predefined algorithm and stores the complete path according to the condition of the links at the particular request time of the day (courtesy of the EXIS data analyser). The complete course of links from the origin to the destination is computed.

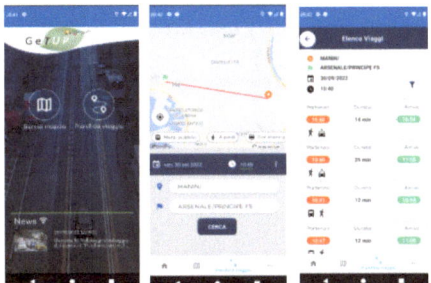

Fig. 3. User interface for GeTUP application (screenshots from actual application)

Later, the detailed path is passed to the sub-algorithm inside the server which after here will be termed as high-level information algorithm (HLI- algorithm). The HLI works on an importance index factor to skim out the links to be traversed by the user according to two major factors; first, the amount of information the GeTUP user requested and second the importance ranking of the links involved in the detailed shortest path found earlier. Once the importance ranking is applied the most important links are skimmed out of the path and given as a result. This information is returned as a response message to the API which then moves it to the GeTUP application as a final result for the user. All the process goes on in a blink of an eye.

2.2 Interface

The GeTUP application allows user to plan their trip with a travel planning system and through interactive information on the modes of transport in the Genoa metropolitan area powered by NA algorithm, also including public and shared transport (Car Sharing and Bike Sharing). The GeTUP application is implemented on Samsung S21 (Android 13) based on an open-source GPS navigator system. In the application, a customizable map interface is used that follows the user's liking and ease of use as shown in Fig. 3. For each request from the user given the origin and destination address; a set of alternative paths are provided with a combination of various modes including private mode, car-sharing, public transport and walking. With every change of request the NA algorithm together with the EXIS traffic analyser the link conditions and predicted traffic flows are updated consequently revealing new alternative paths as per users' preferences.

3   Tests and Evaluation

The GeTUP application system has been implemented as mentioned earlier for validation and evaluation through multi-device field testing with several origin and destination points over different periods throughout the weekday. The details of tests are not reported here due to space constraints; instead, the evaluation and analysis of the application in a real-world environment for the city of Genoa, Italy is carried out to see: *"How accurate is GeTUP in considering the users' needs to provide high-level path information and how does it benefit the traffic network and demand distribution systems in terms of reduced travel times and link loads as compared to the traditional turn-by-turn navigation applications?"*.

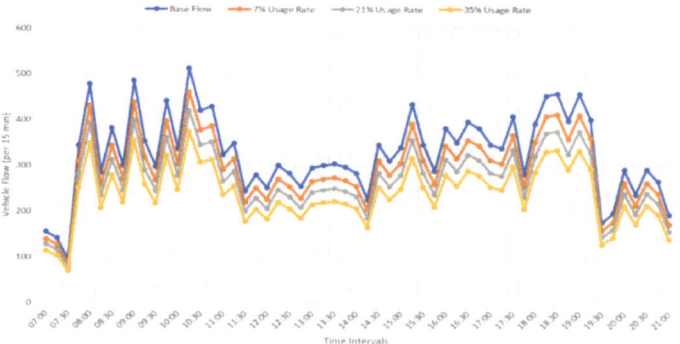

Fig. 4. Traffic flow for various usage rates of GeTUP

For evaluation purposes on a real network, a simulation model for a part of the city of Genoa is set up as in [11]. 17 OD pairs are considered for the simulation from 7 a.m. to 9 p.m. with 15-min intervals for travel time and flow monitoring. Travel-time and flow relationships are used from [9]. It is known that the level of service for any road network is demonstrated by its resistance to congestion. Hence, it is the first thing to analyse on employing any MaaS application. As depicted in Fig. 4, the traffic distribution for an OD pair 1 (Corso italia 44 to Stadio Luigi Ferraris) is compared for three different usage rates of GeTUP for private mode trips (7%, 21% and 35%). The comparison of traditional navigation algorithm simulative results with that of GeTUP implemented balanced re-routing results exhibited that deployment of GeTUP distributes traffic better on every usage rate as compared to the traditional methods for path search. Upon steady increment in the usage of GeTUP for path search, the traffic distribution and relative flow reduction on congested links also elevates steadily. This concludes that the usage rate of balanced path search is linearly proportional to the reduction in traffic flows on saturated links for a certain OD pair.

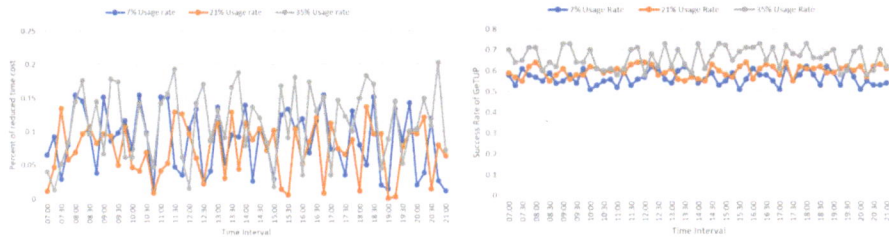

Fig. 5. a) Percent of reduced travel time. **b)** Success rate for various usage rates.

Similarly, travel time cost for a certain OD pair is analysed against different usage rates of GeTUP. Although a general perception is that if a user takes less crowded but longer paths to reach the destination instead of shorter paths it will increase the travel time cost. Albeit, the GeTUP implementation and its evaluation tells a different story. An index of *success rate* (number of trips suggested by GeTUP faster than those suggested by the traditional path search) and *percent of reduced time* (percent of travel time saved by using GeTUP as compared to the traditional navigators) are used to anlyse the travel time cost savings. Witnessing Fig. 5a, we can observe that for any period of the day specially the peak hours (morning and evening peak) among all the usage rates of GeTUP the travel times for a certain OD pair is reduced up to 21% for the usage rate of 35%. However, the reduction in travel time costs is lower for normal traffic hours. Figure 5b shows us a similar trend for the success rate index, it reveals that over 45% of the paths suggested by the GeTUP routing search cost less travel time as compared to the traditional navigators. This saving rises well above 58% on a steady increase in the usage rate of GeTUP. In fact, even for the lower usage rate GeTUP path suggestion is a great motivation to shift from the traditional shorter path to reach the destination with a lesser time cost.

4   Conclusions

This article presents a deliverable product of Green MaaS for the Adaptive Urban Planning project of the Liguria region in Italy in the form of a high-level navigation application based on Internet of Things architecture. GeTUP application is based on a powerful navigator aggregato algorithm designed to meet the user's navigation request in a sophisticated way following the user's travel profile and providing an option for customized percentage level of path information. Based on four executable entities and two APIs responsible not only for providing real-time path information for any OD pair to the user but also for updating the traffic conditions in the database to help the traffic analyser server entertain the next routing request efficiently. Successful implementation of the beta version of the application on Android devices reveals that even as low as 7% usage of this system can reduce the immense traffic flows on the vulnerable links. Also, it alleviates the congestion situation by reducing the load on links involved in the shortest paths by inviting users to take unpopular links saving travel time costs to reach their destinations. Since our evaluations depicts that implementation of GeTUP casts a positive impact on the overall network situation the municipal authorities can take a step up by rewarding the users on participating and using the application as a gift for contributing towards sustainable ecosystem.

References

1. Yamashita, T., Izumi, K., Kurumatani, K.: Car navigation with route information sharing for improvement of traffic efficiency. Intell. Transp. Syst. Proc., 465–470, Oct 2004
2. Greenshields, B., Bibbins, J., Channing, W., Miller, H.: A study of traffic capacity. National Research Council (USA) Highway Research Board, vol. 1935 (1935)
3. Wilkie, D., van den Berg, J.P., Lin, M.C., Manocha, D.: Self-aware traffic route planning. In: AAAI (2011)
4. Van Den Bosch, B., Van Arem, M.M., Misener, J.: Reducing time delays on congested road networks using social navigation. In: Integrated and Sustainable Transportation System (FISTS), IEEE Forum, pp. 26–31 (2011)
5. Lim, S., Rus, D.: Stochastic distributed multi-agent planning and applications to traffic. Robot. Automation (ICRA), pp. 2873–2879 (2012)
6. Traffic assignment manual for application with a large high-speed computer. Technical Report (1964)
7. Aslam, J., Lim, S., Rus, D.: Congestion-aware traffic routing system using sensor data. Intell. Transp. Syst. (ITSC), 1006–1013 (2012)
8. Pan, J., Khan, M.A., Popa, I.S., Zeitouni, K., Borcea, C.: Proactive vehicle re-routing strategies for congestion avoidance. In: Distributed Computing in Sensor Systems (DCOSS), pp. 265–272 (2012)
9. Bilal, M.T., Giglio, D.: Evaluation of macroscopic fundamental diagram characteristics for a quantified penetration rate of autonomous vehicles. Eur. Transp. Res. Rev. **15**, 10 (2023). https://doi.org/10.1186/s12544-023-00579-0
10. Bilal, M.T., Giglio, D.: Realization of the penetration rate for autonomous vehicles in multi-vehicle assignment models. Transp. Res. Procedia **62**, 171–180 (2022). https://doi.org/10.1016/j.trpro.2022.02.022
11. Bilal, M.T., Giglio, D.: Analysing inequity in land use and transportation models by genetic algorithm for realistically quantified penetration rate of Advanced Driving System Equipped Vehicles. Transp. Res. Interdisciplinary Perspectives **20**, 100841 (2023). ISSN 2590-1982

Open Access This chapter is licensed under the terms of the Creative Commons Attribution 4.0 International License (http://creativecommons.org/licenses/by/4.0/), which permits use, sharing, adaptation, distribution and reproduction in any medium or format, as long as you give appropriate credit to the original author(s) and the source, provide a link to the Creative Commons license and indicate if changes were made.

The images or other third party material in this chapter are included in the chapter's Creative Commons license, unless indicated otherwise in a credit line to the material. If material is not included in the chapter's Creative Commons license and your intended use is not permitted by statutory regulation or exceeds the permitted use, you will need to obtain permission directly from the copyright holder.

How Can MaaS Be Further Developed to Encourage a Greater Number of People to Adopt Sustainable Mobility in Austria?

A First Outline of the Austrian Lighthouse Project MUST (Multimodal Traffic Control Through a Combination of Innovative Communication Channels)

Melanie Juppe[1]([✉]), Wolfgang Schildorfer[1], Reinhard Tockner[1], and Alexander Hausmann[2]

[1] Department Logistikum, University of Applied Sciences Upper Austria, Wehrgrabengasse 1-3, 4400 Steyr, Austria
melanie.juppe@fh-steyr.at

[2] ASFINAG Motorway and Expressway Financing Joint-Stock Company, Schnirchgasse 17, 1030 Vienna, Austria

Abstract. In recent decades, countries have adopted and tried a number of measures aimed at reducing car use and increasing the use of public transport (PT). Mobility-as-a-Service (MaaS) systems could open up new opportunities to meet both sustainability goals and traveler needs. The fusion of digital climate and environmentally friendly capacity and traffic management, complemented by the associated digital information infrastructure, represents a great opportunity here for the promotion of sustainable mobility. This integration leverages extensive expertise from operators of traffic routes and means of transport, augmented by real-time situational awareness within their respective networks. The MUST (*multimodal traffic control through a combination of innovative communication channels*) project encompasses a consortium composed of transport operators, the largest Austrian automobile association, industry leaders, research institutions, and the Austrian national public broadcaster. The primary objective of the MUST project encompasses an assessment of the current information and communication channels within the consortium, while simultaneously conducting a prospective evaluation for both enhancing existing channels and introducing entirely new ones.

Keywords: Mobility-as-a-Service (MaaS) · Information and Communication Technology (ICT) · Sustainability · Multimodal traffic control · communication channels

1 Introduction

Numerous cities have been designed or altered to accommodate extensive car usage, necessitating significant efforts to reverse this trend. The present challenge lies in persuading travelers to opt for more sustainable modes of transportation. Heikkilä [1]

© The Author(s) 2025
C. McNally et al. (Eds.): TRAconference 2024, LNMOB, pp. 276–281, 2025.
https://doi.org/10.1007/978-3-031-85578-8_35

defined MaaS as: "a system, which provides a comprehensive range of mobility services to customers by various mobility operators". The Mobility-as-a-Service (MaaS) concept, conceived in the last decade, combines both public and private mobility options to facilitate more sustainable on-demand transportation. [2, 3]. Also Soteropoulos et al. suggested that automated vehicles and MaaS could potentially reduce the need for private car ownership and provide more efficient and sustainable transportation options [4]. Recent research has emphasized how MaaS plays a crucial role in enhancing efficiency for both transportation systems and individual users [5]. Appropriate communication support for potential users is therefore necessary and can only be achieved through communication channels adapted to this. The upcoming Austrian lighthouse project MUST *(multimodal traffic control by combining innovative communication channels)* which has started in October 2023, aims to initiate directly, and investigate the most effective communication channels tailored for reaching and informing specific user groups. This will be tested in various contexts during a pilot phase. One of the mentioned information channels is the smartphone app DOMINO OÖ. MUST uses this smart, intermodal mobility service which was developed in the past four years in the completed DOMINO project [6]. As background, here is some basic information about the app: it is an offer for commuters in the central region of Upper Austria. The primary features of the application include a departure monitor, a route organizer, and a ridesharing platform that allows users to either offer or find rides. A total of 1212 available rides were offered in the pilot period from the beginning of March to the end of September 2022, 109 of which were booked. Most of the trips that were offered had a distance of less than 50 km. Utilizing the intermodal DOMINO OÖ app as an integral part of various information channels, MUST endeavors to enhance traffic management in Austria moving forward.

Fig. 1. Screenshot of the DOMINO OÖ app (2021). DOMINO OÖ is an open-source software developed by Fluidlife, Vienna, Austria. Available at www.domino-ooe.at.

2 State-of-the-Art

Researchers have shown significant interest in MaaS to achieve the objective of offering a unified platform that combines diverse travel modes and services. This streamlines the travel planning, booking, and payment process, providing travelers with greater convenience [7]. Most MaaS projects and research studies predominantly focus on European nations, with f.e. Sweden [8, 9] and the Netherlands [10, 11] being a notable example. Parallel to this, Information and Communication Technology (ICT) has revolutionized urban mobility, with smartphones offering users valuable, real-time information for navigating cities. One of the significant benefits of Information and Communication Technologies (ICTs) is their ability to deliver timely, up-to-date information to users whenever they request it, whether through a website or a mobile application [12]. This capability has emerged as a critical necessity for users of transportation systems, requiring information to be not only real-time but also dependable, because in today's world, individuals are extremely conscious of the value of their time. Since transportation is a derivative activity linked to others, travelers have become less patient with any time that is perceived as wasted [13]. This is where we come in with the MUST project - in the coming three years, our goal is to pinpoint and establish information channels that cater to the diverse needs of both the project consortium and various user groups.

3 Research Context

The integration of climate and environmentally friendly capacity and traffic management and a corresponding digital information system is a promising contribution to the promotion of sustainable mobility. By bringing together the comprehensive know-how of transport infrastructure and transport operators as well as real-time situation awareness within their respective networks, the MUST project contributes to creating a fundamental framework for future intermodal and comprehensive traffic management. In this interaction with other Austrian initiatives such as SAM-AT, the cooperative role of MUST is important in that it puts users at the center. The aim is to use the different information channels to put climate and environmentally friendly transport management strategies into practice. This framework relies on the dissemination of service-related information and an increased level of compliance on the part of road users, leading to optimized use of existing transport infrastructure and services. The main objective of the MUST project includes an assessment of the current information and communication channels within the consortium, while at the same time conducting a forward-looking assessment for both the improvement of existing channels and the introduction of entirely new ones.

Within the framework of the MUST project, the primary objective is to investigate the optimization of information channels in to enhance their efficacy. This entails a comprehensive analysis of both national and international traffic information channels to discern their unique attributes. These findings will be faced with user preferences, as outlined in the research question: "How do users presently gather information for mobility decisions, and how would they prefer to receive such information?" Subsequently, this data will inform the development of tailored information offerings for diverse user demographics, with a focus on fostering ecologically sustainable mobility behavior.

The Logistikum Steyr at Upper Austria University of Applied Sciences is the first port of call and now Austria's largest research institution for questions relating to logistics and supply chain networks. Together with international research and cooperation partners, Logistikum Steyr as a center of excellence provides concrete contributions in the competence areas of supply chain management and transport logistics - especially transport logistics and mobility. The main tasks in the project therefore lie primarily in work package 4 - Customer Journey in Practice. We will coordinate the pilot corridor Upper Austria. The customer journeys "Routinised commute (local knowledge, fixed/changing travel times)" and "non-routinised leisure traffic (local knowledge /changing travel times)" will be processed. Linz as a daily traffic jam hotspot will include the daily commuter issues as well as leisure time traffic challenges in the area of event mobility (e.g., major sporting events around the football stadiums in Linz). A close exchange with the State of Upper Austria (Regional Management Upper Austria), the City of Linz, the Upper Austrian Transport Association, the ÖAMTC Upper Austria (call center operation of DOMINO Upper Austria - return channel for traffic information) and the technology partner Fluidtime (app operation) is carried out by the Upper Austrian University of Applied Sciences by means of regular coordination meetings.

The implementation objective of MUST is, on the one hand, an analysis of the information and communication channels available in the consortium and, on the other hand, an analysis of the potential for both the expansion of existing channels and the establishment of completely new channels. Based on this, MUST evaluates possible control options regarding increasingly multimodal mobility offers and combined route chains, considering the behavioral habits of customers, and addresses the following questions:

- When do road users - considering gender and information types - need which information via which channels to change their transport behavior?
- How can we harmonize the degree of information via the channels and the multitude of applications in such a way that certain steering effects occur or are made possible?
- What information is not yet available, but still interesting for behavioral change?
- How must cooperation with the large service providers (international navigation and map providers) be designed so that they also provide consistent information?
- How do we reach the different needs of the target groups? What information do men and women in different life situations, commuters and leisure traffic need and how can this contribute to making the transport system more sustainable, ecological, efficient, and safe?
- Are all the technical channels needed for this already available or do we need to develop and implement new information channels to reach road users?

These questions will be investigated, clarified, and applied in the MUST project.

4 Methodological Approach

A detailed analysis of traffic management measures beyond the use of a specific app (DOMINO OÖ) will be carried out to provide a broader approach to managing and optimizing traffic flow in different environments. A key component here would be, for example, the inclusion of real time information about public transport timetables. User stories are also an important part of the agile development process. They are precise descriptions of a specific functionality or feature from an end-user's point of view. They help to formulate requirements and expectations in a clear and understandable way. In the MUST project, the perspective of commuters and the use of information services for leisure routes at events are being researched. The assessment involves thoroughly examining the current mobility behavior and the way each specific target group utilizes information within their respective use cases. The test case setup involves configuring various messages and channels for testing purposes. This includes preparing multiple messages and selecting the appropriate channels to be used during the testing process. Furthermore, mobility behavior assessment uses test cases to evaluate the movement patterns, capabilities and performance of a person or system in specific scenarios or under specific conditions. The aim is to gather data and insights to make informed decisions, improve systems or take tailored actions. Mobility change behavior assessment usually refers to the evaluation of changes or shifts in a person's ability to move, whether these are due to physical, psychological, or environmental factors. This assessment is crucial for understanding and addressing mobility-related challenges, which are explored in MUST.

5   Expected Results and Discussion

Evidence-based mobility change entails leveraging empirical evidence and data to shape and alter individuals' transportation preferences and habits. This methodology acknowledges that furnishing transparent, trustworthy, and pertinent information can foster more eco-friendly and streamlined mobility practices. When evidence-based mobility strategies encompass various information dissemination channels, such as personalization and relevance, they prove highly effective. The evolution of traffic management owes much of its progress to insights gained from the utilization of these information channels. These insights have spurred the creation of more intelligent, resourceful, and data-informed traffic management approaches. It is planned to present and discuss the first learned lessons of the MUST project during the TRA conference.

References

1. Heikkilä, S.: Mobility as a Service - A Proposal for Action for the Public Administration, Case Helsinki (2014)
2. Guyader, H., Friman, M., Olsson, L.E.: Shared mobility: evolving practices for sustainability. Sustainability **13**(21), 12148 (2021)
3. Mitropoulos, L., Kortsari, A., Ayfantopoulou, G.: Factors affecting drivers to participate in a carpooling to public transport service. Sustainability **13**(16), 9129 (2021)
4. Teo, T.W., Choy, B.H.: in. In: Tan, O.S., Low, E.L., Tay, E.G., Yan, Y.K. (eds.) Singapore Math and Science Education Innovation. ETLPPSIP, vol. 1, pp. 43–59. Springer, Singapore (2021). https://doi.org/10.1007/978-981-16-1357-9_3
5. Jittrapirom, P., Caiati, V., Feneri, A.M., Ebrahimigharehbaghi, S., Alonso-González, M.J., Narayan, J.: Mobility as a service: a critical review of definitions, assessments of schemes, and key challenges. Urban Plan **2017**(2), 13–25 (2017)

6. DOMINO Project Homepage. https://www.domino-maas.at/de/projekt-domino. Accessed 21 Oct 2023
7. Kamargianni, M., Matyas, M., Li, W., Muscat, J.: Londoners' Attitudes Towards Car-Ownership and Mobility-as-a-Service: Impact Assessment and Opportunities That Lie Ahead (2018)
8. Hesselgren, M., Sjöman, M., Pernestål, A.: Understanding user practices in mobility service systems: results from studying large scale corporate MaaS in practice. Travel Behav. Soc. **21**, 318–327 (2020)
9. Karlsson, I., et al.: Development and implementation of mobility-as-a-service – a qualitative study of barriers and enabling factors. Transp. Res. Part A Policy Practice **131**, 283–295 (2020)
10. Meurs, H., Sharmeen, F., Marchau, V., van der Heijden, R.: Organizing integrated services in mobility-as-a-service systems: principles of alliance formation applied to a MaaS-Pilot in the Netherlands. Transp. Res. Part A: Policy Practice **131**, 178–195 (2020)
11. Mobility-as-a Service in the Netherlands: The Implementation of a Stated Choice Experiment to Examine Travel Behavior Adaptations (2019)
12. Zhao, X., Vaddadi, B., Sjöman, M., Hesselgren, M., Pernestål, A.: Key Barriers in MaaS Development and Implementation: Lessons Learned from Testing Corporate MaaS (CMaaS). 2590–1982, Vol. 8, p. 100227 (2020)
13. Casquero, D., Monzon, A., García, M., Martínez, O.: Key elements of mobility apps for improving urban travel patterns: a literature review. Future Transp. **2**(1), 1–23 (2022)

Open Access This chapter is licensed under the terms of the Creative Commons Attribution 4.0 International License (http://creativecommons.org/licenses/by/4.0/), which permits use, sharing, adaptation, distribution and reproduction in any medium or format, as long as you give appropriate credit to the original author(s) and the source, provide a link to the Creative Commons license and indicate if changes were made.

The images or other third party material in this chapter are included in the chapter's Creative Commons license, unless indicated otherwise in a credit line to the material. If material is not included in the chapter's Creative Commons license and your intended use is not permitted by statutory regulation or exceeds the permitted use, you will need to obtain permission directly from the copyright holder.

Scenario-Based Analysis of Automated Mobility Service Costs in Urban Areas – The Case of Oslo

Charly Beye[1,2(✉)], Nicole van den Boom[1,2], Adrian Boos[1], and Guy Fournier[1,2]

[1] Pforzheim University, Tiefenbronnerstr. 65, 75175 Pforzheim, Germany
charly.beye@hs-pforzheim.de
[2] Laboratoire Génie Industriel, CentraleSupélec, Université Paris-Saclay, 9 rue Joliot Curie, 91190 Gif-sur-Yvette, France

Abstract. Automated Vehicle (AV) services hold significant potential for revolutionizing public transportation networks across Europe, promising to elevate public service quality and enable more sustainable, citizen-centric mobility. However, to successfully integrate these innovative mobility options, it is crucial to understand their economic impacts on public transport systems. This study aims to provide a comprehensive cost analysis of AV mobility services, particularly automated minibuses, through a scenario-based approach using a developed Fleet and Cost Calculator. Employing a sequential exploratory research design, we collected qualitative and quantitative data to clarify the pivotal criteria influencing AV service costs. As part of the Horizon ULTIMO Project, we calculated and analysed AV service costs within the Oslo-Groruddalen (Norway) demonstration site across two different mobility scenarios. The analysis not only highlights the economic feasibility of integrating AVs but also provides recommendations for the use of automated mobility services, focusing on cost efficiency. In terms of economic feasibility, the analysis showed that AV service costs can decrease between 2,5% and 56% by adapting certain operational factors, such as vehicle supervision, energy costs, and maintenance costs.

Keywords: Automated Driving · Automated Mobility Services · Automated Mobility Costs · Public Transport · Urban Mobility

1 Introduction

Automated vehicles (AV) have considerable potential to change the current mobility paradigm by transforming (sub-) urban and rural regions and creating transportation systems with a common economic approach, which benefits the entire population of the region [1]. From the perspective of different scenarios, positive effects are associated with an increase in transport efficiency, reduced use of private cars and the related saving of public space, improvement of travel time and higher accessibility of mobility services [2]. Substantial improvements in the efficiency of transport network systems can additionally reduce travel costs, as AV technology might also contribute to more frequent use of vehicle- or ridesharing [3]. By making AVs available, this technology can offer travellers the opportunity to circumvent the elevated costs associated with driving only

© The Author(s) 2025
C. McNally et al. (Eds.): TRAconference 2024, LNMOB, pp. 282–287, 2025.
https://doi.org/10.1007/978-3-031-85578-8_36

with their personal cars [4]. Concerning the costs of AVs, the impact of electrification and automation must also be considered in certain areas, such as maintenance and repair [5] or energy consumption [6]. As part of the Horizon ULTIMO Project, this study examined the AV service costs in the suburban area Oslo-Groruddalen (Norway) with an implemented Fleet and Cost Calculator, considering different mobility scenarios. To analyse service costs, we considered an automated minibus (AM) from Navya (now Gama) [7]. These results may support the Public Transport Operator/Authority (PTO&PTA) in making informed decisions regarding the implementation of AV services.

2 Methodology

A sequential exploratory research design [8] was used to develop a cost calculation model and interpret service costs. This design involved both qualitative and quantitative research methods to establish a Fleet and Cost Calculator (FCC). The qualitative phase involved collecting and analysing data through two key methods: reviewing cost factors for automated mobility services and scenarios and consulting with the local PTA to understand key criteria affecting AV service costs. In the quantitative phase, the FCC was used to estimate the required fleet sizes (RFS) and service costs. The framework for calculating AV service costs comprises four crucial steps, as illustrated in Fig. 1.

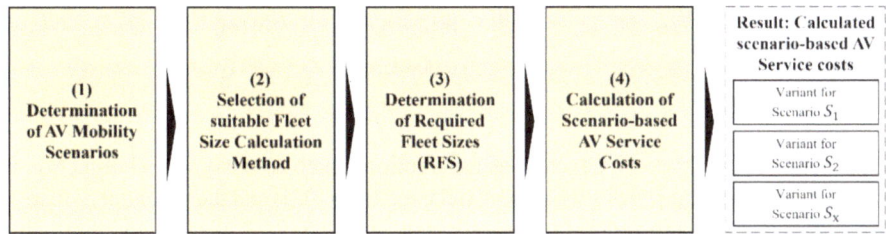

Fig. 1. General Conceptual Framework of scenario-based AV service cost calculation

For each scenario, we determined the daily Aggregated Total Costs (ATC), including the specific cost breakdown and the costs per passenger kilometer Travelled (PKT). Scenario 1 "AV as bus line" implies the substitution of internal combustion engine buses with automated minibuses (AMs) used along a 5.2 km circular bus route (spatial dimension). Concerning the temporal aspect, the AV serves a fixed schedule. In contrast, Scenario 2 "AV in MaaS" aims to seamlessly integrate AVs into the existing Public Transport infrastructure through a Mobility-as-a-Service (MaaS) system [9]. Temporally, the AMs operate on-demand, acting as a feeder service. Spatially, a door-to-door service is offered, covering an estimated area of 19.25 km^2, with a hub (metro station) at the border of the service area. Despite these distinctions, both scenarios feature shared specifications, including a vehicle speed of 25 km/h, a vehicle capacity of 15 passengers and an assumed evenly distributed daily demand of 10,000 passengers. For Scenario 1 "AV as bus line" the fleet size calculation follows the method in [10], while for Scenario 2 "AV in MaaS", it adheres to the approach in [11]. Concerning the costs outlined in Table 1, the vehicle acquisition and commissioning costs were converted to the present values [12], with a discount rate of 3% in Norway (as of April 2023) [13]. The commissioning costs of 22,500€ (rounded up) were assumed based on the previous AVENUE project [14]. Due to a lack of data, we posited that cleaning costs would amount to twice the average minimum wage in the European Union [15]; concurrently, we incorporated a monthly workload of 120 h to determine the hourly costs. For the maintenance costs, we applied the assumption for electric buses from [16]. The operational hours are designated as t_O (16 h/day).

Table 1. Cost Factors for AV Service Cost Calculation

Cost Factors	Value	Source
c_A : Vehicle Acquisition	5.09 €/h	[7]
c_{OFB} : Off-board Supervisor Salary	18.50 €/h	[17]
c_{OB}: On-board Driver Salary	18.50 €/h	[17]
c_{CL} : Cleaning costs	15 €/h	[15]
c_{IP} : Insurance premium	0.097 €/h	[18]
c_{EN} : Energy costs	1.88 €/h	[19, 20]
c_{FE} : Fees	1.7 €/h	Consulted by PTA
c_{MA} : Maintenance Costs	0.17 €/km	[16]
c_{TA} : Tax	0.008 €/h	[21]

We incorporated adaptive factors as assumptions, such as the potential reduction in maintenance costs (r_{MA}) from 3% to 30% (adapted from [22]) and the potential reduction in energy costs (r_{EN}) from 5% to 50% (adapted from [23, 24]). We estimated monthly cleaning cycles (q_{CL}) from 1 to 10 times, assuming a 1-h duration per cycle and a maximum of 480 cleaning hours per month. We considered an adaptable number of off-board supervisors (SPV); with a minimum of 2 and maximum value dependent on RFS. The formula for the daily aggregated costs (1) is as follows:

$$ATC = \left[c_A + c_{IP} + c_{EN} \cdot (1 - r_{EN}) + c_{CL} \cdot \frac{q_{CL}}{480h} + c_{TA} + c_{OB} + c_{OFB} \cdot \frac{1}{SPV} \right]$$
$$\cdot t_O \cdot RFS + c_{FE} \cdot t_O + c_{MA} \cdot (1 - r_{MA}) \cdot PKT \qquad (1)$$

In (1), energy costs are related to r_{EN}, maintenance costs to r_{MA} (offset against travelled distance), cleaning costs to q_{CL} and off-board supervisor salary to the inverse value of SPV. Consulted by the PTA, the fees were generalised to the entire fleet.

3 Results

Figure 2 presents the key results (including the cost breakdown) for both scenarios. In Scenario 1, RFS is 11 AMs, covering a daily distance of 4,400 km. This service constellation results in daily aggregated costs (ATC) of 5281.83€. In Scenario 2, RFS is 2 AMs covering a daily distance of 611 km. Thus, the daily ATC are 950.37€.

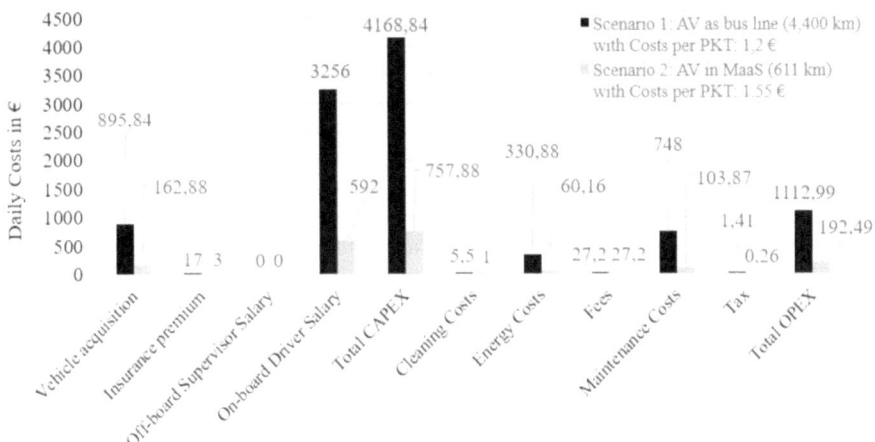

Fig. 2. Service Characteristics and Results of Cost Calculation (Baseline)

The driving cost factors include vehicle acquisition, on-board driver salary, energy, and maintenance costs. The sensitivity analysis indicates: Examining the potential of energy cost reductions, ranging from the minimum to maximum values, reveals a consistent trend with a cost decline of 2.58% in the "AV in MaaS" scenario and 3.33% in the "AV as bus line" scenario. When considering potential maintenance cost reductions, there is an average cost decline of 3.23% (AV in MaaS) and 4.17% (AV as bus line). This finding is notable because maintenance costs account for a higher percentage of ATC than energy costs. Moreover, variations in cleaning frequencies have a minimal cost impact. Focusing on the distinction between on-board drivers and off-board supervisors, substantial changes in cost structures compared to the baseline (Fig. 2) are evident. By introducing off-board supervisors, daily costs decrease by 55.83% (down to 2,322€) in scenario "AV as bus line" and by 30.97% (down to 654€) in scenario "AV in MaaS". These findings underline the cost efficiency of off-board supervision.

4 Discussion and Conclusion

Labour costs, encompassing both on-board drivers and off-board supervisors, emerged as significant contributors to daily expenses in the AV scenarios examined. Potential energy and maintenance cost reductions led to a slight decline in the overall costs. Scenario 1, with a larger fleet size, indicated higher daily costs compared to more flexible scenario 2. Scenario 1 aims to optimise services for the PTO, whereas Scenario 2 extends the benefits to citizens and the city, emphasizing the importance of considering diverse stakeholders in urban mobility. Not having included external costs and the impact of changing demand through various offerings, which could affect modal share, further represents a notable limitation. Moreover, the impact of competition with public transport needs to be examined as it could substantially impact research outcomes. Finally, additional learnings from different locations (location-specific demand and cost patterns etc.) could influence the model. In conclusion, research on the economic impact of AVs must be further conducted. Defining potential AV mobility services is crucial for the thorough calculation of service costs, implementation support, identification of required input data, and conclusive interpretation of results. Therefore, this process is critical to strategically enhance the cost efficiency and accessibility of AV services.

References

Legacy, C., Ashmore, D., Scheurer, J., Stone, J., Curtis, C.: Planning the driverless city. Transp. Rev. (2019). https://www.semanticscholar.org/paper/Planning-the-driverless-city-Legacy-Ash more/7b8f782bce7f69620ad0e0fa5992b40bab85288d

Papa, E., Ferreira, A.: Sustainable Accessibility and the Implementation of Automated Vehicles: Identifying Critical Decisions (2018). https://www.semanticscholar.org/paper/Sustainable-Accessibility-and-the-Implementation-of-Papa-Ferreira/7434624089bb316ca491c1e466ff22 97e104ce1f

Saghir, C., Sands, G.: Realizing the Potential of Autonomous Vehicles. Planning Practice & Research (2020). https://www.semanticscholar.org/paper/Realizing-the-Potential-of-Autono mous-Vehicles-Saghir-Sands/1fe30d9829439b96b4a26dc0257a744554ea9775

Clements, L.M., Kockelman, K.M.: Economic effects of automated vehicles. Transp. Res. Rec. J. Transp. Res. Board **2606**(1), 106–114 (2017). https://doi.org/10.3141/2606-14

Alonso Raposo, M., Grosso, M., Mourtzouchou, A., Krause, J., Duboz, A., Ciuffo, B.: Economic implications of a connected and automated mobility in Europe. Res. Transp. Econ. **92**, 101072 (2022). https://doi.org/10.1016/j.retrec.2021.101072

Bauer, G.S., Greenblatt, J.B., Gerke, B.F.: Cost, energy, and environmental impact of automated electric taxi fleets in Manhattan. Environ. Sci. Technol. **52**(8), 4920–4928 (2018). https://doi. org/10.1021/acs.est.7b04732

Electric Motor Engineering, The electric bus without driver is arriving in Bari - Electric Motor Engineering. https://www.electricmotorengineering.com/the-electric-bus-without-dri ver-is-arriving-in-bari/. Accessed 2 Oct 2023

Creswell, J.W., Plano Clark, V.L.: Designing and conducting mixed methods research. Thousand Oaks, California: SAGE (2017)

Pathak, A., Scheuermann, S., Ongel, A., Lienkamp, M.: Conceptual design optimiza- tion of autonomous electric buses in public transportation. World Electric Vehicle J. (2021). https://www.semanticscholar.org/paper/Conceptual-Design-Optimization-of-Autono mous-Buses-Pathak-Scheuermann/9bc7340fc46fe553340f72845db6848866f0fd82

Jara-Díaz, S., Fielbaum, A., Gschwender, A.: Optimal fleet size, frequencies and vehicle capacities considering peak and off-peak periods in public transport. Transp. Res. Part A: Policy Practice **106**, 65–74 (2017). https://doi.org/10.1016/j.tra.2017.09.005

Badia, H., Jenelius, E.: Design and operation of feeder systems in the era of automated and electric buses. Transp. Res. Part A: Policy Practice **152**, 146–172 (2021). https://doi.org/10.1016/j.tra. 2021.07.015

Fagnant, D.J., Kockelman, K.M.: "Dynamic ride-sharing and fleet sizing for a system of shared autonomous vehicles in Austin, Texas", (in En;en). Transportation **45**(1), 143–158 (2018). https://doi.org/10.1007/s11116-016-9729-z

Trading Economics, Norway Interest Rate. https://tradingeconomics.com/norway/interest-rate. Accessed 2 Oct 2023

Antonialli, F., Mira-Bonnardel, S., Bulteau, J.: Economic assessment of services with intelligent autonomous vehicles: EASI-AV. In: The Robomobility Revolution of Urban Public Transport: Springer, Cham, 2021, pp. 85–105 (2021). https://link.springer.com/chapter/https://doi.org/10.1007/978-3-030-72976-9_4

Eurostat, Monthly minimum wages - bi-annual data. https://ec.europa.eu/eurostat/databrowser/view/EARN_MW_CUR/bookmark/table?lang=en&bookmarkId=d05d73c3-70cb-4135-bc87-7a351ee38718. Accessed 13 Jul 2023

Thorne, R.J., Hovi, I.B., Figenbaum, E., Pinchasik, D.R., Amundsen, A.H., Hagman, R.: Facilitating adoption of electric buses through policy: Learnings from a trial in Norway. Energy Policy **155**, 112310 (2021). https://doi.org/10.1016/j.enpol.2021.112310

ERI Economic Research Institute, Bus Driver Salary in Norway. https://www.erieri.com/salary/job/bus-driver/norway. Accessed 2 Oct 2023

Statista, Norway: average insurance premium price by type|Statista. https://www.statista.com/statistics/1075912/average-insurance-premium-price-in-norway-by-type/. Accessed 13 Jul 2023

Statista, Norway: monthly electricity prices 2023 | Statista. https://www.statista.com/statistics/1271469/norway-monthly-wholesale-electricity-price/. accessed 2 Oct 2023

Navya, "Brochure-Autonom-Shuttle-Evo-EN," vol. 2020 (2020). https://navya.tech/wp-content/uploads/documents/Brochure-Autonom-Shuttle-Evo-EN.pdf

European Commission, Incentives and Legislation | European Alternative Fuels Observatory. https://alternative-fuels-observatory.ec.europa.eu/transport-mode/road/norway/incentives-legislations. Accessed 2 Oct 2023

Grosso, M., et al.: How will vehicle automation and electrification affect the automotive maintenance, repair sector? Transportation Research Interdisciplinary Perspectives, vol. 12, None (2021). https://doi.org/10.1016/j.trip.2021.100495

Zhong, S., et al.: Energy and environmental impacts of shared autonomous vehicles under different pricing strategies. (in En;en), npj Urban Sustain, vol. 3, no. 1, pp. 1–10 (2023). https://doi.org/10.1038/s42949-023-00092-2

Taiebat, M., Brown, A.L., Safford, H.R., Qu, S., Xu, M.: A review on energy, environmental, and sustainability implications of connected and automated vehicles. Environ. Sci. Technol. **52**(20), 11449–11465 (2018). https://doi.org/10.1021/acs.est.8b00127

Open Access This chapter is licensed under the terms of the Creative Commons Attribution 4.0 International License (http://creativecommons.org/licenses/by/4.0/), which permits use, sharing, adaptation, distribution and reproduction in any medium or format, as long as you give appropriate credit to the original author(s) and the source, provide a link to the Creative Commons license and indicate if changes were made.

The images or other third party material in this chapter are included in the chapter's Creative Commons license, unless indicated otherwise in a credit line to the material. If material is not included in the chapter's Creative Commons license and your intended use is not permitted by statutory regulation or exceeds the permitted use, you will need to obtain permission directly from the copyright holder.

Governance of Automated Vehicles in Urban Transport System: A Case Study of the Oslo Region, Norway

Wale Arowolo[1](✉), Magnus Larsson[2], and Isabelle Nicolai[1]

[1] Sustainable Economy Research Group, Department of Industrial Engineering,
CentraleSupélec (Paris Saclay University), Gif-sur-Yvette, France
adewale.arowolo@centralesupelec.fr
[2] Ruter, Oslo, Norway

Abstract. Shared automated electric vehicles (SAEV) can increase safety, reduce congestion, and provide environmental benefits to the urban transport system. Nonetheless, SAEV's governance in the urban transport system is challenging. This paper proposes a novel interdisciplinary methodology and contributes to the governance debate from the 'policy' and 'polity' dimensions. We attempt to draw insight from the Norwegian cultural context and a willingness to use SAEV survey in Oslo, Norway. We analyse informal institutions from the cultural viewpoint using the Hofstede 6D model vis-à-vis a quantitative willingness to use SAEV survey. Then, we combine insight from the cultural indicators and survey to draw insight on the appropriate governance approaches from the 'policy' dimension of governance to promote SAEV deployment in Norway.

Keywords: SAEV · governance · urban transport system

1 Introduction

Autonomous vehicle (AV) represents a potentially disruptive yet beneficial change to the transportation system with the potential to improve vehicle safety, reduce emissions and congestion, facilitate efficient use of travel time and change travel behavior [1-4]. The World Health Organization estimates 1.35 million road accident deaths and 50 million injuries annually, with human error responsible for around 95 percent of cases [5]. AVs' most revolutionary impact is integrating them into existing public transport systems to improve performance and fulfill current and latent mobility needs [6]. Shared autonomous electric vehicles (SAEV) are electric AVs in shared mode, such as robotaxis and automated shuttles. SAEV can provide different services, such as on-demand ride services within a fixed route with fixed stops, a fixed route with on-demand stops, or door-to-door on-demand services. SAEV can provide unimodal or multimodal door-to-door trips [7]. SAEV can also reduce vehicle ownership, fleet size requirements, parking demand, labour costs, and cold-start emissions [2]. Nonetheless, if not carefully managed, SAEV may have adverse environmental impacts such as sprawling land use,

© The Author(s) 2025
C. McNally et al. (Eds.): TRAconference 2024, LNMOB, pp. 288–298, 2025.
https://doi.org/10.1007/978-3-031-85578-8_37

increased vehicle miles travelled associated with congestion, and increased travel time consequences [3, 8, 9]. Thus, appropriate governance frameworks to align public and private interests are required to capture the potential benefits, minimise costs, and support the integration of SAEV in the urban transport system [10].

This paper attempts to contribute to the governance of SAEV in the urban transport system debate. Governance can be in the 'political' dimension (the actor constellation - range of actors involved in the process of policymaking), 'policy' dimension (political steering - the nature and character of steering instruments in use) and 'polity' dimension (the institutional landscape - in which the actors operate based on a system of rules that shape their actions) [11]. Institutions are the humanly devised constraints that structure political, economic, and social interactions. Institutions consist of both informal constraints and formal rules in an interdependent manner and consequent political and economic organisations [12]. Informal institutions cannot be fully understood without considering culture, and understanding culture presumes insight into institutions. Culture is the collective mind programming that distinguishes the members of a group of people from others [13]. Culture matters because it affects attitude, behaviour, and choices, providing an indicative road map to distinguish between proper and improper behaviour in a society [14]. Culture is a primary factor influencing the adoption or acceptance of new technologies and innovation [15, 16].

The willingness to use SAEV is the strength of people's intention to use SAEV [17]. Willingness to use is impacted by culture in a society and can be supported by a workable context-specific governance approach. We attempt to contribute to the governance debate to increase willingness to use SAEV from the 'polity' (focusing on informal institutions from the cultural viewpoint using the Hofstede 6D model) and the 'policy' dimension by adapting the framework developed in [18]. We attempt to answer the following questions:

- How does Norwegian culture impact the willingness to use SAEV in Norway?
- What is the workable governance approach to support SAEV deployment in Norway?

  The original contribution of our paper is as follows:

- We propose a novel interdisciplinary methodology to analyse the governance of SAEV in the urban transport
- We provide insight into the appropriate governance approach to support SAEV deployment in Norway

  This paper proceeds as follows: Sect. 2 is the literature review. Section 3 is the methodology. Section 4 presents our results. Section 5 is the conclusion.

2   Literature Review

This section presents three streams of related literature to our research: i. SAEV[1] integration in the urban transport system ii. Willingness to use SAEV iii. Governance of SAEV in the urban transport system.

---

[1] AV, SAEV and self-driving vehicles are used interchangeably in this paper.

2.1 SAEV in the Urban Transport System

Several recent studies have shown the impact of SAEV integration in the urban transport system. [19] found that SAEV fleet could lower negative externalities and generate cost benefits for commuters. [20] found that rapid reduction in technology costs, a rise in people's willingness to pay and supporting policies are required to promote AV adoption. [2] found that SAEV could pose additional challenges, such as increased waiting time and the associated inconvenience(s), increased vehicle miles travelled, and could increase congestion due to empty miles. A study in Oslo, Norway, to understand the impact of integrating SAEV in urban transport shows that in a scenario where all the existing ICEV users switch to SAEV, there will be drastic traffic reduction in Oslo as about 7% of the current vehicle fleet will be required to meet transport demand during the peak period [21]. Related studies show similar results, such as the OECD International Transport Forum's Lisbon study (3%), Helsinki (4%), Dublin (2%), Auckland (7%) and Stuttgart (7%) [22-26].

2.2 Willingness to Use SAEV

The results of the willingness to use SAEV survey in the literature seem heterogeneous. The results are demographically diverse, context-specific, and a function of the researchers' data and adopted methodology. [27] found that people willing to use AV are most likely early adopters. [28] found that around half of their survey respondents in different cities are willing to use self-driving vehicles. [29] found that nearly two-thirds of their survey respondents consider they will be late adopters of AV technology. [30] found reduced willingness to use AV due to extreme concern about AV technology. [31] found that users with lower anxiety and increased self-confidence were more open toward AVs. [32] found that privacy concerns and perceived safety significantly affect general concerns about the willingness to use AV. [18] found that the more positive the attitude toward AV, the higher the willingness to use AV.

2.3 Governance of SAEV in the Urban Transport System

This sub-section reviews the literature on the governance of SAEV in the two dimensions applied in this paper: i. steering instruments (policy) and ii. Institutional properties (polity) [33].

'Policy' Dimension of the Governance of SAEV

[34] proposed 'a no-response strategy, prevention-oriented (avert)', 'control-oriented (regulation)', 'tolerance-oriented (reform)', and 'adaptation-oriented (manage policy uncertainty)' approach for SAEV governance. Building upon and combining the works of [35] and the multi-level perspective [20, 36] proposed five governance frameworks (governing by doing, governing by enabling, laissez-faire, self-governing and governing by authority) for SAEV integrated in Mobility-as-a-Service (MaaS) in the urban transport system. [37] proposed a comparative analysis framework for governance based on the assumed roles of the technology, identified domains and mechanisms of governance, and the assumed actors responsible for steering the development process.

'Polity' Dimension of the Governance of SAEV

[38] found that existing regulatory standards and industry self-regulation shape AV regulatory governance, and international policies and conventions co-exist simultaneously in Sweden and Norway. [39] argue that existing regulatory institutions affect the regulatory response to emerging technologies and are responsible for different trajectories of regulatory development across countries. [40, 41] found that informal institutions can complement formal institutions to facilitate coordination and achieve effective governance in the European public transport context. [42] argue that governance strategies that aim to tap the potential of AVs in supporting sustainable urban mobility should explicitly consider institutional dynamics.

3   Methodology

We discuss in this section the adopted methodology in this paper.

3.1   Willingness to Use Survey

We conduct a willingness-to-use survey in the Oslo region, Norway. The questionnaires were designed based on the following determinants: (i) personal technology innovativeness[2], (ii) perceived usefulness/benefit, (iii) safety perception (perceived safety and perceived risk) and (iv) price perception. The determinants are among the empirically proven significant determinants of willingness to use SAEV in the literature [19]. The respondents were asked questions on their attitude to technology adoption, willingness to share a vehicle with strangers if it results in a lower price for the journey, perception of self-driving vehicles as a solution to meet their future transport needs, and perception of safety and security of self-driving vehicles to gain insight into their willingness to use SAEV. The 5-point Likert scale was used from 'totally disagree' to 'totally agree'. The respondents were contacted on their mobile phone numbers for the interview. The survey was conducted for all citizens over 15 years old in the Oslo region (old Akershus county (Viken), Groruddalen and Ovrig) from Tuesday to Saturday every week in September 2022. There were 1,738 respondents: Groruddalen (615), Viken County (604), and Ovrig (519).

3.2   Informal Institution (Norwegian Cultural Dimension)

We draw insight from Hofstede's 6-D model to discuss Norwegian culture. In Hofstede's 6-D model, culture is described in six dimensions: Power Distance, Individualism, Tough/Tender, Uncertainty Avoidance, Long-Term/ Short Term Orientation, and Indulgence. Hofstede's model has been used extensively during recent decades in the theoretical and empirical literature in different social science fields. Hofstede's model can be regarded as a grounded approach for describing culture [16].

---

[2] Personal technology innovativeness is the willingness of a person to try out a new technology.

3.3 Governance (Policy Dimension)

Building upon the framework proposed by [37] on public authorities' approaches for governing cities and climate change, [20] proposed five governance approaches for governing the development of MaaS in public transport: i. Governing by authority (employing traditional top-down mechanisms to govern) ii. Governing by enabling (facilitating and encouraging actions with non-public actors through partnerships and incentives development) iii. Governing by doing (public authorities take care of the entire service production and delivery and avoid collaborating with private actors) iv. Self-governing: governing by 'showing the way', and v. laissez-faire (public authorities allow the network of actors to reach a stable state without getting involved). We adopt the framework of [20] to develop the corresponding actions in developing SAEV for the five governance approaches.

Figure 1 shows the schematic representation of our methodology discussed in Sects. 3.1 to 3.3. We focus on informal institutions and culture from the 'polity' dimension of governance. Then, we analyse informal institutions from the cultural viewpoint vis-à-vis the quantitative willingness to use SAEV survey. Then, we draw insight on the appropriate governance approach from the 'policy' dimension of governance.

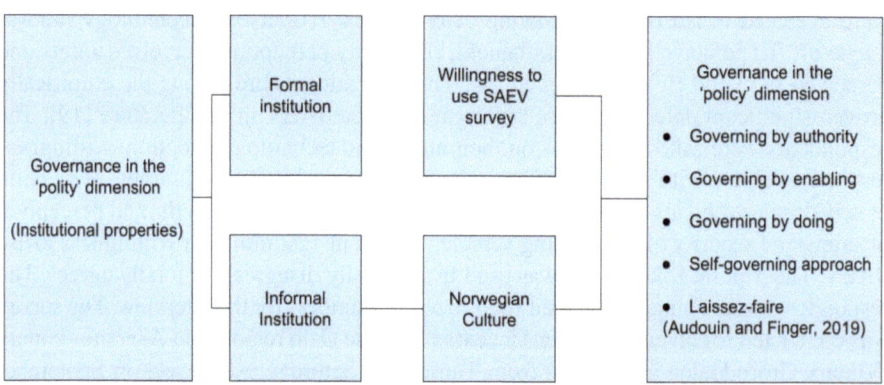

Fig. 1. Schematic diagram of the methodological framework

4   Results

4.1 Willingness to Use SAEV

We discuss the results of our questions for the willingness-to-use SAEV survey in the Oslo region, Norway: i. Attitude to new technology ii. Willingness to share a vehicle with strangers if it results in a lower price for the journey iii. Believe that self-driving vehicles can solve many of the transport needs of the future. iv. Believe that self-driving vehicle is safe and secure. We found that 27 out of every 100 respondents partly or totally believe SAEV is safe and secure, 41 out of every 100 believe that SAEV can meet their future transport needs, 30 out of every 100 respondents partially or completely agree to share SAEV and 46 out of every 100 respondents like new technology and are willing to be among the early adopters. We infer that, on average, 36 out of every 100 are partly or totally willing to use SAEV.

We summarise the insights from the Norwegian cultural indicators vis-à-vis the willingness to use SAEV survey and workable governance framework in Table 1.

Table 1. Summary of Norwegian cultural dimensions, indicators, and proposed governance framework

Cultural dimension	Norwegian Cultural Indicators [43]	Impact of cultural indicators on the willingness to use SAEV	Proposed workable governance approach to mitigate willingness to use barriers based on the Norwegian cultural context
High individualism	• The right to privacy is important and respected	Privacy loss concerns could reflect the low perception of the safety and security of SAEV and the willingness to share SAEV with strangers in public transport → Reduce willingness to use SAEV	'Governing by doing' could help increase the perception of safety and security and increase willingness to use SAEV: •Developing SAEV projects in-house in a closed manner could increase the perception that the government is thorough to ensure that all safety and security concerns are addressed •Minimising collaboration with third parties could help allay the fear of unnecessary private data exposure to many third parties (such as private sector institutions) or corporate entities
High short-term orientation	• Strong concern for establishing the truth • Normative thinking with great respect for traditions • Preference to maintain time-honoured traditions and norms • Viewing societal change with suspicion	Lack of full understanding of SAEV, disruption of traditional public transport paradigm, and suspicion of societal change could reflect a relatively low interest in new technology, willingness to be early adopters, and believe that SAEV can meet future transport needs → Reduce willingness to use SAEV	'Governing by doing' could help address the challenges with the traditional Norwegian culture and pessimistic view of societal change: Developing SAEV projects in-house in a closed manner with minimum collaboration with third parties could help address the perception of change in tradition, societal norms and public transport disruption

(*continued*)

Table 1. (*continued*)

Cultural dimension	Norwegian Cultural Indicators [43]	Impact of cultural indicators on the willingness to use SAEV	Proposed workable governance approach to mitigate willingness to use barriers based on the Norwegian cultural context
Moderate uncertainty avoidance Moderate restraint	• Moderate structure and planning are required • Moderate perception that social norms restrain actions	Due to uncertainty concerns, citizens should be assured that SAEV is carefully planned and structured. The perception that social norms restrain actions could affect the willingness to use SAEV since the innovation is new and its impact is not fully understood. The indicators could also reflect on the willingness to share SAEV with strangers in public transport. → Reduce willingness to use SAEV	Combining governing by doing and self-governing by: Developing solutions in-house in a closed manner, minimising collaboration with third parties, and providing all government employees with a SAEV solution could allay uncertainty fears and 'restraint' constraints. Combining the governance approaches could serve as a springboard to promote SAEV, increase awareness, encourage sharing with strangers, and integrating SAEV in the public transport system
Tender Culture Power Distance	• Levelling with others, consensus, cooperation • Taking care of the environment and societal solidarity • Incentivisation • Dialogue • Decision-making is achieved by involving people • consensus-oriented communication • Management facilitates and empowers • Decentralised power	These indicators could serve as enablers to increase awareness and willingness to use SAEV	'Governing by enabling' actions appear to align with the 'tender' and 'power distance' Norwegian cultural indicators: • Initiating public-private partnerships → Levelling with others, consensus, cooperation, dialogue, consensus-oriented communication, and decision-making • Providing funding → Incentivisation • Influencing negotiations in favour of SAEV → Incentivisation, societal solidarity • Leveraging SAEV opponents using horizontal network governance → decentralised power, dialogue, management facilitates and empowers and consensus-oriented communication

Based on the short-term orientation, privacy loss and the combination of moderate uncertainty avoidance and restraint concerns that appear influential on the survey results, we suggest combining governing by doing, self-governing and features of governing by enabling (Table 2).

Table 2. Proposed governance approaches to support SAEV deployment in Norway

Governance approaches for SAEV deployment in Norway	'Governing by doing', 'self-governing' and 'governing by enabling' • Develop a SAEV solution in-house in a closed manner • Minimise collaboration with third parties • Provide government employees with a SAEV solution to show citizens examples to follow • Provide funding • Initiate public-private interactions • Define vision with strong quantitative objectives • Influence negotiations in favour of SAEV • Leverage SAEV opponents using horizontal network governance

As the willingness to use SAEV increases, a 'governing by enabling' approach could be fully adopted in relation to Norway's tender, power distance and individualism cultural indicators. 'Governing by enabling' could promote end-user acceptance and awareness of the environmental benefits of SAEV based on the tender cultural indicator of Norway.

5   Conclusions

This paper attempts to provide insight into the governance of SAEV integrated in the Norwegian urban transport system using insight from the Norwegian cultural context and willingness-to-use survey. The survey result suggests that less than half of Norwegians are willing to use SAEV in the Oslo region. We then suggest that a workable governance approach for SAEV integration in the Norwegian public transport system should allay fears of privacy concerns (high individualism) while making people feel safe and secure, allay uncertainty fears, manage pessimism to technological innovation (high short-term orientation), regard time-honoured traditions and address the moderate 'uncertainty avoidance' and 'restraint' indicators. Also, the cultural indicators should align with the Norwegian culture's 'tender' and 'power distance' indicators, such as consensus, cooperation, societal solidarity, incentivisation, dialogue, facilitating management and decentralised power. Therefore, we argue that 'governing by doing', 'self-governing' and 'governing by enabling' should be combined to increase the willingness to use and promote SAEV deployment in Norway.

Acknowledgement. This research received funding from the Horizon Europe research and innovation programme under Grant Agreement No. 101077587 (The ULTIMO project).

References

1. Fagnant Daniel and Kockelman Kara: Preparing a nation for autonomous vehicles: opportunities, barriers and policy recommendations. Transp. Res. Part A **77**(2015), 167–181 (2015)
2. Harprinderjot, S., Kavianipour, M., Ghamami, M., Zockaie, A.: Adoption of autonomous and electric vehicles in private and shared mobility systems. Transp. Res. Part D **115**(2023), 103561 (2023)
3. Alexandros, N., Thomopoulos, N., Milakis, D.: The environmental and resource dimensions of automated transport: a nexus for enabling vehicle automation to support sustainable urban mobility. Annu. Rev. Environ. Resour. **46**, 167–192 (2021)
4. Kelsey, O., Isaksson, K.: Governance arrangements shaping driverless shuttles in public transport: the case of Barkarbystaden. Stockholm. Cities **113**(2021), 103146 (2021)
5. KPMG (2020). Autonomous Vehicles Readiness Index. Assessing the preparedness of 30 countries and jurisdictions in the race for autonomous vehicles. KPMG, 2020. Retrieved from https://smart-cities-marketplace.ec.europa.eu/sites/default/files/avri.pdf. Assessed 30 Jan 2023
6. Dorien, K., Naderer, G., Fournier, G.: The potential of automated minibuses in the socio-technical transformation of the transport system. Transp. Res. Procedia **00**(2022), 000 (2022)
7. Horschutz, N.E., Issaoui, R., Korbee, D., Jaroudi, I., Fournier, G.: How to measure the impacts of shared automated electric vehicles on urban mobility. Transp. Res. Part D **93**(2021), 102766 (2021)
8. Shelly, E., et al.: Modelling cross-national differences in automated vehicle acceptance. Sustainability **2020**(12), 9765 (2020)
9. Yonah, F., Hudson, A., Zhao, J.: Policies for autonomy: how american cities envision regulating automated vehicles. Urban Science **2020**(4), 55 (2020)
10. Iain, D., Marsden, G., Anable, J.: The governance of smart mobility. Transp. Res. Part A **115**(2018), 114–125 (2018)
11. Oliver, T., Bähr, H., Falkner, G.: Modes of Governance: A note towards conceptual clarification. European Governance Papers (EUROGOV) No.N-05-02 (2005). http://www.connex-network.org/eurogov/pdf/egp-newgov-N-05-02.pdf
12. Douglass, N.: Institutions. Journal of Economic Perspectives **5**(1), 97–112 (1991)
13. Gert, H., Jan Hofstede, G.: Cultures and organisations: Software of the mind. Second Edition, McGraw Hill Publishing 2005 (2005)
14. Anneli, K., Luca, A.: Determinants of institutional trust: the role of cultural context. Journal of Institutional Economics (2022), 18, 45–65 Cambridge University Press
15. Yun Yongdeok, O., Hyungseok, M.R.: Statistical modelling of cultural differences in adopting autonomous vehicles. Appl. Sci. **2021**(11), 9030 (2021)
16. Jhanghiz, S., Gyulavári, T., Jászberényi, M., Ásványi, K., Kökény, L., Chairy, C.: Surrendering personal control to automation: Appalling or appealing? Transp. Res. Part F **80**(2021), 90–103 (2021)
17. Tamara, K.: Behavioural intention to use autonomous vehicles: systematic review and empirical extension. Transp. Res. Part C **119**(2020), 102732 (2020)
18. Audouin Maxime and Finger Matthias: Empower or Thwart? Insights from Vienna and Helsinki regarding the role of public authorities in developing MaaS schemes. Transp. Res. Procedia **41**(2019), 6–16 (2019)
19. Guy, F., Adrian, B., Ralf, W., Jaroudi, I., Inna, M., Nemoto, E.H.: Substituting individual mobility by mobility on demand using autonomous vehicles – a sustainable assessment simulation of Berlin and Stuttgart. Int. J. Automot. Technol. Manage. **20**(4), 369–407 (2020)

20. Prateek, B., Kockelman, K.M.: Forecasting Americans' long-term adoption of connected and autonomous vehicle technologies. Transp. Res. Part A **95**(2017), 49–63 (2017)
21. Ruter: The Oslo Study: How autonomous cars may change transport in cities report? Ruter, 2019 (2019)
22. ITF (2016). Shared Mobility. Innovation for liveable cities International Transport Forum, OECD, 2016. https://www.itf-oecd.org/sites/default/files/docs/shared-mobility-liveable-cit ies.pdf
23. Helsinki (2017), Shared Mobility Simulations for Helsinki, International Transport Forum, OECD, 2017. https://www.itf-oecd.org/sites/default/files/docs/shared-mobility-simulations-helsinki.pdf
24. Dublin (2018): Shared Mobility Simulations for Dublin, International Transport Forum, OECD, 2018. https://www.itf-oecd.org/sites/default/files/docs/shared-mobility-simulations-dublin.pdf
25. Auckland (2017), Shared Mobility Simulations for Auckland, International Transport Forum, OECD, 2017. https://www.itf-oecd.org/sites/default/files/docs/shared-mobility-simulations-auckland.pdf
26. Markus, F., Hartl, M., Magg, C.: A modelling approach for matching ridesharing trips within macroscopic travel demand models. Transportation, Springer, vol. 45(6), 1639–1653, November 2018
27. Timo, L., Heikki, L., Markus, P.: Attitudes and concerns on automated vehicles. Transportation Research Part F. Traffic Psychol. Behav. 2018, **59**, 24–44 (2018)
28. Ipek, S.N., Johanna, Z., Thomas, W.: Measures of baseline intent to use automated vehicles: A case study of Texas cities. Transportation Research Part F. Traffic Psychology and Behaviour **62**, 66–77 (2019)
29. Ilias, P., George, D.: An empirical investigation on consumers' intentions towards autonomous driving. Transp. Res. Part C. Emerging Technol. **95**, 773–784 (2018)
30. Sophie, W., Muhammad, A., Sebastian, K.: Are we ready to ride autonomous vehicles? a pilot study on austrian consumers' perspective. Logistics **2019**(3), 20 (2019)
31. Franziska, H., Peter, F., Hudecek, M.: Identification of user groups of autonomous shuttle buses: A latent profile analysis (2023). https://osf.io/t4y9d/. Assessed May 29 2023
32. Klara, L., Groznik, A.: The role played by social factors and privacy concerns in autonomous vehicle adoption. Transport Policy **132**, 1–15 (2023)
33. Oliver, T., Bähr, H., Falkner, G.: Modes of governance: Towards a conceptual clarification. J. Eur. Publ. Policy **14**(1), 1–20 (2007)
34. Tan Si Ying and Taeihagh Araz: Adaptive governance of autonomous vehicles: Accelerating the adoption of disruptive technologies in Singapore. Gov. Inf. Q. **38**(2021), 101546 (2021)
35. Harriet, B., Kristine, K.: Local government and the governing of climate change in Germany and the UK. Urban Studies **43**(12), 2237–2259 (2006)
36. Frank, G.: Technological transitions as evolutionary reconfiguration processes: a multi-level perspective and a case-study. Res. Policy **31**(2002), 1257–1274 (2002)
37. Milos, M., Dominic, S., Dimitris, M., Kate, P., Moshe, G.: Governance cultures and sociotechnical imaginaries of self-driving vehicle technology: Comparative analysis of Finland, UK and Germany. In: Milakis, D., Thomopoulos, N., van Wee, B. (eds.) Policy Implications of Autonomous Vehicles. Advances in Transport Policy and Planning. Volume 5, 2020, pp. 235–262 (2020)
38. Lisa, H.: Regulatory governance in emerging technologies: The case of autonomous vehicles in Sweden and Norway. Res. Transp. Econ. **83**(2020), 100967 (2020)
39. Alberto, A., Inna, K.: Uncertainty, institutions and regulatory responses to emerging technologies. Regulation Governance **15**, 1111–1127 (2021)

40. Robert, H., Monios, J., Rye, T., Isaksson, K., Scholten, C.: The interplay of formal and informal institutions between local and regional authorities when creating well-functioning public transport systems. Int. J. Sustainable Transp. **11**(8), 611–622 (2017)
41. Tom, R., Monios, J., Hrelja, R., Isaksson, K.: The relationship between formal and informal institutions for governance of public transport. J. Transport Geography **69**, 196–206 (2018)
42. Jens, S., Truffer, B., Fleischer, T.: Potential impacts of institutional dynamics on the development of automated vehicles: Towards sustainable mobility? Transp. Res. Interdisciplinary Perspectives **14**(2022), 100587 (2022)
43. Hofstede Insight: Cultural dimensions values for Norway (2023). https://www.hofstede-insights.com/country-comparison/norway/. Assessed 15 Feb 2023

Open Access This chapter is licensed under the terms of the Creative Commons Attribution 4.0 International License (http://creativecommons.org/licenses/by/4.0/), which permits use, sharing, adaptation, distribution and reproduction in any medium or format, as long as you give appropriate credit to the original author(s) and the source, provide a link to the Creative Commons license and indicate if changes were made.

The images or other third party material in this chapter are included in the chapter's Creative Commons license, unless indicated otherwise in a credit line to the material. If material is not included in the chapter's Creative Commons license and your intended use is not permitted by statutory regulation or exceeds the permitted use, you will need to obtain permission directly from the copyright holder.

How Do Tourists Use Bike-Sharing? Analyzing Origin-Destination Data of Tourists' Bike-Sharing Usage in Budapest, Hungary

Dávid Földes$^{(\boxtimes)}$ (ID) and Ali Maktabifard (ID)

Department of Transport Technology and Economics, Faculty of Transportation Engineering and Vehicle Engineering, Budapest University of Technology and Economics, Budapest, Hungary
`foldes.david@kjk.bme.hu`, `a.maktabifard@edu.bme.hu`

Abstract. The theoretical goal of bike-sharing services in metropolises is to serve short-distance direct trips and provide access to or egress from the closest high-capacity public transport stop. Besides citizens, bike-sharing can be popular among tourists if the service area covers the most important tourist attractions. This study aims to reveal the differences between tourists' and non-tourists' bike-sharing usage habits. The origin-destination data of the station-based bike-sharing system in Budapest, Hungary, were analyzed using over 300,000 rentals from September 2022. It was estimated that users were tourists if the country code of their billing address was not Hungary. Accordingly, 3.7% of the rentals were made by tourists. We found that there are significant differences in bike-sharing usage habits between tourists and non-tourists. Tourists' renting time is longer, but the distance covered is the same as for non-tourists. Moreover, tourists mostly use the bike-sharing system in the afternoon and mostly (3/4 of their rentals) in the downtown area where the most popular tourist attractions are located. Decision-makers can use the results of this study to improve bike-sharing services and operators to optimize the reallocation.

Keywords: Bike-sharing · Tourist transport · Tourism · Tourist attraction

1 Introduction

Bike-sharing (BS) services are operating in app. 1600 cities in 90 countries [1]. BS aims to provide a sustainable transport option, reducing pollution [2, 3], promoting healthier lifestyles, and achieving efficient land use. BS can be used mainly for short-distance and covering the first or last kilometers to/from a public transport stop. The purpose of BS use can be daily commuting or even recreational (including touristic purposes) [4]. Usually, BS services are operated downtown where most tourist attractions are located. BS is attractive for tourists as part of a short urban holiday, accordingly, cities that consider expanding their attraction as tourist destinations could benefit from BS services as an integral part of the touristic experience [5].

Our aim was to reveal the differences in BS use between tourists and non-tourists. Accordingly, OD pairs covered by tourists and non-tourists were analyzed in Budapest.

© The Author(s) 2025

C. McNally et al. (Eds.): TRAconference 2024, LNMOB, pp. 299–306, 2025.
https://doi.org/10.1007/978-3-031-85578-8_38

We hypothesized that the average renting time and distance by tourists is longer, tourists use the BS downtown, and the number of round-trips is higher. The result can be useful for operators planning BS service deployment or improvement.

The structure of the paper is as follows. The literature is reviewed in Sect. 2. The case study, data collection, and analysis are described in Sect. 3. The results are discussed in Sect. 4. Finally, Sect. 5 contains concluding remarks.

2 Literature Review

The literature on tourist-related BS can be categorized as whether the tourists are (i) the main focus or (ii) they are only one aspect of the analysis in the research.

The purpose of using BS as a tourist is to have an environmentally friendly vacation and save money and time [5]. Tourists' psychological benefits accounted for the highest proportion of tourists' satisfaction, followed by easy access to bikes, and environmental awareness [6]. Perceived usefulness has the highest influence on using the BS system. Accordingly, adopting specific strategies to enhance the quality of recreational trips, establishing promotional policies, and promoting BS is necessary if the aim is to increase BS use among tourists [7]. Results show that tourists' satisfaction with attractions and the popularity of an attraction are influenced positively by the available BS stations nearby [8, 9]. Specifically, the quality of bikes and accessibility explain destination satisfaction [10]. The opening of a BS station within 200 m of a given attraction led to an increase its demand [11].

Compared to tourist-related and commuting-related trips, weekends attract more occasional users than weekdays. They are more influenced by weather, and day type [12]. The number of BS users is generally increasing in tourist seasons (e.g. summer time) [13]. Commuters ride a bike directly from the start to the end station while the tourists stop at different points-of-interest locations along the way. Accordingly, the renting time for tourists is higher [14, 15]. The highest BS demand is associated with tourist attractions, dormitories, and high-capacity public transport stops [16].

Based on the literature, we have concluded that mostly the perceived usefulness and attitude towards BS are measured among tourists; only a minority of the papers analyzed the differences between tourist and non-tourist users' habits.

3 Applied Methods

3.1 Study Site: Budapest's Bike-Sharing Service (BuBi)

A public and station-based bike-sharing service has been operating in Budapest since 2014 by the Centre for Budapest Transport. After the renewal of the system (lighter bikes, new mobile application, and renting process) in 2021, the number of rents increased dramatically (2015: 651,592; 2022: 2,906,00). This study considered the total operational area in September 2022 (>2000 bikes, 172 stations covering app. 35 km^2). The number of stations per square kilometer is about 5, which is lower than in Paris (33) and New York (23) but the same as in Soul (6.5). The BS tariff system is affordable. The monthly pass costs 1.3 EUR, including free first 30 min for each ride. The most popular tourist attractions are located downtown; in this study, this area, as the area of tourist interest, was considered (10.3 km^2, 69 stations).

3.2 Data

This study uses OD-level BS renting data, users' nationality, and distance between BS stations. Monthly BS usage data from September 2022 was retrieved. Retrieved data are anonymized and contain the origin and destination stations, rental and return timestamps, and the country code of the user's registered bank account. Whether the user is a tourist can only be assumed as the travel purpose is not known. Accordingly, we considered that users with foreign bank accounts are tourists. However, this is a limitation as Hungarian inland tourists cannot be considered in this way.

The data were cleaned (Table 1). The rent was deleted if the origin or destination was not at a BS station, or the two stations were the same and the duration was less than 5 min. Leaving the bike not in a station or defective renting (failure of the bike) caused these problems. The total number of rides retrieved is 344,572, the considered is 303,979; thus, the rate of total defected data is 11.78%. The number of non-tourist-related rides is 292,694 and the tourist-related is 11,285 (3.7%).

Table 1. Result of data cleaning.

	Number of defective data	Rate of defective data [%]
Origin is not a BS station	10,951	3.18
Destination is not a BS station	8,498	2.47
Origin and destination are not a BS station	2,013	0.58
Same origin and destination and renting time <5 min	19,111	5.55
Origin or destination is a maintenance station	20	0.006

3.3 Distance Matrix and Analyzing Methods

We assumed that BS users prefer bicycle-friendly infrastructure. As the exact routes ridden are not available, the shortest possible distance on bicycle-friendly roads between OD pairs was considered. *Open Route Service API* was used to estimate the distance between stations. The 'cycling-regular' profile available in the API was used as this profile prefers bicycle-friendly infrastructure use and avoids high car traffic. The result is an asymmetric OD matrix; distances in the two directions may differ. If the origin and destination station were the same, the distance covered was neglected.

The rides are analyzed using descriptive statistics. Raw renting data were filtered and grouped according to date, origin, and destination. Count and rate were used to describe the occurrence and frequency of filtered data. The center of the filtered data set was determined by mean and median values; the dispersion of the filtered data set was described using the standard deviation.

4   Results and Discussion

The average distance between BS stations is 707 m. The closest BS stations are 136 m, and the furthest stations are 14.6 km from each other. The average riding distance for non-tourists and tourists is practically the same, 2.4 km. There is no relevant difference in the minimum distance: 0.14 km for non-tourists and 0.19 km for tourists. However, the maximum distance is significantly different: 13.1 km for non-tourists and 8.42 km for tourists. Table 2 summarizes the main descriptive statistics regarding distances. Figure 1 depicts the distances by the frequency of occurrence on days. 50% of the rides between stations are lower than 2.0 km for non-tourists and lower than 2.22 km for tourists. Accordingly, tourists ride a slightly longer distance than non-tourists.

Table 2. Descriptive statistics of riding distance.

[km]	Average	Standard deviation	Modus
Non-tourists	2.38	1.52	0.99
Tourists	2.39	1.25	2.22

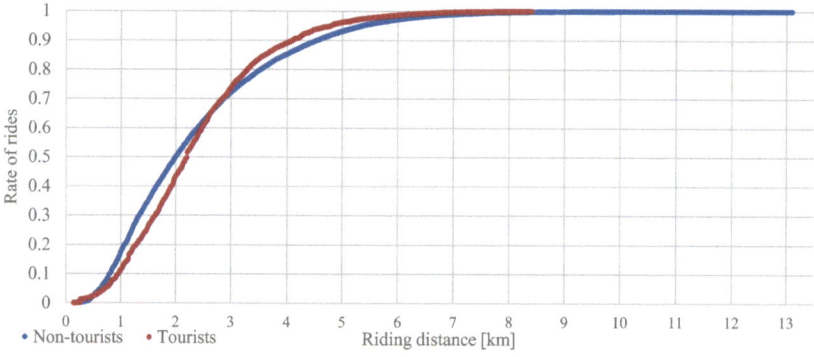

Fig. 1. Riding distance frequency between stations

There are 29,584 possible OD pairs, including round trips. Tourists covered only 3,546 (11.99%) OD pairs, but non-tourists covered 22,601 (76.4%). Accordingly, tourists use only a restricted part of the service area; tourists are concentrated in the tourist area. The 3/4 of rides done by tourists were located inside this area. The tourist area covers most of the downtown; 2/3 of the non-tourist-related rides are also connected to this area; however, 1/3 of the rides did not even involve this area. The round trips may refer to the recreational motivation. The rate of round trips is doubled among tourists; 7.87% compared to non-tourists 3.87%. Moreover, the average renting time for round trips is 27.48 min for non-tourists and 49.02 min for tourists.

Regarding TOP10 OD paris, Toursts' 8 OD pairs are located in Margaret Island, a car-free recreational zone. Another popular OD pair is the route between tourist attractions Market Hall and the gate of the Buda Castle. These rides are less than 2.4 km long. The

TOP10 OD pairs for non-tourists are mostly connected to universities. There are five OD pairs between university buildings or dormitories (<1 km) outside the tourist area. Similarly to [16], it refers to the working-related motivation and BS popularity among students. Moreover, downtown-related short (<600 m) rides and recreational rides in Margaret Island are popular. However, the TOP10 OD pairs are more determining for tourists than for non-tourists. 7.68% of the total rides is done between the TOP10 OD pairs by tourists, but only 1.41% is done by non-tourists.

The average renting time is 14 min for non-tourists and 28.05 min for tourists. Though the average distance covered by tourists is not significantly longer than the distance covered by non-tourists, the average renting time is twice as long. The possible reason for this fact is the different motivations. Tourists may use BS as a sightseeing tool, stop several times along the trip (checking the attraction), and the general speed is lower. The same conclusion was given by [14, 15].

Fig. 2. Number of daily rentals

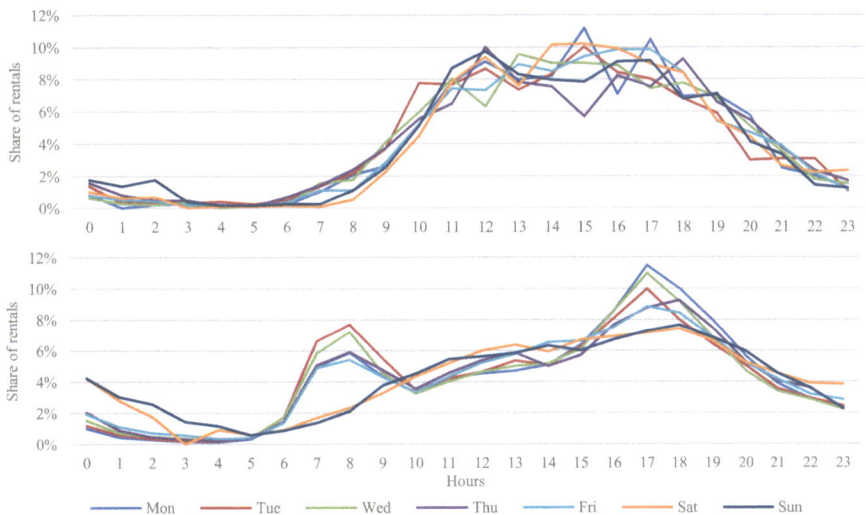

Fig. 3. Hourly share of rentals according to days (up: tourists, down: non-tourists)

Figure 2 shows the number of daily rentals. Tourists and non-tourists use BS differently. Non-tourists use BS more on weekdays, which refers to the commuting motivation. Thus, tourists use BS more on weekends; it may result from more tourists being in the city during weekends. However, the influence of weather is high; 17–18 and 25 September, the weather was rainy, the number of rentals was low.

The starting time of the rentals is aggregated according to days in the week and hours for tourists and non-tourists in Fig. 3. Among the tourists, there is a daily plateau; the share of the rentals is the highest and mostly consistent daytime between 10 am and 6 pm. There is a small surplus on Sunday nights, presumably because of the leisure activity on Saturday evening. For non-tourists, there are morning and afternoon peaks in the share of weekday rentals, connecting to commuting and work-related traffic. On Saturdays and Sundays, the share of rentals is similar to that for tourists; there is a plateau in the afternoon. Moreover, there is a remarkable share on Saturday and Sunday nights. Accordingly, considering the share of rentals in a day, there is no significant difference among the days for tourists, but there is a significant difference among workdays and weekends for non-tourists. Moreover, leisure-related night traffic is visible mostly among non-tourists on Saturday and Sunday.

5   Conclusion

Understanding the usage habits of tourists and non-tourists may support BS service development. In this paper, we analyzed OD pairs of the station-based BS in Budapest to reveal the differences in usage habits between tourists and non-tourists.

We found that the average distance covered by BS is not significantly different, but the average renting time difference is remarkable; tourists may spend more time seeing the city during a ride (hypothesis 1 is not entirely proven). Moreover, ¾ of the rents by tourists are located in the inner city, which covers the most popular destinations (hypothesis 2 is confirmed). Contrarily, non-tourists use the BS system in a more spatially balanced way. The most popular OD pairs connect to a recreational zone for tourists and connect to university campuses for non-tourists. The share of round-trips is doubled for tourists than for non-tourists; round-trips may refer to recreational purpose (hypothesis 3 is proven). Non-tourists use BS more on weekdays, which refers to commuting motivation, but tourists use BS more on weekends. Typically, on weekdays, the rentals are done during the day by tourists and in the morning and afternoon peaks for non-tourists. The rents are mostly done in the afternoon, but night-related rides are also significant on weekends.

There is a limitation to this study. We consider only the origin and destination data of BS without knowing the origin and destination of the total travel chain. Furthermore, the shortest bikeable route was considered without knowing the exact route covered by a line. Our further goal is to overcome the limitations and analyze the data deeply using machine learning and big data analysis techniques.

Acknowledgment. Supported by the ÚNKP-23-5-BME-1 New National Excellence Program of the Ministry for Culture and Innovation from the source of the National Research, Development and Innovation Fund.

References

1. Meddin: The Meddin Bike-sharing World Map Report (2022). https://bikesharingworldmap.com/reports/bswm_mid2022report.pdf
2. Cheng, B., Li, J., Su, H., Lu, K., Chen, H., Huang, J.: Life cycle assessment of greenhouse gas emission reduction through bike-sharing for sustainable cities. Sustain. Energy Technol. Assess. **53**, 102789 (2022). https://doi.org/10.1016/j.seta.2022.102789
3. D'Almeida, L., Rye, T., Pomponi, F.: Emissions assessment of bike sharing schemes: The case of Just Eat Cycles in Edinburgh, UK. Sustain. Cities Soc. **71**, 103012 (2021). https://doi.org/10.1016/j.scs.2021.103012
4. 2016 Capital Bikeshare Member Survey Yields Top Reasons For Usage, Capital Bikeshare. Accessed 06 Oct 2023. http://ride.capitalbikeshare.com
5. Kaplan, S., Manca, F., Nielsen, T.A.S., Prato, C.G.: Intentions to use bike-sharing for holiday cycling: an application of the Theory of Planned Behavior. Tour. Manag. **47**, 34–46 (2015). https://doi.org/10.1016/j.tourman.2014.08.017
6. Zhou, B., Liu, T., Ryan, C., Wang, L., Zhang, D.: The satisfaction of tourists using bicycle sharing: a structural equation model - the case of Hangzhou, China. J. Sustain. Tour. **28**(7), 1063–1082 (2020). https://doi.org/10.1080/09669582.2020.1720697
7. Khajehshahkoohi, M., Davoodi, S.R., Shaaban, K.: Factors affecting the behavioral intention of tourists on the use of bike sharing in tourism areas. Res. Transp. Bus. Manag. **43**, 100742 (2022). https://doi.org/10.1016/j.rtbm.2021.100742
8. Yang, Y., Jiang, L., Zhang, Z.: Tourists on shared bikes: can bike-sharing boost attraction demand? Tour. Manag. **86**, 104328 (2021). https://doi.org/10.1016/j.tourman.2021.104328
9. Wiyoso, D., Pramitasari, D.: Urban tourism space based on bike-sharing tourist in Yogyakarta city. Jurnal Teknosains, vol. 11, no. 1, Art. no. 1 (2021). https://doi.org/10.22146/teknosains.56079
10. Lin, J.-J., Chen, W.-T.: Bike-sharing systems and destination satisfaction in overseas tourists. Asian Transp. Stud. **5**(2), 423–435 (2018). https://doi.org/10.11175/eastsats.5.423
11. Zhang, Z., Yang, Y., Yang, S., Zhang, Z.: Can bike-sharing availability improve tourist satisfaction? evidence in Chicago. Tour. Manag. Perspect **48**, 101164 (2023). https://doi.org/10.1016/j.tmp.2023.101164
12. Jaber, A., Csonka, B.: Investigating the temporal differences among bike-sharing users through comparative analysis based on count, time series, and data mining models. Alex. Eng. J. **77**, 1–13 (2023). https://doi.org/10.1016/j.aej.2023.06.087
13. Maas, S., Nikolaou, P., Attard, M., Dimitriou, L.: Examining spatio-temporal trip patterns of bicycle sharing systems in Southern European island cities. Res. Transp. Econ. **86**, 100992 (May 2021). https://doi.org/10.1016/j.retrec.2020.100992
14. Kou, Z., Cai, H.: Understanding bike sharing travel patterns: an analysis of trip data from eight cities. Phys. A: Stat. Mech. Appl. **515**, 785–797 (2019). https://doi.org/10.1016/j.physa.2018.09.123

15. Banet, K.: Using data on bike-sharing system user stopovers in smart tourism: a case study. Commun. Sci. Lett. Univ. Zilina. **23**(2), G1–G12 (2021). https://doi.org/10.26552/com.C. 2021.2.G1-G12

16. Zhu, R., Zhang, X., Kondor, D., Santi, P., Ratti, C.: Understanding spatio-temporal heterogeneity of bike-sharing and scooter-sharing mobility. Comput. Environ. Urban Syst. **81**, 101483 (2020). https://doi.org/10.1016/j.compenvurbsys.2020.101483

Open Access This chapter is licensed under the terms of the Creative Commons Attribution 4.0 International License (http://creativecommons.org/licenses/by/4.0/), which permits use, sharing, adaptation, distribution and reproduction in any medium or format, as long as you give appropriate credit to the original author(s) and the source, provide a link to the Creative Commons license and indicate if changes were made.

The images or other third party material in this chapter are included in the chapter's Creative Commons license, unless indicated otherwise in a credit line to the material. If material is not included in the chapter's Creative Commons license and your intended use is not permitted by statutory regulation or exceeds the permitted use, you will need to obtain permission directly from the copyright holder.

Public Transport Attractiveness Issues in Lithuania's Major Cities

Aldona Jarašūnienė$^{(\boxtimes)}$ ⓘ and Rytis Engelaitis ⓘ

Faculty of Transport Engineering, Vilnius Tech, Vilnius, Lithuania
aldona.jarasuniene@vilniustech.lt,
rytis.engelaitis@stud.vilniustech.lt

Abstract. The article analyses the attractiveness problems of public transit in the largest Lithuanian cities. The analytical-methodical part of the article discusses the concept of attractive public transport system. In the research section, the expert analysis that evaluates the public transport systems of the large Lithuanian cities is presented, and ways to improve the quality in this area are considered. Additionally, a qualitative method is applied to study the way to increase the attractiveness of public transportation services in the field of sustainable development by focusing on existing public transportation problems. The causal connections between expert evaluation and qualitative analysis are examined. Working methods and tools include analysis of methodological and scientific literature, quantitative research using questionnaires. In conclusion, the article presents a proposed solution that includes transit model, which will allow to promote and increase the attractiveness of public transport (PT) itself towards commuters.

Keywords: Public Transport · Attractiveness · Flows · Sustainable Development

1 Introduction

The current state of PT attractiveness is in major decline while the car is the most attractive mode of transport because people believe that it comes with comfort, convenience and sense of freedom and speed. This implies that PT service needs to be adjusted to get the same attractiveness [1]. Today's society is linked to rapidly evolving transport services [2]. The growing number of cars every year, not only in Lithuania but also in all European Union countries, is leading to increasing urban congestion, traffic intensity and pollution [3]. The aim of this research is to investigate the attractiveness issues of public transport in major Lithuanian cities and to propose an innovative model as a solution. The first part of the paper analyses the literature on the attractiveness and quality of public transport and discusses the classification of strategies to increase the use of public transport. The second part deals with the analysis of the methods of decision-making based on Expert Assessment and the determination of the compatibility of the Expert Group's approaches to the problems of attractiveness of public transport in the major Lithuanian cities. The third part presents the result a model for promoting the attractiveness of urban public transport system based on motivating and required actions focused on sustainable mobility and multimodality principles.

© The Author(s) 2025
C. McNally et al. (Eds.): TRAconference 2024, LNMOB, pp. 307–313, 2025.
https://doi.org/10.1007/978-3-031-85578-8_39

2   Analysis of Scientific Literature on the Attractiveness and Service Quality of Public Transportation

The fundamental objective of the urban transportation system is to provide complete freedom of movement and quality of goods and services by combining aspects [4]. To achieve such a result, it is necessary to study problem areas and implement the necessary proposals in a complex way [5]. A real turnaround in transportation is not feasible without strengthening local public transportation [6]. For people to take the bus instead of their own car, bus transportation must become more attractive, i.e., it must better meet people's needs [7]. All urban residents benefit from high-quality public transport, including motorists who experience less traffic and parking congestion, accident risk and pollution exposure [8]. The Green Deal emphasizes the compatibility of growth prospects with climate change action. An overall goal of sustainable transport solutions is improving the accessibility of a region and quality of life of residents [9]. Combining different modes of transport, creating convenience and a basis for their use can increase the impact of sustainable transport and the efficiency of the whole transport system [10, 11]. The quality of public passenger transport services can significantly impact road safety. Improving the quality of passenger transport is one of the most critical areas of urban transport system development [12]. The management of the city's public transport is an important area of urban life whose objective is to satisfy the users' needs for passenger transport in a punctual, high-quality, and complete manner [13]. High quality transit can attract discretionary travelers who would otherwise drive, which reduces traffic problems including congestion, parking costs, accidents, and pollution emissions [14]. Stimulating PT-use and car-habit disrupting techniques are meant to motivate habitual car users to give public transportation a try [15]. The positive changes in public transport are possible if the process involves not only state and municipal institutions but also infrastructure planners, passenger carriers, and residents themselves [16]. Some networks experimenting with free or reduced pricing have created a tailwind of political support [17]. Solutions to problematic issues need to be based not only on solving the specific transport problem and the foreseeable prospects for urban development, but also on the achievement of strategic objectives, high safety standards, environmental protection, and economic benefits [18]. There is a lack of scientific literature analyses and studies on the attractiveness issues related to public transport.

3   Research on the Compatibility of Experts' Opinions

To find out about the relevance of public transport problems, it is necessary to consult specialists and conduct an expert survey. Therefore, this paper uses an expert survey approach. Expert surveys offer more reliable information than other approaches, such as general public polls. This is because specialists are far more likely to have the understanding and ability to answer the questions presented. In this paper, the Kendall's rank correlation and Pearson's correlation methods are used to determine the consistency of the opinions of a group of experts on the problems of public transport attractiveness in major Lithuanian cities. Then, the assigned ranks are transformed into weights using the average rank transformation into weight method (ARTIW) [19]. The significance of the concordance coefficient can be determined using Pearson's Chi-square test [20–22]. It is important to conduct qualitative research – an expert survey – based on the evaluations of specialists regarding the problems pointed out related to the attractiveness of public transportation. A questionnaire was developed based on the main problems in metropolitan public transport. The questions were based on the situations and issues that affect the attractiveness of the public transport service towards commuters. The questionnaires were handed out to a group of 9 public transport experts. The ranks assigned to the criteria by the group of experts and the results of the calculations carried out are shown in Table 1.

Table 1. Ranks assigned to the Experts Group criteria and results of the calculations.

Expert No. ($i = 1, 2, ..., n$)	Public transport attractiveness problems in Lithuania's major cities ($j = 1, 2, ..., m$)										
	A	B	C	D	E	F	G	H	I	J	Sum
E_1	4	2	5	4	4	4	4	5	2	4	38
E_2	4	5	5	4	4	4	4	5	4	2	41
E_3	2	5	4	4	2	4	2	4	5	3	35
E_4	4	4	4	2	4	2	4	4	2	3	33
E_5	4	4	5	2	4	2	4	5	5	5	40
E_6	4	4	5	4	3	5	4	2	4	4	39
E_7	4	4	4	4	4	5	4	2	5	5	41
E_8	4	4	4	2	5	2	5	4	5	2	37
E_9	4	4	4	4	2	2	4	4	5	4	37
$R_j = \sum_{i=1}^{n} R_{ij}$	34	36	40	30	32	30	35	35	37	32	341
$\bar{R}_j = \frac{\sum_{i=1}^{n} R_{ij}}{n}$	3.8	4	4.5	3.3	3.6	3.3	3.9	3.9	4.1	3.6	38
$\sum_{i=1}^{n} R_{ij} - \frac{1}{2}n(m+1)$	−4	−2	2	−8	−6	−8	−3	−3	−1	−6	−31
$\left[\sum_{i=1}^{n} R_{ij} - \frac{1}{2}n(m+1)\right]^2$	16	4	4	64	36	64	9	9	1	36	243

The statistical Pearson correlation coefficient is determined according to formula:

$$\chi^2 = \frac{6 \cdot 243}{10 \cdot 5 \cdot (5+1)} = 4.86.$$

The empirical value $\chi^2 = 4.86$ is equal to the critical value $\chi^2_{v,\alpha} = 4.86$. Therefore, the experts' evaluations are consistent. The concordance coefficient, which indicates the consistency of the experts' opinions, is then determined using formula:

$$W = \frac{12 \cdot 243}{9^2 \cdot (5^3 + 5)} = 0.3.$$

At last, the value of W_{min} is calculated:

$$W_{min} = \frac{4.86}{9 \cdot (5-1)} = 0.135.$$

The estimated (empirical) concordance coefficient $W = 0.3$ is higher than its minimum value $W_{min} = 0.135$, which suggests that the experts' opinions are consistent.

4 Results: A Model for Promoting the Attractiveness of PT

The proposed model reflects the current situation in every city and metropolis; therefore, it is relevant not only for Lithuania's major cities but also for larger cities of Europe. What is evident is that public transportation may be made more attractive by implementing actions (Fig. 1) that consider all the factors described in the model. The motivating actions to promote the use of public transport are intended to improve the quality of the service for the commuters, making it easy and understandable at the same time – modern and innovative. Required actions are the laws that should be carried out by the government or municipalities. The first step would be to implement the European Green Deal. Also banning the cars from the city centres is a method that has been used by some cities all over Europe. With these actions in mind, the PT priority in traffic should be extended although this measure could need some infrastructural changes.

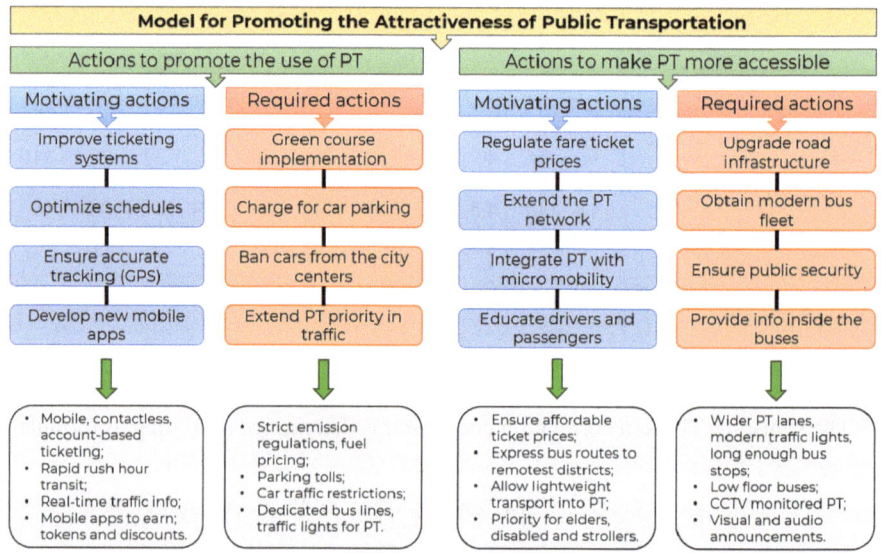

Fig. 1. The proposed scientific model for promoting the attractiveness of public transportation (developed by authors).

It must be noted that it's not enough to promote the public transportation to attract commuters to public transport because there is a part of society who doesn't have a different choice but to use PT, so it is important to take actions to make public transport more accessible and safer. The Multi-Criteria Decision Analysis method could be used to assign scores to safety and accessibility factors and weight them based on importance then the final score would reflect overall attractiveness. It should be noted that some of the actions presented in the model are mandatory in EU countries. However, the action plan on encouraging modal shift to public transport and the action plan on improving accessibility to public transport will help to create an attractive public transit system. Non-EU countries could adapt the model in the same way if they would agree to join the Green Deal and Vision Zero policy and withdraw from car-centric thinking.

5   Conclusions

1. Literature analysis shows that traffic congestion remains a major problem on the roads and suburban motorways of major Lithuanian cities.
2. To identify the problem areas of unattractiveness of public transport, a qualitative study was carried out by interviewing experts in the field. The estimated concordance coefficient is higher than its minimum value, which suggests that the experts' opinions are consistent.
3. The proposed model reflects the current situation in each city and metropolitan area and is therefore relevant not only for Lithuania's major cities, but also for larger European cities.

References

1. Dimoula, K.: The perceived attractiveness of public transportation. Master thesis. Erasmus Universiteit Rotterdam, Erasmus school of economics (2018). https://thesis.eur.nl/pub/47810/Dimoula-K-469624-MA-thesis.pdf
2. Juškevičius, P., Valeika, V.: Transport systems of Lithuanian cities [Lietuvos miestų susisiekimo sistemos]. Vilnius: Technika (2019). https://doi.org/10.20334/2019-041-M
3. Samašonok, K., Jarašūnienė, A., Išoraitė, M.: A study of the satisfaction of the population of major Lithuanian cities with public transport services. Bus. Theory Practice **22**(2), 392–405 (2021). https://doi.org/10.3846/btp.2021.14708
4. Bazaras, D., Jarašūnienė, A., Norkūnas, M.: Improvement of the urban transport system by developing the platform "Park and Ride" in Vilnius City. In: Prentkovskis, O.Y. (eds.) TRANSBALTICA XII: Transportation Science and Technology. TRANSBALTICA 2019. Lecture Notes in Intelligent Transportation and Infrastructure. Springer, Cham (2022). https://doi.org/10.1007/978-3-030-94774-3_62
5. Jarašūnienė, A., Česnulaitis, D.: Improving the efficiency of the Vilnius City transport system in the context of sustainable mobility and multimodality. Baltic J. Road Bridge Eng. **16**(3), 31–46 (2021). https://doi.org/10.7250/bjrbe.2021-16.531
6. Verbraucherzentrale Bundesverband e.V. Verbrauchern den Umstieg in den öffentlichen Verkehr erleichtern. Berlin. (2020, 19 Oktober). Retrieved from vzbv: https://www.vzbv.de/sites/default/files/downloads/2020/11/17/positionspapier_offentlicher_verkehr_verbandsthema_allgemeiner_teil.pdf

7. Klingenhöfer, F., Huber, B.: Blick in die Zukunft „Mobilitätswende 2030" Vom Linienbus zur öffentlichen Mobilität der Zukunft. Fraunhofer IESE – Institut für Experimentelles Software Engineering (2022). https://www.iese.fraunhofer.de/content/dam/iese/dokumente/media/stu dien/mobilitaetswende_2030-dt-fraunhofer_iese.pdf

8. Saeidi, O.: Investigation and Analysis of Barriers to Public Transport Development in Ahvaz Metropoli. . Iran: Shahid Chamran University of Ahvaz. https://vtpi.org/Omid_Ahvaz_P ublic_Transit_Summary.pdf

9. Commission for the Environment, Climate Change and Energy. Implementing the European Green Deal: Handbook for Local and Regional Government (2022). https://doi.org/10.2863/ 359336

10. Singal, B.: Public Transport Planning: A Viable Integrated Multimodal Citywide Network. Copal Publisching Group, India (2019)

11. Macioszek, E., Kurek, A.: The use of a park and ride system a case study based on the City of Cracow (Poland). Energies **13**(13), 3473 (2020). https://doi.org/10.3390/en13133473

12. Ogryzek, M., Kmie, D.A., Klimach, A.: Sustainable transport: an efficient transportation network - case study. Sustainability **12**(19), 8274 (2020). https://doi.org/10.3390/su12198274

13. Campos-Alba, C.M., Prior, D., Pérez-López, G., Zafra-Gómez, J.L.: Long-term cost efficiency of alternative management forms for urban public transport from the public sector perspective. Transp. Policy **88**, 16–23 (2020). https://doi.org/10.1016/j.tranpol.2020.01.014

14. Litman, T.: Evaluating Public Transit Benefits and Costs: Best Practices Guidebook. Retrieved from Victoria Transport Policy Institute, 9 May 2023. https://www.vtpi.org/tranben.pdf

15. Zarabi, Z., Taniguchi, A., Waygood, E.O.D.: Shifting to public transport: The influence of soft interventions. Retrieved from École Polytechnique de Montréal (2021). https://www.researchgate.net/publication/357132178_Shifting_to_public_tra nsport_The_influence_of_soft_interventions

16. Ranceva, J., Ušpalytė-Vitkūnienė, R.: Models of public transport organization and management in Lithuania and foreign countries. Mokslas – Lietuvos Ateitis / Science – Future of Lithuania **13**, 1–10 (2021). https://doi.org/10.3846/mla.2021.15168

17. Future of public transport: Report. KPMG (2022). https://assets.kpmg.com/content/dam/ kpmg/xx/pdf/2022/11/future-of-public-transport-report.pdf

18. Kuhn, B.: Transportation Engineering: A Practical Approach to Highway Design, Traffic Analysis and Systems Operations. McGraw-Hill Education (2019)

19. Maskeliūnaitė, L., Sivilevičius, H.: Identifying the importance of criteria for passenger choice of sustainable travel by train using ARTIW and IHAMCI methods. Appl. Sci. **11**(23), 11503 (2021). https://doi.org/10.3390/app112311503

20. Bolboacă, S.D., Jäntschi, L., Sestraş, A.F., Sestraş, R.E., Pamfil, D.C.: Pearson-Fisher Chi-Square Statistic Revisited. Information **2**(3), 528–545 (2011). https://doi.org/10.3390/info20 30528

21. Turhan, N.: Karl Pearson's chi-square tests. Full Length Research Paper. Educ. Res. Rev. **15**(9), 575–580 (2020). https://doi.org/10.5897/ERR2019.3817

22. Sivilevičius, H.: Application of expert evaluation method to determine the importance of operating asphalt mixing plant quality criteria and rank correlation. Baltic J. Road Bridge Eng. **6**(1), 48–58 (2011). https://doi.org/10.3846/bjrbe.2011.07

Open Access This chapter is licensed under the terms of the Creative Commons Attribution 4.0 International License (http://creativecommons.org/licenses/by/4.0/), which permits use, sharing, adaptation, distribution and reproduction in any medium or format, as long as you give appropriate credit to the original author(s) and the source, provide a link to the Creative Commons license and indicate if changes were made.

The images or other third party material in this chapter are included in the chapter's Creative Commons license, unless indicated otherwise in a credit line to the material. If material is not included in the chapter's Creative Commons license and your intended use is not permitted by statutory regulation or exceeds the permitted use, you will need to obtain permission directly from the copyright holder.

Monitoring of Acoustic Performance
of Noise Barriers Along Roads in Ireland

Andreas Fuchs[1]([✉]), Marco Conter[1], Maureen Marsden[2], and John Cullen[2]

[1] AIT Austrian Institute of Technology GmbH, Vienna, Austria
andreas.fuchs@ait.ac.at
[2] Fehily Timoney and Company, Dublin, Ireland

Abstract. Noise barriers are an integral part of noise abatement measures for road and rail traffic noise. To assess the quality of installed noise barriers in-situ, two standardized methods are used, which assess the intrinsic characteristics of sound reflection (EN 1793-5) and airborne sound insulation (EN 1793-6) under direct sound field conditions on roads. In a long running monitoring project funded by TII (Transport Infrastructure Ireland), over 120 noise barrier measurements have been performed in Ireland in the last 5 years. All noise barriers are timber noise barriers and are categorized as reflective and absorptive. The overall performance of typical Irish noise barriers is shown, based on the datasets available. Due to the local climate, not all measurements could be performed with the noise barrier in a dry condition. In the comparison of the results, the measured moisture content of the noise barrier is explicitly analysed and the effect of the intrinsic characteristics of sound reflection and airborne sound insulation under direct sound field conditions is shown. Finally, the obtained results are compared to the typical performance of European noise barriers obtained from the SOPRANOISE project.

Keywords: noise barrier · monitoring · sound reflection · airborne sound insulation

1 Introduction

The National Roads Network in Ireland totals 5,306 km. There is approximately 220 km of Environmental Noise Barriers (ENBs) installed along the national road network (excluding earth berms and walls) and the majority of these ENBs were installed between 2004 and 2017. At the beginning of 2019 Transport Infrastructure Ireland (TII) funded a five-year monitoring project to assess the intrinsic characteristics for sound reflection and airborne sound insulation of installed noise barriers over the years. During this project Fehily Timoney & Company and the AIT Austrian Institute of Technology GmbH have completed in-situ measurements of 20 noise barriers each year according to EN 1793-5 [1] for sound reflection and EN 1793-6 [2] for airborne sound insulation properties under direct

© The Author(s) 2025
C. McNally et al. (Eds.): TRAconference 2024, LNMOB, pp. 314–320, 2025.
https://doi.org/10.1007/978-3-031-85578-8_40

sound field conditions. Most of the noise barriers on the Irish road network are of timber construction and have seldom developed a design identity of their own. These barriers are generally considered as reflective barriers; however, on some barriers an absorptive material is fixed to the barrier surface.

2   Measurement Campaign

The measurement campaign of the monitoring project started in 2019 and was completed over five consecutive years. The noise barriers were selected by TII and all measurements were performed by Fehily Timoney and Company. The noise barriers tested are timber barriers, and are categorized into two Categories - *reflective* and *absorptive*.

For the five years, 20 airborne sound insulation measurements and between five to eight sound reflection measurements were performed each year, giving a total of 100 airborne sound insulation on absorptive and reflective barriers and 31 reflection measurements on absorptive barriers. The manufacturer information (supplied by TII where available) and the noise barrier locations have been anonymized for this publication and each noise barrier is identified with a three digit code: The first digit identifies the product type (A for absorptive, R for reflective). The second digit identifies the manufacturer (X, Y, Z or U for Unknown) and the third digit is an integer index (for barrier 1 to 9). For example, the noise barrier AX3 is an absorptive barrier of the manufacturer X with the index 3. For the noise barriers with unknown manufacturer no unambiguous identification of the manufacturer and type was possible.

Table 1 shows all 26 measured noise barriers with the height of the barrier and the year of installation, where known. Where tests were repeated on a barrier, they were in different years. For noise barriers with heights below 4 m, the Adrienne temporal window must be reduced, resulting in a higher lower frequency limit f_{min}. This must be considered when comparing results of the single-number ratings, as averaging could be performed over a different frequency range.

The reflective barriers are a single leaf construction, with vertical timber elements. The joins of the timber vertical elements on one side of the barrier have overlapping timber elements. All, but one barrier had I-section steel posts-a single barrier had timber posts. The fixing mechanism around the post varied-some barriers had very deep posts with a large timber packer, and some had narrower steel posts. The top of the timber barriers had a wooden cap. Many of the barriers had horizontal timber elements to the rear of the barrier. The barriers (RZ1 to RZ5) had a concrete base. Some timber barriers were constructed of tongue and groove vertical timber elements (AU2, AU3, RU6 and RU7). Eight barriers had joins in the panel, where the barrier was composed of two panels. Six of the barriers had joins within the area covered by the microphone grid.

The absorptive barriers tested were of a timber construction, with similar construction to the reflective barriers. Most had overlapping elements over the vertical boards. The absorptive material was mostly protected by a weather proof membrane comprised of different materials, such as polyethylene or woven

Table 1. Number of Measurements for each noise barrier with the height of the noise barrier, the year of installation and the minimum third-octave band frequency f_{min} for the airborne sound insulation index SI_j and the sound reflection index RI_j

Barrier	AX1	AX2	AX3	AY1	AU1	AU2	AU3	AU4	AU5	RX1	RX2	RX3	RZ1
# Measurements	3	5	5	4	2	5	5	1	1	5	5	3	4
Installation year	'17	'17	'17	'09	-	-	'10	-	-	'17	'17	'17	'19
height (m)	3.00	2.01	3.53	3.02	2.02	3.02	2.48	3.55	3.03	1.99	3.02	3.04	3.06
f_{min} for SI_j (Hz)	315	630	315	315	630	315	400	315	315	630	315	315	315
f_{min} for RI_j (Hz)	400	1600	315	400	1600	400	630	315	400	-	-	-	-

Barrier	RZ2	RZ3	RZ4	RZ5	RU1	RU2	RU3	RU4	RU5	RU6	RU7	RU8	RU9
# Measurements	5	5	5	5	2	3	3	4	4	1	5	5	5
Installation year	'19	'19	'19	'19	-	-	-	'08	-	-	-	-	'16
height (m)	3.05	3.05	2.52	3.57	1.97	3.51	3.69	3.00	2.58	2.10	2.24	2.50	2.97
f_{min} for SI_j (Hz)	315	315	400	315	630	315	315	315	400	630	630	400	315

membranes. There were vertical wooden elements at regular intervals on the absorbing side of the barrier.

During the measurements all the meterological conditions required by the standards were recorded (moisture content, wind speed and air temperature). The moisture content was measured with a pin-type meter measuring the electric resistance between two electrodes. Air humidity was not recorded, but as the standard requires a dry surface of test object, measurements were not performed during rain or foggy weather and the humidity can be estimated to be <90 %.

3   Results Sound Reflection

Figure 1 shows the single-number rating DL_{RI} for sound reflection for all barriers with the measurement year encoded in the marker colour. The three barriers for

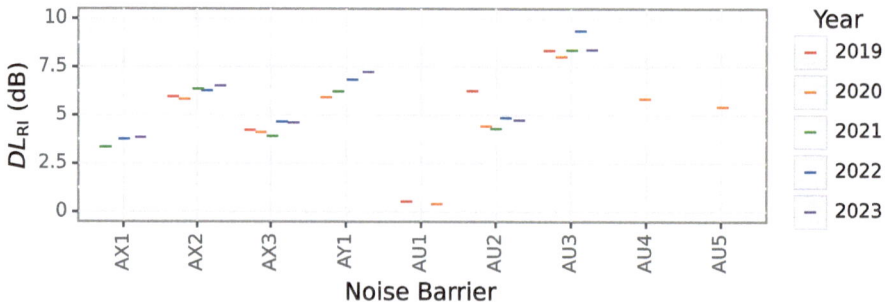

Fig. 1. Overview of the single-number rating DL_{RI} for sound reflection. Please note the different valid minimum frequency f_{min} from Table 1.

manufacturer X (AX1-AX3) show very different results for the different barriers, although the repeatability over time is quite consistent. Nevertheless, the barrier AX2 with a height of approx. 2 m shows significant higher values for the single-number rating. This may be explained by the fact that AX2 has rounded vertical slats, which generates a more diffuse sound field. AX1 and AX3 have both flat vertical wooden slats.

However, the Adrienne temporal window for the barrier AX2 was shortened excessively to values down to 1.3 ms (from 6.0 ms), which may lead to excessive cutting of the impulse responses. This could also be true for the sample AU3 with a height of approx. 2.5 m. The single-number rating for AU1 was calculated from above 1600 Hz, where the polyethylene weather membrane seems to be very reflective, resulting in quite low sound reflection performance.

Figure 2 shows the sound reflection index for two barriers, where both barriers showed differences in the high frequency behaviour for the repeated measurements. For AY1, the sound reflection performance increased over the monitoring period for third-octave band values above 1 kHz. This barrier had no weather proof lining at start and end of the measurement period. The absorbing material was more damaged with time (and had lots of dust from the road), which might lead to increased sound absorption due to an improved impedance match between absorptive material and air. Nevertheless, AU2 had a significant change in the sound reflection index above 2 kHz between the measurements in 2019 and 2020. Tests after 2020 indicate a significantly decreased performance in the upper frequency range, where most of the sound energy is reflected.

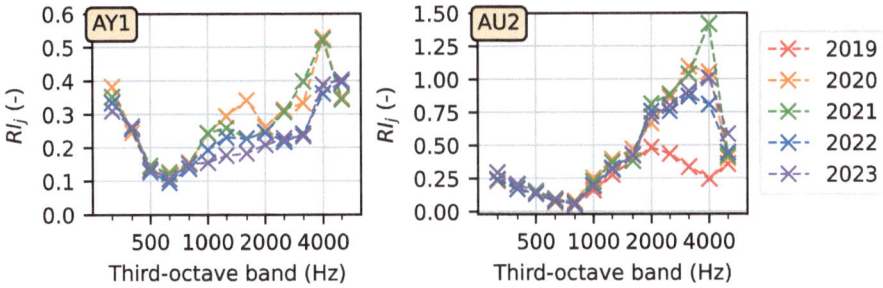

Fig. 2. Third-octave bands of the sound reflection index RI_j for the barriers AY1 (left) and AU2 (right).

In general, the obtained values for sound reflection (although most of them from low-height barriers) are in good agreement with the values of typical timber noise barriers (4.5 dB to 6.5 dB) from the SOPRANOISE database [4]. No relationship between the sound reflection properties and the moisture content could be found.

4   Results Airborne Sound Insulation

Figure 3 shows the single-number rating $DL_{SI,E}$ for airborne sound insulation in front of the acoustic element with the measurement year encoded as marker colour. The most obvious difference is between the absorptive and reflective barriers with the exception of RU7, which has a high variation of the individual measurements as well as values above 30 dB. The high sound insulation can be attributed to an increased thickness of 15.5 cm and the low height of 2 m and the reduced Adrienne time window. In general, the absorptive barriers have a better sound insulation performance, with values between 20 and 35 dB. The reflective noise barriers generally have values between 15 to 22 dB. In the SOPRANOISE database [4] many measurements on timber noise barriers are evaluated on a reduced frequency range. Here, the range of values for the $DL_{SI,E}$ is from 15 to 45 dB. The absorptive barriers can be considered in the centre of this range, where nearly all of the reflective noise barriers are at the lower limit.

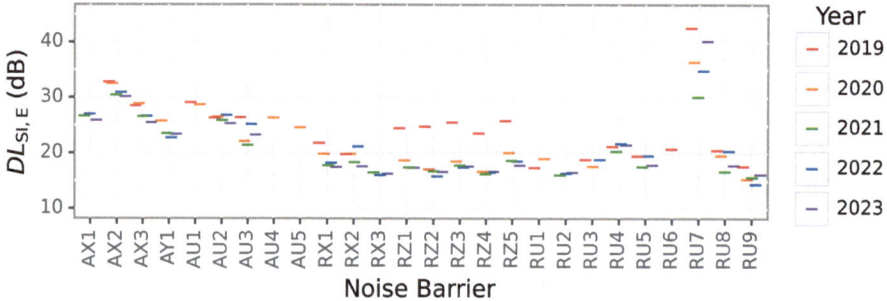

Fig. 3. Overview of the single-number rating $DL_{SI,E}$ for airborne sound insulation. Please note the different valid minimum frequency f_{min} from Table 1.

For the five reflective noise barriers of manufacturer Z, the measurements in the year of installation (2019) show a value around 25 dB. In the following years the airborne sound insulation performance of these barriers is significantly reduced (where the reduction happens across the whole frequency range). The measured performance in later years of RZ1 to RZ5 is similar to other reflective barriers, whose year of installation is unknown or at least two years before the first measurement. Nevertheless, it must be considered, that the measurements in the first year of these noise barriers were performed in December in cold and damp weather. Also, the wood of these noise barriers was unweathered and a yellow/pine colour in the first year. Most of the measured reflective barriers and also RZ1-RZ5 in the following years had weathered to a silver/grey colour and presumable had lost the moisture that newer wood contains. According to Morgan [3] the performance of timber noise barriers might decrease in the 25 d after installation, as the increased moisture content due to pressure treatment decreases. The measurements of RZ1-RZ5 in the year 2019 were performed at minimum one month after installation.

Figure 4 shows a comparison of the $DL_{\mathrm{SI,E}}$ and the moisture content of the reflective noise barriers. The points in the upper right corner correspond to the barriers RZ1-RZ5 with a moisture content of around 20%. This gives an indication that the results might be influenced by an increased moisture content of the elements, as the airborne sound insulation performance of timber noise barriers decreases after installation due to loss of moisture content and may also react to increased air humidity and rain.

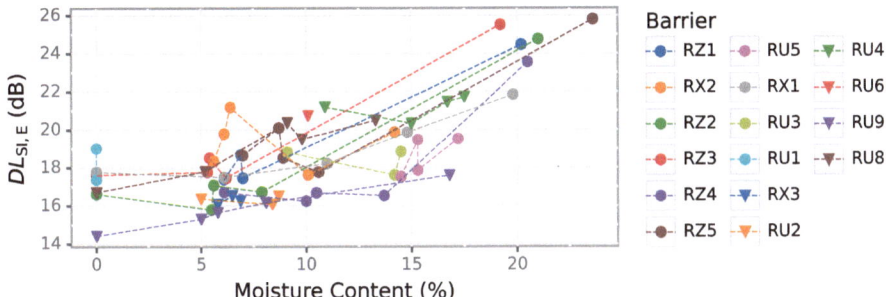

Fig. 4. Comparison between $DL_{\mathrm{SI,E}}$ and the moisture content of the reflective noise barriers

In the analysis of the performed measurements for airborne sound insulation at the post, no new information could be found.

5 Summary

This paper gives an short overview of the monitoring project of the acoustic performance of timber noise barriers in Ireland. Due to the small height of many of the observed noise barriers, results can not easily compared between the barriers, but monitoring of the performance of the individual noise barriers is still possible. For sound reflection the performance was mostly constant, one barrier performance steadily increased, and another decreased. Both effects can be attributed to degradation due the environmental influences. For airborne sound insulation five similar newly built reflective timber noise barriers showed a significant decrease of 6–7 dB one year after installation and highlights the importance of in-situ testing and monitoring of installed noise barriers.

References

1. EN 1793-5:2016+AC:2018, Road traffic noise reducing devices - Test method for determining the acoustic performance - Part 5: Intrinsic characteristics - In situ values of sound reflection under direct sound field conditions (2018)
2. EN 1793-6:2018, Road traffic noise reducing devices - Test method for determining the acoustic performance - Part 6: In situ values of airborne sound insulation under direct sound field conditions (2018)
3. Morgan, P.A.: The acoustic durability of timber noise barriers on England's strategic road network. Published Project Report 490, Transport Research Laboratory, Wokingham UK (2010)
4. Conter, M., Fuchs, A.: D2.2 - Final report on the main results of WP2 - Acoustic assessment of the intrinsic performances of noise barriers. Project report SOPRA-NOISE (2021)

Open Access This chapter is licensed under the terms of the Creative Commons Attribution 4.0 International License (http://creativecommons.org/licenses/by/4.0/), which permits use, sharing, adaptation, distribution and reproduction in any medium or format, as long as you give appropriate credit to the original author(s) and the source, provide a link to the Creative Commons license and indicate if changes were made.

The images or other third party material in this chapter are included in the chapter's Creative Commons license, unless indicated otherwise in a credit line to the material. If material is not included in the chapter's Creative Commons license and your intended use is not permitted by statutory regulation or exceeds the permitted use, you will need to obtain permission directly from the copyright holder.

Trends and Scenarios of Urban Logistics in EU: A Survey in 21 Greater Cities

Fabio Cartolano[1](✉), Carlo Vaghi[1] (iD), and Maria Rodrigues[2]

[1] FIT Consulting Srl, Via Merulana 272, 00185 Roma, Italy
cartolano@fitconsulting.it
[2] Panteia BV, Bredewater 26, 2715 CA Zoetermeer, The Netherlands

Abstract. The knowledge of urban logistics in European cities is fragmented and data is not harmonised due to lack of systematic methodological approaches, along with the reluctance of targeted operators to share information in a highly competitive market. A recent survey targeted on urban logistics has been commissioned by the European Commission, within the "Study on new Mobility Patterns in European Cities", to collect data and make available indicators related to economic, environmental, and usage aspects of goods distribution and other logistics movements in EU cities. The survey was developed in 21 EU greater cities of 12 Member States plus UK, among 1590 logistics operators, between April 2021 and February 2022.

Additionally, a systematic review of the main previous studies on urban logistics in the targeted cities was achieved in order to have a methodological appraisal and quantitative references to be used as background information in the selected cities, so as to validate the results of the survey and analyse any discrepancies found in the data collected by the survey with statistics from previous studies. This review revealed extreme differences in the methodologies and identified limitations in the comparisons of data collection and analysis, especially concerning the definition of indicators and consistency of targets.

The survey represents a novel achievement and a credible baseline towards understanding urban logistics dimensions and benchmarking relevant indicators. As such, the outcome of the survey can serve as baseline for future data collection and analysis with the goal to develop sustainable urban logistics policies.

Keyword: Urban logistics · City logistics · deliveries · sustainable freight transport

1 Introduction

Urban logistics deliveries are one of the most complex and least efficient segments of freight transport, being responsible for a significant share of traffic congestion and emissions in EU cities. According to the European Commission, light commercial vehicles') are responsible for around 2.5% of total EU emissions of carbon dioxide (CO_2) in EU [7]. Despite the assessment is influenced by the heterogeneity of sources, urban goods movements account for 20 to 30% of all vehicles [16] and freight vehicles account up

© The Author(s) 2025
C. McNally et al. (Eds.): TRAconference 2024, LNMOB, pp. 321–334, 2025.
https://doi.org/10.1007/978-3-031-85578-8_41

to about 25% of CO_2 emissions in specific cities [18]. At same time, urban logistics is a fundamental service for citizens and comprises a substantial part of all commercial activities contributing to local economic development.

Moreover, cities have become increasingly concerned with the impact of urban logistics in terms of traffic, noise, pollution, land use, road accidents. With the European Green Deal [9] and the Sustainable and Smart Mobility Strategy (SSMS) [10], the EU is striving to achieve the 90% cut in emissions by 2050, delivered by a smart, competitive, safe, accessible and affordable transport system, as committed by the Climate Law [11]. The SSMS aims to make all modes of transport more sustainable by identifying specific actions such as Making interurban and urban mobility healthy and sustainable and Greening freight transport. One way of achieving this is to provide well-organized, safe and environmentally friendly transport widely. To this purpose, Sustainable Urban Logistics Plans (SULP) are considered a crucial policy instrument to bring the freight dimension into urban planning processes and to accelerate the deployment of available zero-emission solutions within the freight sector. However, it is also important to take into account the various factors influencing logistics actors' choices to adopt more sustainable procedures, including operational efficiency, environmental footprint and economic sustainability.

A survey targeted on urban logistics has been recently commissioned by the European Commission, within the "Study on new Mobility Patterns in European Cities" (NMP)[1], to collect data and make available indicators related to economic, environmental, and usage aspects of distribution of goods and other logistics movements in EU cities.

The main objective of the survey was to provide analytic insights from statistical elaborations of the surveyed variables in order to establish a basis for a better understanding of the urban logistics market and actors while simultaneously promoting standardized frameworks for further in-depth surveys at city level.

The study has assessed the status of the urban logistics market in a selection of 21 European "greater cities", covering key indicators like vehicle fleet dimension and type, transport activity indicators (number of deliveries, vehicle-km, tonne-km, load factor, carbon footprint, etc.) and additional characteristics concerning delivery modes (including bike and powered two-wheelers) and current/future actions prepared by operators to reduce the carbon footprint of their urban logistics activities.

1.1 A Review of Urban Logistics Surveys

The knowledge on urban logistics in Europe is fragmented and data is not harmonized due to lack of systematic methodological approaches for collecting data. This hampers the opportunity to benchmark data from analyses conducted in different cities and regions. The gap is augmented by the reluctance of targeted operators to share information in a highly competitive and remunerative market. The survey undertaken for this study, the findings of which are presented in this document, aims at providing analytic insights from statistical elaborations of the surveyed variables in order to establish a basis for

---

[1] European Commission - Directorate General for Mobility and Transport (DG MOVE), Service Contract MOVE/A3/SER/2019–401/SI2.814412.

a better understanding of the urban logistics market and actors while simultaneously promoting standards for further in-depth surveys at city level.

Prior to the field survey a systematic review of urban logistics studies and surveys in main EU cities was performed among 26 EU city surveys, to have a methodological appraisal and quantitative references to be used as background information. This allowed the comparison of the field survey results with data from previous studies.

It should be noted that urban logistics surveys are usually conducted to set the basis for developing sound transport policies in a specific area or city-wide; Sustainable Urban Logistics Plans (SULPs), often part of Sustainable Urban Mobility Plans (SUMPs), are policy tools used by local authorities to assess and monitor logistics performances, to identify specific problematic locations, such as areas with traffic congestion, queues for delivery areas, or noise, and to identify suitable solutions (e.g. number of deliveries, peak hours volume, need for urban consolidation centres or access restrictions).

The quality of data on urban freight transport has been extensively dealt in the "Final Report - Indicators and data collection methods on urban freight distribution" of "non-binding guidance documents on urban logistics" (N° 6/6), [20]. The main findings from this analysis are that there is a lack of publicly available urban logistics data in European countries. According to the report, freight transport in many surveys and models is often neglected. Where freight data is collected, this is in most cases done for freight transport in general and not specifically at the urban level. Moreover, the methodologies used to collect data hamper the benchmark between different data.

Main Conclusion from the Literature Survey. The extensive overview of studies conducted in the selected cities revealed significant differences in the methodologies: urban logistics surveys are methodologically heterogeneous and not easily comparable. Among the main limitations encountered in the comparison of data collection:

- The definitions of indicators across studies are not always consistent (e.g. Freight movements should stand for deliveries but there is often ambiguity with trips),
- Some studies use classification systems of enterprises different than NACE [7], as a result of which the meanings of urban logistics are not aligned, sometimes including activities different from goods transportation;
- Some studies compute reference populations for indicators using inference methods, reporting extrapolated values without reporting related accuracy,
- The definitions of urban area used are sometimes generic, mentioning "inner city", "city centre" as well as "urban area" with no exact correspondence to the EC-OECD definition of the "greater city"[2] and its boundaries.

Despite these limitations, for a few cities data collected can help validating the findings of the field survey, such as fleet composition (Heavy Goods Vehicles-HGV and Light Goods Vehicles-LGV).

Other indicators were also investigated for LGV and HGV aggregated. The most significant results are the following:

---

[2] https://ec.europa.eu/eurostat/statistics-explained/index.php?title=Territorial_typologies_manual_-_cities,_commuting_zones_and_functional_urban_areas.

Table 1. Fleet composition (own elaboration from various sources).

City	% HGV	% LGV	Source	Limitations
Budapest	30%	70%	[6, 13]	Data collected in Buda Castle and related to traffic patterns
Milan	16%	84%	[153]	Data related to registered vehicles in Milan and Rome Metropolitan Areas
Rome	17%	83%		
Paris	46%	54%	[4]	
Madrid	6%	94%	[5]	Data related to registered vehicles in region
Lisbon	46%	54%	[19]	Data related to traffic patterns in city centre
Brussels	45%	55%	[15]	LGV: vehicles < 3 tonnes. Others: HGV

- Euro class: prevalence of Euro 4-5-6 classes in the range of 90% of the total circulating fleets (Hamburg, [15], and Rome, [17]), compatible with age classes, considering that Euro 5 standard was introduced from 2009, which is in line with a 12–13 year of lifecycle for most of commercial vehicles [1].
- Fuel: prevalence of Diesel (over 80% in Rome, 2019 and Madrid, 2013 [5]), although this share is going to shift towards the transition to electric vehicles.
- Deliveries: the number of deliveries is often calculated by asking shopkeepers and local business number (or frequency) of supplies and then deriving traffic flows and volumes of transported freights. In Lisbon (2015, [2]) most of local businesses (76%) declare to receive freight every working day. This indicator, however, excludes home delivery services, very relevant during the last years.
- Business model of enterprises: own account is generally found to be prevalent in cities reaching up to 50% (Paris, 2016 [20]).
- Vehicle-Km: distance is covered mainly by LGV (about 80% in Madrid, 2013 and London, 2020, [21])
- Tonne-Km: this indicator is mainly collected with direct surveys to logistics operators or intermodal hubs (i.e. ports). In some cases the indicator is provided per specific categories of goods (e.g. in Brussels palletized flows were analysed, in London a 45 Mtonne-km/day flow was estimated)
- Bicycle and powered two-wheelers: in many EU cities there are several initiatives at pioneering level to introduce bicycle and powered two-wheelers logistics (for last mile or home delivery), but no systematic studies have been found on their incidence in the urban logistics landscape. This lack of data on such initiatives is a gap that is going to be covered in the next future research.
- Pollutant emissions: freight vehicles contribute for about 25% of CO_2 emissions in the city (Brussels, 2015). Measurements are derived by elaboration of traffic flows combined with emission/km with no information about the accuracy of the estimations.

Finally, for some cities, additional studies (suitable for London, Brussels, Milan and Paris) were reviewed to retrieve the total number of tonne-km, deliveries or trips per day and to use such indicators to attempt bringing data collected in the survey up to the dimension of the reference population. However, the accuracy and the geographical coverage of the primary data were not indicated and, while comparing data normalized per population, discrepancies among various cities have been found.

2 Methods of the Field Survey

The field survey performed in NMP study has been developed in 21 EU greater cities of 12 Member States plus UK, among logistics operators providing delivery services (on own account or for third parties). The sample comprised the performance of 1590 interviewed companies. Operators interviewed provided features of 2800 reference delivery trips, developed over 11703 vehicle-days between April 2021 and February 2022. The translated NMP questionnaire was submitted in CATI and CAWI formats.

The target population was all freight deliveries by HGVs and LGVs, travelling in to, out of and within the boundaries of the 21 greater cities, each having a resident population of at least 1 million: Barcelona, Berlin/Munich, Brussels/Antwerp, Bucharest, Budapest, Hamburg/Bremen, Lisbon, London, Madrid, Milan, Paris, Prague, Rome, Rotterdam/Amsterdam, Sofia, Stockholm/Gothenburg.

The panel consisted in enterprises in the NACE categories G.46 (wholesale trade, except of motor vehicles and motorcycles), H.49.41 (freight transport by road), H.52.24 (cargo handling), H.52.29 (other transportation support activities), H.53 (postal and courier activities) with a defined ZIP code of the designated urban areas. For each city a sample quota was created in the data collection application to keep track of the sample disposition for each city in the conduction of the telephonic interviews.

The average refusal rate of 75% was higher than expected compared to similar B2B surveys conducted by the provider in the same period. Thus some remedial actions were set out to make a panel of 119,940 enterprises available for contacts, randomly selected for interviewers. At the end of the survey, 6,782 companies were contacted, with a survey response rate of 23.44%, leading to the final outcome of 1,590 interviews.

Survey Quality Indicators. In order to focus on most significant indicators surveyed by the questionnaire, some "survey quality indicators" were selected as most statistically significant. Those were calculated according to Eurostat Guidelines on passenger mobility statistics [13], considered valid for freight transport as well.

Statistically, the net sample of 1,590 interviews is greater than 384 (5% of the total sample), which is the minimum threshold to sit in the 95% of confidence interval and 5% of error. Therefore, results are largely within the range of the margin of error.

As concerns economic data of the enterprise and fleet size and future plans for low emission logistics, indicators retained were (i) composition of fleet owned by the enterprise, (ii) number of city deliveries by bikes, (iii) number of city deliveries by powered two-wheelers, (iv) share of companies who have a plan for low emission logistics, (v) share of companies who do not have a plan for low emission logistics.

As concerns "Activity and traffic data", survey quality indicators are (a) number of deliveries, (b) vehicle-km, (c) tonne-km, (d) average load factor, (e) pollutant emissions. For these indicators the reference population is the number of trips performed by a vehicle in a day. The analysis has not included indicators that exceeded – by city – a 5% Margin of Error (MoE), threshold above which data are considered not reliable.

Table 2. NMP field survey performance (Source: NMP)

City	Completed interviews	Reference Trips	Completed Vehicle Days	Trips per Interview	Vehicle days/Trip	Vehicle days per Interview
Rotterdam	43	72	428	1.7	5.9	10.0
Amsterdam	27	49	279	1.8	5.7	10.3
London	101	155	711	1.5	4.6	7.0
Sofia	170	177	706	1.0	4.0	4.2
Prague	123	215	789	1.7	3.7	6.4
Lisbon	66	165	709	2.5	4.3	10.7
Bucharest	124	275	763	2.2	2.8	6.2
Barcelona	87	155	712	1.8	4.6	8.2
Madrid	88	167	781	1.9	4.7	8.9
Paris	122	171	809	1.4	4.7	6.6
Hamburg	52	87	399	1.7	4.6	7.7
Bremen	18	42	312	2.3	7.4	17.3
Budapest	158	293	737	1.9	2.5	4.7
Rome	105	176	712	1.7	4.0	6.8
Milan	88	161	705	1.8	4.4	8.0
Stockholm	38	98	597	2.6	6.1	15.7
Goteborg	16	28	111	1.8	4.0	6.9
Brussels	60	137	664	2.3	4.8	11.1
Antwerp	10	15	45	1.5	3.0	4.5
Berlin	53	93	467	1.8	5.0	8.8
Munich	41	69	267	1.7	3.9	6.5
Total	**1590**	**2800**	**11703**	**1.8**	**4.2**	**7.4**

Table 2 reports the detail of interviews and related targets achieved. The total performance of the survey was 1590 interviews, surveying 2800 "reference trips" for deliveries and pick-ups of goods, deployed over 11708 vehicle-days. Where:

- *Trip (or "delivery trip")* is defined as the movement made by a vehicle deployed to perform deliveries and pick-ups of goods during a day, from the start of the day operations to its end. More than one trip could be performed during the working day;
- *Reference trip* defines one or more routes usually covered by one or more vehicles of the company fleet to perform deliveries in the city;
- *Completed vehicle days* define the total transport duration of reference trips performed by for the net sample size in each city.

3   Results and Discussion

According to the response to the questionnaire during the 1590 interviews performed, the present section describes a sample of results achieved by the NMP survey on (i) economic data of the enterprise and fleet size, (ii) activity and traffic data.

Fleet Composition. Enterprises performing urban distribution are made by a large share of LGV (73%), whereas 16% of interviewed enterprises declare to own both LGV and HGV. Cities with the highest share of LGV are Brussels and Prague with over 90%. Exceptions are recorded in Barcelona, with an equal distribution between LGV and HGV, and Bremen with a larger share of HGV. The overall distribution confirms the literature review findings, with some cities increasing the share of LGV over the years (Paris, Lisbon and Brussels) (Figs. 1 and 2).

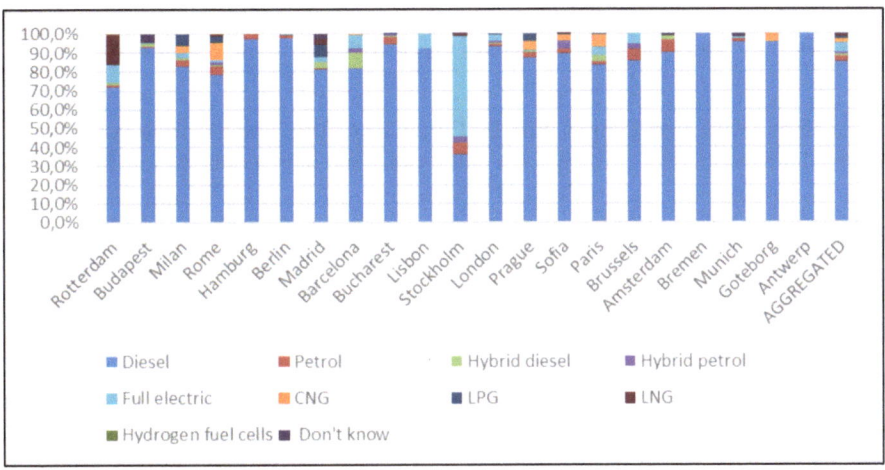

Fig. 1. Fleet composition - performed trips by type of fuel LGV (Source: NMP)

As evident in the figures, diesel is predominant both types of vehicles, with diesel HGV exceeding 90%. Other types of fuel are CNG/LNG (respondents may have used indifferently one or the other definition for both types of vehicles). For LGV there is a limited share of petrol (more than 3%) and electric (more than 2%). At city level, the distribution is similar with the exceptions of:

- Full electric for LGV, which penetration in Brussels, Lisbon, Paris and Rotterdam exceeds 6% and Stockholm, where the share is over 53%;
- Full electric for HGV in Rotterdam with a share of 14%;
- Hybrid diesel for LGV, whose share in Paris and Madrid is now above 3% and Barcelona, where the share is 8%.

These exceptions show that policies aiming at restricting the circulation of more pollutant vehicles produced a modal shift toward cleaner vehicles.

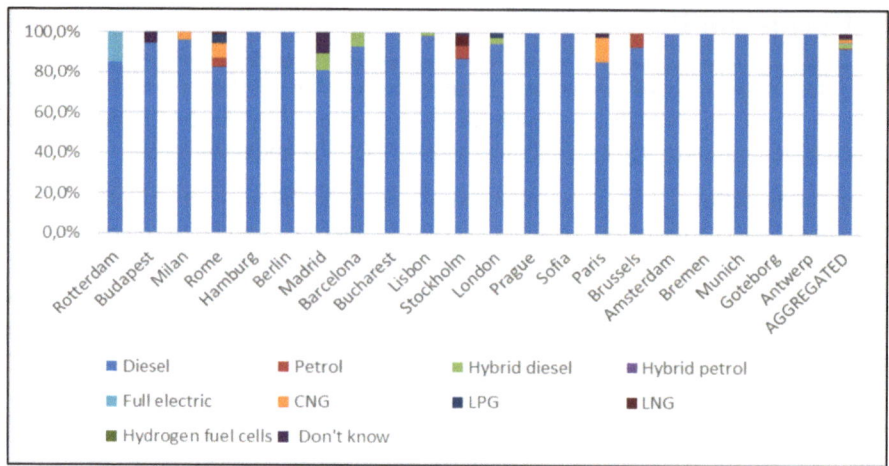

Fig. 2. Fleet composition - performed trips by type of fuel HGV (Source: NMP)

Looking at the fleet composition by Euro Class, Euro 5 and Euro 6 comprise the largest share in all cities. Still, in some cities, older and more pollutant vehicles (Euro 4 and less) have a relevant share: Madrid records almost 40% of LGV Euro 4 or less (20% Euro 0), Sofia 37%. Stockholm on the contrary records 86% of Euro 6.

Business Models. Transport on "Own account" is the prevalent category with 50% of share as expected and in line with literature review. Barcelona and Bremen record the lowest shares of enterprises providing delivery services on own account (respectively 20% and 28%). Stockholm is the city with the highest share of deliveries performed on behalf of other logistics operators (39%). This feature is usually present in well interconnected logistics ecosystems, with high level of cooperation among operators, thus increased efficiency.

Deliveries by Bike. Deliveries by bike and powered two wheelers are very limited in all cities. 2,58% of enterprises perform deliveries by bike and 3.65% of enterprises perform deliveries by powered two-wheelers. Bicycle logistics is not very developed nor monitored and this data confirm the limited share of logistics operators using such mode. Due to the limited number of respondents, additional dimensions per city, NACE category or other variables that, for the collected sample, show equal distributions.

Number of Deliveries. This very relevant indicator shows the different degree of efficiency of delivery trips, provided that the number of deliveries for a *full loaded* vehicle may vary across cities due to the peculiarity of cities (e.g. port cities having a larger share of container traffic) and businesses. The average number of deliveries per trip range from 4 to 10. Antwerp and Goteborg were excluded from the calculation, because of unreliable results due to limited data collected.

Expectedly, LGV perform the large majority of deliveries among all cities (80% of the total sample), except the peculiar case of Munich, where 57% are performed by HGV. The same correspondence with data on fleet composition is showed for deliveries by type of fuel, with strong prevalence of Diesel and some exceptions for electric vehicles (Stockholm) and LPG (Milan), or CNG/LNG (Rome, Sofia).

Concerning the distribution of deliveries by Euro class, there is a general prevalence of Euro 5 and Euro 6 and the distribution reflects what was presented in the fleet composition, with the exception of Madrid where LGV Euro 0 vehicles are used to perform more deliveries with a share of 31% (against 20% of the fleet composition). A similar situation applies to Barcelona for HGVs, where the number of deliveries with Euro 2 is 10%, against 7% of the fleet composition, highlighting more intense activity performed by enterprises owning older vehicles (Fig. 3).

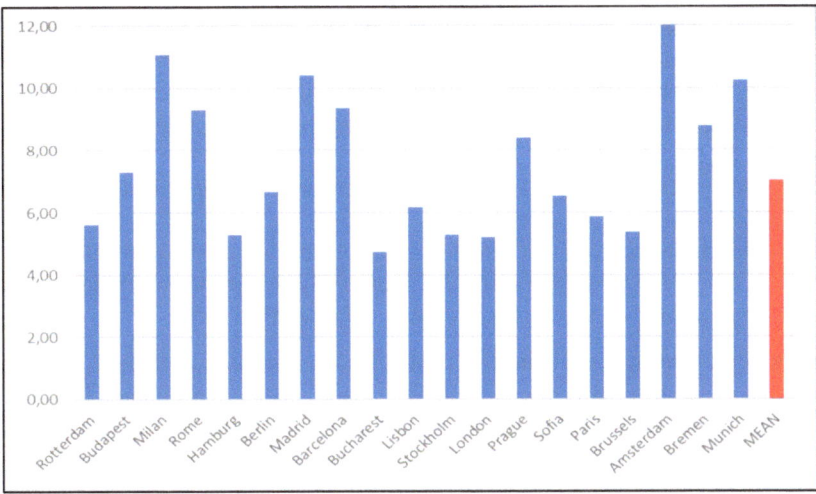

Fig. 3. Number of deliveries per trip per city (Source: NMP)

Vehicle-Kilometres. Vehicle-km is the reference indicator for traffic flows and provides a picture of the intensity of urban logistics trips. The indicator shows trends similar to fleet composition, namely a prevalence of LGV. However there are some deviations in London, where about 43% of Vehicle-Km are performed by LGV, against 65% in the fleet composition, and in Rotterdam (56% of Vehicle-Km by LGV, against 70% in the fleet composition). These results indicate an average greater distance driven by HGV in those cities, confirmed also at aggregate level (68% of vehicle-km for LGV, against 75% of the fleet composition for this type of vehicle).

The distribution by type of fuel does not show deviations when compared to the fleet composition, with Diesel as the predominant fuel. In terms of distribution by Euro class, it is interesting to note that lower Euro classes are used less intensively. For LGV, Madrid records 23% of vehicle-km with Euro 4 or less (representing 40% of the fleet), Sofia 17% (37% of the fleet) and Barcelona 8% of Euro 0 (10% of the fleet). For HGV, Sofia records 17% of vehicle-Km with Euro 4 or less (representing 35% of the fleet).

These results indicate that the fleets of vehicles with lower Euro class are used for limited distances in cities, due to circulation restrictions.

Tonne-Kilometres. Tonne-km represents the volume of freight goods delivered/collected in a specific area. As underlined in the literature review, a clear methodology to capture this indicator is lacking. Data collected in this survey can be considered as a baseline to collect indicators applied to urban logistics, and to investigate their evolution over time, together with vehicle-km. Such developments could be relevant to understand increasing home delivery practices, supposed to contribute to fragment the last mile distribution and therefore expected to increase vehicle-km (i.e. traffic flows in cities) more than tonne-km. The impact of home delivery also depends on drop density, speed, frequency of delivery and the return policy.

Concerning the distribution by fuel, in addition to the expected prevalence of Diesel fueled vehicles, the share of electric vehicles in Stockholm is remarkably lower for this indicator (6%) than the share for vehicle-km (36%): e-vehicles are mostly used for last mile deliveries and for transporting smaller volumes., the share of vehicles with lower Euro class (Euro 4 or less) in some cities deviates from the vehicle-km distribution. For LGV, Madrid has 76% of Tonne-km performed by Euro 4 or less (against 23% of vehicle-km for these classes) Sofia has 40% (against 17%) and Barcelona 28% of Euro 0 (against 8% of vehicle-km). For HGV, Barcelona records 28% of Euro 4 or less (against 21% of vehicle-km for these classes).

These deviations confirm that vehicles with lower classes are used to transport larger volumes of freight but for shorter distances, most likely heading to consolidation centres and avoiding last mile deliveries because of circulation restrictions.

Average Load Factor. This indicator is calculated to identify the efficiency of delivery activities in a specific urban environments. The load capacity of a commercial vehicle is dependent from a combination of space (goods with light weight but high volumes, i.e. toilet paper) and weight (less volume higher weight, i.e. liquid wash machine soap). For the scope of this study and according to the level of quality of the information collected, this has been calculated as a ratio between the weight of parcels and load capacity of the vehicle (Fig. 4).

At the aggregate level, the average indicator is 50% for both LGV and HGV. Cities with local logistic operators presenting the lowest loading factors are Barcelona (30%), Rotterdam (32%) and Amsterdam (28%). The highest loading factors are reported in Madrid (74%), Lisbon (78%), Stockholm (72%) and Antwerp (82%).

The collected data does not show a clear correlation with other variables but represents a baseline that will be useful for future comparative studies. Analysing the evolution over time in each city could be relevant to understand the efficiency of logistics processes and could offer suitable information to decision makers for designing urban logistics processes. In this regard, the average load factor is a relevant proxy indicator for estimating the overall urban logistic market efficiency. Such indicator is very prone to be influenced by external factors, such as policies aimed at increasing it (e.g. introducing space or temporal incentives or restrictions) that could results in distortions like detours and extra-stops, impacting traffic congestion or usage of public spaces.

Greenhouse Gases (GHG) Emissions. The emission impact was calculated upon distances driven and by using standard Copert data [7] applied per type of vehicle, maximum allowed mass, Euro class, rigid/articulated (for HGV only), load factor (ratio between total weight and maximum allowed mass).

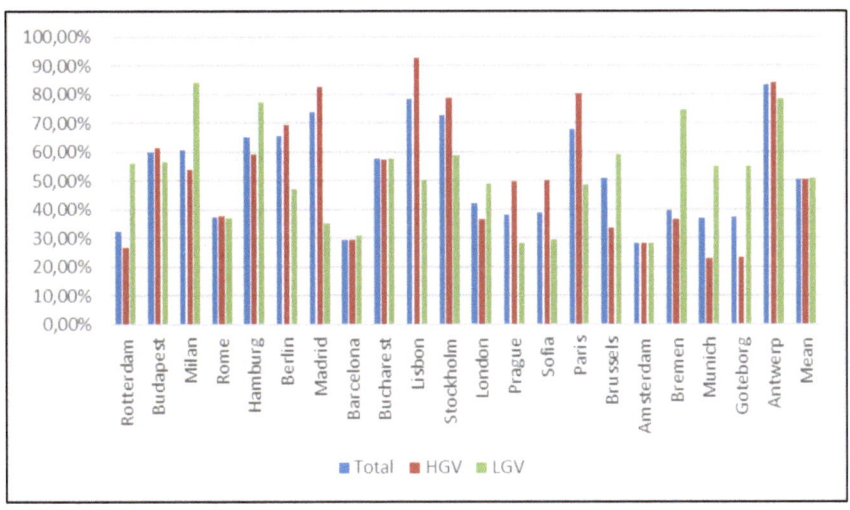

Fig. 4. Average loading factor (Source: NMP)

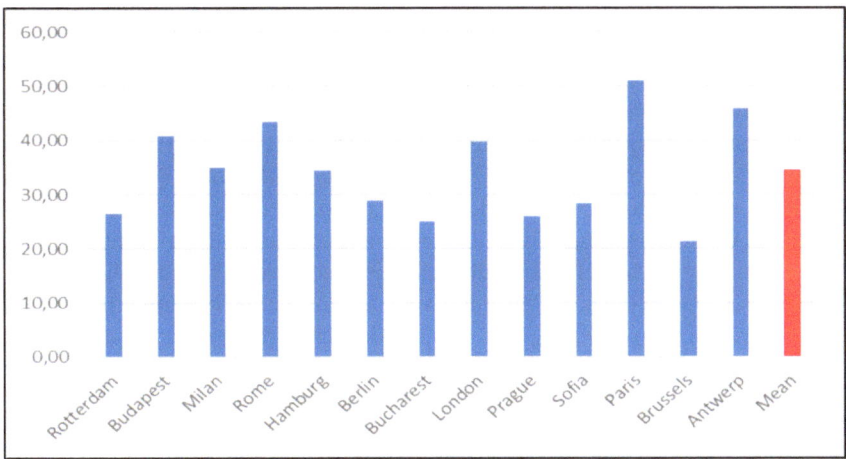

Fig. 5. CO_2 emissions per trip (Source: NMP)

Figure 5 shows the CO2 emissions per trip. The computation of this indicator for some cities revealed high margins of error and related data, not reported in the figure, have been considered unreliable. Estimations for pollutant emissions in cities are usually reported as share in the total measured emissions and by considering traffic patterns for freight vehicles. There were not found comparable studies reporting the CO_2 emissions per trip, given the current mix of vehicles used as assessed by this study and reported in the following figure. The average value at aggregate level is **34 kg/trip**, which could be a baseline for computing overall emissions at city level, knowing the number of vehicles accessing the urban boundaries and the average distance driven per trip.

4 Conclusions

The field survey performed within NMP study commissioned by the European Commission provide an overall picture of urban logistics markets and figures in a sample of European cities, assessing the performance with several indicators.

A review of the latest surveys in European cities showed limited availability of data especially at urban level, differences in the methodologies to collect data, and scarce analysis of collected datasets, making it difficult to compare results with statistical reliability. Nevertheless, the systematic review that was conducted in the surveyed cities, allowed us to collect information useful to validate data collected from the survey and to corroborate results of the analysis.

The NMP field survey was based on CATI and CAWI interviews, delivered to a target sample of all freight deliveries by heavy goods vehicles and light commercial vehicles, owned or working on behalf of logistics companies, travelling in to, out of and within the boundaries of greater cities in the biggest urban agglomerations of 12 Member States plus UK, with a resident population above 1 million. The most relevant results from the data analysis can be summarised as follows:

- The fleet composition of enterprises performing urban deliveries presents - as expected due to access control zone regulations in place in several EU cities - greater share of LGV than HGV. In terms of business model, own account operators are prevalent with an average share of 50% at aggregate level.
- Diesel powered engines are largely popular across Europe with exceptions in few cities. These had a remarkable share of Hybrid Diesel fuel used (Barcelona 8%) and full electric (Stockholm, 53% for LGV, and Rotterdam, 14% for HGV), demonstrating that local policies promoting cleaner vehicles have produced a certain modal shift. However, when comparing vehicle-km and tonne-km, it became evident that electric vehicles are mostly used for last mile deliveries and smaller freight volumes.
- Euro 5 and Euro 6 (aged > 10 years) vehicles are the most common. Vehicles of lower Euro class are used to perform more deliveries but for shorter distances and to transport larger volumes of freight, likely heading to consolidation centres and avoiding last mile deliveries due to circulation restrictions.
- Overall, the median age of vehicles is between 3 and 4 years at aggregate level, with younger fleets recorded in Bremen, Antwerp, Rotterdam, Berlin and Stockholm and older fleets recorded in Amsterdam, Bucharest, Sofia.
- City deliveries performed by bike or powered two-wheelers have a very limited market share. In addition, they are not frequently mentioned as a possible solution by logistics operators when asked about policies and measures to reduce CO2 emissions.

NMP study provides comparable indicators and therefore represents a valuable step towards understanding specific indicators related to urban logistics, the status of specific markets and figures in EU cities. As such the data collected through the study can serve as baseline for future data collection and analysis, with a systematic and standardised approach, and to support transport planners and public administrations in development of sustainable urban logistics policies.

List of Abbreviations

CATI	Computer-assisted telephone interview
CAWI	Computer-assisted web interview
EC	European Commission
EU	European Union
EUROSTAT	Statistical Office of the European Union
HGV	Heavy Goods Vehicle
LGV	Light Goods Vehicle
MS	Member State
NACE	Statistical Classification of Economic Activities in the European Community for the French term "Nomenclature statistique des Activités économiques dans la Communauté Européenne"
NMP	New Mobility Patterns

Acknowledgements. This publication was made possible thanks to the contributions provided authors and affiliated organizations involved in the "Study on new Mobility Patterns in European Cities" (NMP) (European Commission - Directorate General for Mobility and Transport (DG MOVE), Service Contract MOVE/A3/SER/2019–401/SI2.814412), commissioned by the European Commission, all of whom are credited for their respective contributions.)

Declarations.

Availability of Data and Material: Data and material are available upon request.

Funding:. The study upon which the paper is based has been commissioned by the European Commission.

References

1. ACEA, Average age of the EU vehicle fleet, by country. https://www.acea.auto/figure/ave rage-age-of-eu-vehicle-fleet-by-country/. Accessed 12 Dec 2021
2. Alho, A.R., Silva, J.d.: Lisbon's Establishment-based Freight Survey: revealing retail establishments' characteristics, goods ordering and delivery processes. Eur. Transp. Res. Rev. **7**, 16 (2015). https://doi.org/10.1007/s12544-015-0163-7
3. ANFIA Homepage. http://www.anfia.it/en. Accessed 8 Jan 2022
4. APUR - Atelier parisien d'urbanisme, Les outils au service de la logistique urbaine dans les PLU - L'exemple Du Plan Local D'urbanisme De Paris, NOTE n°170, (2020). https://www.apur.org/fr/file/56124/download?token=C0yWAvhS. Accessed 10 Feb 2022
5. CITET and PiperLab, Análisis Madrid Central (2019), sourced from Madrid City Council. https://www.citet.es/es/clean-air-app. Accessed 10 Feb 2022
6. Create Value Kft., Városkutatás Kft, Budapest Citylogisztikai Célkitűzéseinek Egységes Koncepció Javaslata, Budapesti Közlekedési Központ, Budapest (2014)
7. EMISIA, COPERT database. https://www.emisia.com/utilities/copert-data/. 12/12/2021
8. European Commission, CO_2 emission performance standards for cars and vans. https://climate.ec.europa.eu/eu-action/transport/road-transport-reducing-co2-emissions-vehicles/co2-emission-performance-standards-cars-and-vans_en. Accessed 11 Oct 2023

9. European Commission, The European Green Deal. https://commission.europa.eu/strategy-and-policy/priorities-2019-2024/european-green-deal_en
10. European Commission, EU Mobility Strategy. https://transport.ec.europa.eu/transport-themes/mobility-strategy_en
11. European Commission, European Climate Law. https://climate.ec.europa.eu/eu-action/european-climate-law_en
12. Eurostat, NACE Rev.2 - Statistical classification of economic activities in the European Community, Luxembourg (2008)
13. Eurostat, Guidelines on Passenger Mobility Statistics, December 2018. Accessed Oct 2023
14. Gabor, P.: Városi területek megújítása - különös tekintettel a szabadterekre, https://doi.org/10.14267/phd.2014012, Corvinus, Budapest (2013)
15. Locklair, A.: Erhebung der Lkw-Euronormklassen im Umfeld des Hamburger Hafens. Hamburg Port Authority (2018)
16. Macharis, C., Melo, S. (eds.): City Distribution and Urban Freight Transport, pp. 13–14. Edward Edgar Publishing Inc., UK (2011)
17. Roma Servizi Mobilità: Il Piano Urbano della Mobilità Sostenibile. PUMS), Rome (2019)
18. Strale, M., Lebeau, P., Wayens, B., Hubert, M., Macharis, C.: Goederentransport en logistiek in Brussel: stand van zaken en vooruitzichten. Brussel Mobiliteit, Brussels (2015)
19. TIS.PT. Estudo de Logística Urbana para a Zona Piloto da Baixa de Lisboa, Lisbon (2012)
20. Toilier, F., et al.: How can Urban Goods Movements be Surveyed in a Megacity? The Case of the Paris Region, Transportation Research Procedia, Volume 12, 2016, pp. 570–583, ISSN 2352–1465, https://doi.org/10.1016/j.trpro.2016.02.012. (2016)
21. UK department of Transport - Road traffic statistics London. https://roadtraffic.dft.gov.uk/regions/6. Accessed 12 Dec 2021
22. Van den Bossche, M., Maes. J., et al.: Non-binding guidance documents on urban logistics N° 6/6, Final report: Indicators and data collection methods on urban freight distribution, European Commission, Brussels (2017)

Open Access This chapter is licensed under the terms of the Creative Commons Attribution 4.0 International License (http://creativecommons.org/licenses/by/4.0/), which permits use, sharing, adaptation, distribution and reproduction in any medium or format, as long as you give appropriate credit to the original author(s) and the source, provide a link to the Creative Commons license and indicate if changes were made.

The images or other third party material in this chapter are included in the chapter's Creative Commons license, unless indicated otherwise in a credit line to the material. If material is not included in the chapter's Creative Commons license and your intended use is not permitted by statutory regulation or exceeds the permitted use, you will need to obtain permission directly from the copyright holder.

Spectral Characterization of the Rail Surface in Urban Environments Using in-Service Vehicles

Benjamin Baasch[1](✉), Judith Heusel[2], Alexander Lähns[1], Michael Roth[2], and Jörn Groos[2]

[1] German Aerospace Center (DLR), Rutherfordstr. 2, 12489 Berlin, Germany
benjamin.baasch@dlr.de
[2] German Aerospace Center (DLR), Lilienthalplatz 7, 38108 Braunschweig, Germany

Abstract. Rail monitoring using in-service vehicles enables the fast detection of surface defects, which are often responsible for high noise emission. In this paper a processing sequence is presented that converts axle box accelerations into rail condition indicators based on spectral characteristics of the rail surface. The methodology is exemplified with data acquired with a shunter locomotive operating at an inland harbour in the city of Braunschweig, Germany.

Keywords: railway · condition monitoring · rail roughness · axle box acceleration · rolling noise

1 Introduction

Railways play an important role for urban transportation and mobility. However, railway noise is a considerable challenge in urban environments and its minimization is an important task. The continuous monitoring of the rail surface can help to identify track segments that are potential sources of increased noise emissions. Traditional methods of rail condition monitoring are based on visual inspections and manually operated measurement equipment. These are accurate and reliable but relatively expensive and slow. They cannot be carried out during railway operations and thus are performed only at dedicated time intervals. Vehicle-based condition monitoring (VBCM) in contrast is fast and cost efficient, especially if carried out with in-service vehicles. It provides data of entire track networks continuously. Therefore, there is a growing interest in VBCM with in-service vehicles for urban railway networks. However, urban railway operations present unique challenges for VBCM compared to mainline railways [4]. Highly variable vehicle speeds, short travel intervals and frequent stops complicate vehicle positioning and data analysis. This results in a special demand for dedicated data processing and analysis algorithms. In this paper we present a combination of signal processing, data fusion and machine learning techniques for the

© The Author(s) 2025
C. McNally et al. (Eds.): TRAconference 2024, LNMOB, pp. 335–341, 2025.
https://doi.org/10.1007/978-3-031-85578-8_42

spectral characterisation of the rail surface in urban environments. Specifically, georeferenced axle-box accelerations (ABA) are analysed and decomposed in different spectral components. The use of accurate speed information from vehicle positioning allows to transform the data from a time-frequency representation to an equivalent representation in the spatial domain, which reveals speed independent information on the wavenumber spectrum of the rail surface. In this spectral domain unsupervised feature extraction is performed. The features can then be linked to rail surface defects such as corrugation.

2   Materials and Methods

The goal of the VBCM methods described here is to find and extract spectral patterns that can be linked to characteristics of rail surface irregularities. The following sections describe the complete process chain from data recording through signal processing to the extraction of characteristic features for rail condition monitoring.

2.1   Data Acquisition

Onboard data have been acquired with a shunter locomotive operating at the Braunschweig (Germany) Harbour. Condition monitoring of the track is carried out by using analogue broadband three-component accelerometers, which measure the ABA with a working frequency band of 0.8-8,000 Hz. The accelerometers are mounted on the axle boxes on the left and right side of the shunter's front axle. The resulting six ABA channels are digitised by an analogue-to-digital converter and sampled with 20.625 Hz. A central data processing unit is used to collect and process the data. The multi-sensor system further comprises GNSS (global navigation satellite system) receiver and antenna and an IMU (inertial measurement unit) for vehicle positioning tasks.

2.2   Georeferencing

In this context georeferencing refers to the association of actual locations in the track network to the ABA recordings. It facilitates the track-dependent analysis of monitoring data from repeated runs and is crucial for tracking the development of rail defects over time. Furthermore, accurate velocity estimates are important.

The actual georeferencing is performed using an advanced processing pipeline that is based on [6]. It employs map data and Kalman filters. In brief, the following steps are performed. First, the GNSS and IMU data of entire measurement days (sessions) are processed in a Kalman filter and smoother to provide accurate estimates of the vehicle velocity, longitudinal acceleration, and the IMU acceleration bias. Based on these results, the session is divided into single journeys from vehicle start to stop (without changes of direction). For each journey the driven path in the network is then found from a graph of the network and comparing different path hypothesis with the GNSS measurements. Finally, the

GNSS and IMU are re-processed using an on-path Kalman filter and smoother that encodes the vehicle position as a one-dimensional on-path distance. The georeferencing results comprise position and velocity estimates with covariance matrices. The 100 Hz output rate translates to a spatial resolution of ca. 0.14 m at 50 km/h.

2.3 Signal Processing

When a train travels along a track with a defect characterized by a specific wavelength λ at a constant speed v, it undergoes a vertical movement, and the excitation frequency of the resulting vibration is given by $f = v/\lambda$ [7].

Time-frequency representations are powerful tools to analyse these frequency patterns in the ABA data [2,5]. The task of the signal processing here is to transform the time domain data to a space-wavenumber representation that facilitates the extraction of characteristic wavelength patterns of rail surface irregularities.

First, the data is transformed from the time domain to a time-frequency representation using a Short-Time-Fourier-Transform (STFT). The discrete STFT of the signal $y[n]$ with the window $w[n]$ can be expressed as

$$\mathbf{STFT}\{y[n]\}[m, f] \equiv Y[m, f] = \sum_{n=1}^{n=N} y[n]w[n - m]e^{-i2\pi fn}, \tag{1}$$

where n, m and f are discrete time and frequency steps, respectively.

In the time-frequency domain the frequency response of the wheel-rail system is removed employing log-spectral averaging as follows:

The wheel-rail interaction can be modelled by a linear time-invariant system in form of the convolution of the source (rail roughness) function $r(t)$ and the impulse response of the rail-wheel system $s(t)$:

$$y(t) = s(t) * r(t). \tag{2}$$

The logarithmic amplitude of the STFT of $y(t)$ can then be expressed as

$$\log |Y(m, f)| = \log |S(m, f)| + \log |R(m, f)|. \tag{3}$$

If the rail roughness function is considered spatially non-stationary and the rail-wheel response stationary, by averaging the log-spectra of the overlapping segment of the STFT, the log-spectra of the rail roughness will average out and the resulting estimate of the log-spectrum of the rail-wheel system

$$\log \left|\hat{S}(m, f)\right| = 1/M \sum_{m=1}^{m=M} \log |Y(m, f)| \tag{4}$$

can then be subtracted from $\log |Y(m, f)|$, which yields an estimate of the logarithmic amplitude of the STFT of the rail roughness:

$$\log \left|\hat{R}(m, f)\right| = \log |Y(m, f)| - \log \left|\hat{S}(m, f)\right|. \tag{5}$$

Finally, the time-frequency representation of the rail roughness function is transformed to a distance(x)-wavenumber(k) representation using vehicle speed obtained from georeferencing with $x = tv$ and $k = f/v = 1/\lambda$.

This representation serves as input to the feature extraction via machine learning, which is explained in Sect. 2.4. Figure 1 shows the raw onboard data from one journey in the time-domain (top) and after signal processing as distance-wavenumber representation (bottom).

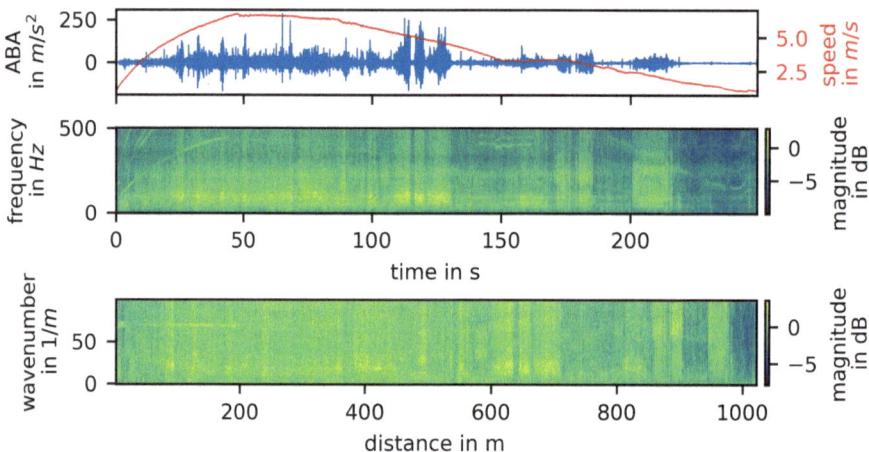

Fig. 1. ABA data in the time domain (top), in the time-frequency domain (middle) and distance-wavenumber domain (bottom).

2.4   Machine Learning

An unsupervised machine learning approach is used to extract characteristic spectral patterns from the distance-wavenumber representation of the ABA data. Specifically, an undercomplete sparse dictionary is learned that represents the actual input data in a lower-dimensional space. The atoms of the dictionary represent characteristic spectra of the rail surface. The weights of the atoms can then be used to find and describe rail surface defects. The dictionary H and the weights W can be found via the following optimization problem:

$$\arg \min_{W,H} \frac{1}{2} \|Y - WH\|_{\text{Fro}}^2 + \alpha \|W\|_{1,1}, \tag{6}$$

where $\|.\|_{\text{Fro}}^2$ stands for the Frobenius norm and $\|.\|_{1,1}$ stands for the element-wise matrix norm, namely the sum of the absolute values of all the entries in the matrix.

3   Results

Onboard data acquired with a shunter locomotive as described in Sect. 2.1 are used to exemplify the results of the proposed methodology. In Fig. 1 the time frequency representation of the raw data is compared with the distance-wavenumber representation of the data after signal processing. It can be seen how, after signal processing, the rail-wheel system's response and spectral components of rotating parts of the wheelset and engine are suppressed, whereas the spectral components associated with the irregularities of the rail surface stand out. In the distance-wavenumber domain, characteristic spectra are extracted by sparse dictionary learning. Eight atoms representing wavenumber spectra of the rail surface are found (Fig. 2). The low wavenumber atoms (dark blue lines) represent track geometry anomalies with longer wavelength, whereas atoms in high wavenumber ranges (light-green and yellow lines) represent short-wavelength

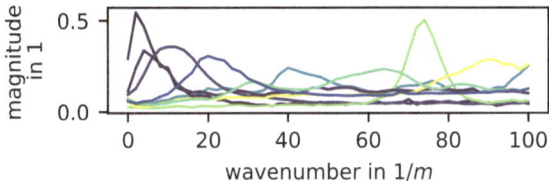

Fig. 2. Different atoms of the learned dictionary. Brightness of lines increases with increasing wavenumber of the maximum magnitude of the different atoms.

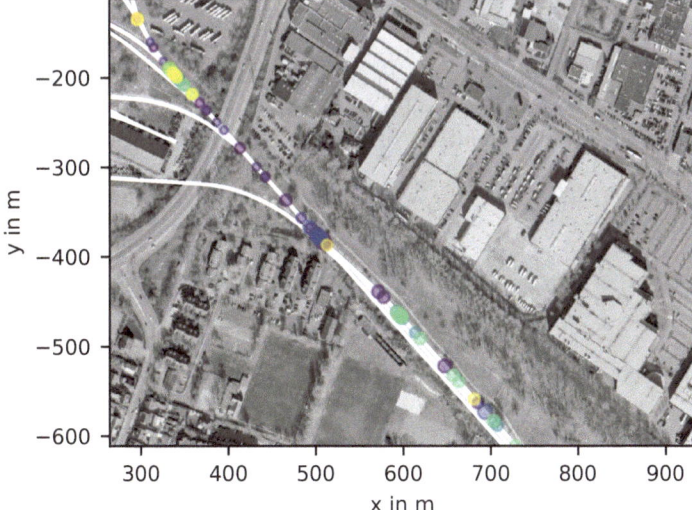

Fig. 3. Map of the railway network [1]. Dots indicate rail surface anomalies. The color represents the atom according to Fig. 2 that best approximates the wavenumber spectrum of an anomaly. The size represents the magnitude of the corresponding weight.

rail irregularities. The sparsity constraint on the weights ensures that mainly frequency components from relatively strong track and rail irregularties are considered. The weights of these atoms can be mapped on the railway network (Fig. 3). The high-wavenumber anomalies indicated by yellow and light-green dots at a rail segment in the north-west of the map reflect known corrugation defects.

4   Conclusions

A methodology to monitor the rail surface condition using in-service vehicles has been presented. It includes sensor fusion, signal processing and machine learning approaches that are suitable to extract information from ABA data in challenging environments. Real-world data from a shunter locomotive operating at an inland harbour in the city of Braunschweig, Germany were presented and analysed. The results indicate the great potential of the presented methodology to detect and describe rail surface irregularities with in-service vehicles. The conditions of the shunting operations pose typical challenges that are shared by other urban railway systems such as light rails and trams. Therefore, the methodologies presented here are readily applicable to other kinds of urban rail transportation systems.

Acknowledgements. This research is part of the OnboardEU [3] project funded by mFund, an innovation initiative for digital data-based applications for the mobility of the future of the German Federal Ministry for Digital and Transport (BMDV).

References

1. Aerial photograph: City of Braunschweig (Department of Geoinformation)
2. Baasch, B., et al.: Detecting singular track defects by time-frequency signal separation of axle-box acceleration data. In: WCRR 2019. https://elib.dlr.de/121517/
3. DLR Verkehr. OnboardEU - mit KI Schäden am Gleis automatisch erkennen. Ed. by DLR Verkehr (2022). https://verkehrsforschung.dlr.de/de/news/onboardeu
4. Heusel, J., et al.: Detecting corrugation defects in harbour railway networks using axle-box acceleration data. In: Insight - Non-Destructive Testing and Condition Monitoring 64.7 (2022), pp. 404-410. issn: 1354–2575. https://doi.org/10.1784/insi.2022.64.7.404.
5. Niebling, J., et al.: Analysis of Railway Track Irregularities with Convolutional Autoencoders and Clustering Algorithms. In: Bernardi, S., et al. Dependable Computing - EDCC 2020 Workshops. EDCC 2020. Communications in Computer and Information Science, vol 1279. Springer, Cham (2020). https://doi.org/10.1007/978-3-030-58462-7_7
6. Roth, M., et al.: Map-supported positioning enables in-service condition monitoring of railway tracks. In: 21st International Conference on Information Fusion (FUSION), pp. 2346-2353, July 2018. https://doi.org/10.23919/ICIF.2018.8455377
7. Salvador, P., et al.: Axlebox accelerations: Their acquisition and time- frequency characterisation for railway track monitoring purposes. In: Measurement 82 (2016), pp. 301–312. issn: 02632241. https://doi.org/10.1016/j.measurement.2016.01.012

Open Access This chapter is licensed under the terms of the Creative Commons Attribution 4.0 International License (http://creativecommons.org/licenses/by/4.0/), which permits use, sharing, adaptation, distribution and reproduction in any medium or format, as long as you give appropriate credit to the original author(s) and the source, provide a link to the Creative Commons license and indicate if changes were made.

The images or other third party material in this chapter are included in the chapter's Creative Commons license, unless indicated otherwise in a credit line to the material. If material is not included in the chapter's Creative Commons license and your intended use is not permitted by statutory regulation or exceeds the permitted use, you will need to obtain permission directly from the copyright holder.

Transitioning Towards Area-Oriented Approaches in Transport Infrastructure Planning: An Analysis of European Best Practices

Cheyenne Raskeyn[1] ⓘ, Jos Arts[1,2](✉) ⓘ, Tertius Hanekamp[3], and Sjaak van der Werf[4]

[1] Department of Planning, Faculty of Spatial Sciences, University of Groningen, PO Box 800, 9700AV Groningen, The Netherlands
{c.a.raskeyn,jos.arts}@rug.nl

[2] Environmental Sciences and Management, North-West University, Potchefstroom, South Africa

[3] TEMAH, Zandweg 90, 3454JX De Meern, The Netherlands
th@temah.nl

[4] Rijkswaterstaat, Ministry of Infrastructure and Water Management, PO Box 2232, 3500GE Utrecht, The Netherlands
sjaak.vander.werf@rws.nl

Abstract. Investments in transport infrastructure severely impact the regions in which they are planned. These impacts can be both positive (e.g., connectivity, accessibility) and negative (e.g., environmental pollution). To advance more inclusive and sustainable development of transport infrastructure, a variety of European studies were conducted about the potential of area-oriented planning approaches in which infrastructure and spatial development are integrated. Although area-oriented approaches are promising, they are not standard practice. Barriers such as political dynamics, organizational structure and limited scope of mandate restrain National Road Authorities in becoming area partners. The successes of area-oriented projects are found in communicative and collaborative skills but are not fully developed to advance the transition to a new area-oriented planning regime. To accelerate the transition to area-oriented transport infrastructure planning, a European agenda for further action is presented.

Keywords: Integrated Infrastructure and Spatial Development · Area-Oriented Planning · Collaborative Planning · Multi-Level Perspective on Transitions

1 Introduction

Throughout Europe, large investments are made in transport infrastructure – roads, railways, waterways, pipelines. These investments have serious consequences for the urban and rural regions in which they are planned: positive impacts (such as connectivity, accessibility, spatial-economic development) and negative impacts (on such issues as nature, noise, water, air quality, health, safety, visibility). In addition, the same regions

© The Author(s) 2025
C. McNally et al. (Eds.): TRAconference 2024, LNMOB, pp. 342–348, 2025.
https://doi.org/10.1007/978-3-031-85578-8_43

already face huge challenges, such as climate adaptation, energy transition, sustainable development and social inclusion. To advance more inclusive and sustainable development of transport infrastructure and their surrounding areas (addressing the impacts on, and challenges and needs of regions) various European studies have been carried out recently, including: the EU-Horizon2020 project *Vital Nodes* (https://vitalnodes.eu), the CEDR-projects on Freight and logistics in a multimodal context (*FLUXNET*) and on Collaborative Planning of Infrastructure and Spatial Development (*SPINdesign* – https://www.cedr.eu/). these studies have provided a rich source of valuable insights and best practice cases. In several European countries there is a beginning of a change from a sectoral towards a more inclusive approach [1]. However, this is still not standard practice throughout Europe. Often infrastructure planning is still rather sectoral and linear project-oriented. Many infrastructure projects still experience cost and time overruns, and limited stakeholder satisfaction. Therefore, an important question is how to create a transition from a traditional, narrow-scoped, sectoral, unimodal, project-oriented infrastructure planning towards a more integrated, inclusive, multi-modal, area-oriented planning of infrastructure and spatial development that is oriented at sustainable transport and liveable urban and rural regions [2].

This paper examines how a transition towards an area-oriented planning of integrated infrastructure and spatial development can be achieved, and which issues are urgent to address with such approach. To this end, the pilot cases of the abovementioned European studies were revisited to examine the barriers, success factors and conditions for achieving area-oriented infrastructure planning. This was done by document analysis, interviews with key persons and workshops [3]. In line with the earlier European studies, the main focus is on road infrastructure and the role of National Road Authorities (NRAs). However, the findings may provide useful insights for all types of transport infrastructure – not only roads, but also railways, waterways, pipelines – as well as other actors in the planning process – such as infrastructure providers, (spatial) planning authorities (at national, regional, local level). The paper presents the outcomes of this study and an agenda for further research and action for a transition towards area-oriented approaches in transport infrastructure planning.

2   Transition Theory

To understand the transition from the traditional sectoral approach to an area-oriented approach, insights from transition theory are used to analyse why – despite the huge *'landscape'* pressures on infrastructure planning authorities (such as policy ambitions for sustainable and inclusive development and transportation) – the *'niches'* of innovative insights and best practice cases of area-oriented planning are often not upscaled and mainstreamed into the *'regime'* of current infrastructure planning.

These three analytical levels of the *'Multi-Level Perspective'* interact [4]: niches are embedded in regimes, and regimes are embedded in the landscape (nested character). Change processes are initiated when niche innovations are upscaled to the regime level. Consequently, niche innovations become more influential and develop momentum for alterations in the regime. The success of upscaling does not only depend on the niche innovations but is also influenced by developments in the landscape, and the adaptiveness, openness and stability of the regime. To exploit change opportunities pressure on the landscape is necessary [5]. Thus, a transition is the result of interaction between all three levels (see Fig. 1).

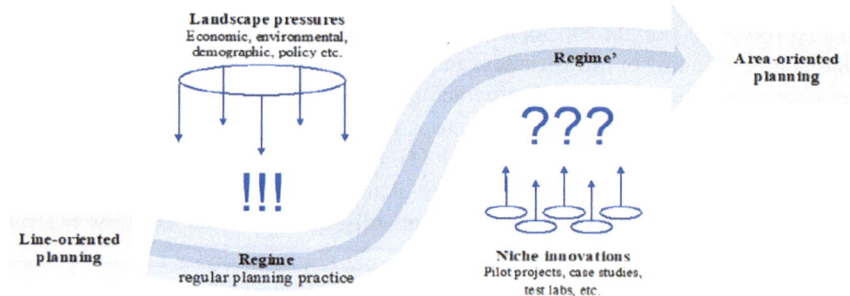

Fig. 1. From line- to area-oriented planning from the Multi-Level Perspective on transitions.

3   Findings: Barriers, Success Factors and the Role of NRAs

The 17 case studies or pilots from the three European research programs (SPINdesign, FLUXNET and Vital Nodes) all experimented with niche innovations regarding area-oriented approaches. These pilots were all executed in collaboration with the respective NRAs with the aim to develop toolboxes for area-oriented planning that would provide guidance to those NRAs. Although the NRAs experienced promising results from such area-oriented perspective, this new approach proves to be not (yet) institutionalized in their regular planning practice [3]. Therefore, we investigated in-depth four cases of the research programs to identify barriers and success factors for an area-oriented approach, and the role of NRAs within area-oriented projects (see Table 1).

Table 1. Barriers, success factors and the role of NRAs in four European cases.

Case Study	Barriers	Success factor	Role of NRA
Rotterdam, NL	Sectoral organization top-down decision making	Overarching 'Topcorridors' program	Area partner
Linz, AT	Limited involvement, limited scope of mandate	Regional collaboration, partial involvement	Partner/ co-financer
Oslo, NO	Political dynamics, exclusion from National Transport Plan	Public-public collaboration/relationships	Coordinator
Norrköping, SE	Political dynamics, communicative conflict with Linköping	Public-public collaboration/relationships	Coordinator

3.1 Barriers

In the process of upscaling area-oriented practices to regular planning practices, a major barrier for NRAs proves to be the lack of capacities needed. This includes insufficient financial resources, knowledge, experience, personnel and time. Area-oriented planning requires broader knowledge and collaboration across multiple sectors, thereby demanding more diverse personnel and resources than line-oriented approaches. Next to this, political dynamics from national and local levels have negatively influenced the progress in the Norrköping and Oslo cases. In the case of the former, the development of a high-speed rail connection was put on hold by the national government due to the high costs. In the case of Oslo, the planning of a tunnel was stopped as result of political disagreement about the project scope. The Norwegian government decreased funding of road construction due to environmental concerns and thus the Norwegian NRA (Statens Vegvesen) excluded the project from their prioritization. And it is unlikely that the project is incorporated in the revised National Transport Plan. Furthermore, the organizational structure of NRAs proves to function as a barrier. In the Rotterdam case, for instance, the NRA (Rijkswaterstaat) is sectorally organized with a project-oriented planning approach. As a consequence, relationships with other actors are usually aimed at lower levels of involvement, such as information and at most consultation, but no co-creation. Top-down decision-making characterizes the standard approach of the Dutch NRA, which complicates the use of area-oriented approaches.

3.2 Success Factors

A major overall success factor in area-oriented planning lies in communication and collaboration. Long-standing and continuous relationships and collaborations are crucial for the success of a project. The extent to which this was done differed per case, however all NRAs engaged in regional collaboration and acknowledged the importance of relationship management for successful area-oriented planning. This was done most successfully in the case of Rotterdam, where the project became a part of a larger scale cross-border and integral program ('Topcorridors'). In this program, multiple national and regional stakeholders are connected and linked to each other to share responsibilities concerning the development and maintenance of the corridor. Through such a sense of joint ownership and joint collaboration, NRAs can establish the right environment for upscaling area-oriented planning practices.

3.3 The Role of NRAs

NRAs fulfil a vital role in the upscaling process of niche innovations. As large-scale and government-driven organizations, NRAs are in the position to be change agents and policy entrepreneurs. To accelerate the transition process, NRAs need to position themselves as area partners within collaborative structures. In the Rotterdam case, the Dutch NRA (Rijkswaterstaat) provided a platform for the cross-border and integral Topcorridors program. The program allowed governmental organizations to collaborate beyond their own administrative borders with the goal of joint corridor development and maintenance, while simultaneously encouraging knowledge sharing [6]. However, positioning a NRA as an area partner may be restricted by the mandate of an organization. For instance, the Austrian NRA (ASFiNAG) was deprived of the opportunity to remain involved in the Linz case of a multimodal junction south of the city since the project no longer required road infrastructural measures, which is the sole responsibility of ASFiNAG. Although the project still has a large impact on the road network of ASFiNAG, their narrow mandate limited them in engaging in the project whit its area perspective. Although NRAs often may have developed the right knowledge and know-how, they are prone to significant barriers, limiting the success of upscaling.

4 Discussion and Conclusion

The findings show that transitioning from the traditional sectoral, project-oriented approach towards an area-oriented planning approach is complex. There are substantial landscape pressures (need for climate adaptation, energy transition, sustainable development and social inclusion), that stimulate the development of niches (pilots and experiments) providing innovation. Eventually, these niche innovations can initiate change processes at the regime level and the area-oriented approach can be institutionalized. For this, however, the area-oriented approach needs to be retained in the regular planning regime. Thus far, such institutionalization does not happen, despite the huge landscape pressures (the much-talked challenges) and good practice cases of the European programs (providing relevant niche experiences). NRAs incrementally learn the essence and necessity of area-oriented approaches from the pilots, but the knowledge gained is not actively practiced. There is no system change (yet). This implies that the landscape pressure on the regime is still experienced too low (awareness) and/or that the regime is insufficiently adaptive and open (inertia). Revisiting the studied cases showed that the European NRAs remain at the starting phase of the transition needed.

Another explanation for the difficulty in retaining niche innovations into the planning regime relates to differences in the planning context for pilot cases as opposed to the regular planning arena (*'pilot syndrome'*). During experimentation with pilots there is often more room for innovation and/or additional resources, resulting in elevated success rates compared to projects in a regular planning environment. Also, the *institutional memory* of NRAs seems to be limited. The toolboxes developed in the European programs, aiming to provide guidance in area-oriented approaches, were little known and used. Next to this, documentation on the pilot projects proved to be hard to find. Combined with (meanwhile) unavailable websites of research programs, this slows down learning processes of NRAs and their partners.

Finally, NRAs play a key role in the upscaling and institutionalization of area-oriented approaches into regular planning practice. Barriers such as political dynamics from the national and local level, a sectoral or project-oriented organizational structure and the limited scope of mandate, restrain NRAs in becoming area partners and prevent the transition towards area-oriented approaches. Furthermore, the successes found in communication and collaboration are not (yet) fully exploited and thus not used to their full potential. To accelerate the transition towards area-oriented approaches in transport infrastructure planning, an agenda for further research and actions is prepared.

5 Outlook: Towards a European Working Agenda

The European Association of Road Administrations (CEDR) sees rapid changes the context in which their road networks are embedded and functioning. CEDR is benchmarking its course in its CEDR Compass. Due to a higher degree of interdependencies and uncertainties it is expected that area-oriented planning approaches (*'Collaborative Planning'*) becomes even more significant. In order to guide implementation and development of existing area-oriented planning (toolboxes), a CEDR Working Group on Collaborative Planning was established. The group is founded by Sweden, Norway, Finland, Estonia, Poland, the United Kingdom, Austria and the Netherlands, however other European countries are invited to participate. In line with the CEDR Compass, the Working Group develops a *working agenda* in which the toolbox will be enriched with experiences focussing on 5 emerging themes in the 'landscape' of NRA practice:

A. *New economic patterns:* Developments – such as circular economy, re-shoring, an EU that is self-sufficient and less vulnerable for geo politics – are creating new (multimodal) network logics and driving new investments.
B. *A social inclusive approach towards mobility infrastructure:* Equality in opportunities for citizens is important ('broad prosperity'). NRAs' 'social licence to operate' is based on an economic narrative (efficiency, productivity). NRAs need to develop new narratives that are socially inclusive to keep their licence to operate.
C. *Ecology inclusive asset management:* Much road infrastructure was developed in an era with limited attention for impacts on biodiversity, water and soil. As the environment is under pressure, and because of EU policies and regulations (e.g., Natura 2000), NRAs have to adopt an ecologically inclusive approach.
D. *Vital infrastructure development:* New infrastructure projects need a collaborative approach to meet high ambitions and requirements concerning: multimodal network development, mobility shift in urban areas, inclusive socio-economic development, climate adaptation, landscape and ecology.
E. *Mature infrastructures in a dynamic context:* In many European countries the road network has reached the maturity phase of network development, creating a shift from focussing on expanding to maintaining the network. Parallel, the context in which the network is operating is changing due to climate adaptation, energy transition, urbanisation, multi-modality, etc. Therefore, the mature networks are in need of a re-design within the current spatial footprint.

In 2024, the CEDR Working Group will implement this agenda and further develop practices for area-oriented planning in the context of these emerging themes. This to stimulate mainstreaming of (the toolboxes for) area-oriented planning, thereby renewing Europe's NRAs' license to operate and collaborative plan for a sustainable future.

Acknowledgement. This study was supported by the Working Group on 'Collaborative Planning of Infrastructure and Spatial Development', of CEDR.

References

1. Arts, J., Hanekamp, T., Linssen, R., Snippe, J.: Benchmarking integrated infrastructure planning across Europe. Transp. Res. Procedia **14**, 303–312 (2016)
2. Arts, J., Leendertse, W., Tillema, T.: Road infrastructure: planning, impact and management. In: Vickerman, R. (ed.) International Encyclopedia of Transportation, vol. 5, pp. 360–372. Elsevier, UK (2021)
3. Raskeyn, C.: Area-oriented approaches in transport infrastructure planning: A cross-case analysis of European best Practices. University of Groningen, Groningen (2023)
4. Geels, F.W., Schot, J.: Typology of sociotechnical transition pathways. Res. Policy **36**, 399–417 (2007)
5. Raven, R.P.J.M., van den Bosch, S., Weterings, R.: Transitions and strategic niche management: towards a competence kit for practitioners. Int. J. Technol. Manage. **51**(1), 57–74 (2010)
6. Faith-Ell, C., Kalle, H., Arts, J.: Towards sustainable and inclusive corridors: rethinking planning of infrastructure and spatial development. Transport Transitions: Advancing Sustain. Inclusive Mobility **2**(43), xx – yy (2025)

Open Access This chapter is licensed under the terms of the Creative Commons Attribution 4.0 International License (http://creativecommons.org/licenses/by/4.0/), which permits use, sharing, adaptation, distribution and reproduction in any medium or format, as long as you give appropriate credit to the original author(s) and the source, provide a link to the Creative Commons license and indicate if changes were made.

The images or other third party material in this chapter are included in the chapter's Creative Commons license, unless indicated otherwise in a credit line to the material. If material is not included in the chapter's Creative Commons license and your intended use is not permitted by statutory regulation or exceeds the permitted use, you will need to obtain permission directly from the copyright holder.

Evaluating the Ability to Telecommute by the Potential Female Telecommuters by Logistic Regression Model

Mootaz M. Jaff[1]([⊠]) and Abdul Azeez Kadar Hamsa[1,2]

[1] International Islamic University Malaysia, 26300 Gambang, Malaysia
mootaz_munjid@iium.edu.my
[2] International Islamic University Malaysia, 50728 Kuala Lumpur, Malaysia

Abstract. This paper evaluates the ability of potential female employees to telecommute in the selected business organizations in Kuala Lumpur, Malaysia. The ability to telecommute by the potential female employees was determined by targeting few employment types namely clerical, professional, associate professional and manager through a pre-designed questionnaire survey. A binary logistic regression model was developed to evaluate the effects of the explanatory variables on the ability to telecommute by controlling and not controlling the "employing industries" and "employment categories". The results of the model reveal that 'having prior experience working from home' and 'increase in the frequency of telephone usage' almost doubles the ability to telecommute, whereas 'spending considerable time working with others' and the 'frequent use of the photocopier' reduces it. The respondents employed in the 'financial intermediation' and 'real estate industries' were approximately 2.5 times more likely to be able to telecommute as compared to those employed in the 'telecommunication industry'. Managers, associate professionals, professionals were 3.5, 2.3 and 1.5 times respectively more likely to be able to telecommute more frequently than clerical workers.

Keywords: Telecommuting · Female Workforce · Binary Logistic Model · Ordinal Logistic Model · Kuala Lumpur · Malaysia

1 Introduction

In Malaysia, the female workforce in various types of employment is high. The increase in the penetration of telecommuting depends on the ability of the female employees and their willingness to adopt it. Despite employees are able to telecommute, nonetheless they may not willing to adopt it due to various reasons. As a result, it paves the way to investigate the ability and willingness of the female employees to telecommute if they are given the choice to do so. In this paper, however, only the female employees' ability to telecommute were investigated using logistic regression models. Two models were developed to predict the ability of the respondents to telecommute. The first model is a binary logistic regression model predicting the ability to telecommute in the form of a

© The Author(s) 2025
C. McNally et al. (Eds.): TRAconference 2024, LNMOB, pp. 349–355, 2025.
https://doi.org/10.1007/978-3-031-85578-8_44

dichotomous (able vs unable) dependent variable. The second model was developed to assess the influence of the predictor variables on the ability to telecommute by including industries and employment.

2 Literature Review

A considerable number of studies and pilot projects has highlighted both the potential and actual impacts of substituting the "daily commute" with a "commute" in cyberspace [1]. A 2008 Bloomberg Business week report has suggested that the number of American employees working away from the office at least once a month had been growing at 10% annually for the past several years. However, empirical evidence suggests that the percentage of American employees working from home at least once a week grew by nearly 10.3% between 1997 and 2005, and grew by a further 17.9% from 2005 to 2010. Recent statistics from USA indicate that the percentage of American employers allowing employees to telecommute occasionally almost doubled from 34% to 63%. Between 2005 and 2012. However, the percentage of employers allowing workers to telecommute regularly had increased only at 2% from 31% in 2005 to 33% in 2012 [2].

More and more women in the labor force had increased the income level of households while the time available for performing household chores decreased. This phenomenon has resulted in women needing to buy services such as childcare and meals from outside and as a result increasing the number of non-work trips [3]. A number of research has dwelled on the unique travel patterns that men and women develop [3–7]. These studies found significant differences along gender lines. Women were found to make more but shorter trips (in both time and distance) and they were more inclined to perform more child and home-oriented travel and trip-chain [8]. Due to the "gender gap" in travel behavior and the existing two-fold responsibilities shouldered by working women, they are more likely than men to view telecommuting as a potential solution to these extra pressures [9]. This is especially applicable in the Asian context where women bear greater responsibilities in spending considerable amount of time on childcare and caring other dependents in the household whereas men, on the other hand, were predominantly spent time on office and office-related works. This favor telecommuting especially for women because it allows them to fulfill household commitments while still remaining in the labor force.

3 Study Methodology

This study had targeted 400 female employees working as "clerical workers", "professionals", "associate professionals", and "managers" from a total population of 168,300 (N). The sample size (n) was determined using $n = N/(1 + Ne^2)$ with a precision level 'e' assumed as 5%. The sample size was then distributed to each employment type by applying stratified sampling technique. These female employees were employed at four main telecommuting compatible industries namely "financial intermediation", "real estate renting", "transportation and communication" and "education" [10] in Kuala Lumpur, Malaysia.

A questionnaire survey was administered to collect data covering four main sections namely "employment characteristics", "perceptions towards telecommuting", "household characteristics" and "personal characteristics". Additionally, a "place-based" travel diary was also administered to collect data on "travel characteristics" of the targeted respondents. More than 600 questionnaires were distributed to the targeted samples in stages but only 524 questionnaires were eventually received representing a response rate of 87%. However, only 454 valid questionnaires were used in the analysis due to incomplete data. Both the data and the analysis were administered during pre-COVID period.

Two models were developed in this paper. The first model is a binary logistic regression model predicting the ability to telecommute having a dichotomous (able vs. unable) dependent variable. A second model was developed to assess the influence of the predictor variables on the ability to telecommute by including "industries" and "employment". In these two regression models, a forced entry method was applied because of the limited sample size.

4 Modeling the Ability of Female Employees to Telecommute

About 58% of the respondents stated their ability to telecommute given their job scope. Only the independent variables related to employment-related category were included in the model because ability to telecommute can only be influenced by such variables. The selection of five independent variables in the model was purely based on the correlation results with the dependent variable.

The coefficients (ß-values) of the independent variables in the regression model represent the change in the logit of the outcome variable associated with one unit change in the predictor variable [11]. At this stage, it is important to examine the Wald statistic of each independent variable included in the model. This statistic allows to assess whether the coefficients (ß-values) of each independent variable are significantly different from zero which eventually considered safe to assume that the variables are contributing significantly to the predictive power of the model. The results of Wald statistics and its significance confirm all of the variables included in the model contribute significantly to the model's predictive power ($p<0.01$).

The coefficients of the variables such as 'prior experience of home-working', 'amount of time spent working alone', and 'telephone usage' were positive indicating a one-level increase of these variables would result in a net increase in the logit of the ability to telecommute. On the other hand, the coefficients of the variables such as 'time spent working face-to-face with others', and 'photocopier usage' were negative indicating a one-level increase of these variables would result in a net decrease in the logit of the ability to telecommute.

A unit increase in the respondents' "experience working from home" has resulted in an almost doubling the odds of being able to telecommute (OR = 1.95). Likewise, a unit increase in "telephone usage" has resulted in an almost identical change in the odds of being able to telecommute (OR = 1.93). A one-level increase in the "time spent working alone per week" has increased the odds of being able to telecommute by 1.35 times. On the other hand, a one-level increase in both the 'time spent working face-to-face with

Table 1. Results of the Binary Logistic Regression Model (n = 398)

Dependent variable: Ability to telecommute (able vs. unable)					
Independent variables	Coefficient ß	SE	Wald	p-Value	Exp(B)
Constant	.098	.650	.023	.880	1.103
Experience home-working	.667	.148	20.406	.000	1.949
Time spent working face-to-face	−.391	.096	16.636	.000	.677
Amount of time working alone	.297	.103	8.283	.004	1.346
Frequency of using the phone	.657	.169	15.193	.000	1.929
Frequency using the photocopier	−.690	.169	16.616	.000	.501
-2Log likelihood	501.189				
Model Chi-square	79.721			.000	
Cox & Snell R Square	0.171				
Nagelkerk R Square	0.229				
Tjur R Square	0.373				

others', and 'photocopier usage' has resulted in low ability to telecommute (OR = .68 and .50 respectively). These results are shown in Table 1.

5 Assessing the Influence of Respondents' Industry and Employment Category on the Ability to Telecommute

A second model predicting the ability to telecommute was developed by including two additional variables namely "industries" and "occupation type". The reference category for both variables was the last category such as "telecommunication and transport industry", and the "clerical employees". The beta coefficients for other categories will be interpreted with reference to the reference category.

The inclusion of the variable 'industry' has made significant contributions to the model's predictive power. However, it is not in the case of the variable "employment". The chi-square value of the overall model was statistically significant indicating that the model is a good predictor for predicting the ability to telecommute. Furthermore, the inclusion of the two additional variables has resulted in improving model's pseudo R^2 value which is .202 (Cox & Snell R^2) and .271 (Nagelkerke R^2). An improvement in the coefficient of discrimination (Tjur R^2) which is 0.400 was also noticed. The results are shown in Table 2.

The results show only two industries namely "financial intermediation" and "real estate" have identical positive beta coefficient with a statistically significant Wald statistics. By examining the odd ratios of these two beta coefficients, it reveals that the respondents employed in these two industries were 2.5 more likely to be able to telecommute than those employed in the telecommunication industry (OR = 2.45 and 2.52 respectively).

Table 2. Results of the Binary Logistic Regression Model with industry and employment

Dependent variable: Ability to telecommute (able vs. unable)

Independent variables	Coefficient ß	SE	Wald	p-Value	Exp(B)
Constant	−.306	.817	.140	.708	.736
Industry			8.364	.039	
Industry (1)	.896	.360	6.191	.013	2.449
Industry (2)	.924	.401	5.313	.021	2.519
Industry (3)	.419	.418	1.002	.317	1.520
Occupation			2.774	.428	
Occupation (1)	.674	.508	1.758	.185	1.961
Occupation (2)	.172	.312	.303	.582	1.187
Occupation (3)	.557	.465	1.437	.231	1.746
Experience home-working	.769	.163	22.176	.000	2.158
Amount of time working face-to-face	-.393	.108	13.113	.000	.675
Amount of time working alone	.309	.111	7.702	.006	1.362
Frequency of using the phone	.567	.188	9.131	.003	1.762
Frequency using the photocopier	−.797	.201	15.733	.000	.451
-2Log likelihood	476.996				
Model Chi-square	94.353			.000	
Cox & Snell R Square	0.202				
Nagelkerk R Square	0.271				
Tjur R Square	0.400				

Finally, it is important to mention that the coefficients of the original five variables that were included in the first model were not found to be varied considerably across the two models. In other words, the association between the five original variables and the ability to telecommute remains largely unchanged even after controlling the influence of respondents' industry and employment type. In fact, the coefficients of the two variables 'amount of time spent working alone' and 'amount of time spent working face-to-face with others' were identical in both models indicating that their influence are exactly the same regardless of the respondents' employment and industry. The effects of 'having home-working experience', and 'photocopier usage' were slightly more pronounced after controlling respondents' industry and employment type, while the effect of the 'telephone usage' was slightly reduced. However, all the changes remain marginal.

6 Conclusions

The results of the model reveal that 'prior experience working from home' almost doubles the likelihood of being able to telecommute. Also, the 'increased frequency of telephone usage' had an almost identical influence on the ability to telecommute. On the other hand, 'time spent working with others' and the 'usage of the photocopier' were found to reduce the likelihood of being able to telecommute. The respondents employed in the 'financial intermediation' and 'real estate industries' were approximately 2.5 times more likely to be able to telecommute than those employed in the 'telecommunication industry'. Managers, Associate Professionals, Professionals were found to be 3.5, 2.3 and 1.5 times more likely to be able to telecommute than Clerical workers. The models were found to be satisfied with the minimum sample size requirements. Finally, the models are considered useful for predicting the group membership of the dependent variable for entire groups of employees and not individual respondents where noticeable overlaps between the predicted probabilities and observed group membership were detected.

References

1. Nilles, J.M.: Telecommuting and urban sprawl: mitigator or inciter? Transp. Res. **18**, 411–432 (1991)
2. Mateyka, P.J., Rapino, M.A., Landivar, L.Ch.: Home-based workers in the United States: 2010, US Census Bureau Report Number P70–132 (2012). https://www.census.gov/prod/2012pubs/p70-132.pdf
3. Levinson, D.: How travel pattern changed from '68 to '88. Review (Journal) Institute of Transportation Studies 20(2) (1997)
4. Mauch, M., Taylor, B.: Gender, race and travel behavior: analysis of household-serving travel and commuting in San Francisco Bay area. Transp. Res. Record: J. Transp. Res. Board, 147–153 (1997)
5. Taylor, B.: Beyond the gender gap: An array of social, ethnic and economic factors influence travel behavior. Review (Journal) Institute of Transportation Studies 20(2) (1997)
6. Wachs, M.: The gender gap: How Men and Women develop different travel patterns. Review (Journal) Institute of Transportation Studies 20(2), (1997)
7. White, P., Christodoulou, G. and Mackett, R.: The role of teleworking in Britain: Its implications for the transport system and economic evaluation. In: Proceedings of the European Transport Conference, The Netherlands (2010)
8. Crane, R.: Is there a quiet revolution in Women's travel? Revisiting the gender gap in commuting. J. Am. Plann. Assoc. **73**(3) (2007)
9. Mokhtarian, P.L., Michael, N.B., Hulse, L., Salomon, I.: The influence of gender and occupation on individual perceptions of individual perceptions of teleworking. Transportation Centre, the University of California, Berkeley, CA 94720 (1997)
10. Ng, C.: Teleworking and development in Malaysia. Prevalence survey report 3 (1999)
11. Field, A.: Discovering statistics using SPSS, 2nd edn. SAGE Publications Ltd., London (2009)

Open Access This chapter is licensed under the terms of the Creative Commons Attribution 4.0 International License (http://creativecommons.org/licenses/by/4.0/), which permits use, sharing, adaptation, distribution and reproduction in any medium or format, as long as you give appropriate credit to the original author(s) and the source, provide a link to the Creative Commons license and indicate if changes were made.

The images or other third party material in this chapter are included in the chapter's Creative Commons license, unless indicated otherwise in a credit line to the material. If material is not included in the chapter's Creative Commons license and your intended use is not permitted by statutory regulation or exceeds the permitted use, you will need to obtain permission directly from the copyright holder.

Predicting Demand and Supply in a Real-Time Traffic Management Framework

Athina Tympakianaki[1]([✉]) [iD], Mohammadmahdi Rahimiasl[2] [iD],
Charis Chalkiadakis[3] [iD], Monica Dominguez[1] [iD], Ynte Vanderhoydonc[2] [iD],
Jordi Casas[1] [iD], Eleni I. Vlahogianni[3] [iD], and Siegfried Mercelis[2] [iD]

[1] Aimsun SLU, Ronda Universitat 22B, 08007 Barcelona, Spain
athina.tympakianaki@aimsun.com
[2] University of Antwerp – imec, IDLab - Faculty of Applied Engineering, Sint-Pietersvliet 7,
2000 Antwerp, Belgium
[3] National Technical University of Athens, 5 Iroon Polytechniou Street, Zografou Campus,
15773 Athens, Greece

Abstract. Predicting the future supply and demand of a transport network are challenging and important problems in real-time traffic management systems that are essential to enhance the decision-making process for deploying adequate traffic strategies under different conditions (e.g., road works, accidents). In the context of the TANGENT H2020 project, simulation-based and data-driven methodologies are developed focusing on the real-time demand and supply prediction problems. This paper focuses on the development and integration of the demand and supply models as well as incident detection methods into traffic simulation environments for network-wide traffic predictions. The role of each component of the framework and their interoperability is explained in the paper, using as testbed the network of Athens, Greece.

Keywords: Intelligent Transport Systems · Traffic simulation · Demand and Supply Traffic predictions · Anomaly detection · Deep learning

1 Introduction

Traffic prediction has more than 30 years of history, yet a challenging problem, fragmented in terms of methodologies and tools implemented for real-time traffic management systems. Traditional parametrical approaches for traffic prediction present limitations when capturing the spatial and temporal relationships of traffic patterns, hence, there is a need to develop more advanced data-driven methods for predicting the future traffic supply and demand. Furthermore, the role of traffic simulation is very important as it captures the demand and supply interactions within the transport ecosystem. This work focuses on the development and integration of demand and supply methods into traffic simulation environments for traffic predictions.

© The Author(s) 2025
C. McNally et al. (Eds.): TRAconference 2024, LNMOB, pp. 356–362, 2025.
https://doi.org/10.1007/978-3-031-85578-8_45

2   Framework for Real-Time Traffic Monitoring and Forecasting

Figure 1 depicts the integration process and workflow between the demand and supply models with the simulation environment. The first step involves the collection and preparation of the traffic data that are fed to the demand and supply components. The supply component consists of data-driven traffic prediction models as well as traffic simulation models. The selection of the most adequate supply prediction model and their interaction depend on the scope of the application. For instance, data-driven models can be applied for local traffic predictions where historical and real-time traffic measurements are available for specific locations. Simulations can be performed for predicting the network-wide traffic conditions, covering also locations for which real traffic observations are not available. The demand is a key input to transport simulation models, along with information on the traffic supply. In offline transport applications (e.g., design and evaluation of response plans), the estimation of a base demand, representing average historical travel patterns and specific circumstances (e.g., recurrent congestion, special event), is needed to simulate and assess the network-wide conditions. In online applications, data-driven models can be trained for predicting the future demand, adapting the base demand to real-time traffic conditions. The output of the real-time local traffic prediction models can be used as input into the demand prediction component. Subsequently, network-wide simulation can be performed to predict adequate network performance metrics (travel times, speeds, flows, etc.). Besides the demand input, the simulation can incorporate changes in the network supply, such as detected traffic-related incidents or road closures due to an event, to evaluate their impact on the road network. The simulation output can be further fed into the data-driven prediction models by providing additional information to enhance the accuracy and reliability of the local traffic predictions (e.g., when an incident is identified and its impact is simulated).

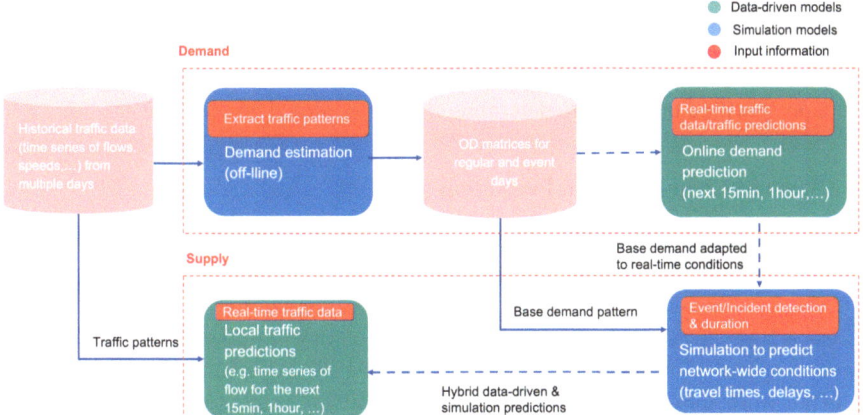

Fig. 1. Framework for real-time traffic monitoring and forecasting.

2.1 Data-Driven Supply Predictions Methods

Traffic Supply Forecasting
Traffic supply forecasting contributes significantly to anticipating and preventing traffic congestion. The supply side of a transport network can be defined by traffic measurements such as traffic speed, travel time, density, and flow. Traffic supply forecasting has been tackled from different modeling perspectives during the last few years. Among various approaches, data-driven approaches have shown excellent results with the rise of data collection and computational resources. The state-of-the-art (SotA) models in deep learning utilize Graph Neural Networks (GNNs), Recurrent Neural Networks (RNNs), Convolutional Neural Networks (CNNs) and Attention Mechanisms. By benchmarking different SotA models, optimizing their hyper-parameters, and performing an in-depth fusion of multiple models for the examined case study (described in Sect. 3), Multivariate Time Series Forecasting with GNN (MTGNN) [1] is shown to be one of the best models for this task. The method uses graph learning layer, Graph Convolutional Networks and Temporal CNNs to capture latent dependencies in spatial and temporal contexts. The entire framework is trained in an end-to-end manner.

Incident Detection
Incident detection is defined as the detection of extreme traffic conditions caused by, e.g., an accident or nonrecurrent high traffic volumes. Early detection is crucial for timely activation of adequate response plans aiming to alleviate the impacts and returning the traffic conditions back to normal. Incident detection methods detect nonrecurrent congestion by comparing the current traffic state with the predictions. When the measurements deviate significantly from the predicted values, an anomaly is detected.

The deviation is measured through different statistics (such as absolute or squared error, z-scores). In the next step, isolation forest [2], which is an unsupervised machine learning model, is used to aggregate all those anomaly scores into a unified anomaly score. The algorithm works by isolating observations by randomly selecting one initially calculated anomaly score, and then randomly selecting a split value between the maximum and minimum values of the selected one. The unified anomaly score is compared with a prespecified threshold to determine if it's considered an anomaly.

2.2 Demand Estimation and Prediction Methods

Simulation-Based Demand Estimation
Demand estimation refers to the process of adjusting the Origin-Destination (OD) trip matrices of a given transport model, usually using simulation-based approaches. This process is formulated as an optimization problem, aiming to find an OD matrix that best reproduces a set of traffic measurements when assigned to a given network.

In this work, the existing gradient-based algorithm [3], in the Aimsun Next simulation software [4], is used to perform the offline estimation of OD matrices. The approach requires the calculation of the assignment matrix, which maps the contributions of OD flows on links with available traffic counts in the network. Its main advantage is the analytical computation of the gradient of the objective function, which is expected to result in robust solutions. Nevertheless, it is limited to using only traffic counts (due to the assumption about their linear relationship with the OD matrix), which can be problematic for congested networks as this assumption does not hold.

Data-Driven Demand Predictions
Online demand prediction methods are used in the context of real-time traffic management and route guidance, aiming at providing fast estimates for recent time intervals with predictions for future time intervals.

In this work, a preliminary attempt is made in using a GNN architecture to predict aggregated demand generated at origins (in the OD matrix). The proposed approach adapts the work proposed in [5], which spatial-temporal graph attention network (ST-GAT) architecture for traffic supply forecasting to the demand prediction problem. Supervised experiments are carried out to predict the demand at time t (at all origins generating vehicles) using traffic data from the previous hour (t-1 h) from all sections aggregated in 15-min intervals. The spatial-temporal components of the ST-GAT architecture allow for modelling both the geometry of the network (spatial) as well as the evolution of flow as a vector of time series from the previous hour in 15-min intervals (i.e., t-60 min, t-45 min, t-30 min, t-15 min as a feature vector and t as the class to be predicted) across the whole network. Synthetic simulated data has been used to test several configurations in a controlled experimental machine learning setting.

3   Application of the Framework to a Case Study

Athens Case Study
The Athens case study is one of the TANGENT project case studies, which is used as testbed to apply the proposed framework. In particular, the inner-ring urban transportation network of Athens, Greece (Fig. 2) is used, which is calibrated and validated in Aimsun Next using hourly traffic data (counts and speeds) from 2023. The network consists of 860 nodes and 1,856 sections as well as 237 loop detectors. With respect to the calibrated demand, approximately 48,000 vehicles are inserted in the network during morning peak hour (08:00–09:00) for a typical day.

Specifically, traffic data was collected for two weeks in 2023; a week with normal operation of the network (February 20th – February 26th) and a week with a known (severe) flooding event (September 11th – September 17th). The collected data was further used as input to the simulation model to perform a series of demand calibration experiments to derive representative OD matrices for both the normal week and the week with the disturbance. The same data set was further used to train the data-driven traffic supply forecasting as well as the incident detection models. Following the framework workflow (Fig. 1), the adjusted OD matrices and traffic predictions can be used to train the data-driven demand model, which in a real-time context can provide the future demand as input to simulation for deriving the network-wide traffic conditions.

In this work, due to space limitation, preliminary results are shown from the application of the data-driven traffic supply forecasting (local predictions), incident detections and the demand calibrations for different traffic patterns.

Fig. 2. Athens simulation testbed.

3.1 Preliminary Results

Traffic Predictions and Anomaly Detection Results
The trained model, which considers 1-h predictions and a test set including both normal and event days, has an accuracy MAE of 5.57 km/h. Usually, a MAE of around 3 km/h is expected for 1-h predictions, but since half of the testing is done on a period where the network was not functioning normally, the achieved performance is expected. For every sensor, a graph with anomaly scores is constructed. Figure 3 depicts anomaly scores for the different select days described in the previous section. Red points are anomalies, while blue points are normal scores. The contamination threshold is set at 5%. This implies that the thresholds are determined such that 5% of the data is classified as anomalous. However, certain anomalies, specifically those where the predicted traffic speed is lower than the observed traffic speed, are not of interest to our analysis. Consequently, these specific anomalies are not flagged, resulting in a final anomaly percentage that is less than the initial 5% threshold. The figure shows that the red dots only occurred during event days. As explained in Sect. 1, anomalies are detected by comparing the predicted with the observed traffic speed (Fig. 4).

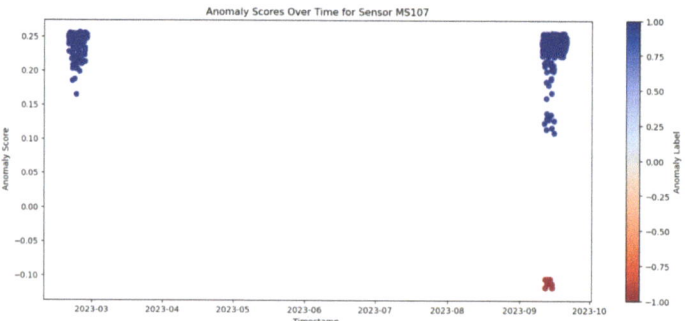

Fig. 3. Anomaly scores over days for a specific sensor for the Athens case study.

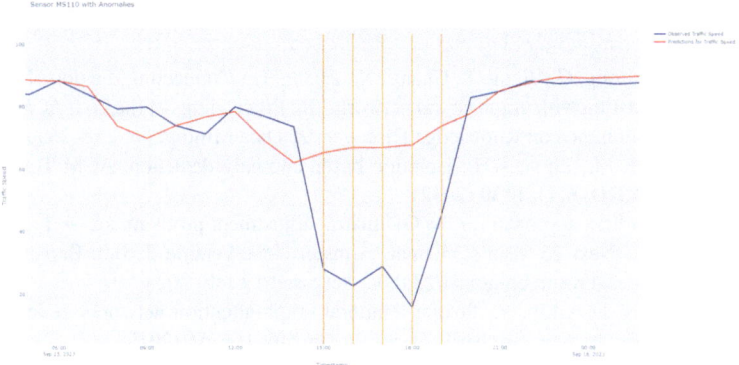

Fig. 4. Predicted versus observed values for a specific sensor. Yellow lines are anomalies.

Demand Estimation Results

For the demand estimations, simulation-based experiments were executed for the normal week and the week with the disturbance. For each day of both weeks, extensive morning (06:00–12:00) and afternoon (13:00–19:00) peak hours are taken into consideration. As a result, 84 hourly OD matrices are adjusted for each week. In terms of the accuracy of the final output, for the normal week, the Root Mean Square Percentage Error (RMSPE) varies between [0.23, 0.34]. For the week with the disturbance, RMSPE varies between [13, 15]. The high RMSPE values, in the latter, are expected considering the fact that a network-wide disturbance (flooding event) is examined, hence, a high value quantifies the error of the adjustment in relation to the baseline values.

4   Conclusions

This paper outlines the developments of various data-driven and simulation-based methods for network supply and demand predictions, integrated into a framework for real-time traffic management. Preliminary results and insights are provided through the case study of Athens, focusing on data-driven traffic (local) predictions and incident detection. Ongoing work continues on training and testing a demand prediction model, for enhancing the demand input to the simulation model for network-wide predictions. Hybrid data-driven and simulation predictions are also examined, by performing simulations that reproduce identified incidents and providing the simulation outputs as additional information to the data-driven traffic predictions under non-recurrent events.

Acknowledgements. This work has received funding from the European Union's Horizon 2020 research and innovation programme under grant agreement No 955273 (TANGENT).

References

1. Wu, Z., Pan, S., Long, G., Jiang, J., Chang, X., Zhang, C.: Connecting the dots: Multivariate time series forecasting with graph neural networks. In: Proceedings of the 26th ACM SIGKDD International Conference on Knowledge Discovery & Data Mining, pp. 753–763 (2020)
2. Liu, F.T., Ting, K.M., Zhou, Z.H.: Isolation-based anomaly detection. ACM Trans. Knowl. Discov. Data (TKDD) **6**(1), 1–39 (2012)
3. Spiess, H.: A gradient approach for the OD matrix adjustment problem. a∈ ˆA **1**, 2 (1990)
4. Aimsun. Aimsun Next 23 User's Manual, Aimsun Next Version 23.0.0, Barcelona, Spain (2023). https://docs.aimsun.com/next/23.0.0/. Accessed 19 July 2023
5. Zhang, C., James, J.Q., Liu, Y.: Spatial-temporal graph attention networks: a deep learning approach for traffic forecasting. IEEE Access **7**, 166246–166256 (2019)

Open Access This chapter is licensed under the terms of the Creative Commons Attribution 4.0 International License (http://creativecommons.org/licenses/by/4.0/), which permits use, sharing, adaptation, distribution and reproduction in any medium or format, as long as you give appropriate credit to the original author(s) and the source, provide a link to the Creative Commons license and indicate if changes were made.

The images or other third party material in this chapter are included in the chapter's Creative Commons license, unless indicated otherwise in a credit line to the material. If material is not included in the chapter's Creative Commons license and your intended use is not permitted by statutory regulation or exceeds the permitted use, you will need to obtain permission directly from the copyright holder.

Sustainable Future Flight Business Models: Motivations and Barriers

Chenyi Liao[1] , Christopher J. Parker[1](✉) , Graham Parkhurst[2] ,
Magdalena Oldziejewska[3] , Mohammad Uddin[3] , and Ram Ramanathan[3]

[1] Loughborough University, Loughborough, Leicestershire, UK
c.parker@lboro.ac.uk
[2] University of the West of England, Bristol, UK
[3] The University of Essex, Southend-on-Sea, Essex, UK

Abstract. This paper investigates stakeholders' motivations and barriers within emerging Future Flight business models. Aviation is vital for developing economies, and the urgent need for a transition towards more sustainable practices is gaining prominence. Hence, understanding the factors shaping Future Flight technology's adoption is crucial. Drawing on ten interviews with pioneering Future Flight-related technologists, business leaders, social entrepreneurs, and policy-makers, we employed the Technology-Organisation-Environment framework and Transaction Cost Economics theory to analyse critical factors influencing Future Flight business models. We show participants are concerned about sustainable aviation fuel availability and Future Flight technologies' readiness. We emphasise the importance of technology maturity and commercial viability for successful Future Flight implementation. Smaller start-ups are poised to lead such development because of their nimbleness and sustainability focus. Concurrently, larger companies face challenges transitioning from traditional business models. We identified regulatory frameworks, social acceptance, and public demand as key drivers. Finally, we show how entrepreneurs desire standardised global regulations to support sustainable aviation practices. We offer insights into the complex dynamics of Future Flight adoption, highlighting companies' need to evaluate their cultural and human resource strategies while emphasising global regulatory standards' importance – as part of The CoFFEE Project's (www.coffeefutureflight.com) broader research programme.

Keywords: Urban Air Mobility (UAM) · FlyDrive · Unmanned Aerial Systems (UAS)

1 Introduction

Aviation is increasingly critical in developing economies. Aviation enabled cities - including Denmark's Legoland (in otherwise secluded Billund) - to become 'experience' destinations [1] and is essential for some people in maintaining social relationships concerning spatially defused familial and friendship networks. However, aviation's high energy demands - and low-carbon technologies' limited development stage - raise important questions about environmental impacts.

© The Author(s) 2025
C. McNally et al. (Eds.): TRAconference 2024, LNMOB, pp. 363–369, 2025.
https://doi.org/10.1007/978-3-031-85578-8_46

FF developments have focused on the engineering challenges of building Advanced Air Mobility (AAM) – comprising Door-to-Door Regional Air Mobility (FlyDrive, aka. Flying cars travelling >50k) and Urban Air Mobility (UAM, aka. flying taxis travelling <50k) - and Unmanned Aerial Systems (UAS, aka. drones) powered by electricity or hydrogen. Engineers assume businesses and social entrepreneurs will someday develop interconnected social networks and capabilities ('ecosystems') to make their mechanisms viable. Nevertheless, FF can only enhance society and strengthen the economy when individuals, groups of users and non-users, innovation ecosystem stakeholders, and local communities adopt these new technologies and aviation forms.

Understanding the emerging innovation ecosystems is critical as the UK Research and Innovation Agency's (UKRI) Vision and Roadmap for FF [2] proposes geographically distributed service integration to be achieved by 2028. FF requires rethinking critical business theory, as current business models require reconsidering sustainable development and circular economy. Without understanding the emerging innovation ecosystem, complex stakeholder networks, and technology implementation process, FF will likely fail.

This paper aims to determine the motivations and barriers of aerospace industry technologists, business, social entrepreneurs, and supply chains within the evolving FF scenarios and how they influence the emergent FF business models. We achieve this aim by reporting on ten interviews with pioneering Future Flight-related technologists, business leaders, social entrepreneurs, and policymakers while employing the Technology-Organisation-Environment framework and Transaction Cost Economics theory.

2 Methodology

2.1 Setting and Sample

We recruited participants through professional networks, social media posts, and snowballing. Purposive sampling against key attributes (e.g., company size) determined participant selection, including AAM/UAS manufacturers, transport business users and suppliers, and social entrepreneurs who understand FF's social and economic impact. We also applied snowball sampling to identify cases of interest during the interview.

We interviewed 10 participants (M = 6, F = 4), including entrepreneurs (n = 4), professional publics (n = 3), technologists (n = 2), and policymakers (n = 1) with expertise and experience in economics, logistics, aviation engineering, and occupational psychology.

2.2 Data Collection

We developed an interview protocol to explore the technological, organisational, and environmental factors influencing FF business models. Technological factors describe the characteristics of FF's technology. Organisational factors include the company's scale and reach in the market. Environmental factors are those outside of the organisation's control.

Questions elicited participants' opinions on the closest technologies they expect to be developed in the next five to ten years, the social, economic, and environmental impact of FF, and their intention to engage with the circular economy. Individual semi-structured interviews (c.60 min) were recorded, professionally transcribed, and imported into NVivo for analysis.

2.3 Data Analysis

Influenced by the Technology-Organisation-Environment (TOE) framework [3] and Transaction Cost Economics (TCE) theory [4], broad initial codes were set up to guide the coding process as the top layer codes in the catalogue. After familiarisation with the data, the first reading mapped out the contextual information about each participant and their organisation/company. The second reading was a data-driven coding process, cross-checking codes with contextual information to determine potential patterns. From the participants' responses, we determined the reasoning behind their predictions regarding trends, technologies, and potential impacts. This followed Wiltshire and Ronkainen's [5] inductive and abductive thinking.

3 Results

3.1 Technological Factors

Fuel Availability
Most participants (7/10) were aware that decarbonisation's scope surpasses superficial rhetoric. In particular, there was a high awareness of the importance of a 'green' production process and physical infrastructure to achieve 'net zero'. Some participants addressed the problem of batteries for their limited circular economy attributes to date (p11), distance limitation (p4, p9), and the reliability of hydrogen as a fuel (p7). Some participants were, however, optimistic about Sustainable Aviation Fuel's (SAF) development, considering it the *"most promising"* technological potential and expecting an *"acceleration of SAF"* (p3). Critically, large companies' motivation to invest in SAF and hydrogen shall influence their business models' evolution.

Readiness
It is plausible that the readiness of FF technologies takes longer to mature than the UKRI roadmap proposes. Many participants spoke about the reality of the timeline regarding AAM and UAS. It is plausible that an unhealthy equity investor culture leads to over-hyped timelines and unrealistic expectations. Participants confirmed the potential for advances with low-carbon fuels, navigation, and autonomy; however, their commercial viability, efficiency at scale, and usability will determine the readiness of FF technologies.

3.2 Organisational Factors

Company Culture
In part - because the participants' motivations for creating a positive impact align with the

company/organisation they are part of - many participants are sensitive to sustainability. Some participants intend to engage with a Circular Economy (CE) approach. However, CE is still a niche concept in the industry because of the technological limitation of recyclable composite materials (p4). One participant (p1) even criticised CE for not advocating for reduction as it risks increasing consumption.

Human Resources

It is plausible that companies in the FF industry are having difficulty recruiting appropriate talents. Eight of ten participants expressed concerns about the industry's talent and skills shortage. Two participants described a *"war of talent"* with other industrial sectors in need of artificial intelligence talent (p5), identifying a *"massive shortage of engineers"* (p11). This shortage also relates to public bodies charged with regulating FF links with the difficulty of offering competitive salaries to skilled workers (p11).

The data also suggest that staff expertise is vital in shaping a company's products/services (p9, p11). Many participants believe the shortage can be mitigated by identifying industry skill requirements (p5, p11) and developing new training protocols (p2, p4, p8). One participant suggested that a diversified workforce could benefit FF. For example, recruiting female engineers (p11). Younger workers' emphasis on desiring to contribute to sustainability could positively change a company's culture.

Inertia to Change

It is plausible that smaller start-up companies with new business models will change the FF industry. There is inertia for big companies to continue making a profit without changing business models. Some participants explained that *"big companies' business models are planned around planes designed 25 years ago"* (p4). Small to Medium sized Enterprises (SMEs) also are reluctant to change their business models *"because they are locked into the supply chain"* (p4).

In contrast, smaller start-ups honed their resources on niche products/services. They built a competitive advantage over multinational companies by obtaining patents and securing their supply chains (p7). Conversely, participants saw big companies as being more motivated to invest in SAF and retrofit their existing planes. However, smaller start-ups certify early in their development process to avoid later expenses.

3.3 Environmental Factors

Regulations and Government

All participants stressed the crucial role of regulations. Many participants identified the delay and gaps in regulations to support the FF industry's development. The industry's growth created demands for certifying vehicles and operational safety clearances (p11). The companies sought for the regulators to establish *"a level playing field"* (p8) and address the disparity in global standards (p8). Considering big companies' inertia to change, governmental regulations supporting sustainable fuel development (i.e., tax on traditional fuel) will drive the industry significantly. Currently, the lack of regulatory changes resulted from the limited regulatory capacity and shortage of skills (p9, p11), which is believed to be the biggest challenge.

Social Norms

Many participants acknowledged the public as a critical FF stakeholder. Public acceptance of FF products/services will likely be shaped by social norms and the potential impact of social disparity. Some participants suggested the high likeness of the more affluent, smaller population benefiting from FF technologies. One participant (p9) indicated that the Fly Drive market will probably diffuse in society before the UAM market because the public already accepts flying outside of urban environments.

4 Discussion and Conclusion

This paper aims to determine the motivations and barriers of aerospace industry technologists, business, social entrepreneurs, and supply chains within the evolving FF scenarios and how they influence the emergent FF business models. In pursuing this aim, we identified seven key factors that will likely impact the emergent business models from the perspectives of technology, organisation, and environment. We determined that readiness and fuel availability are two critical barriers in the technological context. Other research suggests an unlikelihood of aviation businesses transitioning to carbon-zero air mobility because of a lack of alternative fuels. The costs incurred from adopting new technologies when technology readiness and scalability are uncertain would lead to low or negative profitability for airlines [6].

Our findings suggest that start-ups will likely lead the FF industry's development, addressing the less attractive niches to established aviation businesses and focussing on those niches requiring moderate capital investment. In the organisational context, start-ups possess a company culture that drives the design of niche products and services intending to improve sustainability performance. Unlike multinational aviation businesses and SMEs with, to date, insufficient financial incentive to push for 'jet zero', start-ups actively engage with upcoming FF technologies to carve out niche markets.

In the environmental context, we identified concerns over accessibility, privacy, and mental health (noise pollution). While research noticed the impact of the increasing popularity of private aeromobility on sustainability [7], our findings suggest that social disparity is a critical barrier to public acceptance. Moreover, entrepreneurs are aware of the crucial role of the public. They are optimistic about changing social norms that demand sustainable air travel, which can drive the industry to invest in sustainable fuels and new business models. Like the public's preference for a central government FF regulation [8], entrepreneurs demand regulation to standardise the industry globally and roll out regulations more quickly to accommodate FF development.

Theoretically, the TOE framework can be mobilised to identify factors that support implementing a circular economy approach to businesses. We progress this theory by showing that environmental factors are critical to start-ups, SMEs, and multinational companies. In contrast, organisational and technical factors' importance varies depending on the organisation's scale of operation, market segment, and culture.

In conclusion, businesses aiming to establish themselves in the industry should evaluate their organisational culture and human resources. Companies should revise their strategic business models to offer a more significant sustainability commitment. Businesses should diversify their hiring to gain broader talent (gender, age), work ethos, and focus on sustainability. Government and regulatory bodies must establish industry standards globally to accommodate the FF industry's development. In addition, regulatory bodies should address their skills shortages and increase their capacities to fulfil the industry's certification and safety clearance needs.

4.1 Limitations and Future Research

This paper's underpinning TOE framework is limited in capturing the complexity of the adoption and implementation process and the rapidly changing external environment for novel industries such as future aviation. To mitigate this limitation, the UKRI-funded umbrella project (CoFFEE Project: https://coffeefutureflight.com) shall continue to collect more interview data from a diversified body within the Future Flight Community. In particular, a more significant sample of policymakers and technologists. Doing so shall allow us to extend our findings' generalisability into policymakers' perspectives.

References

1. Lassen, C., Smink, C.K., Smidt-Jensen, S.: Experience spaces, (aero)mobilities and environmental impacts. Eur. Plan. Stud. **17**(6), 887–903 (2009). https://doi.org/10.1080/096543109 02794034
2. UKRI: Future Flight Vision and Roadmap. https://www.ukri.org/publications/future-flight-vis ion-and-roadmap/. Accessed 02 Oct 2023
3. Tornatzky, L.G., Fleischer, M.: The Process of Technology Innovation. D.C. Heath & Company, Lexington (1990)
4. Ketokivi, M., Mahoney, J.T.: Transaction cost economics as a theory of the firm, management, and governance. In: Oxford Research Encyclopedia of Business and Management. Oxford University Press (2017). https://doi.org/10.1093/acrefore/9780190224851.013.6
5. Wiltshire, G., Ronkainen, N.: A realist approach to thematic analysis: making sense of qualitative data through experiential, inferential and dispositional themes. J. Crit. Realism **20**(2), 159–180 (2021). https://doi.org/10.1080/14767430.2021.1894909
6. Gössling, S., Humpe, A.: Net-zero aviation: time for a new business model? J. Air Transp. Manag. **107**, 102353 (2023). https://doi.org/10.1016/j.jairtraman.2022.102353
7. Cohen, M.J.: Sustainable mobility transitions and the challenge of countervailing trends: the case of personal aeromobility. Technol. Anal. Strateg. Manag. **21**(2), 249–265 (2009). https://doi.org/10.1080/09537320802625330
8. Camilleri, E., Gisborne, J., Mackie, M., Patel, R., Reynolds, M.: Future Flight Challenge – Mini Public Dialogue (2022)

Open Access This chapter is licensed under the terms of the Creative Commons Attribution 4.0 International License (http://creativecommons.org/licenses/by/4.0/), which permits use, sharing, adaptation, distribution and reproduction in any medium or format, as long as you give appropriate credit to the original author(s) and the source, provide a link to the Creative Commons license and indicate if changes were made.

The images or other third party material in this chapter are included in the chapter's Creative Commons license, unless indicated otherwise in a credit line to the material. If material is not included in the chapter's Creative Commons license and your intended use is not permitted by statutory regulation or exceeds the permitted use, you will need to obtain permission directly from the copyright holder.

Smart and Innovative Solutions for Maximizing Public Transport Ridership and Passengers' Satisfaction: Case Study of the City of Žilina

Ghadir Pourhashem[(⊠)] and Tatiana Kováčiková

Department of International Research Projects (ERAdiate+), University of Žilina, Žilina, Slovakia
{Ghadir.Pourhashem,Tatiana.Kovacikova}@uniza.sk

Abstract. Public transport and shared mobility services are widely acknowledged as the backbone of sustainable urban transportation in terms of reduction of traffic congestion, air pollution and noise and increasing traffic safety. These services play pivotal roles in achieving the objectives of EU Cities Mission and the European Green Deal ambition of net zero greenhouse gas emission by 2050.

Alongside this, the evolution of emerging mobility concepts is also transforming the future of mobility landscape, which impose a paradigm shift towards smarter carbon-neutral mobility services. Therefore, development of smart sustainable mobility solutions and ensuring improvement of customized mobility services for all citizens are key challenges for the cities of tomorrow. All of these require experimentation besides nudging people's mobility behavior towards sustainable and climate-friendly choice of transport modes especially in Central and East European cities with high tendency for car ownership and use of private car for daily commuting.

With a twofold objective of raising public transport ridership and passengers' level of satisfaction with collective and public transport systems, this paper focuses on presenting various mobility-related solutions and measures and the implementation of the most promising ones through utilizing co-creation process for improvement of public transport service features and their smart integration with existing shared mobility services. The case study is provided for the city of Žilina as one of the twining cities in the Horizon Europe SPINE "Smart Public transport Initiatives for Climate-Neutral cities in Europe" Innovation Action project.

Keywords: Public transport · passengers' satisfaction · awareness raising · co-creation · smart parking management · travel pattern change

1 Introduction

Increasing the use of public transport is essential for reducing traffic congestion, improving air quality, and promoting sustainable urban development. In recent decades, multitudinous methods and measures (objective and subjective) to encourage people to opt

© The Author(s) 2025
C. McNally et al. (Eds.): TRAconference 2024, LNMOB, pp. 370–376, 2025.
https://doi.org/10.1007/978-3-031-85578-8_47

for public transport (PT) while nudging behavior change in favor of PT and uptake of shared mobility services as a promising solution for sustainable and carbon-natural urban development have been studied in diverse contexts [1]. Thus, achieving the objective of maximizing PT ridership and making PT a viable choice for travelers requires a multi-faceted approach that focuses on improving the overall user experience and satisfaction [2–4]. In this vein, combining and utilizing various strategies and smart and people-centric solutions would catalyze cities and regions capacities to effectively increase the use of PT and ensure travelers satisfaction, leading to a more sustainable and efficient transportation system. Hence, the main objective of this paper is presenting various mobility-related solutions and measures and the implementation of the most promising ones through utilizing co-creation process for improvement of PT service features and their smart integration with existing shared mobility services in the city of Žilina as one of twinning cities in the Horizon Europe SPINE project.

The remainder of this paper is divided as follows. Section 2 presents the SPINE methodology which will be employed in the SPINE living labs (LLs). SPINE digital solutions as nudge interventions for behavioral change among trip makers towards more use of PT and shared mobility services in Žilina are presented in Sect. 3. Conclusion together with some expected impacts are addressed in Sect. 4.

2 SPINE Methodology

SPINE develops and applies an equity centered design thinking approach that spans across all its steps, starting with empathizing with the users of the solutions, proceeding to defining the plausible set of solutions, collectively co-creating and refining the solutions, implementing them, testing and assessing their impact and finally amplifying successful solutions to cities across Europe. All solutions (push and pull measures), are centered around improving the offer of PT system, addressing the diverse needs of existing and potential PT users, concentrating, and emphasizing on axes such as equity, accessibility, affordability, and inclusiveness. At the same time, a diverse multitude of co-creators (e.g., citizens, vulnerable groups, decision-makers, transport sector stakeholders, urban planners, local influencers and ambassadors) is assembled in order to acknowledge the existing condition in the local contexts, come to terms with existing barriers and potential personal biases and identify key drivers to overcome them.

3 SPINE Digital Enablers for Triggering Behavior Change in Favor of Public Transport in Žilina

City of Žilina is the fourth largest city of Slovakia with an approximate population of 85,000 inhabitants. It is an important industrial hub and transport corridor in Slovakia; therefore, it regularly faces traffic congestion and all the detrimental impacts associated with it. Although Žilina public transport is a cornerstone of urban transport system, private cars still have a high share in the city and region's modal split mostly for daily commuting to Žilina for work and education. As such, the city of Žilina as a twinning city in the SPINE project strive to unlock the true potential of PT system for increasing its share by 30% compared to the baseline and users' satisfaction by 25% considering the needs of diverse user groups in the years of 2023 to 2027. To do so, an ambitious implementation plan has been set for the co-design, development and deployment of push and pull measures. Furthermore, integration of smart solutions and utilizing the co-creation efforts within the Žilina LL has been envisaged to reinforce existing PT offerings and integrate them with new smart mobility services which will increase users' satisfaction and will eventually contribute to the achievement of the modal shift and decarbonization targets. It is pertinent to note that for the successive learning regarding inclusive user-centered design of measures and accelerating uptake of offered solutions, the city of Žilina will also take the advantage of SPINE collaborative network of LLs fostering transferability and replicability of gained knowledge and experience across lead and twinning cities horizontally ("cross-pollination" activities) and vertically from lead city to twin cities ("twinning" activities). The implementation progress and the impact of all envisaged solutions for the city of Žilina as described in the following sections will be measured and monitored using a set of 7 predefined key performance indicators: *modal share, PT ridership, users' satisfaction, access to mobility services, congestion, air pollution and noise pollution.*

3.1   Real-Time Information for Public Transport Riders

Digitalization has become a pervasive strategy for the cost-effective management of services by public authorities while neglecting sometimes the mobility needs of non-digital travelers, people with low digital skills and city visitors and tourists. Given the fact that accurate, timely passenger information is crucial for enhancing passengers' satisfaction with PT services, this solution is aiming at accelerating the equity among all social groups and boosting passenger satisfaction in regard with providing the real time information of bus schedules (e.g. arrivals, delay), service disruptions and occupancy levels.

Real-time information for PT passengers displayed on digital screens as a relatively simple technical solution is expected to contribute to making PT more attractive for various social groups through changing citizens perception of four fundamental operational features of PT services: reliability, short waiting time, speed, and capacity [5–7], nudging green travel choice and making more informed decisions [8]. As a part of this solution, digital screens will be installed in areas with various land-use in Žilina: main PT hubs (i.e., places with significant commuter traffic where passengers can exchange between different PT lines), main hall of rail-way station and shopping malls with high travel demand. To monitor progress and the opportunities brought by this solution, the collected data on delay and passengers' occupancy together with results from the conducted quantitative and qualitative satisfaction surveys will be processed and analyzed to further explore the required areas of intervention besides the factors influencing PT users' level of satisfaction and travel experience.

3.2   Multimodal Journey Planner

Multimodal journey planners with comprehensive door-to-door information received from sensors and PT onboarding data loggers also proven a substantial impact as a digital enabler on travel modal shift and increasing PT patrons' perception of seamless journey in particular coping with the first mile/last mile issues, usually faced by commuters from the sub-urban areas [9]. The multimodal journey planner proposed in SPINE project is an integrated, dynamic journey planner which provides the passenger with real-time and scheduled information for several mobility services (PT, sharing modes, micro-mobility, etc.) incorporating multimodal routing and optimal travel planning. Utilizing dynamic vehicle routing for on-demand mobility and real-time traffic information, the journey planner ensures that passengers can make informed decisions for their travel routes, while creating a seamless and efficient mobility experience, and optimizing the travel process. The planner will be designed to respond to user requests for a route by presenting a curated list of routes, considering factors such as quantified CO_2 emission (carbon saving) compared to car, fastest, cheapest and more suitable options. In addition, the multimodal journey planner customizes its suggestions to meet the unique needs of each traveler by analyzing user preferences and behavioral clustering. The Žilina multimodal journey planner is expected to further incentivize citizens to use PT for their entire journey, reduce traffic congestion in the central area of city and most importantly increase the use of PT through unleashing the potential of Bike & Ride as a complementary relationship between bike-sharing and public transport [10] that enables passengers to use sharing bikes in conjunction with PT system for first and last mile access in a journey.

3.3 Smart Parking Management and Low Emission Zone Action Plan

Implementing a novel smart parking management system in urban areas with mixed land-uses can significantly enhance the parking experience for users of private cars (e.g., travel time and cost reduction), reduce traffic congestion in the urban road network, and contribute to a more sustainable and efficient urban environment. On the other hand, integrating smart parking management with other modes of transportation, such as PT and bike-sharing systems using multimodal journey planner will promote seamless connections and multimodal travel between different transportation options comparing the cost, travel time and level of CO_2 emission of each transport offer for the selected trip destination to encourage people choose sustainable and efficient ways of reaching commercial areas located in the Žilina city center.

To augment the adaptation of the new measures in parking management policy, a low emission zone (LEZ) feasibility study will also be carried out in the 2nd city circuit (broader city center) close to the smart parking management implementation area. The environmental impact assessment of the LEZ study as a complementary part of smart parking management, would play an essential role in preparation of a strategic document for triggering behavior change in favor of PT and bike-sharing services. This multifaceted solution is expected to contribute to changing people's travel behavior towards multimodal trips and more use of PT and bike-sharing services in the central area of the city.

3.4 Smart City Platform for Transport and Mobility Planning

The SPINE smart city platform (SCP) as a Software as a Service (SaaS) is aimed to build and implement sustainable and resilient mobility concepts and strategies, as well as manage their operations and assets, through the combination of big data and predictive analytics. Žilina smart city platform will be comprising of a simulation engine and tools for testing SPINE innovations, assessing their impact using quantitative key performance indicators (KPIs) and launching open-dialogue digital hub (i.e., platform for enabling participants to respond to specific solution/s and new mobility services, comment on each other posts and engage in group discussions). A stakeholders' dashboard will also provide the city authorities, policymakers, and transport operators with a series of interactive screens with the main goal of monitoring the deployed innovative solutions and monitoring the impact of combined push/pull measures. It is important to note that the SCP itself does not directly contribute to increasing passenger satisfaction or increasing the number of PT passengers. While it is supposed to increase the city's capacity by gaining a better overview of the situation in different transport sections with major implications such as contributing to more data-driven decision making and strategic planning, for an increase in PT ridership. In short, the expected output of the Žilina SCP will be a data integration space and a web-based platform providing a set of different functionalities and dashboards (e.g., goal view, modules view, KPIs and analytics view). This tool will allow the city authorities to efficiently monitor the SPINE-related mobility measures through 4 sets of identified KPIs for the city of Žilina: *I) PT quality of services* (e.g. Ridership volume, Passengers' satisfaction, Travel time reliability and Accessibility),*II) On-street parking management* (e.g. Parking capacity and Parking search time), *III) Traffic information*(e.g. Modal share and Congestion indexes) and *IV) Environmental impact* (e.g. Air pollution and Noise pollution levels).

3.5 Rebranding Public Transport Campaigns

To raise awareness and emphasize on the environmental benefits of using PT, among diverse social group of people, a series of pop-up exhibition supported by city of Žilina, Žilina public transport company and other relevant stakeholders will be organized. The main aim of the pop-up exhibitions is to encourage critical thinking among participants, observe the interactions of the participants with the various displays of open questions and proposed solutions and to gather quantitative and qualitative data focused on key drivers of modal shift to PT and satisfaction from PT services and other shared mobility services [11]. To further foster the communication between the city and citizens, promote the mobility solutions and push and pull measures, the SPINE Citizens App will be also developed within the project. The Citizens App will serve as a citizens engagement platform with features such as travel logging, offering tailored incentives per user group based on user preferences, previous experience. Incentives, offers, and promotions will also be discussed and presented during the Living Lab (LL) and pop-up exhibitions organized in Žilina.

4   Conclusion

Public transport system is the robust and collective backbone of urban mobility, playing the central role in the desired future of transportation. Considering that most European major cities are undergoing to experience urban population growth by 2030, hereupon urbanization trend is envisaged to resonate demands for smart technology initiatives to make PT more appealing and accommodate seamless and inclusive mobility needs of citizens. Therefore, the co-designed technological and non-technological mobility solutions jointly developed in the cities' LLs would empower transportation systems to become resilient and able to deal with exceptional situations. The proposed measures, thanks to deployment of design thinking and co-creation methods, are expected to unlock opportunities in the city of Žilina by bringing together public and private interests in the evolution of PT and sustainable and green mobility leading to an increase of the share of PT ridership by approximately 30% and the user satisfaction with PT by 25% compared to the baseline, and cover different market/customer segments providing a better social inclusion and healthier lifestyles. In conclusion, it is essential to acknowledge the inherent limitations posed by the absence of primary data. The forthcoming research will be dedicated to collecting quantitative and qualitative data that will enable us to enhance the depth and breadth of our analysis, leading to more robust conclusions and practical implications. This study serves as a steppingstone toward a more comprehensive understanding of digital enablers' role for triggering behavior change in favor of PT and its implications.

Funding. This project has received funding from the European Union's Horizon Europe research and innovation programme under grant agreement No. 101091456.

References

1. Rasca, S., Saeed, N.: Exploring the factors influencing the use of public transport by commuters living in networks of small cities and towns. Travel Behav. Soc. **28**, 249–263 (2022)
2. Walker, J.: Human transit: How clearer thinking about public transit can enrich our communities and our lives. Island Press, Washington, D.C. (2012)
3. Brakewood, C., Macfarlane, G.S., Watkins, K.: The impact of real-time information on bus ridership in New York City. Transp. Res. Part C Emerg. Technol. **53**, 59–75 (2015)
4. Lunke, E.B.: Commuters' satisfaction with public transport. J. Transp. Health **16**, 100842 (2020)
5. Watkins, K.E., Ferris, B., Borning, A., Rutherford, G.S., Layton, D.: Where is my bus? Impact of mobile real-time information on the perceived and actual wait time of transit riders. Transp. Res. Part A Policy Pract. **45**(8), 839–848 (2011)
6. Redman, L., Friman, M., Gärling, T., Hartig, T.: Quality attributes of public transport that attract car users: a research review. Transp. Policy **25**, 119–127 (2013)
7. Soza-Parra, J., Raveau, S., Muñoz, J.C., Cats, O.: The underlying effect of public transport reliability on users' satisfaction. Transp. Res. Part A Policy Pract. **126**, 83–93 (2019)
8. Maclean, S., Dailey, D.: Measuring the utility of a real-time transit information system. In: Proceedings the IEEE 5th International Conference on Intelligent Transportation Systems, pp. 846–850 (2002)
9. Reed, T., Biermann, S.: Journey planner usage analysis. Final Report, iMOVE Project, Planning and Transport Research Centre, University of Western Australia (2020)
10. Martens, K.: The bicycle as a feedering mode: experiences from three European countries. Transp. Res. Part D: Transp. Environ. **9**(4), 281–294 (2004)

11. Pourhashem, G., Malichova, E., Kováčiková, T.: The role of participation behavior and information in nudging citizens sustainable mobility behavior: a case study of Bratislava region. In: 19th International Conference on Emerging eLearning Technologies and Applications (ICETA) Proceeding, Košice, Slovakia, pp. 300–306. IEEE (2021)

Open Access This chapter is licensed under the terms of the Creative Commons Attribution 4.0 International License (http://creativecommons.org/licenses/by/4.0/), which permits use, sharing, adaptation, distribution and reproduction in any medium or format, as long as you give appropriate credit to the original author(s) and the source, provide a link to the Creative Commons license and indicate if changes were made.

The images or other third party material in this chapter are included in the chapter's Creative Commons license, unless indicated otherwise in a credit line to the material. If material is not included in the chapter's Creative Commons license and your intended use is not permitted by statutory regulation or exceeds the permitted use, you will need to obtain permission directly from the copyright holder.

Transition to Multimodal Digital Mobility Services (MDMS) and User-Friendly Multimodal Transport Networks

Lambros Mitropoulos$^{(\boxtimes)}$ ⓘ, Aikaterini Maria Fotiou ⓘ, Annie Kortsari ⓘ,
Andreas Nikiforiadis ⓘ, and Georgia Ayfantopoulou ⓘ

Department of Infrastructure, Networks, Mobility and Logistics,
Centre for Research and Technology Hellas, Hellenic Institute of Transport, Athens, Greece
lmit@certh.gr

Abstract. Multimodal Digital Mobility Services (MDMS) are defined as "systems providing information about the location of transport facilities, schedules, availability and fares, of more than one transport provider, to make reservations, payments or issue tickets". Public transport (PT) is the backbone of a multimodal transport network and although, several MDMS solutions exist, few of them consider the traveller needs at disaggregated level, while public transport predictive data are not integrated within provided services. In addition, the Mobility as a Right (MaaR) concept that may be considered the keystone of an inclusive multimodal transport network is barely covered in literature. The MaaR concept is omitted in the majority of the MDMS and MaaS definitions, and little attention is given on incorporating it in innovative measures. This study builds on the principles of MDMS and MaaR to develop a conceptual framework for forming MDMS- and MaaR - based public transport measures to support the transition to a user-friendly multimodal transport network. To set the foundations for a structured discussion on the scarcely studied terms, first the definitions of MaaS, MDMS and MaaR are explored, to highlight key aspects and determine the potential relationship among them.

Keywords: Multimodal Digital Mobility Services (MDMS) · Mobility as a Service (MaaS) · Mobility as a Right (MaaR) · Public transport

1 Introduction

With the growing range of mobility options, travelers encounter daily difficulties related to accessibility, mobile applications, ticketing, and trip planning [1, 2]. Innovative technology-enabled services such as ride-hailing, ride-sharing, car-sharing, bike-sharing, and e-scooter-sharing, have become popular [3, 4] and the need for a single, user-friendly platform that integrates all services has enabled the concept of Mobility as a Service (MaaS) [1, 5]. MaaS is a user-focused approach that integrates digital solutions and various transport modes into a single platform, making it easier for users to plan, book, and pay for their journeys. Public transport plays a significant role in MaaS, but the concept extends to encompass a wide range of mobility options.

© The Author(s) 2025
C. McNally et al. (Eds.): TRAconference 2024, LNMOB, pp. 377–384, 2025.
https://doi.org/10.1007/978-3-031-85578-8_48

Aligned with its vision of harnessing the capabilities of digital solutions to enhance transport sustainability, the European Commission has engaged in the development of a framework for Multimodal Digital Mobility Services (MDMS) [6] to promote MDMS deployment in a coherent way across the European Union (EU). The provision of personalized trip suggestions, although is mentioned in several MaaS definitions [7], it is not claimed as core MaaS characteristic [8]. On the contrary, the concept of Mobility as a Right (MaaR) advocates for transport as a fundamental right that provides equitable access to mobility services for all individuals, regardless of their socioeconomic status, location, or physical abilities. However, clear considerations and directions on how to incorporate MaaR into MDMS, and thus MaaS, are not presented in the Sustainable and Smart Mobility Strategy (SSMS) [9].

2 Review of MaaS, MDMS and MaaR

2.1 Mobility as a Service (MaaS)

Considering that MaaS is an evolving mobility concept, its definition is regularly updated according to research studies [4, 5, 8, 10] and it is often interpreted in various ways. Planning a trip, booking and paying for a ticket are now considered core elements of the MaaS [7, 11], while integration of personalized bundled mobility packages according to users' preferences are also considered one of its components by some studies [5]. Table 1 presents a sample of MaaS definitions provided by EU studies.

Table 1. Sample of EU MaaS definitions.

Reference	Definition
Maas (2022) [8]	MaaS integrates different types of transport services into a single mobility service. This service is accessible on demand via a single platform. The platform provides booking, payment, and the provision of access (e.g., tickets) to the end customer via one digital interface. Public transport has a central role in all MaaS concepts
Interreg (2022) [12]	MaaS uses digital platforms to integrate different forms of transport – public transport, vehicle sharing, taxis, and rental offers – into a single service, which can be accessed on demand by the user. The service is accessible via a single platform, and has a single payment/ticketing system, calculating routes and costs so users can decide on best option

(continued)

Table 1. (*continued*)

Reference	Definition
UITP [7]	MaaS is the integration of, and access to, different transport services (such as public transport, ride-sharing, car-sharing, bike-sharing, scooter-sharing, taxi, car rental, ride-hailing and so on) in one single digital mobility offer, with active mobility and an efficient public transport system as its basis. This tailor-made service suggests the most suitable solutions based on the user's travel needs. MaaS is available anytime and offers integrated planning, booking and payment, as well as en route information to provide mobility without having to own a car
International Road Transport Union - IRU (2019) [13]	The key concept behind MaaS is to put the user at the core of transport services, offering them tailor-made mobility solutions based on their individual needs. MaaS is the integration of various forms of transport modes into a single mobility service accessible on demand. It combines all possible transport modes, enabling users to access services through a single application and purchase
European Metropolitan Transport Authorities – EMTA [7]	With MaaS, customers fulfil and manage all their mobility needs on demand, based on their general preferences and journey-specific needs. The service is based on the seamless integration of all different public and commercial modes of transport and is delivered via a digital interface. The service must enable multimodal travel possibilities and thus allow for the planning and booking of multimodal journeys, support on the go and payment as well as alteration of the planned journey

Understanding and synthesizing the definitions one may conclude that MaaS transcends the role of an application that simply allows users to plan and pay for their trip. Brokering travel with suppliers, repackaging, and reselling it as a bundled package, it is a distinguishing characteristic of MaaS [14].

2.2 Multimodal Digital Mobility Services (MDMS)

An official MDMS term does not exist, and for this reason MDMS is defined vaguely within literature [6, 15]. Table 2 provides a collection of MDMS definitions within literature.

Table 2. EU MDMS definitions.

Reference	Definition
European Commission (2023) [15]	MDMS are "services providing information on traffic and travel data such as location of transport facilities, schedules, availability or fares for more than one transport mode, which may include features enabling the making of reservations, bookings or payments or the issuing of tickets." It could still be amended by the co-legislators
Sorensen (n.d.) [16]	MDMS delivers the booking and payment services on top of the planning, pricing and navigation services as defined in the EU regulation for Multimodal Travel Information Services
European Commission—Mobility and transport (2023) [17]	MDMS are defined as "systems providing information about, inter alia, the location of transport facilities, schedules, availability and fares, of more than one transport provider, with or without facilities to make reservations, payments or issue tickets"
Mobility Talks episode 14 (2022) [18]	MDMS make shared mobility more accessible to users by bundling different modes and mobility providers in a single booking
Railway PRO (2021) [19]	MDMS, such as MaaS applications, route-planners or ticket vendors, help to compare different travel options, choices and prices, and in some cases facilitate the purchase of mobility products

MaaS and MDMS are two separate definitions that although complete each other, it is theoretically possible to deploy a MaaS without MDMS, but the effectiveness, convenience, and user-friendliness of such a system would be significantly limited. MDMS are a critical enabler of MaaS, and they play a central role in achieving the core objectives of MaaS. Technically, it would be possible to implement some aspects of MaaS without MDMS (e.g., through traditional ticketing systems and manual planning), but it would not align with the vision and goals of MaaS to create a seamless, user-centric, and efficient transportation system.

2.3 Mobility as a Right (MaaR)

The Mobility as a Right (MaaR) is a recent term used within transport [20, 21] that is related to the rights of people to move freely. MaaR is a strategic vision aiming at defining an inclusive mobility system where the needs of all citizens are taken into consideration, regardless of their capacities or their status in society [22].

Distributional justice addresses how people with disabilities are given less physical access to urban spaces through differential investment in disability fixtures and fittings within the built environment [23]. In this sense, MaaR encompasses both mobility infrastructure (e.g., wheelchair ramp) and smart applications that enable trip planning. Lately, Information and Communications Technology (ICT) have enabled mobility applications and facilitated access to public transport and other modes of transport. However, it becomes essential that these apps are aligned to the MaaR concept by involving users in the overall mobility decision chain. The MaaR concept is omitted in the majority of the MDMS and MaaS definitions, while infrastructure and vehicle characteristics are not considered when providing travel suggestions to travelers through travel solutions. This study builds on the principles of MDMS and MaaR to develop a conceptual framework for building MDMS- and MaaR - based public transport measures.

3   MDMS and MaaR Integration Framework

A well-functioning public transport system relies on the synergy among three components: the infrastructure (e.g., stations and stops, terminals, roadways, ticketing systems, information displays, etc.), the vehicle (e.g., buses, trains, trams, metro, ferries, etc.) and the users (e.g., commuters, tourists, families, people with special needs, etc.). Infrastructure should support the efficient operation of vehicles, which, in turn, should provide a comfortable and reliable service for all users. User preferences and needs should inform the planning and improvement of both infrastructure and vehicle design. A pyramid may be used to depict the relationship among these three components (Fig. 1). One of the latest EU funded projects on public transport (PT), the UPPER [21], aims to strengthen the role of public transport as the cornerstone of sustainable and innovative mobility by implementing a combination of measures looking to push people out of private cars and to pull them closer to public transport in cities across Europe. Since planned measures may engage any of the three PT components, we propose developed measures to focus on: 1) the development of public transport-oriented infrastructure, 2) enhancing the efficiency, reliability, and appeal of public transport, and 3) encouraging a shift in user's behavior towards greater utilization of public transport.

At the same time, incorporating the MaaR concept in the proposed measures, promotes the notion of equity for an inclusive mobility system where all citizens are taken into consideration. A "sieve" is used in Fig. 1, through which the three PT components with their characteristics (i.e., infrastructure, vehicle and users) penetrate to acquire the MDMS features and comply with the MaaR concept. The MaaR concept ensures universal access for all users and integration of various transport modes, including public transit, cycling, walking, and shared mobility services, to offer users a range of options based on their needs. It is noted that a "transport mode" includes both the infrastructure and the vehicle.

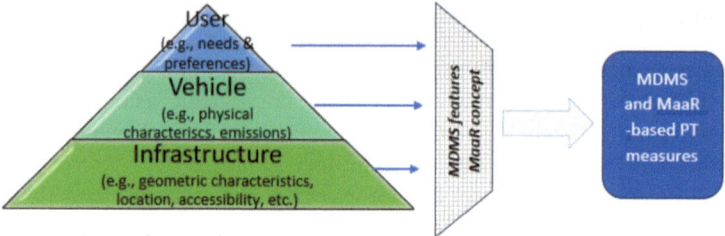

Fig. 1. Development of MDMS and MaaR measures.

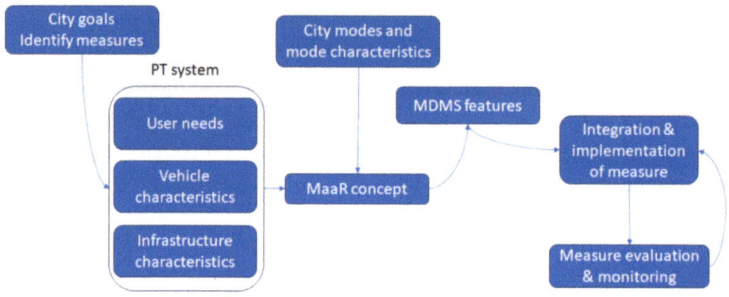

Fig. 2. Flowchart for planning MDMS-MaaR based public transport.

Utilizing the conceptual components from the previous section, and the need to incorporate the MaaR concept within MDMS, it could be concluded that along with the PT system, the transition to user-friendly multimodal transport network requires considering available modes' characteristics to achieve city goals. The city goals are formed by considering regional factors, legislation, policy directions and citizens' needs. The PT system is affected by the set goals, and when combining the PT system characteristics, the MaaR concept and existing transport mode characteristics, the developed MDMS features are expected to provide an inclusive transport system. The supporting elements of monitoring and evaluating the measure implementation, through dedicated Key Performance Indicators (KPIs), play an important role to sustain the integrated environment. A diagrammatic representation is shown in Fig. 2. The MDMS and the MaaR are related in that MDMS can be a practical implementation of the principles outlined in Mobility as a Right. MDMS platforms have the potential to enhance accessibility, affordability, and user-centricity in transport services, all of which are key aspects of MaaR. By integrating these principles into the design and operation of MDMS, cities and regions can work towards providing equitable transport options for their citizens. The MaaR concept dictates the consideration of user characteristics, and their matching with MDMS features to optimize user experience in public transport.

References

1. Mitropoulos, L., Kortsari, A., Mizaras, V., Ayfantopoulou, G.: Mobility as a service (MaaS) planning and implementation: challenges and lessons learned. Future Transp. **3**, 498–518 (2023). https://doi.org/10.3390/futuretransp3020029
2. Mitropoulos, L., Kortsari, A., Apostolopoulou, E., Ayfantopoulou, G., Deloukas, A.: Multimodal traveling with rail and ride-sharing: lessons learned during planning and demonstrating a pilot study. Sustainability **15**, 13755 (2023). https://doi.org/10.3390/su151813755
3. Mitropoulos, L., Kortsari, A., Ayfantopoulou, G.: A systematic literature review of ride-sharing platforms, user factors and barriers. Eur. Transp. Res. Rev. **13**, 61 (2021). https://doi.org/10.1186/s12544-021-00522-1
4. World Bank Blogs. https://blogs.worldbank.org/transport/mobility-as-a-service-can-help-developing-cities-make-most-complex-urban-transport-systems-if-they-implement-it-right. Accessed 27 Sept 2023
5. Arias-Molinares, D., García-Palomares, J.C.: The Ws of MaaS: understanding mobility as a service from a literature review. IATSS Res. **44**, 253–263 (2020). https://doi.org/10.1016/j.iatssr.2020.02.001
6. MOVE EU. https://www.move-eu.eu/our-positions/multimodal-digital-mobility-services-mdms. Accessed 27 Sept 2023
7. Ertico Mobility as a Service (MAAS) and Sustainable Urban Mobility Planning (2019)
8. Maas, B.: Literature review of mobility as a service. Sustainability **14**, 8962 (2022). https://doi.org/10.3390/su14148962
9. EUR-Lex (2020)
10. CIVITAS ECCENTRIC project Guidelines on How to Implement MaaS in Local Contexts (2020)
11. Deloitte the Rise of Mobility as a Service (2017)
12. Interreg Europe A Policy Brief from the Policy Learning Platform on Low-Carbon Economy (2020)
13. IRU. https://www.iru.org/who-we-are/where-we-work/europe/maas-mobility-service. Accessed 27 Sept 2023
14. Forward, M.: Mobility on Demand (MOD) and Mobility as a Service (MaaS) How Are They Similar and Different? Move Forward Blog (2019)
15. European Commission Multimodal Passenger Mobility Forum - Report from the Expert Group (2023)
16. Sorensen, S.: MDMS, an EU MaaS Regulation in the Making? Green Deal Flagship 6 – Making Connected and Automated Multimodal Mobility a Reality
17. European Commission - Mobility and Transport. https://transport.ec.europa.eu/index_en. Accessed 27 Sept 2023
18. Mobility Talks Episode 14: Multimodal Digital Mobility Services. EIT Urban Mobility (2022)
19. Railway PRO EC Begins Consultation for Multimodal Digital Mobility Services. Railway PRO (2021)
20. Smart Public Transport Initiatives for Climate-Neutral Cities in Europe | Spine Project | Fact Sheet | Horizon. https://cordis.europa.eu/project/id/101096664. Accessed 27 Sept 2023
21. Upper Project EU – Home. https://www.upperprojecteu.eu/. Accessed 3 Oct 2023
22. Ilieva, L.: "Mobility as a Right" Concept for Ruse City. Presented at the CIVITAS FORUM 2017, Torres Vedras, Portugal (2017)
23. Waitt, G., Harada, T.: Towards a relational spatial mobility justice of disability as territory. Mobilities **18**, 115–131 (2023). https://doi.org/10.1080/17450101.2022.2099753

Open Access This chapter is licensed under the terms of the Creative Commons Attribution 4.0 International License (http://creativecommons.org/licenses/by/4.0/), which permits use, sharing, adaptation, distribution and reproduction in any medium or format, as long as you give appropriate credit to the original author(s) and the source, provide a link to the Creative Commons license and indicate if changes were made.

The images or other third party material in this chapter are included in the chapter's Creative Commons license, unless indicated otherwise in a credit line to the material. If material is not included in the chapter's Creative Commons license and your intended use is not permitted by statutory regulation or exceeds the permitted use, you will need to obtain permission directly from the copyright holder.

Integrated Approach for Convoy Dispatching and Passenger Routing at Railway Stations with Variable Composition Trains

Federico Gallo$^{(\boxtimes)}$ ⓘ, Alice Consilvio ⓘ, and Nicola Sacco ⓘ

Department of Mechanical, Energy, Management and Transport Engineering,
University of Genoa, Via Montallegro 1, 16145 Genoa, Italy
federico.gallo@unige.it

Abstract. This paper addresses a railway scenario with variable composition trains, and focuses on the management and simulation of the passenger flows and trains at stations. In particular, it proposes a management system aimed at optimally routing the passengers from their entrance in the station to the correct platform segment according the train they have to board, and scheduling the train service in terms of convoy composition, capacity and destinations. The resulting system is then modelled as a discrete event system (DES) and simulated, to evaluate its performance on the basis of indicators like the queue length, the unsatisfied demand, the passenger travel time in the station, and the passenger density at the platform. The proposed station management system is then tested on a mixed real/synthetic numerical example aimed at proving the feasibility and functioning of the proposed approach.

Keywords: Automated mass transit · Travel guidance and route planning · Mobility information for citizens

1 Introduction

This work addresses a passenger railway scenario where trains can change their composition along their path, that is, trains can be composed of carriages with different destinations that split or merge together at some stations of a railway line. While this procedure is sometimes already done in current railway systems [1], it become more widely applicable with modern technologies such as automated train operations and wireless communication systems, which combined together form the basis for the *virtual coupling* concept. Virtual coupling can be considered an extension of ERTMS level 3 [2] where trains, automated and connected through wireless communication systems, can split and merge dynamically and automatically. This concept is increasingly recognized as capable of delivering the railway sector with many advantages such as an increased capacity [3], since trains along the same line could be grouped into a convoy, thus overcoming the concept of fixed or moving blocks, and an increased flexibility [4, 5], since trains could split to reach different destinations or could have a number of

© The Author(s) 2025
C. McNally et al. (Eds.): TRAconference 2024, LNMOB, pp. 385–391, 2025.
https://doi.org/10.1007/978-3-031-85578-8_49

carriages variable along the line depending on the transport demand (e.g. more carriages on the central segment of a metro line). For these reasons, virtual coupling is regarded among the most relevant innovations to be studied within the European Horizon 2020 Shift2Rail Joint Undertaking [6].

However, while the above advantages are accepted now, operating with convoys of trains (or carriages) with different destinations leads to some problems at the stations, where trains must be dispatched appropriately considering the available resources and the transport demand, and passengers must reach the most suitable platform segment according to the carriage they have to board. This work addresses this problem, by proposing an integrated approach for the combined management of passenger and trains at the stations. To avoid confusion, hereafter the term *convoy* will refer to a group of virtually coupled trains, whereas the term *train* will refer to a self-propelled rolling stock made up of one or more mechanically-coupled carriages.

The main contribution of this work is an integrated approach for the management and simulation of trains and passengers along a railway line and, in particular, at stations. In more details, the approach presented in this paper is made up of two modules, suitably joined to represent the whole system:

- The *train model,* which starts from the optimal train scheduling provided by the solution of an optimisation problem and verifies and assesses whether the number of trains in a convoy and their destination is able to satisfy the transport demand;
- The *passenger model*, which is aimed at assigning passengers to trains and at routing them in the station from their entrance to the most suitable platform segment. This model allows to verify whether passengers are routed towards a suitable train for their destination, and if an equal distribution of passengers is achieved, both on board (no overcrowded carriages and empty carriages) and at the platforms.

2   Methodology

The considered scenario and the two modules of the proposed station management system are modelled as a discrete event system (DES), where the events are related to train dispatching (e.g., train arrival and departure) and passenger behaviour (e.g., passenger boarding a train). To model the passenger actions and routes in the stations, the latter are spatially discretised and represented by a graph where each arc models an area of the station that can be crossed by a pedestrian flow.

Figure 1 shows the main components of the station management system: starting from the input data, the railway service is first scheduled by solving the optimisation problem described in [7]. The output consists of the convoy composition, number of trains/carriages and their destination. Considering the resulting convoy composition and the demand data, passengers are then routed in the station by solving a routing problem, which finds the best path for the passengers according to their entrance in the station and the position of the train they have to board. More details about the passenger routing and simulation can be found in [8], which describes a similar approach (although with a different aim) to the one followed in the passenger model of the proposed station management system.

After the optimisation phase, the DES resulting from the solution found is simulated, and the solution consists of the values of different indicators and performance measures,

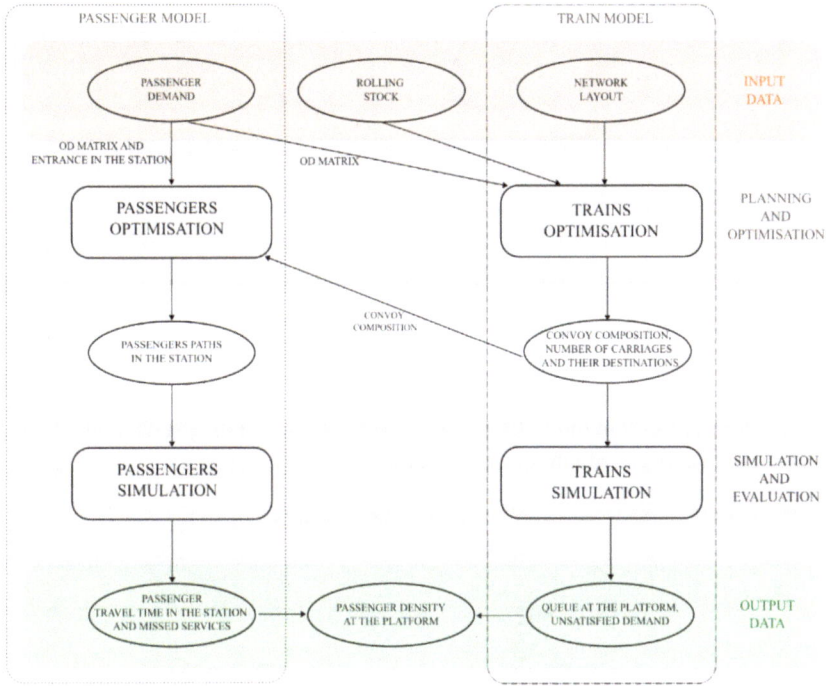

Fig. 1. Scheme of the proposed station management system.

such as the passenger queue at the platform and at some areas of the station (e.g., at the elevators), the passenger Level of Service both in the station and on board, the passenger travel time in the station and the level of unsatisfied demand for each convoy. This latter represents the number of passengers who do not find a suitable train for their destination at a certain time or do not find space in it due to crowding, and thus they have to wait for the following one.

3   Results

This section presents an application of the proposed station management system to a case study made up of both real data and generated data. In particular, the real layout of the Brignole metro station of the Italian city of Genoa is considered, whereas the demand data (OD matrix and passenger entrances) have been generated randomly. For the train service scheduling, the East direction of an Y-shape line (Fig. 2a) with 8 stations is considered, with Brignole being the junction station: though simplified, this layout can represent the Genoa metro network, once the currently planned new lines will be built (Fig. 2b). With such layout, the generated OD matrix has dimension 8x8; in addition, the trains can take two different directions just after Brignole, and passengers must be routed towards the right train.

Figure 3 and Fig. 4 show, respectively, the spatial discretisation and the related graph of the upper level of the Brignole metro station: the green nodes represent each location

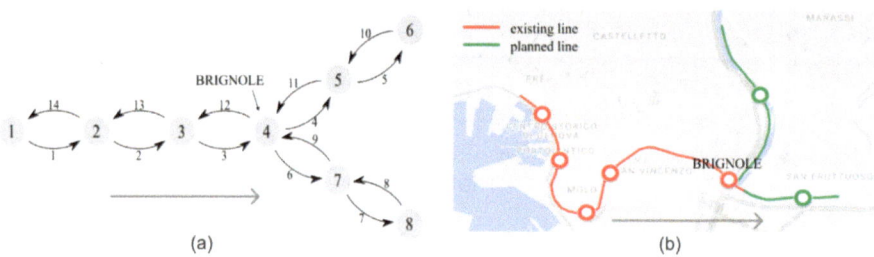

Fig. 2. Scheme of the considered railway layout (a) and of portion of the future Genoa metro line (b). The gray arrow represents the considered travel direction. Maps Data: ©2023 Google.

(e.g., atrium and platform segments) where people can move from one area to another, where the arrows represent the links between them (e.g., corridors, stairs and elevators). Each link has an associated length and travel time.

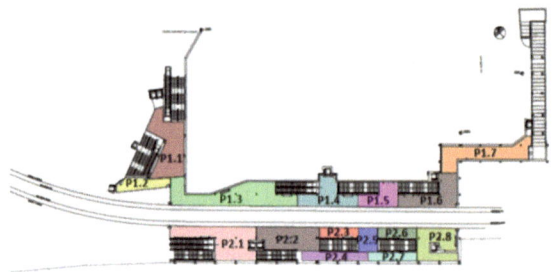

Fig. 3. Discretisation of Brignole metro station.

Fig. 4. Graph of Brignole metro station.

The optimisation problem has been solved with IBM-Ilog Cplex®, and the simulations have been performed with ExtendSim®. 10 simulations lasting 180 min each have been conducted. The passenger arrival rates have been deduced from the OD matrix; the capacity of each train was set to 70 pax. The minimum headway of the convoys was 3 min, and the available number of trains in the rolling stock was 20.

Figure 5 shows the passenger queue (blue curve) at the platform: the queue increases linearly according to the passenger arrival rate, which was computed from the OD matrix considering all the destinations reachable by the considered station; once a convoy arrives at the station, the queue drops of a value equal to the capacity of the trains in the convoy (considering the destinations of the passengers and of the trains). If the passenger queue does not drop to 0, the difference represents the unsatisfied demand of passengers that have to wait for the next convoy. As it can be seen, when the unsatisfied demand increases too much, more capacity is added, and average value of the queue (red curve) remains constant over all the simulation.

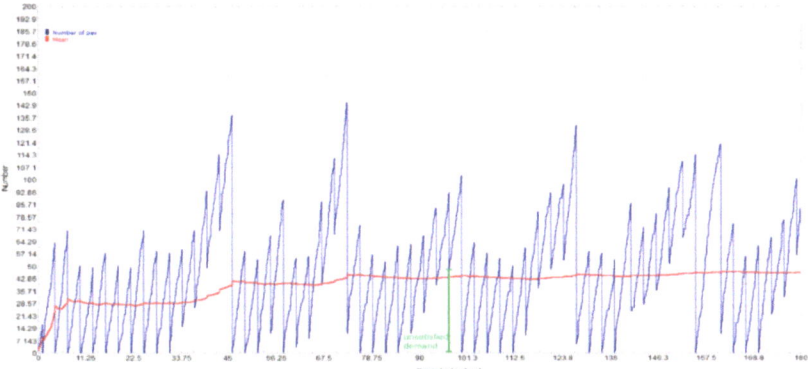

Fig. 5. Passenger queue at the platform (simulation 1).

The average passenger density over all the simulations was 0.1460 pax/mq. The associated level of service was A according to the Highway Capacity Manual [9] scale, which corresponds to a situation where pedestrians move using their desired paths and are not influenced by other people.

Figure 6 shows the passenger travel time in the station for the destination P1E (see Fig. 4) as a function of the passenger entrance time. The travel time is kept at a constant value for quite all the passengers. The highest travel times are due to passengers who lost their train and, as a result, where redirected to another platform segment where waiting for a suitable train in the following convoy.

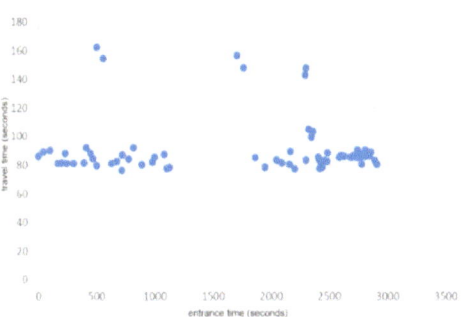

Fig. 6. Passenger travel time in the station (first hour of simulation 1).

4　Conclusions and Future Works

This work proposes a combined approach for managing and simulating both trains and passengers of a railway line where, thanks to virtual coupling, convoys can be composed of trains with different destinations. Results allow to evaluate the performance of the system, both in terms of demand satisfaction and passenger management in the stations, and to test possible real-time or long-term reactive actions (e.g., increase the number of available vehicles in the rolling stock).

Future extensions of the proposed approach include the study of real-time automatic reactions that may be realised in case of worsening of the performance indicators (e.g., schedule additional trains when the passenger density at the platform exceeds a given threshold value), and the consideration of specific user needs by the routing model (e.g., people with reduced mobility should be routed towards an elevator).

References

1. Fioole, P.-J., Kroon, L., Maróti, G., Schrijver, A.: A rolling stock circulation model for combining and splitting of passenger trains. Eur. J. Oper. Res. **174**(2), 1281–1297 (2006)
2. Di Meo, C., Di Vaio, M., Flammini, F., Nardone, R., Santini, S., Vittorini, V.: ERTMS/ETCS virtual coupling: proof of concept and numerical analysis. IEEE Trans. Intell. Transp. Syst. **21**(6), 2545–2556 (2020)
3. Felez, J., Kim, Y., Borrelli, F.: A model predictive control approach for virtual coupling in railways. IEEE Trans. Intell. Transp. Syst. **20**(7), 2728–2739 (2019)
4. Gallo, F., Febbraro, A.D., Giglio, D., Sacco, N.: A mathematical programming model for the management of carriages in virtually coupled trains. In: 2020 IEEE 23rd International Conference on Intelligent Transportation Systems (ITSC), Rhodes, Greece, pp. 1–6 (2020)
5. Gallo, F., Febbraro, A.D., Giglio, D., Sacco, N.: Planning and optimization of passenger railway services with virtually coupled trains. In: 2021 7th International Conference on Models and Technologies for Intelligent Transportation Systems (MT-ITS), Heraklion, Greece, pp. 1–6 (2021)
6. Marrone, S., Nardone, R., Petrillo, A., Santini, S., Vittorini, V.: Towards railway virtual coupling. In: 2018 IEEE International Conference on Electrical Systems for Aircraft, Railway, Ship Propulsion and Road Vehicles and International Transportation Electrification Conference (ESARS-ITEC), Nottingham, UK, pp. 1–6 (2018)
7. Gallo, F., Febbraro, A.D., Giglio, D., Sacco, N.: Global sensitivity analysis for the evaluation of the effects of uncertainty of transport demand and passenger behavior on planning railway services with variable train composition. In: 2021 IEEE 24th International Conference on Intelligent Transportation Systems (ITSC), Indianapolis, US, pp. 2298–2305 (2021)
8. Calcagno, P., Consilvio, A., Febbraro, A.D., Sacco, N.: Reshaping metro station spaces to improve social distancing during COVID-19 pandemic. In: 2021 7th International Conference on Models and Technologies for Intelligent Transportation Systems (MT-ITS), Heraklion, Greece, pp. 1–6 (2021)
9. HCM 2010: Highway Capacity Manual, Washington, D.C.: Transportation Research Board (2010)

Open Access This chapter is licensed under the terms of the Creative Commons Attribution 4.0 International License (http://creativecommons.org/licenses/by/4.0/), which permits use, sharing, adaptation, distribution and reproduction in any medium or format, as long as you give appropriate credit to the original author(s) and the source, provide a link to the Creative Commons license and indicate if changes were made.

The images or other third party material in this chapter are included in the chapter's Creative Commons license, unless indicated otherwise in a credit line to the material. If material is not included in the chapter's Creative Commons license and your intended use is not permitted by statutory regulation or exceeds the permitted use, you will need to obtain permission directly from the copyright holder.

The Organisation and Regulation of Door-to-Door Journeys Involving Public Transport: Lessons from Four Countries

Russell Cannon[1,2]([⊠]), Lena Winslott Hiselius[1,2], Chunli Zhao[1,2],
and Rosalia Camporeale[1,2]

[1] Transport and Roads, Lund University, Box 118, 22100 Lund, Sweden
russell.cannon@tft.lth.se
[2] K2, Sweden's National Knowledge Centre for Research and Education on Public Transport,
Bruksgatan 8, 222 35 Lund, Sweden

Abstract. Enabling journeys from door-to-door is an essential part of attracting people towards public transport and away from the private car. Facilitating such journeys requires comprehensive and integrated planning across modal and spatial boundaries. Public transport governance, however, is often constrained by fragmented, multi-level structures and regulatory barriers that limit the ability of responsible organisations to provide a seamless user experience. Sharing experiences between cities in different countries can accelerate learning processes and facilitate the development of more effective organisational structures and regulatory frameworks. The aim of this paper is to analyse how organisation and regulation can support (or hinder) door-to-door mobility. This is done by mapping the roles, responsibilities and relationships between actors who coordinate and collaborate on the design of public transport networks and tariff systems in the context of multimodal networks in Malmö/Lund in Sweden, Copenhagen in Denmark, Berlin in Germany and Trondheim in Norway. Information is gathered through stakeholder workshops in each country. Key regulatory and organisational challenges are identified and national differences between cases are discussed. Examples of structures and mechanisms that support multimodal integration and coordination in each country are also provided.

Keywords: Seamless travel · Organisation · Regulation · Public transport · Active modes

1 Introduction

The ability of the private car to provide a door-to-door journey is recognised as an important part of its appeal. It is therefore important to think in terms of door-to-door mobility in order to make public transport more competitive (Tovey et al., 2017). Indeed, efforts to "mimic the door-to-door efficiency and flexibility of the private car through increasing seamlessness" have been described as central to facilitating a shift towards reduced car use and a simultaneous uptake of walking, cycling and public transport (Tønnesen et al., 2021, p. 2).

© The Author(s) 2025
C. McNally et al. (Eds.): TRAconference 2024, LNMOB, pp. 392–397, 2025.
https://doi.org/10.1007/978-3-031-85578-8_50

The governance of public transport, however, is characterised by fragmented, multi-level structures that increase the complexity of providing seamless intermodal journeys (Kumar & Agarwal, 2013). Fragmentation in the organisation of public transport can result in a lack of focus on the whole consumer journey (Merkert et al., 2020). Divided responsibilities, access to finance and silo mentalities can limit the ability of responsible bodies to implement measures that support multimodal integration (Mulley & Yen, 2020). Sharing experiences between cities in different countries can accelerate learning and aid the development of organisational structures and regulatory frameworks that can more effectively support sustainable door-to-door mobility.

This study is part of the EASIER project, funded by the Urban Accessibility and Connectivity (ENUAC) call within ERA-NET; it focuses on how organisation and regulation can support (or hinder) the implementation of policies aiming for seamless door-to-door journeys involving public transport.

The aim of the study is to analyse organisational and regulatory aspects of door-to-door journeys by mapping the roles, responsibilities and relationships between stakeholders who coordinate and collaborate on public transport network design and tariff systems in the context of multimodal networks in city regions in four countries: Malmö/Lund in Sweden, Copenhagen in Denmark, Berlin in Germany and Trondheim in Norway.

2 Method and Data

Data was collected through workshops with stakeholders representing organisations involved in the operation and governance of public transport in the four city regions. The aim of the workshops was to map stakeholder responsibilities and relationships and to identify challenges and examples of good practice in relation to seamless intermodal travel. The workshops were designed around a selected sample journey within each study area. The sample journey was designed to provide a realistic scope for the workshop and to draw out important elements of coordination across modal, spatial and organisational boundaries. Each workshop was conducted digitally and lasted approximately two hours. Typically, 4–5 participants attended each workshop. The cases used in the study are as follows:

Berlin: The selected journey for Berlin is from the eastern side of the city, Berlin-Lichtenberg, to Potsdam, a neighbouring city located in the federal state of Brandenburg. The journey represents a common commuting route between Berlin and Potsdam and combines walking, cycling, tram and train. Organisations represented at workshop: BVG (public transport authority in Berlin), Jelbi (shared mobility provider), the district authority of Berlin-Lichtenberg, and VBB (regional public transport authority) (Fig. 1).

Copenhagen: The example journey is from a home in Amagerbro to a work location at Fabriksparken. The journey involves cycling or walking as the first and last mode of the trip and a combination of Metro, S-train and bus for the public transport part. Organisations represented at workshop: Movia (public transport agency), DSB (national train operating company), and the Capital Region of Denmark (regional authority) (Fig. 2).

Trondheim: The example journey is from Stjördal to the Tiller area of Trondheim. The origin and destination are situated in different municipalities and belong to the Trøndelag

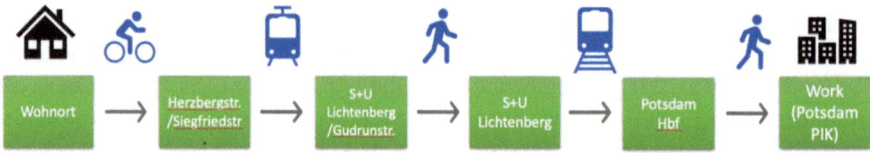

Fig. 1. Example journey for Berlin, Germany.

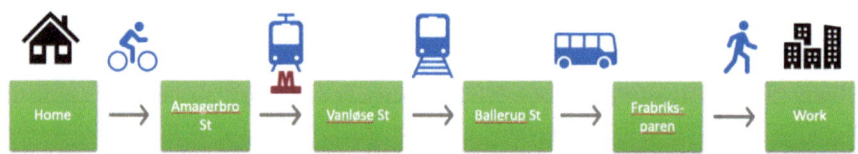

Fig. 2. Example journey for Copenhagen, Denmark.

region. There are high commuting flows between these areas. The example journey is a mix of walking and cycling, as well as train and bus travel. Organisations represented at workshop: Stjørdal Kommune (municipality), Trøndelag Fylke (regional authority), AtB (regional public transport company) and the Norwegian Railway Directorate (government agency) (Fig. 3).

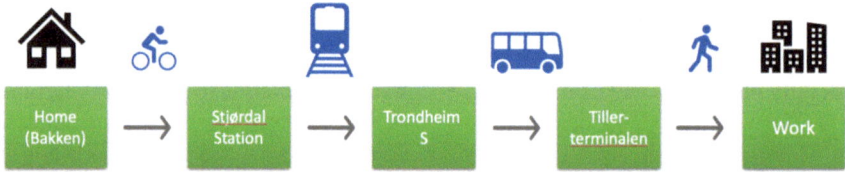

Fig. 3. Example journey for Trondheim, Norway.

Malmö/Lund: The example journey is between two cities in the region of Skåne: a residential area in Malmö to the university in Lund. There are high commuting flows between these cities. The example journey incorporates a combination of walking, cycling, bus, train and tram. Organisations represented at workshop: SJ (national train operating company), Skånetrafiken (regional public transport authority), Region Skåne (regional authority), Lund municipality and the City of Malmö (Fig. 4).

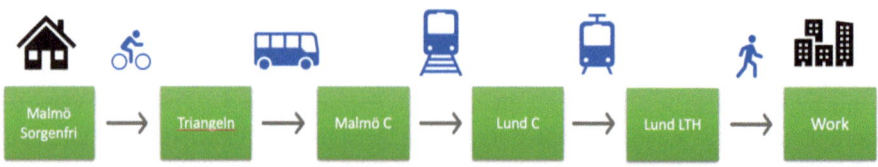

Fig. 4. Example journey for Malmö/Lund, Sweden.

3 Results

The results of the workshops highlight a variety of issues related to the organisation and regulation of door-to-door, intermodal journeys involving public transport. This includes a significant number of challenges and a smaller set of positive examples of seamless transport coordination and collaboration between stakeholders in the studied case cities.

The challenges include unclear delineation of responsibilities between organisations (both horizontally and vertically) and the difficulty of working collaboratively across organisational boundaries. A common challenge among the cases is the vertical alignment of goals between organisations at different levels of government (i.e., municipal, regional and national) and conflicting prioritisation of modes, particularly where limited street space is an issue. Such issues are particularly prevalent at railway terminals and surrounding areas, which typically involve a large and complex network of stakeholders. Differing organisational objectives regarding public transport access and egress via active and shared modes such as e-scooters can make it difficult to agree on joint actions to support door-to-door mobility.

A challenge identified in the Berlin case is a lack of collaboration and clear division of responsibilities between the district, the senate and the transport operators. There is a lack of collaboration between the local authorities (city, districts) responsible for the built environment around the stations or the walk/cycle path from one station to another (or from the surrounding neighbourhood to the station) and the transport operators.

The regulatory requirements and legal responsibilities of public organisations can limit their ability to support seamless door-to-door journeys across organisational boundaries. One example is the procurement of shared bike systems in Lund/Malmö, for which stakeholders describe being required to run separate procurements for the sake of proper competition, thus making it harder to integrate schemes.

One means of supporting service integration that was discussed is the establishment of formal partnerships. For example, in Denmark, DOT is a collaboration between the public transport operators DSB, Movia and the Metro that supports multimodal journey planning and enables use of the same ticket (Rejsekortet), between train, bus and metro. While it can be challenging to reach mutual solutions which suit each party, DOT was described in the workshop as a benchmark for cooperation. Stakeholders in Malmo/Lund highlight Resplus, a ticketing collaboration for booking journeys across different modes and operators, which is operated via Samtrafiken, a company formed by Sweden's regional public transport authorities to support coordination.

Funding schemes can also provide an impetus for collaboration. In the Trondheim case, there is a collaboration between and across levels of government (e.g., between region-municipality; or between municipalities) facilitated by a national funding scheme, Miljøpakken. Miljøpakken is designed to facilitate increased cycling, walking and use of public transport. Among other effects, Miljøpakken has enabled integration of travel data and the adjustment of the fare system to encourage multimodal journeys. While the need for agreements between politicians, for example at city and regional level, can slow down the process, the scheme has acted as a stimulus for joint action that supports integrated sustainable mobility.

In general, the cases showed that multi-actor governance poses a challenge to the development of integrated tariff systems, due to a lack of standardisation, different pricing policies and the legal requirements of different organisations. An example of good practice comes from the tariff system in Berlin-Brandenburg, established through collaboration between BVG and VBB. The resulting zone-based tariff supports journeys across municipal boundaries. Germany's recently introduced 49€-ticket, which enables travel with all public transport modes in all federal districts of Germany, was also highlighted within the workshop as a positive step for intermodal travel, in part due to its impact in unifying regulations and thus simplifying travel for users.

4 Conclusion

The results of this study highlight the complexity of multi-actor public transport governance and the critical importance of collaboration between different authorities, transport operators and agencies to create a cohesive and efficient transport network that enables users to make seamless, intermodal door-to-door journeys.

Seamless travel often involves the integration of different transport modes and services, each potentially operating under different regulatory frameworks and with different objectives. Harmonising regulations and aligning objectives across different entities can be a way forward to promote seamless travel experiences.

The results show that there are contextual differences between the cases studied. For example, Berlin and Copenhagen, as capital cities, have more complex administrative structures with multiple layers of governance. On the other hand, Trondheim and Malmö/Lund, as smaller cities, have simpler governance models with more localised decision-making processes. Despite these differences, common challenges can be observed. The most important of these is the need for a clear delineation of responsibilities between organisations and the alignment of objectives between organisations at different levels of government.

To achieve this, the study points to the establishment of more integrated transport authorities, partnership organisations or national funding schemes, such as Miljøpakken in Norway, that support integration. Such structures can play a key role in coordinating planning, policy-making and service provision across different transport modes and administrative boundaries, thereby promoting integration and seamless connectivity.

References

Kumar, A., Agarwal, O.P.: Institutional Labyrinth: Designing a Way Out for Improving Urban Transport Services: Lessons from Current Practice. World Bank, Washington, DC (2013)

Merkert, R., Bushell, J., Beck, M.J.: Collaboration as a service (CaaS) to fully integrate public transportation – lessons from long distance travel to reimagine mobility as a service. Transp. Res. Part A Policy Pract. **131**, 267–282 (2020). https://doi.org/10.1016/j.tra.2019.09.025

Mulley, C., Yen, B.T.: Workshop 6 report: better service delivery through modal integration. Res. Transp. Econ. **83**, 100913 (2020). https://doi.org/10.1016/j.retrec.2020.100913

Tønnesen, A., Knapskog, M., Uteng, T.P., Øksenholt, K.V.: The integration of active travel and public transport in Norwegian policy packages: a study on 'access, egress and transfer' and their positioning in two multilevel contractual agreements. Res. Transp. Bus. Manag. **40**, 100546 (2021). https://doi.org/10.1016/j.rtbm.2020.100546

Tovey, M., Woodcock, A., Osmond, J.: Designing Mobility and Transport Services: Developing Traveller Experience Tools. Routledge (2017)

Open Access This chapter is licensed under the terms of the Creative Commons Attribution 4.0 International License (http://creativecommons.org/licenses/by/4.0/), which permits use, sharing, adaptation, distribution and reproduction in any medium or format, as long as you give appropriate credit to the original author(s) and the source, provide a link to the Creative Commons license and indicate if changes were made.

The images or other third party material in this chapter are included in the chapter's Creative Commons license, unless indicated otherwise in a credit line to the material. If material is not included in the chapter's Creative Commons license and your intended use is not permitted by statutory regulation or exceeds the permitted use, you will need to obtain permission directly from the copyright holder.

Bridging the Green Gap: Understanding the Intention-Behavior Gap in Mobility and Urban Mobility

Christian Brock[1], Jeanne Lallement[2(✉)], Lars Findeisen[1], and Ulysse Soulat[2]

[1] University of Rostock, Ulmenstr. 69, 18057 Rostock, Germany
[2] La Rochelle Université, 23 Avenue Albert Einstein, BP 33060, 17031 La Rochelle, France
`jeanne.lallement@univ-lr.fr`

Abstract. Against the backdrop of the global challenge of climate change, the concept of sustainable mobility has gained significant attention. Governments, organizations and individuals are increasingly adopting green transportation practices to reduce carbon emissions and create greener and smarter cities. Despite the growing awareness and intention to sustainable mobility, a significant gap remains between individuals' intentions and their actual behavior; this phenomenon is well known as the intention-behavior or green gap. However, an up to date analysis that evaluates existing evidence to develop overarching approaches for further research is still lacking. Against this background we conduct a systematic literature review to explore the reasons for the green gap in the (urban) mobility context. Therefore, we have included research from different disciplines in our analysis. Based on our analysis, we propose measures to reduce the gap and develop a research agenda to stimulate future research in this area.

Keywords: mobility · green gap · sustainability · intention-behavior gap · mobility

1 Introduction

Against the backdrop of the global challenge of climate change, the concept of sustainable mobility has gained significant attention [18]. Governments, organizations, and individuals are increasingly adopting green transportation practices to reduce carbon emissions and create greener and smarter cities. However, despite the growing awareness and intention to sustainable mobility, a significant gap remains between individuals' intentions and their actual behavior; this phenomenon is well known as the intention-behavior or green gap [6].

In the context of mobility, the intention-behavior gap refers to the discrepancy between individuals' intention towards sustainable mobility and their actual behaviors. While many people express concern for the environment and acknowledge the importance of sustainable transportation, their actions often do not align with these intentions. For example, individuals may express support for public transportation or cycling, yet

© The Author(s) 2025
C. McNally et al. (Eds.): TRAconference 2024, LNMOB, pp. 398–402, 2025.
https://doi.org/10.1007/978-3-031-85578-8_51

continue to rely heavily on private cars for their daily commute. This discrepancy between intention and behavior can be attributed to a variety of factors. In recent years, numerous research studies have been published on this topic. However, an up to date analysis that evaluates existing evidence to develop overarching avenues for further research is still lacking. Against this background we conduct a systematic literature review to explore the reasons for the green gap in the (urban) mobility context. We thus make two contributions: First, we propose measures to reduce the gap, and second, we develop a research agenda to stimulate future research in this area.

2 Conceptual Background

The intention behavior gap has a long tradition across a broad range of disciplines. In recent years, research has been conducted primarily in the field of sustainability e.g. environmentally friendly behavior [8], ethical consumption [9], and sustainable tourism [11]. Further, research has also increased in the field of individual mobility, since the gap between intention and behavior also plays a major role in this context. Hence, the gap also exists in everyday mobility, which is sometimes environmentally friendly, and leisure mobility, which is more individual and carbon-intensive [2]. For instance, even though I intend to cycle to work more often, I still use the car for a variety of reasons.

The focus and findings of previous research are quite diverse. Some studies focus on the gap, [7], while other research downplays or even refutes it [13]. The reasons for the different results are also diverse: for example, different methodological approaches in different contexts play a crucial role. Therefore, a condensed review of previous findings is essential.

Most research exploring user reasons for choosing mobility modes is based on the theory of planned behavior, which comprises the three key components of attitude, subjective norms, and perceived behavioral control, that determine individuals' behavioral intentions [1]. Behavioral intention is the core antecedent of behavior. However, numerous meta-analyses show that there is still a gap between intention and behavior. Therefore, we aimed to identify the underlying reasons for this gap in the field of individual mobility.

3 Methodological Approach and Initial Findings

Based on an extant literature search, we identified several hundred studies. The identified articles, from databases such as Google Scholar, ABI/INFORM, and EBSCO (Business Source Premier) contain keywords such as "mobility" and "behavioral gap". Each article was reviewed by two co-authors separately and approximately one hundred passed the screening process.

Based on our analysis, we identified key overarching factors that could foster the green gap: First, customer and user related antecedents comprise traits and predispositions such as competence/familiarity, knowledge, convenience, and habit. Second, infrastructural factors such as accessibility to sustainable transportation options and information about sustainable mobility. Third, contextual factors such as weather conditions or journey characteristics (e.g. distance, travel reasons).

Most studies in our review focus on **customer and user related antecedents**, thus on individual factors. For instance, prior research has shown that competence and knowledge of customers play a crucial role for reducing the intention-behavior-gap [6]. We also found evidence of the usefulness of self-measured information in raising awareness of the CO2 impact of mobility [3].

Studies concerning individual factors are divided into two main streams. One the one hand, Du *et al.* [4] indicate that "low-carbon knowledge" can increase the low-carbon behavioral intention of behavior. Previous studies have used different methods to explain the awareness gap. One the other hand, many studies [4] focus on psychological factors which influence mobility behaviors, and in particular, the subjective variables described in the theory of planned behavior. Attitudes (an individual's positive or negative evaluation of the behavior), subjective norm (a person's assessment of social pressure to perform or not to perform the behavior) and perceived behavioral control (a person's perception of the difficulty of performing the behavior), are determinants of the individual's choice of mobility mode. The literature describes different psychological factors whose weaker effects explain the behavior of less carbon-intensive mobility. Some of the articles emphasize the symbolic value of mobility products. Cars, in particular, can be a symbol of one's identity and status. Electric vehicles can also convey social values and self-image motives because they are environmentally friendly, intelligent, and socially desirable [15]. **Infrastructural factors** are well documented in the literature; however, these factors are less connected to intention-behavioral gap in the context of daily mobility. The Covid pandemic, for example, had a major impact on infrastructure. Hence, during the pandemic, some public transport services were restricted or even closed, while others, especially cycling infrastructure, were promoted [16]. Among **the contextual factors**, there are certain mobility products that lead to more independence. For instance, car use, which is independent of public transport departure times, more comfortable in bad weather, and also saves considerably more time. Furthermore, research shows that the purpose of the trip also influences the choice of transportation mode [14, 17]. In light of this, Campbell's paradigm shows that the gap between attitude and behavior is related to behavioral costs [12]. Many soft mobility situations have contextual costs. Contextual factors include temporal, spatial, and weather-related factors associated with these type of costs [5].

4   Implications and Avenues for Further Research

Based on our findings, we develop valuable guidance for management – mobility providers and policy makers – as well as for scholars. Policy makers, should, for example, focus on the knowledge and competence of potential customers [10]. Targeted information campaigns for customers/users who already have a positive intention of sustainability can improve actual behavior.

The second contribution of our research is the development of a research agenda to stimulate future research in this area. Based on the three key factors identified, Table 1 below provides an overview for further research.

For instance, Geng et al. [7] and van Rijnsoever et al. [18] suggest that measures to close the gap between intention and behavior should consider the nature of this inconsistency. Since this research is still in progress, we will address this recommendation,

Table 1. Research Agenda (selection)

Theme/issue	Selected recommendations/research questions
Customer and user related antecedents	What could a customer/user segmentation look like in order to specifically align corresponding measures? What methods should be applied to map dynamic and multi stage effects? Do other psychological determinants exist that have been neglected so far but contribute (e.g. self-licensing)? What role do mobility experience and emotions play in related research (e.g. emotions connected with autonomous vehicles or SUV, flight shaming)?
Infrastructural factors	Which infrastructural measures have a positive influence on green mobility behavior (e.g. car parking fees; ease of access; connectivity)? How do new technologies influence the use of a green mobility provider (e.g. green travel apps)?
Contextual factors	What measures can be taken to reduce perceived contextual barriers? How can contextual factors be differentiated or better specified?

among other. We will also delve deeper into the literature to more precisely specify the factors and recommendations for further research.

References

1. Ajzen, I.: The theory of planned behavior. Organ. Behav. Hum. Decis. Process. **50**, 179–211 (1991)
2. Barr, S., Prillwitz, J.: Green travellers? Exploring the spatial context of sustainable mobility styles. Appl. Geogr. **32**, 798–809 (2012)
3. Chambon, T., Soulat, U., Lallement, J., Guillaume, J-L.: The effect of visual information complexity on urban mobility intention and behavior. In: Nurcan, S., Opdahl, A.L., Mouratidis, H., Tsohou, A. (eds.) Research Challenges in Information Science: Information Science and the Connected World. RCIS 2023, vol. 476 (2023)
4. Du, H., Liu, D., Sovacool, B.K., Wang, Y., Ma, S., Li, R.Y.M.: Who buys new energy vehicles in Chuan? Assessing social-psychological predictors of purchasing awareness, intention and policy. Transp. Res. Part F **58**, 56–69 (2018)
5. Elmashhara, M.G., Silva, J., Sá, E., Carvalho, A., Rezazadeh, A.: Factors influencing user behaviour in micromobility sharing systems: a systematic literature review and research directions. Travel Behav. Soc. **27**, 1–25 (2022)
6. Frank, P., Brock, C.: Bridging the intention–behavior gap among organic grocery customers: the crucial role of point-of-sale information. Psychol. Mark. **35**(8), 586–602 (2018)
7. Geng, J., Long, R., Chen, H., Li, W.: Exploring the motivation-behavior gap un urban resident'green travel behavior: a theoretical and empirical study. Resour. Conserv. Recycl. **125**, 282–292 (2017)
8. Grimer, M., Miles, M.: With the best of intentions: a large sample test of the intenyion-behavior gap in pro-environmental consumer behavior. Int. J. Consum. Stud. **41**(1), 2–10 (2017)

9. Hassan, L.M., Shiu, E., Shaw, D.: Who says there is an intention–behaviour gap? Assessing the empirical evidence of an intention–behaviour gap in ethical consumption. J. Bus. Ethics **136**, 219–236 (2016)

10. Hiselius, L.W., Rosqvist, L.S.: Mobility Management campaigns as part of the transition towards changing social norms on sustainbale travel behavior. J. Clean. Prod. **123**, 34–41 (2016)

11. Juvan, E., Dolnicar, S.: The attitude–behaviour gap in sustainable tourism. Ann. Tour. Res. **48**, 76–95 (2014)

12. Kaiser, F.G., Byrka, K., Hartig, T.: Reviving Campbell's paradigm for attitude research. Pers. Soc. Psychol. Rev. **14**(4), 351–367 (2010)

13. Kopplin, C.S., Brand, B.M., Reichenberger, Y.: Consumer acceptance of shared e-scooters for urban and short-distance mobility. Transp. Res. Part D **91**, 1–14 (2021)

14. Lanzini, P., Khan, S.A.: Shedding light on the psychological and behavioral determinants of travel mode choice: a meta-analysis. Transport. Res. F: Traffic Psychol. Behav. **48**, 13–27 (2017)

15. Li, L., Wang, Z., Gong, Y., Liu, S.: Self-image motives for electric vehicle adoption: evidence from China. Transp. Res. Part D **109**, 1–13 (2022)

16. Möllers, A., Specht, S., Wessel, J.: The impact of the Covid-19 pandemic and government intervention on active mobility. Transp. Res. Part A Policy Pract. **165**, 356–375 (2022)

17. Shibayama, T., Emberger, G.: Ensuring sustainable mobility in urban periphery, rural areas and remote regions. Eur. Transp. Res. Rev. **15**(1), 11 (2023)

18. Van Rijnsoever, F., Frala, J., Dijst, M.: Consumer car preferences and information search channels. Transp. Res. Part D **14**, 334–342 (2009)

Open Access This chapter is licensed under the terms of the Creative Commons Attribution 4.0 International License (http://creativecommons.org/licenses/by/4.0/), which permits use, sharing, adaptation, distribution and reproduction in any medium or format, as long as you give appropriate credit to the original author(s) and the source, provide a link to the Creative Commons license and indicate if changes were made.

The images or other third party material in this chapter are included in the chapter's Creative Commons license, unless indicated otherwise in a credit line to the material. If material is not included in the chapter's Creative Commons license and your intended use is not permitted by statutory regulation or exceeds the permitted use, you will need to obtain permission directly from the copyright holder.

Seamless and Attractive Transfers to and Within Public Transport Stations: Responsibilities, Characteristics and New Visions

Heike Marquart$^{(\boxtimes)}$, Benjamin Heldt , Christian Wolf, and Rita Cyganski

Institute of Transport Research, German Aerospace Center (DLR), Rutherfordstraße 2, 12489 Berlin, Germany
heike.marquart@dlr.de

Abstract. For tackling societal, health and environmental challenges, it is desirable to increase public transport usage and make transfers to public transport stations attractive and seamless. The study explores which factors are important for an attractive station environment and who is responsible for planning attractive transfers. It focuses on the walk to or from the station and its forecourt. First, stakeholder workshops with practitioners were conducted. Second, an online survey about the station surroundings and transfers was conducted amongst users of two stations in Berlin, Germany. Third, interviews with users applying virtual reality were carried out. Users experience and discuss 'new visions' of the station's forecourt with virtual reality, applying 'digital Tactical Urbanism'. The survey data was analyzed by using descriptive analysis, the interview data by applying thematic coding. The findings show that cycling-/pedestrian-friendliness is not adequately addressed by practitioners. It is not clear who is responsible to provide attractive, seamless and safe station environments. Wayfinding, quality of the walking environment and safety are important issues of the transfer situation. The results help to understand the needs of users and stakeholders and deliver recommendations to strengthen the combination of walking and public transport by introducing the approach of 'digital Tactical Urbanism'.

Keywords: public transport · walking · virtual reality · planning

1 Introduction

For tackling societal, health and environmental challenges, it is desirable to increase the share of people who use public transport (PT) combined with walking or cycling. Thus, it is important to make transfers to and within public transport stations attractive and seamless. To ensure that individuals enjoy walking, a pedestrian-friendly environment is an important prerequisite. Taking public transport in urban areas can be challenging: Complex wayfinding and environmental features in and around stations can cause an unpleasant trip experience. Previous work has shown that the design of the environment at a bus stop has an influence on the extent to which journeys to and from the stop are made on foot or by bicycle, and what distance to the stop or waiting times are

© The Author(s) 2025
C. McNally et al. (Eds.): TRAconference 2024, LNMOB, pp. 403–409, 2025.
https://doi.org/10.1007/978-3-031-85578-8_52

considered acceptable [1]. Accordingly, it is desirable that practitioners take the user's needs regarding the PT station environment into consideration. Thereby, a seamless and comfortable public transport experience can be provided.

This paper delivers insights into how the surrounding of PT stations is planned from the decision makers' point of view and which factors are important for an attractive PT station from the users' point of view. In particular, the forecourt of PT stations is explored and new visions for PT stations are made perceptible by virtual reality. The objective of this paper is firstly, to introduce the concept of (digital) Tactical Urbanism (TU) and explore its possibilities to enhance PT station's surroundings and secondly, to investigate potential TU measures and their influence on a perceived seamless and attractive transfer to/from PT stations.

2   State of the Art

About 53% of the total PT travel time takes place inside of a public transport vehicle, the remaining 47% comprises walking to, at or from the stop and changing or waiting at the station [1]. Hence, for improving the attractiveness of PT stations, analyzing the walking environment to/from or within the station is essential. In transport research, walkability has been studied for several years, traditionally with a focus on measurable mesoscale indicators: accessibility or proximity, mixed land use, density, aesthetic design, sidewalks, street connectivity, safety or neighborhood type [2]. In recent years, research has also focused on the microlevel, the perceived walkability and the quality of the walking environment, which often lacks attention in traditional walkability research [3, 4]. The quality of the walking environment is interlinked with the sojourn quality of the neighbourhood, comprising factors such as protection (e.g. traffic or social safety), comfort (e.g. sights, social interaction, places to rest or walk) and joy (e.g. urban design adapted to human scale, positive sensory experiences) [5]. Factors regarding the quality of the walking environment can increase the motivation to walk [3].

Besides land use mix or density, the access and proximity to a public transport stop is often used as an indicator for walkability [6]. Yet, the walking environment around PT stops is often underexplored, as walkability quality analysis is usually based on macroscale indicators. A lack of wayfinding signs, complex entrances and traffic situation around PT stops can cause confusion and stress, which influences wellbeing and health [7]. Moreover, informal or hidden paths and shortcuts, which are often used when walking for transport purposes, are difficult to consider when using macroscale datasets [8]. Yet, informal shortcuts are important route sections that improve the trip experience [9] and can be considered by investigating the microlevel.

In transport-related policy and planning, walking as part of intermodal trips is still underrepresented and also lacks evaluation in national traffic surveys [10, 11]. In Germany, many transport authorities do not adequately address walking as a mode of transport, hence, financial and personal resources are missing [10]. Recently, European countries have applied walking policies, including safe and convenient walking infrastructure, short trip distances, speed limits or traffic calming [11].

One option to call attention to shortcomings regarding walkability is Tactical Urbanism (TU). TU is a tool to draw attention to perceived deficiencies in the physical design of urban spaces and guide policy principles. TU has been used in urban development, often motivated by the goal of improving cities on a small scale "from bottom up" [12]. TU is often described as "resident-generated, low-cost, often temporary" interventions in urban space [12, p. 135]. It targets long-term change through short-term interventions. Lydon & Garcia (2015) [13] refer to such initiatives as tactical because they use targeted, easy-to-apply means to achieve predetermined goals while providing flexibility in the planning and implementation processes.

3   Methods

The research design comprises different methods in three subsequent stages, focusing both on decision-makers' and public transport users' view (Fig. 1).

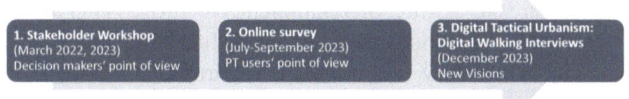

Fig. 1. Study design: methods, time frame and research aims

The study investigates two structurally different PT stations in Berlin, Germany: station 1 Berlin-Lichtenberg and station 2 Berlin-Haselhorst (Fig. 2). Both stations feature a mobility hub ('Jelbi') nearby with carsharing, e-kickscooters and bikesharing.

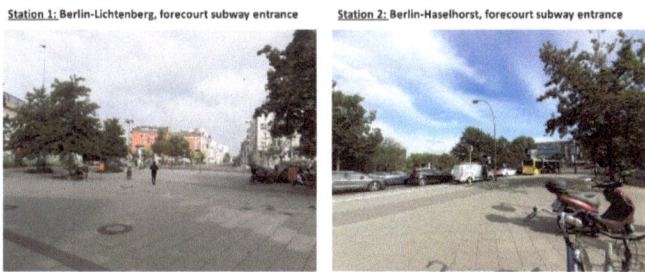

Fig. 2. View from the subway entrance to the forecourt (own image)

First, (1) stakeholder workshops were conducted online to discuss TU interventions. 14 representatives participated, e.g. district administration for climate protection and for impaired people, mobility providers, an architect, lobby groups for pedestrians and public transport, and science. Participants discussed strengths and weaknesses of the station's surroundings and evaluated potential measures. Although the workshop results represented a starting point for TU interventions, subsequent discussions found that getting the necessary legal approvals to realize interventions were not feasible due to current

German traffic law regulations: (a) every single measure must be approved separately, (b) traffic data is needed to argue why measures are necessary but could not be retrieved in scope and time frame of the project, (c) many departments need to be involved. Thus, we decided to develop a new methodology that does not require as much effort and legal approval (see step 2 and 3).

Next (2), the needs and requirements of public transport users with regard to the two stations under research were identified by conducting an online survey (n = 333) from mid-July until mid-September 2023. For recruitment, non-random sampling methods were used and posters in the PT stations, newsletters, online platforms and flyers in the areas announced the survey, making sure that people knowing the stations take part. Descriptive analysis was used to analyze the data. Based on the results of the online survey, a virtual reality environment for station 1 was jointly developed by DLR Institute of Transport Research, DLR Earth Observation Center (EOC) Science Communication and Visualization department and an architect (Katja Pfeiffer). Virtual reality (VR) is a type of extended reality (XR) technology, with which a person can explore and interact with a virtual world in a 360° setting. Silvennionen et al. (2022) [14] used VR to explore the effect of urban design on user's perceived walkability, showing participants different VR simulations [14]. While they used a complete virtual environment, we decided to use 360° pictures of the PT station 1 forecourt, so that the participants can relate to the station. In these 360° pictures, 3D-elements of urban design features based on the results of the online survey were included, e.g. benches, food truck, drinking fountain, greenery and wayfinding signs (Fig. 3).

Fig. 3. Graphical representation of two example views from the forecourt with VR elements in Berlin-Lichtenberg (image created by DLR EOC)

Finally, (3) Interviews were conducted with 19 participants who wore the VR glasses und explored the PT station with the digital TU features. The interviews took between 27 and 45 min. The participants were recruited from the online survey as well as neighborhood networks and social media. Based on the principle of Tactical Urbanism, the participants could turn different features in their VR world on or off, e.g. take a look how the forecourt looks with or without wayfinding signs, what influence a crowded place has on their perceived walkability or if they prefer the forecourt with a food truck. Meanwhile, the interviewers applied a semi-structured interview guideline while the interviewee moved through the virtual world to explore (a) perceptions and emotions, (b) environmental complexity and wayfinding and (c) environmental familiarity and activities. Hence, digital Walking Interviews were conducted. This intervention, based on principles of TU, can be referred to as 'Digital Tactical Urbanism'. The interview data was analyzed applying an inductive-deductive approach with thematic coding.

4   Results and Discussion

The findings from the survey data show that a PT stations' surrounding should be clean and safe and have a high quality of the walking environment: features such as greenery, seating options, aesthetics, security personnel, light, toilets, drinking fountains and good shops/gastronomy were desired the most. Furthermore, a safe and short access from the station to other modes of transport (e.g. bus or sharing options) and to the neighborhoods was desired. This is in line with the results from the stakeholder workshops. The stakeholders emphasized the importance of the signage or visibility of the station, proposing pictograms on the sidewalks, as well as seating options and safety through lighting. Generally, the results from the stakeholder workshops showed that the forecourts in front of the station's entrance are rarely considered in planning processes. By investigating who is responsible for planning PT forecourts, it became apparent that the responsibilities are unclear: neither the transport providers nor the municipality seem to be in charge of transfers to/from PT stations. Hence, transfers are lacking attention. This is also reflected in the online survey respondents' answers, which highlighted that the station's surrounding needed improvement regarding safety, wayfinding, aesthetics, pleasantness and comfort.

Applying the digital Walking Interviews using the digital TU intervention showed similar results as the online survey, but gave in-depths insights into the benefits and barriers of TU measures from the user's point of view. The 19 interviewees emphasized that greenery and wooden benches improved the trip experience at the PT station. Seeing people interact and rest at the forecourt can improve perceived safety and influence how often participants would choose to visit the PT station. The digital Walking Interviews using the VR environment improved the immersive experience of potential TU measures, without the need to physically implement them. Participants described their impressions and potential usage of the redesigned station. Tactical Urbanism is a participatory, low-cost instrument to intervene in urban space. Thus, it can contribute to urban transformation processes by including local actors and questioning current planning principles. As the stakeholder workshop has shown, it can be difficult to physically intervene in urban space and change urban environments. As a consequence, this study applied virtual reality as a tool to create digital TU measures by introducing 'digital Tactical Urbanism'. In areas with unclear planning responsibilities and hard-to-reach authorities involving long-lasting decision processes, digital Tactical Urbanism can be an option to make changes in the urban environment tangible, both for laypersons and for decision-makers, in a low-cost and fast way. By taking surveys or interviews beforehand as bottom-up approaches, the design elements for the virtual world are shaped by the users' needs and desires, thus meeting the demand of Tactical Urbanism.

5 Conclusion

This study highlights the access and transfer situation at PT stations. By integrating decisions makers' and public transport users' views, the quality of the walking environment was explored. Digital Tactical Urbanism made new visions tangible and served as a basis for discussing improvements and shortcomings. Hence, we identified ways of making PT stations more attractive and strengthen their role as cornerstones for a sustainable urban mobility system.

References

1. Hillnhütter, H.: Pedestrian Access to Public Transport. Stavanger, University of Stavanger. Dissertation (2016)
2. Saelens, B.E., Handy, S.L.: Built environment correlates of walking: a review. Med. Sci. Sports Exerc. **40**(7 Suppl.), S550–566 (2008)
3. Bozovic, T., Hinckson, E., Smith, M.: Why do people walk? Role of the built environment and state of development of a social model of walkability. Travel Behav. Soc. **20**, 181–191 (2020)
4. Guzman, L.A., Arellana, J., Castro, W.F.: Desirable streets for pedestrians: using a street-level index to assess walkability. Transp. Res. Part D Transp. Environ. **111**, 103462 (2022)
5. Gehl, J.: Städte für Menschen. Jovis, Berlin (2010)
6. Arellana, J., Saltarín, M., Larrañaga, A.M., Alvarez, V., Henao, C.A.: Urban walkability considering pedestrians' perceptions of the built environment: a 10-year review and a case study in a medium-sized city in Latin America. Transp. Rev. **40**(2), 183–203 (2019)
7. Merzoug, S., Jarass, J.: Emotionen beim Zufußgehen im urbanen Raum – Einflüsse eines alltäglichen Fußwegs auf das mentale Wohlbefinden in Berlin. Internationales Verkehrswesen. **4**, 64–68 (2021)
8. Sun, G., Oreskovic, N.M., Lin, H.: How do changes to the built environment influence walking behaviors? A longitudinal study within a university campus in Hong Kong. Int. J. Health Geogr. **13**, 28 (2014)
9. Marquart, H., Stark, K., Jarass, J.: How are air pollution and noise perceived en route? Investigating cyclists' and pedestrians' personal exposure, wellbeing and practices during commute. J. Transp. Health **24**, 101325 (2022)
10. UBA (Umweltbundesamt, German Environmental Agency): Geht doch! Grundzüge einer bundesweiten Fußverkehrsstrategie. Umweltbundesamt: Dessau-Roßlau (2018)
11. Buehler, R., Pucher, J.: Overview of walking rates, walking safety, and government policies to encourage more and safer walking in Europe and North America. Sustainability **15**(7) (2023)
12. Talen, E.: Do-it-yourself urbanism: a history. J. Plan. Hist. **14**(2), 135–148 (2015)
13. Lydon, M. and Garcia, A,: Tactical Urbanism: Short-term Action for Long-term Change. 2015, Washington, D.C.: Island Press (2015)
14. Silvennoinen, H., Kuliga, S., Herthogs, P., Recchia, D.R., Tunçer, B.: Effects of Gehl's urban design guidelines on walkability: A virtual reality experiment in Singaporean public housing estates. Environment and Planning B: Urban Analytics and City Science. **49**(9), 2409–2428 (2022)

Open Access This chapter is licensed under the terms of the Creative Commons Attribution 4.0 International License (http://creativecommons.org/licenses/by/4.0/), which permits use, sharing, adaptation, distribution and reproduction in any medium or format, as long as you give appropriate credit to the original author(s) and the source, provide a link to the Creative Commons license and indicate if changes were made.

The images or other third party material in this chapter are included in the chapter's Creative Commons license, unless indicated otherwise in a credit line to the material. If material is not included in the chapter's Creative Commons license and your intended use is not permitted by statutory regulation or exceeds the permitted use, you will need to obtain permission directly from the copyright holder.

Local Transport Plans: An Integrative Approach to Local Plan Making

David Clements[(✉)]

National Transport Authority, Dublin, Ireland
david.clements@nationaltransport.ie

Abstract. Over the past number of years, Ireland's National Transport Authority (the "NTA") have been funding and overseeing the preparation of Local Transport Plans (LTPs) across the country. Such plans are prepared for regional growth centres and key towns identified in the Regional Spatial and Economic Strategies, as well as other settlements identified by local authorities, and are based on the Area Based Transport Assessment (ABTA) guidance that was previously prepared by the NTA and Transport Infrastructure Ireland.

In overseeing this programme, the NTA have been to the forefront of transport planning in Ireland at the local level and as such have developed a unique insight into local plan making and the relationship between national, regional and local policy. This paper summarises the approach the NTA has taken to the preparation of these plans, the key issues which have emerged, and the challenges of achieving modal shift across Ireland and the associated objective to reduce emissions from transport.

Keywords: Transport Policy · Sustainable Development · Strategic Planning · Active Travel · Public Transport · Placemaking

1 Introduction

1.1 Definition of a Local Transport Plan (LTP)

LTPs represent the application of national, regional and metropolitan transport policies and objectives at the level of the individual settlement. They can be undertaken for towns of varying scales and suburban areas of larger conurbations.

The precise content of an individual LTP is dependent on a number of factors. These include the scale of the settlement; the potential for investment in sustainable transport; and the extent to which transport planning in the location may be influenced by individual major schemes or issues.

In general, the LTP will comprise a clearly prioritised framework for investment in transport, with a focus on sustainable modes, for a settlement over a 12–20 year timeframe.

© The Author(s) 2025

C. McNally et al. (Eds.): TRAconference 2024, LNMOB, pp. 410–416, 2025.
https://doi.org/10.1007/978-3-031-85578-8_53

1.2 International Context

The NTA Local Transport Plan programme could be viewed in the context of the Sustainable Urban Mobility Plan (SUMP) approach [1]. The 8 principles of the SUMP are applied in the making of LTPs, in particular the requirements of the functional urban areas and cooperation across institutional boundaries. In general, however, LTPs are prepared for towns with forecast populations of up to 50,000, rather than for Irish cities or Metropolitan Areas, for which Regional or Metropolitan Area Transport Strategies are prepared.

1.3 Irish Policy Context

LTPs sit at the bottom of the hierarchy of transport plans and programmes in Ireland and, as such, from the perspective of local communities, comprise the most detailed and perhaps, most tangible expression of transport policy, other than individual transport schemes. Figure 1 illustrates this hierarchy:

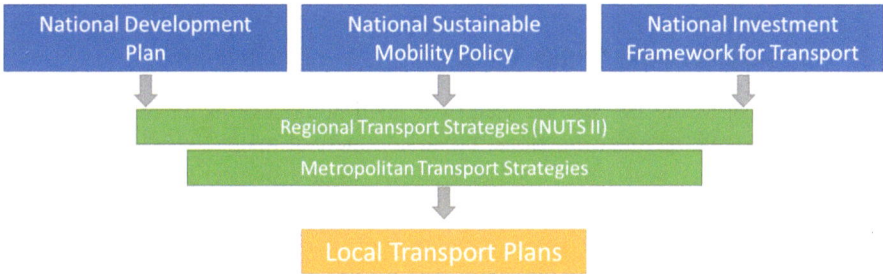

Fig. 1. Transport Planning Hierarchy in Ireland

2 The Legacy Challenge

2.1 Existing Development Patterns

Ireland has a challenging settlement and population distribution pattern for the provision of efficient, effective and sustainable transport. While it ranks 7th in the EU for percentage of population living in rural areas, at 36% [2], this statistic masks the extent to which this population is also dispersed. In 2022, 20% of all dwellings granted permission was in the form of one-off housing, equating to 40% of all applications for houses.

In addition to the dispersal of population into rural areas, there has also been significant growth in the form of edge-of-town suburban development across towns in Ireland, in particular in leinster, which throughout the period known as the "Celtic Tiger" became the commuter belt for the dublin Metropolitan Area [3]. Vast swathes of suburban development were appended onto towns of all sizes up to 100 km from workplaces in the capital.

Allied to this, at the settlement level, retail moved to the edges in the form of out-of-town developments, and commonly, new schools were developed on peripheral greenfield sites. In some cases, central schools were closed and consolidated into single strategic campuses away from the existing population centres.

The combination of these land use trends, and their perseverance into the era of climate change and post-Covid behavioural change, has presented serious challenges for the fostering of sustainable transport culture in Irish towns. Car-dependency has been locked-in via settlement patterns and urban forms which do not support public transport, walking or cycling.

2.2 Functional Urban Areas of Settlements

Allied to the population distribution challenge is the associated extension of the functional urban areas of Irish settlements into the hinterland. While this phenomenon is not unusual, it is the proportion of the population which is rural and dispersed and their associated impact on the urban fabric of settlements due to their need to drive and park within the urban area which provides a distinct challenge for sustainable transport.

2.3 Transport Networks Within Towns

In general, Irish towns are characterised by a small number of major routes, many of them with strategic national functions; broad commercial main streets; collections of narrower side streets; back lanes; suburban housing estates; and peripheral employment zones.

This format presents a number of challenges as follows. The strategic routes tend to be single carriageway with limited potential for the introduction of public transport priority or high-quality segregated cycle tracks. The main streets tend to be similar but with the added constraint of the perceived need for on-street parking to serve retail and other services. The narrower side streets can often accommodate an enhanced public realm, including fully pedestrianised areas, but only do so by significantly reducing circulation options for all modes to one-way streets etc.

As such, the potential for improving the sustainable transport offer in many Irish towns is limited without radical interventions in the traffic circulation regimes to accommodate roadspace reallocation.

2.4 Car Dependency and Political Will

The above factors have combined to create a high level of car dependency across major urban areas in Ireland. This car dependency has fostered a very car-centric culture which manifests itself by creating a challenging environment for the progress of sustainable transport schemes which, by definition, are required to reduce the amount of roadspace dedicated to the car. Through the local democratic process, this has placed another significant barrier to the implementation of LTPs.

3 Approach to Local Transport Plans

3.1 Relationship to Local Area Plan

The NTA has encouraged local authorities to undertake Local Transport Plans concurrently, and as a formal element of, their statutory Local Area Plans (LAP). The latter comprise the land use plans for the settlements and the iteration and cooperation between the two is fundamental to the success of the non-statutory LTP. It is regarded as essential, that the key objectives and policies of the LTP are included in the LAP.

3.2 Establishing Context

In line with steps 2 and 3 of the SUMP guidance, the establishment of the baseline or context is a vital first step in the making of an LTP. This consists, generally, of the following tasks:

- Review of existing local transport planning context;
- SWOT analysis of settlement /study area or similar;
- Establish the objectives of the plan; and
- Determine forecast demand for travel for horizon year;

These tasks may require close consultation with various departments within the local authorities in order to gain a clear understanding of planning, public realm, heritage, etc. considerations, and will require consultation with the NTA, Transport Infrastructure Ireland (TII) and transport operators.

A key element of the SWOT analysis is the use of the NTA's Accessibility to Opportunities and Services (ATOS) connectivity tool to identify locations where pedestrian and cyclist accessibility may be sub-optimal. In larger settlements / study areas, the Public Transport Accessibility Tool (PTAL) may also be appropriate to use.

At the end of this stage, the plan team should be in a position whereby they clearly understand what they are planning for and can then consider what potential measures could address the weaknesses uncovered in this stage, and meet the objectives for the plan.

3.3 Developing Measures and the Draft Plan

Based on the established context, the plan team will then develop the planned transport networks for each mode, ensuring that the key trips generators and attractors are connected by sustainable modes. From these networks, and in an iterative manner, potential measures can be identified. These are generally set out according to transport mode and, depending on the scope of the plan, may contain specific land use measures for assessment as part of the associated and concurrent LAP-making process.

The potential measures are assessed against the objectives of the plan, qualitatively and quantitatively, using transport modelling tools if necessary, and a draft preferred local transport plan is prepared.

3.4   Iteration and Finalisation

In advance of publication of the draft LTP, the NTA will seek to ensure that the associated draft LAP provides for its implementation both by including its policies and objectives and by proposing a land use pattern which supports sustainable transport.

In most cases, the consultation on the draft LTP is undertaken alongside the statutory LAP, with the associated Strategic Environmental Assessment and Appropriate Assessment requirements.

4   Key Deliverables

4.1   Networks for Each Mode

The critical outputs for each LTP are the networks for each transport mode identified during the plan preparation, and from which the potential measures are defined. The following principles apply for each:

- Walking – all of the major trip generators and attractors in the study area should be connected by safe and convenient pedestrian facilities, including crossings;
- Cycling – similar to walking, but with consideration for interactions with the public transport network and traffic in order to determine the most appropriate routeings for cycle trips;
- Public transport – focussed on the centre with high-quality stops and facilities. In settlements with existing or proposed dedicated town bus services, the objectives of the NTA in relation to their introduction or expansion will be accounted for;
- Traffic – In line with the road user hierarchy, private car traffic circulation will be determined once the plan team is satisfied that safe and convenient networks for walking, cycling and public transport have been set down.

4.2   Priority Investment Areas

When illustrating the transport networks by mode, the prioritisation of individual schemes or areas of investment is required. In the past, there has been a tendency to measures success in terms of kilometres of infrastructure delivered. While this remains a key metric, it can be inappropriately applied and can result in less effective, less central, easier routes being delivered. An example of this would be a cycle track on a ring road which does not serve any cycling demand to schools or the town centre.

As such, it is essential that the LTP prioritises investment in infrastructure with a clear path set out to meeting the key transport demands by sustainable modes. A "centre-first" approach is generally appropriate here.

4.3   Travel to Schools

Schools are a major source of trip demand in towns and, due to their peak-time nature, contribute significantly to congestion. As such, LTPs require to clearly identify how travel to school by sustainable modes is being catered for and how front-of-school environments will be managed in the study area to facilitate movement by pedestrians and cyclists.

5   Guidance Available

5.1   Area Based Transport Assessment Updated

The NTA and TII published a suite of guidance documents in 2018 under the "Area-Based Transport Assessment" (ABTA) label for use in the delivery of Local Transport Plans. These guidance documents are being re-examined with a view to their updating in light of the experience gained to date by the NTA, local authorities and industry practitioners during the making of the initial tranche of LTPs.

6   Main Lessons Learned to Date

The lessons learned, according to which the ABTA guidance will be updated, can be summarised as follows:

- The need for comprehensive iteration between the LTP and LAP;
- The need to only use transport modelling where required;
- The need for early stakeholder engagement;
- The advantage of high-quality presentation of materials; and
- The prioritisation of measures.

The NTA is currently managing 32 LTPs as part of the programme. By the end of 2024, it is anticipated that all of the Regional Growth Centres and Key Towns will have undertaken an LTP and will have a clear framework for promoting and facilitating sustainable transport infrastructure and services, and a clear path for achieving the required reduction in emissions from transport.

References

1. Rupprecht Consult (ed), Guidelines for Developing and Implementing a Sustainable Urban Mobility Plan, Second Edition, European Platform on Sustainable Urban Mobility Plans, Cologne (2019)
2. The Global Economy website. www.theglobaleconomy.com/rankings/rural_population_perc ent/European-union/. Accessed 25 Sept 2023
3. Pope, C., Major growth in urban sprawl around Dublin, The Irish Times web-page (2004). https://www.irishtimes.com/news/major-growth-in-urban-sprawl-around-dublin-1.977210. Accessed 25 Sept 2023

Open Access This chapter is licensed under the terms of the Creative Commons Attribution 4.0 International License (http://creativecommons.org/licenses/by/4.0/), which permits use, sharing, adaptation, distribution and reproduction in any medium or format, as long as you give appropriate credit to the original author(s) and the source, provide a link to the Creative Commons license and indicate if changes were made.

The images or other third party material in this chapter are included in the chapter's Creative Commons license, unless indicated otherwise in a credit line to the material. If material is not included in the chapter's Creative Commons license and your intended use is not permitted by statutory regulation or exceeds the permitted use, you will need to obtain permission directly from the copyright holder.

Traffic Restriction Scheme in Downtown Athens: Is It Effective?

Ioanna Berntoufi$^{(\boxtimes)}$ and Ioanna Spyrospoulou

National Technical University of Athens, 9 Iroon Polytechniou Street, Zografou, 15780 Athens, Greece
ioannaberntoufi@gmail.com

Abstract. In this study we develop, the first to our knowledge, multiple linear regression model to estimate the effectiveness of the Athens' traffic restriction scheme called "daktylios". The utilized data involves traffic flow obtained from 50 measurement stations on road segments inside and outside but close to the restriction zone limits. Multiple linear regression models were designed to explore the parameters affecting traffic flow and traffic flow difference as a result of the scheme's operation. Results provided insights about the variation of the traffic conditions during the implementation time of the scheme. They demonstrated a slight decrease of traffic flow for sections located inside the restriction zone, indicating the ineffectiveness of the scheme. Furthermore, specific circumstances indicated a variation of drivers' obedience. For example, the effectiveness of the scheme varied with trip direction (from, to or within the zone), time of day and retail sales. The parameters affecting traffic flow inside the restriction zone were similar throughout the years indicating stabilized driver behavior, with the only exception being the year 2021–2022 during which traffic flow increased probably as a result of travelers' reluctance to use public transport in fear of COVID-19. The results of the study indicate that the "daktylios" scheme needs to be revisited and probably replaced by more contemporary measures, which address the present needs and behavior of the travelers, and have the potential of being more effective.

Keywords: traffic restriction · congestion · sustainable mobility

1 Introduction

Evidently, traffic congestion has been a major problem in almost every metropolitan area since the rise of urbanization. And while several measures are implemented worldwide towards reducing the use of private vehicles including green mode promotion and private car restrictions, cities still suffer from this phenomenon, with Athens not being the exception. Athens is striving to achieve shift from the passenger car, a transport mode that is highly popular among its inhabitants, with car-ownership rates reaching the profoundly high value of 799/1,000 persons, to more sustainable models. As a result, several traffic management strategies have been designed and implemented throughout the years; some being successful and others ineffective.

© The Author(s) 2025
C. McNally et al. (Eds.): TRAconference 2024, LNMOB, pp. 417–423, 2025.
https://doi.org/10.1007/978-3-031-85578-8_54

The oldest such scheme, "daktylios", was introduced in 1982 as a countermeasure towards reducing energy consumption and mitigating the increased environmental pollution in the centre of Athens as a result of heavy traffic flows and traffic congestion. "daktylios" scheme involves the restriction of private car use, in a specific zone in central Athens, based on the last digit of the vehicle's number plate i.e. on odd dates only vehicles with odd number plates are allowed to enter the zone, while several exemptions also apply (e.g. residents, green cars, emergency vehicles). Although several cities have adopted more modern schemes of vehicle restrictions, the "daktylios" scheme is still in operation today, notwithstanding that its effectiveness has never been assessed.

Schemes similar to "daktylios" are still in operation in cities worldwide. In Hangzhou, China, Yao et al. (2018) noted a reduction of traffic flow and increase of mean speed as a result of the adoption of a traffic restriction scheme, based on vehicle number plates, which classified vehicles into restricted circulation, free circulation, vehicles from other cities and taxis. Traveller response to the scheme included modification of trip departure times, from peak to off-peak hours, and shift to different modes. However, restrictions where more horizontal compared to the Athenian scheme. Guerra and Millard-Ball (2017) assessed a similar scheme to "daktylios", which is in operation in Mexico City, and data analysis identified four different types of traveller responses: use of a second car, change of trip timings, shift to private car as a result of anticipated reduction of congestion and scheme violating behaviour. Results indicated that overall, traffic conditions in Mexico City were not affected by the scheme. On the other hand, a small improvement i.e. 5.5% reduction of traffic flow and 3.1% increase in metro ridership was observed in the city of Santiago, Chile, (De Grange and Troncoso 2011) as a result of the adoption of a similar traffic restriction scheme. Assessment results of such schemes from international literature highlight the minimal effectiveness or even ineffectiveness of the schemes, as well as, the adoption of scheme violating behaviours by travellers (Liu et al. 2016, Bonilla 2019).

The objective of this research is to assess the effectiveness of the "daktylios" scheme utilizing traffic data in the city of Athens, with the design of suitable general linear models. Section 2 describes the collected data and adopted methodology, while in Sect. 3 the performed analysis is presented. Policy implications are discussed based on study's findings discussed in Sect. 4.

2 Data and Methodology

The present study assesses the effect of "daktylios" on traveller behaviour, through the metric of traffic flow. It is based on the analysis of historical traffic data, and in particular traffic flow data collected from 50 detectors located at dedicated locations of the Athenian road network (see Fig. 1). The spatial and temporal characteristics of the collected data should consider several elements including traveller habits, road network characteristics and the scheme's characteristics – "daktylios" is in operation between September/October and July, on weekdays between 07.00 and 20.00 (with the exception of Friday when it ends at 15.00).

Considering their spatial distribution, the selected detectors involved both locations inside (31 detectors) and outside but close (19 detectors) to the "daktylios" operation zone. Considering their temporal distribution two types of datasets were designed.

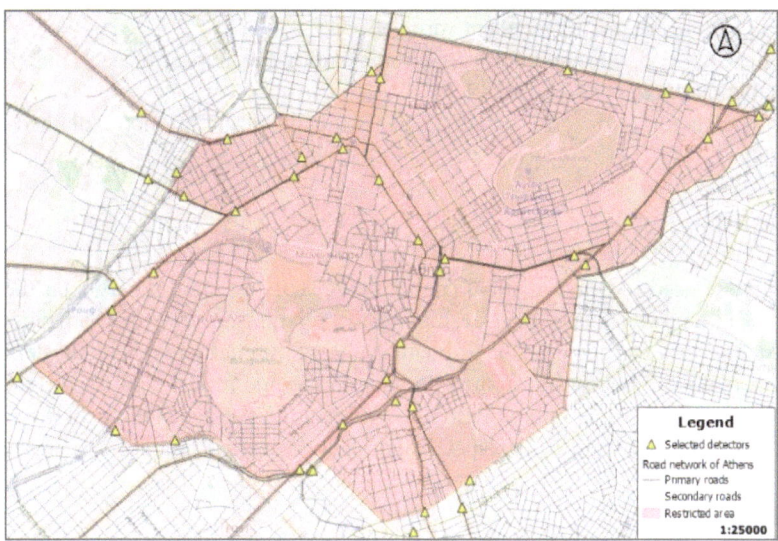

Fig. 1. Selected detectors for data collection

The first involved only scheme operational months including both pre-COVID and post-COVID periods, and the second involved a period of two weeks before and two weeks after the commencement of the scheme. The first dataset was opted to estimate the variation of traffic flow within the implementation period of the scheme, as well as to detect possible changes considering the traffic demand throughout the recent years. The second dataset was selected to support the investigation of the effect of "daktylios" operation on traffic flow. It should be noted that "daktylios" is in operation shortly after the observation of high traffic flows (close to school start dates) in the Athenian road network, while the measure is uplifted every year when traffic flow declines (during summer holidays). In order to explore the actual effect of the scheme on traveller behaviour, the before and after periods should involve periods where inhabitants' travel needs are similar. As such the before and after investigation period, was selected to have a duration of two weeks.

Multiple linear regression models were designed to model the relationship between the parameters affecting traffic flow for the first dataset and traffic flow difference as a result of the scheme's operation for the second dataset. The objective was to build the best-fitting linear equation that explains the variation in traffic flow and traffic flow difference based on the values of the independent values. Factors that could potentially affect traffic flow vary and mostly refer to road and trip characteristics. The equation used for the multiple linear regression was the typically represented one, showed as follows:

$$y = b_0 + b_1 x_1 + b_2 x_2 + \cdots + b_n x_n + \varepsilon \tag{1}$$

where:

- y is the dependent variable traffic flow or traffic flow difference.

- x_1, x_2, ..., x_n are the independent variables.
- b_0, b_1, b_2, ..., b_n are the coefficients representing the impact of each independent variable on the dependent variable.
- ε represents the error term, which accounts for the variability not explained by the model.

3 Results

The software used for the statistical analysis is R-statistics, and in particular the glm2 library. The analysis focused on evaluating the relationship between dependent and independent variables based on the estimated value of the coefficient, with the dependent variables being the hourly traffic flow per lane in the first model, and the difference in hourly traffic flows (after minus before) as a result of the scheme's operation in the second for years 2022 and 2021 respectively. The two designed models are presented in Tables 1 and 2.

Table 1. Multiple linear model for hourly traffic flow per lane of year 2022

Variable	Coefficient	Std. Error	t-value
Intercept	453.236	10.214	44.372
Time interval [08]	−34.212	7.074	−4.836
Time interval [10]	−35.757	7.057	−5.067
Time interval [11]	−62.859	7.093	−8.862
Time interval [12]	−74.132	7.092	−10.453
Time interval [13]	−77.743	7.08	−10.98
Time interval [14]	−73.759	7.145	−10.323
Time interval [15]	−52.446	7.1	−7.387
Time interval [16]	−39.201	7.096	−5.524
Time interval [17]	−33.29	7.121	−4.675
Time interval [18]	—31.735	7.116	−4.46
Time interval [19]	−40.859	7.105	−5.751
Time interval [20]	−70.136	7.169	−9.784
Day [Wednesday]	7.657	3.659	2.093
Day [Thursday]	9.944	3.845	2.586
Day [Friday]	12.127	4.051	2.994
Restriction area [inside]	−64.38	3.766	−17.097
Road [main branch]	349.124	14.047	24.854
Road [primary]	338.618	9.33	36.295
Lanes [2]	−43.159	4.871	−8.86

(continued)

Table 1. (*continued*)

Variable	Coefficient	Std. Error	t-value
Lanes [3]	−21.264	4.938	−4.306
Lanes [4]	37.089	11.733	3.161
Direction [to city center]	−196.856	4.807	−40.949
Direction [out of city center]	−154.21	4.44	−34.734
R–squared	**0.1528937**	AIC	**236503**

Table 2. Multiple linear model for hourly traffic flow difference of year 2021

Variable	Coefficient	Std. Error	t–value
Intercept	75.917	4.64	16.363
Time interval [09]	−9.706	3.03	−3.204
Time interval [10]	−8.794	3.047	−2.886
Time interval [11]	−5.62	3.019	−1.861
Time interval [12]	−11.292	3.013	−3.748
Time interval [13]	−7.501	3.019	−2.484
Time interval [16]	7.763	3.058	2.539
Time interval [17]	7.66	3.058	2.505
Time interval [20]	−8.784	3.033	−2.896
Day [Monday]	−50.676	2.915	−17.387
Day [Tuesday]	−58.745	3.784	−15.525
Day [Wednesday]	−50.429	2.92	−17.269
Day [Friday]	−29.349	2.724	−10.776
Trip [Shoppimg]	4.964	2.159	2.3
Restriction area [inside]	−10.825	2.873	−3.767
Direction [to city center]	−8.926	2.774	−3.218
Direction [within city center]	−9.524	2.257	−4.22
Road [secondary]	−9.871	2.63	−3.754
Lanes [2]	−15.353	2.77	—5.543
Lanes [3]	−11.816	2.692	−4.389
Lanes [4]	−14.354	3.868	−3.711
R–squared	**0.2419003**	AIC	**51078**

Traffic flow, as the first model indicates, is affected by several spatial and temporal parameters including time of day and day of week, and type of road, which represent traveller mobility habits. Considering the examined topic, lower traffic flows are observed at the inner road sections compared to the outer road sections (considering the restriction zone), indicating an effectiveness of the scheme. At the same time, traffic flow in road sections with direction to and from the city centre exhibit lower flows compared to road sections already inside the city centre. The reduction in flows although evident, is not as high.

The second model presents attributes affecting the difference of traffic flow following the operation of the measure. The positive intercept indicates an increase of traffic flow following the operation of "daktylios", this however can be explained by the selected reference levels of the model, as almost all the variable coefficients have negative values. The road sections inside the restriction zone, exhibit a small reduction, indicating a small effectiveness of the measure. Once more as expected, the traffic flow change following the operation of the scheme is affected by several spatiotemporal variables. Time of the day, especially peak-hours, appear to influence the traffic flow difference indicating higher differences during these hours. Trip purposes also influence traffic flow difference, as an increase in traffic flow difference during high retail days is observed. Furthermore, the decrease in traffic flow difference for trips directed towards the city center and those within it may suggest some effectiveness of the scheme.

4 Discussion

The present study explores the effect of the operation of an old traffic restriction measure in the centre of Athens, through the design of dedicated general linear models utilizing traffic data, and in particular traffic flow. Results indicate a limited effect of the measure, being in agreement with findings from international literature. Questionnaire surveys could be employed as an alternative method to assess the effectiveness of the scheme, and could provide more in-depth information. It is interesting, however, that studies employing stated preference questionnaires instead of revealed ones, indicate a substantially high improvement as a result of the adoption of similar vehicle restriction schemes. Reality however, as this research and other similar studies report, actually differs. Congestion pricing comprises a more "modern" and effective traffic restriction measure that has been adopted in several cities worldwide, with positive impact (Ramos et al. 2017). Congestion pricing was discussed in Athens in the past, but the initiation of these discussions coincided with the financial crisis, and the scheme was abandoned. Parking pricing schemes and infrastructure changes such as pedestrianisation comprise other successful policies towards improving traffic congestion, through traffic restrictions. The "daktylios" scheme should be revisited, as its impact is minimal, while at the same time it establishes violating behaviours due to minimal enforcement.

At present, although there are efforts towards promoting green transport modes, Athenians still depend greatly on their private car contributing to congestion and air pollution. Urban planners, transport engineers and decision makers should consider more modern schemes that have the potential to transform the city of Athens into a sustainable and livable city.

References

Bonilla, J.A.: The more stringent, the better? rationing car Use in Bogotá with moderate and drastic restrictions. World Bank Econ. Rev. **33**(2), 516–534 (2019). https://doi.org/10.1093/wber/lhw053

de Grange, L., Troncoso, R.: Impacts of vehicle restrictions on urban transport flows: the case of Santiago, Chile. Transp. Policy, S0967070X11000825 (2011). https://doi.org/10.1016/j.tranpol.2011.06.001

Guerra, E., Millard-Ball, A.: Getting around a license-plate ban: behavioral responses to Mexico City's driving restriction. Transport. Res. Part D: Transp. Environ. **55**, 113–126 (2017). https://doi.org/10.1016/j.trd.2017.06.027

Liu, Y., Hong, Z., Liu, Y.: Do driving restriction policies effectively motivate commuters to use public transportation? Energy Policy **90**, 253–261 (2016). https://doi.org/10.1016/j.enpol.2015.12.038

Ramos, R., et al.: From restricting the use of cars by license plate numbers to congestion charging: analysis for Medellin, Colombia. Transp. Policy **60**, 119–130 (2017). https://doi.org/10.1016/j.tranpol.2017.09.012

Yao, W., et al.: 'Analysis of cars' commuting behavior under license plate restriction policy: a case study in Hangzhou, China. In: 2018 21st International Conference on Intelligent Transportation Systems (ITSC), pp. 236–241. IEEE, Maui (2018). https://doi.org/10.1109/ITSC.2018.8569742

Open Access This chapter is licensed under the terms of the Creative Commons Attribution 4.0 International License (http://creativecommons.org/licenses/by/4.0/), which permits use, sharing, adaptation, distribution and reproduction in any medium or format, as long as you give appropriate credit to the original author(s) and the source, provide a link to the Creative Commons license and indicate if changes were made.

The images or other third party material in this chapter are included in the chapter's Creative Commons license, unless indicated otherwise in a credit line to the material. If material is not included in the chapter's Creative Commons license and your intended use is not permitted by statutory regulation or exceeds the permitted use, you will need to obtain permission directly from the copyright holder.

European Research and Innovation in Support for Sustainable and Smart Urban Mobility

Konstantinos Gkoumas[1] , Marcin Stępniak[1](✉) , Ilias Cheimariotis[1] ,
Fabio Marques dos Santos[1] , Monica Grosso[1], Ferenc Pekar[1] , Chiara Lodi[2] ,
and Alessandro Marotta[1]

[1] European Commission, Joint Research Centre, Via E. Fermi 2479, 21027 Ispra, VA, Italy
marcin.stepniak@ec.europa.eu
[2] Piksel S.r.l., 20126 Milan, Italy

Abstract. The European Green Deal and its target to achieve a 90% reduction in transport-related greenhouse gas emissions by 2050 together with the new European Union Urban Mobility Framework put urban mobility and logistics in the spotlight of European mobility policies. Research and innovation are paramount to respond to the challenges and to further improve mobility and transport systems in cities, while fully tackling the negative impacts of transport. This study provides a review of recent trends, challenges and achievements of European research and innovation initiatives in urban mobility and logistics and examines their alignment with the overarching transport and mobility policy objectives. The paper identifies relevant projects that focus on urban mobility, using the Transport Research and Innovation Monitoring and Information System (TRIMIS) database. It identifies the main trends in European research and discusses main areas of development and key achievements. It also presents recommendations for future research priorities and initiatives.

Keywords: Urban Mobility · Urban Logistics · Research and Innovation · Transport policy

1 Introduction

The new European Urban Mobility Framework [1] is built upon the emissions reduction target, as defined in the European Green Deal, and on the mobility development pathways established in the EU Sustainable and Smart Mobility Strategy. The document emphasizes the importance of public transport, active mobility options and efficient zero-emission urban logistics, including last mile delivery. It also stresses a need for more effective sustainable urban mobility planning and urban mobility data.

The principal objective of this study is to assess the extent to which European Research and Innovation (R&I) potentially contributes to achieve main policy aims, as defined in the New Urban Mobility Framework. For each of the identified thematic areas, we provide an overview of the R&I activities and achieved progress. We conclude with recommendations on the focus of future R&I initiatives, having in mind key policy challenges and identified research gaps.

© The Author(s) 2025
C. McNally et al. (Eds.): TRAconference 2024, LNMOB, pp. 424–429, 2025.
https://doi.org/10.1007/978-3-031-85578-8_55

2 Methods

The study uses the Transport Research and Innovation Monitoring and Information System (TRIMIS) database [2] to identify relevant projects which focus on urban mobility and logistics. TRIMIS was founded in 2017 by the European Commission and is operated by the Joint Research Centre [3]. The main product of TRIMIS is its original and open database of transport R&I projects and programmes. In addition to the latest Framework Programmes (i.e. Horizon Europe, Horizon 2020 and FP7), it gathers data on other relevant European projects (e.g. Interreg, Connecting Europe Facility etc.) as well as projects funded by Member States. Currently, the database consists of nearly 9000 projects and is constantly updated and extended.

In this study, the focus is on the H2020 Framework Programme for R&I projects that target urban mobility and logistics. An automatic search for projects in the TRIMIS database was applied, using pre-selected keywords that focused on: active modes, micromobility, urban rail, last mile delivery, logistics, low emission transport, electromobility, connected and automated mobility, Mobility as a Service (MaaS), urban transformation, inclusive and accessible urban mobility planning, and mobility management. In a second step, a manual verification of the list of projects was carried out, to ensure the selected ones were relevant to urban mobility and logistics.

For the thorough analysis of the identified projects, they have been clustered into subthemes to help to organise the review process and to detect main trends. The subthemes are derived from the action areas of the new Urban Mobility Framework, to facilitate assessment of the progress towards aims listed in the policy document. The full list of analysed thematic areas includes the following:

- Digitalisation – which covers new mobility services including MaaS, connected and automated mobility, Urban Air Mobility together with ICT, data processes, and digital solutions and services.
- Sustainable Urban Mobility Plans (SUMP) – including urban mobility and logistic planning and management plans, stakeholder involvement, Sustainable Urban Logistic Plans integration.
- Climate-neutral cities – including efficient urban transport, recharging and alternative fuels refuelling infrastructure, clean fuels, electromobility, hydrogen.
- Active mobility – including walking, cycling and micromobility, modal shift towards active modes, no car solutions, e-bikes and e-cargo bikes, cycling infrastructure, road safety.
- Trans-European Transport Network (TEN-T) urban nodes – including multimodal hubs (passenger and freight), park & ride solutions; (with the special attention to vulnerable road users).
- City freight logistics and last-mile delivery – including development and implementation of zero-emission freight urban logistics, last-mile delivery, freight urban logistics infrastructure, new distribution models.
- Monitoring progress – sustainable urban mobility indicators including initiatives on harmonised mobility indicators, development and implementation of Sustainable Urban Mobility Indicators benchmarking tools etc.

3   European R&I Supporting Sustainable and Smart Urban Mobility

3.1   Overview of Selected Projects

In total, 331 projects were considered [4, 5]. Apart from H2020 projects (302 projects), we included also projects supported by Interreg programme and Urban Innovative Actions (UIA). The total EU contribution in those projects exceeded EUR 1.5 billion against a total budget of almost EUR 2 billion (Fig. 1).

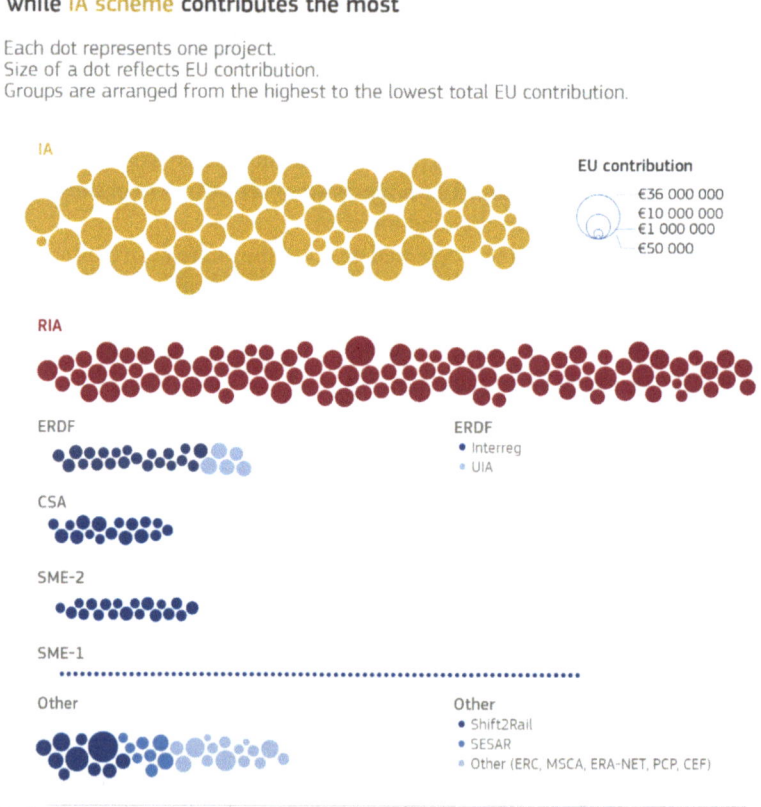

Fig. 1. Urban mobility and logistics research and innovation projects by source of funding.

Further, we assessed the effort directed towards the main identified subareas of research (Fig. 2). The highest number of projects (137) focus on digitalisation, followed by number of projects working on SUMP (106) and Climate neutral (75) and Active mobility (48). SUMP has higher synergies with other subthemes, which means that projects within this category also contributed to other topics (mostly digitalisation).

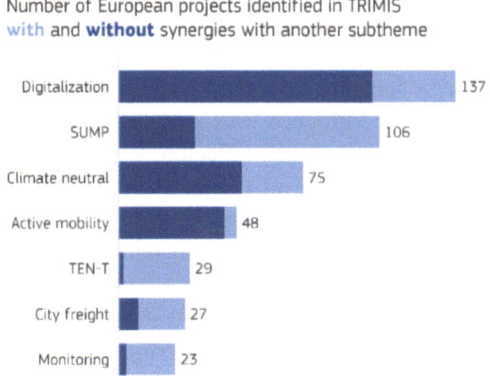

Fig. 2. Research and innovation projects by subtheme.

3.2   Key Achievements of R&I Projects and Future Challenges

Digitalisation, Innovation and New Mobility Services. The observed progress cover further developments related to cooperative, connected and autonomous mobility and in urban environment, MaaS, shared mobility services or urban air mobility. Projects worked on digital tools for transport planning, e.g. cloud-based cooperative system, including traffic prediction and multimodal routing system. With technologies reaching higher maturity, the challenge will be to integrate them into operational aspects, test them and implement in real life conditions.

SUMP and Mobility Management Plans. The projects assessed intend to help authorities design SUMPs, from development of models and software tools to several collaboration actions. They also introduce new technological developments into SUMPs and take advantage of Big Data analytics capacities. On logistics, the concept of Sustainable Urban Logistics Plans (SULPs) has been introduced. While SUMPs, and SULPs to a lesser degree, can be described as a mature concept, their uptake is still subject to barriers and disruptions (for example the COVID-19 crisis). Therefore, continuous effort and support for local authorities is necessary to implement, monitor and continuously refine and adapt SUMPs (and SULPs) according to the evolution of urban environments, their mobility and logistics needs and solutions in place.

Towards Climate-Neutral Cities: Resilient, Environmentally Friendly and Energy-Efficient Urban Transport. R&I activities focused on technologies related to electric vehicle charging, grid infrastructure and technology as well as development of fuel cells with higher lifespan. In order to facilitate the transition to climate neutrality, urban mobility will require the fusion and upscaling of best practices in all action areas of the new Urban Mobility Framework, while ensuring synergies with other areas such as renewable energy production and storage. The projects cover a broad range of topics from detailed technological solutions (e.g., fast chargers, charging pads, real-time data for checking availability of charging stations etc.), through grid optimisation or integration of electric vehicles into urban transport systems, to new business models. Further works are still necessary to develop and test new solutions as well as to push forward existing ones, to scale them up and reduce installation and operational costs.

Healthier and Safer Mobility: A Renewed focus on walking, Cycling and Micromobility. Previous European R&I projects have achieved positive results related to active mobility promotion and collecting good practices in different urban settings. Progress has been made in the area of technology development for cycling and micromobility, including innovative solutions focusing on safety, self-charging bikes and parking infrastructure, four-wheel, lightweight vehicles or micro-vehicles for older adults or people with reduced mobility. The safety of vulnerable road users remains one of the main concerns and requires further technical as well as regulatory progress.

Reinforced Approach to TEN-T Urban Nodes. The conducted works on intermodal connection between long-distance and last-mile freight logistics are essential, yet the topic requires further studies. The recent projects address governance of multimodal passenger and freight zero emission mobility hubs. They require further work, similarly to the multimodality of urban nodes and joint consideration of passenger and freight flows to optimise the utilisation of the network and means of transport. Those solutions need to be further embedded into a broader urban transport planning perspective and they would require further technological advances.

Zero-Emission City Freight Logistics and Last-Mile Delivery. The recent progress includes improvement of multi-level governance models and multi-stakeholder cooperation as well as new schemes for horizontal collaboration, urban space management and re-localisation of logistics activities in multi-modal hubs. The use of cargo bicycles, Light Commercial Vehicles (LCVs) and novel small sustainable cargo vehicles helped to improve last-mile delivery in urban settings. Future research should include detailed methodologies for demand and flow characterisation to be able to dispatch efficiently across multimodal networks.

Monitoring Progress – Sustainable Urban Mobility Indicators. Recent R&I activities managed to define and test a set of relevant indicators covering key aspects of urban mobility (e.g., access to mobility services, congestion, road safety, GHG emissions, air quality). The pilot study covered indicators for 46 cooperating cities. Further works are necessary for a widespread use of the tool among all European cities, as it proposes a common approach to data collection and monitoring. Furthermore, indicators reinforce the evidence base for preparing and implementing SUMPs, by tracking changes towards sustainable urban mobility objectives in each city. Future initiatives should include relevant data dissemination tools and related data governance models.

4 Conclusions

Research and innovation contribute to policy objectives and societal needs in Urban Mobility, regarding innovation in transport technologies, improvement of planning tools or evidence gathering. R&I activities foster cross-stakeholder collaboration and citizen engagement. Focusing mainly on relevant EU funded projects, this study presents an analysis of R&I in urban mobility and logistics in Europe in the last years against the aims of the New Urban Mobility Framework.

Future initiatives should build upon the existing experience and achievements outlined in this study, while introducing and building momentum in novel ideas. Effective policy implementation will build upon cross-stakeholder planning tools and collaboration networks. Then it will increasingly rely upon knowledge management and the governance and use of mobility data and indicators, and a mix of living labs and virtual experimentation to project in the future. Knowledge management and data governance and analytics are the transversal methods adapted to the systemic complexity of mobility systems and will need to be further relied upon for both effective policy implementation and solutions in the field. Improving the sustainability, safety and quality of mobility and life in general in cities and neighbourhoods will rely on a mix of technological, planning, infrastructure, and participative solutions, with an increased emphasis on social and behavioural aspects. Finally, the role of research and innovation is crucial to anticipate future opportunities and challenges.

Acknowledgements. The views expressed here are purely those of the authors and may not, under any circumstances, be regarded as an official position of the European Commission.

References

1. European Commission: The New European Urban Mobility Framework, COM(2021) 811 final. European Commission, Brussels, Belgium (2021)
2. European Commission: TRIMIS Transport and Research and Innovation Monitoring and Information System. https://trimis.ec.europa.eu
3. Tsakalidis, A., Gkoumas, K., Grosso, M., Pekár, F.: TRIMIS: modular development of an integrated policy-support tool for forward-oriented transport research and innovation analysis. Sustainability. **12**, 10194 (2020). https://doi.org/10.3390/su122310194
4. Marques Dos Santos, F.L., Gkoumas, K., Stepniak, M., Cheimariotis, I., Grosso, M., Pekar, F.: Research and Innovation in Urban Mobility and Logistics in Europe - list of projects (2022)
5. Gkoumas, K., Cheimariotis, I., Pekár, F., Stepniak, M., Marques dos Santos, F., Grosso, M.: Urban mobility and logistics. Publications Office of the European Union. LU (2022)

Open Access This chapter is licensed under the terms of the Creative Commons Attribution 4.0 International License (http://creativecommons.org/licenses/by/4.0/), which permits use, sharing, adaptation, distribution and reproduction in any medium or format, as long as you give appropriate credit to the original author(s) and the source, provide a link to the Creative Commons license and indicate if changes were made.

The images or other third party material in this chapter are included in the chapter's Creative Commons license, unless indicated otherwise in a credit line to the material. If material is not included in the chapter's Creative Commons license and your intended use is not permitted by statutory regulation or exceeds the permitted use, you will need to obtain permission directly from the copyright holder.

Land Use Planning for a Better Bus: Ireland's Approach to Orientating Development Towards Public Transport

Michelle Poyourow[1][✉] and Eoin Farrell[2]

[1] Jarrett Walker + Associates, Dublin, Ireland
michelle@jarrettwalker.com
[2] National Transport Authority, Dublin, Ireland

Abstract. Planning for "transit oriented development" has long been discussed, mostly in relation to rail and metro lines. However, modern Ireland has developed around its road network, and public transport on roads primarily means buses.

Ireland's national government is making major investments in bus operations, infrastructure, fleet and technology. For these investments to have maximum impact, the way that Ireland uses land and develops new urban areas must change.

To that end, the National Transport Authority of Ireland (NTA) and Jarrett Walker + Associates (JWA) have co-authored an illustrated guidance document for planners and developers, showing how land use decisions affect the ultimate cost, usefulness and patronage of bus services.

This guidance document is contributing to greater coordination among national plans, national investments, local authorities' development decisions and private development. It provides advice that is relevant in any country and that is urgently needed to address the climate crisis.

Keywords: Land Use · Spatial Planning · Development Planning · Transit Oriented Development · Bus · Public Transport

1 Introduction

Since 2018 the national government of Ireland has been increasing investment in bus services across the country, to accommodate population growth and maintain quality of life for residents whilst reducing energy consumption. However, public transport's future success will depend not only on the level of public transport investment, but also on land uses and the built environment. An amenable built environment multiplies public transport's usefulness and value, while a hostile one can reduce it to near zero.

Land use planning has long been incorporated into planning for railway services. A rail station is widely-understood to be an investment whose return depends on the surrounding land uses and built environment. For this reason, rail planning is accompanied by development in the areas around stations. These station-area plans contribute to the expected benefits of the line and help to justify its construction.

© The Author(s) 2025
C. McNally et al. (Eds.): TRAconference 2024, LNMOB, pp. 430–436, 2025.
https://doi.org/10.1007/978-3-031-85578-8_56

Given the scale of new investment in Ireland's bus services, a similar focus on land use and the built environment is needed around bus services just as it has been around rail stations. However, there is little guidance available for land use and urban planners who wish to shape growth for the future success of bus services.[1] To that end, the NTA and JWA have co-authored an illustrated guide for planners, engineers and developers. The guide shows how decisions about land use and the built environment affect the ultimate cost, usefulness and patronage of bus services.

2   What is Useful Public Transport?

Most people's time is limited. Within a limited amount of time, there is a certain geographic area a person can reach by public transport. The sum of the destinations in this area – the schools, jobs, etc. – describe a person's access to opportunity.

Although many factors affect a person's decision to use public transport, travel time is crucial. Most people work or attend school, which means they have busy days and need to use their time efficiently. If a trip can't be made on public transport in an amount of time that feels reasonable, they are more likely to use a car. Access to opportunity – also called accessibility – is a good way to measure this benefit of public transport.

Patronage has many causes and goes up and down for reasons completely external to public transport, such as economic conditions, pandemics or the cost of parking a car. When we increase access we increase the likelihood that any person, looking to make a trip, will find the public transport journey time reasonable. This is why higher access is correlated with both higher public transport patronage and mode share [2, 3].

Public transport is useful if people can reach their destinations in a reasonable amount of time. Access is the best objective measure of public transport's usefulness.

What types of public transport service support high access?

2.1   Frequent

High frequency service improves access by reducing waiting time and thereby shortening overall journeys. High frequency is correlated with high patronage, at the level of the network, the route and the bus stop [2, 4–8]. High frequency on a line is determined by three factors: how many vehicles are assigned to the line, the length of the line, and the speed of the vehicles.

Frequency is often under-emphasised because it is invisible. You can't take a photo of the fact that the next bus is coming soon. However, frequency is what makes a theoretical line into a service someone can use whenever they want to, just as a motorist can use a road whenever they want to.

2.2   Connected

In a well-connected network, each line serves not only the people and destinations along it, but also the people and destinations near *other lines* in the network.

Connectivity between public transport lines can be achieved in two ways:

- By coordinated timing of vehicle arrivals and departures, to ensure short interchanges. However this can only be done in one or two places per line, and is vulnerable to even minor disruptions.
- By high frequencies, which make interchanges fast and reliable at any time.

2.3 Fast and Reliable

Speed affects how far someone can travel, and therefore how many opportunities they can access, once they've boarded a public transport vehicle. Services that are very slow shrink the area that people can access in a reasonable amount of time.

Reliability describes how consistently a service operates at its scheduled speed, and therefore how consistently it provides access to opportunities.

The fact that priority measures make passengers' journeys faster and more reliable is obvious. Less obvious is the fact that they save public funds and make investments in bus service more effective.

Slow service is not just unattractive to passengers. It is also more expensive to operate. The number of buses and drivers needed to operate a route at a certain frequency depends on how long it takes the buses to drive the line out and back. Within any limited budget for service, a bus line that is slowing down will have a degrading frequency as well as a degrading speed.

The design and operation of roads, and the distances bus lines are expected to cover, therefore affect the level of frequency that can be provided.

3 Land Use Planning that Supports High Access

Public transport does not deliver access by itself. High access arises from the interaction between the public transport service, the land use pattern and the built environment.

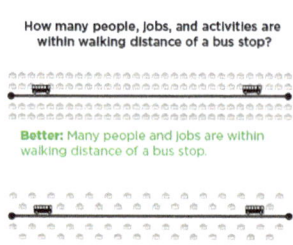

How many people, jobs, and activities are within walking distance of a bus stop?

Better: Many people and jobs are within walking distance of a bus stop.

3.1 Density

High density means that there are more people, jobs, and activities in the area around a public transport stop. This increases the number of residents and opportunities within a short walk of public transport, and thereby shortens journey times and increases access to those opportunities.

Worse: Fewer people and jobs are within walking distance of a bus stop.

3.2 Walkability

There may be a certain number of people or jobs in the area around a bus stop, but not everyone in that area can actually reach the bus stop. The connectedness of the street or path network determines people's walking distance to the stop.

Is the walk to the bus stop direct and comfortable?

Better: In a connected street network, most nearby places are a short distance away by foot.

Worse: In a disconnected street network, walks to nearby places are long and circuitous.

Better: For people to use a bus service both ways, it must be safe to cross the road near the stop.

On a disconnected street network, the walk to the stop might be much longer than the "as the crow flies" distance. Walkability is undermined by:

- Disconnected street networks that require pedestrians to walk an indirect path.
- Fences or walls separating adjacent developments.
- Lack of consistent footpaths along roads.
- Lack of lighting on footpaths, at junctions and at other road crossings.
- Lack of a safe place to cross the street near a bus stop.
- Building orientation that puts front doors far from the street.
- Dual carriageways or train tracks without regularly-spaced pedestrian crossings.

More walking time to and from public transport stops reduces people's access to opportunities within a reasonable journey time. Time isn't the only issue: if the walk is unsafe or uncomfortable, then many people won't walk it at all. This reduces patronage in the area, making it harder to justify high levels of service there [9].

Poor walkability can also affect the directness and linearity of services, if bus routes must deviate into certain areas because people in those areas can't walk to a stop.

Are buses well-used in both directions, at many times of day and week?

Better: A mix of land uses means buses are used in both directions during weekday rush-hours, and throughout the day and week

3.3 Mix of Uses

A mix of uses has a particular value for public transport. In a single-centred urban development, most travellers go into the centre in the morning and out in the evening. This one-directionality is inefficient for public transport, because buses

Worse: Buses serving purely residential areas tend to be used mostly during rush-hours, and mostly in one direction.

or trains may be full in one direction but must return empty in the other.

A mix of housing, retail and employment uses along corridors is much more efficient to serve with public transport. The operating cost of returning buses to the centre in the morning results in additional passenger journeys, rather than empty seats.

3.4 Continuity

Continuity of development reduces travel time by all modes. However, the effect on public transport of longer distances between developments is unique among the travel modes. This is because public transport requires an ongoing **operating cost**, in addition to capital cost.

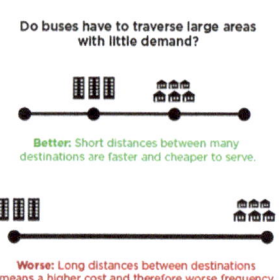

The longer the distance public transport vehicles must cover, the less frequency can be provided, within any given operating budget for service. For public transport, distance therefore trades-off against frequency. For passengers, longer journeys on the vehicle and worse frequencies both reduce access.

Continuous urban development allows people to make shorter journeys and allows operators to provide better frequencies, both of which improve access.

3.5 Linearity

In most areas built before private cars were widespread, buildings are within a short walk of roads that provide direct travel to many other places. This was a natural pattern of development when nearly all travel was by foot, bicycle or shared transport. The result was linear main roads, except where topography required otherwise.

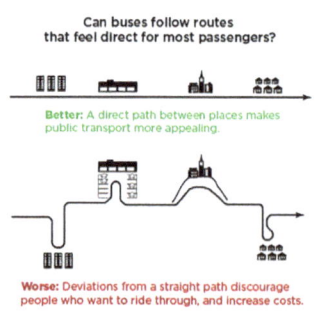

With the advent of the private car, developments can now be put in disconnected street networks and at the ends of cul-de-sacs. When an important destination for public transport is at the end of a cul-de-sac, a bus must deviate from a direct path to serve it, adding time for everyone traveling through.

Deviations don't just cost passengers more time – they also cost the operator. Deviations lengthen the lin. The longer a line, the less frequency can be provided within any given operating budget, undermining access and patronage.

Of the five principles listed here, linearity is least likely to be mentioned in general discussions of planning, because it is uniquely an issue for public transport. When using the individual modes (walking, bicycling or driving a car) each person can travel to the end of their cul-de-sac without inconveniencing anyone else. But for public transport operators and passengers alike, deviations are a problem.

New developments should **"Be on the way!"** so as to benefit from buses going past them, in a linear fashion, to other dense areas and destinations.

4   Conclusion: Three Essential Actions for Planners

The guidance note recommends three essential actions that all land use planners should take to shape growth around successful bus networks.

Make bus Operations Efficient and Direct.   Public transport requires continuous operating cost, so it is important to use it efficiently. When public transport is inefficient, the limited budget for service must be divided across more routes and more kilometres, worsening frequency and shortening hours or days of service. The result of this dilution is less access to opportunity, and fewer people using public transport. Planning urban and suburban growth in continuous, linear patterns makes efficient bus operations possible and makes bus service more useful to more people.

Make it Easy to Get to the Bus.   There are several ways to get to public transport, but walking should be the focus in planning. Walking is the foundation because it requires very little land and no operating cost and it affords people the greatest freedom and spontaneity. However it is extremely sensitive to decisions about land use and the built environment.

Organise Development Around a Frequent Network.   The highest-access public transport networks include a connected network of frequent lines. Dense development and important social destinations should be located *near those lines*. Rather than expecting public transport to chase development to wherever it is permitted, development should be shaped and located such that public transport can service it efficiently.

References

1. There are few published guides on how land use and built environment should be arranged to make bus transport successful. However, a very good English language example is "Buses in Urban Development" by the Chartered Institution of Highways & Transportation (CIHT) of the United Kingdom, published in 2018 (2018)
2. Diab, E., DeWeese, J., Chaloux, N., et al.: Adjusting the service? understanding the factors affecting bus ridership over time at the route level in Montréal, Canada. Transportation **48**, 2765–2786 (2021). https://doi.org/10.1007/s11116-020-10147-3
3. Cui, B., Boisjoly, G., Miranda-Moreno, L., et al.: Accessibility matters: exploring the determinants of public transport mode share across income groups in Canadian cities. Transp. Res. Part D: Transp. Environ. **80**, 102276 (2020). https://doi.org/10.1016/j.trd.2020.102276
4. Berrebi, S., Joshi, S., Watkins, K.: On bus ridership and frequency. Transport. Res. Part A: Policy Pract. **148**, 140–154 (2021)
5. Dill, J., Schlossberg, M., Ma, L., et al: Predicting transit ridership at stop level: role of service and urban form. In: Transportation Research Board 92nd Annual Meeting Compendium of Papers, 13-4693 (2013)
6. Gregory, T., Brown, J.: Explaining variation in transit ridership in U.S. metropolitan areas between 1990 and 2000: multivariate analysis. Transport. Res. Rec. **1986**(1), 172–181 (2006)

7. Stewart, A., Attanucci, J., Wilson, N.: Ridership response to incremental Bus Rapid Transit upgrades in North America. Transport. Res. Rec. J. Transport. Res. Board **2538**, 37–43 (2015)
8. Balcombe, R., Mackett, R., Paulley, J., et al.: The demand for public transport: A practical guide. Transportation Research Laboratory. Report 593 (2004)
9. Ewing, R., Cervero, R.: Travel and the built environment: a synthesis. Transport. Res. Rec. **1780**(1), 87–114 (2001)

Open Access This chapter is licensed under the terms of the Creative Commons Attribution 4.0 International License (http://creativecommons.org/licenses/by/4.0/), which permits use, sharing, adaptation, distribution and reproduction in any medium or format, as long as you give appropriate credit to the original author(s) and the source, provide a link to the Creative Commons license and indicate if changes were made.

The images or other third party material in this chapter are included in the chapter's Creative Commons license, unless indicated otherwise in a credit line to the material. If material is not included in the chapter's Creative Commons license and your intended use is not permitted by statutory regulation or exceeds the permitted use, you will need to obtain permission directly from the copyright holder.

Developing Connectivity Tools: Improving Urban Spaces with the assistance of Connectivity Analysis

Deborah John[1](✉) and Barry O'Neill[2]

[1] National Transport Authority, Dublin, Ireland
deborah.john@nationaltransport.ie
[2] Compass Informatics, Dublin, Ireland
boneill@compass.ie

Abstract. Since 2022, Ireland's National Transport Authority (NTA) have been developing web-based tools that are capable of undertaking connectivity assessments of settlements throughout Ireland. These tools have the opportunity to greatly enhance land use and transport planning decision making processes, and it is the intention of the NTA to introduce them as a standardised approach to connectivity analysis for use by Local Authorities and Government Bodies.

The tools highlight opportunities and constraints for sustainable modes within existing transport networks, while also offering the potential to explore improved options. The application of these tools in the development of Local Transport Plans assists in the improved design of urban spaces by bringing active modes and permeability to the forefront of the planning process.

These toolkits are now widely used amongst Local Authorities during development of their Local Transport Plans and have also attracted significant interest from Government Departments as they look to establish better ways to promote sustainable development methodologies.

Keywords: Connectivity · Sustainable Transport · Web based Assessment · Public Realm · Local Transport Plans

1 Introduction

The NTA has for many years recognized the importance of measuring connectivity to both Public Transport (PT) services and local opportunities and services. Until recently this analysis has been carried out on a relatively ad-hoc basis. The NTA came to realize that to gain the maximum benefit from this type of analysis, a standardized method was required; one that could be used by a wide range of organizations, and that had at its core centrally managed base data.

Transport for London's 'Assessing transport connectivity in London' [1] provided a sound basis for the methodology and the NTA have adapted two of the toolkits for the Irish market. The NTA wished to use these tools both for analysis within Metropolitan areas but also more widely, in smaller regional towns.

© The Author(s) 2025
C. McNally et al. (Eds.): TRAconference 2024, LNMOB, pp. 437–443, 2025.
https://doi.org/10.1007/978-3-031-85578-8_57

The first tool developed was the Public Transport Accessibility Level (PTAL) tool. This tool measures the distance to a local PT stop and scores that stop based on its walk distance combined with the level of service at the stop. Combining the results of this tool with additional land use data can highlight areas that are either in need of additional Public Transport services or would benefit from alternative land use zoning.

The second tool to be developed was ATOS (Access to Opportunities and Services). The ATOS tool measures an area's connectivity to services such as Employment, Education, Health Services, Retail and Open Spaces. It calculates the journey time from grid squares to local services. Each grid is then scored relative to the average journey time for all grid squares within the study area. The NTA have, in the first instance, concentrated on walking and cycling connectivity. As with the PTAL tool, combining the results of the ATOS tool with other data such as housing locations can quickly identify areas that are outside an acceptable walk distance to local services.

2   How the Tools Work

Both the PTAL and ATOS tools were initially developed as ArcGIS Python (ArcPy) script toolboxes and deployed within Esri's ArcMap and ArcPro desktop GIS applications, with later versions being deployed as ArcGIS geoprocessing services in ArcGIS Server.

2.1   Public Transport Accessibility Level (PTAL)

A PTAL is a measure of the accessibility of a location to the public transport network, considering walk access time and service availability. For any selected location, PTAL suggests how well that location is connected to public transport services. It does not consider accessibility by car. PTAL values are simple by design, ranging from zero to six, with the highest value (6b) representing the best connectivity (Fig. 1).

PTAL	Access Index range	Map colour
0 (worst)	0	
1a	0.01 – 2.50	
1b	2.51 – 5.0	
2	5.01 – 10.0	
3	10.01 – 15.0	
4	15.01 – 20.0	
5	20.01 – 25.0	
6a	25.01 – 40.0	
6b (best)	40.01+	

Fig. 1. PTAL score range.

In general a location will have a higher PTAL if it is a short walking distance to the nearest stops or Service Access Points (SAP); waiting times at the nearest stations or stops are short; more services pass at the nearest stations or stops; and here are major rail/bus stations nearby.

The NTA's PTAL methodology focuses on analysis of 100 m grid squares. Wait times, walk-times, access times, and service frequencies are calculated in each grid square. PTAL formulas are applied to generate Accessibility Indexes which are then summed to generate the final PTAL value for a grid square. The NTA's ArcGIS PTAL toolbox is comprised of two tools: Tool 1 and Tool 2 which are responsible for the import, transformation, analysis, and visualization of GTFS data, and subsequent calculation of PTAL results.

PTAL Analysis Results
The following image shows the results of a PTAL analysis for the greater Dublin metropolitan area. Hotter (red) areas are those with better relative access to public transport (Fig. 2).

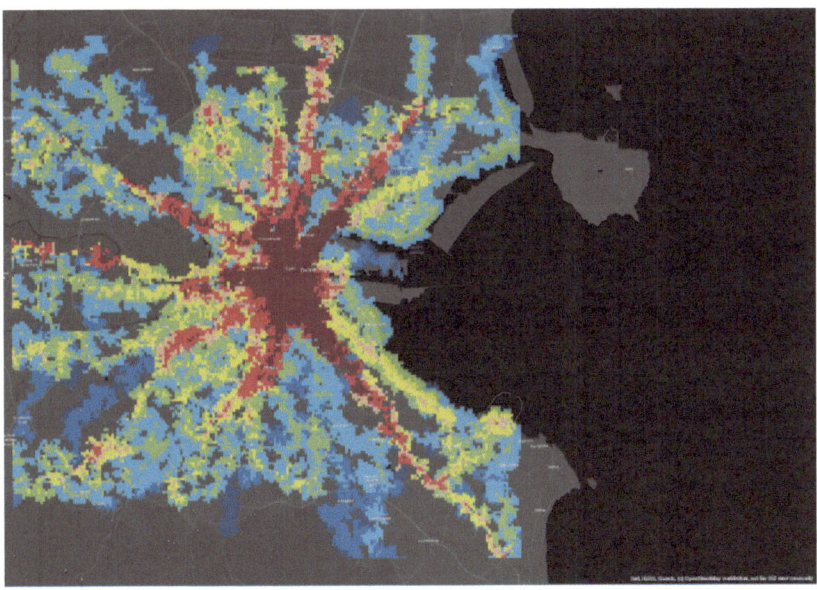

Fig. 2. Dublin PTAL analysis results.

2.2 Access to Opportunities and Services (ATOS)

Access to Opportunities and Services (ATOS) is a metric which serves as an indicator of a location's connectivity to essential key services and opportunities. ATOS scores range between A and E, with A representing the highest level of connectivity (Fig. 3).

The ATOS measure plays a crucial role in public transport planning by assisting in the identification of areas in need of connectivity improvements and guiding decisions on the optimal locations for introduction of essential services such as healthcare and education. This makes it an indispensable tool in the planning process.

ATOS Score	Map colour
A	
B	
C	
D	
E	
NULL	

Fig. 3. ATOS score range.

In general an area will have a more favourable ATOS score if there are many of a given service type within close proximity, and the transport network in the vicinity is dense and well developed since this should reduce travel time to service locations.

The NTA's ATOS methodology focuses on the analysis of 100 m grid squares. The ATOS scoring methodology differs from PTAL in that the score for an individual grid square is dependent upon how travel times to the nearest relevant destinations (for the specific type of service) compare to the average travel times across all selected grid squares. In this manner an ATOS score provides a realistic representation of the relative connectivity of a grid square to a given service within the entire selected grid area.

There are two distinct analysis types available within the ATOS toolbox:

1. The nearest {X} services within {Y} minutes travel time from origin point
   This analysis assesses the relative connectivity to a given number of services {X} within a pre-defined travel time {Y}.
2. Total number of jobs within {Y} minutes travel time from origin point.
   This analysis assesses the relative employment available within **{Y}** minutes travel time from a location.

ATOS Analysis Results

The following image shows the results of an ATOS open spaces analysis for the greater Dublin metropolitan area. Hotter (red) areas are those with better relative access to open spaces (Fig. 4).

3 How We Adapted them for Ireland

Both the ATOS and PTAL toolboxes are closely based upon the algorithms developed by Transport for London (TfL). However several adaptions to the TfL algorithms, along with additional innovations were added to the toolboxes to make them applicable to Irish Public Transport.

The PTAL toolbox enables automatic repair of repair of incorrect stop coordinates and route types; customized PTAL analyses limited to specific services, service dates, and service times; Service Proportional Weighting, a custom weighting used to highlight scoring differences when a service does not operate on all days within the "Mon-Fri" range; and customizable symbology.

Fig. 4. Dublin ATOS Open Spaces analysis results.

The ATOS toolbox focuses on walking and cycling; allows the user to customize the number of service types and travel time cut-offs for analysis; applies negative weightings to illustrate that not all {**X**} services were reachable within the allowed travel time; allows both positive and negative weightings to be applied if the travel times are greater or less than a specified travel time; and enables customizable symbology.

4   How We Made them On-Line Tools

To maximize interactivity and accessibility, and to enable local authorities and authorized third parties to execute their own ATOS and PTAL analyses, the toolboxes were converted to on-line tools by publishing the ArcGIS Python script tools to ArcGIS Server as geoprocessing services. Safe Software's FME Flow Server Apps are used to provide secure web-accessible user-interfaces for both the ATOS and PTAL analyses. The ATOS and PTAL toolboxes along with other components of the NTA's connectivity toolkit are made available via a secure NTA Data Analysis web-portal.

5   Applying the Tools to Local Transport Plans

Both toolkits are now widely used in developing Local Transport Plans (LTPs).

5.1   Using PTAL in Local Transport Plans

At the local level it is possible to use PTAL for more in-depth analysis. A poor PTAL score or a sudden change from a high to a low score could indicate where there is opportunity to improve permeability in the local network or where the frequency level could be improved. It is a relatively easy task to edit the underlying network to include a new walk link through a previously sealed off cul-de-sac, the PTAL tool is run again to see what impact the new walk link has had. For existing residential areas, this improvement can be converted to the population that is now within walking distance of PT.

Introducing a new PT route or improving existing services is often the outcome of a Local Transport Plan. The impact of these can be assessed using the PTAL calculation. A before and after analysis can show if the new services or routes have improved the connectivity in the area. The NTA are currently investigating the simplest way to incorporate new / test services into the PTAL calculation.

5.2 Using ATOS in Local Transport Plans

Like PTAL, ATOS is a useful tool for measuring the permeability of a walk or cycle network. Poor ATOS scores can indicate where permeability measures may help to improve overall connectivity. A walk network can be easily edited to demonstrate the before and after scenario. The images below show that making an existing informal walk link across some park land would open access to a larger area (Fig. 5).

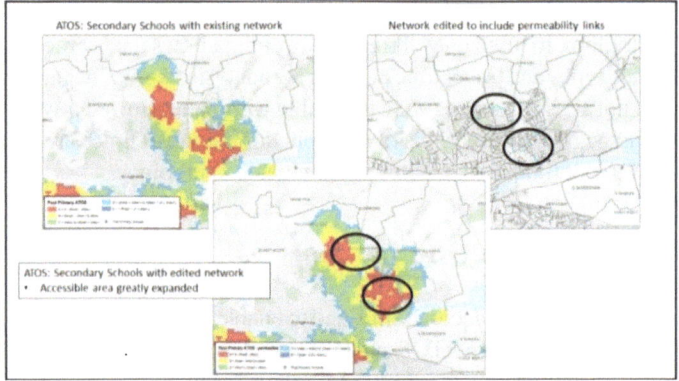

Fig. 5. Network changes improve an ATOS score.

Before and after scenarios can be carried out for the re-location of existing services, or to aid in the site selection for new services. The service location layer required by the ATOS calculation is a point feature class, so it is easy to edit. Calculations can be made using the existing network or if the site selection process requires it, the underlying network can be edited.

6 Conclusion

In every use case it is the intention to bring the integration of Land Use and Transport to the fore when approaching areas for development. Using a simple score such as PTAL or ATOS can be quickly and easily adopted and understood.

References

1. Transport for London: Assessing transport connectivity in London (2015)

Open Access This chapter is licensed under the terms of the Creative Commons Attribution 4.0 International License (http://creativecommons.org/licenses/by/4.0/), which permits use, sharing, adaptation, distribution and reproduction in any medium or format, as long as you give appropriate credit to the original author(s) and the source, provide a link to the Creative Commons license and indicate if changes were made.

The images or other third party material in this chapter are included in the chapter's Creative Commons license, unless indicated otherwise in a credit line to the material. If material is not included in the chapter's Creative Commons license and your intended use is not permitted by statutory regulation or exceeds the permitted use, you will need to obtain permission directly from the copyright holder.

Shared Solutions for Active, Collective, and Inclusive Neighbourhoods

Astrid Bjørgen[(⊠)], Marianne Ryghaug, Hampus Karlsson, and Claudia Moscoso

SINTEF Community, S.P. Andersens Veg 5, 7034 Trondheim, Norway
astrid.bjorgen@sintef.no

Abstract. Urban planning, either from a city or regional perspective, includes the study of land use, mobility, and transport, both of people, goods and services. In recent years, new shared service solutions have been considered to improve transport and mobility by reducing the share of privately owned vehicles. Shared solutions are also seen as a desired solution for urban planning in creating more sustainable neighbourhoods. This paper presents a study designed to map the relationship between urban planning approaches that aim to stimulate shared and active mobility and travel and shopping habits. The findings highlight that introducing several measures to stimulate more sustainable mobility had positive effects, such as reduced car ownership, and as such it should be considered when designing future urban spaces. The paper builds on a case study of Vestre Billingstad community close to Oslo, Norway.

Keywords: Mobility · Sharing · Urban planning · Urban space · Neighbourhoods

1   Introduction

Mobility entails the transport of people, goods, and services, and is a key component of spatial planning in communities, cities or regions. When the focus is to achieve sustainable and healthy neighbourhoods and cities, transport becomes even more important for urban planning, as road transport constitutes around 25% of greenhouse gas emissions in European cities and is a sector where emissions are still increasing [1]. Also transport infrastructure and mobility solutions demand space which is a scarce resource in urban environments. This has lately become an even more pertinent problem because of the huge increase in e-commerce and home delivery solutions spurring an increase of last mile deliveries and consequently the number of light duty vehicles [2, 3]. As a result, it has become more important to secure enough space for last mile deliveries. At the same time, shared space must be ´protected´ to provide good living areas. For instance, families with children want green areas, walkable communities and safe, high-quality corridors that can provide more freedom of movement facilitating active travel modes and simple everyday logistics [4]. This requires more focus on the relationship between mobility and urban design, environment, health and well-being [5, 6], more attention to combined and shared solutions for both people and goods [7, 8], and better last mile solutions for e-commerce and home delivery.

© The Author(s) 2025
C. McNally et al. (Eds.): TRAconference 2024, LNMOB, pp. 444–449, 2025.
https://doi.org/10.1007/978-3-031-85578-8_58

When planning urban areas, local municipalities are responsible for the overall development through public planning, such as the Planning and Building Act [9]. Municipalities, do however, also have responsibility for reaching different political goals such as climate emission reductions and promoting healthy environments. This means that municipalities often have the difficult task of weighing different concerns and goals against each other. Previous studies have shown that different planning approaches shape the integration of urban freight transport in city planning, and that early integration of stakeholders is key to secure proper urban mobility and logistics [10]. A more efficient use of urban spaces can be achieved through integrated planning, dynamic regulation, sharing [11] and co-use of space [12]. Not surprisingly, in cities that have designed for and invested in infrastructure for active transportation such as cycling, there is also a higher probability that more people cycle [13]. This demonstrates that infrastructure and urban planning can be powerful tools that, if properly considered, may lead to multiple benefits such as more liveable cities [5, 10].

This study aims to provide new knowledge about the role of urban planning in designing sustainable, healthy, and desirable urban environments. The study pays particular attention to the importance of including mobility and logistics planning as part of spatial and urban development, which opens more shared solutions and services. The paper thereby contributes to the literature on shared mobility solutions in urban planning, discussing strategies for developing sharing solutions with neighbours, increased coordination between public and private stakeholders and co-creation with citizens.

2 Method

The analysis is based on a case study of a community called Vestre Billingstad (VB), located in Asker municipality, close to Oslo, the capitol in Norway. The VB area is to be developed over the next 10–15 years and will have more than 1600 homes, including kindergartens and other services when completed in 2035. In 2020 the first residents moved in, and so far in 2023, 400 homes are settled. The study was designed to explore how urban planning may impact the possibilities for developing shared mobility solutions, and how these may affect travel and shopping habits. To this end, a document analysis, interviews with representatives from the municipality, developers and mobility providers, and a web-survey with VB inhabitants were conducted.

The survey was carried out in March 2023 at the end of the work week to cover the travel and shopping habits of respondents from the last three days. A total of 158 respondents (23–83 years old) answered the survey. The majority (86%) live in apartments, while the rest live in rowhouses. The households consist of couples without children, individuals living alone, and couples with one child. Most of the respondents reported to have a driver's license, to own their private car and to have access to sharing cars and bikes.

The survey was divided into seven sections: demographics; travel and shopping habits; motivation for moving to VB; experience with services offered at VB; changes in everyday life related to moving to VB; and other. The survey questions are available from the authors upon request. In this paper we focus on the way urban planning may impact the possibilities for developing shared mobility solutions. The objective of the study is two-fold: i) Explore how the sharing mobility service and restricted parking access have contributed to reducing car ownership; and ii) explore how the possibilities for safe bicycle parking and infrastructure for walking and cycling affect residents' choice of means of transport.

3 Results

3.1 Planning Tools to Achieve Intentions and Innovations

There are two key components that proved important of the choices taken and solutions implemented, while maintaining a long-term perspective for the development of VB: i) the joint area plan and ii) the development agreement between the developers and municipality. The joint area plan for VB was initiated by the municipality due to developers submitting different proposals on how to develop their own areas. Due to limited resources, the municipality allowed developers to prepare the joint area plan, as long as the municipality followed the process and could decide the guidelines for the area. As a result, the four developers used two planning consultants to prepare the area plan which included the overarching principles of how VB should be developed with regards to land use, mobility and sharing solutions. The area plan cut across ownership to land use structures and has common requirements for traffic. This has helped to ensure a comprehensive planning of the area throughout the development period. From a developer's perspective, the area plan provided clear conditions in the early phase, while at the same time enabling the development to be realised more quickly. The development agreement, which was legally binding, was initiated by the developers and worked to secure and shed light on the municipality's intentions for VB and to give the developers clear requirements to adhere to in the planning phase. Standing together as a developer group reduced the risk of being innovative, such as trying out new mobility services. Additionally, the developer group created trust in the process of consensus building.

In sum, the requirements stated in both documents put limitations on what is allowed to build and what should be offered to the residents. The requirements included parking restrictions, limits to private car movement within the area and provisioning of shared mobility solutions and infrastructures for active travel. For instance, the apartments did not come with parking lots. Parking had to be purchased in addition to the apartment, to make the cost visible. The parking norm only allowed for owning one parking space per apartment which is stricter than the usual norm for this kind of community, and a location for a self-served pick-up point for home delivery service was included. These measures were implemented to encourage reduced private car ownership and traffic, while at the same time providing mobility alternatives such as car- and bike sharing. Also, safe and attractive infrastructures for walking and cycling were prioritised.

3.2 Mobility Services, Car Ownership and Travel and Shopping Habits

The respondents were asked to report the number of trips made to and from home in the last three days and their main modes of transport for different purposes such as work, leisure, and others. The results show that 32% of the respondents reported to have driven a private car, while 30% of the respondents walked to and from home and 14% used public transport. The rest of the responses were distributed between other means of transport, e.g. bicycle and train. Only 5% of the respondents reported to have used a car sharing service, either as a driver or as a passenger, which demonstrates that the available car sharing service seemed to have a small impact on the travel habits of the residents. At the time of being surveyed, 7% of the respondents reported to have reduced their car ownership and an additional 2,5% were considering selling their car. It is however, interesting to note that private car ownership is lower for VB than the average for the surrounding municipality. Despite of better facilities for walking and cycling, only a small portion of the respondents reported to walk or bike more often. Although these numbers might appear small, they still show positive effects in the right direction, i.e., a group of people walk and/or use their bicycle more often, and there is a small (but positive) reduction of car ownership.

The survey showed that the access to a local self-service pick-up point did, according to the respondents, reduce the use of home-delivery, pick-up in store and "click-and collect" service. But at the same time, contributed to a slight increase in online shopping. This is also in line with the analysis done by the freight forwarder responsible for the pick-up point, both self-service and in store. With deliveries of online shopping to the self-service pick up point there is of course, some transport associated. However, mapping of these movements demonstrated that the number of trips was very low. Thus, VB seem to go against the negative trends discussed in the introduction with a large increase in last mile and light duty vehicles impacting the residential area.

4   Discussion and Conclusion

The study focuses on examining how urban planning affects the possibility of developing shared mobility solutions and active travel modes. In addition, how such shared mobility solutions can affect travel and shopping habits of residents in a newly developed urban residential area in Norway. The results suggest that to support greener and more attractive neighbourhoods, three actions are crucial: i) implementation of planning approaches that prioritize shared solutions, ii) measures targeting private car ownership and use, and iii) implementation of enhanced active mode infrastructure. Specifically, the results showed that:

- The area designated to parking influenced how citizens owned and used cars and bicycles.
- Internal pedestrian and bicycle paths contributed positively to increase the proportion of trips by foot and/or bicycle.
- Residents were satisfied with the access to shared car services.

Given the restrictions on car use and car ownership in VB, a 7% reduction of car ownership might seem meagre. However, when compared to the general trend in Norway this is a rather impressive result, as private car ownership has increased the last years (87% of Norwegian households own a private car [14]). Yet, the fact that the VB area probably attracted residents with different characteristics compared to the population should be taken into consideration when discussing net-effects of each measure. For instance, if the residents of VB already had lower car ownership, biked, and walked more than the whole population, and perhaps was attracted to the area exactly because VB fitted their preferences on these issues, effects will tend to be lower. As we do not know the contrafactual here, this should still be considered as a potential explanation of why numbers may seem low, thus warranting more research. In general, these are promising findings that suggest that urban planning related to transport and mobility can stimulate more active and shared mobility and reduce car ownership.

Furthermore, the fact that strategic use of pick-up points seemed to reduce traffic in the area, is an interesting finding. The role of shared mobility solutions may also warrant further research, as it might have played a larger role than what the reported numbers show. Car sharing solutions may play a role as a safety net solution for those who do not have a car or choose to reduce the number of cars in their household and may have impacted those that now consider selling their cars [15]. This may be seen similar to how fast charging was important for the early adoption of EVs in Norway [16]. To have a better picture of the implications of the car sharing solutions, it will be important to study how having access to shared solutions impact practices and attitudes towards car ownership in the long run.

Shaping spaces according to local needs and supporting communities by rethinking the use of urban areas may be low-hanging fruits in practical planning. This applies in particular to municipal sub-plans, zoning plans and street use associated with downtown development, to ensure flexibility in land use and solutions that can satisfy multiple purposes and multiple user groups. Based on the analysis of this paper we conclude that to innovate with regards to the planning processes is in itself a tool that warrants further focus when striving to develop sustainable and inclusive neighbourhoods. New ways of planning may also serve to build consensus towards increased area flexibility where the needs of the residents may change. In this project, forging a public-private cooperation to produce a joint area plan and developers' agreement, facilitated new ways of implementing shared mobility solutions and the encouragement of active mobility and reduced car ownership. This resulted in some smaller, albeit positive, impacts, which suggests that one could use similar planning approaches in future area development. However, to have larger impacts, stricter regulations of car ownership and more promotion of shared mobility solutions is probably needed to create sustainable neighbourhoods.

Acknowledgement. The authors wish to acknowledge the research project Spatial Planning and Mobility at Vestre Billingstad (project number 332772), funded by the Regional Research Viken (2020–2022).

References

1. Statista. Distribution of carbon dioxide emissions in the European Union in 2021, by key source [Internet]. Statista (2021). https://www.statista.com/statistics/999398/carbon-dioxide-emissions-sources-european-union-eu/. Accessed 2 Oct 2023
2. Allen, J., et al.: Understanding the impact of e-commerce on last-mile light goods vehicle activity in urban areas: the case of London. Transport. Res. Part D: Transp. Environ. **61**(Part B), 325–338 (2018)

3. Bjørgen, A., Bjerkan, K.Y., Hjelkrem, O.A.: E-groceries: sustainable last mile distribution in city planning. Res. Transport. Econ. **87** (2021)
4. Bø, L.A., et al.: Barn i byen. Gode oppvekstmiljøer for barn i sentrale bydeler i Trondheim [Internet]. SINTEF, Trondheim, Norway (2023). https://hdl.handle.net/11250/3046984
5. Nieuwenhuijsen, M.J.: Urban and transport planning, environmental exposures and heath-new concepts, methods and tools to improve health in cities. Environ. Health **15**, 161–171 (2016)
6. Nieuwenhuijsen, M.J.: Urban and transport planning pathways to carbon neutral, liveable and healthy cities: a review of the current evidence. Environ. Int. **140**, 105661 (2020)
7. Lyons, G., Hammond, P., Mackay, K.: The importance of user persective in the evolution of MaaS. Transport. Res. Part A: Policy Pract. **121**, 22–36 (2019)
8. Riemens, R., Nast, C., Pelzer, P., van den Hurk, M.: An assessment framework for safeguarding public values on mobility platforms. Urban Transf. **3**(1), 1–26 (2021)
9. Kommunal- og distriktsdepartementet. Lov om planlegging og byggesaksbehandling (Plan-og bygningsloven). LOV 2008–06–27 nr 71 Jun 27 (2008). https://lovdata.no/dokument/NL/lov/2008-06-27-71
10. Bjørgen, A., Ryghaug, M.: Integration of urban freight transport in city planning: lessons learnd. Transport. Res. Part D: Transp. Environ. **107**, 103310 (2022)
11. Rokseth, L.S., Heinen, E., Hauglin, E.A., Nordström, T., Manum, B.: Reducing private car demand, fact or fiction? a study mapping changes in accesibility to grocery stores in Norway. Eur. Transp. Res. Rev. **13**(39) (2021)
12. Baker, D., et al.: Urban freight logistics and land use planning education: trends and gaps though the lens of literature. Transport. Res. Interdisc. Perspect. **17**, 100731 (2023)
13. Lunke, E.B., Aarhaug, J., De Jong, T., Fyhri, A.: Cycling in Oslo, Bergen, Stavanger and Trondheim, p. 122. Institute of Transport Economics, Oslo, Norway. Report No.: TØI report 1667/2018 (2018)
14. Wangsness PB, Ciccone A, Nenseth V. To what degree can carsharing substitute car ownership? Self-reported evidence from carsharing users and the general population [Internet], p. 57. Institute of Transport Economics, Oslo, Norway. Report No.: 1940/2023 (2023). https://www.toi.no/getfile.php?mmfileid=75535
15. Grue B, Landa-Mata I, Flotve BL. Den nasjonale reisevaneundersøkelsen 2018/19. Institute of Transport Economics. Report No.: 1835/2021 (2021). https://www.toi.no/getfile.php?mmfileid=71405
16. Ingeborgrud, L., Ryghaug, M.: The role of practical, cognitive and symbolic factors in the successful implementation of battery electric vehicles in Norway. Transport. Res. Part A: Policy Pract. **130**, 507–516 (2019)

Open Access This chapter is licensed under the terms of the Creative Commons Attribution 4.0 International License (http://creativecommons.org/licenses/by/4.0/), which permits use, sharing, adaptation, distribution and reproduction in any medium or format, as long as you give appropriate credit to the original author(s) and the source, provide a link to the Creative Commons license and indicate if changes were made.

The images or other third party material in this chapter are included in the chapter's Creative Commons license, unless indicated otherwise in a credit line to the material. If material is not included in the chapter's Creative Commons license and your intended use is not permitted by statutory regulation or exceeds the permitted use, you will need to obtain permission directly from the copyright holder.

A Case Study for Selecting Suitable Mobility Measures for Workplaces

Domokos Esztergár-Kiss$^{(\boxtimes)}$ and Conrado Braga Zagabria

Budapest University of Technology and Economics, Budapest, Hungary
esztergar@mail.bme.hu

Abstract. There is a change from traditional planning approaches providing a shift toward more sustainable transport options. To support these policies, Workplace Travel Plans are applied that encourage the usage of active transport modes and sustainable behavior change. This research considers such mobility plans, where our approach provides a framework of connecting employee requirements, employer willingness, and site-specific options resulting in a list of suitable measures for a specific workplace. The utility of the measure is a value calculated for each measure, which enables the ranking of all measures. The case study was conducted an institution based in Budapest, Hungary. As a result, the most highly ranked measures are related to active modes, while the other top-ranked measures are related to different strategies, such as carpooling walking, and the traffic calming.

Keywords: mobility planning · demand management · travel behavior

1 Introduction

Sustainable urban mobility has been receiving increased attention, and cities are trying to reduce car traffic and increase the share of sustainable transport modes [1]. Usually, hard measures are a typical a solution for problems, which require considerable investments in infrastructure. However, soft measures use information provision, best practice dissemination, and persuasion to support sustainable transport modes [2]. This represents a change from traditional planning approaches providing a shift toward more sustainable transport options [3].

The journey for work is often in the focus of attention since it accounts for a significant proportion of all traffic. On the one hand, commuters are interested in reducing the negative aspects of their commute, but at the same time, employers are challenged to improve the workplace environment and employee wellbeing [4]. To support these policies, Transportation Demand Management strategies can be applied that are connected to Workplace Travel Plans, which encourage the usage of active transport modes and sustainable behavior change [5].

This research considers the importance of creating workplace mobility plans to encourage sustainable employee commuting behavior change by introducing a method for selecting and ranking the most efficient measures for specific locations.

© The Author(s) 2025

C. McNally et al. (Eds.): TRAconference 2024, LNMOB, pp. 450–456, 2025.
https://doi.org/10.1007/978-3-031-85578-8_59

2   Literature Review

Workplace Travel Plans provide a package of measures implemented by an organization to encourage commuters to use more sustainable transport modes [6]. They have to include several stakeholders to assess the problems of mobility planning. Several articles provide an analysis of travel plans. Sprumont et al. [7] assessed the workplace commuting pattern in Luxembourg and analyzed the impacts of office relocation and the effectiveness of measures in peripheral workplace location. Petrunoff et al. [8] evaluated the effects of a three-year workplace travel plan intervention on increasing active travel modes. They concluded that a plan which only included strategies to encourage active travel to work achieved a small but consistent increase in active trips during the implementation period. Cairns et al. [9] analyzing 20 workplace travel plans in the UK, concluded that considerable behavioral change could be achieved in a variety of contexts. However, employers usually need an overall strategy that addresses car parking, in addition to improving alternative travel modes. Vanoutrive et al. [10] conducted an assessment on the situation in Belgium about the implementation of employee mobility management measures. They indicated that employers regularly choose to implement a set of sustainable commuting measures. Petrunoff et al. [11] compared two travel plans and concluded that a plan including parking management complementing strategies to encourage active forms of travel is more effective than a travel plan that includes only encouragement strategies.

Measures in the context of Workplace Travel Plans are the interventions or actions performed by the employer to reach a goal of encouraging employees to travel differently. Measures can be wide-ranging, examples include bicycle parking, discounted public transport tickets, car parking supply restrictions, car-sharing facilities, and customized information on local transport options [12]. Our approach provides a framework of employee requirements, employer willingness, and site-specific options, resulting in a list of measures for a specific workplace.

3   Method

A quantitative method has been developed that analyses a range of sustainable workplace travel plan measures and ranks them based on a set of collected data. The method assesses the employee mobility patterns and preferences, the employer goals and preferences, and features of the physical workplace environment to calculate which measures would be most suitable for the specific workplace.

The first step of the process was to collect relevant measures cited in the literature, suggested by guidelines, mentioned in whitepapers, used in real developments, and measures related to new and emerging transport modes. The identification of measures includes the creation of an identification number, a measure name, and a measure description for every measure.

The categorization of measures contributes to the next phase of the method. The goal of the categorization process is to increase the amount of information about the individual measures and to provide the opportunity of filtering them.

- Mode type: is the primary transport mode, related to or encouraged by the measure. One measure can relate to more than one mode at the same time.
- Strategy cluster: is about how the measure tackles a problem or how it encourages sustainable commuting. The strategy clusters relate to the goal of the measure.
- Approach: relates to the method of action and can also be named 'measure type'. The approach groups the measures that have similar methods of implementation.
- Financial demand: is an overview of the financial aspects of the measure implementation. It considers the cost and investment aspects.
- Time frame: is an estimation of how long the implementation phase lasts. This category does not consider the time of approval of a measure, only execution time.

The Measure Sustainability Index is a tool created to evaluate the measures based on their sustainability impact. It uses indicators of three sustainability domains.

- Environmental domain: This indicator evaluates the effect on the reduction of emissions. For this reason, a scale was designed, where the measures are ranked based on their potential for emissions savings.
- Social domain: Wellbeing is a multidimensional concept that may be measured both objectively and subjectively. Subjective wellbeing is aimed at capturing wellbeing perceived by individuals.
- Economic domain: This indicator was created to evaluate the economic sustainability of measures. The financial factor relates to how the measure is economically sustainable to the employee as well as the employer.

The mobility questionnaires are the instruments used to retrieve the input data used in the method aiming the biggest influencers.

- Employer Questionnaire: The survey applied to the employer is intended to assess organization policies and practices affecting staff travel behavior. It also aims to collect employer preferences over modes and strategies and understand its goals.
- Site Audit: The questionnaire applied to the site of the workplace aims to look for constraints, barriers, and opportunities for measure implementation. It is an evaluation of the physical aspects of the worksite, both inside and in its surroundings.
- Employee Questionnaire: The survey applied to the employee is the basis of a travel plan. It provides numerical data on current travel patterns of staff, reasons why they travel the way they do, and what would be needed to travel differently.

All mobility questionnaires have a connection with the measures. The process starts with three factor tables of the three questionnaires and the measures, where initially all the factors receive the value of 1. This means that all measures are enabled for implementation unless the connection changes it. The next step is to find which measures have connections to which questions and define the factors. If the measure already exists in terms of infrastructure or policy, it does not need extra support.

The utility of the measure is a value calculated for each measure, which enables the ranking of all measures. It is composed of two parts: the mobility questionnaires calculation and the Measure Sustainability Index calculation. The simple average of the utility values provides a final utility value.

4 Results

The case study was conducted at KTI Institute of Transport Science, which is a non-profit, state-owned institution based in Budapest, Hungary, focused on the single worksite where around 500 employees work. The research started with data collection from the employer by conducting the employer survey and the site audit. The second phase was dedicated to collecting commuting data from the employees.

The Employer Questionnaire addresses policies in place regarding the daily commuting of employees. KTI currently does not have a strict set of policies to encourage commuting behavior, although it has noticed policies concerning public transport and strategies of reducing the need for travel. In terms of the Site Audit, KTI office stands next to a transportation hub, well located in terms of public transport. The workplace has a private car parking facility, and public car parking places are available around the worksite. The workplace has sufficient showers and lockers, as well as a bike parking place. The external cyclist infrastructure is well assessed with bike routes. However, the pedestrian infrastructure surrounding the workplace is considered weak.

The results of the Employee Questionnaire provide an analysis of current commuting patterns. KTI modal share is composed of 75% of sustainable commuting modes. Public transport represents 47%, followed by car (25%) and walking (13%), while active modes have a combined rate of 19%. The share of active modes can be further increased as stated by the employer as one of the preferences.

Figure 1 explores the flexibility regarding working hours. In the employer questionnaire, it was stated that flexible times are in place as a measure. It relates to the arrival and departure hours, which are flexible as long as the weekly work hours, are filled. The high adoption is expressed by 78% of the commuters who use it. Staggered shifts are when the starting and finishing hours are set differently for different days, and it differs from flexible times where staggered shifts have a fixed amount of hours per day. Almost half of the employees possess staggered shifts, but the other 40% do not have and would not like to have. These two approaches are useful to fight the congestion in peak hours both in the city center and in the surroundings of the workplace. Compressed workweeks promote working more hours than a typical day but work fewer days a week, which is an excellent strategy to reduce the need for travel. This measure, which most of the employees do not have, has the highest rejection rate (46%) but still a reasonably high demand rate (26%).

Figure 2 addresses the reasons behind the choice of each transport mode. Although a stratification by mode would help understand reasons for commuter, the chart shows the less travel time (25%) and more comfort (22%) as the leading options.

When selecting and ranking the most suitable measures for the workplace, the utility values were calculated. As a result, top-ranked measures are presented in Table 1. The highly ranked measures are related to active modes. The first three measures of supporting bicycle purchase, reimbursement of cycle mileage, and providing accessories have financial incentives in their composition. In contrast, the fourth works with a program-based approach by creating bicycle commuter groups, and the fifth is an improvement of the infrastructure in place, where a bike repair center is installed. The other top-ranked measures are related to different strategies. Measures providing financial benefits for Park

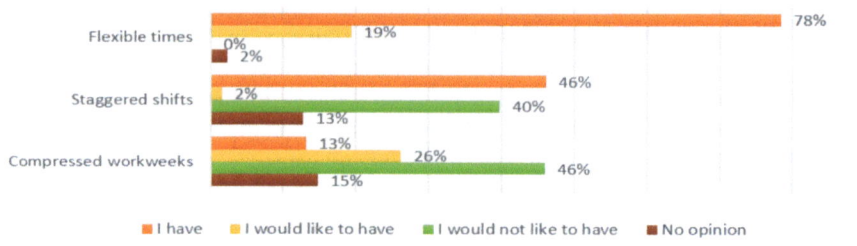

Fig. 1. Flexibility of work hours.

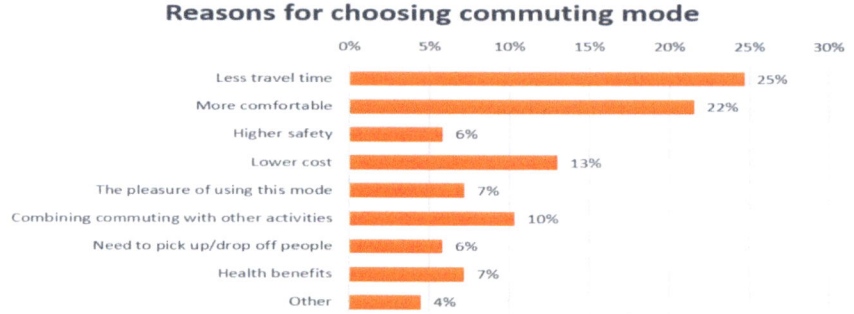

Fig. 2. Reasons for choosing the commute mode.

and Ride and designated parking spaces for carpooling are related to car and public transport. Finally, the measure providing new employees an induction kit is an information approach measure designated to transport modes, while creating a walking commuter group relates directly to walking, and the traffic calming measure relates to active modes, as it encourages both walking and cycling.

Table 1. Top-ranked measures for the case study.

Ranking	Measure
1	Free or low-interest loans for employees' bicycle purchase
2	Reimbursement of cycle mileage for commuting trips made by bicycle
3	Financial contributions for cycling-related accessories and services
4	Create a bicycle commuter/user group to advocate for cyclists
5	Install a bike repair center or partner with nearby bike services shop
6	Create a walking commuters' group

(*continued*)

Table 1. (*continued*)

Ranking	Measure
7	New employee induction kit
8	Financial benefits for Park and Ride
9	Implement traffic calming measures in the surroundings of the workplace
10	Designated parking spaces for carpooling vehicles in the parking garage

5   Conclusion

The goal of the elaborated method was to support the selection and ranking of measures of Workplace Travel Plan. The measures were collected and categorized. The list of measures and the mobility questionnaires created the basis of the development, while the elaborated concept provided a score for the measures. Considering the impacts of mobility surveys and sustainability impacts, the measures could be ranked.

References

1. Hickman, R., Hall, P., Banister, D.: Planning more for sustainable mobility. J. Transp. Geogr. **33**, 210–219 (2013). https://doi.org/10.1016/j.jtrangeo.2013.07.004
2. Bamberg, S., Fujii, S., Friman, M., Gärling, T.: Behaviour theory and soft transport policy measures. Transp. Policy **18**(1), 228–235 (2011). https://doi.org/10.1016/j.tranpol.2010.08.006
3. Arsenio, E., Martens, K., Di Ciommo, F.: Sustainable urban mobility plans: bridging climate change and equity targets? Res. Transp. Econ. **55**, 30–39 (2016). https://doi.org/10.1016/j.retrec.2016.04.008
4. Chatterjee, K., et al.: Commuting and wellbeing: a critical overview of the literature with implications for policy and future research. Transp. Rev. **40**(1), 5–34 (2019). https://doi.org/10.1080/01441647.2019.1649317
5. Enoch, M.: Sustainable transport, mobility management and travel plans. Routledge, London (2016). ISBN: 9781315611563. https://doi.org/10.4324/9781315611563
6. Esztergár-Kiss, D., Tettamanti, T.: Stakeholder engagement in mobility planning, autonomous vehicles and future mobility. Elsevier, Amsterdam (2019). ISBN:9780128176962. https://doi.org/10.1016/B978-0-12-817696-2.00009-3
7. Sprumont, F., Viti, F., Caruso, G., König, A.: Workplace relocation and mobility changes in a transnational metropolitan area: the case of the university of Luxembourg. Transport. Res. Procedia **4**, 286–299 (2014). https://doi.org/10.1016/j.trpro.2014.11.022
8. Petrunoff, N., Wen, L.M., Rissel, C.: Effects of a workplace travel plan intervention encouraging active travel to work: outcomes from a three-year time-series study. Public Health **135**, 38–47 (2016). https://doi.org/10.1016/j.puhe.2016.02.012
9. Cairns, S., Newson, C., Davis, A.: Understanding successful workplace travel initiatives in the UK. Transport. Res. Part A: Policy Pract. **44**(7), 473–494 (2010). https://doi.org/10.1016/j.tra.2010.03.010
10. Vanoutrive, T., Van Malderen, L., Jourquin, B., Thomas, I., Verhetsel, A., Witlox, F.: Mobility management measures by employers: overview and exploratory analysis for Belgium. Eur. J. Transp. Infrastruct. Res. **10**(2), 121–141 (2010). https://doi.org/10.18757/ejtir.2010.10.2.2878

11. Petrunoff, N., Rissel, C., Wen, L.M., Martin, J.: Carrots and sticks vs carrots: comparing approaches to workplace travel plans using disincentives for driving and incentives for active travel. J. Transp. Health **2**(4), 563–567 (2015). https://doi.org/10.1016/j.jth.2015.06.007
12. Gruyter, C.D., Rose, G., Currie, G., Rye, T., van de Graaf, E.: Travel plans for new developments: a global review. Transp. Rev. **38**(2), 142–161 (2018). https://doi.org/10.1080/01441647.2017.1322643

Open Access This chapter is licensed under the terms of the Creative Commons Attribution 4.0 International License (http://creativecommons.org/licenses/by/4.0/), which permits use, sharing, adaptation, distribution and reproduction in any medium or format, as long as you give appropriate credit to the original author(s) and the source, provide a link to the Creative Commons license and indicate if changes were made.

The images or other third party material in this chapter are included in the chapter's Creative Commons license, unless indicated otherwise in a credit line to the material. If material is not included in the chapter's Creative Commons license and your intended use is not permitted by statutory regulation or exceeds the permitted use, you will need to obtain permission directly from the copyright holder.

Understanding Mobility in Sparsely Populated Areas: Between Vulnerability and Asserted Choice?

Sophie Hasiak$^{(\boxtimes)}$ ⓘ and Pénélope Mazari

MATRiS Laboratory, Cerema Hauts-de-France, 59019 Lille, France
`sophie.hasiak@cerema.fr`

Abstract. This paper aims to provide an in-depth knowledge of mobility in sparsely populated areas based on a quantitative and statistical approach. First, it analyses mobility practices in sparsely populated areas and compares them with those in urban areas. It then proposes an assessment of car dependency of population living in rural areas. It points out that this dependency results from either a constraint or an asserted choice, depending on existing alternatives of car use. It also analyses the degree of dependency by having a focus on multimodal practices of people living in these areas. It then concludes with the challenge of shifting populations living in sparsely populated areas from constraint mobility to chosen mobility, by offering mobility services adapted to their needs and, above all, by encouraging them to consider new practices that go beyond their routines.

Keywords: Mobility · sparsely populated areas · car dependency

1 Background and Challenges

Mobility in sparsely populated areas is at the heart of public policy concerns. It represents a major societal issue that is part of climate emergency and ecological challenges. France has been particularly involved since the social movement of the "Yellow Vests" at the end of 2018, a social protest linked to the refusal of an increase in fuel taxation [1, 2]. This challenge has been amplified by the current economic and energy crisis. Public stakeholders have acted the specificities of mobility policies of these territories, given both their mobility vulnerability due to a higher car dependency of population [1] and their environmental and energy vulnerability because of the lack of credible mobility alternatives to solo-driving. Considering the challenge of decarbonized mobility policies, there is an urgent need for an ecological transition in these territories [3]. However, this will be difficult because the economic and social cost of this transition is particularly high and mainly borne by households [4].

In France as in other European countries, public policy stakeholders are mobilized to propose more sustainable mobility services adapted to these areas [5–7]. The scientific community is also working on the evaluation of local initiatives to set up mobility services in sparsely populated areas [1, 8]. To achieve this, stakeholders need to rely

© The Author(s) 2025
C. McNally et al. (Eds.): TRAconference 2024, LNMOB, pp. 457–463, 2025.
https://doi.org/10.1007/978-3-031-85578-8_60

on a better understanding of mobility needs and current mobility practices. However, this knowledge remains partial. The key question is then how to characterize the car dependency of people living in sparsely populated areas? This is the main issue that this paper attempts to answer.

2   Literature Review

Even if residents of sparsely populated areas have the same day-to-day activities program (work, shopping, leisure) as those living in densely areas, there are some specificities in their mobility: longer travel distances due to both the dispersion of housing and the distance from services and employment, poor public transport services due to low density, and predominance of car use [3]. There is a consensus to speak about a population that is captive to the car for its daily journeys, about an involuntary or constrained mobility [3] and, more broadly, about a car dependency [3, 8–11]. The car then appears to be the easiest mode to travel in these territories, and sometimes the only one that exists or is practicable [12]. This feeling is widely expressed by people living in rural areas [2, 3]. As a result, the car ownership rate is among the highest [8, 13], and car use is more akin to solo-driving use [1].

The French scientific literature therefore mainly highlights car practices in rural areas [11]. However, this finding is primarily based on national statistical data from the population census, which tends to limit the assessment to motorization rates and car use for only home-work journeys. This is due to the real difficulty of setting up statistical databases in rural areas, leading research work to rely on qualitative surveys [8, 14]. Although these latter enable researchers to identify major trends, the lack of quantified data and the challenges of these areas encourage research work that will provide a better understanding of mobility needs in rural areas [15].

3   Methodology

3.1   Definition of Sparsely Populated Areas

The term of sparsely populated or rural areas covers a range of geographical realities, from peripheral areas of metropolises or main conurbations, to isolated rural areas or villages [3]. There is no precise definition [7], which means that different methods are used to define them [3, 10, 11, 16, 17]. We have opted for the objective definition of the French statistics institute (INSEE), which characterizes the different degrees of urbanization of the French territory based on the result of European studies [18]. It identifies two classes of sparsely populated territories, the sparsely populated municipalities and the very sparsely ones, which differ from the proportion of half of the population living in communes with a density of less than 300 inhabitants or 25 inhabitants per km2. Considering this definition, 90% of all French municipalities are rural ones. They gather 35% of the French population, above the European average rate (24%).

3.2   Specific Use of Two Databases

Our work is based on some analyses of two databases on daily mobility. The first database corresponds to the gathering of different household travel surveys lead in France. These surveys, called "EMC2 surveys" and based on a national methodology, collect weekday journeys made by people the day before the survey. The database brings together 69 surveys carried out in large and medium-sized conurbations between 2011 and 2022. It covers 24% of French sparsely populated communes and correspond to a sample of 211,234 people surveyed. The database totals about 1765,000 gross trips, of which 26% are made by people living in sparsely populated areas. The second database is the national personal mobility survey, called the "EMP survey", conducted by the French government in 2019. It records both the daily journeys made by people and longer-distance ones. Its sample is composed of 13,825 individuals surveyed, including 4,842 people living in low-density areas. This represents a total of 36,088 gross weekday trips, including 12,687 made in low-density areas. The results of these two databases are not directly comparable, given the size of the sample of sparsely populated communes and the different modalities of some of the data collected. As a consequence, we invite any reader to consider the results as orders of magnitude.

4   Main Results

4.1   Differences in Mobility Practices in Sparsely Populated Areas?

There is a same level of mobility on a weekday. Whether living in a sparsely or densely populated area, an individual makes the same number of average daily trips, more than 3 trips per day. Overall, he almost spends the same time travel budget (between 83 and 87 min). However, trips by residents of sparsely populated areas are longer, averaging 11.8 km per trip versus 7 km in densely populated areas. Moreover, the rate of immobility is higher in sparsely populated areas. It reaches between 11% and 16%, depending on the day and the survey, compared with 10% to 11.1% in densely populated areas.

Moreover, there is no difference in the hierarchy of transport modes use. We can say that density of areas doesn't alter the hierarchy. Whatever the type of area, the car remains the primary mode of transport, walking the second and public transport the third. However, it is the intensity of use that differs. The modal share of the car is much higher in sparsely populated areas than in densely ones (see Table 1). This greater car use in rural areas leads to less use of walking and public transport.

Table 1. Main modal shares according to the two databases (EMC2 base/EMP base).

Mode	Dense area	Intermediate-density area	Low-density area
Car	44.9/43.2	66.0/69.4	71.4/77.2
Walking	35.2/33.2	24.1/21.5	19.2/14.2
Public transport	15.2/17.5	6.1/5.9	5.3/4.1
Cycling	2.3/3.4	1.6/2.2	1.3/2.5

Nota: the first number relates to the result of EMC2 base/the second one of the EMP base.

4.2   An Asserted or Chosen Car Dependency?

The scientific literature points out a car dependency of people living in rural areas. Does this notion of dependence indirectly conceal the behavior of individuals who systematically take the car without thinking about using other transport modes such as cycling and walking for short trips? This refers to a reflex or non-reflexive use [8], especially in a context where it's easy to travel by car due to the fluidity of traffic [10]. Is this dependency chosen or constrained?

We consider here the average trip, which is around 12 km in sparsely populated areas. This distance rules out the alternative of walking. Bicycles, especially electric ones, could be a possible alternative, subject to safe development of country roads. We therefore focus on the public transport alternative, in order to measure the type of dependence in relation to an average journey. The literature highlights the fact that low-density areas are less well covered and served by public transport. Only 25% of low-density communes are part of an urban transport perimeter of large or medium-sized agglomerations with some public transport services. Nearly 40% of sparsely populated communes are part of small intermunicipal structures that are still thinking about setting up mobility services adapted to their territory. If we consider regional transport, it is not possible to measure the geographical coverage of rural areas by regional coach services due to the lack of national data. We can just say that these types of service are intended to serve these territories in a fine-tuned way. On the contrary, we can assess their coverage by the rail network on the basis of a radius of 3km around the railway stations. This leads to a proportion of 30% of rural communes and 45% of the population living in rural areas under the influence of a rail service. These results confirm the diversity of sparsely populated areas seen here from the angle of their accessibility to public transport.

4.3   A Multimodal Resident of Sparsely Populated Areas?

Car dependency is here characterized from another angle, linked to the evaluation of the degree of dependency. In terms of their mobility practices, are people living in rural areas exclusive car users, or, on the contrary, are they multi-modal, using the car for some of their journeys and more virtuous modes for others? The approach of multimodality in travel practices is analyzed from the self-reported frequency of mode use, a question asked in EMC2 surveys. In sparsely populated areas, 89% of people are not exclusively dependent on car use, and use other transport modes (see Table 2). Nearly one in four of them never use the car as a driver. The preferred mode of transport is then walking and car as a passenger, followed by public transport. Finally, less than 1% of multimodal people never use the car, either as a driver or a passenger. In this case, they walk or take public transport to a lesser degree.

Two multimodal patterns stand out among multimodal individuals (see Table 3). Four out of ten people travel by car or on foot and three out of ten people travel by car or on foot or by bicycle. As a result, almost 70% of multimodal people living in rural areas use car and active modes. Public transport ranks third, always associated with car and active modes. There are differences with multimodal people living in densely populated areas. Indeed, the first two multimodal models combine the car with public transport (65% of people). The use of active modes and car represents the third and fourth practices of these people (28%).

Table 2. Proportion of people according to the frequency of modes use by multimodal people

Mode	Regular use	Occasional use	No use
Car passenger	38.0	37.1	24.9
Car driver	76.8	2.9	20.3
Walking	68.1	25.2	6.7
Public Transport	9.0	17.5	73.5
Cycling	12.5	32.4	55.1

Table 3. Being a multimodal people in sparsely and densely populated areas means using…

Modes	% of people living un sparsely populated areas	% of people living in densely populated areas
Car, walking	40.7	18.8
Car, walking and cycling	28.7	9.3
Car, walking, cycling, public transport	11.9	25.0
Car, walking, public transport	11.6	39.8

5   Conclusion

This paper aims to quantify the phenomenon of car dependency among people living in sparsely populated areas. Their aspirations to have alternatives to the car have been affirmed through the national debate led by the French government in 2019 following the "yellow vest" movements. Some households have begun to reconsider their solo-driving use because of their travel budget or because of pro-environmental values [11]. French local stakeholders are already working to implement mobility solutions adapted to sparsely populated areas (shared mobility, rural mobility as a service, self-service bicycles,….), and to revitalize small railroad lines. But will this lead to a lessening of car dependency in sparsely populated areas? Won't habits and social norms be stronger, leading to the continued predominant use of solo-driving car? Even if reducing car use could be perceived as a higher difficulty in rural areas [2], the transformation of mobility aims to move from subjugated mobility to chosen mobility. Beyond technology and suffer, we need to bring about sociological change..

In addition to this issue, local stakeholders are also faced with another big challenge. This concerns the mobility of vulnerable population living in these areas, people with financial and/or physical constraints (elderly, unemployed young adults, people with modest incomes). Their mobility needs are specific, and so the mobility solutions to be provided will not be the same. This diversity in the vulnerability of these areas calls on stakeholders to work together mobility solutions with these different types of population, considering that there will be no single model of solutions for these areas. The challenge is a complex one, relating to implementing a package of mobility services within a reasonable financial budget.

References

1. Flipo A., Sallustio M., Ortar N., Senil N., K. Cariou K.: Projet RE-ACTEURS: Réseaux d'acteurs, innovation et gouvernance de la mobilité. La transition mobilitaire en territoire peu dense. Research report, Ademe (2022)
2. Colard J., de Lapasse B., Clément C.: Mobilités dans les espaces périphériques et peu denses: pour un territoire plus accessible? France Stratégie, Working paper, n. 2021–02 (2021)
3. Jacquin O.: Les mobilités dans les espaces peu denses en 2040: un défi à relever dès aujourd'hui. Information report of the French Senate (2021)
4. Ortar N.: Dealing with energy crises: working and living arrangements in peri-urban France. Transp. Policy 65(C), 72–78 (2018)
5. ATEC, ITF France: Zones peu denses: quels territoires? Quelles solutions? Quels impacts? Roadmap report (2019)
6. European Network for Rural Development: smart villages and rural mobility. Leaflet (2021)
7. UITP: The rural mobility challenge for public transport: how combined mobility can help. Knowledge brief (2022)
8. Baptiste H., Busnot-Richard F., Carrière J.P., Huyghe M., Mattei M.: Quelles mobilités en milieu rural à faible densité? Final research report, Région centre and Ademe (2013)
9. Dupuy G.: La dépendance à l'égard de l'automobile, Rapport de recherche PREDIT, La Documentation française, p. 93 (2006)
10. Hubert J.-P., Pistre P., Madre J.-L.: L'utilisation de l'automobile par les ménages dans les territoires peu denses : analyse croisée par les enquêtes sur la mobilité et le recensement de la population. Économie et Statistiques (483–484–485), 179–203 (2016)
11. Huyghe M.: Habiter les territoires ruraux. Comprendre les dynamiques spatiales et sociales à l'oeuvre et évaluer les perspectives d'évolution des pratiques de mobilité des ménages. Thesis report (2015)
12. Moati P., Cabaud F.: Qui sont les Gilets jaunes, leurs soutiens et leurs opposants? Analysis report, l'ObSoCo (2019)
13. Department For Transport: Travel in Urban and Rural areas. Factsheet, UK Government (2010)
14. Hernja G., Mergier A., Laboratoire de la mobilité Inclusive: La mobilité des jeunes dans un territoire rural, Report summary (2020)
15. International Transport Forum: Innovations for better rural mobility. Research report (2021)
16. OECD: OECD regional typology. Working paper (2011)
17. Census bureau of the United States: Understanding and using American community survey data. What users of data for rural areas need to know. U.S. Government Publishing Office, Washington, DC, (2020)
18. Eurostat: Applying the degree of urbanization. A methodological manual to define cities, towns and rural areas for international comparisons. Manuals and Guidelines collection (2021)

Open Access This chapter is licensed under the terms of the Creative Commons Attribution 4.0 International License (http://creativecommons.org/licenses/by/4.0/), which permits use, sharing, adaptation, distribution and reproduction in any medium or format, as long as you give appropriate credit to the original author(s) and the source, provide a link to the Creative Commons license and indicate if changes were made.

The images or other third party material in this chapter are included in the chapter's Creative Commons license, unless indicated otherwise in a credit line to the material. If material is not included in the chapter's Creative Commons license and your intended use is not permitted by statutory regulation or exceeds the permitted use, you will need to obtain permission directly from the copyright holder.

Decarbonization and Rural On-Demand Transport

Beate Kubitz[1] and Xuefei Wang[2(✉)]

[1] Beate Kubitz Associates Ltd., Hebden Bridge Town Hall, Hebden Bridge HX7 7BY, UK
[2] Padam Mobility, 11 Rue Tronchet, 75008 Paris, France
xuefei@padam.io

Abstract. A reduction in CO_2 and other greenhouse gas emissions related to transport requires a reduction in total km driven by private vehicles. This is difficult to achieve in rural areas because public transport provision is sparse. A redesign of public transport is required to provide people with alternatives to driving. This paper assesses the improvement to the availability and quality of transport options in two areas in the UK and France where demand responsive transport (DRT) is introduced. It calculates the increase in population served, the increase in service speeds, frequency, and span in order to assess the potential for car trip replacement and associated emissions reductions.

Keywords: regional mobility · rural areas · decarbonization · DRT · bus

1 Context

1.1 Carbon, Cars and Public Transport

The transport sector, particularly private car travel, is responsible for nearly a quarter of Europe's greenhouse gas emissions (GHG). Of this, road transport constitutes the highest proportion of overall transport emissions – in 2020 it emitted 77% of all EU transport GHGs. Cars and taxis represent a significant proportion of this (for example in the UK, cars and taxis are responsible for 52% of the emissions from domestic transport [1]. Reduction of GHG emissions requires a reduction in km driven by private cars [2]. Higher levels of driving correlate with poorer provision of public transport and higher levels of public transport provision correlate with lower transport carbon footprints [3]. A reduction in GHG emissions therefore requires the availability of practical alternatives to driving. One potential intervention is to improve bus services. However, achieving useful services in low density areas is problematic.

To be useful, transit must exist in both space and time. It must run not just where we need it but also when we need it. Unless it does both, it doesn't exist for us at all. [4]. Where populations are sparse and travel patterns variable, achieving this is extremely rare. A 'critical mass' of people using specific routes is unlikely, and services are difficult to fund. Indeed, in many territories rural routes have been cut, and rural areas experience transport desertification [5]. However, behaviour change cannot be expected without practical alternatives to car travel.

© The Author(s) 2025
C. McNally et al. (Eds.): TRAconference 2024, LNMOB, pp. 464–471, 2025.
https://doi.org/10.1007/978-3-031-85578-8_61

2 The Areas and Services

This paper looks at designing public transport in sparsely populated areas. It analyses two rural territories with on-demand public transport services (DRT) in France and the UK using data from the technology platform, Padam Mobility. For each service we analyse the following metrics:

- **service accessibility:** the percentage of the population served by the service, within a short walk of the bus.
- **service utility:** the service span (hours of operation) when people can access public transport and the frequency of services.

Fig. 1. Area of Pays de Limours TAD route 39-18 (orange) (Île-de-France Mobilités)

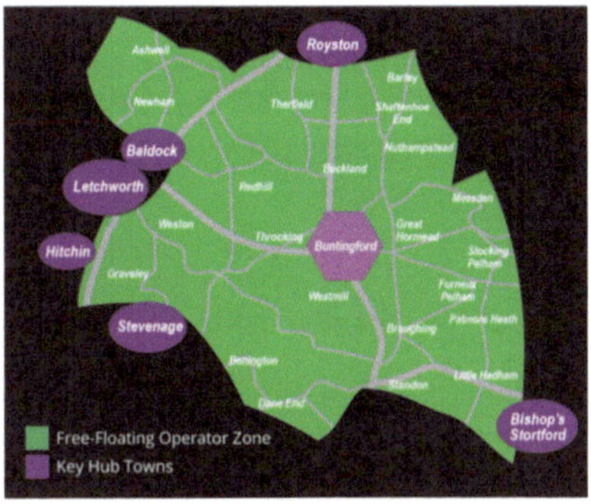

Fig. 2. Hertslynx area showing key hub towns (Hertfordshire County Council)

We compare the availability and quality of public transport before and after the introduction of DRT and in comparison with travelling by private car in order to assess whether DRT provides a practical alternative to car travel (Figs. 1 and 2).

2.1 Pays de Limours

Area: Pays de Limours is a rural community consisting of 14 small towns and villages 30 km to the south-west of Paris. It is situated in the Essonne department of the Île-le-de-France. The population of the area is approximately 23,000 inhabitants, with an average population density of 230 people per km^2. The villages in the area have populations from a few hundred to 6,700 people and are between 2 and 10 km apart, with fields and green space bounding them [6].

Transport: Transport is organised by the Île-le-de-France Mobilités. There is a key fixed line bus route, the 39–18, which traverses the zone connecting Limours with Arpajon via a bus station situated on the main A10 arterial route into Paris in Briis-sous-Forges. This service is at least hourly. There are a number of sub-routes (eg the 39–28) which augment this frequency in term time and cover more of the zone. In addition, a network of three lines (61, 62 and 63) serves schools in term times. This makes the timetable quite complex. The bus station connects with onward services connecting to RER stations with some express services. This complex and dense network appears to serve a high number of people. However, the timetables are at their most frequent around the start and end of school days.

DRT: The DRT scheme (TAD in French) was launched to provide better connectivity within the zone. It covers the entire area and provides different services at peak and off-peak times. During the peak hours (0610 to 0930 and 1630 to 1945), passengers can request a bus to take them to the bus station or regional rail station for rapid onward travel. In off peak hours (0930 to 1630) the service can be used to connect people travelling between any two villages within the zone. It is designed to enable people to access the rapid bus which runs north east/south west to Orsay (in the south of Paris) which operates hourly between 0600 and 2200 and the local fixed line 39–18. The zone is 9 miles (14.4 km) north to south and 7 miles (11.2 km) east to west, giving an approximate area of 63 square miles (176 km^2). Four buses serve the area. Journeys can be booked in real-time or in advance. Fares are equivalent to a single Île-de-France-Mobilités bus fare - €2.10 per trip.

2.2 North East Hertfordshire

Area: North and East Hertfordshire lies north of London, and contains a number of conurbations (including Stevenage, Letchworth, Hitchin, Baldock, Royston and Bishop's Stortford) but with a rural zone lying between them. Around 50% of the population lives outside the towns. In total the population is 300,000 people, with 50,000 of them living in the area bounded by Royston in the north, Stevenage in the west and Bishop's Stortford in the east outwith the towns. They live in isolated dwellings, small hamlets and villages with only one small market town, Buntingford (population 6,844).

Transport: The area is heavily car dependent. Public transport stops are sparsely spread across the area and the least frequent services coincide with the areas where people drive more. Hertfordshire County Council identified this area as including 4,000 people who had no access to bus on any metric prior to 2021. In addition, an estimated 40,000 of the 50,000 people in the area live beyond a 15-min walk from a bus stop.

DRT: The local authority introduced a DRT service, HertsLynx in 2021. It covers an area between 7 and 9 miles (11.2 and 14.4 km) in each direction from Buntingford, serving a total area of around 150 miles^2/400 km^2. HertsLynx launched with three 16-seater minibuses each with one space for a wheelchair user and increased to 4 vehicles in 2022. The service operates 0700–1900 Mondays to Saturdays and 1000–1600 on Sundays and Public Holidays. Journeys can be booked in real-time or in advance. A £2 standard fare supported by the UK government applies. Passengers can use HertsLynx for travel anywhere in the Free-Floating Operator Zone. There are no fixed routes on this service, instead passengers can be picked up and dropped off at a vast number of stops within the zone. They are also able to travel from the Free-Floating Operator Zone to designated locations in the Key Hub Towns. Travel is permitted between Key Hub Towns but is not available for journeys between points within one Key Hub Town.

The service was commissioned by Hertfordshire County Council, funded by the Rural Mobility Fund of the UK Department for Transport. The service is operated by Uno Bus and is managed using the Padam Mobility DRT platform.

3   DRT and Fixed Services: Who is Served and How?

3.1   Population

The population served by fixed line services and DRT was calculated in the following way. Each area was divided into hexagons with 500 m diameter edges, approximating a reasonable walking distance to access public transport. Population within those cells with fixed line stops were counted as having access to public transport prior to DRT. The DRT zone was overlaid and those people in areas with DRT virtual stops counted as having access to public transport. Cells were outlined (in black) where where people have access to DRT who did not previously have access to fixed line buses.

Hertfordshire: The addition of the HertsLynx virtual bus stops increases the number of people able to access public transport by 40–50,000 people. An estimated additional 40–50,000 people (Figs. 3 and 4).

Pays de Limours: The addition of DRT increases the number of people able to access public transport by 5,000 (Figs. 5 and 6).

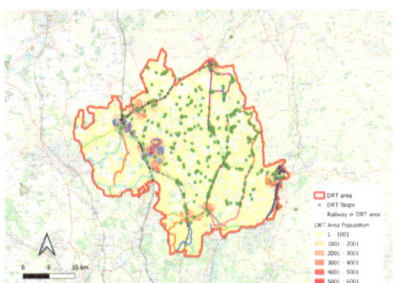

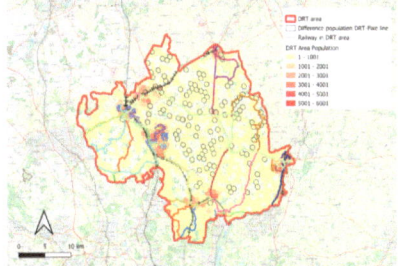

Fig. 3. Hertslynx zone showing virtual DRT stops

Fig. 4. HertsLynx zone showing additional population served by DRT (created with QGIS & OpenStreetMap)

3.2   How? Frequency, Speed and Span

DRT services enable us to derive a set of 'transport desire lines' from the journeys people have requested. These have the advantage that they reflect the trips people want to make rather than those they are forced to make by the available fixed routes. These desire lines are the basis for our analysis. For this part of the analysis we compare car travel, fixed route lines and DRT for a set of top origins and destinations for each area. We look at three metrics: journey times, the span of services (earliest service to last service

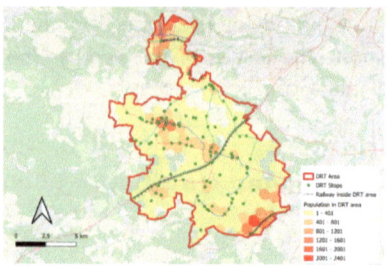

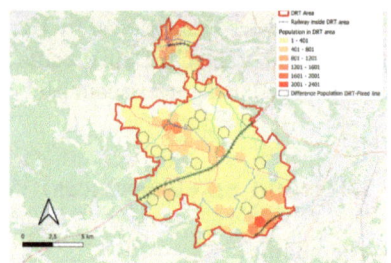

Fig. 5. Pays de Limours showing DRT virtual stops

Fig. 6. Pays de Limours showing additional population served by DRT (created with QGIS & OpenStreetMap)

daily) and the frequency of services. We used the Google Maps route planner across a number of sample times of day for car and fixed line transport timetables and used Padam Mobility data for DRT.

3.3 Hertfordshire

Table 1. Four typical desired journeys to Royston station

Journey	Car	Fixed line bus	DRT
Paddock Road Buntingford	17 min	43–53 min	30 min
Hare Street Buntingford	20 min	84–704 min	32 min
High Street Ashwell	14 min	27 min - overnight	27 min
High Street Walkern	24 min	79 min - overnight	41–50 min

Span: The maximum service span for fixed line routes across these destinations is 0632 to1757 (11 h 25 min) however Royston station to High Street Ashwell is only served between 0730 and 1222 (4 h 52 min). The maximum service span across these destinations for DRT is 0700 to 1900 (12 h) in 2023 (Tables 1 and 2).

3.4 Pays de Limours

See Tables 3 and 4.

Table 2. Frequency of fixed bus routes (weekdays)

Journey	Journey description
Royston station to Paddock Road Buntingford	Bus service 18 every 2–3 h from 0749 to 1708 (5 times per day)
Royston station to Hare Street Buntingford	Bus service 18 every 2–3 h from 0749 to 1708 (5 times per day) plus change onto bus service 331 every 1–3 h from 0632 to 1757 (8 times per day)
Royston station to High Street Ashwell	Bus service 90 at 0927 (one per day) or bus service 91 0730 or 1222 (2 times per day) = effective service 3 times between 0730 and 1222. If the train (Royston to Ashwell Station) is combined with bus, this increases to 4 times per day
Royston station High Street Walkern	The only public transport options between Royston station to High Street Walkern go via rail to Stevenage station and then require a bus (the 384 service) which runs 6 times per day

Table 3. Journey times for 4 typical journeys to Gare Autoroutiere Briis (bus station):

Destination	Car	Fixed line bus	DRT
Mairie – Vaugrigneuse	3 min	24 min	6 min 12s
Lycee Jules Verne – Limours	8 min	25–33 min	12 min 18s
La Fontaine aux Cossons – Vaugrigneuse	5 min	33–46 min	12 min 42s
La Touche - Saint Maurice	8 min	25–58 min	12 min 18s

Table 4. Summary of frequency of the fixed route services. Numerous lines serve the destination from Gare Autoroutiere Briis before/after the school day.

Destination	Timetable description
Mairie - Vaugrigneuse	morning and evening only
Lycee Jules Verne – Limours	morning and evening only (one midday service via a 19 min walk)
La Fontaine aux Cossons – Vaugrigneuse	morning and evening - may include a 18 min walk
La Touche - Saint Maurice	morning and evening - requires a change or a 15 min walk

3.5 Analysis

Population: In both territories the on-demand services clearly increases the number of people served both absolutely and outside of peak hours.

Speed: DRT matches or improves fixed line bus journey times in both territories. It significantly improves journey time where it is replacing a journey that can only be made with a connection.

Span: The span of DRT services and fixed route services is largely similar.

Frequency: The availability of DRT services throughout the day increases access to public transport for trips outside the peak hours so that occasional or irregular trips can be made and people can depend on services for trips that don't follow school or standard working schedules.

When compared with car journeys which are faster and available throughout the day and night, DRT marks a significant improvement on fixed line buses. The smaller difference in journey times between car journeys and DRT also makes DRT attractive as it does not involve parking. DRT in these areas ensures that there is a genuine mode choice for people who have a private car and access to transport for those who don't.

4 Implications for Car Use

The DRT systems studied increased the availability of public transport within a short walk of people's homes. This is a necessary condition for behaviour change. Usage data of the systems studied demonstrate trip patterns that reveal that there is demand beyond the transport corridors and across the day. It enables those who do not drive or have access to a car to make everyday journeys to access education, employment, services and for social reasons.

Behaviour change cannot take place without options between which people can choose. DRT successfully provides an alternative to car travel which is a fundamental condition for people to be able to choose between alternative modes. In this way DRT enables behaviour change. Whilst there is some evidence of spontaneous effects in some areas, the effects on behaviour change are stronger where incentives and regulations encourage people to review their modal choices. We anticipate that extensive behaviour change will require such incentives and regulations, however without the provision of alternatives such as DRT they will not be effective.

References

1. Office for National Statistics (UK). Climate change insights, business and transport, UK: February 2023 (2023). https://www.ons.gov.uk/economy/environmentalaccounts/articles/cli matechangeinsightsuk/february2023#:~:text=In%202020%2C%20over%20half%20(52,Tra nsport%20(DfT)%20measure
2. Marsden, G.: Reverse gear: the reality and implications of national transport emission reduction policies. CREDS (2023)
3. Morgan, M., Anable, J., Lucas, K.: A place-based carbon calculator for England. Cardiff, Wales. Presented at the 29th Annual GIS Research UK Conference (GISRUK), Cardiff, Wales, UK (Online), Zenodo (2021). https://doi.org/10.5281/zenodo.4665852
4. Walker, J.: Human transit: how clearer thinking about public transit can enrich our communities and our lives (2021)

5. Hinchcliffe, C., Taylor, I.: Every village, every hour: a comprehensive bus network for rural England. Report for CPRE based on research and modelling by Transport for Quality of Life. CPRE (2021)
6. Communaute de Communes du Pays de Limours (n.d.). www.cc-paysdelimours.fr

Open Access This chapter is licensed under the terms of the Creative Commons Attribution 4.0 International License (http://creativecommons.org/licenses/by/4.0/), which permits use, sharing, adaptation, distribution and reproduction in any medium or format, as long as you give appropriate credit to the original author(s) and the source, provide a link to the Creative Commons license and indicate if changes were made.

The images or other third party material in this chapter are included in the chapter's Creative Commons license, unless indicated otherwise in a credit line to the material. If material is not included in the chapter's Creative Commons license and your intended use is not permitted by statutory regulation or exceeds the permitted use, you will need to obtain permission directly from the copyright holder.

Developing Pan-European Capacity at Local Level for a New Era of Sustainable Rural Mobility: SMARTA-NET

Andrea Lorenzini[1](✉) and Brendan Finn[2]

[1] MemEx Srl, P.zza Benamozegh, 57123 Livorno, Italy
andrea.lorenzini@memexitaly.it
[2] ETTS, Dublin, Ireland

Abstract. People in rural areas must travel more often and further due to reductions in local opportunities and amenities but have poor mobility options other than the private car. The underlying cause of this is the fact that a specific rural mobility policy is absent in almost all European countries. This led to a lack of goals, obligations, and funds, and ultimately to a lack of mobility supply. A wide range of rural mobility solutions (such as, demand responsive transport, carpooling, and car sharing) already exists. These can be packaged to meet the local mobility needs, and connected to the regular public transport. Nevertheless, the implementation context for rural areas is very different from that of urban areas, not only in scale but also in the nature of the implementers, who usually do not have the skills to implement such services. Capacity building, know-how transfer, peer-to-peer networking, and exchange of experience can be essential enablers for local shared mobility implementers. As part of the response to these challenges, the SMARTA-NET project is establishing the first European network on rural mobility, aimed initially at rural municipalities and others who can act at the local level. Guidance documents are being prepared on four key topics designed to foster the capacity of municipalities to develop and implement sustainable, inclusive, and integrated mobility solutions in their territories. Guidance is then imparted through training events in 14 participating countries. Preliminary Results are a high level of interest across Europe to participate in the network and training and to exchange experience.

Keywords: Sustainable Rural Mobility · Ecotourism · Urban and Rural links · Rural Mobility Network · Training · Capacity building

1 The Rural Mobility Challenge

Rural mobility is one of the fundamental enablers for the stated European Long-Term Vision for Rural Areas – *"stronger, connected, resilient and more prosperous rural areas by 2040"* [1]. It is about whether those places can thrive and continue to retain their population, including their families and especially their youth and future generations; or, whether the 21st Century exacerbates patterns of the 19th and 20th centuries, of decline and decay in which rural areas fall ever further behind urban areas.

© The Author(s) 2025

C. McNally et al. (Eds.): TRAconference 2024, LNMOB, pp. 472–478, 2025.
https://doi.org/10.1007/978-3-031-85578-8_62

The 21st Century reality is that many forms of employment, education, shopping, services, healthcare, public administration and amenities have been disengaged from rural areas, now even from smaller towns and county hubs [2]. This is not going to change any time soon. People in rural areas, now even in rural towns, have become increasingly obliged to travel more often and longer distances for the most basic requirements of a normal life. This is not going to change any time soon either.

When the essentials of life have been removed from the locality, it follows that an acceptable quality of life is only possible when people can travel to the places where what they need is now available. However, public transport and shared mobility services are minimal or non-existent in most rural areas throughout Europe. This places extreme limitation on where a person can go without a car. As a result, in rural Europe, most people have little or no choice about how they travel. Their only choice is whether they can travel at all. Almost all travel is made by personal car, other than what can be accessed nearby on foot. Households own multiple cars, from sheer necessity. Those without cars are dependent on others for lifts, or they must forego travel and their lives become limited.

2   The Relevance of Policy in Rural Mobility

The SMARTA Rural Transport Areas Project (SMARTA) examined the Frameworks within which rural mobility sits in each of the 27 European Union (EU) countries. This identified that virtually all European countries lack any explicit policy on rural mobility, lack targets or explicit obligations on levels of mobility provision, and that such policy and implementation/funding mechanisms as exist are fragmented across agencies (general public transport, education, health, …). Further, regulatory frameworks either do not make provision for Demand Responsive Transport Services (DRTs) or for many emerging forms of shared mobility, or severely constrain what may be provided [3, 4].

What's more, there is no common methodology for extending a metropolitan Sustainable Urban Mobility Plan (SUMP) to its hinterland, and no methodology at all for developing a SUMP-equivalent for a predominantly rural area.

At the transport provision level, most rural areas in Europe have limited public transport at best. Many areas have no public transport at all, except where they are fortunate to have inter-urban services passing through. However, these often have limited stopping places and may not serve many communities along their route. Where local public transport is available in rural areas, frequency is usually low. When only available at commuting times, it is impractical for daytime purposes such as attending social services, healthcare, shopping, training, leisure and cultural activities, socialising, etc. What's more, the impacts of the COVID-19 pandemic on the overall transportation system have further exacerbated the situation by reducing the sense of safety and comfort in sharing trips with others [5, 6].

3   Rural Shared Mobility Solutions and their Connections to the Regular Public Transport

A wide range of rural mobility solutions, which can be packaged to meet the local needs, and connected to the regular public transport, already exists in different rural EU contexts [7]. These solutions, demonstrated in multiple European projects including SMARTA, SMARTA2, LAST MILE, MAMBA, INCLUSION, MARA, MELINDA, Hi-Reach, can be clustered in three main types, as shown in Fig. 1: i) Flexible Transport Services, including on-demand transport; ii) Ride sharing services, such as, carpooling and shared taxi services; iii) Asset sharing services, including car and bike sharing. Other types of initiatives include school services of general access, mobility hubs, non-emergency medical transport, mobility in support of rural tourism, community lift-giving, autonomous shuttles. Rural shared mobility solutions offer a range of distinctive features that cater specifically to the needs of rural communities. Long-lasting durability, community engagement, innovative technology, are key elements that make or break these initiatives. By leveraging these features, rural shared mobility solutions have the potential to enhance accessibility, connectivity, and overall quality of life for residents in rural areas. These become much more effective when they are coordinated or, integrated with the fixed-route bus and rail services. They can extend the coverage of the conventional public transport network, reaching additional areas and offering higher levels of service than would be feasible or affordable with larger vehicles.

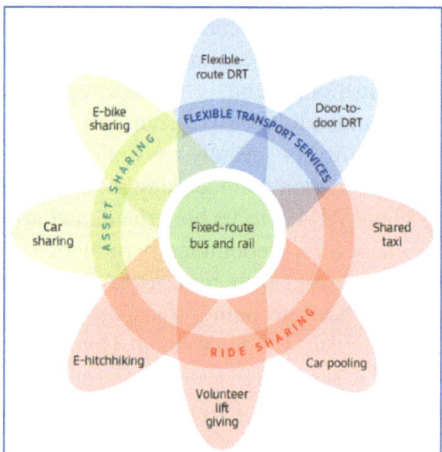

Fig. 1. Rural Shared Mobility solutions. Source: SMARTA

4   The Implementation Context and the Urban and Rural Divide

The implementation context for rural areas is very different from that of urban areas, not only in scale but also in the nature of the implementers. As of today, the development and growth of shared mobility have almost exclusively addressed large urban and metropolitan areas. However, in the case of ever-growing social and economic divisions in Europe in terms of income and wealth, the labour market, education and skills, health, and migration, new shared mobility solutions have a potentially greater role to play particularly in small-sized cities, and their functionally connected peripheral, extra-urban and rural areas. In such environments, the take up of new shared mobility solutions is held back due to the difficulties of operators in adopting suitable business models and cost-effective mobility services. As a result, private operators are not interested in gaining new markets and the transport service provision is weak.

Furthermore, cities have the required structured functions, capacities, and digital tools to deliver a seamless, multimodal, and reliable mobility system, representing a convenient alternative to a private car. By contrast, rural municipality employees have to play a multitasking role, usually burdened with many different tasks. Frequently, a transport department is missing, and there are few/no professionals with the skills required to plan and organise a comprehensive package of mobility solutions [8].

5   Boosting Factors for Local Implementers

Implementation of new generations of rural mobility schemes throughout Europe could be done rapidly if the enabling conditions are there. It requires very little infrastructure or capital financing. Communities generally have a good capacity to mobilize and implement locally. The key gaps right now are sector-specific know-how, access to funding for operations, and administrative/regulatory space for local actors to act. Enablers that would make a major difference and allow rapid deployment are networking, capacity building, know-how transfer, peer-to-peer discussions, and exchange of experiences; these are essential for local implementers, both from the public and private sectors.

Similar to what has been gained for urban mobility, where the consolidation of good practices and the sharing of experiences among cities is a well-established practice (just think of the several urban mobility networks – CIVITAS, POLIS, EUROCITIES, EMTA, UITP, etc.), good progress can be achieved for rural mobility. The National Rural Transport Assistance Program (RTAP) in the USA offers an interesting model, spanning three main clusters – Networking, Capacity Building, and Funding, that Europe could consider [9].

6   Setting up the First Pan-European Network on Rural Mobility: SMARTA-NET

SMARTA-NET is an initiative of the European Commission, under the Directorate-General for Mobility and Transport (DG-MOVE) supported by the Directorate-General for Agriculture and Rural Development (DG AGRI), dealing with rural mobility, with additional emphasis on ecotourism. SMARTA-NET is helping to develop pan-European capacity at local level to respond to these challenges [10].

SMARTA-NET project is establishing the European Rural Mobility Network (ERMN), i.e., the first pan-European network on rural mobility, aimed initially at rural municipalities and others who can act at the local level. The ERMN consists of two main groups:

- The Network of Rural Municipalities
- Rural Mobility Multipliers (regions and organisations that have the mandate and capacity to mobilise and support mobility services in rural municipalities

The ERMN aims to enable the sharing of knowledge and experiences among local authorities, practitioners, and experts on rural mobility.

7   Building Capacity

SMARTA-NET is helping to develop pan-European capacity at local level to respond to rural mobility challenges. In doing so, the project is preparing guidance documents on four key topics, designed to assist local implementers. The Guidance, fed by in-depth research, supported by data collection and interviews and focus groups with relevant stakeholders, are addressing: i) design and implementation of shared mobility services, and their integration with the conventional public transport; ii) Connecting touristic destinations to sustainable mobility networks; iii) integration of aspects of rural mobility into the sustainable urban mobility planning process; and iv) funding and financing schemes for rural transport services.

The knowledge assembled in the Guidance is being imparted through training events in 14 participating countries. More specifically, SMARTA-NET has developed a training programme focused on providing know-how and tools to address mobility and transport issues and sustain them over time and designed for capacity-building and empowerment of rural communities to cope with the mobility challenges and to build their resilience by leveraging ecotourism. The capacity-building is intended to be a participatory process that involve the local stakeholders inspired by SUMP principles (it is codesigned with the ERMN municipalities, hence they are involved since the preparation stage). The Training programme consists of 4 training sessions on the Guidelines in each of the member countries, in own country and language, for the ERMN members and other interested stakeholders (Fig. 2).

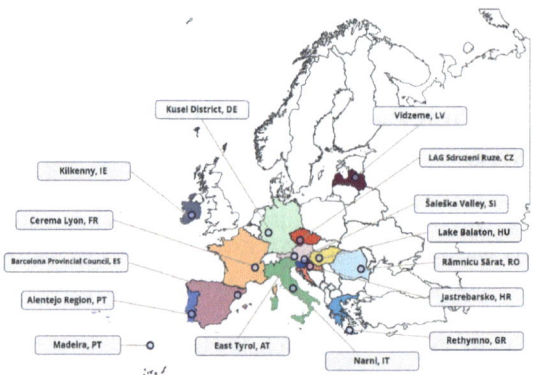

Fig. 2. The SMARTA-NET Lighthouse Sites. Source: SMARTA-NET

8 Conclusion

In the first semester of 2023, the Consortium engaged with one "Lighthouse Site" (LS) in each of 14 target countries - Austria, Croatia, Czech Republic, France, Germany, Greece, Hungary, Ireland, Italy, Latvia, Portugal (mainland plus the special case of Madeira), Romania, Slovenia and Spain. LS are places with a range of tourism interests in their area, and either have an active rural mobility scheme or are actively seeking to develop one. Working with the SMARTA-NET team, they have assessed various aspects of transport behaviour and how this can be improved, contributing to setting up the ERMN. A total of 16 LS from 15 EU countries are participating in SMARTA-NET. This provides good geographic coverage of Europe, of the typology of rural areas and of rural tourism. With the support of the LS, SMARTA-NET mobilized the process of engagement of target municipalities. Some 50 municipalities participated in the first in-person meeting of the ERMN, in October 2023 in Bingen am Rhein, Germany. This highlights a high level of interest across Europe to participate in the network and training, and to exchange experience.

Additional municipalities will be engaged in late 2023 and beginning of 2024. SMARTA-NET will plan the future of the ERMN beyond the project, to be shaped by the Members themselves during 2024. SMARTA-NET aims to widen the membership scope and scale to be the leading voice of the sector, and develop policy and other recommendations to influence European, national and regional frameworks, programs and funding for rural mobility, for a new era of rural transport.

References

1. European Commission, A long term vision for the EU's rural areas. https://ec.europa.eu/info/strategy/priorities-2019-2024/new-push-european-democracy/long-term-vision-rural-areas_en
2. Eurofound,: Bridging the rural–urban divide: Addressing inequalities and empowering communities. Publications Office of the European Union, Luxembourg (2023)
3. Finn, B., Nelson, J., Lorenzini, A.: The importance of Policy, Institutional and Regulatory Frameworks to the provision of DRT and emerging other forms of shared mobility services in rural areas. Paper at Thredbo 16 Conference, Singapore (2019)
4. ITF: Innovations for Better Rural Mobility, ITF Research Reports, OECD Publishing, Paris (2021)
5. Rodrigues, M., et al.: Research for TRAN Committee – Relaunching transport and tourism in the EU after COVID-19, European Parliament, Policy Department for Structural and Cohesion Policies, Brussels (2021)
6. Lorenzini, A., Ercoli, E., et al.: Back to sustainable mobility in times of COVID-19 - strategies for touristic destinations and small islands. CIVITAS DESTINATIONS Project Publication (2021)
7. Lorenzini, A., Ambrosino, G., Finn, B.: Policy Recommendations for Sustainable Shared Mobility and Public Transport in European rural areas. SMARTA Project Publication (2021)

8. Porru, S., Misso, F.E., Pani, F.E., Repetto, C.: Smart mobility and public transport: opportunities and challenges in rural and urban areas. J. Traf. Transp. Eng. (Engl. Ed.) **7**(1), 88–97 (2020)
9. National Rural Transport Assistance Program (RTAP). www.nationalrtap.org
10. SMARTA-NET Project. https://www.smarta-net.eu/

Open Access This chapter is licensed under the terms of the Creative Commons Attribution 4.0 International License (http://creativecommons.org/licenses/by/4.0/), which permits use, sharing, adaptation, distribution and reproduction in any medium or format, as long as you give appropriate credit to the original author(s) and the source, provide a link to the Creative Commons license and indicate if changes were made.

The images or other third party material in this chapter are included in the chapter's Creative Commons license, unless indicated otherwise in a credit line to the material. If material is not included in the chapter's Creative Commons license and your intended use is not permitted by statutory regulation or exceeds the permitted use, you will need to obtain permission directly from the copyright holder.

Decarbonizing Mobility in Minor Cities in Finland

Matti Roine[1], Pekka Leviäkangas[2], and Shahid Hussain[2(✉)]

[1] MH Roine Consulting, 005200 Helsinki, Finland
mattiroine@me.com
[2] Civil Engineering, University of Oulu, Pentti Kaiterankatu 1, 90570 Oulu, Finland
shahid.hussain@oulu.fi

Abstract. Smaller cities struggle to achieve carbon-neutrality goals. In Finland, the goal is to reach carbon neutrality by 2035, which means that by and large the smaller cities need to address this target. Several measures to cut carbon emissions have been analyzed in different cities, but it seems that in most cases the reductions in the mobility sector are far behind the ambitious goals, and hence jeopardize the reaching of the 2035 milestone. However, many other improvements are possible regarding accessibility, equity, health, and affordability of urban mobility in smaller cities that have traditionally relied very heavily on private car use. It seems that one of the keys is the electrification of mobility while shifting energy production from fossil energy to cleaner forms, such as wind, hydro, and solar power in the future but more effective land/use policy in the long term.

Keywords: Mobility · climate change · decarbonizing · carbon neutrality · smart cities · urban form

1 Introduction

In the history of vehicle traffic, it has been blamed for many sins, and not just emissions. It has been claimed that car traffic dominates the urban environments preventing people from moving safely when walking and cycling, taking much of the urban space, localizing activities following the logic of vehicle use, and defining the urban form, i.e. causing urban sprawl. Many research results are available that have addressed the aforementioned issues. For instance, EIT Urban Mobility in the EU (Borgato et al., 2021) published a report in 2021 on the costs and benefits of the possible sustainable urban mobility transition policies and presented the most cost-effective policies in cities. In general, research has identified and addressed the problems well. However, many cities seem to have a tailored strategy to address the Green Deal.

There is an increasing population of citizens who are ready to give up car driving and choose walking, cycling, or e-scooters as their preferred means of mobility. This implies that the accessible activities and services must be within the reach of these mobility patterns. One of the new approaches has been developed mainly by Sorbonnecalled Professor Carlos Moreno 15 min City (Moreno et al., 2021). This idea signifies that all

© The Author(s) 2025
C. McNally et al. (Eds.): TRAconference 2024, LNMOB, pp. 479–486, 2025.
https://doi.org/10.1007/978-3-031-85578-8_63

the necessary daily activities are reachable in 15 min without a motorized vehicle. Some of the main cities in Europe have already supported the new ideas of the 15-min smart cities like Paris and Barcelona. The measures following the 15-min ideology have not yet radically changed the urban forms. However, renewals that improve the situation like walking and cycling conditions have been made, including improvements to have access to necessary services and public transport. The 15-min idea may have different variations in branding, but the basics remain essentially the same – for example the Circular City Vienna includes other aspects in addition to mobility (Circular City of Vienna, 2023).

The body of research results and widely accepted strategies confirm the same paradigm: how to manage carbon emissions in a city by reducing car traffic and use, promoting walking, and cycling, providing charging systems for electric mobility, improving public transport services, enhancing city logistics, and introducing new regulation and policies. Often the main problem in a city is to follow, adopt, and deploy these results so that the well-being of citizens is increased.

This paper presents the preliminary results of a study on the needs and tailored improvements of the mobility services and system in cities under 100,000 in Finland. Mobility and climate action in Finland-

2 Mobility and Climate Actions in Finland

The Finnish Ministry of Transport and Communications is responsible for the transport policy and the development plans. The Ministry introduces frequently The National Transport System Plan covering the whole transport system (Ministry of Transport and Communications, 2021). The Plan is a description of the mobility in Finland as well as in its regions and cities. The Regional Councils are responsible for regional coordination and action, while the cities take care of their mobility planning and implementation of the measures. This division of work applies also to climate action.

The post-pandemic years saw a significant increase in the number of people regularly working from home (Statistics Finland, 2023). The number of public transport users has turned back to the growth track but has not recovered to the pre-pandemic levels. The pandemic radically increased online shopping. By and large, mobility patterns in different cities seem to differ quite much and there is still an incomplete understanding of the local traits and habits. Therefore, successful transition paths and implementations are not that straightforward.

In 2017 The Finnish Climate Change Panel estimated the effects of alternative measures of transport on climate change and emissions (Table 1). At that time energy efficiency measures and alternative fuels were believed to be the best alternatives for reaching climate targets (Liimatainen and Viri, 2017). Since then, many new research projects have been carried out, and more effective measures defined. Also, the cities and municipalities have started to manage and analyze the emissions-cutting actions.

When cities must adapt to climate change and fasten their actions to decarbonization it is of course important to implement profitable and beneficial measures at first. In Finland, good results have been accomplished with technology improvements dealing with the city's electric systems and housing, not only because of the cold and long wintertime. The transport system is more complicated with different users and impacts in the cities e.g. on economy and business.

Table 1. The mitigation of climate change measures in transport by The Finnish Climate Change Panel in 2017 (Liimatainen and Viri, 2017).

Group of measures	Detailed measures	Mitigating emissions compared to the basic scenario (Mtn)
Improvement of energy efficiency of transport system	Development of walking and cycling	0,30
	Development of public transport	0,18
	Improving the utilization rate of cars	0,19
	Energy efficiency of truck transport	0,30
Improvement of energy efficiency of vehicles	Reduction of energy use of new vehicles	0,32
	Electric and gas-powered vehicles	0,14
	Reduction of energy use of trucks	0,13
Substitution of fossil fuels	Renewable fuels in passenger cars	0,90
	Renewable fuels of trucks	0,50
	Liquefied biogas and electric batteries in trucks	0,15

3 Activities in Minor Cities in Finland

The City of Lahti was nominated as the European Green Capital in 2021. The city has been very active in defining and implementing the necessary actions to reduce climate change and decarbonizing measures including mobility actions. Lahti has about 120000 inhabitants and is the regional center taking care of the regional public transport alongside other regional tasks, e.g. of main health care services. The city aims to be carbon-neutral already by 2025. The City of Lahti has been active in many respects and has abandoned coal entirely in its s main energy system. The need for decarbonization concerns all city systems, obviously including transport which causes about 25% of the local carbon emissions.

The city plans in Lahti also include the European SUMP planning for mobility improvements. The main reduction of carbon emissions is due to changes in vehicle traffic and the main trends estimated are electrification of transport and improvements of modal split towards walking and cycling. The fossil CO_2 emissions are estimated

Table 2. Estimated CO2-ekv. Reductions in 2030 in the City of Lahti (City of Lahti, 2021)

Number	Measure	Effect (- Ton CO2 ekv.)
1	Electric vehicles	660
2	Changing in powersystems in public transport	438
3	Electric loading service network	267
4	Traffic arrangements in the center of the city	222
5	Compensating the city staff air travel	145
6	Increasement of cycling	89
7	Enlarging of the cycle network	89
8	New mobility data systems	89
9	City staff mobility plan	82
10	Improvement of public transport trunk lines	62

to be 2030 about 129000 CO2-ekv, which is about -29% less than in 2020. The analysis included more than 30 measures of which only 10–20 proved to be very effective (Table 2).

In Finland, many other cities follow mainly the same procedure as Lahti e.g., the City of Vaasa. They have profoundly analyzed the target towards carbon neutrality by 2030. However, they seemed not to entirely reach the target with different developed scenarios. This approach proved the difficulties cities have with the carbon-neutrality

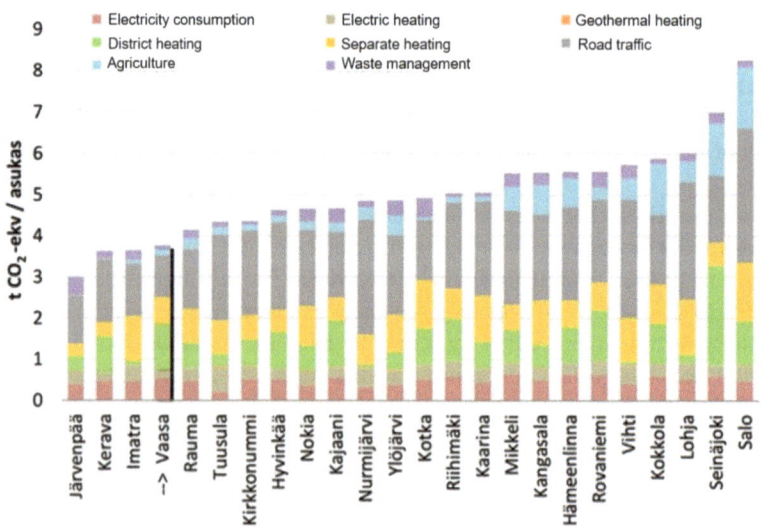

Fig. 1. Different activities and CO2 emissions (CO2 ekv./inhab.) in municipalities having inhabitants 25000–70000 in 2018 (Vaasa, 2020).

target by 2030. The variability of CO_2 emission production in minor municipalities and cities is large (Fig. 1).

4   Methods of Pathways

SUMP has published guidelines on how mobility should be planned in smaller cities (SUMP, 2021). They have suggested many different measures in towns, but they have also considered that many smaller cities have limited financial resources. Then they suggest low-cost measures such as mobility management, organizational and communication measures, improvements of existing infrastructure, measures that are efficient in decreasing the cost of operation, etc.

4.1   Sustainability Urban Mobility Planning in Smaller Cities and Towns

CIVITAS network analyzed in detail the transition pathways of cities and took into account that cities with different sizes should be treated not alike. In many analyses, the pathways can be long-term e.g., to 2050. The short perspectives are implementation plans and are more practical than the long-term plans. Civitas has defined the phases for a long-term plan (Table 3). Most of the plans for carbon neutrality that have been elaborated for minor cities in Finland follow the traditional method starting from state-of-the-art and defining the immediately needed actions to reach the set carbon neutrality targets. They are practical in that sense that they can leave out some variations and unexpected situations that usually arise with long-term planning. The approaches do not usually follow the idea of backcasting.

When reviewing the transport system in some of the minor cities in Finland you can easily find out that they have realized the need for a change towards a more sustainable transport system, and implemented actions to improve walking and cycling by arranging new networks. At the same time, cities have usually improved the public transport system making it more sustainable and accessible to the users, increasing routes, implementing management and information systems, etc. However, cities had to accept the reality that land use has become more dispersed and adapted to business needs.

Table 3.   Recommended process for developing a Transition Pathway. (Smeds et al., 2020)

Review or develop a long-term vision of the desired future city
Define objectives and targets that align with the vision
Identify a policy mix that can achieve objectives and targets
Stress-test the policy mix against alternative future scenarios
Identify milestones for the implementation of the policy mix
Identify enabling actions – institutional capacity and financial resources
Build a timeline visualising interdependencies between policy milestones and enabling action
Bring it all together in a narrative of the Transition Pathway
Translate 30-year Transition pathway into operational 5–10 year implementation strategies

5   Further Activities and Research

A lot of research projects have been going on previously because new technologies e.g. in communication have given possibilities to improve the mobility of citizens. Many research projects have been directed to Intelligent Transport Systems and services especially applications using internet-based services. In many minor cities, the projects have been directed to public transport services ordering, booking, timing, and alternative needs. Also, information systems like bus-stop information have been studied and implemented. An EU-related project is RECIPROCITY (Replication of innovative concepts for peri-urban, rural, or inner-city mobility in 2021–2023 including two minor cities from Finland accelerating the replication of innovative mobility solutions (RECIPROCITY, 2021).

Today, the country-wide service structure reform of social welfare and health care (SOTE) started last year in 2023. The health care services and their availability will be completely changed having a major impact on mobility to services all over the country. This will cause a major change in traveling to hospitals and health centers and traveling in the cities and regions. The second major change is the new law requiring that municipalities make their plan of the activities that are necessary to reach the climate neutrality targets. The results so far show how demanding is to reach the targets especially related to vehicle traffic. The service improvements and new mobility systems have been accepted well. The research activities of the new challenges have just started.

6   Conclusions

Many of the minor cities have already been planning their carbon-neutral reduction process and elaborated the needed action plans. Because of the marked impact of the transport emissions (usually ca. 20–25% of all the city carbon emissions), they have included mobility and transport to all carbon-cutting activities of the city in the attempt to be carbon-neutral already in 2030.

The inevitable conclusion seems to be that major impacts can be achieved only by reducing vehicle mileage and the need for transport. We should acknowledge that the cities' plans with novel and strong measures can be acceptable only by engaging and committing the users to the process of change. It is also clear as the research results show that users' behavioral change is a complex approach - and requires in-depth insight and expertise. The dynamics between land use and mobility transport is a complicated and long-term process. Most of the minor cities have already adapted their land use according to vehicle use and implemented old-fashioned land use plans. Although the authorities are now to react to the devolvement, it will take quite a lot of time, resources, and system re-arrangements until the land use–transport linkage model is changed.

Minor cities seem to be cautious of introducing measures that have effects on vehicle kilometers of travel is one of the main polluters of transport. In Finland for example, the LEZ (Low Emission Zone) concept has not been adopted as in many other European cities. This cautious attitude is understandable because private car use is really necessary to do everyday economic and social activities. Hence such restrictions are politically sensitive. Cities have long been used to rely on private car and commercial vehicle use in their service architecture and supply chains. This has been one of the main directives in their system planning.

It seems that it may not be widely acceptable in Finland, with its small population and long distances, to implement drastic measures without letting the citizens and the economy adapt and accept the by and large inevitable actions to tackle sustainability challenges.

Future research may seek to incorporate the findings into other studies aimed at sustainable development design and implementation, especially as contemporary cities require planning for zero-emission activities. Given the complexity of behavior change, it becomes important for municipalities to find the best way to engage users and participate in the transition process. In addition, the potential impact of incorporating user-centered approaches in the design and execution of future trials toward zero-carbon goals.

References

1. Borgato, S., Fermi, F., Chirico, F., Bosetti, S.: Study on costs and benefits of the sustainable urban mobility transition: D3: Final Report (2021). http://www.eiturbanmobility.eu/
2. Circonnact. Circular Vienna - Circonnact. Circonnact (2023). https://www.circularchange.com/news-cc/2023/7/18/new-circular-insider-in-austria
3. City of Lahti. 2021. The potential of traffic emission reduction and main measures in the City of Lahti. 29.3.2021 (2021)
4. Hussain, S., Ahonen, V., Leviäkangas, P.: Sustainability assessment of smart mobility projects in Finland: a comparative analysis. easychair.org (2023)
5. Smeds, E., Jones, P.: Developing transition pathways towards sustainable mobility in European cities: Conceptual framework and practical. University College London (2020). https://www.diva-portal.org/smash/get/diva2:1791450/FULLTEXT01.pdf
6. Benviroc Oy. Vaasa's greenhouse gas emissions report (2020). https://www.vaasa.fi/app/upl oads/2020/09/b1049193-co2-raportti_vaasa_18022020.pdf
7. Liimatainen, H., Viri, R.: Achieving the transport emission targets 2030 - review of policy measures. Finnish Climate Panel (2017). https://ilmastopaneeli.fi/hallinta/wp-content/upl oads/2024/07/Ilmastopaneeli_Liikenne_2017.pdf
8. Moreno, C., Allam, Z., Chabaud, D., Gall, C., Pratlong, F.: Introducing the "15-Minute City": sustainability, resilience, and place identity in future post-pandemic cities. Smart Cities 4(1), 93–111 (2021)
9. Hussain, S., Ahonen, V., Karasu, T., Leviäkangas, P.: Sustainability of smart rural mobility and tourism: a key performance indicators-based approach. Technol. Soc., 102287 (2023)
10. The Ministry of Transport and Communications. The National Transport System Plan 2021–2032. Finnish Government 2021:75 (2021). http://urn.fi/URN:ISBN:978-952-383-749-2
11. SUMP 2021. European Platform on Sustainable Urban Mobility Plans. Sustainability Urban Mobility Planning in Smaller Cities and Towns (2021). www.eltis.org
12. RECIPROCITY. Deliverable D4.3. Catalogue of mobility best practices (2021). https://recipr ocity-project.eu/communication-mats
13. Statistics Finland. (2023). Quality of work life. https://stat.fi/en/publication/cln0hlj6d8jlh0a vttwdum2g2

Open Access This chapter is licensed under the terms of the Creative Commons Attribution 4.0 International License (http://creativecommons.org/licenses/by/4.0/), which permits use, sharing, adaptation, distribution and reproduction in any medium or format, as long as you give appropriate credit to the original author(s) and the source, provide a link to the Creative Commons license and indicate if changes were made.

The images or other third party material in this chapter are included in the chapter's Creative Commons license, unless indicated otherwise in a credit line to the material. If material is not included in the chapter's Creative Commons license and your intended use is not permitted by statutory regulation or exceeds the permitted use, you will need to obtain permission directly from the copyright holder.

Active Travel and the Common Good: Identifying the Common Good in Public Consultation Reports for Irish Local Authority Active Travel Projects

Mary Noll Venables[✉]

DBFL Consulting Engineers, Ormond House, Ormond Quay, Dublin, Ireland
mary.nollvenables@dbfl.ie

Abstract. This paper examines conceptions of the common good in reports prepared in 2022 by Irish Local Authorities for public consultation on proposed active travel schemes. The proposed active travel schemes form a key component of Irish commitment to reducing transport emissions and represent a public good, which can be used simultaneously by many. The reports are intended to inform the public and local authority elected members about the proposed schemes before elected members vote whether to proceed with the scheme. Approval by elected members of Local Authority projects, including active travel schemes, sits within the Irish land use planning system, which provides for proper planning and sustainable development in the interests of the common good. This study investigates understandings of the common good found within the reports and assesses the degree to which active travel schemes are appreciated as common and public goods that contribute to transformative change in the transport sector. The paper identifies a lack of guidance on shaping reports for public consultation and calls for the development of compelling narratives of how active travel schemes serve the common good and expedite the transport changes envisioned in the Republic of Ireland's Climate Action Plan.

Keywords: Active Travel · Local Authority Development · Planning Processes · Procedural and Teleological Common Good · Public Good · Narratives

1 Introduction

Around the world, governments are looking to reduce greenhouse gas emissions through integrating land use and transport planning [1]. In the Republic of Ireland, the 2023 Climate Action Plan calls for delivering 1,000 km of active travel infrastructure to enable a 20% reduction in vehicle kilometers and a 50% increase in daily active travel journeys by 2025.

Realizing active travel projects in Ireland is not straightforward. Any active travel project, beyond minor schemes, must undergo statutory consultation and be authorized

---

This work contains my personal views and not the views of DBFL Consulting Engineers.

© The Author(s) 2025
C. McNally et al. (Eds.): TRAconference 2024, LNMOB, pp. 487–493, 2025.
https://doi.org/10.1007/978-3-031-85578-8_64

under Sect. 38 of the Roads Traffic Act 1994 or approved by Local Authority (LA) elected members under Part 8 of the Planning and Development Act 2000. For projects that can be approved under the Planning and Development Act, many LAs prepare Part 8 Reports (P8Rs) about the schemes; there is, however, no statutory requirement for P8Rs.

The proposed active travel infrastructure can be considered a quasi-public good that promotes modal shift and delivers co-benefits of better air quality, lower noise levels, improved health, and biodiversity gains. Enhancing public goods is central to pursuing the common good, one of the main goals of the Irish land use planning system [2]. However, active travel proposals are often negatively received by communities, suggesting that not everyone sees a common good or public good in AT schemes [3].

This paper seeks an understanding of the common good in P8Rs. It finds that the P8Rs describe a limited range of procedural and teleological common goods and consequently struggle to overcome the social gap between government policy and local acceptance, and to address prevailing mental models that value private vehicle-based mobility over sustainable accessibility [4]. The common good offered in the P8Rs under review offers a necessary but not compelling understanding of the common good presented by the active travel schemes. These limitations hinder LAs' efforts to construct the active travel infrastructure needed to deliver emissions reductions and other benefits.

2 Theoretical Background

2.1 Planning, Public Goods and the Common Good

The common good rests on a philosophical foundation that stretches back to the ancient Greeks, who recognized that government, a justice system and defense required the work of a community. In its modern context, the common good is found in arrangements 'that are conducive to the rights and welfare of people' and represent the search for a better world [5].

The common good is central to Irish land use planning. In 1963 the first effective planning act began with the preamble 'An Act to make provision, in the interests of the **common good**, for the proper planning and development of cities, towns and other areas'; a similar formulation has been included in every subsequent Planning Act.

The pursuit of the common good includes protecting and enhancing public goods like public health programs, public transportation, police services and other government services like active travel infrastructure that require public support and can be enjoyed simultaneously because they are non-exclusive and non-appropriable [2].

Two approaches to the common good are prevalent within the discipline of planning: a traditional teleological definition based on the results that planning produces, which aligns more closely with providing public goods, and a newer understanding based on the procedural functioning of the planning system [6–8]. Governmental plans to advance a teleological common good or public good often conflict with individual preferences, an incongruence sociologists call the 'social gap' [9]. Others suggest that localized objections to government plans may be better considered as evidence of 'conflicting modes of valuation'—not a rejection of the common good, but a rejection of a particular vision of the common good [10]. In the case of active travel, the public and common good of active travel sometimes conflicts with a different common good, namely an established mentality that values private vehicle-based mobility [4].

3   Methodology

P8Rs were retrieved from each of the 31 Irish LAs' public consultation web portals where available or through Google searches for the name of the Local Authority and Part 8 consultation when the LA did not have a public consultation portal. The search was limited to 2022 because the Covid pandemic had mostly passed, and the national government had greatly increased funding for active travel projects. The definition of an active travel project included walking and cycling routes, cycle lanes, pedestrian crossings, new and improved footpaths, pedestrian and cycle bridges, junction tightening, junction signalization, public lighting, and filtered permeability. A total of 45 P8Rs for active travel schemes were identified.

The reports were analyzed qualitatively, using a grounded theory approach. Key indicators, i.e. descriptions of the common good, were identified through an iterative process, noting conceptions of the common good as they appeared [11]. Because teleological understandings of common good have greater similarity with the provision of public goods, the analysis was attuned to the categories of procedural and teleological common good.

4   Results

4.1   Overview of Part 8 Reports (P8Rs)

Table 1   The percentages of P8Rs that contained depictions of procedural (P) and/or teleological (T) common good

Topic	Procedural (P)/ Teleological (T) Common Good	Included in P8Rs
Safety	T	76%
Policy	P	73%
Design manuals	P/T	62%
Access	T	56%
Submissions procedure	P	53%
Healthier transport modes	T	42%
Modal shift	T	36%
Environmental gains	T	24%
MCA	P	18%
'Common good'	P/T	0%

The proposed active travel schemes were distributed throughout the country in 15 of 31 Irish LAs. The length of the P8Rs ranged from a description on a webpage to 101 pages (mean: 22 pages; median: 19 pages). All P8Rs were written by the LA or an engineering consultancy except for one written by a landscape architect. Multiple types of procedural and teleological common goods were identified in the 45 P8Rs (see Table 1), although the term 'common good' was not found in any of the reports.

4.2 Procedural Common Good

Policy. Almost three-quarters of the P8Rs reference relevant national, regional and local planning policies, have an objective to facilitate planning policy or conclude that because the proposed scheme aligned with policy, it supports 'proper planning and sustainable development'.

Submissions Procedure and Multi-Criteria Assessments (MCA). Over half (53%) of the reports contain instructions on how to make a submission on the proposed scheme or how to challenge environmental screening decisions. 19% of the P8Rs include an MCA in the P8R or mention that an MCA had been performed. Describing submission procedures and MCAs emphasizes that the processes of holding public consultation or selecting a route advance the common good.

Design Manuals. The reference to design manuals in almost two-thirds of the P8Rs combines procedural and teleological common goods. Design manuals set professional standards that scheme designers are obliged to uphold; they also represent a picture of well-designed public goods. The P8Rs reference multiple design manuals. Sometimes P8Rs include an objective to design a scheme in compliance with Design Manual for Urban Roads and Streets or the National Cycle Manual.

Including design guides in P8Rs also suggests that if a scheme is designed according to design guidelines, a teleological common good will be achieved, a sentiment clearly expressed in the P8R for the R510 Improvement Scheme in County Limerick: *"This Scheme aims to deliver improved safety, comfort and security for cyclists, pedestrians and the mobility impaired on the R510. This objective is to be achieved through the delivery of facilities which are designed to comply with the National Cycle Manual, the Design Manual for Urban Roads and Streets, Safe Routes to School Design Guide, Transport Infrastructure Ireland Publications and the Traffic Management Guidelines"* [12].

4.3 Teleological Common Good

Access and Healthier Modes of Transport. Over half of the P8Rs outline that the proposed scheme provides access to local attractions. Almost half of the P8Rs state that the proposed schemes give people the opportunity to choose healthier modes of transport to reach their destinations. The reports frequently mention links and linkages, especially in relation to schools and the cycling network.

The cumulative effect will, as summarized in a P8R from Co. Cork, *'give residents in the area an alternative to using their car for local journeys and allow them to choose a more sustainable, healthier mode of transport.'* [13] Together the mentions of access and healthier transport modes create a picture of a better world in which the public good of active travel makes it easier to reach destinations by foot, bicycle, or scooter.

Modal Shift and Environmental Gains. When more people walk, wheel or cycle to reach their destinations, the modal share of active travel increases and greenhouse gas emissions decrease. Almost a third of the P8Rs mention this potential for modal shift. Even though active travel infrastructure is no guarantee of modal shift, it forms a necessary first step [14].

Roughly a quarter of the P8Rs mention environmental concerns beyond screening for Appropriate Assessment or Environmental Impact Assessment. The reports highlight net tree gain, biodiversity enhancements, pollinator friendly zones, new hedging and street furniture. In these reports, an enhanced environment is a common, public, non-appropriable good for all to share.

Safety. User safety is the most frequently mentioned teleological common good, occurring in over three-quarters of the P8Rs. Frequently schemes were described as improving the 'safety, comfort, security for cyclists.'

The emphasis on safety functions as a double-edged sword. On the one hand, safer infrastructure can ease common fears about cycling or walking [15]. On the other hand, the emphasis on safety could imply that walking, wheeling or cycling are inherently unsafe. Under this interpretation, it is unclear whether the safety of the proposed scheme is a common good after all.

5   Discussion

The P8Rs surveyed describe proposed active travel infrastructure that is intended to help Ireland cut transport emissions by 51% by 2030, as well as provide a host of other societal and personal benefits. They contain limited procedural and teleological understandings of the common good and struggle to address the social gap and conflicting common goods. As a result, they often fail to promote the public and common good of active travel schemes.

The P8Rs do not acknowledge the social gap between local situations and government policy or design manuals. Relying on planning and other governmental policies to make an argument for a proposed scheme assumes knowledge and support for government policies among local communities, both of which may be lacking.

Most importantly, the P8Rs do not challenge the widely shared assumption that a private vehicle is the default mobility mode, and all other forms of transport are optional or recreational. Common objections to active travel schemes are well-known, yet the P8Rs do not address the established mental mode of automobility [16, 17].

There is no legislative requirement for P8Rs; the P8Rs studied here vary from LA to LA and from project to project. The Irish National Transport Authority, the common funding agency of active travel projects, gives guidance on crossings, traffic signals and roundabout redesign. Guidance on how active travel projects are presented is also needed. Giving more attention to the social gap, other understandings of the common good and the public good active travel provides could help increase public acceptance of active travel projects and shift perceptions of active travel as real travel, not just a leisure activity. At a minimum, guidance could recommend preparing a separate document with planning policy and design standards, freeing the P8R to focus on creating a persuasive story that promotes the transformative potential of active travel [18].

References

1. RTPI Research Paper: Net Zero Transport: the role of spatial planning and place-based solutions, RTPI website. http://rtpi.org.uk. Accessed 26 Sept 2023
2. Klosterman, R.E.: Arguments for and against planning. Town Plann. Rev. **56**(1), 5–20 (1985)
3. Timmons, S., Andersson, Y., McGowan, F., Lunn, P.: Using behavioural science to design and implement active travel infrastructure: a narrative review of evidence. ESRI Working Paper No. 745 (2023)
4. OECD: Redesigning Ireland's Transport for Net Zero: towards systems that work for people and the planet (2022)
5. Chomsky, N. Lecture III: what is the common good? J. Philos. **110**(12), 685–700 (2013), quote p. 685
6. Campbell, H., Marshall, R.: Utilitarianism's bad breath? A re-evaluation of the public interest justification for planning. Plann. Theory **1**(2), 163–187 (2002)
7. Lennon, M.: Planning for the common good. London, Routledge (2022)
8. Murphy, E., Fox-Rogers, L.: Perceptions of the common good in planning. Cities **42**, 231–241 (2015)
9. Gonzalez, A., Daly, G., Gleeson, J.: Congested spaces, contested scales—a review of spatial planning for wind energy in Ireland. Landsc. Urban Plan. **145**, 12–20 (2016)
10. Eranti, V.: Re-visiting NIMBY: From conflicting interests to conflicting valuations. Sociol. Rev. **65**(2), 285–301 (2017)
11. Zamani, B., Babaei, E.: A critical review of grounded theory research in urban planning and design. Plan. Pract. Res. **36**(1), 77–90 (2020)
12. Limerick City and County Council: R510 Improvement Scheme, LCC website: https://myp oint.limerick.ie/en/consultation/part-viii-r510-quinns-cross-raheen-roundabout. Accessed 7 Oct 2023
13. Cork County Council: Cork Road Pedestrian Crossing and Active Travel Link, CCC website: https://www.corkcoco.ie/en/resident/planning-and-development/public-consultations/clo sed-part-8-development-consultation/part-8-proposed-r611-cork-road-pedestrian-crossing-active-travel-link-to-cetb-site. Accessed 7 Oct 2023
14. Handy, S.: Thoughts on the meaning of mark Stevens's meta-analysis. J. Am. Plann. Assoc. **83**(1), 26–28 (2017)
15. Central Statistics Office Ireland. Sustainable Mobility and Transport 2021. https://www. cso.ie/en/releasesandpublications/ep/p-smt/sustainablemobilityandtransport2021/cycling/. Accessed 22 Dec 2023
16. Egan, R., Caulfield, B.: Exploring public opposition to active travel planning in Dun Laoghaire-Rathdown. Synthesis Report (2023)
17. Walker, I., Tapp, A., Davis., A.: Motonormativity: how social norms hide a major public health hazard. Int. J. Environ. Health (2023)
18. Freudendal-Pedersen, M.: Sustainable urban future from transportation and planning to networked urban mobilities. Transp. Res. Part D, 82 (2020)

Open Access This chapter is licensed under the terms of the Creative Commons Attribution 4.0 International License (http://creativecommons.org/licenses/by/4.0/), which permits use, sharing, adaptation, distribution and reproduction in any medium or format, as long as you give appropriate credit to the original author(s) and the source, provide a link to the Creative Commons license and indicate if changes were made.

The images or other third party material in this chapter are included in the chapter's Creative Commons license, unless indicated otherwise in a credit line to the material. If material is not included in the chapter's Creative Commons license and your intended use is not permitted by statutory regulation or exceeds the permitted use, you will need to obtain permission directly from the copyright holder.

Taking SUMP to the Streets of Izmir

Geert Koops[✉] and Carolina Ramos

Panteia, Bredewater 26, 2715, CA Zoetermeer, The Netherlands
{g.koops,c.ramos}@panteia.nl

Abstract. The SUMP methodology is characterized by involving stakeholders and citizens using a transparent participatory approach. In this, the SUMP methodology differs from traditional transport master planning, which was primarily led by experts. During the creation of the SUMP for Izmir, which is the first full SUMP in Türkiye, citizens and stakeholders were involved in a manner distinct from the traditional citizen and stakeholder consultation events. Instead of inviting them to the domain of mobility planners, the mobility planners ventured into the streets of Izmir to collect input and feedback for the SUMP. This shift in mindset cannot be achieved overnight. In this paper we elaborate on how we organized the citizen engagement and the preparations made. The preparation comprised of three pillars: a robust communication strategy, capacity building for Izmir municipality staff and proper event planning.

Keywords: SUMP · Izmir · Co-creation · participatory planning

1 Introduction

The citizen engagement is part of a larger project: SUMP Izmir. The purpose of the project is to create a Sustainable Urban Mobility Plan (SUMP) for the Metropolitan Municipality of Izmir (IMM). Izmir is one of the associated cities of the 100 European cities to be climate neutral and smart cities by 2030[1]. Although several successful projects have been implemented in recent years towards achieving that goal, there is a lack of an integrated mobility vision. By implementing this SUMP, IMM wants to establish a comprehensive integrated strategy that moves beyond the traditional transport planning by making sustainability a top priority within its urban mobility planning (Fig. 1).

Izmir is a city in Western Türkiye with approximately 4.5 million inhabitants. With its hilly terrain, hot summers, and C-shaped urban morphology around its bay, active mobility is not a popular modal choice[2]. Therefore, a primary objective in the SUMP is to stimulate citizens to walk and cycle more instead of using their private car. Another challenge in the project is the transition of the informal public transport towards a zero-emission system. Furthermore, being a port city, Izmir faces several challenges with its internal logistic and freight transport which complicates the transition to the climate neutrality[3].

---

[1] https://urbinat.eu/articles/100-climate-neutral-and-smart-cities-by-2030/.

[2] https://izmir.ktb.gov.tr/EN-239196/general-information.html.

[3] Analysis of Izmir's current mobility situation (internal document).

© The Author(s) 2025
C. McNally et al. (Eds.): TRAconference 2024, LNMOB, pp. 494–499, 2025.
https://doi.org/10.1007/978-3-031-85578-8_65

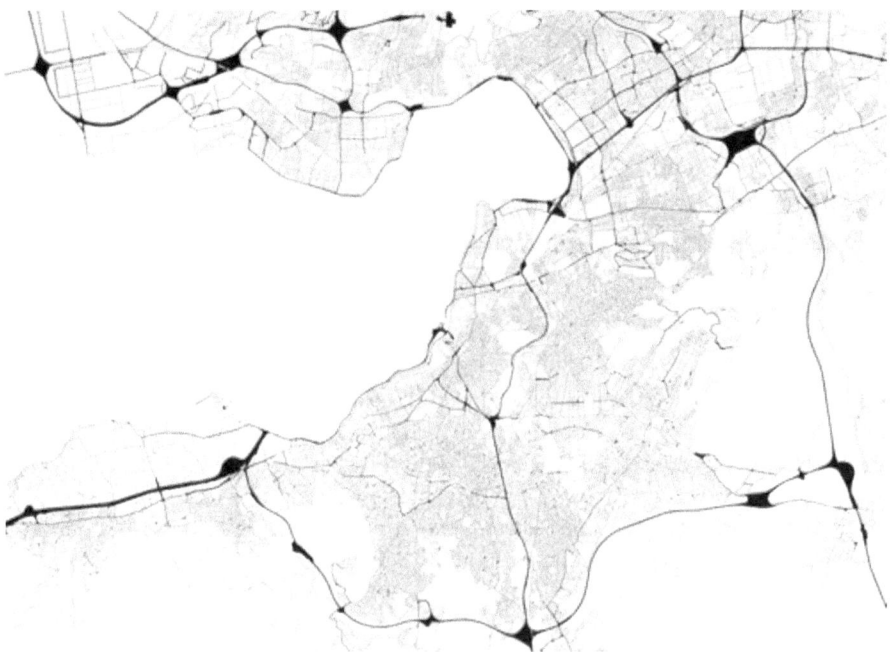

Fig. 1. Map of Izmir. Image courtesy of the SUMP Izmir project, reproduced with permission.

The first milestone of the project was the **Engagement & Communication Plan (ECP)** that described how to engage the public throughout the project, elaborating on this plan is the focus of this paper. The objective of the ECP is that with active citizen co-creation not only to co-create a stronger SUMP-Plan, but also to win the hearts and minds of citizens to use sustainable mobility options more. First we will describe the plan and the engagement activities so far. Secondly, we will examine the success factors of the innovative approach towards citizen engagement. Finally, we will present conclusions and discuss the tangibility of the approach to other cities who are in the process of creating a SUMP.

2 Citizen Engagement Activities

As a base for all engagement activities the SUMP Izmir team developed an **Engagement and Communication Plan (ECP)**. The ECP had the following objectives:

- Establishing a framework and creating the spaces for **collaboration among key stakeholders** to **reach agreements** and a consensual basis to support the change management process in Izmir and the sustainable mobility development - SUMP.
- Enabling the **inclusion of various stakeholders** and participants throughout the planning process. This includes expanding the scope of potential actors that could participate in the process and support the project, also considering especially concerns of groups with less ability to articulate their concerns and groups with specific needs.

- Adapting the integrated engagement approach to the **local conditions of Izmir**
- Supporting **transparency and accountability** to the public throughout the planning process, thereby increasing the overall quality and credibility of decision-making in Izmir.

The engagement process consisted of three stakeholder engagement events and two citizens' forums to interact with the public and co-create a sustainable urban mobility vision and plan with its citizens. At the time of writing, one stakeholder event and two citizens forums were organized.

2.1 Stakeholder Workshop

A first stakeholder workshop was organized to establish an interactive dialogue with stakeholders, to collect their opinions on current mobility issues in Izmir, how they envisage the (ideal) future, and how to shape the city's urban mobility. The engagement during the workshop began with a Mentimeter, where participants could indicate mobility improvements and their priorities towards those improvements. Afterwards a quiz was designed to raise awareness about the mobility situation in Izmir.

During a voting exercise stakeholders could express their preferred vision of the city's future. For instance, participants could choose between a city that prioritises mass-transit and a city that prioritises active travel. Finally, world café[4] breakout sessions were held to discuss how to shape Izmir urban mobility future. The groups engaged in discussions centered on six different topics: (1) Decarbonization and Air Quality, (2) Public Transportation and Shared Mobility Services, (3) Traffic and Increasing Private Car Ownership, (4) Active Mobility, (5) Urban Logistics, and (6) Accessibility of Vulnerable Groups. Moderated by six experts from IMM, each group spent 15 min to discuss the assigned topic before moving to the next group.

The stakeholders workshop gave insights into the perceived mobility issues in the city (qualitative analysis of the mobility situation) and offered first indications for the SUMP vision and measures for Izmir.

2.2 Citizens Forums

During the engagement with citizens the SUMP Izmir decided not to organise the event hotels or offices, but rather on the streets. We believe that in order to reach citizens you can better approach them directly instead of asking them to come to your event. Therefore, stands from the SUMP Izmir team were set up in various locations in the city. At those stands citizens could (1) spin the Awareness Wheel, where participants can gain knowledge via quiz questions about mobility in Izmir in a fun and interactive way, (2) give feedback and suggestions on specific aspects of mobility in Izmir by posting sticky notes on different areas of neighborhood maps, (3) vote on their preferred future mobility modes in Izmir.

The first citizen forum took place in June 2023, the second in September 2023 during the European Mobility Week, a week full of urban mobility outreach. During the second

---

[4] https://theworldcafe.com/key-concepts-resources/world-cafe-method/.

citizen forum the sticky notes activity was replaced by a voting exercise, where citizens could express their priority towards the preliminary SUMP measures that were previously identified by the SUMP Izmir team. Also, the quiz was the second time more focused on what could potentially be good SUMP measures. The citizen engagement provided insights into the issues experienced by citizens and what kind of city they prefer. These results were compared with the stakeholders results and the results of the quantitative analysis of urban mobility in the city, which resulted in a balanced assessment of in which direction the city is evolving.

3   Succes Factors for Citizen Engagement

3.1   Experienced Event Planning Team

A success story has many shoulders to stand on. In the SUMP Izmir project, these are a combination of the experts from IMM, consultants from Germany, Netherlands and Türkiye, and students from Izmir. Via "Young professionals Orientation Program (SUMPYS)", the SUMPYs were involved in the execution of the engagement activities.

The team did not start from scratch. The engagement activities built upon SUMP experience in other cities and previous co-creation in Izmir, called the 'Izmir story'. The key was to bring different expertise together, so that international best practices were combined with the local context.

3.2   Strong Communication Strategy

'Knowing your audience' is the first step in a successful communication strategy. The SUMP Izmir team first mapped the target (Stakeholders)audience for her engagement activities. During these mapping exercises it was concluded that the usual suspects are most likely to engage in the co-creation of the SUMP. To reach the non-usual suspects, going out in the streets is not enough, you need to prepare your audience with a strong communication strategy.

The highlights in the SUMP Izmir communication strategy are the dedicated creative team for social media posts and the use of influencers. A dedicated creative agency was hired to design social media posts, see example below. To amplify our online message to the public, we dedicated a budget to influencers to post about the SUMP Izmir activities & using sustainable modes for travelling around Izmir (Fig. 2).

3.3   Capacity Building of Municipal Staff

To prepare the employees of Izmir metropolitan municipality for a shift in mindset, multiple trainings have been organised. These trainings followed the participatory planning approach. Here employees of IMM, who were not a part of the core team that works on the SUMP, could familiarize themselves with planning a city from multiple perspectives. With participants from different departments of the municipality, urban planners became familiar with incorporating unconventional views in their vision for the city.

SUMP trainings were given on the topics of: (1) creating a walkable city, (2) cycling in your city, (3) introduction of micromobility, (4) Hubs training, (5) Blue-Green Corridor

Fig. 2. Example social media posts. Image courtesy of the SUMP Izmir project, reproduced with permission.

training. During the Hubs training a serious game was played among participants to get a feeling with weighing different interest in your city.

Most crucial in the trainings was that participants could distance themselves from their daily work and were totally free to redesign their city. This freedom gave the participants new views on their city and paved the way for an open-minded mindset to co-creation with citizens.

4   Conclusions and Tangibility

The success of the engagement of citizens in the SUMP Izmir project is not the result of the traditional technical transport planning steps. Its success story is the result of a participatory co-creation process that enabled a behavioural change that reached the mind and soul of the citizens. SUM-Planning demands a different way of thinking, which can be especially challenging in new cultural settings, that are not used to an integrated planning approach. Thus, adapting and understanding the local context is essential as well as the development of innovative communication tools for and reaching the entire community.

When transferring the experience from Izmir to other cities certain conditions should be considered:

- When you take citizen engagement to your streets, organise it in the shoulder months of the summer (May-June, September-October). The sunny weather of Izmir makes organizing on-street activities easier than in other cities.
- Seek external help with organising this new approach of citizen engagement when you are not familiar with the concept yourself. For instance, by hiring a creative agency.
- Ensure your organization is prepared for a shift in mindset towards co-creation. Organising capacity building trainings can facilitate in this transition process.
- Be bold! Step out of your own comfort zone and be open to unexpected interactions with citizens. They are more enriching (for your mobility plan) than you would initially think.

Finally, a meta-study comparing different engagement plans with the SUMP Izmir would be an interesting next step. With this meta-study the success factors per engagement strategy can be defined. A meta-study can also proof which lessons-learned are tangible and which ones are not.

References

URBinAT. 100 Climate Neutral and Smart Cities by 2030 (2022). https://urbinat.eu/articles/100-climate-neutral-and-smart-cities-by-2030/

Izmir Governorship. General Information (2024). https://izmir.ktb.gov.tr/EN-239196/general-information.html

The World Cafe. Key Concepts & Resources: World Cafe Method (2024). https://theworldcafe.com/key-concepts-resources/world-cafe-method/

Panteia. Izmir's Current Mobility Situation. For Internal use only (2024)

Open Access This chapter is licensed under the terms of the Creative Commons Attribution 4.0 International License (http://creativecommons.org/licenses/by/4.0/), which permits use, sharing, adaptation, distribution and reproduction in any medium or format, as long as you give appropriate credit to the original author(s) and the source, provide a link to the Creative Commons license and indicate if changes were made.

The images or other third party material in this chapter are included in the chapter's Creative Commons license, unless indicated otherwise in a credit line to the material. If material is not included in the chapter's Creative Commons license and your intended use is not permitted by statutory regulation or exceeds the permitted use, you will need to obtain permission directly from the copyright holder.

Operational Performance of Passenger Ferry Service

Samruddhi Gujar[✉] and Mohit Dev

School of Planning and Architecture, Bhopal 462030, India
samruddhigujar1704@gmail.com, mohit.dev@spab.ac.in

Abstract. Passenger ferry boats of different capacities operate and are one of the most affordable modes of transportation. This research investigates the effectiveness of passenger ferry services within Mumbai Municipal Boundary, operating in its backwater, creek networks, and western and eastern coastlines. Passenger ferries have the potential to promote sustainable and resilient transportation in the region. The literature review shows an absence of evaluation mechanisms to assess the quality and performance provided for passenger ferries in Indian cities using the Level of Service (LOS) concept. The study aims to assess operational performance by proposing a step-by-step evaluation framework for ferry services based on the LOS. Both passengers and operators' perspectives are considered while suggesting proposals. The outcome is in the form of a procedure that can be adopted for forming and calculating route-wise LOS and recommendations for operators to generate additional revenue. The insights gained from this research will be valuable for policymakers, academics, and stakeholders, enabling them to assess, improve, and allot investments for ferry services in their respective cities as per their unique requirements.

Keywords: Passenger Ferry Service · Level of Service · Evaluation Framework

1 Introduction

Waterways offer advantages such as less resistance to traction and low maintenance costs as the channels are naturally available [1]. Understanding the functions and maintenance of a ferry system is crucial for providing reliable services. Boadu Solomon et al. advise that it is essential to understand an institutional framework and how operators can benefit from efficiently managing the services [2]. In Mumbai, the infrastructure decisions are taken by the Maharashtra Maritime Board, whereas other operational decisions lie within the private operators such as determining fare prices, number of vessels operated, etc. Michael Tanko et al. examine excess travel fares incurred when choosing ferry transport over bus transport between the same origin and destination in Brisbane, Australia. The study highlights consideration of the additional factors beyond travel time when assessing the value of the water transit system [3]. A research gap is identified: the absence of LOS indicators and their stepwise evaluation procedure for water transportation in India. The LOS is a crucial indicator for transportation planning, benchmarking, and comparing systems, assessing safety, convenience, reliability, and efficiency, enabling planners

© The Author(s) 2025
C. McNally et al. (Eds.): TRAconference 2024, LNMOB, pp. 500–507, 2025.
https://doi.org/10.1007/978-3-031-85578-8_66

and managers to identify areas for improvement. C.J. Khisty emphasizes selecting and evaluating eight performance measures (PM) viz, accessibility, transit time, frequency, reliability, cost, public information, passenger comfort, and delay to assess the LOS of Washington State Ferries. The outcome is an approach for monitoring the performance of ferry service on an immediate basis and allocating resources for improvements [4]. Ka Ho Tsoi focuses on the declining importance of ferry transportation in finance. The research is tested in Hong Kong. This study emphasizes the importance of creating diverse opportunities, enhancing passenger experience, and integrating ferry services and pier infrastructure for the long-term viability of ferry transport, thereby increasing its usage [5].

2 Methodology

The Maharashtra State Road Development Corporation Limited maintains 20 ferry routes, eight of which are located inside the Mumbai Municipal Boundary at four different locations: Ferry Wharf, Gateway of India, Marve Jetty, and Gorai Jetty. A questionnaire-based survey is designed, and 157 samples are gathered from daily, non-daily, and tourist users. The user survey is aimed at those who utilize passenger ferry services regularly. No sampling approach is applied. The research then moves to analyzing qualitative data for existing scenarios and quantitative data for forming categories for each LOS parameter. Further, the Analytical Hierarchy Process is adopted to rank, and form composite LOS. The results include strategies to enhance ferry services from the standpoint of operators, as well as review processes for computing location-based LOS.

2.1 Existing Scenario

Mumbai's coastline attracts tourists with major attractions and amusement parks. The users prefer public transport (PT) services like buses, shared cabs, and autos to and from the ferry terminal for less travel time, comfort, and scenic views. During the monsoon season, travelers plan trips based on weather conditions and safety concerns. The operator's survey reveals multiple private operators offering ferry services, with daily trips ranging from 8 to 50. Fare prices are affected due to varying travel times and distances. The schedule is determined depending on the ferry service's total journey time and the ferry terminal's demand. It is also important to identify the profit and loss of the ferry system to implement changes based on the LOS of that identified route. The financial mechanism for ferry operators is tested by considering ticket fare prices. The current profit is Rupees 30,000 per month, but it incurs losses during monsoons, necessitating additional revenue for terminal improvements.

3 Parameter Measures (PMs) for Level of Service (LOS)

The PMs for LOS for the Indian ferry system are determined through a literature review and a user survey questionnaire is designed accordingly. Five PMs are selected for this research and are elaborated further. The LOS is categorized from A (best) to F (worst).

The existing classification of reviewed literature influences the formulation of the LOS categories while adjustments are made to accommodate available data. 31 tourist samples are excluded due to significant opinions as was evident while calculating LOS categories. Tourists primarily use ferry services during the summer and winter seasons, and their user numbers change dramatically. This research study focuses on generating LOS categories by analyzing data from consumers who utilize ferry services regularly.

3.1 Frequency

The frequency is determined by demand at the terminal and hence percentages of users' satisfaction with the current frequency of ferry services is calculated for forming the categories from A to F. Table 1 shows the distribution.

Table 1. Calculated LOS categories for frequency and location-wise LOS.

LOS Category	% of riders satisfied	Location	% of riders satisfied	LOS
A	100%	Ferry Wharf	34.84%	E
B	80% and above	Gateway Of India	50%	D
C	60% and above	Marve Jetty	100%	A
D	40% and above	Gorai Jetty	91.18%	A
E	20% and above			
F	Below 20%			

3.2 Travel Time (TT)

Assessing travel time is crucial for users to choose ferry services as a daily transport mode. So, comparing ferry travel time with PT time helps understand ferry efficiency compared to other transit options, ensuring a meaningful comparison. Here the method adopted is the total journey time using ferry by automobile travel time. The distribution of the LOS category is shown in Table 2.

Table 2. Calculated LOS categories for Travel Time along with location-wise LOS.

LOS Category	Index	Comparison with automobile	Location	Index - Average	LOS
A	<0.20	Ferry TT is less than 20% of automobile TT	Ferry Wharf	0.8	E

(*continued*)

Table 2. (*continued*)

LOS Category	Index	Comparison with automobile	Location	Index - Average	LOS
B	0.20–0.40	Ferry TT is 20 to 40% of automobile TT	Gateway Of India	0.5	C
C	0.40–0.60	Ferry TT is 40 to 60% of automobile TT	Marve Jetty	1.0	F
D	0.60–0.80	Ferry TT is 60 to 80% of automobile TT	Gorai Jetty	1.2	F
E	0.80–1.00	Ferry TT is 80 to 100% of automobile TT			
F	>1.00	Ferry TT is more than 100%			

3.3 Ticket Fare

PT fare significantly influences user willingness, especially for lower-income groups. High ticket fares discourage users. To urge usage, fares should be competitive or lower than driving costs, including fuel, parking, and maintenance. Thus, ferry Passenger Fares (PF) are compared to automobile Operating Costs (OC) for the same origin and destination and LOS is designed in Table 3.

Table 3. LOS categories for Ticket Fare along with location-wise LOS.

LOS Category	Index	Comparison with automobile	Location	Index - Average	LOS
A	<0.20	PF is less than 20% of automobile OC	Ferry Wharf	0.18	A
B	0.20–0.40	PF is 20 to 40% of automobile OC	Gateway Of India	0.23	B
C	0.40–0.60	PF is 40 to 60% of automobile OC	Marve Jetty	0.06	A

(*continued*)

Table 3. (*continued*)

LOS Category	Index	Comparison with automobile	Location	Index - Average	LOS
D	0.60–0.80	PF is 60 to 80% of automobile OC	Gorai Jetty	0.07	A
E	0.80–1.00	PF is 80 to 100% of automobile OC			
F	>1.00	PF is more than 100%			

3.4 Passenger Comfort

Passengers' experiences with PT rely on comfortable sufficient space, especially during peak periods. A clean and hygienic environment is equally vital for both regular commuters and tourists, both on the ferry and around the pier. Thus, the satisfaction levels of five indicators are assessed: cleanliness on the ferry, cleanliness around the pier, seating comfort, operator behavior, and ferry crowdedness. The average of these indicators is obtained for overall satisfaction categories for LOS, as shown in Table 4.

Table 4. LOS categories for Passenger Comfort along with location-wise LOS.

LOS Category	Index	% of passenger comfort	Location	Mean value	LOS
A	>10	100%	Ferry Wharf	8.02	B
B	8–10	80% and above	Gateway Of India	7.36	C
C	6–8	60% and above	Marve Jetty	6.16	C
D	4–6	40% and above	Gorai Jetty	6.98	C
E	2–4	20% and above			
F	0–2	Below 20%			

3.5 Vessel Capacity

Vessel capacity is the number of passengers and vehicles on board a ferry, affecting its functionality and reliability. It is crucial for meeting PT demand and ensuring service availability. The percentage capacity used is calculated based on the number of passengers and vehicles on board relative to the full capacity of the ferry. The distribution of the LOS category is shown in Table 5.

Table 5. LOS categories for Vessel Capacity along with location-wise LOS

LOS Category	% of capacity		Location	% Capacity	LOS
A	<20%	Least Capacity	Ferry Wharf	82.67%	E
B	20% to 40%	Low Capacity	Gateway Of India	100%	E
C	40% to 60%	Medium Capacity	Marve Jetty	170%	F
D	60% to 80%	High Capacity	Gorai Jetty	125%	F
E	80% to 100%	Full Capacity			
F	>100%	More than full capacity			

4 LOS Ranking Using Analytical Hierarchy Process (AHP) and Final Route LOS

The AHP - Multi-criteria decision-making method is adopted, and an expert opinion survey is conducted. The relative importance scale is used to rank all five parameters against each other in the form of a matrix. This formalizes a method for evaluating ferry LOS categories, using a ranking formed of the final PMs. The ranking shown in Table 6 will also help while selecting parameters for the improvement of route LOS.

Table 6. Final ranked parameters and their weights.

Rank	Parameters	Weights	%
1	Frequency	0.43	43%
2	Travel Time	0.21	21%
3	Ticket Fare	0.17	17%
4	Passenger Comfort	0.12	12%
5	Vessel Capacity	0.07	7%

Once calculation for location-wise LOS is completed, individual points are assigned i.e., a 5-point scale is used as A = 5, B = 4, C = 3, D = 2, E = 1, F = 0, and further multiplied by the weights from Table 6 to get the final LOS of that route. An example of the further process and relevant calculations are shown in Table 7.

The above table indicates that the Ferry Wharf location has LOS C, which can be enhanced by selecting the parameter through calculated ranking in Table 6.

Table 7. Final calculated LOS of the individual route.

Location	Parameters	LOS	Points (A)	Weights (B)	A X B	Final route LOS
Ferry Wharf	Frequency	C	3	0.43	1.29	2.9 = Approx 3 = LOS C
	Travel Time	E	1	0.21	0.21	
	Ticket Fare	A	5	0.17	0.85	
	Passenger Comfort	B	4	0.12	0.48	
	Vessel Capacity	E	1	0.07	0.07	
	Grand Total				2.9	

5 Recommendations and Conclusion

This research study aims to offer additional revenue generation recommendations to operators as stated in sub-Sect. 2.1 and proposes an evaluation procedure for forming LOS for any route using given PMs. The subsequent recommendations aim to increase their revenue generation. Charter service provision refers to offering exclusive ferry services for specific purposes or groups. Utilizing ferry terminal space and launches for advertising can generate additional advertising revenue. Collaboration with local tourism organizations can attract more customers and generate extra income. A separate administrative department, if not, should be established to evaluate the efficiency of a city's waterways transportation. Dedicated officials and policymakers of the department can adopt the procedure as described in this study, i.e., by understanding the existing scenario, identifying PMs for LOS assessment based on analysis, collecting data through various surveys for selected PMs, adopting the above methodology as in Sect. 3 for LOS calculation, or altering them according to the city's needs and further ranking them using AHP or another suitable method. By attentively studying customer feedback and preferences, operators and officials can make well-informed choices by adopting this step-by-step proposed framework to improve the passenger experience, ultimately leading to increased customer satisfaction levels for passenger ferry services in the Indian scenario.

References

1. Sriraman, D.: Long-term perspectives on inland water transport in India. Rite J. (2010)
2. Solomon, B., Otoo, E., Boateng, A., Koomson, D.: Inland waterway transportation (IWT) in Ghana: a case study of volta lake transport. Int. J. Transport. Sci. Technol. **10**, 20–33 (2021)
3. Tanko, M., Matthew, I., Yen, B.: Water transit and excess travel: discrete choice modeling of bus and ferry trips in Brisbane, Australia. Transport. Plan. Technol. **42**(3), 244–256 (2019)
4. Khisty, C.J.: Level-of-service measures for ferry systems. Transport. Res. Rec. **1222** (1989)
5. Tsoi, K., Loo, B.: Cutting the loss: international benchmarking of a sustainable ferry business model. Transp. Res. Part A **145**, 167–188 (2021)

Open Access This chapter is licensed under the terms of the Creative Commons Attribution 4.0 International License (http://creativecommons.org/licenses/by/4.0/), which permits use, sharing, adaptation, distribution and reproduction in any medium or format, as long as you give appropriate credit to the original author(s) and the source, provide a link to the Creative Commons license and indicate if changes were made.

The images or other third party material in this chapter are included in the chapter's Creative Commons license, unless indicated otherwise in a credit line to the material. If material is not included in the chapter's Creative Commons license and your intended use is not permitted by statutory regulation or exceeds the permitted use, you will need to obtain permission directly from the copyright holder.

EU-Rail FutuRe Project – Innovative CCS Solutions for Interoperable Regional Lines

José A. Reyes[1], José A. Quintano[1], Michal Matowicki[2], Petr Kačmařík[2], Peter Gurník[2], Noelia Medrano[3], Gabriele Ridolfi[4], Patrick Urassa[5], and Marta García[6(✉)]

[1] CAF Signalling S.L, Alcobendas, Spain
[2] AZD Praha SRO, Praha, Czech Republic
[3] Enclavamientos y Señalización Ferroviaria ENYSE SA, Alcobendas, Spain
[4] Rete Ferroviaria Italiana (RFI), Roma, Italy
[5] Department of Mechanical and Industrial Engineering, Norwegian University of Science and Technology, 7491 Trondheim, Norway
[6] UNIFE – The European Rail Supply Industry Association, Brussels, Belgium
marta.garcia@unife.org

Abstract. With the support of EU's key funding program Horizon Europe, the Europe's Rail Joint Undertaking (EU-Rail) aims to deliver a high-capacity integrated European railway network by eliminating barriers to interoperability, providing solutions for full integration, and achieving faster uptake and deployment of innovation. In this context, EU-Rail FutuRe project aims at tackling the increasing Total Cost of Ownership (TCO) of European interoperable regional lines by means of providing new innovative solutions to make these lines cost-effective and attractive, while meeting safety standards and improving reliability, availability, and capacity of the railway system. Interoperable regional lines are connected to the mainline railway system as part of the Single European Railway Area (SERA) and characterised by low-density passenger traffic, as well as rail freight services.

Keywords: EU-Rail · FutuRe · FRMCS · CCS TSI 2023 · ATO · ERTMS · TMS · ASTP · TIMS · Train Integrity · Train Length

1 Introduction

Regional lines are gradually disappearing. Current economic, social, and environmental conditions negatively impact their survival throughout Europe to the extent of being abandoned. In response, EU-Rail FP6 Project (FutuRe) is born to revitalize them by exploiting leading-edge technologies which lead to a reduction in the Total Cost of Ownership (TCO), while meeting safety standards and improving reliability, availability, and capacity of the regional railway system.

The expected outcomes of FutuRe shall form the basis for a common European regional rail development management framework characterized by green, digital, safe, and cost-efficient solutions where the Command, Control and Signalling (CSS) system plays a crucial role.

© The Author(s) 2025
C. McNally et al. (Eds.): TRAconference 2024, LNMOB, pp. 508–514, 2025.
https://doi.org/10.1007/978-3-031-85578-8_67

In the context of CCS, FutuRe work packages 3 and 8 are leading the assessment for the applicability of several solutions covering an integrated control and command system, which shall first be demonstrated in laboratory conditions targeting the Technology Readiness Level (TRL) 4/5:

- Automatic Train Operation (ATO), up to GoA4
- ERTMS/ETCS level 2, considering both Fixed Virtual Blocks and Moving Block implementations.
- Traffic Management System (TMS)
- Absolute Safe Train Positioning (ASTP)
- Train Integrity and Train Length

As a key enabler for the mentioned technologies, the Future Railway Mobile Communication System (FRMCS) shall bring further radio-communication possibilities, leading to broader digitalization while decoupling radio technologies and applications, thus making easier transitions towards new radio technologies in the future. Additionally, it shall contribute to the main goal of lowering both CAPEX and OPEX by employing Mobile Network Operators (MNO) public radio networks on low-density lines.

Building on the outcomes of previous European initiatives (Shift2Rail), the System Pillar, and other EU-Rail projects, FutuRe work packages 3 and 8 shall assess and demonstrate the suitability of the abovementioned technologies to make European regional lines an attractive and sustainable transport system.

2   Innovative CCS Solutions on Europe's Regional Lines

2.1   ATO Over ETCS L2

The European Railway Traffic Management System (ERTMS), which includes the European Train Control System (ETCS) for automatic train protection, the first baseline of ATO up to GoA2, the Global System for Mobile Communications for Railways (GSM-R) and FRCMS, as introduced in the newly published Technical Specification for Interoperability (CCS TSI 2023/1695), is being implemented all around Europe to ensure interoperability. Improving the safety of national and international train traffic and bringing benefit in terms of TCO of the entire CCS by reducing trackside assets are among the most relevant advantages of ERTMS on regional lines.

Depending on the application level, ETCS trackside and on-board subsystems are characterized by specific equipment on both the track and the train, their interaction, and their functions. The ETCS "levels" establish different uses of ERTMS as a train control system:

- ETCS Level 1: track-to-train spot communications, where a significant and costly amount of trackside elements are required to control the movement of the train. Train detection and train separation are performed by the trackside equipment of the underlying signalling system.
- ETCS Level 2: continuous communications between the train and the track, which enable a reduction in OPEX and CAPEX and increase line capacity. Train detection and train separation can be performed by the Radio Block Centre (in co-operation

with the train which sends position reports and train integrity information) and/or by other trackside equipment [1].

This technology is now being complemented by the Automatic Train Operation (ATO), which provides speed control, higher comfort and ergonomics for passengers, accurate stopping, door opening and closing, and driver-related functions [2].

In comparison with non-automated train operation and being applicable to a wide range of railway applications, among which regional lines are found, ATO over ETCS boosts a reduction in energy consumption, which is a key driver of FutuRe.

From the perspective of regional lines, FutuRe work packages 3 and 8 are conducting research into innovative up-to-GoA4 ATO over ETCS solutions in order to provide these lines with energy-efficient systems and higher punctuality.

2.2 Remote Driving

The introduction of a train system with Grade of Automation 3 or 4 (GoA3/4) on regional lines signifies a shift towards driverless train operation. In such a system, while the train manoeuvres all the operation automatically, staff might still be present for roles unrelated to driving or handling the train in the event of disruption. FutuRe work packages 3 and 8 are developing a remote driving demonstrator characterized mainly by on-board staff for monitoring purposes, use of public networks for communications, and a camera for transmission of live video feeds for control purposes.

The choice to use public networks for communication brings its set of challenges: security is crucial; encrypted and secure communication is essential to avoid potential cyber threats; reliability is another concern; the system has redundant communication paths to ensure uninterrupted connectivity. Furthermore, the real-time video streaming with minimal latency from the field will be displayed in the remote-control monitor for control purposes while the control commands will be relayed back to the train through a customized control system.

2.3 Absolute Safe Train Positioning (ASTP)

The current train location systems, which rely on trackside train detection equipment (e.g., track circuits or axle counters) coupled with on-board odometry and ETCS position reports with reference to the last ETCS relevant balise group encountered by the train, have proven to be effective, but imply inherent limitations being prone to drift-induced inaccuracies and frequent re-calibration and maintenance.

An Absolute Safe Train Positioning system (ASTP) is under development within EU-Rail aiming to combine different sensors and technologies in order to:

- Overcome some of the weak points of current ETCS odometry that have an undesirable impact on operation and traffic optimisation (e.g., increase in confidence interval of train positioning, systematic odometry error, high maintenance cost, and difficult drivability when approaching a target)
- Provide new additional benefits allowing:

o Trackside asset reduction by using virtual reference points, provided a standard Digital Map is available, instead of only physical balise groups, and facilitating the reduction of trackside train detection systems, in combination with the additional functions Train Integrity and Train Length.

o Help to continuously estimate train position along the track in order to minimise the distance run in non-supervised modes at start of mission or after recovery from a failure

- Provide output for different on-board and trackside consumers using position or other train kinematic data.

In a regional context, the most promising expected benefit is related to the reduction of the number of balises and of trackside train detection systems (whose maintenance can be even more critical in rural areas). This has an evident positive impact on the total cost of ownership of the CCS system and on the regularity of operation increasing the attractiveness of the regional rail service.

Through feeding ATO, ASTP can also be a crucial enabler for significant energy savings and optimisation of the use of track by a particular train.

The most promising technologies for ASTP that are considered for development and testing within FutuRe are:

a) **GNSS-based positioning system**: Global Navigation Satellite Systems (GNSS) are commonly used for positioning in a wide range of applications. The fact is that in Europe, GNSS is not utilized as a primary positioning technology in safety-relevant applications. GNSS-based localization functions use signals from satellites to determine the exact position of the train. However, the performance of GNSS can be hampered in areas with dense foliage or in urban areas, where satellite signals might be weak or distorted due to multipath. It is obvious that environments such as tunnels completely deny the use of this technology.

b) **Inertial Navigation Systems (INS)**: INS incorporate kinematic sensors to follow changes in the train position. These systems can function independently of external signals, making them reliable even in environments where GNSS signals might be compromised. However, because of the used physical principles, the synchronization with absolute position is needed for continuous operation in open world. Furthermore, the position quality degrades over time and travelled distance if it is not supported with the source providing absolute position. Therefore, GNSS and INS are usually coupled with the common positioning system to complement each other.

c) **Radio-based systems**: Certain train positioning systems leverage radio signals to triangulate the position of trains. The motivation is to bring the radio navigation signals into the areas where GNSS signals are weak. These systems can provide a position whose performance is comparable with GNSS. However, they require substantial infrastructure investment including the installation of radio towers along the railway network.

d) **Optical sensing and image processing**: In this approach, onboard cameras coupled with advanced image processing algorithms are utilized to precise the train's position based on visual landmarks and their recognition of objects in the database. This approach is relatively newer and might still be in the experimental or early deployment stages in 2026 when FutuRe shall finish.

2.4 Train Integrity and Train Length

Transitioning from trackside fixed blocks to fixed virtual or moving blocks results in an increase in capacity and in a potential reduction of CCS overall TCO by trackside asset reduction, which are the most remarkable advantages brought by ETCS level 2. One of the keystones of this promising implementation of the European ATP is the supervision of the completeness of trains (i.e., train integrity), which must be continuously monitored by the Train Integrity Monitoring System (TIMS) by supervising the train length and reporting it to ERTMS by means of a defined set of variables. Since train distancing is based upon the position of the safe rear end of the train provided by the vehicle itself, when train completeness is guaranteed, the value of the train length assumes a more significant meaning in terms of safety and therefore robust solutions.

2.5 Traffic Management System (TMS)

TMS is a fundamental actor in any signalling system. In the scope of regional lines, its evolution is drawn to follow two streams of development:

- To update the TMS by means of the inclusion of new features and innovative technologies (e.g., ATO, ASTP, etc.)
- To improve the existing TMS functions by adapting its current algorithms with novel and innovative mechanisms, such as the use of AI for conflict detection and resolution management.

By considering the implementation of both streams, the TMS shall remain as a key component of the signalling systems applied to regional lines, keeping its traditional and well proven way of operating while it evolves over time.

2.6 FRMCS: The Successor of GSM-R

GSM-R is one of the key parts in ERTMS as it supports the ETCS signalling system by providing a harmonised and standardised track-to-train communication network. Not only has it been crucial to the European automatic train protection system, but it has also been adopted by rail operators due to its secure operational communication functions. Therefore, it is a crucial part of the railway landscape.

Being one the most notable examples of mass-scale take-ups, it has however passed into obsolescence owing to the rapid development of new telecommunication assets. It is now estimated that GSM-R networks shall be life-expired by 2030, which shall lead to an increase in its maintenance costs.

Succeeding GSM-R, FRMCS comes in as the new railway telecommunication network that shall enable the modernisation of rolling stock and the digitalisation cycle of the entire sector. Technologies such as ATO, train monitoring systems, and remote application shall benefit greatly from its potential.

European lines, which account for half of those fitted with GSM-R worldwide, shall be gradually equipped with FRMCS over a lengthy period of at least 10 years between 2025–2035 [3].

2.7 FutuRe Socio-Economic Objectives

Socio-Economic Objectives (SEO) (Table 1) are strongly linked to the impacts of FutuRe work packages 3 and 8, which are derived from the validation of the demonstration activities (Table 2).

Table 1. Socio-economic objectives of FutuRe work packages 3 and 8

SEO	SEO Description	Expected Improvement
SEO1	Overall TCO reduction of the CCS system	Decrease of 25%
SEO2	CAPEX reduction of radio networks	Decrease of 15%
SEO3	Reliable cost-effective fail-safe on-board train integrity, train length detection, and train positioning	Increase of 15%
SEO4	Optimized energy consumption through ERTMS/ATO targeting GoA4	Decrease of 10%
SEO5	Optimized punctuality	Increase of 15%

Table 2. Demonstration activities of FutuRe work packages 3 and 8

DO	Demonstration objectives (DO)	Targeted TRL
DO1	ERTMS/ATO up to GoA4 to optimize energy consumption and achieve higher punctuality	TRL4/5 by 2025
DO2	ETCS L2 in its different system types by reducing the necessary wayside assets	TRL4/5 by 2025
DO3	TMS functionalities integrated with ATO to optimize traffic regulation and management of disruptions allowing a reduction of the overall energy consumption	TRL4/5 by 2025
DO4	Fail-safe train positioning solutions to reduce CAPEX and OPEX of regional lines	TRL4/5 by 2025
DO5	Cost-effective fail-safe on-board train integrity and train length as necessary component for ETCS Level 2	TRL4/5 by 2025

References

1. Subset 026 4.0.0 System Requirements Specification (2023)
2. Subset 125 0.2.0 System Requirements Specification (2022)
3. UIC webpage. https://uic.org/rail-system/frmcs/

Open Access This chapter is licensed under the terms of the Creative Commons Attribution 4.0 International License (http://creativecommons.org/licenses/by/4.0/), which permits use, sharing, adaptation, distribution and reproduction in any medium or format, as long as you give appropriate credit to the original author(s) and the source, provide a link to the Creative Commons license and indicate if changes were made.

The images or other third party material in this chapter are included in the chapter's Creative Commons license, unless indicated otherwise in a credit line to the material. If material is not included in the chapter's Creative Commons license and your intended use is not permitted by statutory regulation or exceeds the permitted use, you will need to obtain permission directly from the copyright holder.

Impact Assessment of a Rural Tourism Mobility Service

Jenni Vestinen[1]([✉]) [iD], Riina Isola[1], and Minni Haanpää[2] [iD]

[1] VTT Technical Research Centre of Finland Ltd., P.O.Box 1100, 90571 Oulu, Finland
`jenni.vestinen@vtt.fi`
[2] University of Lapland, P.O.Box 122, 96101 Rovaniemi, Finland

Abstract. The purpose of this paper is to present an impact assessment of the new Apukka Shuttle Bus service in Finnish Lapland. The service was created to serve both overnight and activity customers at a remote tourist destination. Tourism mobility in rural areas is challenged e.g. by limited services and insufficient information. The new Apukka Shuttle Bus service was aimed to offer an easy and ecological access to a remote destination. The service was created in collaboration between the destination (Apukka Resort), a transport operator (Kutilan liikenne) and a Mobility as a Service operator (Matkahuolto). Based on the results the service has several benefits on user, organizational and societal levels.

Keywords: mobility service · accessible destination · individual traveller

1 Introduction

The number of foreign individual tourists is increasing, and many tourists in Finland wish to travel to remote destinations for the nature and outdoor experiences. However, in remote areas, accessing nature attractions without your own car is difficult.

Tourism is mostly about creating experiences and memories. Destinations aim to provide valuable tourism experiences, not only through core activities, but by all the related services offered by different stakeholders (Buhalis and Amaranggana 2015). Mobility plays an important role in creating memorable tourism experience and experienced stress during the travel (Kim et al. 2021). The accessibility of remote areas could be improved by new and efficient transport services that integrate different types of transportation and exploit digitalization (Eckhardt et al. 2018).

This paper presents the impacts of a novel tourism mobility service in a remote area implemented in collaboration between tourism and mobility stakeholders.

2 Methodology

The results are based on a case study consisting of both qualitative and quantitative data (Eisenhardt 1989). The case study was conducted in a rural tourist destination in Finnish Lapland. The research is mainly qualitative, and quantitative data includes end-user survey supporting the qualitative analysis. The research methodology includes:

© The Author(s) 2025
C. McNally et al. (Eds.): TRAconference 2024, LNMOB, pp. 515–521, 2025.
https://doi.org/10.1007/978-3-031-85578-8_68

- end-user survey (online) targeted for overnight visitors conducted during March-May 2023 (63 respondents)
- three interviews: destination (Apukka Resort), transport operator (Kutilan liikenne) and digital service provider (Matkahuolto)

3   Impacts of Apukka Shuttle Bus Service

3.1   Case Description

Apukka Resort is a tourist destination in a rural area by a lake, located 18 kms from the city of Rovaniemi, which is the capital of Finnish Lapland. Apukka Resort offers e.g., accommodation, restaurants and various activities such as reindeer, husky, snowmobile, and northern light adventures. Customers consist mainly of foreign individual travellers.

For the winter season 2022–2023, a new Apukka Shuttle Bus service was launched and it operated from November 2022 until early April 2023. The aim was to offer a common service, which is easy-to-use, simple and ecological, for both overnight and activity customs. The personnel of Apukka Resort could also use it for free. Before the Apukka Shuttle Bus, customers mainly used taxis and very limited long-haul public transport to reach the destination. Transportation between Apukka Resort and Rovaniemi was organized only for pre-booked activity customers and was included in the price of the activity. The means of transport varied and also personnel and vehicles of Apukka Resort were utilized.

The Apukka Shuttle Bus operated between the city of Rovaniemi and Apukka Resort, having a stop at the Arctic Circle approximately halfway, five departures both ways daily with the same schedule, 7 days a week. In addition, one of the departures stopped at the Rovaniemi airport. The bus stop located next to Apukka Resort main building and in the city of Rovaniemi in a central location with an easily noticeable Apukka sign. When the parking place of the bus in the city centre needed to be shifted slightly during the season, a guide collected customers at the sign and walked to the bus with the customers. For an easy recognition, the bus has large decals including Apukka Resort text and logo as well as resort related images. The pricing was made easy: 10 euros/adult, 5 euros/child and 25 euros/family regardless of the length of the trip.

Apukka Resort organized the Shuttle Bus service in collaboration with Matkahuolto and Kutilan liikenne. Apukka Resort was in charge of the entity, thus planning, sales, marketing and communication. Matkahuolto Ltd. is a company promoting bus and coach services in Finland and a national Mobility as a Service operator. It has launched national Trips and Tickets mobile application offering multimodal travel planning, ticket purchase and seat reservation. Apukka Shuttle Bus tickets were also available for purchase in Trips and Tickets application and in Matkahuolto's online store. Matkahuolto also provided information services including e.g. real time sales information, travel data, reporting and analytics, tools for drivers and operators, as well as customer service (for Apukka Resort, Kutilan liikenne & end-users). Kutilan liikenne is a local transport operator and was responsible for the operation of the route, as well as verifying travel rights when boarding the bus (online or app purchases, and activity customers) and ticketing & payment in the bus. No cash was accepted and tickets could be purchased by contactless payment in the bus. Kutilan liikenne also participated in the planning of the service with Apukka Resort.

3.2   Survey Results

An end-user survey was conducted as part of the study with 63 respondents. The digital survey form was distributed at the Apukka Resort reception via QR code and via email to overnight customers. Respondents were mainly female (67%) and the most of the respondents were travelling with their family (40%) or their spouse (36%). 5% were travelling alone and 19% with a group, e.g., with friends. The most common channels for finding information about the Apukka Shuttle Bus service were Apukka Resort web page (51%) and Apukka Resort reception (48%).

Passengers were asked if the Apukka Shuttle Bus service had affected their decision to stay at or visit Apukka Resort (Fig. 1). Almost two thirds (62%) said that the service had not affected their decision, whereas 22% answered it had affected their decision, and 16% could not say.

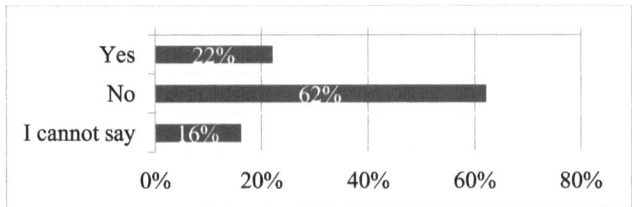

Fig. 1. Did the Apukka Shuttle Bus service affect your decision to stay at / visit Apukka Resort?

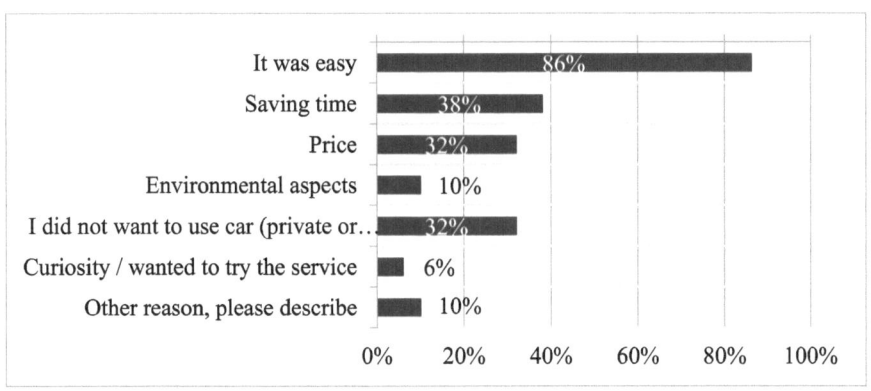

Fig. 2. What made you choose the Apukka Shuttle Bus service?

The respondents were asked what made them choose the Apukka Shuttle Bus service (possible to choose multiple options). The most remarkable reason to use the Apukka Shuttle Bus service (Fig. 2.) was easiness (86%). Saving time (38%), price (32%) and willingness not to use a car (32%) were also common reasons to choose the Apukka Shuttle Bus service. Most of the respondents were very satisfied (48%) or somewhat satisfied (34%) with the Apukka Shuttle Bus service.

Most respondents (85%) had received enough information about the Apukka Shuttle Bus service, 13% had not received enough information. Most respondents (89%) had received enough information about the timetable of the bus, 78% about the location of the bus stops and 75% about the appearance of the bus. The location of bus stops (Rovaniemi City center, Santa Claus village) had caused most unawareness (Table 1).

Table 1. Did you receive enough information about

	Yes	No	I cannot say
Timetable of the bus	88.9%	11.1%	0.0%
Location of the bus stops (Rovaniemi City center or Santa Claus village)	77.8%	19.0%	3.2%
Appearance of the bus	74.6%	14.3%	11.1%

Respondents were asked about the reliability of the Apukka Shuttle Bus service. Most users (76%) rated the service very reliable, 11% somewhat reliable, 8% chose a neutral option and 5% rated the service somewhat unreliable. Respondents who had rated the service unreliable were asked more details about their opinion. The respondents found that there were too few departures. 75% of respondents informed that the Apukka Shuttle Bus service had a positive impact on the image of the Apukka Resort.

Approximately half of the respondents (53%) had purchased the ticket in the bus and another half (47%) online. Nearly all users found the purchasing process easy or somewhat easy. The price was rated expensive or somewhat expensive by 48%, reasonable by 17% and affordable or very affordable by 35% of the respondents.

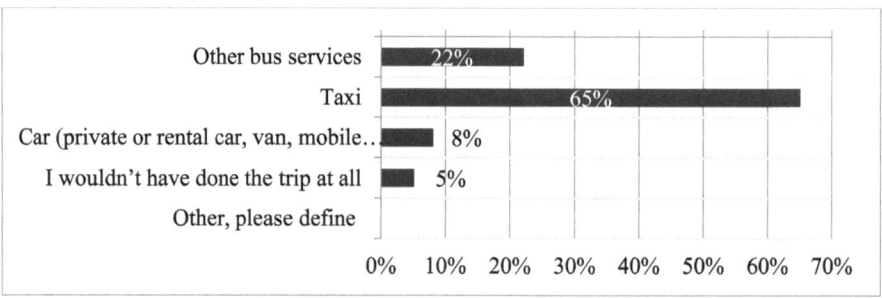

Fig. 3. Which transport mode would you have used if the shuttle bus service wasn't available?

Respondents were also asked, which transport mode they would have used if the Apukka Shuttle Bus service was not available (Fig. 3). Almost two thirds (65%) reported that they would have used taxi, 22% would have used other bus services, 8% would have used a private or rental car and 5% would not have done the trip at all.

3.3 Impacts of the Apukka Shuttle Bus Service

Based on the survey and interviews, the main impacts of the Apukka Shuttle Bus are summarized from user (U), organizational (O) and societal (S) levels (L) in Table 2.

Table 2. The main impacts of Apukka Shuttle Bus service

L	Key performance indicator	Description (all impacts are positive)
U	Perceived accessibility to destination	5% wouldn't have done the trip at all without the service
U	Modal shift	The service shifted 73% from cars (taxis and other cars) to public transport
U	Perceived sufficiency of information	85% received enough information
U	Perceived reliability of the service	87% found the service reliable
U	Satisfaction on the service	82% were satisfied with the service
O	Impact of mobility services on destination selection	The service affected 22% on their decision to visit/stay in destination
O	Number of customers	The service also attracted other day visitors than activity customers
O	Profit/turnover	The service was profitable and increased turnover
O	Image	75% found that the service had a positive impact on the image of the destination
O	Collaboration in value chain	New collaboration was created (Apukka Resort-Matkahuolto)
S	Emissions	The modal shift to buses with high utilization rates decreases emissions
S	Accessibility to services (mobility, tourism)	A new service was created where sufficient public transport did not exist to the destination

4 Discussion and Conclusions

All the involved stakeholders (Apukka Resort, Matkahuolto and Kutilan liikenne) found the Apukka Shuttle Bus service successful. Clear roles and responsibilities, and smooth collaboration between the involved stakeholders can be considered as a success factor of the Apukka Shuttle Bus service. The service was profitable for Apukka Resort and was considered to have a positive impact on the destination recognition and image. Apukka Resort also had a clear concept and branding of the service and implemented it with competent and experienced partners.

The Apukka Shuttle Bus service was popular among the customers, and they were very satisfied with the service. It also affected positively the decision to choose the destination, and brought a new customer group to Apukka Resort: other day visitors than activity customers, e.g. visitors simply enjoying the nature landscape, animals in the yard area and the restaurant with a lake view. The Shuttle Bus service was also utilized for trips between the city of Rovaniemi and the Arctic Circle. The use of digital mobility solutions is becoming more popular in the transport sector. Nearly half of the users bought the Apukka Shuttle Bus ticket online. The lack of cash payment caused challenges only for few.

Along with the Apukka Shuttle Bus service, accessibility of the remotely located destination was improved. Offering bus rides for personnel in a remote location may also attract more employees to the destination, especially those not having a car, such as students and international seasonal workers. Mobility can be a major source of stress for tourists, and easy and reliable service can reduce stress and thus make the experience of the entire travel and destination more positive (Kim et al. 2021; Eckhardt and Haanpää 2022). In addition, the use of a guide was considered to improve the Apukka Shuttle Bus experience.

The modal shift from taxis and other cars to bus was remarkable, and occupancy rates of buses were high, resulting in reduced emissions. In addition, fewer vehicles entering Apukka Resort may improve the safety of the yard.

As a result of the successful experiment the Apukka Shuttle Bus service has been extended to start already in September for the season 2023–2024. Also, the service has been extended to cover transportation between Apukka Resort and Arctic Circle Hiking Area. In addition, the price of the Apukka Shuttle Bus service has decreased by 20%.

To conclude, easy and reliable tourism mobility services based on sustainable public transport have several benefits on user, organizational and societal levels and are thus recommended to be further developed. Digitalization also enables better integration of mobility and tourism services, and the creation of larger service entities is recommended for future studies.

Acknowledgements. The authors gratefully acknowledge Business Finland for enabling and co-financing the FIT ME! (Foreign Individual Travelers' hospitality and Mobility Ecosystem) project, as well as Apukka Resort, Matkahuolto and Kutilan liikenne for their contribution.

References

Buhalis, D., Amaranggana, A.: Smart tourism destinations enhancing tourism experience through personalisation of services. In: Information and Communication Technologies in Tourism, pp. 377–389. Springer, Heidelberg (2015)

Eckhardt, J., Nykänen, L., Aapaoja, A., Niemi, P.: MaaS in rural areas - case Finland. Res. Transp. Bus. Manag. **27**, 75–83 (2018)

Eckhardt, J., Haanpää, M.: Impact assessment framework for tourism mobility services. In: Proceedings of 3rd International Conference on Mobility as a Service (ICOMaaS). Tampere University of Technology, Tampere, Finland (2022)

Eisenhardt, K.M.: Building theories from case study research. Acad. Manag. Rev. **14**(4), 532–550 (1989)

Kim, H., Koo, C., Chung, N.: The role of mobility apps in memorable tourism experiences of Korean tourists: stress-coping theory perspective. J. Hosp. Tour. Manag. **49**, 548–557 (2021)

Open Access This chapter is licensed under the terms of the Creative Commons Attribution 4.0 International License (http://creativecommons.org/licenses/by/4.0/), which permits use, sharing, adaptation, distribution and reproduction in any medium or format, as long as you give appropriate credit to the original author(s) and the source, provide a link to the Creative Commons license and indicate if changes were made.

The images or other third party material in this chapter are included in the chapter's Creative Commons license, unless indicated otherwise in a credit line to the material. If material is not included in the chapter's Creative Commons license and your intended use is not permitted by statutory regulation or exceeds the permitted use, you will need to obtain permission directly from the copyright holder.

Regional Rail Rolling Stock Requirements and Specifications

Rickard Persson[1] (iD), Libor Lochman[2], and Jens König[3](✉) (iD)

[1] KTH Royal Institute of Technology, Rail Vehicles, 100 44 Stockholm, Sweden
[2] Wabtec, Via Pianodardine, 83100 Avellino, Italy
[3] DLR German Aerospace Centre, Pfaffenwaldring 38-40, 70569 Stuttgart, Germany
jens.koenig@dlr.de

Abstract. With the support of EU's key funding program Horizon Europe, the Europe's Rail Joint Undertaking (EU-Rail) aims to deliver faster deployment of innovations in the railway design. The EU-Rail FutuRe project aims at tackling the increasing Total Cost of Ownership (TCO) of European interoperable regional lines by means of providing new innovative solutions to make these lines cost-effective and attractive. Setting the right level of requirements and specification for the regional rail rolling stock plays therefore an important role in meeting the FutuRe objectives. The detailed objectives are to define a vehicle concept for up to 100 passengers or corresponding freight with significant vehicle weight savings compared to existing rolling stock. The key systems for the research are the mechanical arrangement to avoid unnecessary weight and reduce maintenance costs on vehicle and track, the propulsion system which must be emission free despite that these lines often are non-electrified and the control system which may be combined with train protection system and offer autonomous running.

Keywords: Regional railway · rolling stock · low weight · emission free · automatization · cost of ownership

1 Introduction

Regional railway (lines with lower usage or secondary network) plays a crucial role not only in serving European regions, but also as feeders for passenger and freight traffic for the main/core network. Therefore, they have an essential function as an environmentally friendly mode of transport and are an enabler for increasing the transport performance of railways.

Unfortunately, many of these routes were abandoned in the past - mainly due to high costs. Therefore, these rail lines need to be revitalized or even renewed to make them economically, socially and environmentally sustainable and to meet current customer needs and challenges. To achieve these ambitious goals a novel approach is necessary which is in the focus of the FutuRe – the project from the family of the Europe's Rail Joint Undertaking research and innovation activities.

© The Author(s) 2025
C. McNally et al. (Eds.): TRAconference 2024, LNMOB, pp. 522–528, 2025.
https://doi.org/10.1007/978-3-031-85578-8_69

It is the ambition of the European Union to increase the market share of environmentally friendly transport modes. The rail plays the pivotal role in this challenge and the regional railway is an essential part of the public mobility.

Setting the right level of requirements and specification for the regional rail rolling stock plays therefore an important role in meeting the FutuRe objectives. The detailed objectives are to define a vehicle concept for up to 100 passengers or corresponding freight with significant vehicle weight savings compared to existing rolling stock, Fig. 1.

The operation of the vehicles is considered for two groups of lines. The Group 1 Lines are connected with the main line networks. The Group 2 Lines have a limited or no connection with the mainline network.

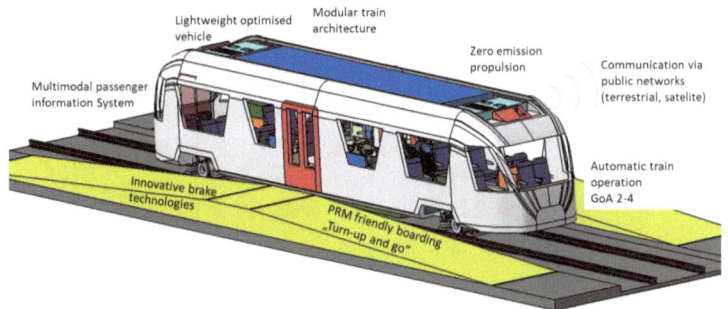

Fig. 1. An illustration of a possible regional vehicle

2   Total Cost of Ownership

The overall objectives of the FutuRe project, part of the Horizon Europe project, are to ensure the long-term viability of the regional railway by reducing the Total Cost of Ownership (TCO), while ensuring high service quality and operational reliability. It also aims to increase customer satisfaction and to make rail an attractive and preferred mode of transport. Therefore, the main objectives of FutuRe are: lowering Capital Expenditures (CAPEX), increasing productivity by lowering Operating Expenses (OPEX), and improving customer satisfaction.

The TCO will here cover the vehicle CAPEX and OPEX and those costs on the infrastructure side that is influenced by decisions for the vehicle. One such case is platform heights, where either the vehicle should be able to adopt to different platform heights or the platform heights should be unified. The platform standard height in Europe is 550 mm, but many regional lines have lower platforms. To allow passengers with reduced mobility to enter the vehicle without support, entrances without steps must be offered. The TCO can here be used as a tool to find the best solution in terms of cost.

$$TCO = CAPEX_{veh} + CAPEX_{infra} + OPEX_{veh} + OPEX_{infra} \qquad (1)$$

The vehicle weight is a threshold to get access to some regional lines as the axle load is limited by the design of the track, but vehicle weight is also important for the operational costs by its influence on the running resistance and the power losses for accelerating the vehicle mass [3].

3   The Vehicle

3.1   General

The foreseen vehicle is in the size of a long distance road bus. The control system, the mechanical arrangement, the propulsion system and a few other systems are identified as critical for the success of a new regional vehicle, Fig. 2. Principal designs will be made bringing Technical Readiness Level (TRL) up to level 3 and for certain systems up to level 5.

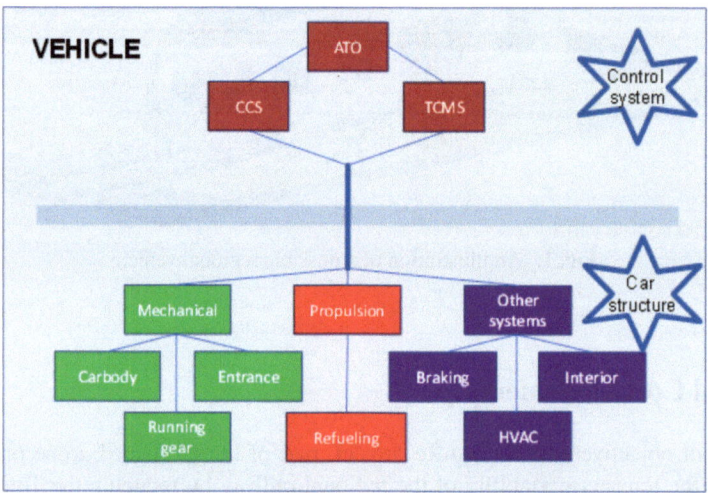

Fig. 2. Block diagram of the vehicle with its systems

3.2   Control System

The Control Command and Signaling (CCS) is a focus area as installing the present Automatic Train Protection (ATP) on low use railways may not be justified and less costly solutions will be looked for. Installing European Rail Traffic Management System (ERTMS) on a railway can even contribute to that passenger rail service is abounded as the cost for the onboard equipment is high. The solution looked for includes increased automatization by high level of Automatic Train Operation (ATO) – GoA 3–4, interacting with the Automatic Train Protection (ATP) and Train Control and Management System (TCMS) on the train. A detailed view is presented in [1].

3.3 Mechanical Arrangement

Lightweight design and costs are crucial aspects here [2]. They are largely determined by the use-case and, related to this, by the corresponding requirements. These in turn result from the line groups which have to be considered for creation a demand-adapted vehicle.

Due to the focused use cases in rural areas with less utilization, small, lightweight rail vehicles with two axles are a suitable solution and address the frame requirements custom fit. The maximum vehicle weight is limited to 32 tons considering the focused axle load of 16 tons to reduce the necessary effort for the refurbishment and maintenance of the superstructure of the lines.

This results in the need for a comprehensive lightweight design approach that includes adaptation to the group of lines. In particular, this is the car body, which is designed for line group 1 in terms of strength and crashworthiness according to the light commuter trains (EN 15227, category C III and EN 12663, category P III). For vehicles for line group 2, even a reduced strength and crashworthiness is possible, since the collision with a 129-ton multiple unit and an 80-ton heavy freight wagon can be ruled out.

The running gear must provide a good ride comfort even with a suboptimal track bed under consideration of the wear and tear of the tracks. Also, here the lightweight design is of particular importance, especially in the area of unsuspended masses.

Above all, cost-efficient design is of great importance which has to be considered for all approaches.

3.4 Propulsion

The modern propulsion system is an enabler for an emission free vehicle and for regional train operation. Many regional lines are unelectrified and electrification cannot be justified by the low passenger demand. Diesel drives are dominating today, but is not considered to be a good choice for the future. Battery based solution are suggested, but it is unclear what design to select, and the choice may be depending on the conditions. Gathering information about user cases, put these in groups and select a representative user case for each group to explore the different propulsion choices.

The base configuration is a traction converter driving one or more motors. The converter is feed by an onboard battery. When accelerating, energy is taken from the battery, converted and fed into the motors. When braking the motor act as generators recovering the kinetic energy to battery power. Hence, the losses are limited to the running resistance and the conversion losses. These losses must be compensated and three solutions have been drafted, see Fig. 3.

1) Charging the battery from an external charger (similar to today's for electrical buses)
2) Charging from catenary (as electrical trains use), the charger must be onboard
3) Charging from hydrogen, a fuel cell and a reservoir must be onboard

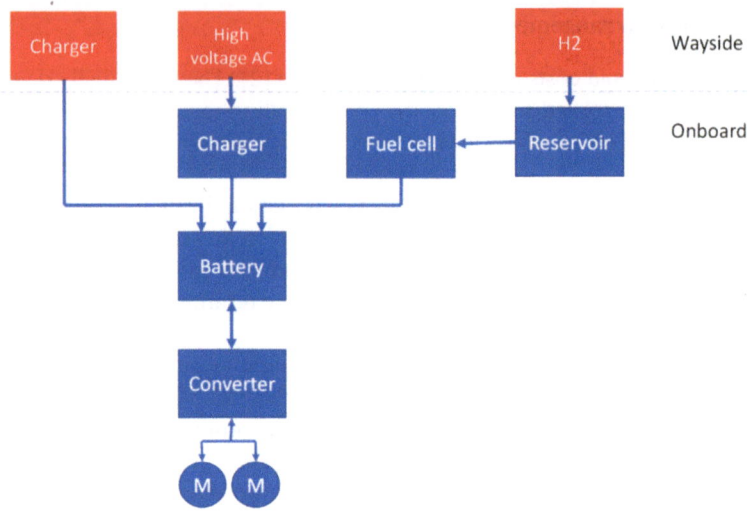

Fig. 3. Possible propulsion setups

3.5 Other Systems

Beside the mechanical and propulsion part of the vehicle also other components and parts must be considered which are in close interactions with the mechanical and traction subsystems architecture.

One focus is on the braking system which has to allow an agile and high braking performance. This is necessary to enable the reduced strength and crashworthiness of the car body compared to conventional trains for mainlines. The reduction of the weight of the braking system plays a relevant role due to the position of the braking system in the running gear and as a part of the at least partial unsuspended mass.

Another focus is on the interior compartment. A main requirement is the easy adaptability to different needs of services which includes passenger and freight services. Especially for the passenger focus group discussions for the recording and classification of user requirements and parameters of the comfort experience are the base for an optimised interior in the intended context of use of the vehicle. An easy and fast adaptable, modular interior concept offers a high potential to fulfill these partially contradictory requirements for the transport of passengers and freight in the same vehicle.

4 Compliance to Standards

An analysis will be made concerning the compliance with the applicable Technical Specifications for Interoperability (TSIs) and/or European standards. This analysis will primarily address the vehicle itself, however in connection with the regional lines architecture requirements, including the aspects of traffic management, control-command and signaling and operational safety.

A detailed study of the applicable legal framework/ safety and regulatory requirements will be made in order to allow developing the necessary basics of a preliminary vehicle concept for up to 100 passengers and transport of goods.

Consequently amendments might be proposed in justified cases, providing an input to standardisation activities to and beyond the TSIs (defined per the European Union Agency for Railways (ERA) Single Programming Document 2022–2024). This includes identification of relevant bodies and entities (ERA TWGs and WPs contributing also to European Sector Railways Forum (SFR) and the Rail Standardisation Coordination Platform (RASCOP), Standards Bodies' committees, and initiatives such as EuroSpec) as well as documents and deliverables (TSIs and other Regulations/Directives, European Standards and International Standards including other deliverables, industrial standards). The need for creating new or updating existing regulations, standards and specifications will be prioritised (i.e., requirements to be added, requirements to be more specific, requirements to be erased), Fig. 4.

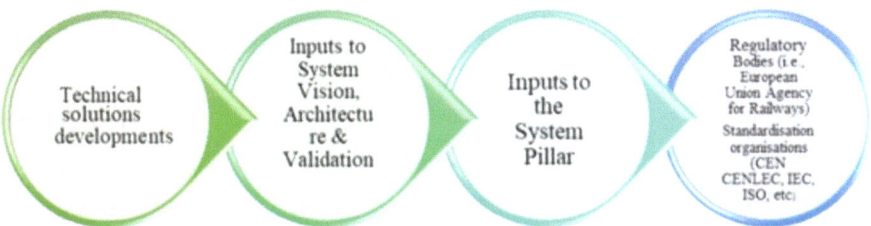

Fig. 4. Road map to suitable regulations

The impact of the Future project is even broader: aiming also at positively influence in the developments of the System Pillar and thus ensuring the proper integration of the vehicle into the overall system architecture.

5   Outlook

The definition of the requirements and specifications of the regional rail rolling stock is the base for the conception of the vehicle in this second part of the FutuRe project ending in 2026. In the subsequent project the detailing of the vehicle is planned which is used for building up first parts of the vehicle as demonstrators. In a further step the realisation of a vehicle demonstrator is planned containing the technologies and approaches achieved in the work before as well as the testing of the vehicle on track.

Acknowledgements. This project has received funding from the Shift2Rail Joint Undertaking under the European Union's Horizon 2020 research and innovation program under grant agreement (No. 101101962). The content of this paper reflects only the author's view and the JU is not responsible for any use that may be made of the information it contains.

References

1. Burro, F., Lochman, L., Mascis, A.: EURAIL – FA6 FutuRe project innovative solutions for G2 regional lines. In: Proceedings of TRA 2024 (2024)
2. Shift2Rail – NextGear: NEXT generation methods, concepts and solutions for the design of robust and sustainable running GEAR, Deliverable 2.2 - Component level demonstrator using fibre reinforced plastic (2021)
3. Persson, R., Stichel, S., Liu, Z., Giossi, R.: Cost reduction with single axle running gears in metro trains. In: Proceedings of TRA 2022 (2022)

Open Access This chapter is licensed under the terms of the Creative Commons Attribution 4.0 International License (http://creativecommons.org/licenses/by/4.0/), which permits use, sharing, adaptation, distribution and reproduction in any medium or format, as long as you give appropriate credit to the original author(s) and the source, provide a link to the Creative Commons license and indicate if changes were made.

The images or other third party material in this chapter are included in the chapter's Creative Commons license, unless indicated otherwise in a credit line to the material. If material is not included in the chapter's Creative Commons license and your intended use is not permitted by statutory regulation or exceeds the permitted use, you will need to obtain permission directly from the copyright holder.

National Roads Network (NR2040) Sustainable Future

Tahel Wexler[1]([✉]), Aoife Hurd[1], and Derek Brady[2]

[1] AECOM Ireland, Dublin, Ireland
tahel.wexler@aecom.com
[2] Transport Infrastructure Ireland, Dublin, Ireland

Abstract. Recently, the focus of Irish transport policy has shifted significantly towards sustainable infrastructure development. Transport Infrastructure Ireland (TII) published National Roads 2040 (NR2040), which responds to policy shift placing greater emphasis on multi-modal solutions, decarbonisation, and sustainability.

This research paper explores the underlying analysis of NR2040, including; (1) how it aligns with Government policies and plans, (2) underlying analysis of future scenarios; functional corridor assessment of national secondary roads; and estimation of road transport emissions, and (3) the NR2040 implementation and investment approach. In respect to methods, research topics 1 and 3 are primarily policy reviews, while the work involved with research topic 2 required extensive analysis and research tools including TII's VISUM-based National Transport Model (NTpM) and the Irish Road Emissions Model (REM).

The development of a comprehensive road transport strategy (NR2040) is a critical part of the national policy effort to reduce negative externalities associated with motorised vehicles and contribute towards the achievement of climate action targets. Thus, NR2040 provides a valuable case study for policy makers and researchers, showcasing a national shift away from the traditional 'predict and provide' model of transport planning in a modern European country with high car dependency and a large rural population.

Keywords: National Roads · Ireland · Sustainable · Policy · Decarbonisation

1 Introduction

In recent years, the focus of Irish transport policy has shifted significantly, focusing less on new roads development and instead seeks to deliver different types of infrastructure that enable more sustainable mobility and cater to all road users. NR2040 seeks to transition from the traditional 'predict & provide' model of transport planning more towards 'decide and provide' with a focus on multi-modal solutions, modal shift, and sustainable transport planning. NR2040 is TII's long term strategy for planning, operating, and maintaining of the National Roads network (NRN) [1].

© The Author(s) 2025
C. McNally et al. (Eds.): TRAconference 2024, LNMOB, pp. 529–536, 2025.
https://doi.org/10.1007/978-3-031-85578-8_70

2 Literature Review

NR2040 ensures the strategic alignment of the National Roads Network (NRN) to recently published Governmental policies and plans, most notably Project Ireland's National Planning Framework (NPF) [2], which sets out the overarching spatial strategy for the country till 2040, and the National Development Plan (NDP) [3], which sets out the investment strategy and budget for the period 2021–2030. The National Investment Framework for Transport in Ireland (NIFTI) [4], prioritises future investment in the land transport network to support the delivery of Project Ireland 2040, and its four investment priorities: Decarbonisation; Protection and Renewal; Mobility of People and Goods in Urban Areas; Enhanced Regional and Rural Connectivity, have been incorporated into NR2040.

The Climate Action Plan (CAP) requires the transport sector to achieve a 50 percent emission reduction by 2030 and become a fully decarbonised sector by 2050 [5]. The National Sustainable Mobility Policy [6] supports CAP and aims for at least 500,000 additional daily active travel and public transport journeys and a 10% reduction in kilometres driven by fossil fuel cars nationally by 2030. NR2040 has been developed in alignment with both.

3 Methodology

The process of developing NR2040 included significant analysis of the existing and future function of National Roads and the services they facilitate for society. In addition, it included development of commitments, that will firmly hold TII true to the values it proclaims in the strategy.

3.1 Planning for Uncertain Futures - Central and Plausible Scenarios

As planning decades ahead requires versatility and flexibility, NR2040 was designed to be adaptable to different futures and has been tested as such in two general steps.

First – TII's VISUM-based National Transport Model (NTpM) - a software representation of Ireland's transport network requiring various model inputs, was used to produce a central demand forecast of road transport activity in 2040 based on Project Ireland 2040 NPF projections. This is the central scenario around which plausible futures pivot.

The second - four plausible demand scenarios, which represent less desirable yet credible outcomes that are more difficult to foresee, were developed based on the relationship between two key variables: changes in journey length and journey frequency, to capture a broad range of plausible futures (see Fig. 1). These variables may be affected by technological developments, changes in energy or travel costs, changes to settlement and employment patterns, and a host of other factors.

This scenario analysis was not executed as a 'predict and provide' method meant to guide investment in capacity increase, but rather as a means for considering alternative future behavioral trends and their implications, given unforeseen factors.

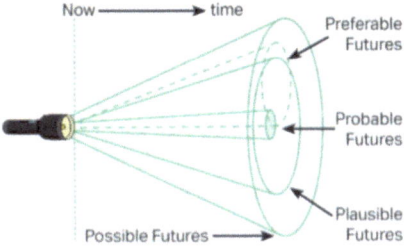

Fig. 1. Planning for Uncertain Futures

3.2 Functional Corridors - Role of National Secondary Roads

NR2040 identified the need that Secondary National Roads sections serve, providing vital access for local communities. The process entailed 'slicing' the National Secondary network into approximately 190 sections, where each was sequentially 'removed' from the road network and run in TII's NTpM, forcing modelled traffic to reroute. Differing levels of impact were recorded. Two separate measures were gathered from this analysis for each section of secondary road, gauged as its National and local importance.

National importance indicates how important each road is in a national context, for the movement of many vehicles at high speeds. While Local importance, of equal stature, determines the impact on localised movements for single users accessing their nearby community facilities, market towns and other regions. Links of high local importance often indicate where topography or geography creates a substantial detour and delays.

This analysis is unique as a national roads authority not only prioritises infrastructure that serves the convenience of the majority, but also recognises the importance of certain 'lifeline routes' for a handful of people, for whom there is no alternative.

3.3 Decarbonisation

In line with national policy and the focus towards decarbonisation of road transport, the development of NR2040 employed TII's Road Emissions Model (REM) in conjunction with the NTpM, to facilitate a strategic analysis of Greenhouse gas (GHG) emissions and non-GHG pollutants from road vehicles. The effect of speed limits on emissions was demonstrated by examining the relationship between travel speed (km/h) and emissions (gram/km) of different vehicle types.

4   Results and Discussion

As part of TII's effort, as a state agency responsible for national roads, to promote a shift in focus towards more sustainable mobility, the development of NR2040 employed TII's NTpM and REM and unique insight was produced for (1) Planning for Uncertain Futures, (2) Role of National Secondary Roads, and (3) Decarbonisation. These results are presented within this section as well as a brief discussion of each.

4.1 Planning for Uncertain Futures

Results of the sensitivity testing impact on travel (see Fig. 2) reflects the extreme boundaries of the plausible scenarios and how they deviate from the Project Ireland 2040 NPF central forecast. The potential effect on vehicle kilometres was calculated to be as much as a 30 percent reduction, in the sustainable & urban communities scenario, to a 21% increase, in the dispersed communities scenario. Interestingly in 2020, during the height of the Covid-19 pandemic and national travel restrictions, National Roads traffic volumes reduced by 40 to 70 percent compared to equivalent period in 2019 [7].

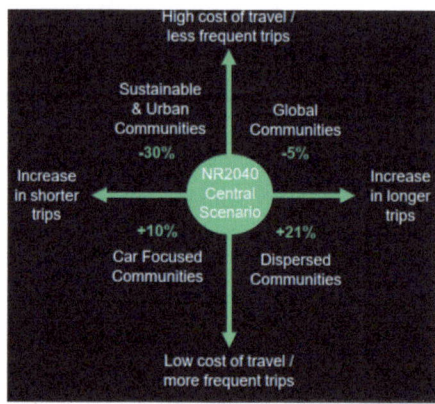

Fig. 2. Plausible Scenario Impact on Travel in National Vehicle Km

4.2 Role of National Secondary Roads

Each National Secondary Road was plotted against National and local importance (see Fig. 3). The blue dots represent sections with high National importance and are categorised as arterial collectors. The red dots represent high local importance and are categorised as Lifelines that should always remain open and accessible. Closure of these roads would impose significant disruptions on communities as alternative routes are sparse, hence they require a high level of resilience. Lastly, the yellow dots represent National Secondary Collectors, where the function of this road type will be targeted for consistency.

Performing the lifeline analysis and determining the functional classifications provides a measure of importance for TII and road managers and appraisers, particularly when it comes to maintenance, improving resilience and ensuring year-round availability and access on lifeline sections.

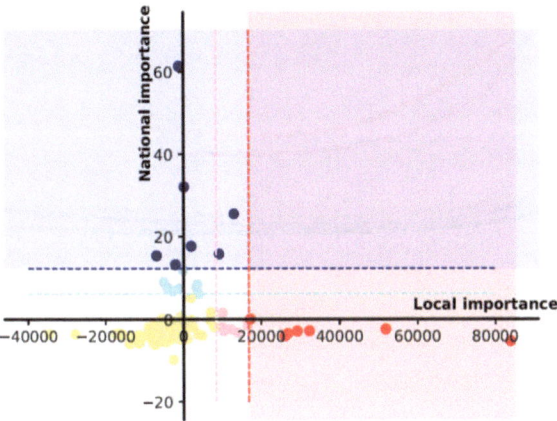

Fig. 3. All categories of National Secondary Roads, divided by the importance and relative importance level.

4.3 Decarbonisation

Results from the REM analysis (see Fig. 4) illustrate the complex relationship between emissions and speed. In general, the optimum speed for minimizing road emissions per km was found to be 50-60kph, with a marginal increase at higher speeds. Below 40kph emissions increase rapidly as speed decreases and stop-start congested conditions present worse emissions than the slowest free-flow speeds. On congested urban road sections where trucks are concentrated, inefficient engine performance at low speeds results in greenhouse gases being emitted on a large scale.

The supporting analysis for NR2040 has identified that while 45 percent of all road travel nationally occurs on the National Roads network, this traffic generates 35 percent of road transport emissions in the state, accounting for 4.1 of 11.6 mega tonnes CO2e national road travel emissions in 2022. Large Goods Vehicles (Articulated) (LGVs) contributed 34 percent (1.4 mega tonnes CO2e) of National Roads emissions in 2022 [8]. Ensuring efficient movement of large goods vehicles is therefore key to reducing carbon emissions.

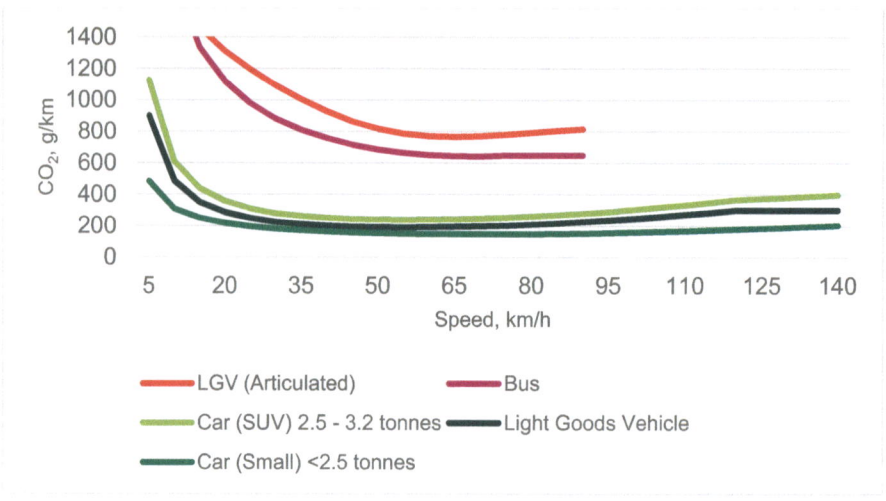

Fig. 4. Relationship between vehicle type, speed, and emissions

5 Conclusion

NR2040 has identified inter-related strategic issues for the future of national roads and its approach to sustainable road transport recognises different user needs and contexts across Ireland. To address the identified strategic issues, the implementation of NR2040 aims to apply to the four investment priorities (see Fig. 5) and to see an evolved NRN with movement of people and goods, safety, and accessibility all central to this evolution. The investment priorities were further reinforced by a series of TII commitments, such as TII's commitment to reduce carbon emissions associated with the development, construction, and operation of new infrastructure, holding itself accountable to the values pronounced in NR2040 through measurable actions.

The development of NR2040 as a comprehensive national roads transport strategy is a critical component of Ireland's national policy effort to reduce the negative externalities associated with motorised vehicles and contribute towards the achievement of climate action targets. Thus, NR2040 provides a valuable case study for policy makers and researchers as a practical example of a national shift away from the traditional 'predict and provide' model of transport planning in a modern European country with high car dependency and a large rural population.

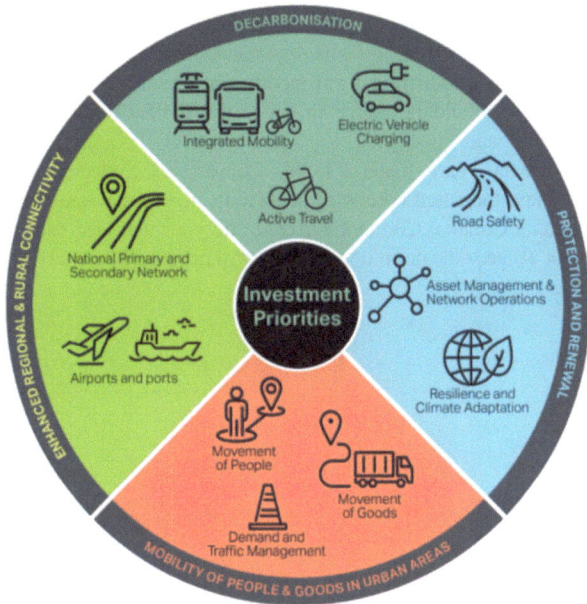

Fig. 5. NR2040 Investment Priorities and Portfolios

References

1. Transport Infrastructure Ireland: National Roads 2040 Final Report (2023)
2. The Department of Housing Planning and Local Government: Project Ireland 2040 National Planning Framework (2018)
3. The Department of Public Expenditure, NDP Delivery and Reform: National Development Plan 2021–2030 (2021)
4. Department of Transport: National Investment Framework for Transport in Ireland (NIFTI) (2021)
5. Department of Transport: National Sustainable Mobility Policy (2022)
6. Department of the Environment, Climate and Communications: Climate Action Plan 2023 (2023)
7. Transport Infrastructure Ireland: National Roads Network Indicators 2021 (2022)
8. Transport Infrastructure Ireland: National Roads Network Indicators 2022 (2023)

Open Access This chapter is licensed under the terms of the Creative Commons Attribution 4.0 International License (http://creativecommons.org/licenses/by/4.0/), which permits use, sharing, adaptation, distribution and reproduction in any medium or format, as long as you give appropriate credit to the original author(s) and the source, provide a link to the Creative Commons license and indicate if changes were made.

The images or other third party material in this chapter are included in the chapter's Creative Commons license, unless indicated otherwise in a credit line to the material. If material is not included in the chapter's Creative Commons license and your intended use is not permitted by statutory regulation or exceeds the permitted use, you will need to obtain permission directly from the copyright holder.

EURAIL – FA6 FutuRe Project
Innovative Solutions for G2 Regional Lines

Fabrizio Burro[1(✉)], Libor Lochman[1], Alessandro Mascis[1], and Marta Garcia[2]

[1] Wabtec Corporation, Rome, Italy
fabrizio.burro@wabtec.com
[2] UNIFE, Brussels, Belgium

Abstract. European regional lines are an essential part of the multimodal mobility scheme. Despite their relevance, they are gradually being abandoned due to the increasing total cost of ownership (TCO). FutuRe FP6 project aims at the development of the traffic management/control/command and signalling system for low density lines that are not functionally/operationally fully connected with the mainline network (ER JU Multi Annual Work Program – [1]) – so called Group 2 Lines (G2), that are operated by passenger and/or freight services that do not usually enter mainline infrastructure. Following the analysis of the current regional railways key cost drivers, we will demonstrate less expensive and more advanced solutions for COTS-based Group 2 lines, having as funding pillars public radio communication & satellites in support to fail-safe highly accurate positioning, intelligent systems on-board, cloud-based control centre for vital fail-safe Interlocking and RBC related functionalities.

Today's available technologies are suitably mature for creating new solutions for low density railway; use of COTS products dramatically reduces the CAPEX and OPEX costs paving the way for long-term viability of these lines.

Keywords: regional · low density · CAPEX · OPEX

1 Introduction

Within the [1] Multi Annual Work Plan (MAWP) of the EJRU, G2 lines have been identified as the ones at major and immediate risk of decommissioning. This is true both for passenger regional and freight services. The pressure of competition (trucks, buses, cars) on the railway operators is intense and growing, requiring more efficiency in terms of performance and energy saving. For those lines very low-cost solutions, without compromising the need for standardization and safety. The signalling/train control world is heading in a clear direction towards minimisation of wayside equipment, wireless communication, integrated intelligence on-board and large national traffic management centres.

© The Author(s) 2025
C. McNally et al. (Eds.): TRAconference 2024, LNMOB, pp. 537–542, 2025.
https://doi.org/10.1007/978-3-031-85578-8_71

1.1 Key Cost Drivers Affecting Regional Low-Density Lines Competitiveness

To reduce the cost of G2 regional low-density lines it is required to consider what in Fig. 1.

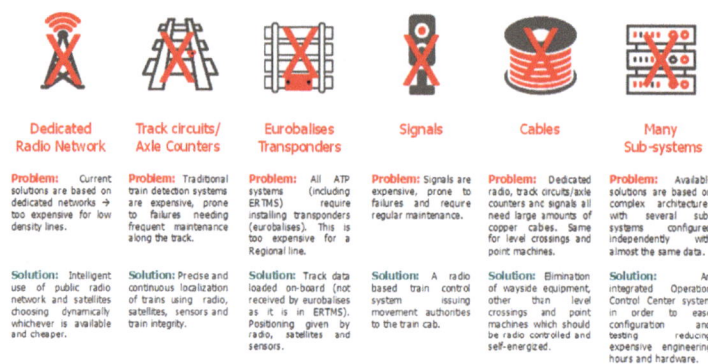

Fig. 1. Key cost drivers currently characterising regional lines

1.2 Pillars Sustaining a Cost Competitive Solution

There is a clear need to define a modern signalling/train control system based on the below described pillars:

1) Significant limitation of wayside installations deploying more **onboard centric solutions** by monitoring the train integrity and determining location through communication systems/sensors (Satellite, Radio).
2) Usage of geographical data in **Onboard Track databases,** similar as to CBTCs, in support to train positioning and detection.
3) Moving wholly to **Moving block** techniques allowing the removal of lineside signals.
4) Moving away from expensive, soon obsolete dedicated radio network solutions in favour of **public networks**.
5) **Rationalisation** of the **Traffic Management & Control Centre**.

There are, worldwide, no Railway lines equipped with such a minimalist solution while introducing low cost high performing technologies.

2 G2 Lines Architecture

A high-level description of main building blocks, internal and external interfaces of the G2 Lines system architecture is provided, in adherence to the [1] and [2] statements to propose innovative solutions for regional lines having *"no or limited connection to mainline traffic"* ([1] sect. 7.6.2.3).

2.1 Rationale

Due to the nature of the traffic targets characterising the G2 Lines, reduced, or not required, interoperability is the reference operative scenario. Similarly, while guaranteeing high-level of the system safety, safety requirements for the G2 Lines subsystems shall result from the review of European and national regulations and proposed amendments to them when need be. The above implies a new concept of traffic management (TMS) and associated services delivery, supporting *"long term viability of regional railways by decreasing the total cost of ownership (TCO), in other words, cost per kilometre both in terms of OPEX and CAPEX, while offering a high quality of service and operational safety"* (ref. [1] sect. 7.6.1.1). The implementation of the G2 Lines requirements should bring to (ref. to [1] 7.6.2.1) *a low-cost technical and operational framework for low density lines to reduce the cost per kilometre both in terms of CAPEX and OPEX as a well as an increase in customer satisfaction, which should be applied as a European solution.*

2.2 Standardization and Interoperability

G2 Lines are expected to (ref. to [1] 7.6.2.3) ensure standard interfaces and standard data protocols (preferably supporting the use of COTS). Intentionally we avoid defining the architecture of each key component: our approach is to define the minimal data to be exchanged among the components and leave to the industry the freedom to develop innovative solutions without being constrained by pre-defined architectures. For example, we do not care about the architecture of the train on-board systems, we care that the train is able to receive and send the data needed to ensure a safe and performant travel. In other words, in our approach trains equipped with different ATP systems can run in the same line until they are able to communicate with the standard protocols and until they are able to safely respect the movement authorities received by the control room. This interoperability among different equipped trains and different control rooms will bring competition and innovation. We do think that in the past the railway market had the tendency to over specify and constrain the architecture of systems with the unexpected result to delay innovation, increase costs and face obsolescence.

2.3 Building Blocks

Following the key drivers, the proposed solution is depicted in Fig. 2 below. The G2 Lines architecture is based on four main segments, substantially the interrelated building blocks on which it can be decomposed from an Operational, Functional and System point of view such as: Communication data network, Train On-board, Wayside, Traffic Management and Control Centre (CCS, including signaling).

Communication Data Network. A dedicated radio network typically represents around 20% to 30% of the overall CAPEX cost of a new signalling systems, not mentioning the OPEX costs for its maintenance. Telecommunication technology evolves faster than railway technology: therefore, it is critical to keep the train radio communication module separated from the Railway applications (i.e. the ATP and ATO). Therefore,

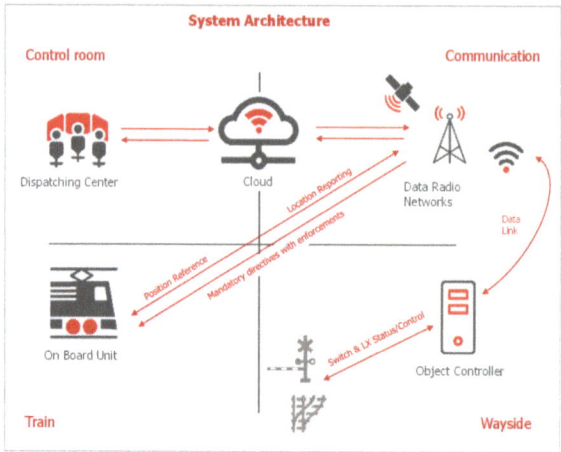

Fig. 2. G2 regional lines high level architecture

it is strategic to use Internet Protocol standards-based messaging, public networks and routing COTS solutions allowing:

1) Transparent access to cellular, wi-fi and satellite communication (and future technologies) with an automatic selection of the available and cheapest one for maximum availability and minimised costs. Data communication, for exchange of train position reference, location reporting, data link with controlling equipment is extensively based on standard Internet Protocol messaging/routing.
2) Cloud and Web based application, either through public services with certified proper level of security or implementing "Cloud-like" concept in legacy/proprietary communication system of infrastructure manager, accessing to most advanced encryption and authentication algorithms.

Based on the above, the G2 Lines communication data network should result in a "telecom technology agnostic" architecture. As per [1] 7.6.2.2, the proposed G2 Lines approach embraces *a cost-effective use of dedicated railway mobile networks or communication infrastructures owned by a third party (e.g., LTE, 4G, 5G, satellite comms and IoT communications) for the innovative ways to exchange of the information between the regional railway subsystems (e.g., Train to Train, Train to Trackside – wayside assets, Trackside to Trackside).*

On-board Train

Train Detection. On-board segment of the G2 Lines architecture recalls the well-known concept of Communication Based Train Control (CBTC) to which positioning based on radio communication technologies can be now integrated. Modern train control solutions are radio based (various CBTC systems for automatic metros, Railways by ERTMS L2/L3 and Positive Train Control): information among traffic control rooms, trains and wayside elements is exchanged wirelessly. Radio and satellites provide (augmented high precision, where required) train positioning to the on-board unit, which can use this to index an onboard track database thus removing the need for wayside balises/transponders along the track. For safety considerations, to remove high-cost track circuits and axle counters the system requires the on-board unit to determine train length and train integrity.

Train Positioning. Positioning technology performance is mature, currently till precision of few centimetres using satellites plus publicly available correction services. Soon, 6G & 7G will make localization cheaper and easier to implement (integrating cellular with satellites technology). Such technologies will evolve quicker than railway technology, therefore that scope must be separated from the railway applications (i.e. the ATP, ATO, train control). With such approach the proposed solution implements the positioning exploiting COTS hardware & software, without changing the railway onboard applications. It is possible to achieve a SIL4 positioning using alternative non SIL certified components without the need of a wayside transponder. The G2 Lines approach will meet the implementation of a cost-effective, interoperable, fail-safe, highly accurate train position based on among others hybrid, multi sensor technologies, digital maps, onboard database, as a mean to reduce balises installations, increase operational efficiency and decrease TCO in the context of regional lines.

Wayside. Minimisation of the wayside equipment deployment is a target for the G2 Lines concept that does not include the integration and exploitation of any wayside based train detection and train positioning technologies, as well as lineside visual signaling equipment and cables. The proposed System Architecture, at wayside level, encompasses the level crossings (wirelessly controlled or autonomous) and point machines (including also self-returning type). Such equipment is essential for safe operation of the infrastructure, then it cannot be removed either for cost reasons or to manage capacity. G2 Lines architecture addresses an upgrade of these elements to make them renewable energy operated (bringing a minimised impact on environment) and radio controlled (not requiring any expensive signaling and power cables). Combination of batteries, solar panels and radio communication allow to align these structural elements to the modern solution proposed becoming an integral part of the overall solution.

Summarising, the proposed G2 Lines approach will meet the need of *infrastructure components and wayside elements for regional railways including signalling, level crossings, switches and track vacancy detection which are energy self-sufficient and/or wireless enabled to reduce costs, cable and power supply and enable remote control or full or partial automation and/or autonomous operation.* ([1] 7.6.2.2).

Integrated Control Room. Existing system architectures are based on several layers to efficiently manage railway traffic, but that results too expensive for much simpler low-density lines. Therefore, low density lines need a fast and easy configurable low-cost Integrated Control Room: one HW unit can manage all functions needed: TMS/CTC; IXL; RBC. A single database will ease the configuration and testing, reduce expensive engineering hours and software maintenance. Such Integrated Control Room can control trains optimizing their journey to increase punctuality, reduce travel time and allow on-demand passenger services encompassing, but not limiting to, maximum safety (SIL4) and non-safety related functions via radio such as the management of interlocking functions, the commanding and control of point machines and level crossing, the optimisation of real-time routing and scheduling of train operations, the remote control of trains and the interface to intermodal traffic management. Finally, cloud-based services provided by third parties can reduce the cost of managing and maintaining servers, applications and database allowing web-based access for remote operators (e.g., maintenance staff on the track side).

3 Conclusions

The proposed innovative approach is a game changer for railway system. Today's available technologies are suitably mature for creating new solutions for low density railway lines based on radio public networks, cloud computing, satellite and COTS products, to reduce at least 30% on CAPEX and 40% on OPEX preserving them from the risk of closure. The key beneficiaries are the public transport authorities in charge of regional mobility, operating passenger low density (G2) lines and freight delivering cargo in remote areas. The presented concept paves the way to the introduction of several advanced services for both passengers and freight lines such as precise forecasting, flexible/on demand service and intermodality.

Acknowledgements. The work presented in this paper has received funding from the Europe's Rail Joint Undertaking, Grant Agreement no. 101101962. The contents of this publication only reflect the authors' views; the Joint Undertaking is not responsible for any use that may be made of the information contained in the paper.

References

1. Europe's Rail Joint Undertaking Multi-Annual Work Programme – 2.0 March 2022
2. Project 101101962 - FutuRe FP6 Grant Agreement
3. www.ertms.net, www.railroad.dot.gov, https://metra.com/positive-train-control-ptc, https://www.railwaypro.com/wp/ertms-the-eu-large-scale-project/
4. https://www.uiprail.org/files/uip/20190206_6th_RMMS_Report and http://www.europarl.europa.eu/RegData/etudes/BRIE/2019/646107/EPRS_BRI(2019)646107_EN.pdf
5. https://uic.org/rail-system/frmcs/ about the migration from GSM-R to modern radio communication
6. https://www.revistaitransporte.com/virtual-balises-for-european-trains/

Open Access This chapter is licensed under the terms of the Creative Commons Attribution 4.0 International License (http://creativecommons.org/licenses/by/4.0/), which permits use, sharing, adaptation, distribution and reproduction in any medium or format, as long as you give appropriate credit to the original author(s) and the source, provide a link to the Creative Commons license and indicate if changes were made.

The images or other third party material in this chapter are included in the chapter's Creative Commons license, unless indicated otherwise in a credit line to the material. If material is not included in the chapter's Creative Commons license and your intended use is not permitted by statutory regulation or exceeds the permitted use, you will need to obtain permission directly from the copyright holder.

Ventilation Fan Development
for Next-Generation Regional Aircraft

Sahan Wasala[1(✉)], Pela Katsapoxaki[1], Werner Gumprich[2], Dominik Christ[2],
Ruben Hernandez[2], and El Hassan Ridouane[1]

[1] Collins Aerospace Applied Research and Technology, Cork, Ireland
sahan.wasala@collins.com
[2] Collins Aerospace Air Management (Nord-Micro), Frankfurt, Germany

Abstract. The Clean Aviation project TheMa4HERA "Thermal Management for Hybrid Electric Regional Aircraft" aims to develop and mature the key technology bricks to enable efficient thermal management on these new aircraft concepts, where the increasing exploitation of electrical power will come with increasing number and entity of heat sources even more widely distributed in the aircraft compartments. Collins Aerospace provides advanced Thermal Management systems and services for commercial, regional, business aviation, military, and government customers. In TheMa4HERA Collins Aerospace is leading the ventilation work package focused on cabin pressure control and ventilation in unpressurized areas. This paper presents the recent findings in the TheMa4HERA project related to the design and optimization of a novel e-fan using state-of-the-art Computational Fluid Dynamics (CFD) modelling techniques. The end goal is to replace conventional bleed-air-driven subsystems in terms of performance, efficiency, and volumetric flow while reducing their weight. This e-fan is expected to be used for the ventilation of unpressurized bay areas during ground operations of HERA.

Keywords: Thermal Management · Electric Aircraft · Computational Fluid Dynamics · Optimization

1 Introduction

The journey to a climate-neutral aviation system through the European Green Deal is ambitious and a formidable opportunity for society and citizens. The answers proposed by Clean Aviation will probably challenge all conventions and classical approaches in bringing forth a real transformation leading to new propulsion solutions and sustainable vehicle configurations and operations. Thermal management is one of the key challenges for the successful realization of the Hybrid Electric Regional Aircraft. Novel propulsion technologies, e.g., Fuel Cells or Electric propulsion, come with challenging thermal management while fulfilling sustainability goals.

An aircraft thermal management system does not consume fuel directly, however, it takes energy from aircraft propulsion and secondary power systems, and as a result, indirectly participates in greenhouse gas emissions. Thermal management is the second

© The Author(s) 2025
C. McNally et al. (Eds.): TRAconference 2024, LNMOB, pp. 543–548, 2025.
https://doi.org/10.1007/978-3-031-85578-8_72

biggest fuel consumer for modern aircraft after propulsion where the pneumatic Environmental Control System (ECS) consumes 75% of non-propulsive power during cruise and from 3% to 5% of engine power [1].

2 TheMa4HERA Project

The objective of the TheMa4HERA project (Thermal Management for Hybrid Electric Regional Aircraft) is to demonstrate the dissipation of required additional heat in the order of 20 to 50 kW for systems and 300 to 1000 kW for power storage and generation in batteries, the APU and fuel cells in future commercial aircraft. Modern-day regional aircraft of 100 passengers generate approximately 10 kW of heat in the cabin. Electrical appliances, in-flight entertainment, light, galleys, etc. may add another 10kW to the cabin. The avionics system generates approximately 15 kW of heat. In total, today's aircraft thus produces an order of 35–50 kW of heat. The TheMa4HERA project aims to design a thermal management system for the HERA aircraft, keeping the heat waste minimum while addressing key performance indicators: weight, energy consumption and drag.

The development of an efficient, airworthy, lightweight, and high-performing thermal management system highly depends on efficient subcomponents that are part of the electric propulsion system. A major subcomponent is the ECS and the development of novel ECS architectures, integrated with new power system cooling is vital to ensure optimal TheMa4HERA architecture, which may comprise batteries, fuel cells and/or direct SAF/H2 burning. In addition to its traditional functions, the new ECS will achieve the latest fresh air flow regulations thanks to the involvement of EASA in the project, and may rely on pressure recovery and innovative cabin pressure control Additionally, the availability of bleed air for non-propulsive uses is limited by the novel hybrid propulsion a/c, therefore no-bleed/less-bleed disruptive solutions are critical and will require forced (electrical) ventilation in unpressurized areas. Collins Aerospace will be focused on developing an e-fan up to Technology Readiness Level 5 (TRL) at the end of this project.

3 Unpressurized Bay Ventilation

Bleed air driven sub-systems are often used for ventilation/cooling of aircraft unpressurized areas such as the belly fairing or tail cone compartments in which fuel vapor may be present. In most cases, the belly fairing ventilation consists of two parallel bleed air-driven turbo fans with their related supply bleed air valves used to control the fan operation in order to cope with the challenging operational environment with stringent requirements, particularly on a bleed-less aircraft, but potentially also as efficiency improvement on bleed aircraft, new technology of electrically driven fan will be introduced to replace the bleed air driven fans and their related supply valve, the electrically driven fan will be capable to operate in the unpressurized areas of the aircraft achieving a better sub-system efficiency, volume and weight than a conventional bleed-air driven subsystem.

Ventilation of the unpressurized area of the HERA aircraft is expected to be performed using a dual electric mix-flow fan that is designed and manufactured by Collins

Aerospace. In the initial design of the fan, a trade off study is being performed in order to shortlist the possible fans and fan configurations. The work that is presented in this paper is based on a study that used for selecting optimum number of stator and rotor blade configurations. The reference mix-flow fan used for this study consists of a stator with 25 blades. Fan performance is typically evaluated using pressure – volume flow rate (PQ) curves [2]. Most importantly PQ curves are also used for validation of the numerical model that will potentially be used in optimization process of a new fan development.

4 Methodology

The purpose of this study was to look at fan performance by varying stator and rotor numbers by means of numerical simulations. All simulations are conducted using the finite-volume commercial solver ANSYS CFX. A domain with a wedge-shaped channel, that take in to account a single rotor blade and a single stator blade leveraging the periodic nature of the domain, is used for simulations in order to minimize computational expense. Four separate domains with three sliding interface boundary conditions are used for modelling of the inlet domain, fan domain, stator domain and outlet domain as depicted in Fig. 1.

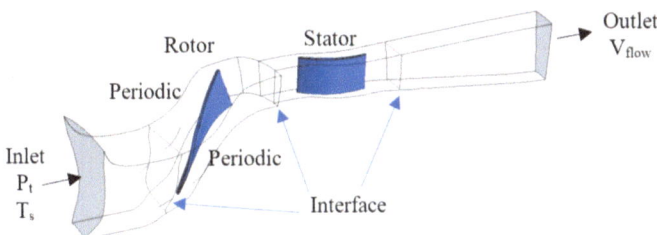

Fig. 1. Boundary conditions of the fan computational domain.

The aerodynamic flow filed of the fan was simulated using the three-dimensional compressible Reynolds-Averaged Navier Stokes equations (RANS) with k-ω SST turbulence model. Inlet and outlet numerical domains were meshed using unstructured tetrahedral cells. Rotor and stator domains were meshed using structured hexahedral cells using ANSYS TurboGrid. Inflation layers have been used near walls with no-slip boundary conditions where y+ is order unity. Final mesh consists of approximately 1.5 million finite volume cells. Inlet boundary is defined with constant total pressure and temperature while the outlet boundary is defined with a constant volume flow rate. The rotating domain is specified with the relevant rotational speed requirement of the fan, while rest of the domains are stationary compared to the stationary reference frame. No-slip boundary at the inner shroud surface of the rotor domain was set as a counter rotating wall. The operating point of the PQ curve is controlled by changing the volumetric flow at the outlet. Initial simulations have been verified based on a grid convergence study, and validated against a similar fan with available experimental data.

A parametric study was conducted in order to investigate the optimum number of stator blades for the fan. Here, five separate simulations were run by changing the number of stator blades. All the other parameters were kept unchanged. The flow field of each configuration is used for justifying the best overall design in terms of performance and weight.

5 Results

Figure 2 show the change of pressure and efficiency of different fans with different number of stator blades namely: 19, 21, 23, 25 and 27. Here the highest-pressure gain and efficiency are observed on the case with 23 stator blades. The efficiency of the fan gradually drop when the number of stator blades are increased of decreased compared to that, and the fan with 27 stator blades shown to be the most inefficient.

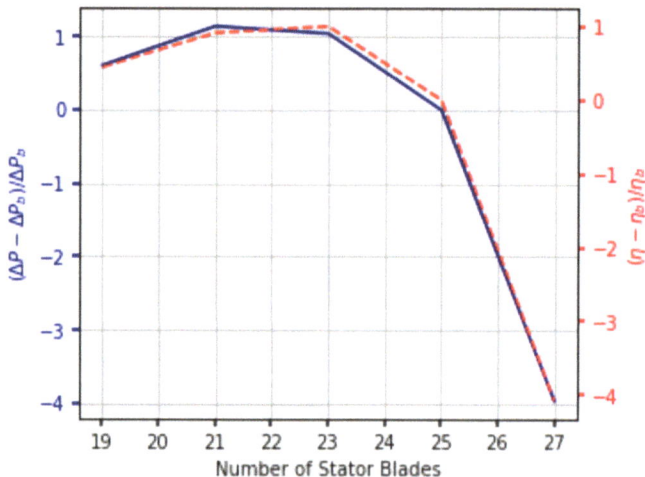

Fig. 2. Pressure and Efficiency of the configurations with different number of stator blades

Figure 3 shows the average values of total pressure distribution across the fan, where cases with stator blades 19, 23, 27 are observed. The total pressure at the upstream of the stator did not show any significant change. This also indicates that the performance of the rotor did not have a significant effect from increased number of stator blades. However, the increased stator number recognizably reduced the total pressure gain towards the fan outlet and thus the fan become less efficient.

Figure 4 show the wall shear stress contours on the stator surface at cases with 19, 23 and 27 stator blades. Here, wall shear stress values reaching zero indicates (i.e., depicted in red) the flow separation. All cases show flow separation near the hub of the fan. On the case with 23 blades, the flow separation area reached minima, which also means that this configuration is aerodynamically more efficient.

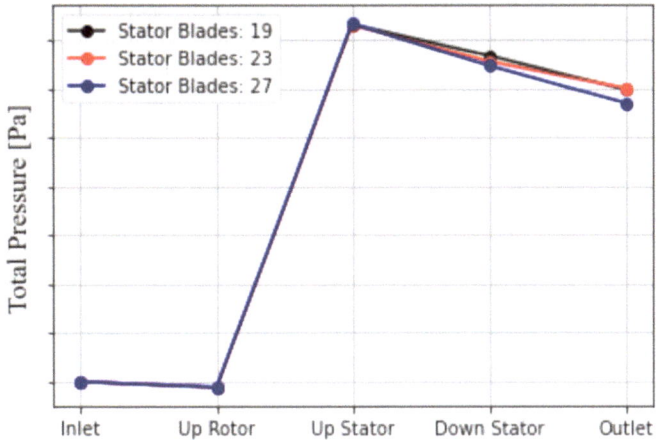

Fig. 3. Average total pressure through the fan

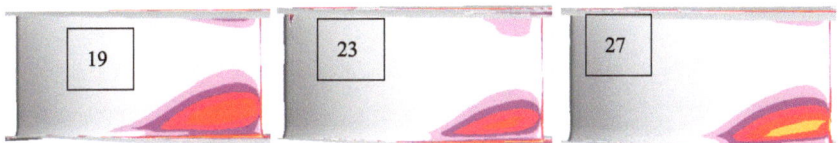

Fig. 4. Wall shear stress contours of the stator blade, where rea indicates $\tau_{wall} \sim 0$

6 Conclusions

This study provides an insight of a stator design of an electric fan which is expected to be used for unpressurized bay ventilation of the next generation hybrid electric regional aircraft. The next step will be to further optimize the stator and rotor blades in order to provide high efficient, light weight and less-noisy fan for the HERA ventilation sub-system.

Acknowledgements. The project is supported by the Clean Aviation Joint Undertaking and its members. Funded by the European Union under Grant Agreement No. 101102008. Views and opinions expressed are however those of the author(s) only and do not necessarily reflect those of the European Union or Clean Aviation Joint Undertaking. Neither the European Union nor the granting authority can be held responsible for them.

References

1. Thermal Management for the Hybrid Electric Regional Aircraft. https://cordis.europa.eu/pro ject/id/101102008. Accessed 25 Sept 2023

2. Wasala, S., Xue, Y., Wiegandt, T., Stevens, L., Persoons, T.: Aeroacoustic noise prediction from a contra-rotating cooling fan used in data center cooling systems. In: AIAA AVIATION FORUM, Virtual (2021)
3. Qu, W., Gong, W., Chen, C., Zhang, T., He, Z.: Optimization design and experimental verification for the mixed-flow fan of a stratospheric airship. Aerospace **10**(2), 107 (2023)

Open Access This chapter is licensed under the terms of the Creative Commons Attribution 4.0 International License (http://creativecommons.org/licenses/by/4.0/), which permits use, sharing, adaptation, distribution and reproduction in any medium or format, as long as you give appropriate credit to the original author(s) and the source, provide a link to the Creative Commons license and indicate if changes were made.

The images or other third party material in this chapter are included in the chapter's Creative Commons license, unless indicated otherwise in a credit line to the material. If material is not included in the chapter's Creative Commons license and your intended use is not permitted by statutory regulation or exceeds the permitted use, you will need to obtain permission directly from the copyright holder.

Exploring the Dynamics of Ride-Hailing Fares in Madrid: A Machine Learning Approach

Tulio Silveira-Santos[1]([✉]) [iD], Anestis Papanikolaou[2] [iD], Thais Rangel[1,3] [iD], and Jose Manuel Vassallo[1] [iD]

[1] Transport Research Center (TRANSyT), Universidad Politécnica de Madrid, 28040 Madrid, Spain
tulio.silveira@upm.es
[2] Volkswagen Data:Lab, Volkswagen AG, 80805 Munich, Germany
[3] Department of Organizational Engineering, Business Administration and Statistics, Universidad Politécnica de Madrid, 28012 Madrid, Spain

Abstract. Ride-hailing apps are getting increasingly common in cities all around the world. However, the major factors that determine how supply and demand interact to determine the ultimate prices are still mostly understood. By using statistical and supervised machine learning techniques (Linear Regression, Decision Tree, and Random Forest), this study aims to comprehend and forecast the behavior of ride-hailing fares. Ten months' worth of data were taken from the Uber Application Programming Interface for the city of Madrid and used to calibrate the model. The results show that the Random Forest model is the most suitable for this kind of prediction due to its superior performance metrics. The unsupervised methodology of cluster analysis (using the k-means clustering method) was also used to examine the variation of the difference between Uber fare forecasts and observed values to better understand prediction error patterns. The investigation found that a tiny percentage of observations (approximately 1.96%) had substantial prediction errors due to unexpected surges caused by supply and demand imbalances, which typically happen during major events, peak hours, weekends, holidays, or when there is a taxi strike. This study assists in the understanding of pricing, service demand, and ride-hailing market pricing structures by policymakers.

Keywords: Ride-Hailing · Dynamic Pricing · Machine Learning · Prediction Error · Clustering Analysis · Transport Policy

1 Overview and Motivation

The transportation landscape has witnessed significant changes in recent years with the advent of new technologies. Ride-hailing services, enabled by mobile applications, have emerged as a popular alternative to traditional modes of transportation [1]. These services, such as Uber and Lyft, offer convenient, on-demand transportation options, transforming the way people move in urban areas [2]. The popularity of these companies can be explained in part by the fact that they often provide cheap, comfortable, on-demand door-to-door transportation options in urban areas [3].

© The Author(s) 2025
C. McNally et al. (Eds.): TRAconference 2024, LNMOB, pp. 549–555, 2025.
https://doi.org/10.1007/978-3-031-85578-8_73

Ride-hailing companies use real-time dynamic algorithms to adjust their fares at any moment, whereas taxi fares are usually fixed and regulated [4]. However, the dynamic pricing mechanism employed by ride-hailing companies, which adjusts fares based on demand and supply, has led to uncertainty and concerns among users, drivers, policy-makers, and regulators. Understanding the behavior of ride-hailing fares is crucial for various stakeholders to make informed decisions, improve policy measures, and enhance the overall transportation system.

This study departs from previous studies and contributes to the prediction of ride-hailing fares in the following ways: (i) total fares were predicted, not just the surge multiplier, as noted by Battifarano and Qian (2019) [5]; (ii) data from a European city were analyzed (most research had previously focused on American cities) and data were collected over an extended period, a total of ten months; (iii) statistical and machine learning models were applied and compared considering the complete dataset, thus working with fixed and dynamic prices, not only dynamic prices, as noted by Rangel et al. (2021) [6]; (iv) a conceptual framework was described that can be adopted by interesting parties to better understand the pricing dynamics of the ride-hailing market; and (v) valuable information and policy recommendations for the ride-hailing market are provided.

To sum up, this study aims to fill the knowledge gap by leveraging machine learning models to gain insights into the behavior of ride-hailing fares, with a specific focus on the city of Madrid, by analyzing a comprehensive dataset obtained from Uber.

2 Methodology, Results, and Main Contributions

The methodology employed in this study involves data collection from Uber, one of the largest ride-hailing companies globally, to gain insights into the behavior of ride-hailing fares. Data were collected through the Uber API and stored at 1 h intervals over 10 months (from September 2018 to June 2019), and 667,051 entries were collected. The dataset comprises a wealth of information, including fare amounts, trip distances, travel times, trip request times, and additional exogenous variables such as weather conditions, month, hour, holidays, peak hours, and taxi strikes. The availability of this comprehensive dataset enables a thorough analysis of ride-hailing fare dynamics and their relationship with various factors.

To collect the data, a custom script leveraging web-scraping techniques was developed to extract ride information from the Uber Application Programming Interface (API). The script was designed to retrieve data from ten selected locations in Madrid, representing both high-demand areas (e.g., airports, public transport stations) and other locations across the city (see Fig. 1). The inclusion of diverse origin and destination points ensures a representative sample for analysis. This study is not intended to compete with existing open tools for fare prediction (such as Uber's Fare Estimator and UberFareFinder), but rather to define a framework to identify the issues of ride-hailing fare prediction and the errors associated with it.

Three supervised machine learning models – Linear Regression, Decision Tree, and Random Forest – were employed to predict ride-hailing fares based on the collected data. These models leverage various features, including trip distance, travel time, delay,

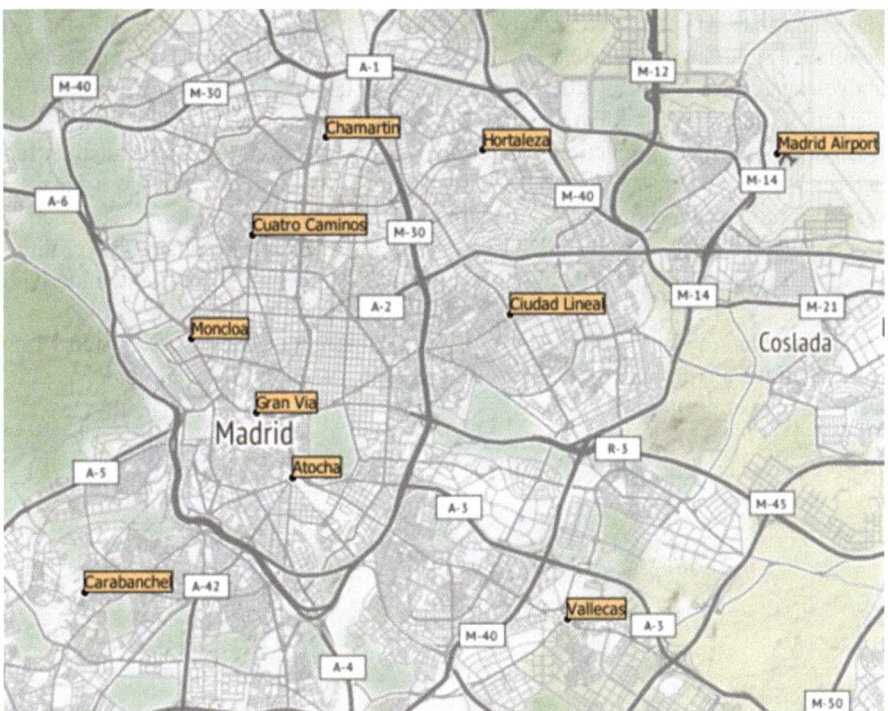

Fig. 1. Madrid city and selection of the ODs of the requested rides. (Source: Map created using QGIS)

weather conditions, month, hour, holidays, peak hours, and taxi strikes, to estimate fare amounts accurately. The objective is to develop robust models that can predict ride-hailing fares with a high degree of accuracy.

In this study, the default hyperparameters for the three machine learning models were used. It is noteworthy that the same random data split of training and testing sets was used for all models (i.e., Train/Test equal to 80/20) for comparison purposes. Predictive accuracy is evaluated and compared across all models (see Table 1).

Table 1. Comparison of performance metrics of the models.

Performance Metrics	Linear Regression	Decision Tree	Random Forest
RMSE (EUR)	3.41	3.85	3.40
MAPE (%)	8.01	6.60	6.32

The results of the machine learning models demonstrate their efficacy in predicting ride-hailing fares. The Random Forest model outperforms the other two models, exhibiting superior performance metrics. The Root Mean Squared Error (RMSE) for the Random Forest model is €3.40, indicating the average prediction error in fare amounts, while the Mean Absolute Percentage Error (MAPE) is 6.32%, providing insights into the relative accuracy of fare predictions. The Random Forest model's exceptional predictive capabilities can be attributed to its ability to capture complex relationships and interactions among the predictor variables, leading to improved fare estimation.

In addition to fare prediction, cluster analysis is employed to identify distinct fare patterns within the dataset. The clustering technique used in this research is k-means, a widely employed algorithm for grouping observations based on similarity. The analysis aims to classify rides into clusters based on fare amounts, trip distances, travel times, and other variables, enabling the identification of different fare profiles and underlying patterns.

The results of the cluster analysis revealed three distinct fare clusters: Low Error, Medium Error, and High Error. Figure 2 shows the number and percentage of observations per cluster.

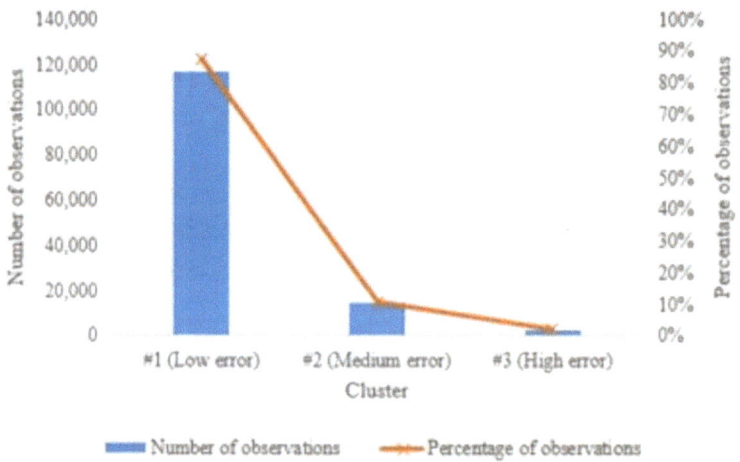

Fig. 2. Number and percentage of observations per cluster.

The number of observations in each cluster also shows that cluster #1 (Low error) contains 87.32% of the observations, followed by cluster #2 (Medium error) with 10.72% and cluster #3 (High error) with 1.96%. All valid observations are included in the clusters. The results show that a high share of the observations have small MAPE prediction errors (87.32%), which shows good accuracy of the machine learning models used to predict the UberX fare (the Random Forest is the machine learning model with the best performance metrics and is also one of the models contributing the most to the group with small MAPE prediction errors). Likewise, it shows a very small share of the observations with high MAPE prediction errors (only 1.96%), which can be caused by unexpected surges due to imbalances between supply and demand, as well as being related to outliers that were not removed and/or other variables.

Table 2 presents a summary of the statistics of the clusters in terms of key continuous variables (e.g., UberX fare, trip distance, travel time, etc.).

Table 2. Relationship between the continuous variables and the formed clusters.

Cluster	Summary Statistics	Fare (EUR)	Distance (Km)	Travel Time (min)	Delay (min)	Precipitation (mm)
#1 (Low error)	Mean	18.57	11.35	18.47	4.60	0.14
	SD	7.29	5.01	5.51	3.30	1.66
#2 (Medium error)	Mean	20.90	9.74	17.02	4.31	0.14
	SD	11.52	5.08	5.17	2.95	1.56
#3 (High error)	Mean	23.38	10.24	17.93	4.62	0.21
	SD	16.38	4.83	5.35	3.51	2.07

Forecasting ride-hailing fares appears more difficult (high MAPE prediction errors) when the supply value of the fare variable and when rain precipitation is higher, but the high standard deviations of the errors overshadow this effect (due to the existence of outliers in all groups). It is also noteworthy there is no clear trend for the variables trip distance, travel time, and delay concerning the errors across clusters, which means the models have already captured the statistical signal from these variables. Table 3 shows the percentage of frequency of observations of the categorical variables within the three clusters.

Table 3. Percentage of frequency of observations of the categorical variables within the formed clusters.

Cluster	Strike (Strike is True)	Business Day (Business Day is True)	Peak Hour (Peak Hour is True)
#1 (Low error)	6.25%	71.06%	16.02%
#2 (Medium error)	6.77%	42.79%	20.77%
#3 (High error)	10.20%	31.79%	20.44%

The results of the percentages of frequency of observations of the categorical variables per cluster show how much they interfere with the MAPE prediction errors, mainly in cluster #3 (High error). The prediction of ride-hailing fares becomes slightly more complex when there is a taxi strike, as well as during peak hours (from 07:00 to 09:00 and from 18:00 to 20:00). The results also show that the forecasts are more accurate on business days, being thus less accurate on weekends and holidays, in which there is a high percentage of frequency of observations in clusters #2 (57.21%) and #3 (68.21%). The previous reasons appear to be related to potential demand peaks that cause an imbalance between supply and demand.

3 Conclusions

This study contributes significantly to the understanding of ride-hailing fare behavior using machine learning models and cluster analysis. By predicting ride-hailing fares and exploring their implications, as well as identifying distinct fare profiles, the research provides valuable insights for various stakeholders in the urban transportation domain.

Public authorities can take advantage of knowing and predicting ride-hailing fares to define and adopt policy measures to set a rational competition and coordination with the taxi industry and across ride-hailing companies. In Spain, for example, taxi services claim that they should be able to establish their prices with the same freedom as ride-hailing services to compete fairly with them (6). Regulatory authorities can use the methodology and result coming out from this study to safeguard fair competition among different ride-hailing operators. The findings can help users know the ride's price in advance, thus facilitating them in choosing the most favorable option for their trips according to their priorities. They can also help drivers keep track of fare rises to secure higher pricing and hence more potential earnings.

Future research directions include: (i) adopting these methods to perform ride-hailing fare prediction using data from different stages of the COVID-19 pandemic; (ii) extending the research methods to other cities (as the ride-hailing market in Spain is restricted and the number of ride-hailing licenses is also limited); (iii) using other robust models for predicting ride-hailing fares, especially to better estimate unexpected surges; and (iv) collecting data from shorter intervals to more accurately capture peak fares.

References

1. Gomez, J., Aguilera-García, Á., Dias, F.F., Bhat, C.R., Vassallo, J.M.: Adoption and frequency of use of ride-hailing services in a European city: the case of Madrid. Transp. Res. Part C Emerg. Technol. **131**, 103359 (2021)
2. Dong, X., Guerra, E., Daziano, R.A.: Impact of TNC on travel behavior and mode choice: a comparative analysis of Boston and Philadelphia. Transportation (2021). (0123456789)
3. Dias, F.F., Lavieri, P.S., Garikapati, V.M., Astroza, S., Pendyala, R.M., Bhat, C.R.: A behavioral choice model of the use of car-sharing and ride-sourcing services. Transportation **44**(6), 1307–1323 (2017)
4. Chen, M.K., Sheldon, M.: Dynamic pricing in a labor market: surge pricing and flexible work on the Uber platform, pp. 1–19 (2015)
5. Battifarano, M., Qian, Z.S.: Predicting real-time surge pricing of ride-sourcing companies. Transp. Res. Part C Emerg. Technol. **107**, 444–462 (2019). https://doi.org/10.1016/j.trc.2019.08.019
6. Rangel, T., Gonzalez, J.N., Gomez, J., Romero, F., Vassallo, J.M.: Exploring ride-hailing fares: an empirical analysis of the case of Madrid. Transportation (2021). (0123456789). https://doi.org/10.1007/s11116-021-10180-w

Open Access This chapter is licensed under the terms of the Creative Commons Attribution 4.0 International License (http://creativecommons.org/licenses/by/4.0/), which permits use, sharing, adaptation, distribution and reproduction in any medium or format, as long as you give appropriate credit to the original author(s) and the source, provide a link to the Creative Commons license and indicate if changes were made.

The images or other third party material in this chapter are included in the chapter's Creative Commons license, unless indicated otherwise in a credit line to the material. If material is not included in the chapter's Creative Commons license and your intended use is not permitted by statutory regulation or exceeds the permitted use, you will need to obtain permission directly from the copyright holder.

Soft Measures Speeding up the Change – Showcases from the City of Turku, Finland

Stella Aaltonen[1](✉) and Katariina Kiviluoto[2]

[1] City of Turku, Yliopistonkatu 27, 20101 Turku, Finland
stella.aaltonen@turku.fi
[2] Turku University of Applied Sciences, Joukahaisenkatu 1, 20520 Turku, Finland
katariina.kiviluoto@turkuamk.fi

Abstract. Mobility Management is a concept to promote sustainable transport and manage the demand for car use by changing attitudes and behaviour through so called "soft" measures (for example, information and service provision, communication or coordination). Mobility management measures do not generally require large financial investments, tend to have a high benefit-cost ratio, and are generally well accepted by citizens. They are often used as complementary activities to other "harder" measures. The City of Turku piloted three soft measures: a digital service map and two nudges. A wide team was included in both measures right from the initial planning phase. Cities can bring together relevant actors ranging from evaluators to service design to complement the work done in topic related departments. Collaboration between different actors ensures that planned actions are not only functionally executed but also designed for diverse target groups and properly evaluated. Wide enough cooperation paves the way for future scaling of successful activities with a deeper impact on travel behaviour.

Keywords: Soft measures · Nudge · Digital

1 Soft Measures

1.1 Mobility Management

Mobility Management is a concept to promote sustainable transport and manage the demand for car use by changing travellers' attitudes and behaviour through so called "soft" measures. These include a range of activities related to information provision and communication, service provision and coordination of different actors. Many times, mobility management measures in comparison to "hard" measures do not necessarily require large financial investments and may have a high benefit-cost ratio [1]. In addition, citizen acceptance of "soft" measures tends to be higher than with measures targeting, for example, infrastructure [2]. In general, the effectiveness of both "soft" and "hard" measures seems to depend on various socio-psychological factors, the type of information used in communicating of the measures as well as the physical urban context within which the measure is carried out [3].

© The Author(s) 2025

C. McNally et al. (Eds.): TRAconference 2024, LNMOB, pp. 556–561, 2025.
https://doi.org/10.1007/978-3-031-85578-8_74

An important part of the soft measures are specific targets and evaluation actions that aim to verify the impact of the action. Research suggests an average 10% decrease in car use when using soft measures [4, 5] and a reasonable cost-benefit ratio. [6] However, due to the complexity of travel behaviour, it may be difficult to verify the actual impact a specific 'soft' measure has on travel behaviour. A review of 141 soft measure related interventions [4], indicates that without proper evaluation, careful study design or precise information on the study design the causality between the intervention and the attributed behaviour change may be difficult to demonstrate. Therefore, careful planning of soft measures is needed to provide a cost-effective alternative or a complementary activity to other actions. In addition, if digital channels of information and communication are used these should be planned together with other intended measures. Digital tools are in many cases developed and evaluated in isolation from the overall mobility management actions. This may lead to isolated platforms, apps and user interfaces and can have a negative impact on the impact of the soft measures.

Very often cities lack to perform valuable evaluation actions in connection with measures. This is in many cases due to the lack of ownership, strategic and systematic approach in actions. There is a clear need in cities to strengthen the role of the soft measures in the actions towards sustainable mobility. It is not enough to create high quality infrastructure and services without understanding their full potential and not nudging citizens towards the behaviour changes needed.

1.2 Circumstances for Change

The city of Turku has passed 200 000 inhabitants in August 2023 being the sixth largest city in Finland [7]. The functional urban area of the Turku has more than 340 000 inhabitants and it's the third largest in Finland [8]. The city of Turku is the oldest city in Finland turning 800 years in 2029. Turku aims to become carbon neutral by the anniversary and increase the share of sustainable transport over 66% by year 2030 [9]. Meeting these goals require big emission cuts in terms of traffic and large behavioural changes.

An important part of this is the large transport reform projects, increasing the amount of bike lanes and investing in walking conditions, as well as tightening parking regulations. These hard measures require a great political will to be implemented and they are not very agile and fast, for example, the planning of traffic reform projects may take years. Fortunately, there are also various so-called "soft measures" to speed up the change in traffic towards a more sustainable direction and to influence on people's behaviour. These soft measures are vital parts of the Sustainable Urban Mobility Plan of the Functional Urban Area of the Turku region.

2 Measures Piloted

2.1 Mobility Map

In the city of Turku, various soft measures have been developed and tested in the EU-funded Scale-up project [10]. In 2022, Turku's mobility service map was launched, from which it is possible to encourage, visualize and monitor various mobility data in the city of Turku and in the Turku region. Through the mobility service map, for example, you can check whether the route of your way to and from work has been ploughed from the snow, which makes it more convenient to cycle in the winter, as it is easier to predict the conditions of the commute route. A lot of focus has been on showcasing bike services, walking routes and real time parking situations. All the shared mobility offers can be found from the map and be combined with service and accessibility of different locations. This acts as incentives for example for tourism locations and interregional travel.

The mobility map is currently the only map that can showcase real time data from variety of sources and offer the content in Swedish and English. The number of users visiting the map has increased every time the mobility map has been showcased in local media. The biggest number of users so far is from February 2023 after media announcement and article on the local newspaper Turun Sanomat [11].

The mobility map is created by open-source code, so it is freely available for others to scale up [12]. Through the embedding tool, the mobility map can be added to any website and this enlarges the usability of the map considerably as all the actors interested can use it freely. One of the most valued features of the mobility map is that all the information is on one place and mobility is not anymore separated from the services.

The mobility map has been created using service design process in several stages of the process. This has led to an increased understanding of the needs of different user groups and how those can be met though mobility map. Evaluating a digital tool is very often based on usage and the number of data sets. One of the most interesting approach for the real impact is, how different actors have used it for their needs and also integrated in their communications campaigns and guidance elements. This often means that the resources are saved in other divisions of the city and within stakeholder organizations as the information needed is already provided for them. This can be considered as the repel effect of creating a common digital platform. At this stage, there is still a need to find a way to create a traceable evaluation for it.

2.2 Nudges

In addition to the mobility service map, mobility nudges are playing an important role, as their purpose is to influence people to change their behaviour in terms of mobility towards the desired direction. Originally developed by Thaler and Sunstein [13], nudging has become a common means to encourage, for example lifestyle changes [14], modal shift [15, 16], or energy saving [17, 18]. There are many ways to categorize nudges, but one simple way is to divide nudges into two types: 'pro-social' and 'pro-self' [19]. 'Pro-self' nudges encourage behaviour that is beneficial for an individual, whereas 'pro-social' nudges aim to help the target group choose behaviours that are beneficial for society. Nudging should not force the target to behave in a certain way but should always make it possible to opt out.

There are numerous examples of successful nudges around the world, and at best they are easy and inexpensive to implement. During the years 2023 and 2024, ten different nudges will be implemented in Turku, focusing on different areas of mobility and traffic. The purpose of these nudges is to support both the transition from own car to sustainable modes of transport and to improve the conditions and usability of sustainable modes of transport, as well as increase the citizen engagement. The nudges will also be aimed at different target groups, such as workers, cyclists, and students.

In the winter of 2023, in Turku, the first nudge was carried out, which was targeted at schools and kindergartens. The purpose of the nudge was to increase active school trips of children by means of simple gamification. The nudge was based on a visually motivating sticker campaign and very cost-effective study design. Alongside the polar bear sticker campaign, a communication campaign was carried out, providing schools, daycares, and parents with practical tips for everyday mobility. A survey for the parents was conducted after the pilot campaign. Despite its low response rate, the survey provided valuable information on how to improve the nudge. Key findings from this experiment highlighted the need for careful pre-planning, incentivizing survey responses, and the importance of age group-based targeting.

Second nudge was launched in October 2023. The regional public transportation card was introduced to citizens as part of the public library system in three different libraries. Individual library users can borrow the public transport card once for two weeks. The aim is to test what kind of effect this can have on changing the perception of using the public transportation and into becoming a regular user. If the nudge is successful, the offer can be enlarged to all the municipalities within the public transportation network.

The intensive development of the nudges has been carried out in wide cooperation with stakeholders in order to create win-win synergies. This has led to a further understanding that there is a need to differentiate that different stages. In some of the nudges a service design of the product needs to be carried out before any intervention can happen, whereas in some nudges it is a matter of targeted marketing of alternative options. Both are directly linked to the evaluation of the effectiveness of the nudge and how this can be carried out.

3   Key Findings of Soft Measures

At this stage of the process, it can be already showcased that a shared common vision of the future is needed as a base for developing effective soft measures. The role of the city in this a crucial as the city acts as a gluer between the different actors and as an enabler creating conditions for all the actors to flourish. Cooperation with different research institutions is one of the win-win situation, as the evaluation actions of the different soft measures is a must, and this knowhow is not yet adequately existing nor resourced in cities. Without the evaluation, the measures could not be adequately scaled up. This is well showcased in the case of the mobility map and in nudges.

The development of a digital mobility map has showcased that integrating data from variety of sources in a user-friendly format creates synergies for many actors. Open data and open-source code create a base for resource efficiency and open the work for others to scale up. The development of the mobility map in design sprints is efficient and enables launching of the new features in stages. This then speeds up the development and increases the interest of the stakeholders creating a dynamic process. The focus on accessibility of the different elements adds up to the process and increases the dialogue.

Lessons learnt from developing and carrying out nudges have pinpointed that tailored target group orientation with suitable evaluation actions is crucial. Carefully planned evaluation together with the overall design may increase the impact of a nudge and may help in avoiding some of the potential pitfalls at the execution phase. Successful nudging requires tight cooperation with different actors. In some cases, the nudging needs a service design phase with different service providers before the actual nudge can take place.

Key ingredients

Digital:

Integrating data in user-friendly format.
Open data and open-source code.
Design sprints with new launches.
Accessibility elements.

Nudges:

Tailored target group orientation needed.
Evaluation planned before action.
Tight cooperation between different actors.
Integration into service design processes.

At the next stage the two elements – the mobility map and nudges will be synchronized, and it is expected that this will increase the speeding effect of the soft measures to the next level. This all is expected to have a significant impact on reaching the high sustainable modal share ambitions of the city of Turku.

References

1. Mobility Management | EPOMM. epomm.eu/about/mobility-management, 9 September 2023
2. Eriksson, L., Garvill, J., Nordlund, A.M.: Acceptability of single and combined transport policy measures: the importance of environmental and policy specific beliefs. Transp. Res. Part A Policy Pract. **42**(8), 1117–1128 (2008). https://doi.org/10.1016/j.tra.2008.03.006
3. Piras, F., Sottile, E., Tuveri, G., Meloni, I.: Does the joint implementation of hard and soft transportation policies lead to travel behavior change? An experimental analysis. Res. Transp. Econ. **95**, 101233 (2022). https://doi.org/10.1016/j.retrec.2022.101233
4. Möser, G., Bamberg, S.: The effectiveness of soft transport policy measures: a critical assessment and meta-analysis of empirical evidence. J. Environ. Psychol. **28**(1), 10–26 (2008). https://doi.org/10.1016/j.jenvp.2007.09.001
5. Semenescu, A., Gavreliuc, A., Sârbescu, P.: 30 Years of soft interventions to reduce car use – a systematic review and meta-analysis. Transp. Res. Part D Transp. Environ. **85**, 102397 (2020). https://doi.org/10.1016/j.trd.2020.102397
6. Taylor, M.A.P.: Voluntary travel behavior change programs in Australia: the carrot rather than the stick in travel demand management. Int. J. Sustain. Transp. **1**(3), 173–192 (2007). https://doi.org/10.1080/15568310601092005
7. https://www.turku.fi/uutinen/2023-08-08_nyt-se-totta-turkulaisia-yli-200000
8. https://yle.fi/a/74-20021918
9. https://issuu.com/turunviestinta/docs/turun_ilmastosuunnitelma_2029
10. https://www.scale-up-project.eu/
11. Article "Aura-auton liikkeitä voi seurata lähes reaaliaikaisesti", Turun sanomat, 22 February 2023

12. GitHub - City-of-Turku/smbackend: Service Map backend, GitHub - City-of-Turku/servicemap-ui: The Service Map UI. github.com/City-of-Turku/servicemap-ui

13. Thaler, R.H., Sunstein, C.R.: Nudge: The Final Edition. Penguin Books, New York. (2021). First published in the United States of America by Yale University Press, 2008

14. Ledderer, L., Kjær, M., Madsen, E.K., Busch, J., Fage-Butler, A.: Nudging in public health lifestyle interventions: a systematic literature review and metasynthesis. Health Educ. Behav. **47**(5), 749–764 (2020). https://doi.org/10.1177/1090198120931788

15. Sina, Z., Thomas, S., Andreas, H., Heiko, G., Helmut, K.: Motivating change in commuters' mobility behaviour: digital nudging for public transportation use. J. Decis. Syst. (2023). https://doi.org/10.1080/12460125.2023.2198056

16. Loidl, M., et al.: Unlocking the potential of digital, situation-aware nudging for promoting sustainable mobility. Sustainability **15**(14), 11149 (2023). https://doi.org/10.3390/su151411149

17. Soomro, A.M., Bharathy, G., Biloria, N., Prasad, M.: A review on motivational nudges for enhancing building energy conservation behaviour. J. Smart Environ. Green Comput. **1**(1), 3–20 (2021). https://doi.org/10.20517/jsegc.2020.03

18. Colasante, A., Idiano, D., Piergiuseppe, M.: Nudging for the increased adoption of solar energy? Evidence from a survey in Italy. Energy Res. Soc. Sci. **74** (2021). https://doi.org/10.1016/j.erss.2021.101978

19. Hagman, W., Andersson, D., Västfjäll, D., Tinghög, G.: Public views on policies involving nudges. Rev. Philos. Psychol. **6**(3), 439–453 (2015). https://doi.org/10.1007/s13164-015-0263-2

Open Access This chapter is licensed under the terms of the Creative Commons Attribution 4.0 International License (http://creativecommons.org/licenses/by/4.0/), which permits use, sharing, adaptation, distribution and reproduction in any medium or format, as long as you give appropriate credit to the original author(s) and the source, provide a link to the Creative Commons license and indicate if changes were made.

The images or other third party material in this chapter are included in the chapter's Creative Commons license, unless indicated otherwise in a credit line to the material. If material is not included in the chapter's Creative Commons license and your intended use is not permitted by statutory regulation or exceeds the permitted use, you will need to obtain permission directly from the copyright holder.

Evolution and Safety Concept of the Modular U-Shift Vehicle for Sustainable Mobility of People and Goods

Manuel Osebek$^{(\boxtimes)}$ ⓘ, Sebastian Scheibe ⓘ, Marco Münster ⓘ, and Tjark Siefkes

DLR - Institute of Vehicle Concepts, Pfaffenwaldring 38-40, 70569 Stuttgart, Germany
{manuel.osebek,sebastian.scheibe,marco.muenster,
tjark.siefkes}@dlr.de

Abstract. At the German Aerospace Center (DLR), new transport and vehicle concepts for sustainable and user-friendly mobility are being researched. The Institute of Vehicle Concepts of the DLR developed a concept study of a modular and fully automated transport system. The so-called U-Shift enables the transport of different capsules as load carriers and changing them during operation without requiring additional infrastructure. This paper describes the vehicle concept and its project landscape developed in cooperation with other partners in order to picture the evolution towards real life testing. The latest prototype, U-Shift IV, was presented in April 2023 as a new research platform. The first active use case of the U-Shift was decided to be within a public area, in which a research operation is implemented over half a year. Testing prototypes in public areas brings new challenges for DLR in applying legal safety requirements as well as following new processes of assigning automated vehicles in Germany. Due to its interfaces and interactive research, the U-Shift concept of the DLR strongly addresses several TRA themes like user-centered transport or transport safety.

Keywords: Automated Driving · Modularity · User-centred Development · Real-life Testing · Prototype · Safety Concept

1 Introduction, Goals and Methodology

Due to the constant increase in commercial traffic in large cities, especially delivery vehicles and small trucks, there is a great load on the traffic system, particularly at peak times. This is accompanied by a pollution of the air and burdening residents through emissions, traffic jams and noise. This is further intensified by additional passenger traffic with buses and other individual transport. This is where U-Shift comes in, by combining these two essential business areas and offering a new transportation solution. At the German Aerospace Center (DLR), transport researchers are developing vehicle concepts with the aim of devising sustainable and user-centric mobility solutions [1, 2] to face European and global development goals. Within the DLR, the Institute of Vehicle Concepts has developed a concept study for a modular and automated transportation system, the U-Shift (see Fig. 1). Specific use cases are public transport, inner-city store

© The Author(s) 2025

C. McNally et al. (Eds.): TRAconference 2024, LNMOB, pp. 562–568, 2025.
https://doi.org/10.1007/978-3-031-85578-8_75

deliveries, parcel logistics or other services. By addressing these main issues and with its zero-emission drivetrain, the automated U-Shift is meant to contribute to reach mobilities sustainability goals and help decarbonating the cities. The overall goal of this research in DLR is to develop and test this modular vehicle concept and find out requirements for user-friendly and cost-effective transport solutions. Therefore, the DLR is facing particular research topics in separate projects (see Sect. 2). By increasing the technology readiness level (TRL) a public testing is made possible (see Sect. 3). This is an enabler for participative development, which is a main goal in this project. To allow the U-Shift prototype for public research, a state-of-the-art safety concept is being developed and introduced in Sect. 4. The results of the research operation itself is not in the scope of this paper.

2 The Vehicle Concept U-Shift and TRL Improvement

The automated *Driveboard* integrates all components and systems necessary for driving. In addition to the electric engines, battery and automation components, it has an integrated lifting system for the simple and quick exchange of different capsules. The capsules are changed "on-the-road", i.e. during operation at designated areas, to enable new transport and mobility services without additional assistance.

Fig. 1. Graphic of Driveboard with sensors and computers for object detection (left side). Different capsule types (right side) for persons (right) or goods (middle) and a bottom structure (left) with same interface structures (Source: DLR).

The U-Shift *capsules* are only equipped with the most necessary technical equipment and can be produced cost-efficiently. This is an important design principle, especially for use in freight transport. U-Shift freight capsules can have a modular design and serve a wide range of applications. The integrated *lifting system* enables the quick and easy exchange of different capsules: First, the Driveboard is sliding under the rail of the capsule bottom. It then raises vertically upwards and lifts the capsule. A locking mechanism centers and locks the capsule. An electrical coupling can be connected to transmit signals or energy. The vehicle is now ready for use.

There are different projects within the U-Shift ecosystem with various research and development goals, covering different development stages and TRLs (see Table 1). The very first version of the vehicle concept was created during the U-Shift I project [3] to define main technical details and build a real scale demonstrator. The U-Shift II project followed and achieves to improve the technology and create an automated Driveboard as a research platform focussing on citizen participation for research questions [4].

Table 1. The table shows an overview over the completed and ongoing U-Shift projects [5]

Project	Objective	Runtime	Funding	Partners
U-Shift I	First demonstrator with remote control	2019–2020	DLR, WM-BW	DLR, KIT, FKFS, UULM
U-Shift II	Drivable automated prototype for research purposes	2020–2024	WM-BW	DLR, KIT, FKFS, UULM
U-Shift III	Remote operated drivable prototype with high performance and ability for MAD	2020–2022	BMWK	DLR, Industry
U-Shift IV	Drivable automated prototype for real-life testing with approval criteria	2022–2025	DLR, WM-BW, SDA	DLR

In the U-Shift III project, DLR and industrial partners built another Driveboard designed to operate remotely and provide a test vehicle for Managed Automated Driving (MAD). Currently, the U-Shift IV project is running with the goal to demonstrate automated driving in simple real-life environments. This real-life testing connects international research projects such as EU project "SHOW". Also technology transfer activities are made to analyze and transfer the knowledge to the industry.

3   Research Operation at BUGA

The U-Shift is used to allow the public transparent insights into current mobility research. To do this, an intensive citizen dialog and citizen surveys are taking place at the National Garden Show (BUGA) in Mannheim, Germany. The DLR also conducts workshops with the prototype with potential users and operators for future deployments of autonomous movers. For example, an on-the-move survey was conducted. This means that guests were observed and interviewed during the trip and then invited to take part in in-depth discussions about their needs and expectations. In total, over 150 visitors of the garden show took part in this survey spontaneously and delivered in a transdisciplinary feedback. The driving operation takes place with the Driveboard and the passenger capsule to allow visitors to board the U-Shift.

The BUGA takes place from April 14 to October 8, 2023 with a total of about 2 million visitors inside the area. For the research operation, a 2,000 m long test track is available on the park site. The U-Shift is exhibited for seven days a week which results in a total exhibition time of about 1,500 h. Approximately 10 journeys can be made per day, which means a total mileage of about 2,000 km with about 10.000 passengers in the entire testings.

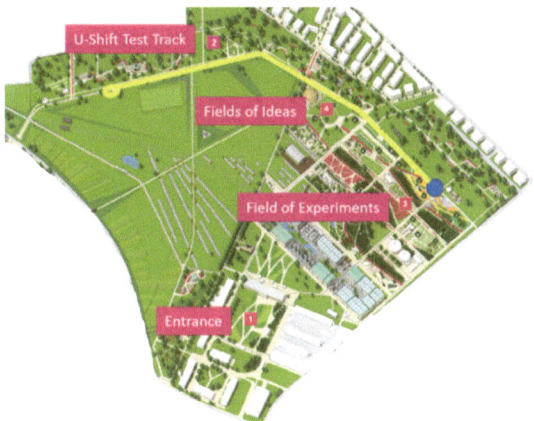

Fig. 2. U-Shift test track at the National Garden Show with start and ending point (blue circle) and the 2.000 m test route (yellow line) inside the park area (Source: https://www.buga23.de/die-parks/spinellipark/. Accessed on 08.01.2024)

The test route is a 4.0–5.0 m wide asphalt road. The terrain is flat, but there is varying vegetation along the road. The exact route can be seen in Fig. 2. It contains several straight sections and curves, as well as a U-turn at the left end of the route. In addition to the U-Shift, Light Electric Vehicles are present on the road, as well as a trackless park train for the visitors. Pedestrians can move or cross the road along the entire track in a short distance, so there is no separate driveway for the U-Shift which is a challenge in forecasting road users' movements. For monitoring such interactions, the sensor system is able to record the field of view and provide information about quality and range of different sensor types. For communicating with pedestrians, the U-Shift is equipped with visual panels and sound systems to inform about its movement. Additionally, a safety operator is always on board to control the vehicle and act in critical situations.

The test track characteristics result in specific requirements for the safety concept and they are part of the DLR research. For example, the bus stop is used to investigate safe boarding and alighting. At intersections, communication between the vehicle and pedestrians as well as passengers can be investigated.

4 Safety Concept for Automated Vehicle Operation

The safety concept created applies exclusively to the single U-Shift IV vehicle, consisting of a Driveboard, a passenger capsule and a multi-use platform. Although the BUGA is a private area with no needs for official approval, the goal was to satisfy all legal safety requirements, especially the StVZO (German road traffic registration ordinance). This is intended to safeguard the demonstration operation at BUGA as well as other test fields and improve the vehicle's suitability for future road use.

The safety assessment follows the AV-Permit of TÜV Süd [6] as seen in Fig. 3. It is based on legal requirements and supplements these with an assessment of operational safety. The safety concept covers all driving modes (automated, remote-controlled and manual) as well as safety measures for the operating environment. The result of the AV-Permit is a comprehensive documentation of the vehicle, training concepts for the vehicle operators and the expert report for submission to insurance companies and authorities for final approval.

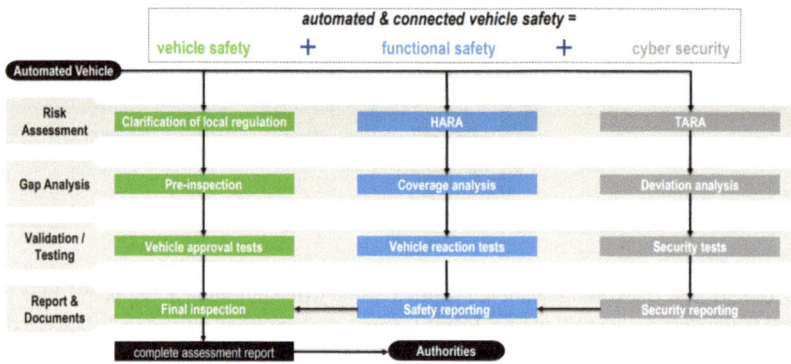

Fig. 3. Framework Details of the AV-Permit for automated and connected vehicles by TÜV Süd. The main assessment fields of Vehicle Safety, Functional Safety and Cyber Security are examined including a risk assessment, gap analysis and validating as well as reporting [6].

The *Vehicle Safety* contains a comprehensive system description of the vehicle. Here, legal requirements according to the German Road Traffic Licensing Regulations (StVZO) are clarified and the vehicle is physically inspected by an independent third-party organization. In *Functional Safety*, a hazard analysis (for the planned operating scenario, HARA), a fault analysis of the technical system (FMEA) and a reliability analysis (Fault Tree Analysis FTA) are carried out (e.g. according to the ISO26262 standard). To minimize errors and misbehavior, prerequisites and qualification measures for operators are defined. The *Cyber Security* includes a Threat Analysis and Risk Assessment (TARA) and a test plan for cybersecurity tests to secure all non-physical interfaces.

The U-Shift IV *Vehicle Safety*, together with the passenger capsule, was successfully approved by the TÜV as a special vehicle. With a few exceptions, U-Shift complies with German regulations, e.g. the vehicle has no steering wheel and no conventional driver's seat. Based on this report, the authorities can issue an individual operating permit and exemption permit according to §21 and §70 of German StVZO.

In the HARA, all potentially dangerous situations occurring at the BUGA were systematically examined for the *Functional Safety*. Corresponding safety requirements and limitations were derived from the analyses such as limiting the maximum speed to 15 km/h along the test route. In designated slow zones, e.g. areas with dense traffic or at intersections the speed is limited to 6 km/h. The FMEA contains a complete system description of the U-Shift and possible (system) fault cases while the FTA shows the effects of simultaneous faults. The critical case here is the brake process. Redundancies ensure that the vehicle is always able to brake to a standstill. An essential part of the safety concept is the driver training to qualify the vehicle operators. Due to the conditions at the BUGA and the prototype status, the safety concept defines the vehicle to be accompanied by a safety driver.

The remote control and the radio emergency stop are particularly relevant in the *Cyber Security*. These components communicate constantly with the vehicle and are a potential gateway for manipulation attempts. This risk was reduced by taking appropriate measures.

5 Conclusion and Outlook

Regarding user-centered transport, the research has shown that it is essential to involve all target groups in the requirements assessment and dialogue. In particular, groups that have doubts about new transportation solutions or that cannot be addressed using conventional methods can be reached by using a participatory approach to define important requirements and hence increase the chances of a successful market launch. In the field of road safety, it has been shown that it is essential to consider vehicle safety, functional safety and cyber security in each specific application. This includes both the operating environment and interactions with other road users or passengers. Unmistakable communication between the vehicle and pedestrians in particular is a prerequisite of safe automated transportation.

Within the U-Shift IV project the DLR successfully assigned the vehicle for the research operation according to legal requirements and the effectiveness of the safety concept was proven at the test area. For a fully road assignment further developments and a comprehensive approval are necessary, which is the next step and targeted for 2024. Further steps are also to research about the vehicle-to-vehicle communication as well as the communication between vehicle and passengers/pedestrians.

At the time of the paper submission, the test phase has not yet been completed and is therefore subject in further publications. The U-Shift IV project and its findings will serve as a blueprint for future DLR projects and further tests of similar vehicle concepts in public areas.

References

1. Münster, M., Kopp, G., Friedrich, H., Siefkes, T.: Autonomes Fahrzeugkonzept für den urbanen Verkehr der Zukunft. ATZ **122**(3), 26–31 (2020)
2. Münster, M., et al.: Fahrzeugkonzepte für die Mobilität von morgen. In: Tagung ATZlive Automatisiertes Fahren, Wiesbaden (2022)
3. Münster, M., et al.: U-shift vehicle concept: modular on the road. In: 21 Internationales Stuttgarter Symposium Automobil- und Motorentechnik, Stuttgart (2021)
4. Münster, M., et al.: U-shift II vision and project goals. In: 22 Internationales Stuttgarter Symposium Automobil- und Motorentechnik, Stuttgart (2022)
5. Münster, M., Osebek, M., Scheibe, S., Siefkes, T.: Modulares Fahrzeugkonzept für die Mobilität von morgen. ATZ **124**(10), 16–21 (2022)

6. TÜV Süd Permit for Operating Autonomous Vehicles on Public Roads. https://www.tuv
   sud.com/en/industries/mobility-and-automotive/automotive-and-oem/autonomous-driving/
   permit-for-operating-autonomous-vehicles. Accessed 27 Sept 2023

Open Access This chapter is licensed under the terms of the Creative Commons Attribution 4.0
International License (http://creativecommons.org/licenses/by/4.0/), which permits use, sharing,
adaptation, distribution and reproduction in any medium or format, as long as you give appropriate
credit to the original author(s) and the source, provide a link to the Creative Commons license and
indicate if changes were made.

The images or other third party material in this chapter are included in the chapter's Creative
Commons license, unless indicated otherwise in a credit line to the material. If material is not
included in the chapter's Creative Commons license and your intended use is not permitted by
statutory regulation or exceeds the permitted use, you will need to obtain permission directly from
the copyright holder.

Integrated Passenger and Freight Transport: Seamless Door-to-Door Mobility and Optimal Use of Resources

Margarita Kostovasili(✉) ⓘ, John Kanellopoulos ⓘ, and Angelos Amditis ⓘ

I-SENSE, Institute of Communication and Computer Systems (ICCS), St. Iroon Polytechneiou 9, 15773 Zografou, Greece
margarita.kostovasili@iccs.gr

Abstract. The provision of safe, resilient and smart mobility services for both passengers and goods is a key priority for the EU transport sector. Even though passenger and freight transport systems share space, infrastructure and challenges, the current governance and regulatory systems deal with them in silos. This paper aims to provide an overview of the current status of passenger and freight transportation, focusing on the specific characteristics, limitations and barriers that prevent the collaboration, as well as the interdependencies and the opportunities for synergies and integrated transport schemes. The latest research related to the adoption of effective approaches enabling integration of passenger and freight transport is also presented, towards interoperability in multimodal transport ecosystem. The concept and results from the EU-funded research project DELPHI are discussed, highlighting the role of regulatory/governance systems and data sharing schemes in the integrated passenger and freight transport. The DELPHI case studies and demonstrators are also presented, along with their expected impact on current mobility and logistics services, and how they can be sustainably implemented, replicated and widely adopted by stakeholders.

Keywords: Integrated Passenger/Freight Transport · Cross-modal Governance · Multimodality · Traffic Management · Federated Platform Network · Sustainable Mobility

1 Introduction

The transport sector is a cornerstone of the global economy and is vital to the fulfillment of the free movement of people, services and goods. At European Union level, transport represents more than 9% of EU gross value added, while employing more than 10 million employees [1]. In recent decades, there has been an increased interest worldwide in transport optimization through sustainable and reliable models that will allow increased mobility at lower cost and reduced environmental footprint. In parallel, growing challenges are identified during the last years in both passenger and freight transport, such as urbanization trends, the increasing complexity of stakeholder scene, e-commerce, rising fragmentation of freight transport [2]. Apart from the various technical challenges and

© The Author(s) 2025
C. McNally et al. (Eds.): TRAconference 2024, LNMOB, pp. 569–575, 2025.
https://doi.org/10.1007/978-3-031-85578-8_76

shortcomings, current governance and regulatory systems have also proven so far inadequate, dealing with the passengers and freight transport systems as independent – and thus disconnected – sectors, while integration between passenger and freight transport has been a subject of debate in the transport community over the last years with a peak coming after the recent Covid-19 crisis [3].

Either transporting people or goods, the ultimate goal is common and is about achieving movement from origin to destination in a quick, easy, safe and reliable way, with the lowest possible cost and environmental footprint. This common approach has already been applied in aviation and rail transportation, where infrastructure is used to move people and goods simultaneously (e.g. airplanes with specially designed spaces for passengers or goods, specially designed wagons for each use, etc.) [4]. However, in road transportation, these two mobility types are still treated as independent systems with different modes of operation and separate infrastructures, resulting in isolated solutions that do not take into account possible interactions and potential opportunities for synergy. A promising solution to this issue could be the definition of a common framework that would enable the integration of passenger and freight transport systems, through the implementation of integrated models satisfying the needs and requirements of both transport types.

2 Integrated Passenger and Freight Transport

In general, an integrated transportation model is based on the concept that people and goods can use common infrastructure, means of transport and services for simultaneous or parallel movement, without one system hindering the other. This integrated approach focuses on the efficiency and reliability of a mixed traffic model, leading to more sustainable operations than the existing model, where passengers and freight are referred to as independent systems. Although the goal is common, there are several functional differences impeding the implementation of a mixed movement model. The main difference is that people are independent decision-making units selecting the transport mode based on subjective criteria, while goods are managed throughout their transport, from origin to destination, based on specific criteria of cost and efficiency. With regards to freight transport, apart from transport means and services, the traffic model shall also take into account processes for (un)loading, management, storage and distribution infrastructure, as well as the corresponding human resources for these procedures.

Another difference is that freight transport follows a predefined pattern, since the routes are known in advance, making it easier to manage transport networks and optimize traffic flows based on available infrastructure. On the other hand, people follow a much looser and irregular framework, which fluctuates significantly. Even though there are some patterns that are followed in passengers' transportation (e.g. commuters), these flows can change dynamically and their modelling is a complex process. At the same time, the two systems are not only treated as independent at an operational level, but also in an institutional and legal framework, since they are governed by different authorities and subject to different rules and guidelines, different contractual and employment structures. An integrated traffic model should be accompanied by consistent policies and coherent planning [5], as the regulatory and regulatory aspects are crucial to the success of such an initiative and are currently the biggest barrier to its spread.

In such an integrated system, passengers and goods simultaneously share vehicles, infrastructure and urban space, increasing the sustainability of the model and reducing service time and energy waste. The central idea of an integrated traffic model lies in the fact that the excess capacity of passenger transport is utilized for the transport of goods, either during the transport of passengers or during periods of inactivity of the transport means. This implies lower direct and generated costs for all parties involved, more care for environmental issues and higher social value. The adoption of a mixed traffic model can bring significant advantages in the transport sector, but also in the wider economy and management of cities. The coexistence of people and goods in shared means of transport can lead to a reduction in the required routes to meet transport needs, leading to a reduction in the traffic load and avoiding traffic congestion, as well as the resulting accidents and emissions.

Furthermore, the combination of such a model with Intelligent Transport Systems can lead to multiple benefits, optimizing the services provided and increasing their efficiency. Especially in cases of urban transport where road transport prevails, combined transport can significantly contribute to the first/last mile management of freight transport, through the utilization of transport means and existing routes [6]. Thus, passenger transport hubs can also be used as hubs for the management and distribution of freight loads. In these cases, the benefit is not only economic or environmental, but also social, since the distribution process of goods is improved and urban space is better used, thus mitigating the impact of transport in the daily life of the inhabitants of a city and increasing social acceptance. Of course, the combined movement of people and goods requires the interconnection of urban freight transport, the urban mobility system and all those involved in the management and economy of the urban ecosystem (public administration, trade associations, residents, etc.).

3   The DELPHI Project

3.1   Concept and Objectives

Recognizing the complexity of stakeholder landscapes, fragmented transportation systems and the need for secure data sharing, the EU-funded research project DELPHI focuses on the strategic dimension of integrating passenger and freight transport in a single system, working towards integrating sectors, harmonizing data, and leveraging advanced methodologies, to transform transportation systems, for a sustainable future. The DELPHI's concept is based on the delivery of technical and governance/regulatory enablers towards a federated network of platforms for multi-modal passenger and freight transport, capable of sharing in a seamless and secure manner, cross-sectoral, multi-modal passenger and freight transport data, as well as traffic management systems information.

The key objectives of DELPHI project include: (i) delivery of novel governance and regulatory schemes and models; (ii) design and development of a "Multimodal Passenger and Freight Transport Network of Platforms" framework; (iii) development of an AI/ML-powered transport network and traffic management optimization framework; (iv) validation of the developed solutions via 4 realistic pilots and simulation-based analysis; and (v) compatibility with EU standards in the freight and passenger mobility domains, as well as promote and contribute to the standardization of multimodal and multi-stakeholder freight and passenger management solutions and information systems. In order to achieve these objectives, DELPHI is expected to deliver an overall architectural design that will enable a secure data federation in a Data Spaces-driven approach, novel governance and regulatory schemes and novel and ultra-efficient methodologies for traffic monitoring, such as Unmanned Aerial System-powered monitoring. Furthermore, it will provide multi-/inter-modal optimization, Artificial Intelligence (AI)/Machine Learning (ML)-powered optimizations and frameworks, and will exploit diverse modes for hybrid passenger and freight transport in different ecosystem types.

3.2 Use Cases and Pilot Demonstrations

DELPHI's digital framework and transversal tools will be validated in the context of 4 pilot demonstrations with complementary requirements and features, in Spain, in Madrid Urban Area, Greece - Athens Metropolitan Area, Greece - Mykonos island, and Romania, city of Cluj-Napoca. In particular, the 1st use case and corresponding pilot demonstration will be implemented in Madrid urban area and targets to create efficient, sustainable, collaborative, faster and safer transport modes for the end-to-end transport process. It focuses on the existing Metro network, avoiding the use of current vans/trucks while it also facilitates remote monitoring and control via drones, as well as mobile applications, providing multiple layers of visualization. Also, crowd-shipping methodologies will be designed and validated, i.e., via the exploitation of e-bikes for sustainable and efficient freight delivery in last-mile delivery scenarios. The platform will enable real-time data exchange thought the use of Blockchain and flexible and dynamic transport planning based on AI/ML-powered predictive analytics, hence leading to reduced emissions and costs for involved stakeholders.

The 2nd use case focuses on multimodal network/traffic management, in the Attica region, examining the possibility of exploiting both the Attica Tollway Road infrastructure and the rapid-transit system. This will be realized focusing on common nodes where both means meet and are commonly located and expand the potential collaboration between the operators for network optimization in an effort to optimize multidimensional congestion, emissions reduction and infrastructure viability and finally cost of transportation minimization. This use case will collate information from transport service providers, as well as end-users and implement an integrated approach and algorithms for delivering goods that are normally trucked on busy roads by using passenger modes such as the Metro exploiting the possibility to be delivered underground. In parallel, non-causal reasoning techniques, as well as AI-enabled decision support systems will be implemented, as well as low-cost, massive mobility data collection via drones and a management system for evaluating and minimizing CO2 and noise emissions.

The 3rd use case deals with the provision of integrated freight and passengers' models and data sharing framework in Mykonos island, focusing on the optimization of the freight transport via exploiting the regular passenger shuttles, as well as the exploitation of underutilized taxi vehicles. This will be realized via the federation of diverse data sources, namely port authority passenger and freight information, as well as Mykonos' shuttle and taxi service real-time schedule and booking information. The shuttle/taxi service routes will be optimized to minimize traffic, operational time and emissions concentration and distribution in the city area, while the resource utilization (human, equipment, vehicles) will be improved based on real-time information and the environmental impact will be reduced.

The 4th use case will be implemented in the city of Cluj-Napoca and it will focus on the provision of Integrated passengers' models and data sharing governance framework in the Cluj-Napoca Metropolitan Area. The use case aims to collate information from public transportation service providers as well as end-users and implement an integrated approach and algorithms for optimizing track routes to minimize traffic and emissions concentration and distribution in the city area, supporting fleet management, along with modeling and optimization of public transport routes. Based on these solutions, the use case will develop a multi-actor governance framework for data exchange and multi-stakeholders' communication for multimodal network and traffic management.

3.3   Expected Outcomes and Impact

Through the development and validation of all aforementioned tools and solutions in the implemented use cases and demonstrators, DELPHI aims to significantly contribute to the improvement of the transport sector and the facilitation of sustainable mobility options for both passengers and goods. The demonstrators are expected to result in improved decision-making process, optimized use of resources, reduced operating costs and environmental impact. In addition, the prediction of transportation network bottlenecks will be enhanced, the mobility flows of passengers and freight will be optimized, the traffic congestion and emissions in urban and sub-urban/rural areas will be reduced and the delivery times in urban areas will be decreased. Moreover, the monitored traffic data will be significantly increased and the collaboration among transport stakeholders will be enhanced. Also, the public transport performance will be increased and the emissions produced by the public transportation vehicles will be reduced.

The key impact areas of DELPHI project include: (i) improved multimodal transport network and traffic management capabilities, facilitating seamless door-to-door mobility for passengers and freight; (ii) effective and resilient network-wide data exchange and new integrated data management systems for dynamic and responsive multimodal network and traffic management; (iii) tested and validated systems for enhanced prediction and resolution of network bottlenecks, substantially increasing safety, security, resilience and overall performance of the entire transport network; (iv) innovative tools and services for optimizing mobility flows of passengers and freight in cities and other operating environments, cutting congestion, journey times and traffic jams across transport modes, and thereby significantly reducing emissions; (v) new governance arrangements for multimodal transport network and traffic management, in view of further regulatory and policy actions; and (vi) high market adoption and transferability of innovations to different ecosystems.

4 Conclusions

Summarizing the above, it is concluded that the adoption of a model of combined movement of people and goods can be a powerful tool to address some of the key problems of the transport sector, such as traffic congestion, environmental burden, limited use of existing infrastructure, extensive spatial coverage etc. Recognizing the interactions and revising the transport approach is crucial for the implementation of such a model, which can be an important tool to address the ever-increasing need to transport people and goods and traffic problems which this entails, through viable and sustainable practices. Towards addressing existing challenges, novel solutions must be introduced, including novel governance and regulatory systems, advanced technical and functional specifications towards interoperability and federation of existing/novel platforms, harmonized data specification and reference information models, as well as common approaches and methodologies towards an integrated approach for passenger and freight transport. DELPHI project will enable the creation of a federated ecosystem and architecture for seamless interaction of diverse freight/passenger transport and traffic management, as well as monitoring platforms, supporting beyond state-of-the-art, multimodal transport optimization models and methods, towards reduced cost, delivery time and emissions.

Acknowledgement. DELPHI project is funded by the European Union, under grant agreement No 101104263. Views and opinions expressed are however those of the author(s) only and do not necessarily reflect those of the European Union or the European Climate, Infrastructure and Environment Executive Agency (CINEA). Neither the European Union nor the granting authority can be held responsible for them.

References

1. European Union – Transport topic. https://european-union.europa.eu/priorities-and-actions/actions-topic/transport_en#:~:text=Transport%20is%20also%20a%20major,employ%20a round%2011%20million%20people. Accessed 10 Oct 2023
2. Bruzzone, F., Cavallaro, F., Nocera, S.: The integration of passenger and freight transport for first-last mile operations. Transp. Policy **100**, 31–48 (2021)
3. World Economic Forum. https://www.weforum.org/agenda/2020/07/covid-19-revealed-the-importance-of-integrating-passenger-and-freight-transport/. Accessed 10 Oct 2023
4. Cavallaro, F., Nocera, S.: Integration of passenger and freight transport: a concept-centric literature review. Res. Transp. Bus. Manag. **43** (2022)
5. Hörsting, L., Cleophas, C.: Scheduling shared passenger and freight transport on a fixed infrastructure. Eur. J. Oper. Res. **306**(3), 1158–1169 (2023)
6. Arvidsson, N., Givoni, M., Woxenius, J.: Exploring last mile synergies in passenger and freight transport. Built Environ. **42**(4), 523–538 (2016)

Open Access This chapter is licensed under the terms of the Creative Commons Attribution 4.0 International License (http://creativecommons.org/licenses/by/4.0/), which permits use, sharing, adaptation, distribution and reproduction in any medium or format, as long as you give appropriate credit to the original author(s) and the source, provide a link to the Creative Commons license and indicate if changes were made.

The images or other third party material in this chapter are included in the chapter's Creative Commons license, unless indicated otherwise in a credit line to the material. If material is not included in the chapter's Creative Commons license and your intended use is not permitted by statutory regulation or exceeds the permitted use, you will need to obtain permission directly from the copyright holder.

Decarbonizing Urban Mobility Through Data-Driven Services: How to Achieve Modal Shift?

Eetu Wallius$^{(\boxtimes)}$, Carlos Lamuela Orta, Toni Lusikka, Olli Pihlajamaa, and Raine Hautala

VTT Technical Research Centre of Finland Ltd., Tampere, Oulu, Espoo, Finland
eetu.wallius@vtt.fi

Abstract. Due to still relying heavily on motorized private vehicles, mobility hinders the sustainable functioning of societies. This has led to an increasing pressure to direct individuals towards sustainable mobility options, mostly including active mobility (i.e., walking, cycling) and public transport. During the recent years, mobility has become increasingly datafied, providing new ways to increase the attractiveness of sustainable options. In this manuscript, we provide a framework to capture the different ways in which data and information can aid in making sustainable modes of mobility more appealing, serving future efforts in the domain. Overall, data and information benefits mobility organizers in their endeavor to create well-functioning mobility systems, as well as users when planning and making day-to-day decisions regarding mobility.

Keywords: Modal shift · datafication · sustainability

1 Introduction

Despite its paramount importance to economy, culture and individual well-being and flourishing, personal mobility is a source of negative externalities that threaten the sustainable functioning of societies. Due to relying heavily on the use of fossil fuels, mobility is a major source of greenhouse gas emissions globally, contributing to the ongoing climate crisis [1]. Additionally, mobility poses local environmental and social issues, such as pollution, congestion, accidents, and the occupation of public space, hindering the well-being of individuals and ecosystems in many areas [2].

The sustainability problems of mobility are particularly pronounced in the case of private motorized vehicles as they have a high energy consumption and occupy parking and driving space that could otherwise be used for improving the livability of areas. The use of these vehicles additionally entails other disbenefits, such as increased accident risk and increased likelihood of health issues, including cardiovascular disease [3]. To tackle these problems, there is an increasing interest in decreasing private car use and directing individuals to use sustainable mobility options, including mostly active mobility (walking, cycling) and public transport.

© The Author(s) 2025

C. McNally et al. (Eds.): TRAconference 2024, LNMOB, pp. 576–581, 2025.
https://doi.org/10.1007/978-3-031-85578-8_77

During the recent years, ICT-based solutions have emerged as a viable tool among other means geared towards encouraging the use public transport and active mobility over privately owned combustion engine vehicles, therefore having potential to contribute to the sustainable modal shift through sustainable mode choices. ICT has become a pervasive part of our everyday lives, and mobility is not left untouched by advances such as artificial intelligence, the internet of things, big data, pervasive computing, and mobile apps [4]. Most notably, mobility systems have become increasingly datafied which allows novel ways to make sustainable modes of mobility more attractive, therefore adding to the repertoire of pull mechanisms for sustainable mobility.

However, the opportunities provided by datafication are multifaceted, and inquiries to holistically make sense of the role that data and technologically mediated information plays in achieving sustainable modal shift are needed to guide research and practitioner endeavors towards sustainable mobile futures. To address this issue, we provide a conceptual framework to broadly capture the different ways in which data and information can aid in making sustainable modes of mobility more appealing.

2 Framework of Information and Data Flows for Sustainable Mobility

The integration of ICT and mobility allows many ways to collect and utilize data for promoting sustainable modal shift. In the framework presented in Fig. 1 and detailed below, we consider two distinct types of actors that benefit from mobility data and information: mobility organizers (i.e., operators, service providers, policymakers, planners, infrastructure managers etc.) and users (i.e., mobile citizens). We conceptualize the ways in which data can support sustainable mode choices to comprise eight distinct approaches that represent different flows of data and information between organizers, users, and the physical elements of the mobility system. Each approach is elaborated through corresponding examples from prior research and existing industry solutions. Overall, data and information can be utilized by mobility organizers in their endeavor to create well-functioning mobility systems, as well as users when planning and making day-to-day decisions regarding mobility.

Data on Mobility Patterns. The widespread adoption of mobile devices equipped with positioning systems has resulted in the availability of individual mobility pattern data. This data provides opportunities for mobility organizers to understand how individuals move in their day-to-day life and leverage this data when planning infrastructure changes, new services or improving and optimizing existing ones to encourage sustainable mode choices. For example, mobility data pertaining to speed and direction obtained through mobile positioning devices can be used to identify risky locations to improve the infrastructure for active mobility, thus encouraging its adoption [5]. Artificial intelligence (AI) provides additional opportunities of utilizing location data by, for example, refining it to information regarding the time used for different phases of a public transport journey (i.e., walking to a stop, waiting at a stop, onboard) that can be further used by mobility organizers to improve passenger experience throughout the entire journey [6].

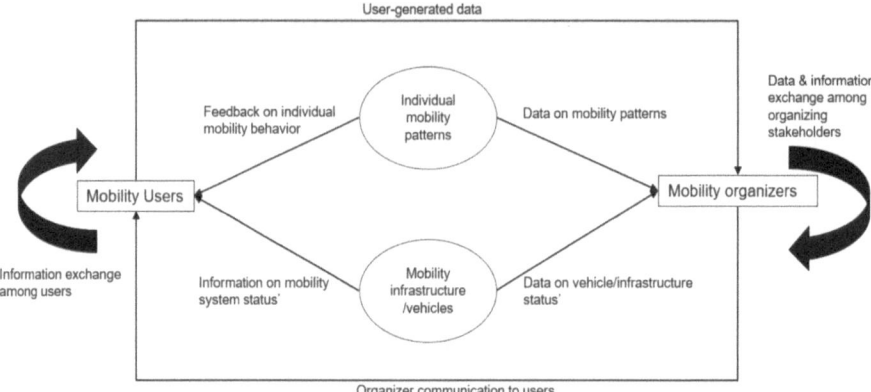

Fig. 1. Framework of data and information flows for sustainable mobility.

Feedback on Individual Mobility Behavior. The individual mobility pattern data not only provides opportunities for improving mobility services and infrastructure but can also be refined to offer mobility users individualized insights and feedback pertaining to their mobility behaviors. This allows nudging users towards more sustainable modes and overall mobility patterns through the provision of health-based information, information based on monetary savings or environmental impacts pertaining to their mobility behavior. For example, by tracking the user, mobility apps can provide information of their CO_2-emissions, or calories burnt due to mobility and therefore make the benefits of using sustainable modes more tangible to encourage their use [7]. Prior research has shown that these nudging approaches can raise awareness of the impacts of different modes of mobility and induce a change in mobility behavior [7].

Data on Vehicle and Infrastructure Status. The sensors embedded within the mobility system enable the gathering of valuable data about infrastructure and vehicles. Mobility organizers can leverage this data to improve their services and optimize their resources. For example, AI-based passenger monitoring systems enable service providers and other stakeholders to understand vehicle occupancy or detect individual positions, and even the activities they engage in inside vehicles and stops [8, 9]. This aids in route and resource optimization, as well as in improving vehicle and infrastructure design for better user experience, therefore having the potential to increase the adoption of sustainable modes.

Information on Mobility System Status. The data of vehicle and infrastructure status can subsequently be transformed into useful information that is then made available to mobility users. For example, real-time information on schedules, routes, or vehicle crowdedness can be provided to the users to encourage sustainable mode choices by easing journey planning and improving the experience of using public transport [10, 11]. Real-time information can additionally be used to direct passenger flows e.g., to less-crowded vehicles through dynamic pricing or other nudges for improved travel experience and accessibility, additionally contributing to inclusive sustainable mobility [10].

User-Generated Data. In addition to the data collected through sensors, ICT developments allow mobility organizers to amass qualitative data pertaining to the user experiences. For example, through crowdsourcing, mobility organizers can swiftly gather user feedback on mobility services, initiatives for future improvements or observations related to infrastructure deficits to improve the attractiveness of sustainable modes. In addition, advances in areas that integrate physical and virtual worlds such as digital twins, and immersive technologies often described under the umbrella concept of the 'Metaverse' [12] offer new opportunities for obtaining user-generated data. These technologies enable constructing immersive virtual models to communicate mobility plans and aspirations to a wider audience of non-expert citizens, while simultaneously eliciting user impressions and feedback regarding these future changes for improved user experience [13].

Organizer Communication to Users. ICT also allows organizers to directly communicate to mobility users. This can include, for example, updates regarding future changes within the mobility system, disruptions within transportation, or other timely information that aid users in planning their journeys using sustainable modes. For example, the public transport app operational in the Finnish capital of Helsinki informs the user of changes in routes and timetables as well as communicates other public transport news to the user [14]. This information provision has the potential to ease the day-to-day use of sustainable modes as well as improve their predictability.

Information and Data Exchange Among Organizing Stakeholders. ICT enables data information exchange between various organizing stakeholders, including both public and private service providers, and policymakers, among others. This is central to approaches such Mobility-as-a-Service, that necessitate the integration of private and public mobility services into a single interface, but also for first and last mile solutions, and overall enhancement of mobility in a given context through data-based planning and operation [15, 16]. These integrated service approaches and joint efforts can aid in making sustainable modes seamlessly integrated, convenient, and flexible options to private motorized vehicles.

Information Exchange Among Users. Finally, ICT provides a means for direct information exchange among mobility users. For example, social networking and media applications and services provide platforms for sharing experiences, tips or other information pertaining to the use of sustainable modes of mobility, thus making their use more convenient. An example of an approach that is dedicated to active mobility is the free-to-use tool Bikemaps.org which uses crowdsourcing to gather information related to bicycle collisions, near misses, hazards, and thefts and presents this information to users in the form of a map, along with specific dates and times when these incidents are more likely to occur, allowing users to proactively avoid and prepare for potentially dangerous areas and circumstances when planning their cycling routes [17].

3 Conclusions and Future Work

Overall, data and ICT-mediated information exchange provide ample opportunities for making sustainable modes of mobility more appealing, and thus foster sustainable modal shift towards active and public transport. This manuscript provided a conceptual framework that broadly outlines the various flows of data and information that aid in this endeavor. Data and information are central to mobility organizers aiming to improve the quality of sustainable mobilities, but also to the day-to-day mobility decisions made by users. The proposed framework aids future efforts by providing a granular understanding of the data needs required for making sustainable modes more attractive, whether by enhancing the quality of sustainable mobility services and relevant infrastructure by more efficient resource management and other improvements or by informing users and providing nudges to encourage sustainable mobility patterns.

In our future work, we seek to utilize the proposed framework to develop data and information driven approaches that will increase the appeal of sustainable modes of mobility in urban settings. To do so, we aim to create collaborative approaches of working with stakeholders in mobility – including cities, research partners, public transport organizations and operators, among others – while at the same time bringing the user to the forefront by revealing the variegated mobility needs and motivations that can be tackled using data and information-based approaches.

Acknowledgements. This work was conducted as a part of Decarbonizing transport and mobility through data-driven modal shift (DeMo) project funded by Business Finland (Dnro 8390/31/2022).

References

1. Singru, N.: Evaluation study independent evaluation department reducing carbon emissions from transport projects (2010)
2. Sheller, M., Urry, J.: The city and the car. Int. J. Urban Reg. Res. **24**(4), 737–757 (2000). https://doi.org/10.1111/1468-2427.00276
3. Sugiyama, T., Chandrabose, M., Homer, A.R., Sugiyama, M., Dunstan, D.W., Owen, N.: Car use and cardiovascular disease risk: systematic review and implications for transport research. J. Transp. Health **19** (2020). https://doi.org/10.1016/j.jth.2020.100930
4. Varghese, V., Chikaraishi, M., Jana, A.: The architecture of complexity in the relationships between information and communication technologies and travel: a review of empirical studies. Transp. Res. Interdiscip. Perspect. **11** (2021). https://doi.org/10.1016/j.trip.2021.100432
5. Yaqoob, S., Cafiso, S., Morabito, G., Pappalardo, G.: Detection of anomalies in cycling behavior with convolutional neural network and deep learning. Eur. Transp. Res. Rev. **15**(1) (2023). https://doi.org/10.1186/s12544-023-00583-4
6. Hosseini, S.H., Gentile, G.: Smartphone-based recognition of access trip phase to public transport stops via machine learning models. Transp. Telecommun. **23**(4), 273–283 (2022). https://doi.org/10.2478/ttj-2022-0022
7. Marquart, H., Schuppan, J.: Promoting sustainable mobility: to what extent is 'health' considered by mobility app studies? A review and a conceptual framework. Sustainability **14**(1) (2022). MDPI. https://doi.org/10.3390/su14010047

8. Abedi, H., Magnier, C., Shaker, G.: Passenger monitoring using AI-powered radar. In: 2021 IEEE 19th International Symposium on Antenna Technology and Applied Electromagnetics, ANTEM 2021. Institute of Electrical and Electronics Engineers Inc., August 2021. https://doi.org/10.1109/ANTEM51107.2021.9518503

9. Capozzi, L., et al.: Toward vehicle occupant-invariant models for activity characterization. IEEE Access **10**, 104215–104225 (2022). https://doi.org/10.1109/ACCESS.2022.3210973

10. Hadas, Y., Tillman, A., Tsadikovich, D., Ozalvo, A.: Assessing public transport passenger attitudes towards a dynamic fare model based on in-vehicle crowdedness levels and additional waiting time. Int. J. Transp. Sci. Technol. (2022). https://doi.org/10.1016/j.ijtst.2022.08.003

11. Meng, M., Memon, A.A., Wong, Y.D., Lam, S.H.: Impact of traveller information on mode choice behavior. Proc. Inst. Civ. Eng. Transp., 11–19 (2018). https://doi.org/10.1680/jtran.16.00058

12. Dwivedi, Y.K., et al.: Metaverse beyond the hype: multidisciplinary perspectives on emerging challenges, opportunities, and agenda for research, practice and policy. Int. J. Inf. Manag. **66** (2022). https://doi.org/10.1016/j.ijinfomgt.2022.102542

13. Christodoulou, N., Papallas, A., Kostic, Z., Nacke, L.E.: Information visualisation, gamification and immersive technologies in participatory planning. In: Conference on Human Factors in Computing Systems - Proceedings, Association for Computing Machinery, April 2018. https://doi.org/10.1145/3170427.3185363

14. HSL, "HSL App". https://www.hsl.fi/en/tickets-and-fares/hsl-app

15. Kamargianni, M., Li, W., Matyas, M., Schäfer, A.: A critical review of new mobility services for urban transport. Transp. Res. Procedia, 3294–3303 (2016). https://doi.org/10.1016/j.trpro.2016.05.277

16. Shaheen, S., Cohen, A.: Mobility on demand (MOD) and mobility as a service (MaaS): early understanding of shared mobility impacts and public transit partnerships. Demand Emerg. Transp. Syst. Model. Adopt. Satisf. Mobil. Patterns, 37–59 (2019). https://doi.org/10.1016/B978-0-12-815018-4.00003-6

17. BikeMaps, "BikeMaps". https://bikemaps.org/

Open Access This chapter is licensed under the terms of the Creative Commons Attribution 4.0 International License (http://creativecommons.org/licenses/by/4.0/), which permits use, sharing, adaptation, distribution and reproduction in any medium or format, as long as you give appropriate credit to the original author(s) and the source, provide a link to the Creative Commons license and indicate if changes were made.

The images or other third party material in this chapter are included in the chapter's Creative Commons license, unless indicated otherwise in a credit line to the material. If material is not included in the chapter's Creative Commons license and your intended use is not permitted by statutory regulation or exceeds the permitted use, you will need to obtain permission directly from the copyright holder.

MASA - Modena Automotive Smart Area as Scalable Living Lab. Experiences and Development from the R-Nord District to the Modena Urban Area

Giulia Tagliazucchi[1], Francesco Pasquale[2], Francesco Gherardini[3],
Alberto Vergnano[3], Gianluca Marchi[1], and Francesco Leali[3]([✉])

[1] Marco Biagi Department of Economics, University of Modena and Reggio Emilia,
Viale Berengario 51, 41121 Modena, Italy
{giulia.tagliazucchi,gianluca.marchi}@unimore.it

[2] Biagio Rossetti Faculty, Department of Architecture, Via Ludovico Ariosto 35,
44121 Ferrara, Italy
francesco.pasquale@unife.it

[3] Enzo Ferrari Department of Engineering, University of Modena and Reggio Emilia,
Via P. Vivarelli 10, 41125 Modena, Italy
{francesco.gherardini,alberto.vergnano,
francesco.leali}@unimore.it

Abstract. The Modena Automotive Smart Area (MASA) is an experimental project, based on a public-private partnership, aiming at the development of an urban living lab to test, evaluate, and implement connected and autonomous vehicles. It is proposed as a case study that contains all the key elements to convey scalability, replicability, and operational flexibility. MASA is an example of public-private partnership with the Municipality of Modena and the University of Modena and Reggio Emilia, strengthened by the endorsement of the Italian Ministry of Infrastructures and Transport and the support of the Emilia-Romagna Region, also involving industrial partners, in the implementation of a triple helix model. The research and experimentation activities concern interactions between vehicles, between vehicle and moving obstacle, between vehicle and city. The present contribution discusses the impact on the complexity of the urban environment both as digital and physical transformations and the possibility of a scalable model from a district area to an urban network.

Keywords: Living Lab · Smart City · Autonomous driving

1 Introduction

Scientific community, public institutions and world automotive industry have been questioning since years about future scenarios related to the transportation of people and goods and their impact on our cities (Fagnant and Kockleman 2015; Dikmen and Burns

© The Author(s) 2025
C. McNally et al. (Eds.): TRAconference 2024, LNMOB, pp. 582–588, 2025.
https://doi.org/10.1007/978-3-031-85578-8_78

2016). The development of new driving technologies is leading to disruptive consequences, rethinking mobility from vehicles, infrastructures and services and reviewing the related legal and insurances issues, up to the adoption and penetration of these new technologies into the market. While connected vehicle and smart city technologies are almost an established reality, the experimentation of connected and autonomous vehicles in urban environments is an increasingly prominent.

The Modena Automotive Smart Area (MASA) is an urban living lab (Nesti 2018), born in Modena in 2017, Emilia Romagna, in the hearth of the Italian Motor Valley. It has been designed to develop and validate both virtually and experimentally innovative solutions for a smart urban mobility. Based on a public-private partnership, the project is the result of the convergence of the strategies of the University of Modena and Reggio Emilia, the Modena Municipality, and numerous industrial partners aimed at promoting science, innovation and technological experimentation (Meissner 2015). MASA is also an example of a triple helix model of innovation (Champenois and Etzkowitz 2018; Tagliazucchi et al. 2023). MASA is in fact characterized by the engagement of intra-institutional stakeholders and led by an interdisciplinary committee of representatives, with the purpose aligning the interest of the partners in creating a collaborative space in an urban area to experiment with technologies and spread knowledge. In particular, the main objective of the MASA project concerns the application of new digital technologies to mobility services in urban (smart city) and extra-urban (smart road) areas and to the automotive industry (connected and autonomous cars and related supply chains). The potential outcomes are: the improvement of health conditions of citizens (health status monitoring, traffic safety, data collection in line with security and respect of privacy); improvement of the quality of city life (more services, a more attractive city, road models adaptable to the conditions of traffic and infrastructures); energy saving strategies (lower emissions and costs); improvement of environmental quality (better air quality and lower urban noise). The role of MASA is now being explored in the light of the recent National Mobility Center, born in the wake of the PNRR, to evaluate its impacts and networks in the creation of a mobility ecosystem to support the city and society.

The present contribution, based on a case study, aims to present the expected results on the application of a variety of vehicles and services in the context of a real urban environment such as the R-Nord district in Modena. The impact on the complexity of the urban environment will be discussed as both digital and physical transformations and will lead to a scalable model from a district area to a structured urban network with spokes and nodes in the city of Modena.

2 The Case Study: MASA

The Modena Automotive Smart Area (MASA) is a living lab designed to develop and experimentally validate innovative solutions for an intelligent urban mobility. Originally based on a public-private partnership, the project was born in Modena in 2017, thanks to the convergence, in a triple helix perspective (Champenois and Etzkowitz 2018), of the public actor at municipal level, the local University, and the industrial partners. Each actor has been involved with specific roles and purposes, developed and refined over time (Tagliazucchi et al. 2023).

More specifically, the University – namely University of Modena and Reggio Emilia (UNIMORE) – conveys its experience in research, training and teaching. It has also strengthened its medium-term development strategy aimed at high-level training and multidisciplinary research with the establishment of the Automotive Academy, an organizational structure designed to coordinate teaching, research and third mission initiatives related to the automotive and mobility sector. The success of UNIMORE strategies is proven by the numerous collaborations with the most prestigious companies in the automotive sector, and – as for the scientific recognition – by the leading role in national and international competitive projects aimed at the development of the most promising automotive technologies, while integrating complementary disciplines, such as engineering, economics, social sciences, medicine. Furthermore, the University of Modena and Reggio Emilia is strongly committed to participating in call for research projects - with outstanding results in competitive calls for research funding. As for the public actor, the Municipality of Modena aims at improving the quality of life of citizens, investing in urban, cultural and technological redevelopment of some peripheral areas of the city, strengthening the infrastructure and digital services offered. More specifically, the Municipality of Modena succeeds in winning the so-called "Call for Peripherals" fund, which is an extraordinary national program of intervention for urban redevelopment and the security of the suburbs of the cities. Thanks to this inflow, the designed area of the urban living lab of MASA has been renovated with the enabling technologies for the experimentation activities. Furthermore, the Municipality of Modena actively promotes the participation of MASA in competitive projects at the national and European level. Finally, the industrial partners – namely AD Consulting and Danisi Engineering, which have strengthened the technological skills of the research group, and Autodromo di Modena – share their industrial research programs aimed at developing solutions for safe and efficient mobility, based on the implementation of connected and cooperative systems. Prototypes of highly-connected vehicles equipped with sensors, such as radars, are then available for closed track and open road testing.

MASA, as a living laboratory, has a tripartite structure (see Fig. 1). The R&D Lab is the University research laboratory equipped with a "Hardware in the loop" type static simulator for carrying out research and development activities. The Smart Dynamic Area is a dedicated area at the Autodromo di Modena equipped with a basic infrastructure for the testing of self-driving vehicles and V2X communication systems in a totally safe and confidential context, as well as for the creation of training activities. The Smart Model Area is an urban area within the city of Modena, equipped with basic infrastructure (cameras; 4.5G communication network; data server; sensors) for testing bidirectional communication between connected vehicles and elements of mobility and city (V2X) and for the testing of vehicles equipped with ADAS devices up to level 3 and 4.

Among the main technical results achieved, several researches concerned Vehicle-to-Vehicle (V2V), Vehicle-to-Infrastructure (V2I) and Vehicle-to-city (V2X) interactions. Autonomous vehicles must be coordinated in order to exploit crossing, turning left (in right-hand drive countries), free parking, whose management is quite challenging. The proposed model is that of a double test: beforehand through the support of the Autodromo di Modena, in order to evaluate the readiness and security of technologies before the urban mobility test, both in the smart model area located within the city and beyond. It is

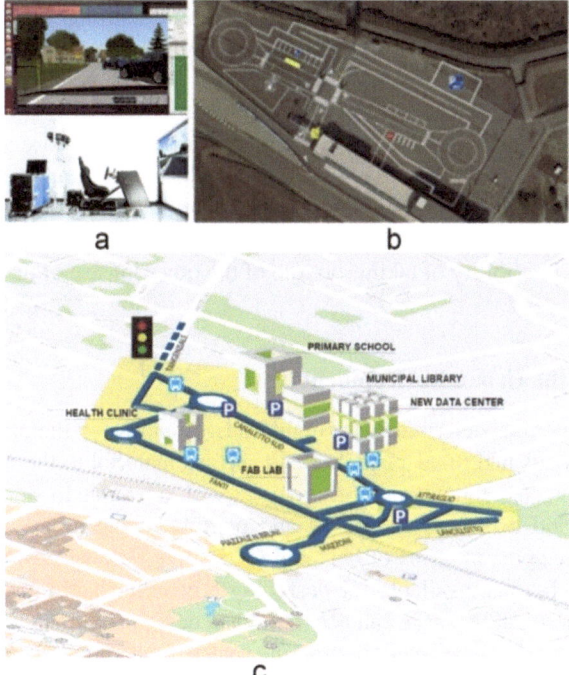

Fig. 1. The Modena Automotive Smart Area: The R&D Lab (a); the smart Dynamic Area at the Autodromo di Modena (b); the Smart Model Area map in the northern part of Modena (c). (Source: https://www.automotivesmartarea.it/)

therefore possible to validate a range of products on roads open to the public. In addition to this, the research groups relating to MASA in the socio-economic and legal fields have explored its characteristics and evolution as a living lab, the application possibilities of road testing in light of the national regulatory framework, the acceptance by consumers and the implications of the spread of self-driving vehicles.

3   Results

3.1   From a PPP to a Living Lab

At their core, public-private partnerships are defined as "working arrangements based on a mutual commitment (over and above that implied in any contract) between a public sector organization with any organization outside of the public sector" (Roberts and Siemiatycki 2015). What makes MASA an exceptional case study is that it overcomes the criticism usually moved to other PPPs, according to which the "structure of the relationship is typically more akin to contracting out than a truly meaningful collaboration between the partners" (Roberts and Siemiatycki 2015). Process management practices were then been activated in order to foster real cooperation and exploit the potential of the PPP collaboration. Within the automotive sector and the transport system, MASA today represents a reference model at a national and international level. Indeed, it goes beyond the traditional purposes of PPPs to "design, build, finance, and/or manage public services and goods" (Roberts and Siemiatycki 2015) by involving the experimentation of cutting-edge technologies in an urban area.

As such, MASA is configured as a laboratory of urban life, whose aim is to develop and test innovative urban solutions (Nesti 2018). As such, it requires a strong collaboration among public, research, and industrial partners (Nesti 2018). The management models are then flexible and involve all the three actors for their roles and skills, also providing for the possibility of taking-on role of the other in case of need (Tagliazucchi et al. 2023).

3.2   Impact on the Urban Environment

Mobility infrastructures determined the main transformations in the urban fabric in the last century, as roads had to adapt in size and surface material to the new standards of automobile. Connected and autonomous vehicles would be able to significantly reduce the amount of land dedicated to driving and parking in the urban areas, which are the most sensitive spots in terms of heat and CO_2 spots. The spaces regained through this process will be then available for designing a healthier and more resilient urban environment, where public space can accommodate design solutions against the effects of climate change, such as vegetation, draining surfaces, shading structures and devices for rainwater collection and reuse. The contribution of AI applications to driving is as well a crucial tool in order to avoid human distractions and errors while driving, which today are responsible for about 85% of road accidents. Self-driving vehicles will also guarantee mobility to all of those users who cannot or have discomfort in driving, whether due to temporary or permanent conditions, in order to make everyone capable of participating in the community and avoiding social isolation.

3.3   Potential of Scalability

As a distinctive element, MASA stands out for its exceptional degree of multidisciplinary, involving different University departments and disciplines. In fact, from the very beginning, some investigation guidelines were identified that guide scientific research and correspond to many multidisciplinary working groups that gather academic researchers, industrial practitioners and public managers. The areas of investigation grouped into MASA are the following: real-time systems; vehicle dynamics; artificial intelligence; automotive cyber security; connectivity and framework; innovative propulsion systems and environmental sustainability; law and ethics, privacy of data; economy, user model, technological adoption and sector analysis; human-machine interaction. Technical experimentation and scientific results are therefore not disconnected from the social, economic and environmental reality, but rather they are evaluated within the context of application that is the urban reality.

It is on this peculiar element that the scalability of the model insists, from a neighborhood in the northern area to the implementation of a smart city. Only thanks to the inter- and intra-disciplinary analysis of the application of new technologies in an extended urban context is an overall assessment of feasibility and impact possible. The global trend toward population concentration in urban areas requires to investigate solutions that are able to connect larger urban areas. The standard paradigms of mobility as they have been developed so far (more people = more roads and more vehicles) will not succeed in this task without the side-effects of a proportional increase in traffic congestion and air pollution. The vision of the so-called "15 min city" can work on a small scale where basic services can be reached by walk or by bike, but on a larger model it can generate physical and social segregation between healthy and disadvantaged areas. Connecting a multi-polar metropolitan area with a sustainable mobility network is the goal to aim for, from an all-around sustainability model perspective.

4 Conclusion

The main objective of MASA concerns the application of new digital technologies to mobility services in urban and extra-urban areas ("smart city" and "smart road") and to the automotive industry (connected car/autonomous car and related supply chains). The potential outcomes are: the improvement of the health conditions of citizens (monitoring of health status, reduction of traffic accidents, data collection in line with security and respect of privacy); improvement of the quality of city life (more services, a more attractive city, road models adaptable to the traffic conditions and to the conditions of the road infrastructure); energy savings (lower emissions and lower costs); improvement of environmental quality (improved air quality and lower urban noise). It proposes, for the near future, an ambitious plan for urban growth and technological development that can only be achieved through the convergence of strategies between the three involved actors: the Municipality, the University and the industrial partners. Its success is ultimately rooted in their actual involvement and participation, which represents a promising case of an urban living lab born according to a triple helix model. The strong point is its multidisciplinary, through which it is possible to analyze the technological development, along with its diffusion and adoption, the juridical-legal constraints on its implementation and, ultimately, its urban impact. This element is configured as the key point to assess the real impact of the living lab and of the technologies implemented in the creation of a mobility ecosystem to support the city and society.

References

Champenois, C., Etzkowitz, H.: From boundary line to boundary space: the creation of hybrid organizations as a Triple Helix micro-foundation. Technovation **76**, 28–39 (2018)

Dikmen, M., Burns, C.M.: Autonomous driving in the real world: experiences with tesla autopilot and summon. In: Proceedings of the 8th International Conference on Automotive User Interfaces and Interactive Vehicular Applications, pp. 225–228 (2016)

Fagnant, D., Kockleman, K.: Preparing a nation for autonomous vehicles: opportunities, barriers and policy recommendations for capitalizing on self-driven vehicles. Transp. Res. Part A **77**, 167–181 (2015)

Meissner, D.: Public-private partnership models for science, technology, and innovation cooperation. J. Knowl. Econ., 1–21 (2015)

Roberts, D.J., Siemiatycki, M.: Fostering meaningful partnerships in public–private partnerships: innovations in partnership design and process management to create value. Eviron. Plan. C. Gov. Policy **33**(4), 780–793 (2015)

Nesti, G.: Co-production for innovation: the urban living lab experience. Policy Soc. **37**(3), 310–325 (2018)

Tagliazucchi, G., Della Santa, S., Gherardini, F.: Design of a living lab for autonomous driving: an investigation under the lens of the triple helix model. J. Technol. Transf., 1–24 (2023)

Open Access This chapter is licensed under the terms of the Creative Commons Attribution 4.0 International License (http://creativecommons.org/licenses/by/4.0/), which permits use, sharing, adaptation, distribution and reproduction in any medium or format, as long as you give appropriate credit to the original author(s) and the source, provide a link to the Creative Commons license and indicate if changes were made.

The images or other third party material in this chapter are included in the chapter's Creative Commons license, unless indicated otherwise in a credit line to the material. If material is not included in the chapter's Creative Commons license and your intended use is not permitted by statutory regulation or exceeds the permitted use, you will need to obtain permission directly from the copyright holder.

Estimating the Impact of Congestion Charging on Traffic Flow Distribution Using *PTV Visum*

David Gruhonjić[1](✉), Valentina Mirović[2], and Marko Šoštarić[3]

[1] Independent Researcher, 21000 Novi Sad, Serbia
davidns99@gmail.com
[2] Faculty of Technical Sciences, University of Novi Sad, 21000 Novi Sad, Serbia
[3] Faculty of Transport and Traffic Sciences, University of Zagreb, 10000 Zagreb, Croatia

Abstract. Various transportation management methods have been designed to tackle traffic congestion and its side effects, and out of these methods, congestion charging distinguishes itself as quite effective. Although its primary goal is to decrease congestion, another important goal is the internalization of externalities – charging polluters an additional fee for the damage caused by their usage of the car. It has proved its effectiveness in London, Stockholm, Singapore, and other cities, by decreasing congestion and improving the modal split. Although congestion charging usually decreases car trips, some users continue using the car. This paper analyses the impact of congestion charging on these users' route choices, using *PTV Visum*. The centerpiece of the paper is the prediction of traffic flow distribution using the *TRIBUT Equilibrium-Lohse* method in *Visum*. The specificity of this method, which is comprehensively explained, is that it considers both the time cost and monetary cost of a trip, while using a unique distribution of the value of time. The paper provides examples and a detailed explanation of necessary functions required for an effective estimation in *Visum*, while highlighting the importance of applying all steps of transportation modelling. It concludes with key findings and recommendations for further research.

Keywords: Congestion Charging · Traffic Flow Distribution · *PTV Visum*

1 Introduction

During the past few decades, the increasing congestion levels in various cities have led to many issues which must be alleviated to improve the sustainability of every urban area. Therefore, transportation experts were forced to find transport management solutions to resolve these problems. One of these solutions is congestion charging, a system that can decrease congestion levels, improve the modal split, and internalize external costs, by charging car users for their mode choice, which causes external costs in terms of pollution, congestion, etc. [1].

If adequately designed, congestion charging can have substantial benefits. For example, the introduction of congestion charging in London has led to a decrease in vehicle-kilometers by 34%, and a decrease in time losses of between 32% and 50% [2, 3].

© The Author(s) 2025
C. McNally et al. (Eds.): TRAconference 2024, LNMOB, pp. 589–595, 2025.
https://doi.org/10.1007/978-3-031-85578-8_79

Simultaneously, public transport ridership has increased, as well as the modal split of active modes of transport. In Stockholm and Gothenburg, congestion levels have been reduced by 22% and 10% respectively, while also reducing travel time variability and increasing public transport ridership [4, 5]. Similar effects were recorded in Singapore [6]. However, such effects could not have been achieved without being coupled with other transport management measures. Also, despite many benefits, congestion charging can have a *rebound effect* – congestion levels can rise on roads which become alternate routes used to avoid the charge, an effect which can be minimized by careful planning [7].

Each congestion charging system has three characteristics which are valid for all systems: the system has a territory or road segment on which it is enforced, it has a period during which users are charged, and the system has locations, also known as *portals*, through which vehicles pass and are charged.

Congestion charging can be categorized according to their territorial organization (enclosed systems, systems with specific portal locations and systems enforced on one specific object), technology (ANPR and RFID), fee variability (fixed fee, variable fee, and dynamic congestion-dependent fee) and charging methodology (daily fee, fee for each passage and specific fee).

This paper firstly lays out the fundamentals of congestion charging and its impact on traditional transport demand models, as well as the importance of the value of time when estimating the effects of the system. After this segment, the paper focuses on the *TRIBUT Equilibrium-Lohse* method in *PTV Visum*. The theoretical background of the method is given, and a practical example is provided, with an explanation of effects and aspects which must be considered when applying this method. Finally, a conclusion is provided with key findings and recommendations for further research.

2   Congestion Charging and Traditional Transport Demand Models

As is known, the traditional transport demand model, also known as the four-step model, consists of four steps: trip generation, trip distribution, mode choice and trip assignment. Congestion charging affects all steps of the four-step model [1].

Introducing the charge can discourage certain citizens from starting a trip to/from the congestion charge zone. This affects trip generation, as some of the previously made trips are no longer done, which reduces the production/attraction of a zone. On the other hand, some zones will see an increase in production/attraction, because there is a chance that the previously mentioned user will ultimately decide to perform her/his trip to an object of the same type and purpose, but in another, non-charged, zone. Also, it must be considered that users cannot give up on or change the origin or destination zone of their work or school trips. Such trips will not change in terms of attraction, production, or distribution, but can change in terms of mode and route choice. Still, the many changed trips affect the O-D matrix, which implies that the second step of the four-step model, trip distribution, also changes as a result.

Congestion charging also affects the modal split and therefore affects the mode choice step of the model. The focus of this paper is the last step of the four-step model, the trip assignment.

In previously published papers and studies, the key variable for calculation was route impedance [5]. Route impedance consists of two elements, or costs: the time cost and the monetary cost. The monetary cost consists of various costs, such as fuel, registration, depreciation, as well as of a cost which is a matter of route choice – the congestion charging fee. The monetary costs of every individual route are similar for all users.

The time cost, on the other hand, is not the same for all users. This cost depends on the value of time for each user, and the value of time mostly depends on the income of these users and the purpose of the trip [8]. The general equation which could be used for the impedance of any given route can be defined as shown in Eq. 1:

$$U = \alpha T + \beta D + C \tag{1}$$

where U represents the route impedance, α represents the value of time, T represents travel time, β represents distance costs, D represents the route distance, and C represents the congestion charge. For a user to choose route A over route B, where route A is charged, and route B is not, the route impedance (U) must be lower for route A than for route B. All variables except α (value of time) are fixed for each route, while α depends on each user, making it the key variable for the route choice of every user.

Equation 1 has been applied while estimating congestion charging effects in Gothenburg, while similar equations were used in other studies [5, 9]. The Gothenburg study has shown that this model, combined with other segments of the transport model of the city, has shown great precision for predicting traffic reduction in peak hours in the charge zone (the mistake was 1%), with a 3% mistake for off-peak hours. When predicting the reduction of travel time on several road sections inside the zone, the model made more mistakes, ranging from 1% to 14% [5].

3 Estimating Congestion Charging Effects on Traffic Flow Distribution Using *PTV Visum*

3.1 Theoretical Background of the *TRIBUT Equilibrium-Lohse* Method

As previously mentioned, an adequate estimate of the impact of congestion charging cannot be achieved without considering the different values of time of various users. Regular methods of traffic flow distribution use the same value of time for all users, and, therefore, another method must be applied to make a correct estimate.

For the estimate, the authors of this paper used the *TRIBUT Equilibrium-Lohse* method found in *PTV Visum*, which has different values of time for all users and all trip purposes. In this method, the value of time is a random variable with a log-normal distribution [10]. When the value of time is distributed among all users, the method calculates the route impedance in the following manner, shown in Eq. 2:

$$Rr = t_r + c_r/VT \tag{2}$$

where Rr represents route impedance, t_r represents route travel time, c_r represents route monetary costs and VT represents the value of time [10]. Equation 2 proves that route impedance is inversely proportional to the value of time. This implies that, for users with a high value of time, paying the charge is more feasible than taking an alternative route with high time cost, while for users with a low value of time, taking an alternative route is more feasible than paying additional costs.

The *TRIBUT Equilibrium-Lohse* method considers all potential routes between two zones, calculates their time and cost impedances, and checks which route is most acceptable for every individual user. As a result of the log-normal distribution of the value of time, one O-D pair of zones will not have only one optimal route choice. Rather, the users will be grouped into different routes – the routes which have the lowest impedance for them. This method belongs to the group of equilibrium methods, in which the system searches for optimal routes in each iteration and determines the share of trips on each route. Multiple iterations, where the software searches for other optimal routes, are done, until no new routes are found. Afterwards, the final share of trips on each route is determined and they are assigned to the network [10].

3.2 Applying *TRIBUT Equilibrium-Lohse* in *PTV Visum*

Congestion charging in *PTV Visum* belongs to the tool named *Restricted traffic areas*. Inside this tool, congestion charging is named *Area toll*. The *Area toll* is inserted using the network editor. First, the shape and area coverage are defined. Afterwards, its characteristics are defined: code, name, and fee for each PrT (Private Transport) group. After this step, the *Procedure sequence* window should be accessed, where the user must choose the right demand segments, as well as the correct PrT Assignment, which is, in this case, *TRIBUT Equilibrium-Lohse*. Running this procedure results in a traffic flow distribution which is then visible in the network editor [10].

For the purpose of this paper, the transport model of the German city of Halle (pop. 240,000) has been used as an example, as all necessary modelling data is made available for this city in *PTV Visum*. The congestion charging zone has been placed in the historic center of the city. According to the categorization from Sect. 1, this system is enclosed, with a fixed fee and with the charging of every passage through the portals. The applied technology was not defined, since the system does not exist in practice, and since this information is irrelevant for application in *PTV Visum*. The fee for each passage was defined as 2 Euros for passenger cars and 5 Euros for heavy goods vehicles. To compare the effects of congestion charging, the PrT Assignment has been run twice – first, using the regular *Equilibrium-Lohse* method, without congestion charging, and later, using *TRIBUT Equilibrium-Lohse*, with the congestion charging zone defined.

After the implementation, the observed reduction of traffic flow on selected links in the charge zone has been between 66% and 95%, with the volume/capacity ratio (V/C) on these links also decreasing by between 20% and 55%. However, two observed links that were not subject to charging, but are in the vicinity of the zone and were identified as potential alternative routes, have seen an increase in traffic flow by 2.4 and 2.8 times, respectively, with their V/C ratio increasing by 19% and 16% respectively. It is important to note that these specifically high values of increase/decrease are a result of the fact that most users previously used the historical city center zone as a transit route, so most of them have simply changed their routes. Had the predominant types of trips been different, the reductions/increases in traffic flow would have been different and, highly likely, lower. Figure 1 shows the difference between the traffic flow in the area before and after the introduction of congestion charging. The green links represent links where traffic flow has been reduced, while the red links represent the links where an increase in traffic flow has been observed. The width of each line represents the intensity of the reduction/increase.

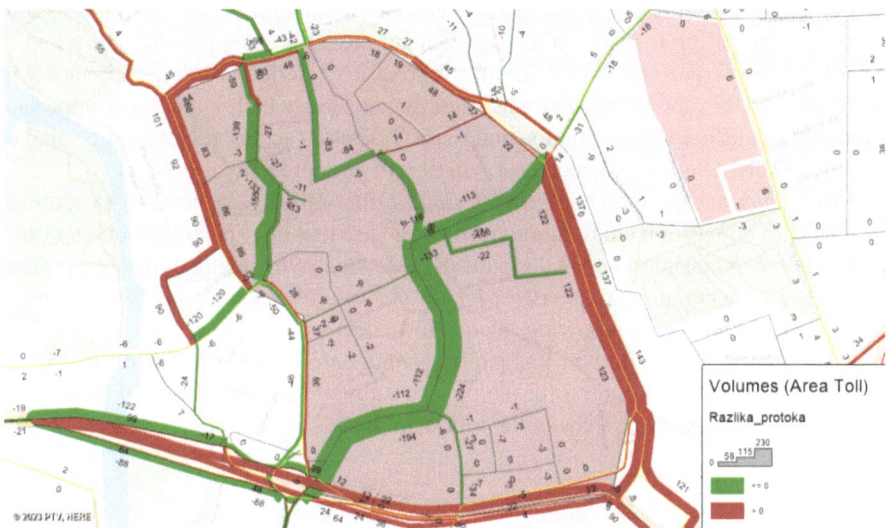

Fig. 1. Difference in traffic flows in two scenarios in the charging zone (made in *PTV Visum*).

Figure 1 confirms the findings stated above. In the figure, the links previously used as transit links, starting from the southwest corner, and ending in the east end of the zone, can be clearly observed, as well as the reduction in traffic flow on these links. Also, the alternative routes and the *rebound effect* can also be clearly observed, those being the links on the south, east and northwest edges of the zone.

The reduction of other parameters, such as vehicle-kilometers and vehicle-hours, has also been observed. The number of vehicle-kilometers inside the zone has decreased by 72%, while vehicle-hours have decreased by 73%. On the links on the edge of the zone, which have served as alternative routes, the number of vehicle-kilometers has increased by 12%, while vehicle-hours have increased by 19%.

Since this paper tackles only the issue of the effects of congestion charging on traffic flow distribution, the authors did not affect the other steps of the four-step model. As a result, the number of vehicle-kilometers and vehicle-hours on the city-wide level have seen a minimal increase of 0.01% and 0.06% respectively. The slight increase is a result of a substantial number of trips which are now avoiding the charge zone. This marginal increase also implies that it is pivotal to apply transportation models which consider the factor of congestion charging for the other steps of the four-step model. If all steps are applied correctly and the system is designed adequately, the prediction should result in a reduction of vehicle-kilometers and vehicle-hours, which is one of the key goals of congestion charging. Additionally, a quality database with all relevant modelling data for the analyzed city must exist, like the database for the city of Halle.

4 Conclusion

Congestion charging has proven to be an effective way of reducing congestion and achieving sustainability goals of urban areas. When estimating its effects, it is vital to affect all steps of the four-step transport demand model in an appropriate way and to consider the different values of time for different users, otherwise, the estimates will be incorrect. The value of time affects route impedance, which is the most important parameter of traffic flow distribution and route choice of each user. Hence, the value of time effectively determines which user will or will not opt to pay the charge.

PTV Visum and its method *TRIBUT Equilibrium-Lohse*, can be used to estimate the effects of congestion charging on traffic flow distribution. The effectiveness of this method lies in its consideration of the value of time, which other methods do not consider, leading to an inaccurate estimate. However, despite its effectiveness, this method cannot serve its purpose without applying adequate models to affect the other steps of the four-step model. The authors of this paper recommend further research of the effects and potential models to be used for the precise estimation of the effects of congestion charging on the other three steps of the four-step model.

References

1. van Amelsfort, D.: Introduction to Congestion Charging. Asian Development Bank. GIZ, Beijing (2015)
2. Transport for London: Central London Congestion Charging: Impacts Monitoring, Fourth Annual Report, London (2006)
3. Transport for London: Congestion Charging: Third Annual Monitoring Report, London (2005)
4. Eliasson, J.: The Stockholm congestion charges: an overview. In: Working Papers in Transport Economics, 2014:7. Centre for Transport Studies Stockholm, Stockholm (2014)
5. West, J., Börjesson, M., Engelson, L.: Forecasting effects of congestion charges. In: Working papers in Transport Economics, 2016:9. Centre for Transport Studies Stockholm, Stockholm (2016)
6. Phang, S.-Y., Toh, R.S.: Road congestion pricing in Singapore: 1975–2003. Transp. J. 43(2), 16–25 (2004)
7. Börjesson, M., Kristoffersson, I.: The Gothenburg congestion charge. Effects, design and politics. Transp. Res. Part A Policy Pract. 75(C), 134–146 (2015)
8. Eliasson, J., Börjesson, M., van Amelsfort, D., et. al.: Accuracy of congestion pricing forecasts. Transp. Res. Part A Policy Pract. 52, 34–36 (2013)
9. Santos, G.: Urban congestion charging: a second-best alternative. JTEP 38(3), 345–369 (2004)
10. PTV Planung Transport Verkehr: PTV Visum 2023 Manual. PTV, Karlsruhe (2022)

Open Access This chapter is licensed under the terms of the Creative Commons Attribution 4.0 International License (http://creativecommons.org/licenses/by/4.0/), which permits use, sharing, adaptation, distribution and reproduction in any medium or format, as long as you give appropriate credit to the original author(s) and the source, provide a link to the Creative Commons license and indicate if changes were made.

The images or other third party material in this chapter are included in the chapter's Creative Commons license, unless indicated otherwise in a credit line to the material. If material is not included in the chapter's Creative Commons license and your intended use is not permitted by statutory regulation or exceeds the permitted use, you will need to obtain permission directly from the copyright holder.

Transport Policy

Towards Sustainable and Inclusive Corridors: Rethinking Planning of Infrastructure and Spatial Development

Charlotta Faith-Ell[1,2](✉) ⓘ, Jos Arts[3,4] ⓘ, and Heikki Kalle[2] ⓘ

[1] Department of Natural Science, Design and Sustainable Development, Mid Sweden University, 831 25 Östersund, Sweden
charlotta.faith-ell@miun.se
[2] Estonian Environment Institute (EKKI), Raekoja plats 8, 51004 Tartu, Estonia
heikki@ekki.ee
[3] Department of Planning, Faculty of Spatial Sciences, University of Groningen, PO Box 800, 9700AV Groningen, The Netherlands
jos.arts@rug.nl
[4] Environmental Sciences and Management, North-West University, Potchefstroom, South Africa

Abstract. The transformation of Europe towards a more sustainable built environment has large implications on the development of transport infrastructure. National and European authorities invest hugely in transport infrastructure focusing usually on improving individual links for single modes in the form of largescale projects. In practice, such planning often results in cost overruns, time delays and limited public support. Often local (land-use) planning issues become the focus instead of the overall system of transport corridors connecting individual places and services. This paper aims at exploring the concept of sustainable and inclusive corridor planning for large infrastructure development that connects with (local) land-use development and (cross) national transport needs. To this end, we examine experiences from three different cases of corridor-oriented planning in Estonia, Sweden and the Netherlands. Our findings suggest the need for a strategic programmatic approach aligning local needs, individual projects and setting overall goals for corridors and networks. This implies developing and applying multi-level governance frameworks with leadership, early and ongoing stakeholder involvement, and joint platforms for monitoring and learning.

Keywords: cross-national corridors · TEN-T network · LUTI (land use - transport interaction) · multi-level governance

1 Introduction

The transition of Europe towards a more sustainable built environment has large implications on the development of transport infrastructure. National authorities invest hugely in infrastructure as it is considered a prerequisite for spatial-economic development

© The Author(s) 2025
C. McNally et al. (Eds.): TRAconference 2024, LNMOB, pp. 599–605, 2025.
https://doi.org/10.1007/978-3-031-85578-8_80

by improving connectivity and accessibility of places and regions [1] – relating to the notion of Land-Use Transport Interaction (LUTI). Many land-use developments (housing, offices, industrial estates) are initiated by local and regional authorities, which interact with policies for large-scale infrastructure projects [2]. While most countries have national infrastructure strategies or plans aiming for optimizing multi-modal infrastructure networks and for achieving sustainability and inclusiveness objectives, their financing practice focuses on granting individual projects for one mode – solving congestion and capacity 'bottlenecks' [1]. At the European level, the TEN-T policy of a system of corridors and networks focuses on improving individual linkages for separate modes aiming especially at connecting peripheral areas to the common market [3].

Therefore, infrastructure planning – both at the European and national level – is dominated by project planning rather than by a network- or corridor-planning approach of integrated infrastructure and spatial development [4, 5]. Also, the resulting practice of large-scale projects meets much local resistance as projects cause impacts on local society (e.g., land take, environmental emissions, barrier impacts) while their benefits to localities remain unclear. To address such issues, usually local (land-use) planning initiatives are integrated in – and paid for by – (large) infrastructure projects. Examples include: modernizing station areas for high-speed railway projects; building tunnels for highway projects through cities creating new space for urban development; or other expensive local mitigation measures. Nevertheless, the planning of infrastructure projects becomes often cumbersome and results in cost overruns, time delays and little public support [6]. Infrastructure projects become arenas for solving local planning issues rather than for addressing systems level connectivity and accessibility – let alone broader goals of sustainability and inclusiveness [7, 8].

Overall, what can be seen is that the multiplicity of transport infrastructure makes its planning very complex, when dealing with functional interrelatedness – i.e., multiple sectors (land-use, transport); and multiple scales (local, regional, (cross)national) – and institutional interdependencies (i.e., multiple actors, and multi-level governance) [9]. To overcome this multiplicity, integrated planning is opted as an approach to deal with the complex circumstances of transport infrastructure [5]. Corridor planning aims to take such integrated approach into account [4, 10].

This paper aims at exploring the concept of sustainable and inclusive corridor planning for large infrastructure development that connects with (local) land-use development and (cross) national transport needs. To this end, we examine experiences gained in three different cases of corridor-oriented planning: Rail Baltic (Estonia), High-Speed Rail (Sweden), and South-East Freight Corridor (the Netherlands).

2 Methods and Cases Selected

This paper builds upon a comparative analysis of the planning and decision-making processes in these cases, and the results of recent European studies. Our study addresses functional interrelatedness (using the LUTI concept) and institutional inter-dependencies (using multi-level governance). The planning and decision-making processes were analyzed on basis of key documents, interviews [see also 7, 8], and the authors' experiences who were directly involved in the cases and all have long-standing experience in infrastructure and land-use planning in their countries. In addition, we build upon European studies we were involved in [see 11]: Vital Nodes (Horizon2020), Freight and Logistics in a Multimodal Context (Fluxnet, CEDR), Collaborative Planning of Infrastructure and Spatial Development (CEDR), Networking for Urban Vitality.

The *Rail Baltic program* in Estonia pertains to a new high-speed rail system (240 km/h) connecting the capitals of the Baltic States with Europe. In Estonia, the 220 km HSR stretch is part of the Estonian National Spatial Plan as well as national transportation strategy. In 2013, the program started with a regional planning process of parallel actions at multiple tiers with various parties and institutions. The program coordinated extensive discussions involving >100 meetings with specific stakeholders as well as the general public. Construction of the railway has started and will finish by 2030.

The *high-speed rail (HSR) program* in Sweden stretches from Stockholm to Malmö and Gothenburg, respectively. The cases in this paper encompass *Jönköping-Malmö* (ca 300 km) and *Linköping-Borås* (ca 217 km). Both projects were carried out by the National Transport Administration at the regional level as so-called 'Strategic Choice of Measures' (SCM) studies in the period 2015–2018. In parallel, a political process called 'National Negotiation on Housing and Infrastructure' was carried out with local cities on the corridors. After the finalization of the SCM studies and the National negotiations, in December 2022, the Swedish government abandoned the corridor approach, deciding to discontinue the planning of new trunk lines for high-speed trains.

The *East/South-East Freight Corridor* program in the Netherlands started in 2017 and comprises the corridor Rotterdam – Rhein-Ruhr (Germany), which has a highly developed network of rail-, water- and highway (and pipeline) connections. This cross-national corridor is not only vital to the Netherlands' transport system and economy, but also to the EU (the TEN-T Rhine-Alpine Corridor). In the Freight Corridor program national government examines how to maintain and optimize this crucial multimodal corridor with the ambition of being a 'topcorridor in 2030'. A coherent package of measures is developed, for which national, regional, local authorities collaborate with private companies. The program focuses explicitly on optimization of 6 major nodes.

3   Results

3.1   Functional Interrelatedness – Land-Use Transport Interaction

Initially, the *Rail Baltic* program involved alignment design of high-speed rail aiming at achieving the optimal technical solution with the least restrictions (nature conservation, cultural heritage, health impacts to settlements etc.). From an urban planning perspective, the study of the Rail Baltic program shows that at the start this program focused mainly on the (detailed) design of stations, including architectural competitions. Only later in the process, the discussion evolved into topics as the role of railways in the city and location criteria for stations. Furthermore, the Rail Baltic served as a steppingstone for adjacent project proposals such as improvements in Tallinn light rail (tram) system and the Finest Link tunnel connecting Tallinn and Helsinki. Consequently, discussions in Tallinn shifted from 'form' (architecture) to 'function' – what sectors and which social groups will benefit?

Initially, the two *Swedish HSR projects* focused mainly on travel time between Stockholm, Malmö and Gothenburg. There was little discussion within the SCM process on the role of the HSR for the cities within the corridors. The cities were instead involved in the National negotiation process. The national negotiation had a very tight coupling between infrastructure and housing, while the SCM process by the Transport Administration process did not. Since the SCM process was carried out from a large-scale perspective, current land-use was considered a prerequisite for the planning of the railway. This means that there was a very limited interaction between land-use planning and transport planning in the SCM process. At the same time, the national negotiation focused very much on planning of new residential areas in the cities that were to become nodes in the HSR system. In this process, the railway functioned as a prerequisite for the negotiation about new housing. This meant that the SCM was planning on basis of current land-uses while the national negotiations at the same time changed the planed land-use, stemming from the assumption that a railway would be present in the future.

The *Dutch Topcorridor program* focused on the optimization of the Dutch part of the TEN-T Rhine-Alpine Corridor and particularly on six major nodes. Measures related to: improving transport flows for different modalities at the corridor (solving bottlenecks); improving intermodal connectivity at nodes (allowing a sustainably modal shift); focusing on service and reliability for end-users (synchro-modality); and looking for sustainability and innovation (renewable energy, alternative fuels). The program strives for economic development with specialization in the six major nodes to prevent unwanted competition between nodes and other cities.

3.2   Institutional Interdependencies – Multi-level Governance

At the (supra)national level, the main institution for the *Rail Baltic program* was the coordinating organization (RBE) overseeing linkages between different plans and governance levels (supra-national collaboration in the Baltic, state, cities). However, RBE focused mostly on engineering; not on regional implications and (land use) planning. At the regional level, there was no permanent organizational form to continue the corridor-level discussions about the meaning of the newly created link, discussing losses and benefits, agreeing on possible roles etc.

The main focus of both the SCM process and the national negotiations in the *Swedish case* was the development of new railway lines for national passenger traffic with few stops. This meant that the current railway system was to become a system for regional traffic and freight traffic the day the HSR system was to be in place. The Transport Administration focused on the regional scale in the SCM process while the National Negotiation focused on the local municipal scale. Also, within the SCM, there was little discussion about the role of the HSR for cities within the corridor. This resulted in two parallel processes and rivalry between cities rather than a reflection on how cities could cooperate and become stronger together. The National Negotiation had a very tight coupling of infrastructure and housing, while the planning process of the Transport Administration did not. Only at the end of the SCM process came the discussion about the function of HSR for cities. This meant that the objectives of the HSR were retrofitted in a separate process after the finalization of the SCM and the National negotiation.

For the *Dutch Topcorridor program*, the program objective (of new economic development with specialization in the major nodes and preventing unwanted competition between nodes and cities) proved to be a complex multi-level governance issue, as it required that: 1) provinces and cities think at corridor/network level (while competing as local terminals and hubs); and 2) that national government links up local/regional spatial-economic development issues (while funding is limited and for national infrastructure). However, the program has increased the awareness about being positioned on a corridor, and about the importance of the corridor to the country, regions, and cities. Under the current institutional framework, the programmatic approach faces difficulties in: implementing multi-level governance (including also the EC), creating multimodality, balancing freight and persons transport, and developing a scope for cross-border issues and measures (to Germany and Flanders).

Table 1. Comparison of the three cases

	Estonia	Sweden	The Netherlands
Land-use transport integration	Yes, rail (freight/persons) and local land uses (and mitigation of impacts)	No, separate processes. SCM focuses on passenger traffic while the National negotiation focused on housing	Yes, multimodal freight transport (road, rail, waterway, pipelines), urban-regional economic development
Spatial scales	Start on EU corridor level developing new rail, later attention for local (land use) issues, and environmental impacts (SEA/EIA)	Start focus on travel time HSR, later on land use in the National Negotiation: (housing) issues for city-regions (only SEA for rail)	Optimizing the overall corridor (for transport) and focusing on 6 specific urban-regional nodes (spatial-economic development) (no SEA/EIA)
Main objectives	(Geopolitical) connectivity as central aim, room for integration (LUTI) locally	No common objective in this period. Separate tracks: municipalities focus on land use, Trafikverket (national) on transport	At program level central policy goals, and alignment with regional/local authorities. Alignment with businesses was missing

(*continued*)

Table 1. (*continued*)

	Estonia	Sweden	The Netherlands
Multi-level governance	(Supra)national level leading (EU funding), rather technical process. Much collaboration with regional and local authorities (via SEA)	Political process, national government leading, other authorities involved through the SCM (regions), and the National negotiation (municipalities)	Nationally driven process (of joint factfinding and will-shaping), with collaboration with provinces, and (major) cities + Port of Rotterdam
Involvement of other parties in the process	All parties (in)directly involved (still court cases). Also, Latvia, Lithuania (and Finland). Businesses late in process	Limited. Main governmental, public and businesses not actively included (National negotiation)	Limited. Mainly governmental parties. Businesses and international (Germany, Belgium and EC) not included
Planning process	No existence of a formal transport planning system. Planning was based on County plans and detailed plans for stations	No existence of a formal planning system for transport corridors. Panning had to be based on SCM. Local implementation still to be done via detailed rail planning	Specific national program, linking up with the existing Planning, Programming Budgeting process for infrastructure. Regional and local implementation still to be done via land-use plans

4 Discussion and Conclusion: Toward a Programmatic Approach

Our analysis (Table 1) shows that in all three cases, there are many institutions at various levels that need to collaborate in the planning of corridors. However, we see imbalances in power between institutions creating barriers for effective planning and decisions. The many dimensions of planning in the cases does not align with current practices of corridor development, as they are often performed on a project base, neglecting the multiplicity of corridors, as an integrated network of nodes and infrastructure [4, 10]. This indicates a need for flexible planning systems that accommodate the challenges but also capture the potentials of planning at corridor level [7].

Our findings suggest the need for rethinking existing practice of large-scale infrastructure project planning, because of the limited added value at both the local and corridor/network level (resulting in missed opportunities and politicization). The cases indicate that a strategic programmatic approach could align local (land use) needs, individual projects and setting overall goals for a corridor and network. This implies developing and applying multi-level governance frameworks with leadership, early and on-going stakeholder involvement, and joint platforms for monitoring and learning.

References

1. Arts, J., Leendertse, W., Tillema T.: Road infrastructure: planning, impact and management. In: Vickerman, R. (ed.) International Encyclopedia of Transportation, pp. 360–372. Elsevier, UK (2021)
2. Bertolini, L., le Clercq, F., Kapoen, L.: Sustainable accessibility: a conceptual framework to integrate transport and land use plan-making. Two test-applications in the Netherlands and a reflection on the way forward. Transp. Policy **12**(3), 207–220 (2005)
3. Witte, P.A.: The Corridor Chronicles - Integrated perspectives on European transport corridor development. Utrecht University (2014)
4. Faith-Ell, C., Kalle, H., Arts, J.: Connecting the dots: rethinking large-scale corridor infrastructure planning. In: 8th Transport Research Arena TRA 2020, Helsinki, Finland (2020)
5. Heeres, N., Tillema, T., Arts, J.: Integration in Dutch planning of motorways: from "line" towards "area-oriented" approaches. Transp. Policy **24**, 148–158 (2012)
6. Arts, J., Hanekamp, T., Linssen, R., Snippe, J.: Benchmarking integrated infrastructure planning across Europe. Transp. Res. Procedia **14**, 303–312 (2016)
7. Vedder, K.: Dealing with infrastructure multiplicity – exploring institutional arrangements for corridor planning. University of Groningen, Groningen (2022)
8. Verhulst, S.: Sustainable Topcorridors – institutional arrangements for sustainable corridor development. University of Groningen, Groningen (2022)
9. Heeres, N.: Towards area-oriented approaches in infrastructure planning - development of national highway networks in a local spatial context. University of Groningen, Groningen (2017)
10. de Vries, J., Priemus, H.: Megacorridors in north-west Europe: issues for transnational spatial governance. J. Transp. Geogr. **11**(3), 225–233 (2003)
11. Raskeyn, C., Arts, J., Hanekamp, T., van der Werf, S.: Transitioning towards area-oriented approaches in transport infrastructure planning. In: 10th Transport Research Arena 2024, Dublin (forthcoming)

Open Access This chapter is licensed under the terms of the Creative Commons Attribution 4.0 International License (http://creativecommons.org/licenses/by/4.0/), which permits use, sharing, adaptation, distribution and reproduction in any medium or format, as long as you give appropriate credit to the original author(s) and the source, provide a link to the Creative Commons license and indicate if changes were made.

The images or other third party material in this chapter are included in the chapter's Creative Commons license, unless indicated otherwise in a credit line to the material. If material is not included in the chapter's Creative Commons license and your intended use is not permitted by statutory regulation or exceeds the permitted use, you will need to obtain permission directly from the copyright holder.

Do People Actually Want Sustainable Urban Mobility Systems? A Residents' Perspective on Policy Measures

Konstantin Krauss[✉] [iD], Dorien Duffner-Korbee[iD], Johanna Flacke, and Claus Doll[iD]

Fraunhofer Institute for Systems and Innovation Research ISI, Breslauer Str. 48, 76135
Karlsruhe, Germany
konstantin.krauss@isi.fraunhofer.de

Abstract. For future transport planning fulfilling climate goals and reduce over-all external costs generated by transportation, further policy measures need to be implemented quickly. Current research is providing an extensive list of measures and instruments to make transportation systems more sustainable. However, many of these proposals fail since single policy measures advantaging the one and dis-advantaging the other group of travelers. We take this challenge as starting point to investigate which policy measures are attractive for which groups of travelers to identify the implementation potential. We do so by presenting residents in 25 large cities in Germany different policy measures. The respondents (n = 2,555) are asked to indicate their consent to the different policy measures in two steps: first, they see the measure only. Second, we present more information about the policy measure and outline advantages and disadvantages of the measure. Doing so enables us to understand which traveler type, i.e. current mode choice and mobility tool behavior, supports which policy measure and how improving infor-mation about it can help to find majority support. We find higher approval after presenting more information and a strong relationship between current mobility behavior and support for measures advocating this behavior.

Keywords: transport policy · urban transport · policy mix · people-centered transport · built environment · sustainable mobility

1 Introduction

To achieve the climate targets for transportation and sustainable transportation systems, all facets from the avoid-shift-improve concept need to be orchestrated successfully [1]. Recently, a great focus has been put on improving existing technology and, thus, making vehicles more climate friendly [2]. However, with this focus, the core idea of the avoid-shift-improve concept is turned upside down: making vehicles more efficient relates to the improve aspect, which is meant to come in last. First of all, transport demand needs to be avoided and this reduced demand shifted to more environmentally friendly modes [1]. Further, there has been a large focus on greenhouse gas (GHG) emissions in latest discussions. While understandable in terms of fulfilling the climate targets, it neglects

© The Author(s) 2025
C. McNally et al. (Eds.): TRAconference 2024, LNMOB, pp. 606–612, 2025.
https://doi.org/10.1007/978-3-031-85578-8_81

the existence of further external effects of transport. These are pollutants, land use, noise, accidents, congestion [3]. However, even these classical components of external cost estimates represent only one piece of the puzzle of peoples' satisfaction with urban mobility [4]. For people, also to understand and follow the decisions made in order to achieve political targets, additional impacts of mobility are crucially decisive. These are related to the external effects of transport and are factors such as the allocation and availability of public space, accessibility, affordability and livability. These are often neglected when assessing transportation systems' status quo.

However, to achieve a sustainable transportation that decreases the external effects generated, the promotion of policy measures towards avoiding or shifting transport demand is crucial. This promotion and implementation requires public consent and needs to fulfill the citizens' needs for livability. While an effective mobility transition requires changes across all dimensions, i.e. built environment, prices and regulations, mobility behavior, and propulsion systems, there is little knowledge about which policy measures residents appreciate or reject beforehand, or could adapt to after measures are implemented. To investigate this phenomenon, we draw on the construct of public acceptability as derived by [5]. The term refers to measures that are yet to be implemented and integrates various issues that are required to make a policy measure publicly acceptable. In this work, we focus on the issues of problem perception, knowledge about the options, and perceived effectiveness and efficiency of the measures.

This is why we pose the following research question: What is the public acceptability of policy measures to restructure urban mobility systems? For the policy measures, we focus on three domains: built environment, regulation, and pricing mechanisms. For the former, we select bicycle expressways. For regulation, we use speed limits. For pricing, we focus on parking fees and a reform of the public transport (PT) ticket pricing system. To answer the research question, we designed and conducted a survey for residents of large cities across Germany and use this data to evaluate the opinion and consent of people towards the aforementioned policy measures.

In this paper we first introduce the methodology we apply, i.e. present the survey conducted and the processing of the data. Second, we show the results for the policy measures and the consent of people towards these. Third, we conclude the paper, derive implications for policy and practice, and propose future research on this topic.

2 Method

For the evaluation of the policy measures, we use a tailored survey amongst respondents living in large cities in Germany (n = 2,555). Large cities are all those with more than 200,000 inhabitants. All residents with an age of at least 18 years were eligible to take part in the survey. For the sampling, we apply fixed quotas for age and gender.

The questionnaire covers four topics: mobility behavior including mobility tools, attitudinal variables, socio-demographic information, and the policy measures. Asking for mobility behavior and tools held provides information about mode choice, usage of shared mobility, number of cars and bikes, and PT pass.

The policy measures cover increases in parking fees, expansion of a bicycle expressway network, a city-wide speed limit of 30 km/h, and free PT. To gain insight in the public acceptability of these measures, four dimensions are taken into account: mobility behavior, mobility tools, and socio-demographics. Mobility behavior in terms of mode choice and general travel frequencies is included, too. As it is known that mobility tools, particularly the number of private cars, affects the overall mobility behavior, these are included as well. Socio-demographics mainly function as control variables. To measure the impact of these four dimensions on the public acceptability of the respective policy measures, factor analysis, clustering, and logistic regression analysis are conducted.

To generate insights into how residents react to different levels and kind of information, we have asked them to evaluate the policy measures twice: first, just by describing the policy measure without additional information. Second, we repeated the same questions after giving the respondents more information about the background of the measure, the reason for it as well as their advantages and disadvantages. Doing so, we aim to better understand which information it takes to inform people and whether this increases their support for measures. We do so for several measures and also ask the respondents to indicate their current mobility behavior. This enables us to analyze the evaluation of different policy measures for different groups of travelers.

The gross sample consisted of 2660 respondents. Data cleaning consisted of three steps. Firstly, all respondents that completed the questionnaire within 240 s (3 min) were excluded (20 respondents). Secondly, respondents that always (30 times) rated the attitudinal variables with the same score, were excluded (35 respondents). Thirdly, we cross-checked whether the city of residence matched the indicated postal code. In 50 cases, these two items did not match. Due to these steps, a total of 105 respondents were excluded from the sample, resulting in a sample of n = 2,555.

3   Results

3.1   Sample Description

From the survey, we can use data from 2,555 respondents. The corresponding socio-demographic information as well as their mobility tool holdings are presented in Table 1. Regarding gender, the sample is rather balanced with slightly more females (52.3%). The age distribution is homogeneous and the majority of respondents works full-time (53.7%). The household net income amounts mainly to 2,000–4,000 EUR (40.2%) with approximately a quarter earning less and another quarter earning more than that. A driver's license is held by 86.5% of the respondents. While the majority has one car in the household (55.9%), a few have at least one electric car (8.1%). 81.4% report at least one bike but only 17.7% report a pedelec in the household. One or more e-scooter is available in 9.2% of all households. With 52.6%, the majority has a PT pass.

Table 1. Socio-demographic and mobility tool description of sample.

Characteristic	Level	Share [%]
Gender[a]	Female	52.3
	Male	47.4
Age	18–40	36.8
	41–60	44.0
	>60	19.1
Employment[a]	Unemployed	23.6
	Yes, part- or full-time	73.9
Household net income[a]	<2,000 EUR	23.9
	2,000 - 4,000 EUR	40.2
	>4,000 EUR	25.4
Driver's license[a]	Yes	86.5
	No	12.3
Number of cars in household	0	21.8
	1	55.9
	≥2	22.3
Number of electric cars in household	0	91.9
	≥1	8.1
Number of bikes in household	0	18.6
	1	29.2
	≥2	52.4
Number of pedelecs in household	0	82.3
	≥1	17.7
Number of e-scooters in household	0	90.8
	≥1	9.3
PT pass	Yes	52.6

[a]Remaining to 100%: Do not want to disclose or "other".

3.2 Evaluation of Policy Measures

We first look at the overall evaluation of respondents regarding the policy measures and, second, at the evaluation based on mobility tools respondents have access to.

Figure 1 shows the results for the overall evaluation of the policy measures, ranginging from 1 "I reject completely" to 7 "I fully support". With sparse information, free PT is the measure evaluated most positive (mean: 5.4) and an increase in parking fees most negative (2.9). With rich information, free PT remains the most supported measure with bicycle expressways following. Although remaining the least supported measure, an increase of parking fees shows the most gain in support with rich information (+0.8).

A speed limit of 30 km/h is the ranging at the mean of the scale. An increase of parking fees is below the mean and the remaining two measures, i.e. bicycle expressways and free PT, are above the scale's mean. Thus, an increase in parking fees is disliked by most on average while bicycle expressways and free PT are liked most on average. Regarding the speed limit of 30 km/h, respondents are on average indifferent.

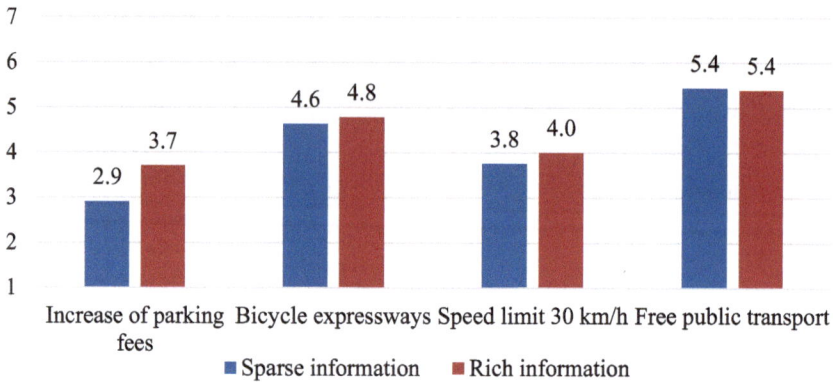

Fig. 1. Means of policy measure evaluation (1: I reject completely, 7: I fully support).

Since the personal mobility behavior might have an impact on the evaluation of policy measures regarding the modes the persons themselves use most, we look at this effect next. Figure 2 shows the distribution of average evaluations of the policy measures depending on the mobility tools people hold in their household. For this, we distinguish between car and bike owners in every combination. While people with a bike but no car evaluate all four policy measures most positive, those with a car but not bike evaluate all measures the most negative. As soon as respondents have the private car in their mobility behavior mix, the evaluation of measures focusing on the car become more

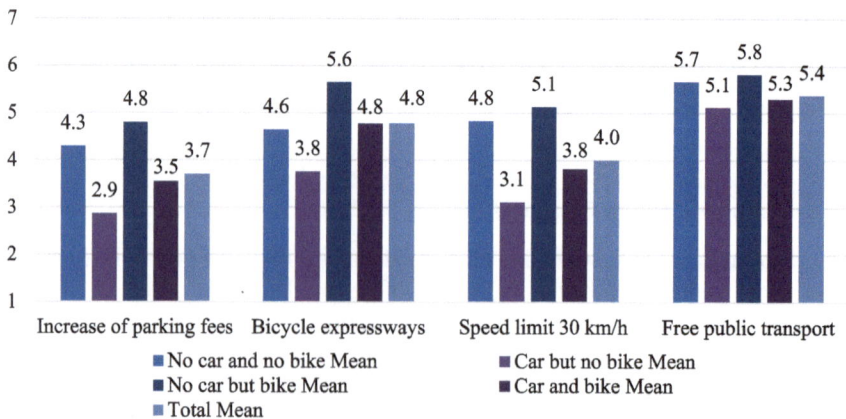

Fig. 2. Policy measure evaluation for different mobility tool groups (1: I reject completely, 7: I fully support).

negative, which can be seen for the group of people with car and bike as well. The least heterogeneity can be observed for those measures that do not negatively intervene with any mode of transport, i.e. expansion of bicycle expressways and free PT. A Kruskal-Wallis H test shows that the distributions of the policy measure evaluations between the groups are statistically significantly different at the 1% significance level.

4 Conclusions

We investigate the public acceptability of policy measures targeted at making transportation more sustainable. Results from a survey in German cities show that free public transport is the most supported measure and an increase in parking fees the least supported one. However, when giving respondents more information, support for an increase in parking fees can be raised substantially. The support for policy measures is linked to the mobility tools respondents have access to. The decisive factor to support car-focussed policy is to not have access to a car in the household. As soon as there is, the evaluations of respective policy measures decrease substantially. This work is of interest to researchers, transport planners, and policymakers. For researchers, it is crucial to understand how different groups behave towards different policy measures. For transport planners, we present insights into which measures to focus on in order to build urban transportation systems that fulfil the residents' needs. Policymakers profit from this work as it shows that informative communication might might alleviate protests against certain policy measures. This communication needs to be specifically targeted at different groups such as households with and without cars. Our work can support in identifying which measures hold the greatest potential in terms of public acceptability.

Acknowledgements. This work is part of the research project MobileCityGame co-funded by the German Federal Ministry for Education and Research (BMBF) and is continued through the EU's Driving Urban Transitions program.

References

1. Dalkmann, H., Brannigan, C.: Transport and Climate Change. Module 5e. Sustainable Transport: A Sourcebook for Policy-makers in Developing Cities. Deutsche Gesellschaft für Technische Zusammenarbeit (GTZ) GmbH, Eschborn (2007)
2. Knobloch, F., et al.: Net emission reductions from electric cars and heat pumps in 59 world regions over time. Nat. Sustain. (2020). https://doi.org/10.1038/s41893-020-0488-7
3. Verhoef, E.: External effects and social costs of road transport. Transp. Res. Part A Policy Pract. (1994). https://doi.org/10.1016/0965-8564(94)90003-5
4. Beckers, T., von Hirschhausen, C., Klatt, J.P., Winter, M.: Effiziente Verkehrspolitik für den Straßensektor in Ballungsräumen. Kapazitätsauslastung, Umweltschutz, Finanzierung. Bundesministerium für Verkehr, Bau und Stadtentwicklung (BMVBS) (2007)
5. Schade, J., Schlag, B.: Acceptability of urban transport pricing strategies. Transp. Res. F Traffic Psychol. Behav. (2003). https://doi.org/10.1016/S1369-8478(02)00046-3

Open Access This chapter is licensed under the terms of the Creative Commons Attribution 4.0 International License (http://creativecommons.org/licenses/by/4.0/), which permits use, sharing, adaptation, distribution and reproduction in any medium or format, as long as you give appropriate credit to the original author(s) and the source, provide a link to the Creative Commons license and indicate if changes were made.

The images or other third party material in this chapter are included in the chapter's Creative Commons license, unless indicated otherwise in a credit line to the material. If material is not included in the chapter's Creative Commons license and your intended use is not permitted by statutory regulation or exceeds the permitted use, you will need to obtain permission directly from the copyright holder.

Possibilities for Supporting Automated Driving by Leveraging Vehicle Data in Road Design and Maintenance

Ane Dalsnes Storsæter[1]([✉]) [ID] and Kelly Pitera[2,3] [ID]

[1] Q-Free, Strindfjordveien 1, 7053 Ranheim, Norway
aneds@q-free.com
[2] OsloMet, Pilestredet 46, 0176 Oslo, Norway
[3] Norwegian University of Science and Technology, Høgskoleringen 1, 7034 Trondheim, Norway

Abstract. Automated driving features from Advanced Driver Assistance Systems (ADAS) to higher levels of automation are shown to have a potential to increase traffic safety. The ability to drive safely is contingent on the automated systems understanding the road infrastructure. However, the road infrastructure has been designed over the course of a century to be optimal for human drivers. Considering the differences between human drivers and sensor-based driver assistance, it is important to ask to what extent is the road infrastructure suitable for automated driving features. Additionally, if there is already a proven safety effect, it is suggested that adapting the road infrastructure to ensure sensor detection will further increase this safety effect. Research into this domain is limited but growing. This paper explores how fundamental changes to the road infrastructure can leverage the sensor capabilities of vehicles of all levels of automation, thus giving an incremental approach to adapting road infrastructure to support Automated Driving Systems (ADS) at all levels of automation.

Keywords: Automated Driving · Automated Freight · Road Infrastructure · Road Maintenance

1 Introduction

1.1 Background

Current road infrastructure has been designed and developed for human drivers over the course of more than a century. From Advanced Driver Assistance Systems (ADAS) to higher levels of automation, Automated Driving Systems (ADS) are becoming more prevalent and in some cases mandatory [1]. These systems have been shown to have the potential to increase traffic safety [2, 3]. Yet, to reap these safety effects, automated driving systems with sensor-based assistance must be able to interpret the road infrastructure correctly.

The automated levels of driving are most ubiquitously explained by the six levels defined by the Society of Automotive Engineers (SAE) [4]. Even the lowest level of

© The Author(s) 2025
C. McNally et al. (Eds.): TRAconference 2024, LNMOB, pp. 613–619, 2025.
https://doi.org/10.1007/978-3-031-85578-8_82

the SAE framework support features that are dependent on the vehicle´s sensor input. Automated emergency braking, blind spot warning, and lane assistance systems (LAS) all require sensors to detect the surroundings followed by the appropriate interpretation and programming to act on it. Advanced Driver Assistance Systems (ADAS) represent lower levels of automation, i.e., level 2 of the SAE definition, and are widely used in today´s transportation system. Additionally, they utilize the same sensors that higher levels of automated driving rely on. Thus, studying the current levels of automation can provide insight into how the road infrastructure is most easily detected by sensors as automation increases. This can result in an understanding of how to provide cost-efficient ways to adapt road infrastructure to support automation and increase safety benefits. Additionally, it serves as a step toward developing road design and maintenance standards for the future of automated transportation.

1.2   Approaches to Adapting Road Infrastructure to ADS

It has been suggested that the "..road infrastructure can play a key role in enabling and supporting automated driving." [3]. Similar to the SAE levels of automated driving, the European Road Transport Research Advisory Council (ERTRAC) has developed a framework to describe this. It is called the Infrastructure Support levels for Automated Driving (ISAD) [5].

The ISAD framework consists of five levels, where the top two levels require the road infrastructure to monitor the road with all road actors and share information about the traffic situation in real-time. This would require sensors to monitor the traffic in all weather conditions and with redundancy. Furthermore, there would need to be a communication infrastructure to support the data transmission. Having road infrastructure that is able to communicate information about road conditions and other users, often referred to as part of Cooperative ITS (C-ITS), is expected to give substantial safety effects [6, 7] and reduce travel times [8].

Traditional road infrastructure is known to have significant road maintenance backlogs both in the EU [9] and in the US [10]. Adding support for automated driving, as suggested in the ISAD framework, would mean additional sensors and the necessary power and communication infrastructure to turn sensor data into real-time traffic information. Providing sensor and communication capability from the roadside could be beneficial for safety effects linked to C-ITS, however, it scales poorly. Therefore, the most likely application due to the required investment and redundancy is likely to be in specific locations, e.g., areas with high traffic volumes.

A contrary approach to the ISAD framework, is to demand that the vehicles can correctly interpret the infrastructure and handle the driving and communication aspects without external sensor and communication support from the infrastructure. For that reason, it is necessary to address how road infrastructure elements, including pavement, markings, barriers, guard rails, and signage, can be optimized to increase detection by automated users. With regards to the infrastructure, the ability to ensure that both human and automated driving systems can correctly sense and interpret their road surroundings has only been explored to a limited extent. This paper explores cost-effective ways to leverage automated driving to increase traffic safety and efficiency. It specifically focuses on automation and ADAS functionality and how infrastructure can be adapted in response to advancements in vehicle technology. First, some main differences between humans and ADS are highlighted. Based on the sensor technology used by ADS, suggestions are made toward how this information can be used to find an incremental path towards optimizing road infrastructure for all road users.

2   Road Design for Automated Driving

In road design for humans, there is a concept called "self-explaining roads," which advocates designing roads so that they are intuitively understood by the driver, e.g., so that they can adapt the speed to fit the road geometry [11]. Likewise, one could envision road design that would consider how automated driving systems sense their surroundings and interpret sensor input to optimize their success rate. That requires understanding how sensors work and finding ways to optimize the road infrastructure for both humans and ADS.

2.1   Sensing the Road Infrastructure

By far, the most common human sense used to drive is vision [12]. Human vision is dependent on electromagnetic radiation to observe the road environment. This is similar to machine vision using cameras. However, there is a difference in the sensitivity range for a human versus a camera. Humans have a best-case range of between 310 and 1100 nm, known as the visible light range [13], while cameras, as well as other sensors, have a much greater range. Cameras, radars, lidars, and ultrasound are the most common sensors used to enable automated driving features [14]. In addition to visible light, some cameras can detect radiation in the lower UV area and the higher IR range. Ultrasound, radars, and lidars are active sensors that transmit their own radiation and detect its reflection. This means they are not dependent on ambient conditions, including light. These differences in how humans versus automated driving systems sense their road environments have not been leveraged in today´s road design.

2.2   The Example of Road Markings

While limited, most of the research into road design for sensor vision has been focused on road markings. Road markings are essential in ADAS features such as lane departure warning systems (i.e., lane assistance systems) which can potentially decrease fatal crashes from 6 to 23% depending on penetration rate, vehicle type, and crash data [15]. Furthermore, lane markings are identified as being an essential traffic control device for automated driving features [16, 17]. As road markings are one of the most widely used and cost-efficient road infrastructure elements, this is an example of using research to find incremental steps towards adapting the road infrastructure to support automated driving.

Within automated driving, road markings are most commonly detected by cameras [18, 19]. Cameras are sensors that, like the human eye, depend on ambient light to function. One of the main quality parameters for humans to be able to see road markings is the retroreflectivity of the roadmarking [20]. Some research has indicated that retrore-flectivity, as well as the contrast between road marking and road surface, are important quality parameters also for camera vision [15]. However, studies comparing a mobile retroreflectometer to a LAS system of a vehicle found that the retroreflection did not affect the detection rate of the LAS [21, 22]. Contrast, on the other hand, is widely indicated as a crucial indicator of whether or not cameras will detect road markings [16, 21–23].

Lidars are also used for lane detection and have the advantage of not being dependent on ambient light conditions [20]. However, lidars are not widely used in commercial vehicles due to a significantly higher cost than cameras [24]. The use of radars has also been tested but requires special lane markings with reflectors for good performance [19, 24].

Road marking detection can fail if the road marking is too worn [25]. Another problem for road marking detection, particularly by camera, is that other road structures with boundaries that appear as longitudinal lines for cameras can be wrongly identified as road markings. This includes highway guardrails and asphalt surface cracks [26, 27]. It would, therefore, be valuable if the different parts of the road infrastructure were easier to separate from each other. The key to being able to do this is to understand how the sensors function.

As an example, qualities of objects, including roughness, color, and reflectivity, affect the success of lidar detection as they affect how much light is reflected [28]. Lighter colors are easier to detect than darker colors. The surface texture also affects lidars' and radars' ability to detect return signals. Higher levels of reflection generally can provide better detection [29]. However, highly reflective surfaces can also be difficult for lidars to register [30]. This implies that the road infrastructure design can make use of differences in color, texture, or patterns to help driver support systems and ADS correctly separate and identify different parts of the road infrastructure. Looking into how colors, textures, and patterns can be used to increase detection of objects likewise applies to vehicles. Some colors and textures are more easily recognized by different sensors such as cameras, radars, and lidars. This implies that traffic safety could be improved also by making small and cost-efficient changes to vehicles.

NordFOU, a cooperation between the national Nordic road and transport administra-tions, released a report in 2023 looking into the visibility of road markings for machine vision [26]. The report points out the challenges in existing quality parameters, for exam-ple, that the contrast between the road marking and road surface is defined in different ways depending on the equipment and method used. Furthermore, the report concludes that it was not possible at this time to find minimum values for performance parameters for machine readability. This work by the NordFOU exemplifies the efforts needed to establish how the greatest safety benefit from ADAS and ADS can be achieved. Fur-thermore, it also highlights the possibility of using the data collected by LAS systems to predict and optimize road maintenance. The data captured by LAS systems can be used to find more complex ways of assessing the quality of road markings and, at the same time, make road maintenance more efficient.

3 Conclusion

ADAS functionality have been shown to provide traffic safety benefits and applies to all levels of automation, but current infrastructure is not designed with considerations of sensor-based driving assistance. Given as an example within this paper, Lane Assistance Systems most commonly use one or several cameras to identify road markings, thus, ensuring that the markings are optimized for machine vision can leverage the safety benefit of these systems. LAS depends on the same road markings designed and optimized for human vision; however, research indicates that the quality parameters used for ensuring human vision are not necessarily the same as for detection by cameras or other sensors. Some research indicates that retroreflectivity may not be the best indicator of LAS success. If this research is validated, it can be used to find more cost-efficient marking products. The LAS systems themselves could also provide the data needed to optimize the maintenance of the road markings.

Optimizing widely used traffic control elements provides an incremental way of leveraging driving support features and automated driving, yet is only starting to receive attention from researchers and ROOs. Further research is needed to identify the correct quality parameters for LAS. Other traffic control elements such as signage and guard rails are other examples that could be optimized for sensor detection.

References

1. European Commission: New rules to improve road safety and enable fully driverless vehicles in the EU, 6 July 2022. https://ec.europa.eu/commission/presscorner/detail/en/ip_22_4312. Accessed 22 Sept 2023
2. AAA Foundation for Traffic Safety: Examining the Safety Benefits of Partial Vehicle Automation Technologies in an Uncertain Future. AAA Foundation for Traffic Safety, Washington, DC 20005 (2023)
3. Erhart, J., Harrer, M., Ruhrup, S., Seebacher, S., Wimmer, Y.: Infrastructure support for automated driving: further enhancements on the ISAD classes in Austria. In: Proceedings of the 8th Transport Research Arena TRA (2020)
4. SAE International: SAE J3016 Surface Vehicle Recommended Practice, Technical report 724. SAE International, Warrendale, PA, USA (2018)
5. ERTRAC: Connected Automated Driving Roadmap. European Road Transport Research Advisory Council, Brussels, Belgium (2019)
6. Tong, J., Nassir, N., Lavieri, P., Sarvi, M.: Putting the connectivity in C-ITS - investigating pathways to accelerate the uptake of road safety and efficiency technologies. iMove Project No: 1-017 Literature Review, Melbourne, Australia (2020)
7. Deveci, M., Gokasar, I., Pamucar, D., Zaidan, A., Wen, X., Gupta, B.: Evaluation of cooperative intelligent transportation system scenarios for resilience in transportation using type-2 neutrosophic fuzzy VIKOR. Transp. Res. Part A Policy Pract. **172** (2023)
8. Agriesti, S., Studer, L., Marchionni, G., Gandini, P., Qu, X.: Roadworks warning—closure of a lane, the impact of C-ITS messages. Infrastructures **5**, 27 (2020)
9. Policy Department Structural and Cohesion Policies: EU Road Surfaces: Economic and Safety Impact of the Lack of Regular Road Maintenance. European Parliament, Brussels (2014)
10. Lehman, M.: The American society of civil engineers' report card on america's infrastructure. In: Layne, P., Tietjen, J.S. (eds.) Women in Infrastructure. Women in Engineering and Science, pp. 5–21. Springer, Cham (2022). https://doi.org/10.1007/978-3-030-92821-6_2

11. Theeuwes, J.: Self explaining roads: subjective categorisation of road environments. In: Vision in Vehicles VI, North-Holland, Netherlands, pp. 279–287 (1998)
12. Macadam, C.: Understanding and modeling the human driver, **40**, 101–134 (2003)
13. Sliney, D.: What is light? The visible spectrum and beyond. Eye **30**, 222–229 (2016)
14. Yeong, D., Velasco-Hernandez, G., Barry, J., Walsh, J.: Sensor and sensor fusion technology in autonomous vehicles: a review. Sensors **21**, 2140 (2021)
15. Babić, D., Babić, D., Fiolić, M., Eichberger, A., Magosi, Z.: Impact of road marking retroreflectivity on machine vision in dry conditions: on-road test. Sensors **22**, 1303 (2022)
16. Hadi, M., Sinha, P.: Effect of pavement marking retroreflectivity on the performance of vision-based lane departure warning systems. J. Intell. Transp. Syst. Technol. Plan. Oper. **15**(1), 42–51 (2011)
17. Hoang, T.: Road lane detection robust to shadows based on a fuzzy system using a visible light camera sensor. Sensors **17**, 11 (2017)
18. Pappalardo, G., Cafiso, S., Di Graziano, A., Severino, A.: Decision tree method to analyze the performance of lane support systems. Sustainability **2**, 13 (2021)
19. Feng, Z., Stolz, M., Li, M., Kunert, M., Wiesbeck, W.: Verification of a lane detection method with automotive radar based on a new type of road marking. In: 2018 IEEE MTT-S International Conference on Microwaves for Intelligent Mobility (ICMIM), Munich, Germany (2018)
20. Burghardt, T.E., Maki, E., Pashkevich, A.: Yellow thermoplastic road markings with high retroreflectivity: demonstration study in Texas **14** (2021)
21. Storsaeter, A., Pitera, K., McCormack, E.: Using ADAS to future-proof roads - comparison of fog line detection from an in-vehicle camera and mobile retroreflectometer. Sensors **5**, 21 (2021)
22. El Krine, A., Girard, J., Redondin, M., Heinkele, C., Stresser, A., et al.: Road marking characterization for ADAS machine vision reliability. In: ESREL 2021 31st European Safety and Reliability Conference, Nantes, September 2021
23. Pike, A., Carlson, P., Barrette, T.: Evaluation of the effects of pavement marking width on detectability by machine vision: 4-inch vs 6-inch markings. American Traffic Safety Services Association, Virginia, USA (2018)
24. Kim, D.-H.: Lane detection method with impulse radio ultra-wideband radar and metal lane reflectors. Sensors, 324 (2020)
25. NordFOU: Machine-readability of road markings in the Nordic countries. NordFOU, Copenhagen, Denmark (2023)
26. Chen, C., Seff, A., Kornhauser, A., Xiao, J.: DeepDriving: learning affordance for direct perception in autonomous driving. In: 15th IEEE International Conference on Computer Vision, Santiago, Chile (2015)
27. Nagarjun Reddy, N., Farah, H., Dekker, T., Huang, Y., van Arem, B.: Road infrastructure requirements for improved performance of lane assistance systems. In: Poster Session Presented at Transportation Research Board Annual Meeting, Washington D.C., USA (2020)
28. Yang, S.-W.: On solving mirror reflection in LIDAR sensing. IEEE/ASME Trans. Mechatron. **16**, 255–265 (2011)
29. Hecht, J.: Lidar for self-driving cars. Opt. Photonics News **29**, 26–33 (2018)
30. Leonard, J., et al.: A perception-driven autonomous urban vehicle. J. Field Robot. **33**, 1–17 (2014)

Open Access This chapter is licensed under the terms of the Creative Commons Attribution 4.0 International License (http://creativecommons.org/licenses/by/4.0/), which permits use, sharing, adaptation, distribution and reproduction in any medium or format, as long as you give appropriate credit to the original author(s) and the source, provide a link to the Creative Commons license and indicate if changes were made.

The images or other third party material in this chapter are included in the chapter's Creative Commons license, unless indicated otherwise in a credit line to the material. If material is not included in the chapter's Creative Commons license and your intended use is not permitted by statutory regulation or exceeds the permitted use, you will need to obtain permission directly from the copyright holder.

Towards a Typology of Arguments for Supporting or Opposing Transport Policy Measures

Wouter Van den Berghe[(⊠)] (iD)

Tilkon Research and Consulting, Poolse-Winglaan 24, 9051 Gent, Belgium
`wouter@tilkon.eu`

Abstract. The level of public support for a potential policy measure is an important consideration for policymakers for whether they will implement such a measure or not – or whether they should try to influence public opinion. It is important to understand what kind of arguments are being used by people when supporting or opposing particular policy measures. Recent research has demonstrated that it is possible to develop a typology of arguments in favour or against policy measures. The typology consists of four groups of arguments in favour and four "mirror" groups of arguments against policy measures. The categorization was developed by decoding and classifying the arguments used by road safety experts from five European countries, who were confronted with several controversial policy measures. Although the typology was developed in the context of road safety, it appears that it can be applied or expanded for use in other policy areas as well.

Keywords: Public support · Policy measures · Transport · Road Safety · Arguments

1 Introduction

Even if an intended policy measure is based on scientific evidence, public support for that measure may be low – which makes policymakers reluctant to implement it. This applies in particular to policy measures that are perceived to restrict freedom, to increase costs or to treat certain people unfairly.

Road safety policy measures are a good illustration of this. Carnis (2015) argues that road safety policymakers are confronted with dilemmas since whatever measure is taken, the benefits tend to differ for different groups of people. Dilemmas may also result from the tension between the benefit/harm of individuals and the overall good for society (Elvebakk, 2015). Examples of road safety measures often perceive as paternalistic are the obligation to wear seatbelts (in cars) or helmets (on motorcycles).

Many arguments for supporting or opposing policy measures are linked to the perceived consequences of the measures. For example, Eby et al. (2017) found that support for lowering the blood alcohol concentration (BAC) limit for drink driving was partially tied to beliefs about the impacts of a change in BAC standards. In relation to making alcohol interlock systems mandatory in the vehicles of people convicted for driving

© The Author(s) 2025

C. McNally et al. (Eds.): TRAconference 2024, LNMOB, pp. 620–626, 2025.
https://doi.org/10.1007/978-3-031-85578-8_83

under the influence of alcohol, McCartt et al. (2010) found that most people favouring the measure thought that it would prevent alcohol-impaired driving, save lives, or prevent crashes. Barton & Pan (2022) showed that using particular arguments when promoting a policy measures significantly increases support for certain mobility policies. Van den Berghe & Christie (2022) found that if people feel safe when using a particular transport mode, they are less in favour for additional or stricter measures affecting their transport mode. They also showed that the perceived restriction of human liberties, fear of discrimination, and resistance to state interventions fuel opposition against measures.

Some studies use theoretical frameworks and models for explaining the support for, or opposition to, policy measures in transport. Eriksson et al. (2006, 2008) developed a model for the acceptability of transport measures, combining the value-belief-norm theory (Stern, 2000) with policy specific beliefs (perceived fairness and perceived effectiveness), problem awareness and personal norms. Huber & Wicki (2021) used a model based on the policy design, trust in government and proximity.

What is absent, however, is a solid categorization of the types and the nature of arguments that are used – not just in transport policies but in other policy areas as well (see e.g., Cannon et al. (2022) and Wicki et al. (2019). It appears useful to understand what type of argument is being used in favour or against the measure. This may help policymakers in finding a compromise between different interests and concerns.

2   Method

2.1   Confronting Experts with Controversial Policy Measures

As part of PhD project (Van den Berghe, 2022) forty European senior transport experts (eight from France, the UK, Greece, Austria and Sweden each) were confronted in an interview with eight 'controversial' policy measures (Table 1). These interviews took place in 2019 and 2020.

The experts were asked whether they considered such measures to be fair and whether they were in favour or against. Subsequently they were asked to provide the main and ancillary arguments to support their position. The characteristics of the interviewees can be summarized as follows:

- about half of the interviewees (19/40) worked for a public authority (11) or were a politician (8)
- eleven worked in a research institute and five in a university
- on average, the interviewees had been 21 years involved in road safety
- half of the interviewees (21/40) had a degree in Engineering or Transport
- one third of the interviewees was female (13/40)
- the median age of the interviewees was 55 years

2.2   Codification of the Arguments

Strauss & Corbin (1998) distinguish between three types of coding practice: open coding, axial coding and selective coding. Open coding was used because it leads to concepts that can be grouped into categories. My approach was mainly data-driven (Boyatzis,

Table 1. List of controversial policy measures in the field of road safety

---

- Zero tolerance for driving under the influence of alcohol (0,0‰ blood alcohol concentration) for all drivers of vehicles (cars, trucks, motorcyclists, cyclists, …).
- In all urban areas and villages the speed limit should be 30 km/h (20 mph) for all vehicles (except on main thoroughfares).
- All people aged 70 or more should be screened on a 5 yearly basis, in order to decide whether they are still allowed to drive a car or not.
- Fines that people have to pay after they have committed a traffic offence should be proportional to their income.
- All cars should be equipped with an alcohol ignition interlock system (which prevents starting and driving the car if the driver's alcohol concentration is above the legal BAC limit).
- All cyclists should wear a helmet.
- Pedestrians should wear retroreflective clothing, shoes or bags when walking in the dark on public roads.
- All cars should be equipped with an Intelligent Speed Assistance (ISA) system that automatically limits the speed of the car to the maximum speed limit.

1998), with codes being drawn inductively from the data itself. The approach adopted resembles the steps in the template analysis approach developed by King and others (Brooks et al., 2015; King et al., 2018).

First, labels were assigned to the arguments and meanings used by three interviewees. These labels were then used for the other interviews. When the existing set of labels was not adequate, new labels were added. The initial argument classification scheme was modified many times as the coding of the transcripts progressed: some of the labels were renamed, others were split into two labels, and some were merged when progressing with the coding.

3   Results

3.1   Overall Typology

The analysis led to a categorisation with eight groups of arguments, four of which are supportive and the other four opposing. Each of these groups can be further broken down into more specific arguments (Table 2). As can be seen, each positive argument area has a mirror area on the negative side – and this is also the case for most of the specific arguments themselves. As an example, the mirror area of the argument area 'Equity' is 'Discrimination' and the mirror argument of the argument 'Strong, clear message' is 'Wrong message'. The 'Pro' and 'Contra' areas are not perfect mirrors. 'Equity' has several negative mirror arguments; this reflects the fact that when someone states that something is equitable it covers all these forms of non-discrimination. Also, some interviewees used the arguments of 'Preserving human liberties', without specifying which liberties. When constructing the typology, no weight was given to any of the arguments, but a label was given only if at least three interviewees had used the argument.

Table 2. Typology of arguments in favour or against policy measures

Supportive arguments	Opposing arguments
Relevance	**Limited added value**
Reduces/avoids harm	Does not reduce/avoid harm
Is effective in meeting its purpose	Ineffective in meeting its purpose
Addresses an important problem	Other problems are more important
Is a good solution to the problem	Other measures are better
Gives the right message	Gives the wrong message
Has positive side effects	Has negative side effects
Regulation is useful	Regulation is not the right approach
Preserving human liberties	**Restricting human liberties**
Proportionate, right, just	Disproportionate
Preserving liberties (general)	
Preserving freedom	Restricting freedom
Preserving mobility	Restricting mobility
Preserving joy in life	Reducing joy in life
Assuming responsibility	Reducing responsibility
Avoiding burden	Increasing burden
Protecting privacy	Reducing privacy
Limited costs for people	Expensive for people
Transparent	Not transparent
Equity	**Discrimination**
Equity (general)	Discrimination (general)
	Discrimination by road user
	Discrimination by age
	Discrimination by gender
	Discrimination by wealth
	Discrimination by group
Difficult to cheat/evade/not comply	Easy to cheat/evade/not comply
Feasibility	**Practical obstacles**
Easy to implement	Complex, difficult to implement
Easy to enforce	Complex, difficult to enforce
Efficient for society	High costs for society
Public support	Public opposition
In agreement with the law	Against the law

3.2   Example of the Meaning of the Labels

Many labels in this table may appear a bit cryptic at first; they are explained in detail in Van den Berghe (2022). For example, the meanings for the groups of arguments related to 'Relevance' are listed below:

- *'Reduces/avoids harm'*: the measure is believed to lead to a (strong) reduction in the number of injuries and fatalities caused by road crashes.
- *'Is effective in meeting its purpose'*: the measure is believed to have the intended impact (e.g., change in behaviour).
- *'Addresses an important problem'*: given the size of the problem (e.g., drunk driving) people tend to support any measure believed to reduce this problem.
- *'Is a good solution to the problem'*: compared to other measures that can address a problem (e.g., speeding), this particular measure is seen as an appropriate/easy/cheap way of solving it or mitigating its consequences.
- *'Gives the right message'*: by implementing the measure you give a clear message about a risky behaviour (e.g., drinking and driving don't go together) or situation.
- *'Has positive side effects'*: in addition to the prime intended effect (e.g., reducing injuries) the measure has positive effects (e.g., making cities more liveable).
- *'Regulation is useful'*: its relevant because without regulation, the situation would not change (e.g., people will continue to engage in risky behaviour).

4   Discussion and Conclusion

It is common sense that arguments in favour or against policy measure are influenced by beliefs on whether the measure is likely to be effective, as well as by ethical or ideological views. This has also been shown in numerous studies. Yet, a sold classification of arguments is lacking. With this research, a first version of a classification scheme has been developed that is useful for categorising the nature of the arguments that people use to support or oppose policy measures – whether they are experts, policymakers or ordinary citizens.

Understanding whether resistance to measures is rooted in the disbelief of its effectiveness, in ethical/ideological concerns (e.g., restriction of freedom, unequal treatment) or concerns about feasibility, may help policymakers in correcting perception errors and/or adapting the measure in order to get broader public support.

As an example, suppose that the main argument against the measure would be its presumed poor effectiveness, then providing evidence for the expected effects should be given high priority in the communication about the measure. Vice-versa, recognising that some of the counterarguments are partially valid may be the right way forward, provided that it can be shown that the positive effects outweigh the possible negative effects. For instance, it may well be that for many drivers a speed limit of 30 km/h will reduce their joy or driving, but the benefits in terms of liveability of cities and reduced safety risks for the population can be argued to outweigh the perceived negative effects.

Although the typology was developed in the context of road safety, it appears that, with minor adaptations, it could also be used in many other policy areas such as transport, environment, education and health. Yet it is recognized that using a more varied set of policy areas, possibly a wider set of argument categories could have been developed.

Another limitation is that the typology was developed based on experts' opinions; using "the general public" might have led to a different categorisation.

There is scope for further research. First, it would be useful to examine the applicability of the scheme – and the need for adaptation – across a wider range of policy measures and policy areas. Secondly, the arguments should also be related to other dimensions and characteristics of people, including the level of expertise/knowledge about the topic and the notion of bounded reality, to what extent the arguments are based on objectives facts (versus subjective judgements), the level of self-interest in the measure being implemented (or not) and ideological views on the matter.

References

Barton, J., Pan, X.: Movin' on up? A survey experiment on mobility enhancing policies. Eur. J. Polit. Econ. **74**, 102172 (2022)

Brooks, J., McCluskey, S., Turley, E., King, N.: The utility of template analysis in qualitative psychology research. Qual. Res. Psychol. **12**(2), 202–222 (2015)

Cannon, J.S., et al.: Perceptions of arguments in support of policies to reduce sugary drink consumption among low-income white, black and Latinx parents of young children. Am. J. Health Promot. **36**(1), 84–93 (2022)

Eby, D.W., et al.: Perceptions of alcohol-impaired driving and the blood alcohol concentration standard in the United States. J. Saf. Res. **63**, 73–81 (2017)

Elvebakk, B.: Paternalism and acceptability in road safety work. Saf. Sci. **79**, 298–304 (2015)

Eriksson, L., Garvill, J., Nordlund, A.M.: Acceptability of travel demand management measures: the importance of problem awareness, personal norm, freedom, and fairness. J. Environ. Psychol. **26**(1), 15–26 (2006)

Eriksson, L., Garvill, J., Nordlund, A.M.: Acceptability of single and combined transport policy measures: the importance of environmental and policy specific beliefs. Transp. Res. Part A: Policy Pract. **42**(8), 1117–1128 (2008)

Huber, R.A., Wicki, M.: What explains citizen support for transport policy? The roles of policy design, trust in government and proximity among Swiss citizens. Energy Res. Soc. Sci. **75**, 101973 (2021)

King, N., Brooks, J., Horrocks, C.: Interviews in Qualitative Research. Sage Publications (2018)

McCartt, A.T., Wells, J.K., Teoh, E.R.: Attitudes toward in-vehicle advanced alcohol detection technology. Traffic Inj. Prev. **11**(2), 156–164 (2010)

Stern, P.C.: Toward a coherent theory of environmentally significant behavior. J. Soc. Issues **56**(3), 407–424 (2000)

Strauss, A.L., Corbin, J.M.: Basics of Qualitative Research: Techniques and procedures for Developing Grounded Theory. Sage Publications (1998)

Van den Berghe, W.: The influence of fairness and ethical trade-offs on public support for road safety measures. An international and intercultural exploration. PhD Thesis University College London (2022)

Van den Berghe, W., Christie, N.: International and intercultural differences in arguments used against road safety policy measures. IATSS Res. **46**(1) (2022)

Wicki, M., Huber, R., Bernauer, T.: Can policy-packaging increase public support for costly policies? Insights from a choice experiment on policies against vehicle emissions. J. Public Policy **40** (2019)

Open Access This chapter is licensed under the terms of the Creative Commons Attribution 4.0 International License (http://creativecommons.org/licenses/by/4.0/), which permits use, sharing, adaptation, distribution and reproduction in any medium or format, as long as you give appropriate credit to the original author(s) and the source, provide a link to the Creative Commons license and indicate if changes were made.

The images or other third party material in this chapter are included in the chapter's Creative Commons license, unless indicated otherwise in a credit line to the material. If material is not included in the chapter's Creative Commons license and your intended use is not permitted by statutory regulation or exceeds the permitted use, you will need to obtain permission directly from the copyright holder.

Sustainable Transport Appraisal: A Literature Review and Implications for Policy Makers

Yeonjung Song[1,3], Warren Whitney[1], Wen Zhang[1,2], Barry Colleary[1,2], Brian Caulfield[2], and Juan Martinez-Covarrubias[1,4(✉)]

[1] National Transport Authority, Dublin, Ireland
juan.martinez@nationaltransport.ie
[2] Trinity College Dublin, Dublin, Ireland
[3] Kobe University, Kobe, Japan
[4] American University, Washington, USA

Abstract. Questions have arisen regarding whether current transport appraisal methodologies and parameters are prejudicial to sustainable transport solutions. With the emergence of the climate change and sustainability agenda, there is wide recognition of the need for modal shift from private-car-based modes to sustainable modes such as public transport and active travel. Despite this policy re-orientation, the use of the traditional appraisal methods and parameters with their inherent assumptions does not always support the policy. Cost-benefit analysis (CBA) is likely to recommend options that do not necessarily align with current sustainable policy goals. To cater for hard-to-monetise benefits, multi-criteria analysis (MCA) is employed in transport appraisal. However, there is a concern that multi-criteria analysis (MCA) is difficult to operate given the need for agreed weightings among stakeholders. This study summarises the challenges of the current appraisal frameworks through a literature review and analysis of the appraisal framework in Ireland. By doing so, this study aims to outline what should be improved for transport appraisal to work under the sustainability policy agenda.

Keywords: Transport appraisal · sustainable transport · cost-benefit analysis · transport investments

1 Introduction

Current well-established frameworks and methods for transport appraisal were first developed more than 50 years ago (Vickerman, 2017) when the socio-technological trends were based on the private-car. Consequently, this was translated to the aim of building fit-for-purpose infrastructure for this mode of transportation. Cost Benefit Analysis (CBA) that focuses on the monetisation of benefits has been a widely used tool in transport appraisal. In the conventional appraisal system using CBA, travel time savings have been a key variable that has a disproportionate impact on the monetised benefits. With the emergence of the climate change and sustainability agenda, the conventional transport appraisal approaches result in a potential systemic failure of having an appraisal framework that can yield results that are not aligned with national policy.

© The Author(s) 2025
C. McNally et al. (Eds.): TRAconference 2024, LNMOB, pp. 627–632, 2025.
https://doi.org/10.1007/978-3-031-85578-8_84

This paper delves into the challenges faced by decision-makers who need fit-for-purpose methods to assess solutions consistent with sustainable policies. We summarise challenges in the current transport appraisal framework from the perspectives of policy alignment and promotion of sustainable transport (Sect. 2). We also review discussions to complement the shortcomings of traditional frameworks (Sect. 3) through the literature review. We then analyse the new appraisal framework for transport projects in Ireland (Sect. 4) and discuss the implications of this analysis (Sect. 5).

2 Challenges in Conventional Transport Appraisal for Sustainable Transport

Although a CBA has been the most widely used tool for a long time, there has been a debate about the appropriateness of CBA as a means of evaluating projects and of choosing between alternatives (Vickerman, 2017). Welde and Volden (2021) argue that the current CBA model has issues including 1) no distribution of effects between groups and regions; 2) negligible impact of environmental consequences on Net Present Value (NPV); 3) discounting negative future effects; 4) uncertainty of input variables and huge variation of appraisal results in individual projects over time; 5) promoting investment in affluent areas; 6) recommendation of projects in conflict with societal goals. While Multi-Criteria Analysis (MCA) is considered an alternative to CBA, MCA can result in a preference ranking which risks either reinforcing the a-priori preferences or being ignored as most stakeholders already had an a-priori preference for a politically sensitive project (te Boveldt et al., 2022).

On the other hand, the value of travel time savings (VTTS) is considered a central notion in transport CBA (Meunier, 2019). The VTTS has been important as a parameter that helps to estimate behavioural choices and as a parameter that helps to determine value for money in allocation of public funds (Goodwin, 2019). However, the critical view that the benefits calculated may give too much weight to time as compared with other criteria of importance to individuals and objectives of transport policy such as sustainability and equity has grown in recent years.

Fosgerau and Jensen (2003) point out that travel time savings account for around 80 per cent of the quantified benefits of new transport infrastructure in CBA. Welch and Williams (1997) find that a substantial part of this benefit may be made up of large numbers of people saving small amounts of time. While the definition of "small amount of time" varies depending on the literature, "small" can be defined as less than 5 min (Welch and Williams, 1997; Mackie et al., 2003). But Welch and Williams (1997) also point out that the threshold of small time could depend on trip length or duration. There are suggestions that a lower or discounted unit value for VTTS could be used for small time savings while the great majority of countries adopt a Constant Unit Value (CUV) approach (Wallis et al., 2015).

3 Alternative Transport Appraisal Approaches

ITF (2022) summarises some modifications to CBA to focus on the accessibility and equity impacts from the literature review: 1) Willingness-To-Pay-based valuations; 2) applying a uniform equity value of time to all users; 3) measure of subjective well-being which considers a greater increase in wellbeing from a given accessibility improvement to low-income group; and 4) use of a range of discount rates that reflect the rate of time preference of the different groups affected by the proposed project. To complement the valuation approach in the dominant style of CBA, some additional methodologies are suggested. These include social choice valuation that values environmental impacts more than travel time savings (Mouter et al., 2019) and Cumulative Effects Assessment as a complementary tool for environmental assessment for projects (Tricker, 2007).

As an alternative form of appraisal from the traditional CBA, quality-based tools have been considered. For example, Transport Quality of Life appraisal on all modes of transport (Carse, 2011), a participatory MCA that involves multiple actors (Hickman and Dean, 2019), a Participatory Value Evaluation that assesses the desirability of government projects (Mouter et al., 2019) and stakeholder-based appraisal method using stakeholder-based impact (te Boveldt et al., 2022) have been proposed in previous research. Thus, there are still questions as to whether these alternative approaches can be widely used by practitioners in terms of robustness and practicality.

4   Appraisal Framework for Transport Projects in Ireland

In Ireland, the Transport Appraisal Framework (TAF) was published in June 2023 to replace the Common Appraisal Framework (CAF) for Transport Projects and Programmes. The CAF was in operation since 2016 before the TAF was published. While the CAF focused on providing guidance on the conventional appraisal methodologies such as CBA, CEA and MCA, the TAF proposes a new appraisal process, the Transport and Accessibility Appraisal (TAA), which replaces MCA as the main qualitative appraisal tool for assessing short-listed options in the transport appraisal process for proposals with costs greater than €30 million. For schemes with estimated costs of up to €30 million, a detailed MCA is sufficient.

The TAA process retains many aspects of the MCA including criteria and impact assessments, but notable differences include the absence of economic criteria and the separation of the previous environment criterion into climate change impact and local environment impact criteria. The TAA intends to capture the impact of the short-listed options across six key criteria: 1) Accessibility, 2) Social, 3) Land Use, 4) Safety, 5) Climate Change, and 6) Local Environment. Scoring of each criterion should be based on impacts to be measured by suggested key performance indicators using potential data sources. Each indicator, sub-criterion and criterion is evaluated using a seven-point Likert scale (high positive, positive, slight positive, neutral, slight negative, negative high negative) and is presented using colours as shown in Fig. 1. Another difference to the MCA is that the scores for each criterion are not intended to lead to a numerical total across all criteria for a given option and should be considered independently of one another.

The TAF also establishes new parameters for journey time reliability, journey quality, electric vehicle fuel consumption, and cycling journey quality, as well as updating parameters to be used in transport CBAs. One of the changes from the CAF is that the TAF discounts travel time savings of 5 min or less. These small amounts of time impacts for transport users of 5 min or less should be excluded from the central scenario in

Option	Accessibility	Social	Land Use	Safety	Climate Change	Local Environment
Option 1	Slight Positive	Slight Positive	Positive	Neutral	Low Positive	Low Positive
Option 2	High Positive	Slight Positive	Positive	Neutral	Positive	Low Positive
Option 3	Neutral	Neutral	High Negative	Neutral	Slight Negative	Neutral
Option 4	Slight Negative	Negative	Neutral	Neutral	Neutral	Neutral

Fig. 1. TAA Summary Reporting Table (Source: TAF Module 7 – Detailed Guidance on Appraisal Techniques)

CBA while these benefits/costs may be applied as part of sensitivity testing. This is a noticeable difference from most other countries' approaches to applying CUV. However, Canada has a similar approach to the TAF which states that savings of less than 5 min per one-way trip should not be included in the overall economic performance results such as Benefit Cost Ratio (BCR) while it uses a CUV approach in valuing these savings. Several countries such as the United States and France that previously used discounted values for small time savings have changed to the CUV approach (Wallis et al., 2015).

5   Discussion

The TAA process can help bring a more qualitative and standardised approach to the assessment of impacts that cannot be sufficiently assessed using existing approaches to MCA and CBA. The use of a standard set of criteria will allow decision-makers to compare projects in their respective investment portfolios using a standard suite of metrics. Also, the standard template for the TAA results table can help in the comparison of impacts across different projects. This may enable TAA to be used as a tool to help guide portfolio investment decisions for state transport agencies, local authorities, and central government. Another likely benefit of the TAA process will be increased assessment of impacts that have not been commonly examined across all transport investment proposals but could support sustainable transport modes. In particular, the TAA impact assessments relating to social impacts and climate adaptation should provide a more detailed understanding of related impacts across a broader range of transport proposals.

However, the TAA process also has some apparent limitations. Firstly, it is likely that not all TAA indicators may be relevant or add value to the appraisal of certain proposals. There is potentially a need to allow for greater flexibility in the selection of indicators that can be used in the TAA process, but this will need to be balanced against the potential benefits of having a standard set of criteria and indicators. Secondly, the TAF does not define a specific refinement approach for reaching a TAA criterion score from multiple indicator results or single indicator score from multiple impact assessment results, such as the quantitative mean and median approaches. Thirdly, there is no guidance on how to weigh certain impacts or indicator results. For example, using the quantitative mean approach to assess an option under the TAA climate change criterion, it is possible for an option to receive both a negative and positive indicator rating that would result in a neutral criterion rating. However, this conceals the possible negative impacts assessed for the option under one or more of the climate change indicators.

The removal of the impact of 5 min or less for the VTTS may reduce the impact of benefits from travel time savings on the economic appraisal. Some road projects, such as a link road to the main carriageway and replacement of the old substandard road, are likely to result in travel time savings of less than 5 min. However, once this applies to traffic volumes through the normal appraisal period, for 30 years, this impact becomes quite large, and the benefit could be overestimated while only an unperceivable change is made to users which would not change any behaviour. Therefore, the removal of the impact of 5 min or less could alleviate the overestimation of the benefit of these projects from an economic perspective and could give more opportunities for sustainable transport projects.

However, this is also likely to be an issue for sustainable transport schemes as these schemes also struggle to generate large journey time savings. In Ireland, travel time savings from public transport schemes and active travel schemes are typically less than 5 min. In addition, travel times by these transport modes are generally less than 30 min. In the Irish context, the threshold of 5 min for small time savings would be high while Wallis et al (2015) find that values for time savings of 3–5 min are typically in the order of half the threshold values through literature review. Therefore, reviewing the travel times and savings of the projects in the past and studying whether Irish users value small travel time differently are recommended to set up a more acceptable threshold and value for the small travel time and to ensure its impact on the sustainable transport modes.

6   Conclusion

Approaches in the TAF attempt to fill the gap between transport appraisal and policy as the TAA highlights qualitative criteria which enable the assessment of broad aspects of projects. The TAA can be a supplement to the CBA results which enhances the information available to decision-makers by shedding light on the social impacts. The removal of the travel time savings of less than 5 min could be considered as a way to reduce the dominance of travel time in a CBA.

However, it still seems that in general the transport appraisal is disconnected from the policy objectives focusing on sustainability and societal benefits of transport projects. CBA still seems to play a dominant role in quantitative approaches and travel time saving represents a large proportion of monetised benefits in the CBA. The benefit from time savings only exists in the short-run while in long-run benefits of investment are seen from changes in land use and spatial distribution (Metz, 2017). Also, transport policy and planning objectives are being fundamentally reconsidered in many countries and a central element of this shift is to focus on accessibility-based perspective (ITF, 2022). In order to respond to the shift of policy objectives, a policy screening prior to the appraisal could be introduced. Reconsidering the valuation of travel time savings in terms of productivity and preference for the faster trip could be also considered to evaluate its impacts considering users' changing concepts of travel time and on-board experience.

Disclaimer.   The views and opinions expressed in this paper are those of the authors and do not necessarily reflect the official position of the National Transport Authority.

References

Carse, A.: Assessment of transport quality of life as an alternative transport appraisal technique. J. Transp. Geogr. **19**, 1037–1045 (2011)

Fosgerau, M., Jensen, T.L.: Economic appraisal methodology — controversial issues and Danish choices. European Transport Conference, 4–6 October, Strasbourg, France (2003)

Goodwin, P.: The Influence of Technologies and Lifestyle on the Value of Time, International Transport Forum Discussion Paper, No. 2019/03, OECD Publishing, Paris (2019)

Hickman, R., Dean, M.: Incomplete cost – incomplete benefit analysis in transport appraisal. Transp. Rev. **38**(6), 689–709 (2018)

International Transport Forum. Broadening Transport Appraisal: Summary and Conclusions, ITF Roundtable Reports, No. 188, OECD Publishing, Paris (2022)

Mackie, P.J., Wadman, M. Fowkes, A.S., Whelan, G., Nellthorp, J., Bates, J.: Values of Travel Time Savings UK, Institute of Transport Studies, University of Leeds, Working Paper 567 (2003)

Meunier, D.: Mobility Practices, Value of Time and Transport Appraisal, International Transport Forum Discussion Paper, No. 2020/01, OECD Publishing, Paris (2019)

Metz,: Valuing transport investments based on travel time saving: Inconsistency with United Kingdom policy objectives. Case Stud. Transp. Policy **5**, 716–721 (2017)

Mouter, N., Cabral, M.O., Dekker, T., van Cranenburgh, S.: The value of travel time, noise pollution, recreation and biodiversity: a social choice valuation perspective. Res. Transp. Econ. **76**, 100733 (2019)

te Boveldt, G., Keseru, I., Macharis, C.: When monetarisation and ranking are not appropriate. A novel stakeholder-based appraisal method. Transp. Res. Part A: Policy Pract. **156**, 192–205 (2022)

Tricker, R.C.: Assessing cumulative environmental effects from major public transport projects. Transp. Policy **14**, 293–305 (2007)

Vickerman, R.: Beyond cost-benefit analysis: the search for a comprehensive evaluation of transport investment. Res. Transp. Econ. **63**, 5–12 (2017)

Wallis, I., Rupp, K., Alban, R.: Travel Time Saving Assessment, NZ Transport Agency Research Report 570 (2015)

Welde, M., Volden, G.H.: Norwegian perspectives on ex post evaluation, Presentation at Broadening the Scope of Transport Appraisal to Capture the Full Impact of Investments Roundtable by ITF (2021). https://www.itf-oecd.org/repository/broadening-scope-transport-appraisal-capture-full-impact-investments-roundtable. Accessed 25 Sep 2023

Welch, M., William, H.: The sensitivity of transport investment benefits to the evaluation of small travel-time savings. J. Transp. Econ. Policy 231–254 (1997)

Open Access This chapter is licensed under the terms of the Creative Commons Attribution 4.0 International License (http://creativecommons.org/licenses/by/4.0/), which permits use, sharing, adaptation, distribution and reproduction in any medium or format, as long as you give appropriate credit to the original author(s) and the source, provide a link to the Creative Commons license and indicate if changes were made.

The images or other third party material in this chapter are included in the chapter's Creative Commons license, unless indicated otherwise in a credit line to the material. If material is not included in the chapter's Creative Commons license and your intended use is not permitted by statutory regulation or exceeds the permitted use, you will need to obtain permission directly from the copyright holder.

Barriers to Adoption of a Fully Electric Vehicle – A Study of Car Users in the Helsinki Region

Risto Öörni[✉]

VTT Technical Research Centre of Finland Ltd, P.O. Box 1000, 02044 VTT Espoo, Finland
risto.oorni@vtt.fi

Abstract. Barriers to purchasing a fully electric vehicle were studied using an online survey aimed at car users located in the Helsinki region in Finland. Opinions on the significance of various barriers were asked from car users who had answered that they would not purchase a fully electric vehicle as their next car. The three barriers that would most affect their decision-making were the high purchase price of an electric vehicle, insufficient operating range on a single charge, and too few charging points in Finland outside the Helsinki and Uusimaa region. The results suggest that the desirability of a fully electric vehicle is affected not only by the availability of charging at home, or in the region where the car user lives, but also in other parts of the country where the user lives.

Keywords: Fully Electric Vehicle · Adoption · Barriers · Finland

1 Introduction

The objective of decarbonizing transport has been set at both European and national levels. In the European regulation, the target level for CO_2 emissions of new passenger cars has been set at 95 g/CO_2/km in the EU area (European Commission 2019). In addition to the European regulation, some countries have set their own targets for reducing CO_2 emissions from transport; for example, Finland aims to reduce them by at least 50% by 2030 from the 2005 level, and to have a fully fossil-free transport system by 2045 (Ministry of Transport and Communications 2021). Deployment of fully electric vehicles contributes to the objectives of decarbonizing transport and reducing the use of fossil fuels.

Deployment of fully electric vehicles (EVs) can be facilitated or hampered by their acceptance among car users. Currently and in the near future, car users can choose between vehicles with an internal combustion engine and fully electric vehicles. Even though the availability of fully electric vehicles and vehicle models is improving, many car users still select a vehicle with an internal combustion engine when buying a new car. It is, therefore, important to study what barriers are perceived by car users to purchasing a fully electric vehicle.

© The Author(s) 2025
C. McNally et al. (Eds.): TRAconference 2024, LNMOB, pp. 633–639, 2025.
https://doi.org/10.1007/978-3-031-85578-8_85

Several factors shaping the adoption of EVs by car users have been identified and fall roughly into four groups: technical factors, contextual factors, cost factors, and individual and social factors (Rezvani, Jansson and Bodin 2015). Other classifications have been described: Zhuge and Shao (2019) identified six factors affecting the decision to purchase an EV: vehicle price, vehicle usage, social influence, environmental awareness, purchase-related policies, and usage-related policies. Hurdles to adopting a fully electric vehicle have been identified as relating, for example, to lack of trust (e.g., in new technology), the inability to convert consumers' interests into actual vehicle sales, poor desirability, and issues related to "living with the technology" such as costs of purchase and ownership and limitations of the technology affecting the user (Krishna 2021). According to a study focused on the Finnish customers of one car brand, the most significant barriers to EV adoption were high purchase price and limited range (Mattila 2019).

While several barriers to EV adoption have been identified, earlier studies do not provide a direct, unambiguous answer to which of these barriers are most relevant in the current situation for car users in Finland, and what is their relative importance to car users.

2 Objectives

The objective of the study was to explore the barriers to purchasing a fully electric vehicle as reported by car users living in the Helsinki region of Finland. As a part of defining the objectives, two main research questions were identified:

1: What is the relative importance of different barriers to purchasing a fully electric vehicle?
2: Is it possible to identify a group or cluster of barriers that would have specific importance for car users' decisions?

3 Methods

3.1 Data Collection

The data set used in the study was collected using an Internet survey in November 2020. The survey was aimed at active car users who had driven at least 1500 km during the last 12 months and who were living in the Helsinki region. The survey was available in Finnish. The data were collected as a part of the CLEMET project (Cleantech Mobility Education for Tomorrow) (CLEMET 2021), and the first results of the study were published in a technical report in 2021 (Lang et al. 2021). The aim of CLEMET was to develop vocational and third-level education on the automotive sector in Finland, to make it ready for new vehicle technologies and vehicles with greener propulsion technologies.

The third part of the questionnaire included a question on the respondent's readiness to purchase a fully electric vehicle as their next car (possible answers: yes, no, I don't know). This question was asked from respondents who had indicated earlier in the questionnaire that they would purchase a new vehicle after their current one. Of the respondents, 464 answered "no" concerning their readiness to purchase a fully electric vehicle as their next car. These respondents were asked to answer an additional question on barriers to purchasing a fully electric vehicle. In addition to the question-naire, demographic information (e.g., age group, gender) was obtained from the panel of the marketing research company which recruited the respondents and implemented the online questionnaire.

The barriers to purchasing a fully electric vehicle were studied by presenting eight statements to the respondents and asking which reasons prevented them from purchasing a fully electric vehicle as their next car, and how much the presented reasons affected that decision. The importance of different barriers was measured using a five-step Likert scale [1: (Very little), 2, 3, 4, 5 (Very much)].

The characteristics of the study participants are shown in Table 1. According to the table, respondents who had turned 60 are overrepresented compared to the target group of the study. For this reason, the main analyses were carried out separately for all respondents, those aged 60 + years, and those younger than 60. Of the respondents, 99% gave their home address as within the municipalities of Helsinki, Espoo, Vantaa or Kauniainen.

Table 1. Characteristics of study participants.

Heading level	N	Share [%]	Cumulative share [%]
Gender	727	64.1	64.1
Male	408	35.9	100.0
Female	1135	100.0	
Age group			
< 30 years	39	3.4	3.4
30–39 years	70	6.2	9.6
40–49 years	126	11.1	20.7
50–59 years	221	19.5	40.2
60–69 years	316	27.8	68.0
70v + years	363	32.0	100.0
Total	1135	100.0	

3.2 Statistical Methods

Mean and standard deviation were calculated for variables representing the importance of different barriers. It was then possible to identify the order of the barriers based on the mean of responses provided in the online survey for all respondents, respondents younger than 60 years, and respondents aged 60 years or older. Kolmogorov-Smirnov and Shapiro-Wilk-tests (Sheskin 2011) were calculated for all statements to find out whether the answers regarding different barriers were normally distributed. The responses were not normally distributed (p < 0.001 for all barriers in both tests) for any of the barriers covered by the survey.

A related-samples Wilcoxon signed-rank test (Sheskin 2011) was performed for all pairs of barriers covered by the study. The test indicates whether the median of differences calculated for a pair of variables (in this case, difference between answers from a single respondent for two barriers to purchasing an electric vehicle) deviates from zero in a statistically significant manner.

4   Results

The distribution of responses regarding the importance of different barriers to purchasing a fully electric vehicle is presented in Table 2, as well as the mean and standard deviation calculated for responses regarding each barrier. Due to rounding to full percentage, the sum of percentages does not equal 100 on all rows in the table.

Table 2. Distribution of responses.

Barrier	Distribution of responses (N = 464)					Mean	Standard deviation
	1	2	3	4	5		
High purchase price of an electric vehicle	24 (5%)	38 (8%)	59 (13%)	116 (25%)	227 (49%)	4.04	1.188
The operating range of an electric vehicle with a single charge is not sufficient for my mobility needs	32 (7%)	35 (8%)	63 (14%)	88 (19%)	246 (53%)	4.04	1.261
Charging points for electric vehicles are too few elsewhere in Finland	37 (8%)	40 (9%)	62 (13%)	114 (25%)	211 (45%)	3.91	1.283
Charging the battery of an electric vehicle takes longer than filling up the tank of a vehicle with an internal combustion engine	48 (10%)	43 (9%)	102 (22%)	105 (23%)	166 (36%)	3.64	1.325

(*continued*)

Table 2. (*continued*)

Barrier	Distribution of responses (N = 464)					Mean	Standard deviation
	1	2	3	4	5		
There is no charging point available at home on the property or in a nearby location	85 (18%)	40 (9%)	62 (13%)	75 (16%)	202 (44%)	3.58	1.546
Charging points for electric vehicles are too few in Helsinki and the Uusimaa region	61 (13%)	61 (13%)	115 (25%)	97 (21%)	130 (28%)	3.38	1.360
A vehicle suitable for my needs cannot be found in the selection of electric vehicle models available	60 (13%)	75 (16%)	129 (28%)	88 (19%)	112 (24%)	3.25	1.332
Electric vehicles are not available as pre-owned vehicles	134 (29%)	92 (20%)	110 (24%)	72 (16%)	56 (12%)	2.62	1.361

The order of importance of different barriers to purchasing a fully electric vehicle was very similar for respondents younger than 60 years, those aged 60 years or older, and all age groups analyzed together. For younger respondents, the three barriers reported to most affect decision-making were high purchase price ($\mu = 4.10$), insufficient operating range on a single charge ($\mu = 3.98$), and insufficient availability of charging spots in other areas of Finland outside Helsinki and the Uusimaa region ($\mu = 4.01$). For respondents aged 60 years or older, the two barriers reported to affect decision-making the most were insufficient operating range on a single charge ($\mu = 4.07$) and high purchase price ($\mu = 4.01$). The third most important barrier for respondents 60 years or older was insufficient availability of charging spots elsewhere in Finland outside Helsinki and the Uusimaa region ($\mu = 3.85$).

The results of the related-samples Wilcoxon signed-rank test did not indicate statistically significant differences between the three most relevant barriers. This result was the same for respondents younger than 60 years, those 60 years or older, and all age groups analyzed together.

5   Discussion of the Results

The results of the study reflect the opinions and perceptions of car users who are hesitant or skeptical towards fully electric vehicles. This is a novel viewpoint, as several earlier studies on barriers to purchase have focused on all car users or buyers of new cars.

Older age groups were overrepresented among the respondents of the study. This limitation in the data was addressed by performing the analysis separately for all respondents, respondents younger than 60 years, and respondents aged 60 years and older. The analysis results also indicated that the three most important barriers were the same when analyzed for all respondents, respondents younger than 60 years, and those aged 60 years or older, but their order was different.

The data set used in the study was collected using an Internet survey, and respondents were recruited by a panel of a marketing research company. This implies that car users not using the Internet were not represented in the data set, and the respondents were at least partially self-selected.

6   Conclusions

The study provided information on the relative importance of different barriers perceived by car users in the Helsinki region in regard to purchasing a fully electric vehicle. The three barriers reported by car users that most affected their decision making were the high purchase price of an electric vehicle, limited availability of charging spots elsewhere in Finland outside Helsinki and the Uusimaa region, and insufficient operating range on a single charge. The results indicate that availability of charging spots outside the home region of the car user in the same country may be perceived as an important barrier to purchasing a fully electric vehicle.

Acknowledgements. The study was carried out as part of the CLEMET project (Cleantech Mobility Education for Tomorrow) (CLEMET 2021). The work of VTT in the CLEMET project was supported by the European Social Fund (project number: S21912). The writing process was supported by the AI4DI project (Artificial Intelligence for Digitizing Industry) of the Horizon 2020 research programme (grant agreement ID: 826060). I wish to express my thanks to Ms. Adelaide Lönnberg for editing the English.

References

1. CLEMET. CLEMET – Cleantech Mobility Education for Tomorrow (2021). https://www.cle met.fi. Accessed 14 Sep 2021
2. European Commission. Consolidated text: Regulation (EU) 2019/631 of the European Parliament and of the Council of 17 April 2019 setting CO2 emission performance standards for new passenger cars and for new light commercial vehicles, and repealing Regulations (EC) No 443/2009 and (EU) No 510/2011 (recast) (Text with EEA relevance)Text with EEA relevance (2019). http://data.europa.eu/eli/reg/2019/631/2021-03-01. Accessed 13 Sep 2021
3. Krishna, G.: Understanding and identifying barriers to electric vehicle adoption through thematic analysis. Transportation Research Interdisciplinary Perspectives, Vol. 10 (2021). https://doi.org/10.1016/j.trip.2021.100364. Accessed 14 Sep 2021

4. Lang, T., Malinen, A., Åman Kyyrö, M. ja Öörni, R. (eds.). Kiihdytyskaistalla tulevaisuuden osaamiseen, Reittejä autoalan koulutuksen ja työelämän kehittämiseen. Metropoliaammattikorkeakoulun julkaisuja, TAITO-sarja 84, Helsinki, Finland, ISBN 978-952-328-317-6 (pdf) ISSN 2669–8021 (pdf) (2021). https://urn.fi/URN:ISBN:978-952-328-317-6. Accessed 19 Jan 2022

5. Mattila, J.: Study on Consumer Demand on Electric Cars: The challenges related to battery electric vehicle adoption in Europe and expectations set by the customer base of Škoda Finland. Bachelor Thesis, Seinäjoki University of Applied Sciences (2019). http://urn.fi/URN:NBN:fi:amk-2019082918061. Accessed 14 Sep 2021

6. Ministry of Transport and Communications. Roadmap to fossil-free transport, Government resolution on reducing domestic transport's greenhouse gas emissions. Publications of the Ministry of Transport and Communications 2021:19 (2021). http://urn.fi/URN:ISBN:978-952-243-604-7 . Accessed 13 Sep 2021

7. Rezvani, Z., Jansson, J., Bodin, J.: Advances in consumer electric vehicle adoption research: a review and research agenda. Transp. Res. Part D **34**, 122–136 (2015)

8. Sheskin, D.J.: Handbook of Parametric and Nonparametric Statistical Procedures. 5th Edition. CRC Press, Boca Raton, USA (2011). ISBN 978-1-4398-5801-1

9. Zhuge, C., Shao, C.: Investigating the factors influencing the uptake of electric vehicles in Beijing, China: Statistical and spatial perspectives. J. Clean. Prod. **213**(2019), 199–216 (2019)

Open Access This chapter is licensed under the terms of the Creative Commons Attribution 4.0 International License (http://creativecommons.org/licenses/by/4.0/), which permits use, sharing, adaptation, distribution and reproduction in any medium or format, as long as you give appropriate credit to the original author(s) and the source, provide a link to the Creative Commons license and indicate if changes were made.

The images or other third party material in this chapter are included in the chapter's Creative Commons license, unless indicated otherwise in a credit line to the material. If material is not included in the chapter's Creative Commons license and your intended use is not permitted by statutory regulation or exceeds the permitted use, you will need to obtain permission directly from the copyright holder.

Exploring Perceptions of Comfort, Security and Safety in Different Modes of Transport: A Comparative Study

Marta Campos Ferreira$^{(\boxtimes)}$ ⓘ and Teresa Galvão Dias ⓘ

INESC TEC, Faculdade de Engenharia, Universidade Do Porto, Rua Dr. Roberto Frias, S/N, 4200-465 Porto, Portugal

{mferreira,tgalvao}@fe.up.pt

Abstract. This study seeks to comprehensively analyze the multidimensional determinants underlying perceptions of safety, security, and comfort in transport mode choice, specifically focusing on private transport, public transport and walking. The research begins with an extensive literature review to identify and delve into the factors influencing perceptions of safety, security, and comfort across various transport modes. This inquiry is further enhanced by organizing two focused group sessions. A total of 35 key factors were identified, forming the basis for subsequent investigation. The study then progressed to the development and administration of a survey aimed at capturing responses from a diverse audience, with the goal of exploring the factors influencing perceptions related to different transport modes. A total of 302 responses were collected and meticulously analyzed to discern the factors impacting various relationships and to identify consistent perceptions across diverse transport modes. Additionally, a factor analysis was conducted to validate the findings derived from the data. The outcomes of this research constitute a significant contribution to the existing literature, offering valuable insights that pave the way for a more holistic understanding of the factors guiding transport mode choices.

Keywords: Mode Choice · Preferences · Multidimensional Analysis · Comfort · Security · Safety

1 Introduction

The choices individuals make regarding transport modes are significantly influenced by their personal perceptions of safety, security and comfort [1]. Traffic safety may be determined by different factors, such as weather, road conditions, built environment, driver tiredness, heavy traffic and vehicle speed [2]. Other studies state that pedestrians' injuries are more likely to occur in areas with high traffic volume and population density [3]. So, by reducing the dominance of the car traffic, there would be a reduction in danger and influence the pedestrians' perception of safety. The existence of lighting is also of great importance, with the purpose of making objects and pedestrians close to the road visible to drivers, helping to reduce the mortality of people at night [4].

© The Author(s) 2025
C. McNally et al. (Eds.): TRAconference 2024, LNMOB, pp. 640–646, 2025.
https://doi.org/10.1007/978-3-031-85578-8_86

Regarding security, although the private vehicle is considered the most secure of the transport modes under study, there are still situations perceived as insecure, such as the possibility of carjacking [5]. Additionally, there are studies that show that people opt for parking the car close to their destination, not only because it is comfortable, but also because it is safer, as people often carry valuables with them after parking [5]. In public transport, several studies have confirmed that personal security is the most important factor, having the potential to discourage individuals from using public transport [6]. It is believed that an individual's perception of their security is influenced by social aspects such as age and purpose of travel [6]. Likewise, street life and the perception of the surroundings influence pedestrians' behavior. Not only the presence of other pedestrians, but the lighting, city arrangement and signs of disorder and vandalism also influence the perception of security [1].

Regarding comfort, individuals tend to consider the private car as a more flexible mode of transport, as it allows the user to decide on schedule, route, speed according to their preferences [7]. However, searching for a place to park may bring stress and discomfort to the driver. Regarding public transport, comfort is affected by several factors, such as travel time, frequency, reliability, accessibility, capacity and seat availability [8]. Regarding the perception of comfort felt by pedestrians, the impact of the surroundings and weather conditions are stressed [9]. Surroundings may include the pavement, cleanliness, noise, travel comfort, traffic conditions and the urban environment [10].

While existing literature offers numerous studies exploring these determinants, they often focus on specific modes of transport and consider a limited number of determinants and dimensions. This research aims to provide a comprehensive examination of the multidimensional determinants underlying perceptions of safety, security, and comfort in transport mode choice. By synthesizing and expanding existing knowledge, it seeks to shed light on the factors that shape individuals' preferences for various transport options, more specifically on private transport, public transport and walking.

The remainder of this article is structured as follows: the next section presents the research approach that was followed. Section 3 presents the main results and discussion. Section 4 draws the main conclusions.

2 Methodology

To achieve the intended objective, two different approaches were adopted, one more qualitative and the other quantitative. First, focus group sessions were organized to explore participants' perceptions of safety, security and comfort in relation to the three modes of transport simultaneously. Then, a survey was carried out with the aim of reaching a larger number of people and understanding whether different factors affect different perceptions and whether these vary according to the modes of transport in question. By reaching a larger audience, it is possible to better understand the phenomena and obtain more tangible relationships and results.

Regarding the focus groups, two sessions were organized with three participants each, from Portugal. The decision to hold sessions with few participants was to avoid them becoming too long and confusing due to the topic covered. These focus groups were held online via the Zoom platform and lasted between 60 and 90 min. To guide participants, the focus groups were divided into five categories: Day, People, Surroundings, Time and Traffic. The purpose of this division was to direct participants to specific factors within each category and, in doing so, obtain more relevant factors.

From the literature review and the focus group sessions, 35 factors were collected that can potentially affect perceptions of safety, security and comfort when using private vehicles, public transport and walking, namely:

- **Day:** day period (morning, afternoon and night), and weather.
- **People:** available seats, crowding, presence of driver, presence of people, presence of security agents, relationship with other drivers, and travel with others.
- **Surroundings:** city arrangement, cleanliness of the street or transport, disorder signs, existence of air ventilation, pavement conditions, proper streets, noise pollution, robbery possibility, stations conditions, street lighting, theft possibility, and user-friendly infrastructures.
- **Time:** accessibility/location of the station, control over schedule, control over speed, driver speed, proximity between the station and initial/final destination, speed limit, travel time, transport frequency, and transport reliability.
- **Traffic:** availability/security of parking spaces, control over route, driving likeability, driver tiredness, and existence of traffic.

Then, a questionnaire was applied to a larger sample to understand the impact of each of the 35 factors identified in each dimension studied. The questionnaire was made available online through the Google Forms platform and was disseminated by email to students and teachers at the university campus in Porto, Portugal, and through personal contacts. A total of 302 responses were collected.

The questionnaire was divided into two parts. The first to characterize the sample in sociodemographic terms and mobility habits and the second with statements for respondents to classify according to their degree of agreement, on a Likert scale from 1 to 5, where 1 meant "completely disagree" and 5 "totally agree". These statements were divided according to the categories mentioned above. Within each category, they were further subdivided according to the dimension studied (safety, security and comfort) and the transport modes (private vehicle, public transport and walking).

Analysis of the survey results included not only a descriptive analysis, but also a factor analysis in order to better understand some of the relations between the different perceptions and the modes of transport. For this the SPSS Statistics and the software Monte Carlo PCA were used.

3 Results and Discussion

3.1 Sample Characterization

Regarding the focus groups, a total of 6 people participated, 4 females and 2 males, aged between 22 and 61. Everyone walked, 5 used public transport and 3 private cars.

As for the survey, a total of 302 answers were collected. From this about 70% were females, and 30% males. The majority was aged between 18 and 25 years old (70%), 3% between 26 and 35, 11% between 26 and 54, 10% between 46 and 55, and 6% above 56 years old. As for the educational level, 4% has the middle school, 35% high school,

39% a bachelor degree, 21% a master degree and 1% a PhD degree. Regarding the use of the modes of transport under study, around 65% use a private car, 57% public transport and 62% walk (in this case, respondents could select more than one option).

3.2   Analysis of the Results

The main results of the analysis carried out regarding the responses to the questionnaires are divided according to the three perceptions under study, namely safety, security and comfort, and considering the three modes of transport, namely private vehicle, public transport and walking.

Safety

The perception of safety of private vehicle users is mainly influenced by weather conditions, where rain makes them feel unsafe. Followed by pavement conditions and driver fatigue. Regarding the latter, 78% of private vehicle users agree with the statement regarding avoiding driving when tired. In this relationship, the importance of well-marked traffic signs in the city's layout stands out.

When using public transport, the perception of safety is influenced by driver speed weather and pavement conditions. Therefore, it is possible to conclude that a sunny day, lower speeds, a well-rested driver and morning or afternoon periods are options that make public transport users feel safer. On the other hand, the pavement condition is one of the most selected, but users do not change their behavior or habits because of it.

For pedestrians, weather conditions, public lighting and pavement conditions are the factors that demonstrate percentages above 75%, as an influential factor and in agreement with the statements. Therefore, sunny days, well-lit places and sidewalks in good condition are the safest choices. The layout of the city and the speed limit are the other two factors that affect the perception of safety regarding walking.

Security

As the concept of security is related to the possibility of intentional harmful events occurring, the possibility of theft, daytime and the possibility of robbery are considered the most influential factors, with more than 75%, for users of private vehicles. Therefore, private vehicle users avoid driving in places they know have high rates of robbery or theft and parking in places with a high incidence of theft. However, 79% of private vehicle users completely agree with the statement about not leaving valuables in the car. Furthermore, nighttime is the period perceived as least safe.

Regarding the perception of security when using public transport, the time of day, the presence of people and the presence of security agents are the three main factors considered most relevant (more than 80% each). As with private vehicle users, night is the time of day when public transport users feel least secure. A large majority agrees that the presence of people is more important at night, as is the presence of security agents, so that the trip is perceived as secure.

For pedestrians, the factors with the highest percentages are daytime, presence of people, possibility of theft and travel accompaniment. Pedestrians appear to avoid places with a high incidence of robberies at night and feel safe with company at night.

Comfort

Regarding the factors that influence the perception of comfort when using a private vehicle, it is possible to verify that weather conditions is the factor that private vehicle users most consider to have the greatest influence on comfort. Sunny days are considered more comfortable. Time travel, time of day and schedule control follow.

As for the public transport users, four factors stand out as more influential: cleanliness of the vehicle, crowding, available seats and presence of people. In this relation, it is possible to witness the effect of an association between two factors, where the availability of seats increases its importance for longer trips.

For pedestrians, comfort seems to be mainly related to weather conditions, time of day and travel time. Therefore, on foot it is comfortable to have a sunny day and avoid traveling at night. Short trips are preferable, however, most pedestrians admit that the feeling of comfort is more important for longer trips, what is expected is that, in case of discomfort, the user will have to deal with it for longer.

3.3 Policy Recommendations

From the analysis carried out it is possible to some draw policy recommendations that promote a safer, more secure, and comfortable transportation environment for all stakeholders involved. For instance, public transport operators and municipal authorities may implement regular cleaning schedules for public transport vehicles, optimize routes to reduce overcrowding, and invest in technology for real-time tracking of vehicle occupancy to manage capacity effectively. Authorities may also launch campaigns to educate private vehicle users about the risks associated with driving in unsafe conditions, encourage adherence to speed limits, and promote strategies to prevent driving while fatigued. Finally, transport planners and policy makers may incorporate perceptions of safety, security, and comfort into transport planning processes through stakeholder consultations, public surveys, and regular evaluations of existing infrastructure to identify areas for improvement.

4   Conclusion

This study provides a comprehensive examination of the multidimensional determinants underlying perceptions of safety, security and comfort in transport mode choice, namely private transport, public transport and walking. It begins with a comprehensive review of the literature that is complemented with the organization of two focus group sessions. As this article addresses three different modes of transport and three perceptions simultaneously, something that has never been done before, the focus group helped to explore the three perceptions simultaneously and draw important conclusions. In total, 35 final factors were selected, which were the subject of further study. The third stage of the study involved the development and implementation of a survey aimed at obtaining responses from a larger audience to explore the factors influencing perceptions and transport modes. A total of 302 responses were collected and then analysed to identify factors influencing various relationships and to identify consistencies in perceptions across different transport modes. Additionally, a factor analysis was conducted to support the conclusions drawn from the data.

Concerning security perception, the night period was perceived as more insecure across transport modes. Consequently, other factors such as the presence of people and security agents gained importance, particularly for public transport. In terms of safety, meteorological conditions, particularly rain, were identified as factors influencing transport modes. Good pavement conditions, proper signage, and low-speed limits were perceived as contributing to safety. Interpreting the perception of comfort proved more challenging, as it yielded higher factor values, especially for private vehicles and public transport. For walking, discomfort was associated with rain and the night period. At night, individuals felt more comfortable when accompanied by others.

The results of this work are an important contribution to the literature and pave the way for a more holistic understanding of the factors that drive transport mode choice. However, it is recognized that there is a need for future work, namely the application of the study to other contexts and broader samples.

References

1. Campos Ferreira, M., et al.: Identifying the determinants and understanding their effect on the perception of safety, security, and comfort by pedestrians and cyclists: a systematic review. Transp. Res. Part F Traffic Psychol. Behav. **91**, 136–163 (2022)
2. Bellizzi, M.G., Eboli, L., Forciniti, C.: Segregation vs interaction in the walkways: an analysis of pedestrians' perceptions. Res. Transp. Bus. Manag. **33** (2019)
3. Ferrer, S., Ruiz, T., Mars, L.: A qualitative study on the role of the built environment for short walking trips. Transp. Res. Part F Traffic Psychol. Behav. **33**, 141–160 (2015)
4. Adkins, A., Barillas-Longoria, G., Martinez, D.N., Ingram, M.: Differences in social and physical dimensions of perceived walkability in Mexican American and non-hispanic white walking environments in Tucson, Arizona. J. Transp. Heal. **14** (2019)
5. Dobbs, L.: Wedded to the car: women, employment and the importance of private transport. Transp. Policy **12**, 266–278 (2005)
6. Friman, M., Lättman, K., Olsson, L.E.: Public transport quality, safety, and perceived accessibility. Sustain. **12** (2020)
7. Arto, O.: Salonen: Passenger's subjective traffic safety, in-vehicle security and emergency management in the driverless shuttle bus in Finland. Transp. Policy **61**, 106–110 (2018)
8. Barone, V., Festa, D.C., Edvige Mongelli, D.W., Tassitani, A.: Comfort Index CI(bus): a methodology to measure the comfort on board. Procedia Comput. Sci. **134**, 439–444 (2018)
9. Landa-Blanco, M., Avila, J.: Factors related to the use of pedestrian bridges in university students of Honduras. Transp. Res. PART F-TRAFFIC Psychol. Behav. **71**, 220–228 (2020)
10. Jamal, S., Mohiuddin, H., Paez, A.: How do the perceptions of neighborhood conditions impact active transportation? A study in Rajshahi, Bangladesh. Transp. Res. Part D Transp. Environ. **87**, 102525 (2020)

Open Access This chapter is licensed under the terms of the Creative Commons Attribution 4.0 International License (http://creativecommons.org/licenses/by/4.0/), which permits use, sharing, adaptation, distribution and reproduction in any medium or format, as long as you give appropriate credit to the original author(s) and the source, provide a link to the Creative Commons license and indicate if changes were made.

The images or other third party material in this chapter are included in the chapter's Creative Commons license, unless indicated otherwise in a credit line to the material. If material is not included in the chapter's Creative Commons license and your intended use is not permitted by statutory regulation or exceeds the permitted use, you will need to obtain permission directly from the copyright holder.

Economic Assessment of Free Public Transport in Athens

Chryssa Vagdatli, Virginia Petraki, Julia Roussou[✉], and George Yannis

Department of Transportation Planning and Engineering,
National Technical University of Athens, Athens, Greece
jroussou@mail.ntua.gr

Abstract. The introduction of Free Public Transport (FPT) is in many cities a potential path to control and reduce the environmental, social and economic problems. FPT can take several forms from the more widely used partial FPT, which includes limitations in its application, to full FPT. This paper aims to investigate the socio-economic feasibility of introducing FPT in Athens. For this purpose, a socio-economic analysis was conducted to assess the economic advantages and disadvantages of the FPT, up to the year 2030. Four Scenarios were examined considering 0% (S0), 50% (S1), 75% (S2) and 100% (S3) fare discount, respectively, on PT tickets. Consequently, a multinomial logistic model was developed to investigate for each of the three discounts the level preferring Athenians the FPT over private car, using data from a stated preference questionnaire survey. For S1-S3 the investment and the operational costs along with the impact on travel time, fuel consumption, road safety and air pollution were estimated and monetized. The socio-economic analysis illustrated that the introduction of FPT in Athens can contribute to social welfare in the medium-term future, mainly due to the modal shift from private cars to PT. More precisely, the examined scheme shows a positive NPV and high IRR in all the Scenarios, indicating its feasibility over time. To be noted that even in extreme price changes of significant input variables, NPV remains positive, ensuring a positive impact on society.

Keywords: Free Public Transport · Socio-economic Analysis · Logistic Model · Net Present Value · Sensitivity

1 Introduction

The introduction of Free Public Transport (FTP) is being explored by many cities as a potential path for the control and reduction of environmental, social and economic problems. FPT can take several forms, from partial FPT, which includes limitations in its application but more widely used, to full FPT. Currently, more than 100 cities around the world offer a form of FPT to their citizens, such as Taihung in Taiwan, Miami in USA and Verenje in Slovenia [16]. The three main benefits of such schemes are to encourage a modal shift from private cars to Public Transport (PT), improve social inclusion, and enhance the urban and economic development of cities. However, a detailed economic assessment has to be made prior to the implementation of FPT schemes to determine

© The Author(s) 2025
C. McNally et al. (Eds.): TRAconference 2024, LNMOB, pp. 647–653, 2025.
https://doi.org/10.1007/978-3-031-85578-8_87

whether they are worth undertaking or not from the social welfare point of view. While FPT is expected to boost ridership [19], it also raises costs for maintenance and financial strain on PT organizations, making it potentially unsustainable.

The most important impact of FPT on society is the mitigation of social inequalities. The main value of reducing or completely abolishing ticket fares lies in establishing simplified use of PT [9], since with FPT everyone can be transported, whenever they want [2]. Abolishing fares has been praised for directly addressing the issue of social exclusion, inequality, and poverty in transport by increasing its accessibility for lower income residents [12].

In this context, the objective of this paper is to investigate the socio-economic feasibility of introducing FPT in Athens. For this purpose, a socio-economic analysis was conducted to assess the economic advantages and disadvantages of the FPT, up to the year 2030. In the framework of the socio-economic analysis, four Scenarios were examined considering 0% (S0), 50% (S1), 75% (S2) and 100% (S3) fare discount, respectively, on PT tickets.

2   Methodology

2.1   Socioeconomic Analysis

Cost Benefit Analysis (CBA) is an analytical tool used to facilitate a more efficient allocation of resources by demonstrating the benefits to society of an intervention [18]. In the framework of the CBA, the socio-economic analysis is developed to evaluate the social value of a project by quantifying the social impacts of the project, with the costs and benefits being comparable in monetary terms considering distortions and limitations in markets [13]. The costs and benefits of the investment scenarios are compared to the do-nothing scenario to determine if the project is worth undertaking from a societal welfare perspective [17]. The costs and benefits arising at different times should be discounted using the Social Discount Rate (SDR).

2.2   Data

The main source of data, for the present study, was the annual report of the activities of the Athens Urban Transport Organization (OASA) [15], which is responsible of the operation and maintenance of all Athens PT, except for the Suburban Railway. These reports include the financial data of the organization, as well as the size of the fleet in use per year for each means of PT, the vehicle-kilometers, the trips, etc. To obtain the necessary traffic data, the OASA traffic simulation model and its results for the year 2018 were used. It is worth mentioning, that the assumption was made that the figures for 2018 are similar to the year 2023, since traffic data of the years 2020–2022 are not considered representative, due to Covid-19 virus. In addition, an annual increase of 1.2% in car and bus trips was considered, due to population growth and increasingly intense urbanization.

2.3 FPT Acceptance Survey

To estimate the modal shift to FPT, data from a stated preference questionnaire survey, were used to investigate the acceptance of commuters in Attica towards FPT [7]. To collect the required data, the method of stated preference in hypothetical scenarios of cost, time and comfort route, were included in a specially designed questionnaire with 234 participants. A multinomial logistic regression statistical model was developed to calculate the probability of shifting from private cars to FPT. Time, cost and convenience were the independent variables. The outcomes of the statistical model showed that the faster and more comfortable the commute, regardless of cost reduction, the more likely respondents are to choose FPT over remaining in their current mode of transportation. For Scenario 1 (50% fare reduction), modal shift from private cars to FPT is approximately 40.1%, for Scenario 2 (72.5% fare reduction) 43.8%, while for Scenario 3 (100% fare reduction), the modal shift percentage reaches 48.5%.

3 Socio-economic Analysis

The socio-economic analysis carried out for the introduction and operation of FPT in Athens is presented for a time horizon until 2030. The analysis was based on finding the costs and benefits that will result from the implementation of FPT in Athens in 4 Scenarios; i.e. 0% (S0), 50% (S1), 75% (S2) and 100% (S3) fare discount, respectively, on PT tickets. It should be noted that scenario S0 represents the do-nothing situation where the ticket fare remains unchanged. For each Scenario the investment and operating costs, and the socio-economic benefits have been calculated in monetary terms.

For the calculation of the impact of FPT on travel time, the annual passenger-hours spent on private cars and PT were taken into account in all the scenarios, as well as the value of the travel time (VOT). In Greece, VOT for work-related car travel is €9.00, for other purposes, it's €4.10 [8]. With 56% other and 44% work trips [5], the average VOT is €6.26/hour. The ratio of this price to the corresponding cost for PT is estimated to be approximately equal to 1.2 [6], i.e. €5.21.

To assess the impact on fuel consumption, the composition of the vehicles by fuel type, the average consumption of vehicles, as well as the price of fuels per year, were identified [20]. According to OASA, 10% of city buses run on compressed natural gas while the rest are diesel-powered, and 3.2% of buses are electric [15]. For passenger cars, the assumption that the entire fleet is gasoline-powered was made, even though approximately 8% use diesel and 1% of these are electric/hybrid in Greece [1]. Every year cars are becoming eco-friendlier and more economical, so a reduction of the average fuel consumption every year, was also considered [20]. On the contrary, the fuel consumption of PT was considered constant until 2030.

To assess the impact of FPT on road safety, the number of road fatalities and injuries in each Scenario, as well as the social cost per fatality and injury were considered. The latest available data from the Hellenic Statistical Authority (EL.STAT) [3] were used to record the road safety indicators. In particular, the minor and serious injuries and deaths in crashes involving private vehicles and PT were collected for the year 2019 as recorded in the Municipality of Athens. To calculate the cost of road safety, the social cost per death (€2,148,034), serious (€273,574) and minor injury (€51,372) in a road crash were used, as calculated for Greece [11]. Finally, it is assumed that road safety improves annually by 2.5%, a percentage that has been derived from the evaluation of road safety data per year for the countries of the European Union.

For the environmental impact assessment, a similar process was followed to calculate the costs of emissions. The cost per ton of CO_2 emissions is 42€/tn in the year 2021 with an annual increase of approximately 2.3%. For nitrogen compounds (NO_x) a fixed cost of 1,900€/tn per vehicle km is considered. The emissions per vehicle type are calculated based on literature for both private cars [10] and PT [14].

The investment cost of each scenario S1-S3 was formed considering two components, the cost of the required study and the purchase of new or used buses. It is worth noting that these two cost categories refer to year 0 (2023), i.e. before the operation of FPT in Athens begins. The cost of the study was considered in all scenarios to be equal to €500,000, which is disbursed once in 2023. The investment of the new bus purchase was found after identifying the number of buses that will be needed to accommodate the new needs for PT travel and achieved by correlating PT trips with PT vehicles [15].

The operating costs of FPT include the operating and maintenance costs of the system, mechanical equipment, as well as costs related to the additional human resources and fuel consumption of the new buses. To find the cost of employing additional human resources, the relationship between the new buses and the required new driver shifts was calculated. For each bus that operates every day, 2.4 drivers [15] were calculated.

4   Summary

4.1   Evaluation of Economic Feasibility

Specific criteria have been used to identify whether a scenario is beneficial [4]. The Net Present Value (NPV) of the investment must be positive, i.e. NPV > 0, the Internal Rate of Return (IRR) should be greater than the SDR, the ratio of benefits to costs should be greater than unity, B/C > 1. Table 1 summarizes the costs and benefits, as well as the economic performance until the year 2030 for Scenario 1.

Table 1. Socio-economic analysis for Scenario 1

Costs and Benefits	Present Value (0.8%)	Implementation 2023	Operation 2024	2025	2030
C1 Investment Cost (mil.€)	−11.43	−11.52	0	0	0
C2 Operating Costs (mil.€)	−83.13	0	−19.48	−18.08	−7.09
Total Costs (mil.€)	**−94.56**	**−11.52**	**−19.48**	**−18.08**	**−7.09**

(continued)

Costs and Benefits	Present Value (0.8%)	Implementation 2023	Operation 2024	2025	2030
B1 Travel time (mil.€)	−237.31	0	−53.35	−54.44	−18.08
B2 Fuel consumption (mil.€)	655.30	0	178.58	164.71	38.28
B3 Road Safety (mil.€)	82.37	0	22.17	21.62	4.49
B4 Emissions (mil.€)	110.11	0	26.29	26.27	7.36
Total Benefits (mil.€)	**610.47**	**0**	**173.69**	**158.16**	**32.05**
NPV (0.8%)	**515.92**	−11.52	154.21	140.08	24.96
IRR	**1,327.6%**				

It was observed that the implementation of FPT in Athens has a significant positive influence on road safety, the environment and society in general. In particular, the considered system presents a positive NPV in all scenarios, from 515 million to 592 million €, IRR > 100% and B/C from 4.5 to 6.4, indicating its economic viability over time. The summarized results of CBA analysis for each scenario are presented in Table 2.

Table 2. Summary results of CBA analysis by scenario.

Scenario	NPV (€)	IRR (%)	B/C
S1 50% fare reduction	515,919,563	>100%	6.46
S2 72.5% fare reduction	540,500,076	>100%	5.24
S3 100% fare reduction	592,985,724	>100%	4.52

5 Discussion

In conclusion, with the introduction and operation of partially and fully FPT in the urban network of Athens, a significant improvement in road safety, the environment and in general the quality of life is expected. Specifically, by the year 2030, a major reduction in road fatalities (by 20 for S1 - 25 for S3), severe (by 38 for S1 - 46 S3) and light injury (by 622 for S1 -733 for S3), as well as improvement in fuel consumption (reduction by 1.3 (S1) − 1.4 (S3) billion liters of gasoline-equivalent fuel), and air pollution (reduction of CO_2 pollutants by 2.5 (S1) − 3.1 (S3) mil. tons) is expected. The suggested intervention in the network of Athens is a socio-economically sustainable investment.

To ensure the successful introduction and operation of FPT in Athens, a pilot operation is proposed as trailhead. This will allow the effectiveness and efficiency of the system to be fully assessed, before being implemented across the whole city. Furthermore, it is proposed to upgrade the existing infrastructure and PT fleet, as well as the level of service of PT, by increasing itineraries, purchasing new environmentally friendly buses and other similar actions to attract more commuters to PT and ensure a smoother transition to the FPT.

Acknowledgements. The present research was carried out within the research project MetaC-Caze – Flexibly adapted Meta Innovations, use cases, collaborative business and governance models to accelerate deployment of smart and shared Zero Emission mobility for passengers and freight", which has received funding from the European Union's Horizon Europe research and innovation programme under grant agreement No 101139678.

References

1. ACEA: Passenger car fleet by fuel type, European Union. https://www.acea.auto/figure/passenger-car-fleet-by-fuel-type/ (2019)
2. Cordier, B.: La gratuité totale des transports collectifs urbains : effets sur la fréquentation et intérêts. Bureau d'études en transports en déplacements. (2007)
3. ELSTAT.: Press Release: Road Traffic Crashes: Year 2020. Hellenic Statistical Authority. https://www.statistics.gr/documents/20181/17776954/NWS_SDT04DTAN2020_30032022_GR.pdf/f9fada24-6252-b269-f95c-4b09213b2bda?t=1648627186870 (2022)
4. Commission, E.: Guide to Cost-Benefit Analysis of Investment Projects. European Commission, Brussels (2014)
5. Eurostat. (2021). Distribution of distance travelled per person per day by travel purpose for urban mobility on all days. Eurostat
6. Fosgerau, M., Hjorth, K., Lyk-Jensen, S.V.: Between-mode-differences in the value of travel time: Self-selection or strategic behaviour? Transp. Res. Part D: Transp. Environ. **15**(7), 370–381 (2010)
7. Goulas, E., Kontaxi, A., Yannis, G.: Free PT in Athens: a stated preference approach. In: Proceedings of the Transport Research Arena TRA 2022 Conference, Lisbon, 14–17 November (2022)
8. Handbook of external costs of transport. Brussels: European Commission (2019)
9. Hodge, D.C., Orrell, J.D., Strauss, T.R.: Free fare policy: costs, impacts on transit service and attainment of transit system goals. Washington State Transportation Center (1994)
10. Holland, M., Pye, S., Watkiss, P., Droste-Franke, B.: Damages per tonne emission of PM2.5, NH3, SO2, NOx and VOCs from each EU25 Member State (excluding Cyprus) and surrounding seas. European Commission DG Environment. https://ec.europa.eu/environment/archives/cafe/activities/pdf/cafe_cba_externalities.pdf (2005)
11. ITF: Covid-19 and Transport: A Compendium. OECD Publishing, Paris (2021)
12. Larrabure, M.: The struggle for the new commons in the Brazilian free transit movement. Studies in Political Economy: A Socialist Review, pp. 175–194. https://doi.org/10.1080/07078552.2016.1211135 (2016)
13. Mouter, N., Dean, M., Koopmans, C., Vassallo, J. M.: Comparing cost-benefit analysis and multi-criteria analysis. In: Advances in Transport Policy and Planning (Vol. 6, pp. 225–254). Academic Press (2020). Author, F.: Article title. Journal 2(5), pp. 99–110 (2016)

14. Nanaki, E.A., Koroneos, C.J., Xydis, G.A., Rovas, D.: Comparative environmental assessment of Athens urban buses—Diesel, CNG and biofuel powered. Transport Policy, pp. 311–318 (2014)
15. OASA: Activity Report 2018. Athens (2019)
16. Rapid Transition Alliance.: Rapid Transition Alliance. https://rapidtransition.org/stories/free-public-transport-the-new-global-initiative-clearing-the-air-roads-and-helping-keep-climate-targets-on-track/ (2021)
17. Sartori, D., Catalano, G., Genco, M., Pancotti, C., Sirtori, E., Vignetti, S., Bo, C.: Guide to CBA of investment projects. Economic appraisal tool for cohesion policy 2014–2020 (2014)
18. Thiedig, J.: An economic cost-benefit analysis of a general speed limit on German highways (2018)
19. Volinski, J.: Implementation and outcomes of fare-free transit systems (No. 101). Transportation Research Board (2012)
20. Yang, Z., Bandivadekar, A.: Light-duty vehicle greenhouse gas and fuel economy standards. ICCT report, p.16 (2017)

Open Access This chapter is licensed under the terms of the Creative Commons Attribution 4.0 International License (http://creativecommons.org/licenses/by/4.0/), which permits use, sharing, adaptation, distribution and reproduction in any medium or format, as long as you give appropriate credit to the original author(s) and the source, provide a link to the Creative Commons license and indicate if changes were made.

The images or other third party material in this chapter are included in the chapter's Creative Commons license, unless indicated otherwise in a credit line to the material. If material is not included in the chapter's Creative Commons license and your intended use is not permitted by statutory regulation or exceeds the permitted use, you will need to obtain permission directly from the copyright holder.

Comparing Access to Services by Public and Private Transport in Ireland to Measure Transport Poverty

Alessandro Giordano[✉][iD], Javier Aycart[iD], and Juan Nicolás Ibáñez[iD]

European Commission, Joint Research Centre, C/Inca Garcilaso 3,
41092 Seville, Spain
alessandro.giordano@ec.europa.eu

Abstract. Transport plays a vital role in social inclusion, enabling people to access essential services and opportunities. Thus, while striving to reduce carbon emissions in the transport sector, the European Union (EU) must balance its environmental targets with "social sustainability" goals and address related challenges. The EU has then proposed establishing a Social Climate Fund to support households at risk of transport poverty and reduce social exclusion risks. However, measures that can help tackle this issue are either poorly defined or depending solely on survey data. This paper aims to provide analytical evidence of transport poverty in the Republic of Ireland, using high-resolution public and private transport networks and identifying where and why people might be at risk of experiencing transport poverty. The contribution to the policy discussion focuses on transport availability as the basis for the analysis. Preliminary results suggest significant differences when comparing access to opportunities, or services, by car or by public transport. With regard to the latter, the number of individuals below a predefined accessibility "sufficiency threshold" varies significantly between rural areas, or near the country border areas, and urban settings. Future work will include a broader variety of destinations and transport poverty dimensions, such as affordability and digital connectivity, population dynamics scenarios, travel demand indicators, and socioeconomic metrics.

Keywords: transport poverty · accessibility · connectivity · transport networks

1   Introduction

The need for a fair transition towards climate neutrality is an integral part of the policy priorities set by the European Commission (EC) in the European Green Deal. Efforts to reduce emissions from passenger mobility can provide a major contribution to achieve these objectives, but because transport plays a significant role in social inclusion, enabling people access to essential services (e.g., education or healthcare) and opportunities, the socioeconomic dimension of the issue cannot be overlooked. To ensure that "no one is left behind", the EU has proposed to establish a Social Climate Fund to support, among others,

© The Author(s) 2025
C. McNally et al. (Eds.): TRAconference 2024, LNMOB, pp. 654–661, 2025.
https://doi.org/10.1007/978-3-031-85578-8_88

households at risk of transport poverty. The design, implementation, monitoring and evaluation of this measures need to be supported by the development of science-based indicators that will help define and quantify transport poverty dimensions and social exclusion risks. However, thus far, these kind of indicators are either vaguely defined or limited in their application scope due to their dependence on location-specific survey data.

This paper aims at producing analytical evidence using both public and private transport high-resolution networks, to identify where and why people might be at risk of experiencing transport poverty. The methodology is applied to the case study of Ireland (IE), and focuses on the "transport availability" dimension of the problem. The assessment is carried out by comparing access to services and to opportunities using either public or private transport, hence measuring people ability to satisfy their accessibility needs. For private transport, we include time-dependent congestion by using average traffic-speeds of road segments on specific time windows. For public transport, we model features like people tolerance for number of transfers or walking distances between stops. Routing tools, able to solve large earliest-arrival time problems with many origins and destinations, will be employed on detailed network representations. The origins are defined as the centroids of populated grid cells in IE, while the destinations are either populated grid cell centroids (as a proxy for opportunities) in IE and Northern Ireland (NI), primary school students, or Public Employment Service facilities (PES). Service data was obtained from institutional sources, following [1,2] and information provided by the Geographic Information System of the Commission (GISCO)[1].

1.1 The Case Study

In recent decades, Ireland has witnessed a rapid depopulation of its rural regions [3]. Responding to this trend, the Irish National Transport Authority has initiated the "Local Link" programme to address the accessibility needs of individuals facing social exclusion in rural settings. Furthermore, it has established 17 "transport coordination units", tasked with identifying the demand for local transport services in rural areas [4]. Thanks to political interest and availability of transport and socioeconomic data at a detailed spatial level, an increasing number of studies have emerged to address transport accessibility concerns.

Carroll et al., (2021) [3] introduce a methodology to identify hotspots of transport disadvantage, and potential forced car ownership risk [5], in IE. The authors achieve this focusing on public transport accessibility employing both statistical approaches and measures of proximity to public transport stops. However, the study does not directly address any accessibility needs of individuals and solely relies on "proximity" to public transport, not considering the quality

---

[1] Sources: www.ec.europa.eu/eurostat/web/gisco (for populated grid cells in 2021 and in 2018); https://ec.europa.eu/eurostat/web/gisco/geodata/basic-services (for basic services in the EU); www.gov.ie/en/collection/primary-schools (primary schools); www.gov.ie (intreo offices, PES).

of service provided. Another highly pertinent study, Kelly et al., (2023) [6], tries to combine spatially-detailed socioeconomic data with accessibility measures, based also on proximity, to build a composite Transport Poverty Risk Index. Expanding on the existing transport poverty literature, and with the aim of enhancing the analytical ground of accessibility analyses in the EU, this study investigates access to services and opportunities.

2 Methods

The methodological section of this paper is divided into two parts, corresponding to the models used in the accessibility analysis: the driving model operating on the road network, and the public transport model running on a multimodal route-based network (including trains, buses and trams) integrating short walks.

Road Network Model: The network includes both IE and NI regions extracted from the commercial mapping dataset TomTom MultiNet (MN) 2021. We include road categories from NET2CLASS[2] $= 0$ (roads of international importance) to 4 (local roads), while filtering out the lower-importance roads, and then run the model for two scenarios: free flow driving conditions and congested traffic conditions. For each segment of the base network, we know the driving time at free flow speed. By using the MN speed profiles representing different degrees of congestion on each segment, we then calculate an increasing factor that represents the additional driving time under congestion conditions. For this analysis, the increasing factor for each segment corresponds to the average congestion factor between 7:00 am and 10:00 am, from Monday to Friday. In those cases where the congestion metrics were not available for specific segments, the congestion factor was set equal to the average congestion factor of the remaining base network segments where data was available.

To represent the last-kilometer routing and overcome the limitations due to the filtering of the minor roads, each origin and destination point is linked to the five nearest junctions in the base network. The length of these artificial edges is equal to the straight-line distance times a circuity factor (i.e., 1.16) that represents the average extra effort of driving through a network compared to a straight-line trip. This 1.16 circuity factor has been pre-computed by taking a random sample of origins and destination junctions in the full MN network for Ireland (all NET2CLASS categories, from major highways to the smallest roads) and calculating the average circuity factor for all the paths between the points in the sample. The factor is hence the ratio between sampled shortest-path distances and their correspondent straight-line distances.

For both scenarios, we use a Dijkstra-based algorithm to solve earliest-arrival time problems from each origin to each destination. We perform these calculation by using the driving time as the cost unit of the problem, so the paths

---

[2] NET2CLASS is the database representation of a classification of roads (and ferries) based on the importance that the constituting roads (and ferries) have in the entire network.

that are found by the algorithm are the fastest under the different traffic conditions (free flow or congestion). To reduce the computational effort, we have only calculated the paths between the origin and destination points that were within a maximum range of 200 km in straight-line distance (i.e., using the projected coordinate system EPSG:3035 for the implementation of the buffering). This figure of 200 km radius from the origin point is the equivalent of driving for two hours in any direction at a constant speed of 100 km/h on a fictitious straight-line road from the origin. The number of total paths (rows in the table) is then equal to the sum of the number of reachable destinations within the 200 km radius range (dwr_i) for each of the n origins i.

Public Transport Model: We use the multimodal networks built from the static public transport timetable schedules provided by the Irish National Access Point[3] and by Translink[4]. in NI, in General Transit Feed Specification (GTFS) format. The optimisation problem is "time-dependent" as public transport services run on specific timetables. To solve this challenge, we use a newly developed schedule-based routing solver (VelociRAPTOR) [7] based on the RAPTOR algorithm, which is a well-known approach for searching for optimal routing options on schedule-based systems [8]. The former extends the latter by focusing on the specific all-pair problem (rather than on one-to-one and one-to-many problems) on large instances. Similarly to the road model, this model follows an earliest-arrival time optimisation. However, in this case the paths can combine any available public transport services and short walks between stations, or origins/destinations and stations, modelled as euclidean distances.

To run the model, we specify routing criteria, such as a maximum walking distance of 1 km between transfers and of 2 km when leaving the origin or arriving at destination, and a walking speed of 5 km/h. Moreover, we define a specific morning time window (7:30 am to 9:30 am), in which it is feasible to board (and remain) on a public transport mode, and set a specific date in the data: which for this case study is Thursday, April 20th, 2023. Selecting a specific date is an essential part of the definition of the network. We aimed to select a weekday that was neither a holiday nor a date with an unusually high density of services relative to the period covered by the data. Finally, we ran scenarios with varying constraints on the number of transfers, ranging from 0 (i.e., no transfers allowed) to 4 (meaning unrestricted number of transfers).

Travel Time and Sufficiency Thresholds: In the next section, we illustrate the results of the simulated scenarios in two different ways. In Fig. 1, we perform a comparison between private and public transport showing results on the map, at a very detailed spatial level, and for a threshold of 40-50 min of travel time, consistent with the considerations made in [9] and based on effects on

---

[3] Irish National Access Point (NAP) https://data.gov.ie.
[4] https://www.translink.co.uk   (accessed   via   https://gtfs.pro/en/uk/Datafeeds/ northern-ireland).

social exclusions of not meeting accessibility needs. In Fig. 2, we want instead to measure transport poverty risk in ten different regions, divided by type of urbanisation, by applying the Foster-Greer-Thorbecke (FGT_2) measure. This measure ranges from 0 to 1, where 0 indicates that every individual is above the poverty line, and it is discussed in detail in [10]. FGT_2 enables a simultaneous assessment of both the magnitude and intensity of poverty. It achieves this by computing the count of individuals below the poverty line, weighted by the extent of the accessibility deficit in relation to the poverty line. To compute this measure, we then need to set sufficiency thresholds (i.e., poverty lines) for the destinations, which should be politically defined by local authorities [11]. We chose to quantify them based on the count of accessible individuals/students/PES, resulting from the population-weighted 50%ile of nationwide public transport accessibility

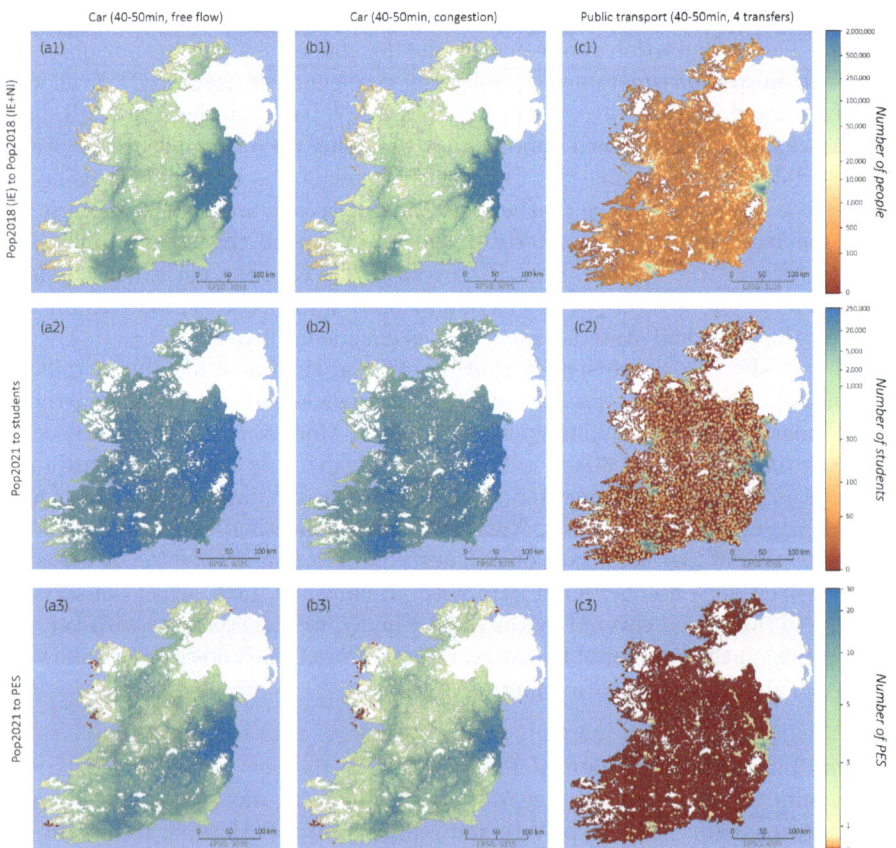

Fig. 1. Comparison of access to people/services, within a 40–50 min travel time threshold: by car in (a) free flow traffic or (b) congested traffic scenarios, and (c) by public transport with unrestricted number of transfers allowed. Own calculations using MATLAB and VelociRAPTOR [7] routing engines for car and public transport cases, respectively, and QGIS to display results.

within 40-50 min, allowing an unrestricted number of transfers. Consequently, we obtained values of 16,520 individuals, 2,500 students, and 1 PES, respectively.

3   Results and Conclusions

Preliminary results for IE suggest differences between access to services and opportunities with public and private transport across the country, with different patterns for people living in very sparsely populated and rural areas, or close to the country border, compared to those living in urban or metropolitan settlements. Differences between car accessibility and public transport accessibility can be readily observed in Fig. 1. The maps also make it possible to distinguish variations in car access under free flow conditions versus congested scenarios.

Figure 2 then illustrates results by region and type of area. For public transport, we provide three scenarios: besides the middle point of 40–50 min travel time threshold with 4 transfers, we provide an upper bound with 40-50 min travel time and 0 transfers, and a lower bound of 55–65 min travel time and 4 transfers. The differences between access by cars, or by public transport in urban environments, and access by public transport in rural environments are dramatic and vary according to regions and scenarios considered.

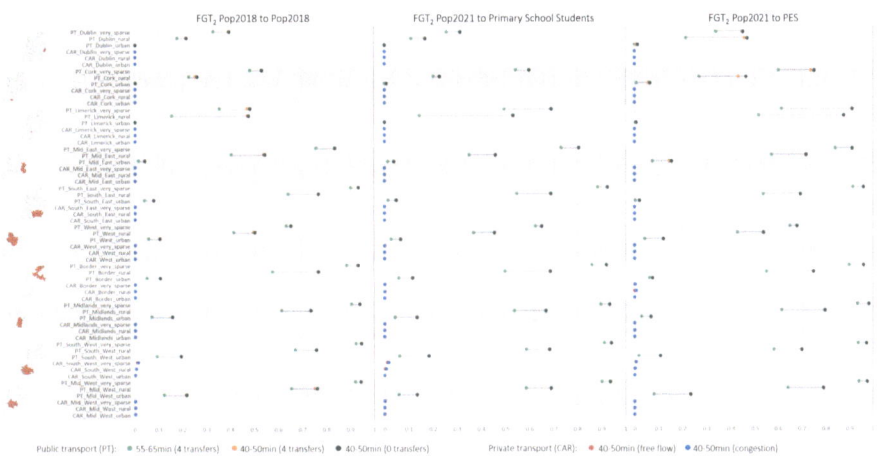

Fig. 2. FGT_2 results for access to people, students and PES in ten regions/ metropolitan areas in Ireland. Results are broken down by settlement type: very sparse, rural, urban.

An important caveat of the results is that, when looking at access to students, school buses have not been included in the Irish NAP datasets yet, and therefore they are not in the analysis. In future work, the accessibility analyses could include these additional transport services along with a broader set of relevant destinations. Further research avenues include investigating population dynamics scenarios, further characterising origins in terms of socioeconomic conditions of the resident population and digital connectivity, and including additional transport poverty dimensions, such as "transport affordability". Finally, these spatial analyses can contribute to the development of composite indicators for transport poverty by complementing and enhancing existing studies and initiatives in this area, such as the one detailed in [6] for Ireland.

References

1. European Commission: Employment and Social Developments in Europe - Addressing labour shortages and skills gaps in the EU (Chapter 3) (2023)
2. Almeida, V., et al.: Geographic inequalities in accessibility of essential services. OECD Social, Employment and Migration Working Papers, 307, OECD Publishing, Paris (2024). https://doi.org/10.1787/12bab9fb-en
3. Carroll, P., Benevenuto, R., Caulfield, B.: Identifying hotspots of transport disadvantage and car dependency in rural Ireland. Transp. Pol. **101**, 46–56 (2021). https://doi.org/10.1016/j.tranpol.2020.11.004
4. Transport For Ireland: Local Link rural transport programme strategic plan 2018 to 2022 (2018)
5. Mattioli, G.: Forced car ownership in the UK and Germany: socio-spatial patterns and potential economic stress Impacts. Env. and Sust. Indicators **5**(4), 147–160 (2017). https://doi.org/10.17645/si.v5i4.1081
6. Kelly, J.A., Kelleher, L., Guo, Y., Deegan, C., Patil, A.: Transport poverty risk - a composite spatial index to support policy design and investment targeting as part of a just climate transition. Env. Sust. Indic. **18**, 100254 (2023). https://doi.org/10.1016/j.indic.2023.100254
7. Tomasi, C., Gualandi, S., Ibáñez, J.N., Duma, D., Fischetti, M., Giordano, A.: VelociRAPTOR: a new algorithm for multimodal all-pairs time-dependent shortest paths at European scale (forthcoming)
8. Delling, D., Pajor, T., Werneck, R.F.: Round-based public transit routing. Transp. Sci. **49**(3), 591–604 (2015). https://doi.org/10.1287/trsc.2014.0534
9. Social Exclusion Unit: Making the connections: final report on transport and social exclusion (2003)
10. Karner, A., Pereira, R.H.M., Farber, S.: An index to measure accessibility poverty risk. Transportation (2024). https://doi.org/10.1007/s11116-023-10460-7
11. Martens, K., Bastiaanssen, J.: Advances and pitfalls in measuring transportation equity. Measuring Transport Equity (Chapter 3), 39–55 (2019). https://doi.org/10.1016/B978-0-12-814818-1.00003-2

Open Access This chapter is licensed under the terms of the Creative Commons Attribution 4.0 International License (http://creativecommons.org/licenses/by/4.0/), which permits use, sharing, adaptation, distribution and reproduction in any medium or format, as long as you give appropriate credit to the original author(s) and the source, provide a link to the Creative Commons license and indicate if changes were made.

The images or other third party material in this chapter are included in the chapter's Creative Commons license, unless indicated otherwise in a credit line to the material. If material is not included in the chapter's Creative Commons license and your intended use is not permitted by statutory regulation or exceeds the permitted use, you will need to obtain permission directly from the copyright holder.

ISO Standards for Urban Planning and Transport in Future Sustainable Medium-Size Cities

Anabela Salgueiro Narciso Ribeiro$^{(\boxtimes)}$ ⬥ and Alexandre Cerejeira

Polo II, Faculty of Sciences and Technology, Department of Civil Engineering,
Coimbra University, Coimbra, Portugal
anabela@dec.uc.pt

Abstract. This work aims to understand the applicability of several recent ISO standards concerning cities' sustainability. Among these standards, we highlight the ones related to sustainable cities and communities, namely ISO 37123:2019 - Indicators for Resilient Cities, ISO 37122:2019 - Indicators for Smart Cities and ISO 37120:2018 - Indicators for City Services and Quality of Life. Furthermore, among the long list of indicators presented in each standard, we are concerned with the ones related to transport and how they can represent a step forward in sustainable mobility promotion. Unfortunately, only a few works have dealt with this validation type. Therefore, using a NUTIII from Portugal - the Coimbra Region, a group of selected indicators concerning transport was applied to a group of municipalities. Unfortunately, the results show that most of the data needed is unavailable, and many indicators cannot be easily calculated. However, when calculated, this group of indicators can be used to compare cities' attractions and advantages, being a precious tool to promote sustainable policies around regional cohesion and equity. Furthermore, this work can also contribute to advice regarding local applications of universal rules.

Keywords: Transport Policies · ISO Standards · Sustainable Cities and Communities · Regional Cohesion

1 Introduction – SDGs Framing

The General UN Assembly adopted the 2030 Agenda for Sustainable Development, with 17 SDGs at its core, at the UN Sustainable Development Summit in September 2015 (https://sdgs.un.org/goals#history). Since then, a consistent body of knowledge has been built around policy discussions and practical implementations for 17 Sustainable development goals. When searching the WoS with the single keyword 'SDGs' there are 166 highly cited papers covering this subject. The biggest increase was observed after 2017, and notably after 2020. Lately, this trend slowed (see Fig. 1)

Among these, the five more cited articles show the most quoted concerns towards Sustainable Development Goals. The two most cited papers (1028 and 830 citations, and from 2018 and 2016, respectively), relate to specific issues related to SDG: the

© The Author(s) 2025
C. McNally et al. (Eds.): TRAconference 2024, LNMOB, pp. 662–667, 2025.
https://doi.org/10.1007/978-3-031-85578-8_89

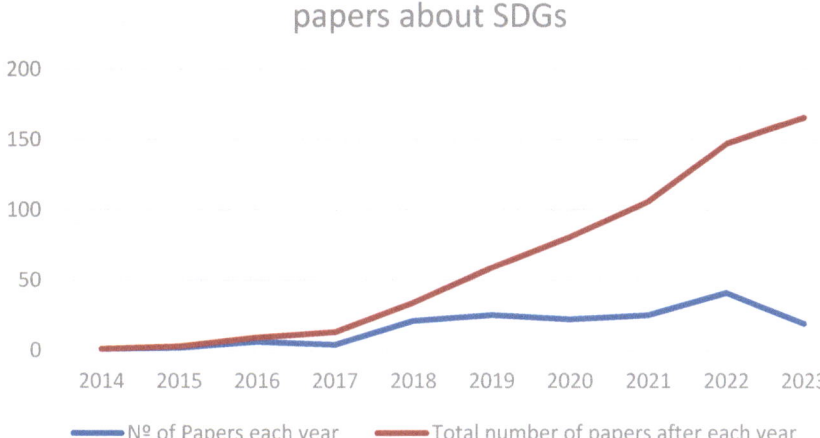

Fig. 1. Evolution in the number of highly cited papers concerning SDGs

first one is related to healt: '*The Lancet Commission on Global Mental Health and Sustainable Development.*'[1] and the second with soil occupancy and climate change '*The significance of soils and soil science towards the realization of the United Nations Sustainable Development Goals*' [2]. The others are mainly related to concerns with integration between SDGs and the efficiency of their application.

Refining the search by using the Keyword 'Cities', and also considered only the highly cited, we find only 14 papers, showing that SDGs are not being discussed so much in what concerns cities and city sustainable development (at least there are not many papers relating research in this thematic area, with SDGs). Among this, one or two papers directly connect with cities and urban development issues, such as the one from [3] called '*Defining and advancing a systems approach for sustainable cities*'. These findings can be further explored, but this initial attempt shows that cities and urban development are not one of the main fields where SDGs are being discussed scientifically.

At the same time, it is also important to note that according to the last report on Sustainable Goals, '*The Sustainable Development Goals Report 2023: Special edition Towards a Rescue Plan for People and Planet*', we are moving back or stagnating, particularly in what concerns cities and urban development [4]:

'Over half of the global population currently resides in urban areas, a rate projected to reach 70 per cent by 2050. Approximately 1.1 billion people live in slums or slum-like conditions in cities, with 2 billion more expected in the next 30 years.

In 2022, only half of the world's urban population had convenient access to public transportation. Urban sprawl, air pollution and limited open public spaces persist in cities. Since 2015, the number of countries with national and local disaster risk reduction strategies has doubled. To achieve Goal 11, efforts must focus on implementing inclusive, resilient and sustainable urban development policies and practices that prioritise access to essential services, affordable housing, efficient transportation and green spaces (see Figure 2).

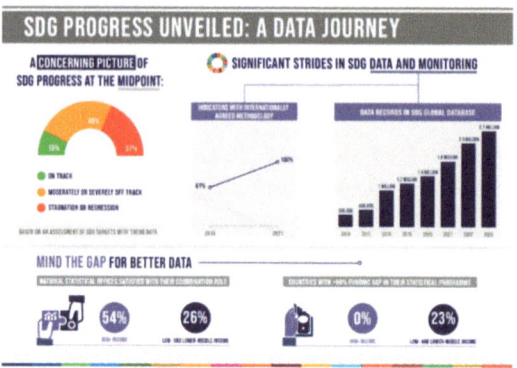

Fig. 2. Global panorama on SDG progress (Source, UN, 2023)

Comparing the progression between all the 17 SGs, Goal 11 aimed at '*Make cities and human settlements inclusive, safe, resilient and sustainable*' is one of the least executed, at the level of 10% in what concerns the criteria 'on track or target met' (see Fig. 3).

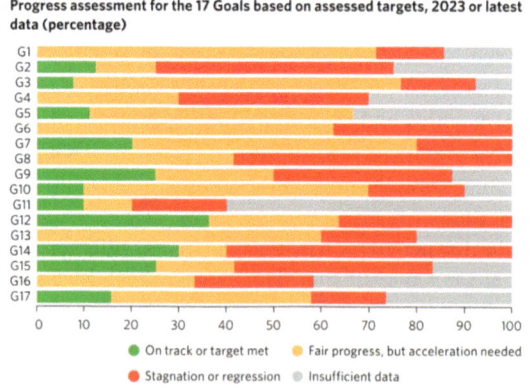

Fig. 3. The small level of execution of SDG11 (Source, UN, 2023)

Looking into what concerns the specific targets that relate more directly to Urban Planning and Transport, we highlight a few ones:

- *11.2 - By 2030, provide access to safe, affordable, accessible and sustainable transport systems for all, improving road safety, notably by expanding public transport, with particular attention to the needs of those in vulnerable situations, women, children, persons with disabilities and older persons;*

- *11.3 - By 2030, enhance inclusive and sustainable urbanisation and capacity for participatory, integrated and sustainable human settlement planning and management in all countries;*
- *11.7 - By 2030, provide universal access to safe, inclusive and accessible green and public spaces, particularly for women and children, older persons and persons with disabilities.*

It is easy to see that these targets compromise policies and actions towards a more sustainable future for cities. One of the observed international actions to promote these policies and actions is the development of ISO standards aimed explicitly at this target's accomplishment.

2   ISO Standards for Sustainable Cities

Adding to the SDGs and to reinforce them, in 2018, some ISO standards were developed related to sustainable cities and communities, namely ISO 37123:2019 - Indicators for Resilient Cities, ISO 37122:2019 - Indicators for Smart Cities and ISO 37120:2018 - Indicators for City Services and Quality of Life. These indicators have yet to be sufficiently explored so far. Searching the WoK and considering the abovementioned ISO standards, we found the following state-of-the-art:

- ISO 37120:2018 - Indicators for City Services and Quality of Life – 10 papers, with the most cited being one about the most suitable indicators to be collected by remote sensing: '*Can ISO-Defined Urban Sustainability Indicators Be Derived from Remote Sensing: An Expert Weighting Approach* [5].
- ISO 37123:2019 - Indicators for Resilient Cities – only one paper, from 2023, entitled '*Standardised indicators for "resilient cities": the folly of devising a technical solution to a political problem*', [6] in which the authors show that

  o *(…) the current standardisation of resilience fails to adequately address the political dimension of disaster risk reduction, reducing resilience to a management tool and missing the opportunity to address the socio-political sources of risks.*

- ISO 37122:2019 - Indicators for Smart Cities – 26 papers showing the dynamic of the ICT sector, with higher business concerns, fashionable trends, and concerns about energy networks. We highlight the paper '*Evaluation of cities' smartness by means indicators for small and medium cities and communities: A methodology for Northern Italy*' [7] and the most recent: '*Smart Cities Maturity Model-A Multicriteria Approach*' [8].

3   Conclusion and Work Being Developed

Overall, the application of the SDGs still needs to be accomplished by developing these new ISO standards regarding the ongoing debate around its utility from a political to a technocratic approach. Therefore, it is fundamental to discuss this relation and foresee the possible utility of these standards by validating it through application in case studies.

The proposed challenge is developing a tool to approach these new tools to real city sustainability, resilience and smartness assessment accuracy, keeping the political targets in focus for 2030. Therefore, developing a detailed and systematic scientific literature discussion to frame the debate. Afterwards, the objective is to use these discussion results to build a model of indicators and weights that can be applied to evaluate if cities are fulfilling these targets. Moreover, it considers comparing different cities to validate this proposed tool while critically assessing the data available to build the proposed indicators.

The case study comprises the municipalities in Portugal's Coimbra Region (see Fig. 4).

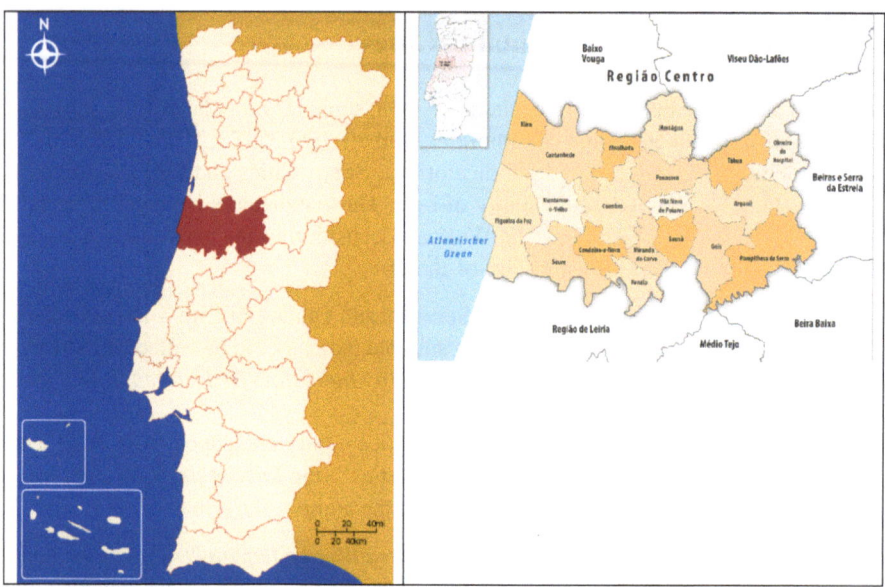

Fig. 4. Portugal - The municipalities of the Coimbra Region, Portugal.

References

1. Patel, V., et al.: The Lancet Commissions The Lancet Commission on global mental health and sustainable development (2007). https://doi.org/10.1016/S0140-6736(18)31612-X
2. World Health Organization. WHO guidelines on physical activity and sedentary behaviour: at a glance. Who, p. 535 (2020)
3. Keesstra, S.D., et al.: The significance of soils and soil science towards realization of the United Nations Sustainable Development Goals. i, 111–128 (2016). https://doi.org/10.5194/soil-2-111-2016
4. Zhao, M., Veeranki, S.P., Magnussen, C.G., Xi, B.: Recommended physical activity and all cause and cause specific mortality in US adults: Prospective cohort study. The BMJ, 1–10 (2020)

5. Bai, X., et al.: ScienceDirect Defining and advancing a systems approach for sustainable cities. 69–78 (2016). https://doi.org/10.1016/j.cosust.2016.11.010
6. (UN, 2023) The-Sustainable-Development-Goals-Report-2023 (2023). ISBN: 978-92-1-101460-0
7. Lehner, A., Erlacher, C., Schlögl, M., Id, T.B., Steinnocher, K.: (n.d.). Can ISO-Defined Urban Sustainability Indicators Be Derived from Remote Sensing: An Expert Weighting Approach. https://doi.org/10.3390/su10041268
8. Chmutina, K., Lizarralde, G., von Meding, J., Bosher, L.: Standardised indicators for "resilient cities": the folly of devising a technical solution to a political problem. International Journal of Disaster Resilience in the Built Environment, Vol. ahead-of-print No. ahead-of-print (2023). https://doi.org/10.1108/IJDRBE-10-2022-0099
9. O, G.D., Bruni, E., Panza, A., Sarto, L., Khayatian, F.: Evaluation of cities' smartness by means of indicators for small and medium cities and communities : a methodology for Northern Italy. Sustain. Cities Soc. **34**(July), 193–202 (2017). https://doi.org/10.1016/j.scs.2017.06.021

Open Access This chapter is licensed under the terms of the Creative Commons Attribution 4.0 International License (http://creativecommons.org/licenses/by/4.0/), which permits use, sharing, adaptation, distribution and reproduction in any medium or format, as long as you give appropriate credit to the original author(s) and the source, provide a link to the Creative Commons license and indicate if changes were made.

The images or other third party material in this chapter are included in the chapter's Creative Commons license, unless indicated otherwise in a credit line to the material. If material is not included in the chapter's Creative Commons license and your intended use is not permitted by statutory regulation or exceeds the permitted use, you will need to obtain permission directly from the copyright holder.

European Regional Airports: Emerging from the Storm or Facing a Gathering Storm?

Noel Hiney[⊠], Marina Efthymiou, and Edgar Morgenroth

Business School, Dublin City University, Dublin, Ireland
noel.hiney2@mail.dcu.ie

Abstract. By the end of 2022, European air passenger numbers had recovered to 78% of 2019 levels. A full recovery to pre-pandemic passenger volumes was expected by the end of 2023, a performance not anticipated by most aviation stakeholders during the crisis. However, capacity issues and the outbreak of war in Ukraine, as economies were emerging from the effects of the pandemic, led to increased levels of uncertainty across aviation and airports. This volatility was further exacerbated by factors such as recruitment/retention challenges, capacity issues in larger airports and the increasing impact of extreme weather events. While smaller European airport volumes recovered more quickly than larger ones, these airports continued to face severe cost and revenue challenges, when they needed to focus on connectivity and repairing balance sheets. Already vital airline relationships and State aid became more critical to the future viability of smaller airports. Aeronautical revenue was increasingly impacted by route deals advantageous to airlines, necessitating a greater airport focus on non-aeronautical revenue. The increasing use of larger airports by low-cost carriers further affected the prospects for sustained regional airport recovery. This paper aims to highlight key influences on European regional airport passenger performance through the uncertain and challenging 2019–2022 period, the impact of increased competition, and future prospects for these airports.

Keywords: Regional Airports · Competition · Revenue Challenges · State Aid · Viability

1 Introduction

Regional airports are a fundamental part of the European transport landscape contributing significantly to air connectivity and regional development. (ACI Europe, 2017; European Commission, 2021; Hiney et al., 2021). The aviation sector's post-pandemic passenger performance has demonstrated its resilience. Despite facing a series of unprecedented challenges over the last two decades, including terrorism, the global financial crisis and, most recently, the COVID-19 pandemic coupled with geopolitical upheavals and their economic repercussions, the industry has demonstrated an ability to adapt and respond to emerging challenges. The airport sector has recovered from the effects of these crises in an agile manner, notwithstanding the continued financial fragility of some smaller

© The Author(s) 2025
C. McNally et al. (Eds.): TRAconference 2024, LNMOB, pp. 668–675, 2025.
https://doi.org/10.1007/978-3-031-85578-8_90

airports. In addition to its assessment of airport passenger performance during challenging times, this paper considers the nature of State aid, i.e., operating and other supports provided to these airports by their respective States, in line with European Commission (EC) State aid guidelines and decisions. Such aid is a critical component of smaller airport business models. Between 2020 and 2022, this support increased significantly and included greater conditionality, especially concerning environmental matters, than was previously the case. Hiney et al. (2023) and Varsamos (2021) highlighted State aid provided to aviation and airports during the pandemic period, through the European Union Temporary State Aid Framework (EU, 2020), as a critical factor for the airports' survival. Research methods followed in this paper comprise observations from a survey of airport managers conducted by the authors during mid-2022 (27 responses), supplemented by an assessment of airport passenger performance using Airports Council International (ACI) data and insights obtained through semi-structured interviews carried out during 2021. Secondary research included a literature review and assessment of contemporary industry and government publications.

2   Recent Passenger Performance

The revival rate of air passenger traffic across Europe has not been consistent, notwithstanding a reasonably uniform post-pandemic removal of travel restrictions across the continent in 2022 and the provision of horizontal and vertical state support to airports during this period (Appendix 1). Many smaller airports in popular tourist locations such as Greece and Spain reached and sometimes exceeded pre-pandemic levels by the end of 2022 (ACI World, 2023). Airports and regions with a higher proportion of flights to Asia and the United States, where restrictions remained for extended periods (Graham et al., 2023), were below 2019 levels, reaching an overall average of approximately 80% of these numbers by the end of 2022. As well as location, airport size and route mix, 2022 passenger trends were also affected by geopolitical events, most notably Russia's invasion of Ukraine, which dramatically impacted passenger traffic primarily in eastern European nations. Ukraine's traffic collapsed due to the war. At the same time, Finland's recovery was affected by the closure of Russian airspace and the slower recovery in Asia, a key hub destination for flights from Helsinki airport. However, the impact of the war in Ukraine was partially offset by activity in Russia's domestic market at a time that its international routes were severely restricted. European airports with less than 10 million passengers annually outperformed larger airports with respect to air traffic recovery, these airports achieving 83–86% of 2019 passenger levels during 2022 (Fig. 1). In its 2023 report on regional airport connectivity, Airports Council International reported also that one third of these airports had fully recovered their 2019 connectivity levels (ACI Europe, 2023).

The low-cost carrier (LCC) share of seats at regional airports increased to 60% in 2022 from 52% in 2019. For many smaller airports, a single LCC's market share can exceed 80% of total traffic, significantly influencing the dependence of the airport on such business and confirming airlines as their most crucial stakeholder. Per-passenger airport costs were also reported to be higher (€15–€17 per passenger). Moreover, non-aeronautical revenue was lower for smaller airports at €4 per passenger for airports with

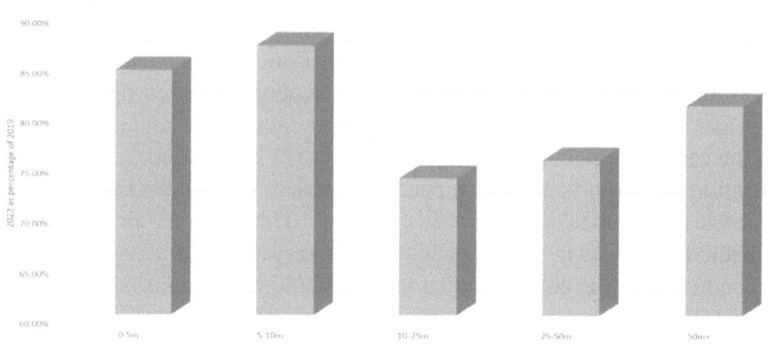

Fig. 1. European Airport Passenger Performance by Passenger Band (2022 vs. 2019). Source: ACI World Intelligence Hub, Authors (2023)

less than 5m passengers per annum. Versus €8–€12 for airports processing more than 25m passengers per annum (ACI Europe, 2023).

One fundamental cost factor for smaller airports is the adverse impact on aeronautical revenue of a keenly priced airline contract, which inevitably reduces income from airport charges and requires the airport to adopt strategies which focus on increasing commercial revenue from its passengers, an income rebalancing approach highlighted by Shin and Roh (2021). The authors note also that smaller airports lack the economies of scale enjoyed by larger ones, given the significant capital costs associated with airport infrastructure.

3   State Aid for European Airports

State aid for regional airports is a critical government investment in each country's economic and social infrastructure, benefiting individual nations and the European Union. The impacts of the COVID-19 pandemic on aviation gave rise to the need for the provision of horizontal and aviation/airport-specific vertical State support for the sector, through the speedy provision of aid in response to 'exceptional circumstances' (European Commission, 2020). The EC approved vertical State aid requests for airports in almost 30 instances to provide for general airport support, including compensation for revenue losses, grants and loan guarantees, subsidised interest rates, and deferral of taxes and charges, e.g., airport concession fees. This support was a vital financial lifeline for most airports receiving it. The authors undertook a survey of 27 airport managers in mid-2022, studying airport stakeholder relationships and selected airport activities. This research found that a higher number of respondent airports were in receipt of, or benefiting, from State aid during the pandemic period than support levels ten years previously for this set of airports, including Air Route Promotion (support provided to 42% of airports in 2022 versus 27% in 2012); Airport Improvements, e.g. safety and security (50% versus 29%); and Airline Support (62% versus 32%). Such airport support was often provided as part of a national-level package of measures (European Commission, 2021). This increased level of airline-specific State aid during the pandemic resulted in a noticeable rise in legal challenges against such Aid. Ryanair took many of these cases, with Bagamery (2021) reporting that the airline was responsible for 16 lawsuits against the EC. These challenges, some of which were upheld, contended that individual States had propped up their 'national champions' at the expense of other carriers who experienced similar challenges (Deasy, 2021).

The existing EC guidelines for the provision of country-level State aid for regional airports (<3m passengers per annum) were due to expire in 2024. In July 2022, the EC opened a consultation process (European Commission, 2022) regarding a proposed extension of its Airport State Aid Framework, suggesting a transition period of three years. Most consultation responses favoured an extension, due mainly to the effects of the pandemic and ongoing financial challenges faced by airports, together with perennial operating aid arguments (essential connectivity, support for business and tourism activity, economic and social benefits). However, some views were expressed opposing the extension, for example, a contention that there were already too many small airports, and that Aid should not be given to unprofitable airports. Underutilisation of the European rail network and environmental factors were also highlighted. Ultimately, the Commission decided to grant a three-year prolongation of its existing Aid framework for European airports to July 2027 (European Commission, 2023).

The difficulties many smaller, unprofitable airports faced and continue to face during increasingly uncertain times have likely resulted in an enduring need for financial support. The future of airport State aid is likely to be determined by EC decisions regarding new post-2027 arrangements, if any, together with associated assessment criteria. In particular, the 'hurdle' for justifying such Aid could be expected to be heightened. Such factors may be more likely to threaten the existence of a more significant number of unviable airports. However, another possibility is that a new structure for airport support might be decided upon, maintaining some elements of current rules. This policy might be interpreted as 'kicking the (airport viability) can further down the road'.

Aid should enable and support fair airport competition. However, broader economic, social, and environmental dimensions will also likely be considered when developing aid policies. Airport resource optimisation at the national level could be a key goal for States. Nonetheless, persistent concerns regarding the provision of such State Aid will be expected to be considered and addressed. State Aid decisions assessed by the EC would not be expected to result in inadequate resource allocation or to support continued airport financial underperformance, i.e., provided to viable airport entities only, so there is no distortive market effect. If new post-2027 policies are not designed and implemented effectively, these market distortions will persist, and new ones may emerge.

4   Conclusions

Regional airports are likely to continue to face severe cost and revenue challenges, exacerbated by geopolitical uncertainty. Intense competition will weigh heavily on recovery prospects. Collaboration with airlines and other stakeholders has become more embedded, and the predominance of the airline role is increasing. The over-reliance of many smaller airports on a single carrier (usually LCCs) is a further risk to future stability. State aid is a key 'third revenue line' for regional airports receiving it.

Steps that regional airports should take to address a more challenging aid framework include diversification of revenue streams, such as increased non-aeronautical revenue, and expansion of commercial activities in and around the airport. An increased investment in sustainable infrastructure should help to reduce costs and enhance future aid eligibility. Stakeholder collaboration will also likely prove beneficial. For national policymakers, strong alignment of regional growth and aviation policies, insofar as they apply to smaller airports, is essential. In addition, airport peripherality and transport substitutability are key factors when considering allocation of aid and selection of Public Service Obligation (PSO) supported routes. Furthermore, some aspects of an airport's activity might be suitable for private investment, which could reduce the level of State Aid required.

In recognition of the continuing importance of State support for regional airports and factors affecting the consideration of the same, the authors have developed a suggested assessment framework (EURAIR) which identifies a set of factors for airports and policymakers when considering, preparing, and assessing airport requests for government support. This framework (summarised in Table 1) can be considered alongside other assessment criteria, for example airport financial performance and future capital requirements.

Table 1. Support for Regional Airports: Selected Assessment Factors for Policymakers (EURAIR). Source: Developed by Authors.

Policy Areas	Points for Consideration (Practitioners and Policymakers)
Connectivity and Accessibility	+ Regional air routes provide essential connectivity in more peripheral and remote areas, also providing community access to essential services in larger national cities - Inefficient / over allocation of airport support could create state of regional route oversupply and reduce individual airport route feasibility
Local/Regional Economic Development and Tourism Activity	+ Substantial economic and tourism activity generated by regional airport route activity, benefiting businesses and citizens - Economic benefits can be overstated; potentially leading to inefficient allocation of financial resources

(continued)

Table 1. (*continued*)

Policy Areas	Points for Consideration (Practitioners and Policymakers)
Employment and Fiscal Support, and Balanced National Development	+ Regional airports are significant taxpayers, and their stakeholders provide major local employment; their presence is a strong support for balanced economic development - In the case of constraints on public funding, State Aid for regional airports needs to be rationalised and well supported
Environmental Sustainability and Innovation	+ Green practices are more straightforward to implement in smaller airports; transformational technology (e.g., electric aircraft) is likely to benefit smaller airports first - Increase in regional air transport (flight and passenger volumes) could threaten ecosystems located in these areas
A competitive and efficient aviation market	+ Aid can improve airport performance through efficiency achievements and greater competitiveness - Efficient airports not receiving Aid will find it more difficult to compete fairly with those (efficient or inefficient) receiving it
Strategic Infrastructure and Inter-modal transport enablement	+ Airports are critical national strategic assets, with their resilience and responsiveness critical during uncertain times - Other transport modes (e.g., rail, maritime) have similar strategic importance and face similar investment challenges

This paper contributes to the airport management discourse by assessing performance and recovery prospects for smaller airports during uncertain and volatile periods. It adds an assessment framework to the body of knowledge, highlighting factors and recommendations for consideration concerning a key exogenous driver, the provision of State Aid, when policymakers and practitioners are contemplating such support.

Appendix 1

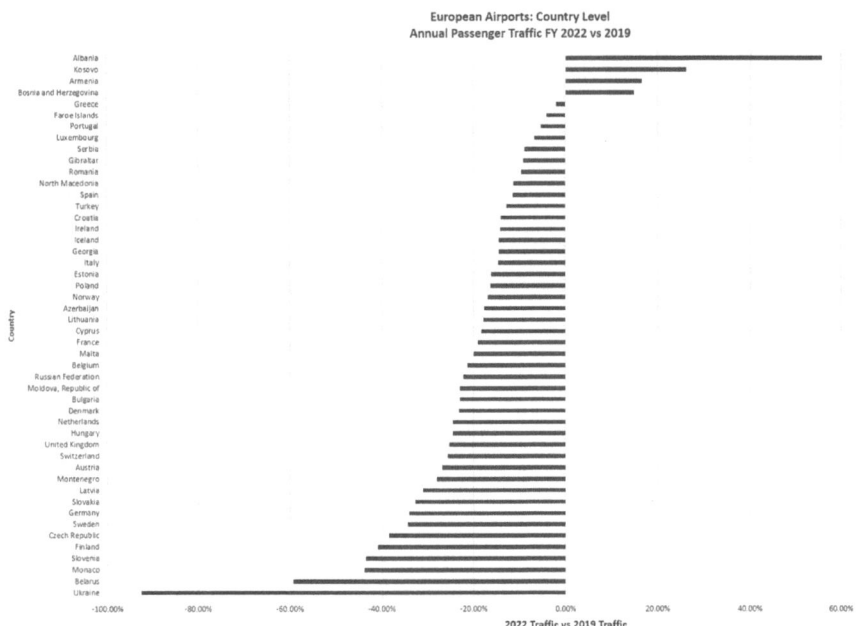

Appendix 1. European Airport Passenger Performance by Country (2022 vs. 2019). Source: Authors based on ACI World Intelligence Hub data (2023).

References

ACI Europe. European Regional Airports, Connecting People, Places and Products. Airports Council International Europe, Brussels (2017)

ACI Europe. European Regional Airports: Sustainably Connecting People, Places and Products (2023). https://www.aci-europe.org/downloads/resources/EUROPEAN%20REGI ONAL%20AIRPORTS%202023.pdf

ACI World. European Airport Passenger Database Intelligence Hub. Airports Council International World (2023)

Bagamery, A.: Ryanair Launches Legal Assault On EU State Aid to Major Airlines. Law.com (2021). https://www.law.com/international-edition/2021/02/19/ryanair-launches-legal-assault-on-eu-state-aid-to-major-airlines/. Accessed 2 Sep 2023

Deasy, G.: European Union and United Kingdom Competition Law Developments in the Aviation Sector: January to June 2021. Air Space Law **46**(4–5), 481–516 (2021)

European Commission. Covid-19 Temporary Framework (2020). https://ec.europa.eu/compet ition/state_aid/what_is_new/sa_covid19_temporary-framework.pdf. Accessed 2 Jul 2023

European Commission. Competition Policy State Aid (2021). https://ec.europa.eu/competition/ elojade/isef/index.cfm?fuseaction=dsp_sa_by_date. Accessed 20 Nov 2022

European Commission. Aviation guidelines – prolongation of operating aid to regional airports (COVID-19 response). 7th July (2022). https://competition-policy.ec.europa.eu/public-consul tations/2023-aviation-guidelines_en#view-the-consultation-document. Accessed 2 Sep 2023

European Commission. Operating aid to certain regional airports (2023). https://ec.europa.eu/com mission/presscorner/detail/en/ip_23_3710. Accessed 3 Sep 2023

Graham, A., Warnock-Smith, D., O'Connell, J.F., Efthymiou, M., Zheng, X.: Market developments on Chinese international air passenger markets in light of COVID-19 policy measures. Sustainability **15**(2), 1525 (2023)

Hiney, N., Efthymiou, M., Morgenroth, E.L.: Regional airport business models: The Shannon Group as a case study (2021)

Hiney, N., Efthymiou, M., Morgenroth, E.: Impact of Covid-19 on Irish airport stakeholder relationships. Ann. Tour. Res. **102**, 103622 (2023)

Shin, T., Roh, T.: Impact of non-aeronautical revenues on airport landing charge in global airports. Transp. Res. Rec. **2675**(10), 667–677 (2021). https://doi.org/10.1177/03611981211012423

Varsamos, S.: State Aid to Airports in the Era of COVID-19. Air Space Law **46**(3) (2021)

Open Access This chapter is licensed under the terms of the Creative Commons Attribution 4.0 International License (http://creativecommons.org/licenses/by/4.0/), which permits use, sharing, adaptation, distribution and reproduction in any medium or format, as long as you give appropriate credit to the original author(s) and the source, provide a link to the Creative Commons license and indicate if changes were made.

The images or other third party material in this chapter are included in the chapter's Creative Commons license, unless indicated otherwise in a credit line to the material. If material is not included in the chapter's Creative Commons license and your intended use is not permitted by statutory regulation or exceeds the permitted use, you will need to obtain permission directly from the copyright holder.

Sustainability and Value Propositions of Smart Mobility Projects in Rural Areas

Valtteri Ahonen(⊠) ⓘ and Shahid Hussain ⓘ

University of Oulu, P.O. Box 8000, 90014 Oulu, Finland
{valtteri.ahonen,shahid.hussain}@oulu.fi

Abstract. The purpose of this paper is to study how smart mobility solutions in rural areas are addressing sustainability in their value propositions. A multi-case study was conducted on four smart mobility projects located in rural areas across Europe. The study evaluates how the value propositions include the dimensions of sustainability. The results show that the value propositions of the projects studied are mostly economic. Social and environmental value propositions are made mainly to end-user customer segments, but not to customer segments that can be regarded as organisational or institutional.

Keywords: smart mobility · sustainability · value proposition · customer segment

1 Introduction

Sustainable development in rural areas has specific challenges, such as population decline, ageing demographics, poor physical connectivity, deteriorating infrastructure, and poor accessibility to services and opportunities [1]. For some of these challenges, smart mobility solutions (e.g. on-demand transport or shared mobility) can offer relief by improving connectivity in rural areas, providing more accessible mobility services, and exploiting data to enhance the efficiency of the services [2].

The financial and economic feasibility of new smart mobility solutions are often limited in rural environments [3]. Prospective value propositions often lack business logic with a clear sight of customer segments, key stakeholders, revenue model, and value capture. Also, projects searching for solutions often concentrate on technology, not on sustainability or viability of the solutions. Yet future smart mobility solutions should be based on sustainable business models.

In a sustainable business model, the value proposition is based equally on sustainability dimensions that are environmental, social, and economic [4]. Moreover, sustainable business models should adopt a multi-stakeholder perspective and not focus only on direct customers [4]. This way the business models would be a more integral part of business ecosystems that recognise also the stakeholder values, business values of the whole supply chain, as well as wider societal value [5].

New mobility solutions are often based on servitisation and product-service integration. As pointed out by Broman & Robert [6], servitisation allows for facilitating the

© The Author(s) 2025
C. McNally et al. (Eds.): TRAconference 2024, LNMOB, pp. 676–682, 2025.
https://doi.org/10.1007/978-3-031-85578-8_91

control of material (for example of the vehicles the solutions use) which makes external costs such as emissions easier to control.

This paper investigates whether the value propositions of four projects, located in rural areas in Europe, are including sustainability dimensions in a balanced manner. Environmental, social, and economic sustainability are the main dimensions evaluated (RQ1) in the business models' value propositions. Also, this study analyses which customer segments are addressed (or overlooked) in the projects' value propositions (RQ2). The aim of RQ2 is to find out how well the studied projects address sustainability in broader terms rather than just focusing on end-customer value.

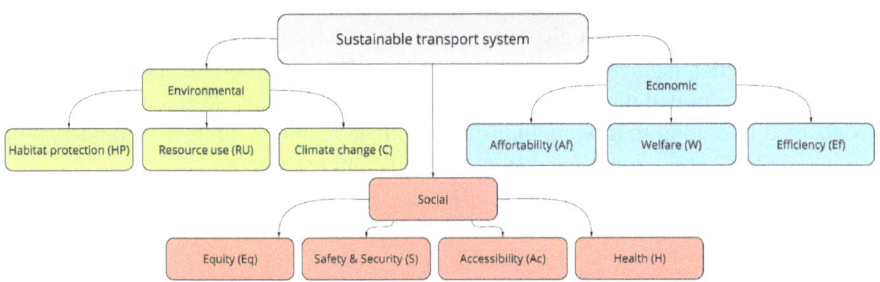

Fig. 1. A framework [7] that is used to evaluate the sustainability of projects' value propositions.

2 Research Approach

All studied projects relate to passenger transport and are part of AURORAL project ongoing until the end of 2024. The projects are: *Digitalisation of the mobility sector* in Hålogaland, Norway (HN), *On-demand public transport* in Alentejo, Portugal (AP), *Shared mobility* in Southern Burgenland, Austria (SBA), and *School transport of children with special* needs in Lapland, Finland (LF). The aforementioned abbreviations are used to identify the projects in the latter sections of this paper.

The data comprises value proposition canvases [8] that were collected from the projects in workshops during 2022 and 2023. This paper focuses on the value propositions (VP) that the projects recognised and used in their value proposition canvases (see Fig. 2). In addition, a project document describing the value propositions of the projects was used [9].

The research design was a multiple-case study in which the content of the data collected was analysed using a framework of sustainable mobility developed by Ahonen et al. [7]. The framework in Fig. 1 includes environmental, social, and economic dimensions of sustainability divided into 10 subdimensions. In this paper, the focus is not on subdimensions but on three main dimensions of sustainability. However, the subdimensions were used to help recognise what dimension of sustainability was addressed in the content under analysis.

3 Analysis

Each VP entity listed in the value proposition canvases were categorised to a certain sustainability dimension and the number of entities addressing different sustainability dimensions was counted (see Fig. 2). VP entities were also investigated separately within the customer segments and different aspects of sustainability were evaluated.

There were eleven customer segments in the projects. Because of the similarities between the customer segments, the authors compiled customer segments into four generalised customer segments. Tourists and citizens (HN), Inhabitants of the localities (AP), End-users (SBA), Private buyer (a parent) (LF), and Private buyer marketplace (LF) were compiled into an *end-users* customer segment. Transport providers from HN and AP, operators from SBA, and logistic companies from LF were compiled into a *transport providers* customer segment. Sectoral firms from SBA and Tourism activity providers from LF were compiled into a *sectoral firms* customer segment. Municipality from LF was decided to belong to a *municipalities* customer segment. Table 1 summarises the most relevant VP entities extracted from each customer segment.

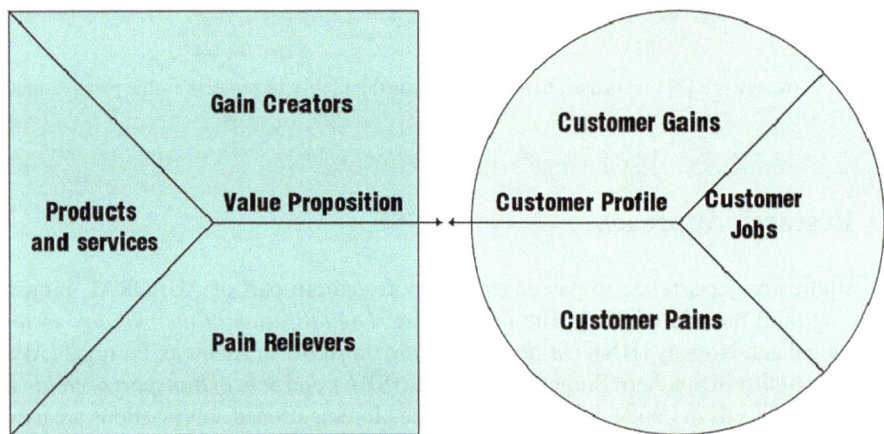

Fig. 2. Value proposition canvas by Osterwalder et al. [8].

4 Findings and Discussion

Regarding customer segments the business models of all projects studied were similar. They all were based on selling mobility services to *end-users* customer segments and organisations that were working with the *end-users* customer segment (*sectoral firms*). In HN and SBA projects the organisations were tourism agencies and for LF project the municipality that was responsible for public school transport. It would have been possible to combine *municipalities* and *sectoral firms* because in LF project the role of the municipality was very similar to other projects' *sectoral firms*. The major difference was that *municipalities* represented a public organisation and sectoral firms were private

Table 1. Value propositions extracted from value proposition entities (customer jobs, customer gains, gain creators, customer pains, and pain relievers).

	End-users (n = 5)	Transport providers (n = 4)	Sectoral firms (n = 2)	Municipalities (n = 1)
Social VPs	- increased skills - increased safety - increased accessibility - increased social cohesion	- guaranteed transport safety - increased digital skills - better social networking	- reduced need for a driving license - better preparation crisis situations	- increased safety - provide socially sustainable services for inhabitants - cope up with language barrier
Environmental VPs	- increased possibilities for environmentally sustainable mobility - lowering carbon footprint - increased access to nature and green spaces	- addressing shared mobility as an overall sustainability objective - reduced emissions - increased environmental sustainability of travel	- reduced emissions	- monitor ecological aspects - reduced emissions - higher ecological standards
Economic VPs	- more services in one application - reduced traffic and parking shortages - decreased costs - saving time - increased livability	- route optimisation - increased profitability - being pioneers in shared mobility - positive reputation - reduction of the number of individual transports - decreased energy consumption - better collaboration between stakeholders - increased availability of skilled personnel - integration of services - developments in infrastructure - better data utilisation	- more complete attraction - collaboration with smart mobility service providers - increased demand - increased marketing channels - long term contracts, more opportunities to R&D activities - increased supply of mobility services - new markets - cost effectiveness - increased availability of skilled personnel	- reduced energy consumption - monitoring cost of transport - enhance public transport network - route optimisation - decreased costs for workforce - increased available workforce

organisations (although some of them were owned by municipalities). For organisations, transport was a key activity they needed in their business environment, which was a challenge, especially in rural areas with large distances, poor existing public transport, and limited availability of transport services.

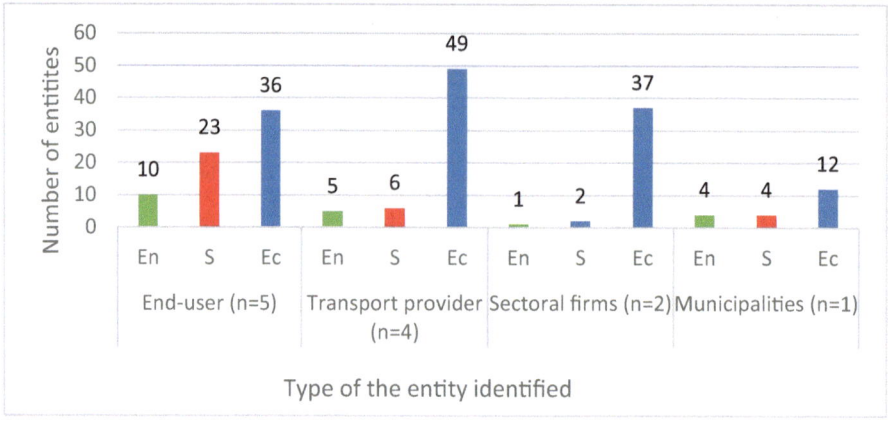

Fig. 3. Identified VP entities (customer jobs, customer pains, customer gains, pain relievers, and gain creators) for the projects where En is *Environmental*, S is *Social,* and Ec is *Economic.*

In all projects, *transport providers* were recognised as a customer segment, not only as a key stakeholder. This was a characteristic that separated these business models from traditional transport operating models where transport service is seen as an out-sourced activity. This kind of business model allows more flexible operations but is more dependent on the market-based local transport services.

In all projects, the value offered was based on the economic benefits emerging from the existence of other customer segments and services provided to *the transport providers.* The existence of these customer segments was necessary for the business models of these projects, otherwise the business models would have been based only on providing service technology for *transport providers.*

One interesting notion was that the AP project had only *end-users* (local inhabitants) and *transport providers* as customer segments. According to the project document [9], municipalities of the regions were recognised as stakeholders, but there were no dis-tinct value propositions for them. The business model could be made more viable by acknowledging municipalities as customer segments. This was done in LF project which allowed the project to offer services to the municipalities directly (e.g. school or health transport). By recognising all stakeholders, the business models can be developed closer to the sustainable business models described by Brozovic [4] where all stakeholders related to the business model are acknowledged.

From Table 1 and Fig. 3 it becomes evident that the projects' value propositions were related more to the economic dimension than other dimensions of sustainability. Social and environmental values were proposed more to *end-users* than to other customer segments. Only LF project had *municipalities* as a customer segment and for that also the economic value was proposed more than other sustainability dimensions.

It can be seen from the results that finding social and environmental value is chal-lenging, especially for organisational customer segments. For the end-users, finding social and environmental value is easier. According to Brozovic [4], value proposi-tions should be based equally on environmental, social, and economic sustainability. Moreover, responsibility for the sustainability of services should not be pushed only to end-users and expect them to find value from social and environmental dimensions over the economic dimension. Environmental and social value propositions should be offered to organisations that have more possibilities to change the system on a larger scale.

All of the projects were based on mobility servitisation but the benefits such as facilitating the control of material use pointed out by Broman and Robert [6] were not acknowledged in value propositions and for that reason were left uncaptured.

5 Conclusion

A multi-case study was conducted on four projects located in rural areas across Europe. The study analysed the customer segments and how the sustainability dimensions in the value propositions were covered. The study revealed that value propositions were mostly based on economic value and not on social or environmental values (RQ1). Moreover, social and environmental values that were proposed were directed to *end-users* customer segments, and not to organisational or institutional customer segments (*sectoral firms, municipalities,* or *transport providers*). Some customer segments such as public service providers (schools, health services etc.) were disregarded in some projects (RQ2).

There are some limitations in this study. First, multi-case study does not provide general knowledge of the phenomenon and must be regarded as observations of a limited number. Second, defining sustainability with three main dimensions is not straightforward due to the different ways that sustainability aspects may be categorised.

References

1. European Commission: A long-term Vision for the EU's Rural Areas – Towards stronger, connected, resilient, and prosperous rural areas by 2040 (2021)
2. Porru, S., Misso, F.E., Pani, F.E., Repette, C.: Smart mobility and public transport: Opportunities and challenges in rural and urban areas. J. Traffic Transp. Eng. 7(1) (2020)
3. Poltimäe, H., Rehema, M., Raun, J., Poom, A.: In search of sustainable and inclusive mobility solutions for rural areas. European Transport Research Review (2022)
4. Borozovic, D.: Business model based on strong sustainability: insight from and empirical study. Bus. Strategy Environ. **29** (2020)
5. Leviäkangas, P., Öörni, R.: From business models to value networks and business ecosystems – what does it mean for the economics and governance of the transport system? Utilities Policy **64** (2020)
6. Broman, G., Robért, K.-H.: A framework for strategic sustainable development. J. Cleaner Prod. **140** (2017)
7. Ahonen, V., Merisalo, V., Hussain, S., Pekkala, V., Leviäkangas, P.: Are smart mobility pilots in Finland fostering sustainability? – An assessment. Transp. Res. Proc. **72** (2023). https://doi.org/10.1016/j.trpro.2023.11.496
8. Osterwalder, A., Pigneur, Y., Bernarda, G., Simth, A., Papadagos, T.: Value Proposition Design: How to Create Products and Services Customers Want (2014)
9. AURORAL: D3.3 PILOT-FOCUSED REQUIREMENT EXTRACTION AND KPIS DEFINITION. https://cordis.europa.eu/project/id/101016854/results

Open Access This chapter is licensed under the terms of the Creative Commons Attribution 4.0 International License (http://creativecommons.org/licenses/by/4.0/), which permits use, sharing, adaptation, distribution and reproduction in any medium or format, as long as you give appropriate credit to the original author(s) and the source, provide a link to the Creative Commons license and indicate if changes were made.

The images or other third party material in this chapter are included in the chapter's Creative Commons license, unless indicated otherwise in a credit line to the material. If material is not included in the chapter's Creative Commons license and your intended use is not permitted by statutory regulation or exceeds the permitted use, you will need to obtain permission directly from the copyright holder.

Enhanced Transport Demand Forecasting and Digitalization

Kevin Cox[✉] and Deborah John

National Transport Authority, Dublin, Ireland
{kevin.cox,deborah.john}@nationaltransport.ie

Abstract. Ireland's National Transport Authority (NTA) has a responsibility for the update and maintenance of the multi-modal National Transport Model which is used in the assessment and analysis of major transport schemes and interventions across the country. Central to the model functionality is the requirement for accurate and consistent land use data inputs, notably in terms of travel demand. To provide a comprehensive demand dataset, in 2023 the NTA undertook a nationwide project to reassess the base year and future year datasets for population and employment. This enhanced forecasting methodology utilized web-based technology to facilitate a collaborative project directly involving all 31 Irish Local Authorities. As part of this project, the NTA assessed how various existing datasets could be utilized together to forecast population and employment variables more accurately. This project worked to improve the consistency and accuracy of data across numerous different organizations and ensured that the data was deliverable at all relevant policy scales, from national and regional to a small area level. Central to this project was the delivery of a new digitized approach, which allowed the coordination of various datasets, firstly to facilitate the use of input data that can improve decision making processes, but also in terms of how the data was collected and controlled for accuracy and consistency. Developing this methodology and working closely with Local Authorities to collate information in an accurate and coordinated manner at a local level, has ensured that the NTA, local authorities and other agencies have robust and consistent base year and future year datasets, which can facilitate evidence-based decisions on Strategic Planning issues and assist in the realization of key sustainable development goals.

Keywords: Connectivity · Sustainable Transport · Web based Assessment · Public Realm · Local Transport Plans

1 Introduction

1.1 Overview

The accuracy of the National Transport Authority's (NTA) national transport models is critical for supporting robust decision-making processes for appraisals of transport projects and the development of transport strategies. Central to the functionality of an effective transport model is the input of accurate and consistent land use data, typically

© The Author(s) 2025
C. McNally et al. (Eds.): TRAconference 2024, LNMOB, pp. 683–688, 2025.
https://doi.org/10.1007/978-3-031-85578-8_92

made up of numerous variables which are key determinants for travel demand. In 2023, the NTA with assistance from KPMG Future Analytics, initiated a major project to enhance forecast year datasets for the NTA National Transport Model through an in-depth review of existing planning sheet data, a comprehensive research programme and a consultation process that involves every Local Authority across the country. This review is focusing on future population, employment and education growth at Census Small Area level with the objective of enhancing the data sets used to forecast growth levels to 2040 and beyond. As a result of this process, the NTA and other state bodies will have a more accurate understanding of likely future travel patterns and as such be better placed to provide the necessary policy, infrastructure and services to cater for future travel demand in a sustainable manner.

This paper will provide an overview of this major project including its key objectives; the methodology used; the primary findings from the research and policy analysis phase and the innovations and improvements this project seeks to bring about.

1.2   Requirement for the Project

Transport modelling, used for a wide range of transport planning processes, is reliant on a comprehensive and accurate set of data that is made up of numerous variables. The NTA has a comprehensive set of transport models including the National Demand Forecasting Model (NDFM) and five Regional Models [4]. The NDFM estimates the overall travel demand on a national scale, while the regional models estimate mode and destination choice, as well as undertaken assignment for road, public transport, walking, and cycling networks. Each of these models are used to make informed decisions and are central to the primary transport planning roles of the NTA including strategic planning, service planning and infrastructure appraisal. Accurate forecasting of population growth as well as growth in other key sectors such as employment and education is critical in ensuring organizations such as the NTA can sustainably plan to provide an adequate level of transport service.

To date, where information has not been made available, the NTA has relied on a standard distribution method for distributing national and regional growth numbers, obtained from the National Planning Framework[1] (NPF) and respective Regional Spatial and Economic Strategies[2], at a sub-settlement level. This approach, whilst useful, is limited in that it assumes areas will consistently grow in the same patterns as before, thereby not factoring in policy objectives, population targets, known development proposals and likely population fluctuations as a result of changing demographics. In this regard the primary aspect of this project involved considerable consultation with local

---

[1] The National Planning Framework, published in 2018 by the Department of Housing, Local Government and Heritage, is Ireland's national strategic plan that seeks to guide development in the country to 2040. Included in the plan are population and employment targets for each region [2].

[2] The three Regional Spatial and Economic Strategies, prepared by the respective Regional Assemblies, function as the key implementation frameworks for overarching national development objectives that are set out in the NPF. Included in each strategy are population projections at a county level to 2031.

authorities across the country to gain a more informed understanding of how population, employment and education are likely to grow in particular settlements and small areas in each county. By identifying how national and regional growth targets are likely to be distributed at the settlement and small area level, the NTA can undertake more refined and robust analysis of likely future travel demand, with confidence that the approach is evidence based and in line with national policy targets.

In addition to the above, it was considered that as a result of new trends in recent years such as remote working and learning, digitalization and changing employment sectors, as well as the emergence of new data sources, the data variables used for the NTA model to derive travel demand were in need of review.

2 Project Methodology

The carefully considered process that is being followed for this project is set out below. Two primary project phases were established: an initial research phase and a second phase for a comprehensive consultation programme with local authorities.

2.1 Phase 1 - Research and Comparative Analysis

A number of research topics were identified that were of interest to the NTA with a view to improving the established approach to data collection and the range of data variables used in the planning sheet. The NTA in collaboration with KPMG Future Analytics undertook a comprehensive programme of research with a view to establishing a list of short-, medium- and long-term actions that could improve the NTA's process for data collection, input and application. A summary of some of the research carried out under each strand is set out below.

Existing NTA Planning Data

The review of the existing NTA planning data (NTA 'Planning Sheet'), comprising 54 variables, noted a number of established weaknesses including the use of base year data to distribute future population growth, when future year growth information was not available. Another major limitation was the difficulty in accurately reflecting the growth data in interim years between census periods. It was noted that through the use of additional data sources that are published on a more regular basis ongoing updates could be made to variables to provide more up to date and reliable forecasts.

Policy

The policy review undertaken as part of the research element of this project was primarily focused on understanding existing and future public sector policy that could potentially materially influence the allocation of future employment and education places. This included a review of policy related to the trends of remote working and learning, and decarbonisation of the economy given the potential impact these trends could have on travel patterns. The review found that whilst there are no quantitative job creation targets for specific employment sectors, there is potential for diverging trends between sectors. It was noted that Government departments and the ESRI are currently developing enhanced analytical capacity to model the macroeconomic implications of Ireland's enhanced climate governance framework and that there is potential to use this macroeconomic model to inform the development of alternative employment allocation scenario(s) in the NTA Planning Sheet [3].

Employment

Employment forecasting has traditionally been challenging to accurately project, given the range of macro-economic variables, as well as the range of land use development options for employment zoned land. Notably, the NTA were interested in considering how to forecast labor force participation at a county and settlement level by age and employment sector as well as considering how employment rates differ at a county level. The research recommended that a structural economic model could help to predict future changes in participation rates for these cohorts and that settlement-level employment ratios are incorporated from the NPF into the NTA planning sheet as a high-level reference for practitioners using NTA models.

This facet of the research also recommended consideration of remote and mobile workers who are not adequately considered as part of the current NTA planning sheet and recommended that these should be accounted for in the updated planning sheet using Census 2022 labor force figures at the county level.

Education

To understand potential alternatives for how educational places could be reconciled against National, Regional and Local scales, research was undertaken to assess potential new methods and data sets that could be used in the NTA planning sheet. The NTA's current method of allocating and projecting school places closely follows population growth and does not account for any future demographic change. It also assumes a consistent ratio of students in primary and post primary level education whereas the Department of Education and Skills publishes projections of full-time enrolment in primary and secondary schools at State and regional level that are based on CSO population forecasts that incorporate regional assumptions on migration, fertility and mortality.

2.2   Phase 2 – Consultation Programme with Local Authorities

The primary task of this project, to update the NTA planning sheet at settlement and small area level, involved an initial consultation phase taking place over a three-month period with every local authority in the country. As part of this process, each Local Authority was asked to review the distribution of population and employment growth for their areas under the established NTA standard distribution method. It is envisaged that this task is primarily undertaken by the planning teams in each local authority and that their input will be primarily based on relevant local planning policy, local authority objectives, knowledge of planned development and the availability of appropriate sites for development.

To facilitate the collection and transfer of this data, a data hub was developed by Compass Informatics on behalf of the NTA, which acted as a central portal for the sharing of data between the NTA and local authorities. This approach facilitated the smooth transfer of data and allowed existing and future growth distributions to be assessed visually through a GIS platform embedded in the portal.

Following completion of the initial consultation, a follow up consultation phase will take place with each local authority. This will consider revised National Forecast figures being developed at National Government level in light of the publication of the final Census 2022 results [1]. This second stage review will allow further refinement and ensure that the final distributions are in line with national policy.

As part of the consultation process, the NTA also regularly engaged with the three Regional Assemblies who assisted in the process and were apart of the overall project steering group that also included the Department of Housing, Heritage and Local Government.

3 Innovations

3.1 Data Hub

The use of a 'Data Hub' accommodated a digitized approach to engagement with Local Authorities and other key stakeholders as part of this collaborative project. This approach allowed for the coordination of various datasets and enabled Local Authorities to review proposed growth distributions on a map-based digital platform that ensured a user-friendly, accurate and consistent method for collecting data. It also allowed for significant time savings from the NTA's perspective in terms of reviewing proposed distribution rates and sharing data with the steering committee. Now that the 'Data Hub' is established, this digital platform can also be used by the NTA in further engagement processes with Local Authorities, enabling closer cooperation on various projects and reducing the amount of time required for inputting data manually. It is envisaged that this platform will enable the NTA to collect planning sheet data on a more regular and consistent basis, thereby improving the overall quality of NTA models.

3.2 Digitization of Planning Sheet

In addition to the production of a Data Hub, the project has also sought to automate data workflows, improve data quality and streamline data exchange between internal and external systems relating to the planning sheet primarily utilizing FME (feature manipulation engine) and SQL database. This has facilitated the automation of the current planning sheet process, replacing the current manual approach. It is considered that this will greatly enhance the way in which data is validated and the overall approach to quality control.

4 Conclusion

Effective and robust transport modeling is based on the input of accurate and consistent land use data, typically made up of numerous variables which are key determinants for travel demand. This paper has detailed the nationwide project undertaken by the NTA to reassess the base year and future year datasets for population, employment and education variables in Ireland based on a collaborative approach with a range of stakeholders. Developing this methodology and working closely with Local Authorities to collate information in an accurate and coordinated manner at a local level, has ensured that the NTA, local authorities and other agencies have robust and consistent base year and future year datasets, which can facilitate evidence-based decisions on Strategic Planning issues and assist in the realization of key sustainable development goals.

References

1. Central Statistics Office. 'Census of Population 2022'. Census of Population 2022 - CSO - Central Statistics Office (2023)
2. Department of Housing, Local Government and Heritage. National Planning Framework. Project Ireland 2040 (2018). The National Planning Framework (npf.ie)
3. Economic & Social Research Institute. The I3E Model. Current Research (2023). The I3E Model|ESRI
4. National Transport Authority. 'Transport Modelling' (2023). Transport Modelling - National Transport

Open Access This chapter is licensed under the terms of the Creative Commons Attribution 4.0 International License (http://creativecommons.org/licenses/by/4.0/), which permits use, sharing, adaptation, distribution and reproduction in any medium or format, as long as you give appropriate credit to the original author(s) and the source, provide a link to the Creative Commons license and indicate if changes were made.

The images or other third party material in this chapter are included in the chapter's Creative Commons license, unless indicated otherwise in a credit line to the material. If material is not included in the chapter's Creative Commons license and your intended use is not permitted by statutory regulation or exceeds the permitted use, you will need to obtain permission directly from the copyright holder.

The Governance and Regulations of Unmanned Aircraft Systems in the European Union – A Comparative Framework

Anssi Rauhala$^{(\boxtimes)}$ ⓘ, Pekka Leviäkangas ⓘ, and Anne Tuomela ⓘ

Civil Engineering, University of Oulu, P.O Box 4200, 90014 Oulu, Finland
anssi.rauhala@oulu.fi

Abstract. Unmanned aircraft systems (UAS) have witnessed a radical increase in a variety of applications. While delivering obvious benefits, there are also hazards and risks regarding, for example, physical safety, people's privacy, and cyber security. The regulatory framework on UAS was considerably changed due to the adoption of Regulation (EU) 2018/1139, which, as a centralised, top-down regulation, covers nearly all UAS, regardless of their weight and size. Although the ensuing regulations and guidelines from the European Commission and European Aviation Safety Agency (EASA) give detailed provisions for things like the operation and design of UAS, the implementation of said rules and the necessary procedures involved are up to the member states and their competent authorities. This paper compares current differences in selected member states considering the implementation of new UAS regulations. The results imply that there is a wide variability between member states in how they cover different aspects of the new regulation. This paper also draws preliminary conclusions on the maturity of the governance and regulatory framework.

Keywords: UAS · EU · governance

1 Introduction

The use of unmanned aircraft systems (UASs) has seen massive growth in civil applications over the past two decades. As with all new technology, UASs can also pose risks to, for example, safety, security, and privacy. To address the potential risks and concerns about privacy, the first national regulations began to appear in 2002–2003 [1]. At the level of the European Union (EU), Regulation (EC) 216/2008 applied only to UASs with a maximum take-off mass (MTOM) of over 150 kg, thus leaving much of the regulatory powers regarding lighter UASs in the hands of EU member states (MSs). Consequently, this led to a fragmented and very heterogenous set of national level rules with varying levels of implementation [1].

The fragmented, dual approach of Reg. (EC) 2008/216 and national regulations were replaced in 2018 by the centralized, top-down framework of Regulation (EU) 2018/1139, which set down a risk-based approach to ensure security and safety for aircraft operation

© The Author(s) 2025
C. McNally et al. (Eds.): TRAconference 2024, LNMOB, pp. 689–694, 2025.
https://doi.org/10.1007/978-3-031-85578-8_93

in the Single European Sky [2, 3]. This new 'Basic regulation' in the field of civil aviation gave the EU Commission a further mandate to adopt delegated and implementing acts to regulate the UAS sector. The European Aviation Safety Agency (EASA), which includes all 27 EU countries plus Iceland, Liechtenstein, Switzerland, and Norway as members, is mandated to develop and issue technical regulations upon request by the European Commission. EASA also has soft powers to introduce guidance materials, acceptable means of compliance, to develop standards on UASs, and to coordinate the actions of MSs [3].

The adoption of new UAS regulations has not been without issues, including suffering multiple delays [4]. Furthermore, real-world experiences and data are required to validate the risk-based approach applied to UAS regulations. With regard to the U-space concept, EASA has acknowledged that there is a lack of data for thorough safety risk assessments. There is also a lack of common data exchange infrastructure [3]. Thus, there is an apparent need for standardized incident and occurrence reporting and a data exchange system that can support proactive identification of potential risks; there is also a need to raise awareness among UAS operators of the importance of reporting occurrences [5].

The OECD [6] has underlined the importance of open government data (OGD) as an essential element of good public governance. UAS regulation transparency, accessibility and reliability highlights how successfully different countries have been able to provide information and data based on the principles of good governance. We will discuss our findings in the comparative analysis in the context of the principles of OGD and good governance. The OECD lists three main pillars for assessing OGD principles: data availability, data accessibility and government support for data reuse.

2 Overview of Current UAS Regulations in the European Union

Unlike the preceding Regulation (EC) 2008/216, the new Reg. (EU) 2018/1139 on common rules in the field of civil aviation applies to all civil applications of UAS regardless of size and weight, with only a few exceptions [4]. The new regulation also shifts the main regulatory powers to the European Commission and EASA. Some exceptions are still admitted, such as the possibility for MSs to lay down national rules for, for example, public security reasons (e.g., defining prohibited/restricted zones for UAS operations); they can also grant time-limited exemptions from non-critical requirements due to urgent circumstances or amend implementing or delegated acts when it offers advantages and greater efficiency [2].

The Implementing Regulation (EU) 2019/947 introduces detailed provisions for the operation of UASs, including for personnel use and organisations, whereby the Delegated Regulation (EU) 2019/945 is more technical, laying down requirements for the design and types of UAS intended to be operated as defined by Reg. (EU) 2019/947. The Delegated Regulation (EU) 2019/945 also place rules on third-country operators when they conduct operations within the Single European Sky airspace. The regulations further categorise civil UAS operations into "open", "specific" and "certified" categories, and set boundaries for the operational contexts of each category [4].

The open category specifies low-risk UAS use cases and sets no requirements for special permits or operational authorisations [4]. However, there is in most cases a requirement for operator registration, and online pilot training and an examination. The open category's general requirements which must be met include a max. Attainable height of 120 m, a MTOM of less than 25 kg, and operations only being carried out on the visual line of sight, keeping a safe distance from uninvolved people. The open category UAS operations are divided into sub-categories (from A1 to A3) which introduce further limitations depending on the MTOM [4]. If the open category terms cannot be met, a risk assessment and possibly an operational authorisation from the national competent authority are required. The certified category UAS operations are considered to be high risk and comparable to manned aviation, and the UAS can be operated only by certified and licensed operators.

As the new basic regulation 2018/1139 applies to all civil UAS regardless of weight and size, amendments were made to the Reg. (EU) 996/2010 on the investigation and prevention of accidents and incidents in civil aviation, and to the Reg. (EU) 376/2014 on the reporting, analysis, and follow-up of occurrences in civil aviation. With respect to UAS-related incidents, the amendments to 996/2010 limit obligatory investigations to UASs certified by EASA or to events where fatalities, serious injuries, or manned aircraft were involved [5]. Amendments to 376/2014 limit the obligatory reporting of occurrences (includes less serious incidents that may affect aviation safety) to only those UAS-related occurrences that result in fatal or serious injuries, or occurrences involving manned aviation [7]. Although reporting accidents is mandatory, high numbers of accidents are likely to go unreported [5].

The U-space regulatory package (i.e., Reg. 2021/664, 2021/665, 2021/666) from 2021 further sets rules which affect, for example, the design, manufacturing, and use of certain UAS types [8]. It is noteworthy that the 2019 and 2021 Acts by the Commission devolve several powers back to the MSs, again running the risk of legal fragmentation, not only regarding the rules of civil aviation but also their complex interplay with other sectors of regulation [8].

3   Comparative Framework

As a first level of analysis, the focus was placed on open category operations as the information and processes for this category are most easily accessible. National differences in adoption and processes related to the new regulations were compared by analysing the relevant sections of the competent authorities' websites, to which links are provided in the "Civil drones" section of the EASA website. In the few instances where the links were missing (e.g., Greece) or the links were not correct (e.g., Romania), the appropriate websites were located via an internet search engine.

A key aspect is the availability of information and necessary procedures in English. The lack of English is an obvious obstacle not only from a research perspective, but also from the user perspective since according to Reg. (EU) 2019/945, third-country operators need to register in the first EASA member state in which they conduct UAS operations. The availability of UAS-related information, pilot registration procedures and exams in English were analysed by comparing the native language website of each competent authority against the English language website (if any). As Fig. 1 highlights, only seven of the EASA MSs seem to have all UAS related information available in

English, two of which (Ireland and Malta) have English as an official language. Five MSs seem to have no UAS-related information in English at all. The other 19 MSs have at least some information available in English, although the scope varies widely, ranging from a short summary to nearly all the available content having been translated.

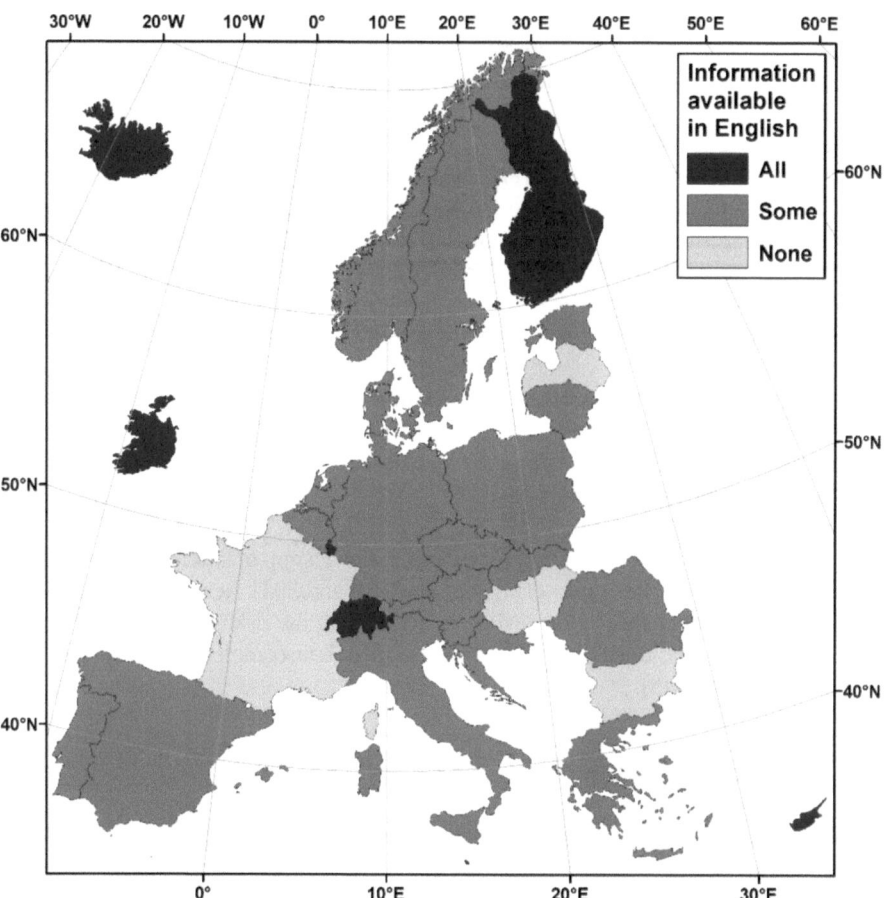

Fig. 1. Map of EASA member states highlighting whether the competent authority website has UAS information and registration procedures available in English as of August 28th, 2023. Note that of the two European microstates that are members of EASA, Malta is in the "All" category and Liechtenstein is in the "None" category.

Another important aspect is the open availability of information and training materials with no need to register, especially if it requires identity verification. The "Civil drones" section of the EASA website warns against fraudulent websites that offer pilot licenses for a price and without the need to take an exam. It is not unexpected for some users to fall victim to fraudulent services if the website of the competent authority offers little information without the user being required to register first. Of the seven countries that seem to have all materials available in English, only one seems to provide all the training materials and only one has the A1/A3 category training materials openly available. Of the 19 states that have at least some information available in English, seven have either all or at least the A1/A3 materials available, eight require registration before access, and, in four cases, the information available is unclear due to translation issues or materials seemingly not existing online. On multiple occasions, the openly available training materials are only in the official language of the state.

As mentioned, proactive identification of safety risks requires sufficient UAS-related occurrence reporting. Of the seven countries providing all information in English, only three seem to include links to proper channels for reporting of the mentioned incidents or occurrences; although, two of the sites do not clearly specify which occurrences UAS operators must report. It should be mentioned that the comparison focused specifically on the UAS-related section of the websites, and it is possible that the reporting information is provided in other sections. Of the states that have at least some information available in English, ten provide information and links for reporting, while four do not seem to have either links or information, and five provide unclear information due to language/translation issues. Of the ten sites which mention reporting, six provide clear instructions on what must be reported, three provide general regulations but no UAS-specific guidelines, and one provides neither the regulations nor the UAS-related guidelines.

The research also revealed some interesting anecdotes. For example, the Croatian Civil Aviation Agency's website directs non-Croatian speakers to the website of the Luxembourg Directorate of Civil Aviation, which has freely available online training and examination for open category A1/A3 in collaboration with Eurocontrol. Regarding the additional open category A2 theoretical examination for non-Croatian speakers, the website notes that "A2 exam from any EU member state will be valid in Croatia". It is true that only the operator registration should be tied to the state of residence or the first state in which a third-country operator conducts UAS operations, while pilot competencies can be acquired in and are valid from any EASA MS. However, it may not be in the spirit of the law nor very user-friendly to direct operators into a different state to acquire competencies.

The degree to which MSs have fully embraced EASA guidelines or taken advantage of the ability to set different geographical zones and, for example, limit certain operations varies. For example, Iceland, as a non-EU country, is yet to implement the EASA guidelines but is expected to do so soon. As of August 28th, 2023, UAS operator registration is not mandatory in Iceland and only commercial UAS registration is required. Spain, on the other hand, still regulates the use of airspace based on Royal Decrees 1036/2017 and 1180/2018 which set out different requirements for recreational and professional activities, although, new national legislation for the use of airspace and UAS geographical zones is expected to be approved in the future.

4   Conclusions

The regulatory landscape of UAS operations in the European Union has changed dramatically in the past few years. The adoption of new UAS regulations seems to take quite different paths in different member states, and there is clearly no agreed paced and synchronised implementation. On the one hand, this is understandable looking at different countries' institutional maturity regarding governance and policymaking. While on the other hand, European citizens and Single Market players have every right to expect better. In some countries, the OECD recommendations for good governance seem to be realised in full, whereas in others, the work is still in progress.

Acknowledgements. This work has been supported by the SPADE project, funded by the European Union's Horizon Europe Research and Innovation programme within HORIZON-CL6-2021-GOVERNANCE-01 under Grant Agreement no. 101060778.

References

1. Stöcker, C., Bennett, R., Nex, F., Gerke, M., Zevenbergen, J.: Review of the current state of UAV regulations. Remote Sens. **9**, 33–35 (2017). https://doi.org/10.3390/rs9050459
2. Bassi, E.: From here to 2023: civil drones operations and the setting of new legal rules for the European single sky. J. Intell. Robot. Syst. Theory Appl. **100**, 493–503 (2020). https://doi.org/10.1007/s10846-020-01185-1
3. Pagallo, U., Bassi, E.: The governance of unmanned aircraft systems (UAS): aviation law, human rights, and the free movement of data in the EU. Minds Mach. **30**, 439–455 (2020). https://doi.org/10.1007/s11023-020-09541-8
4. Rauhala, A., Tuomela, A., Leviäkangas, P.: An overview of unmanned aircraft systems (UAS) governance and regulatory frameworks in the European Union (EU). In: Bochtis, D., Tagarakis, A.C., and Kateris, D. (eds.) Unmanned Aerial Systems in Agriculture, pp. 269–285. Academic Press, London (2023). https://doi.org/10.1016/B978-0-323-91940-1.00012-8
5. Kasprzyk, P.J., Konert, A.: Reporting and Investigation of Unmanned Aircraft Systems (UAS) Accidents and Serious Incidents. Regulatory Perspective. J. Intell. Robot. Syst. **103**, 3 (2021). https://doi.org/10.1007/s10846-021-01447-6
6. OECD: Open Government Data Report: Enhancing Policy Maturity for Sustainable Impact. OECD Digital Government Studies, Paris (2018). https://doi.org/10.1787/9789264305847-en
7. Konert, A., Kasprzyk, P.: UAS safety operation – legal issues on reporting UAS incidents. J. Intell. Robot. Syst. **103**, 51 (2021). https://doi.org/10.1007/s10846-021-01448-5
8. Bassi, E., Pagallo, U.: A Guide to EU Drones Law, and its Work in Progress. SSRN Electron. J. (2022). https://doi.org/10.2139/ssrn.4273048

Open Access This chapter is licensed under the terms of the Creative Commons Attribution 4.0 International License (http://creativecommons.org/licenses/by/4.0/), which permits use, sharing, adaptation, distribution and reproduction in any medium or format, as long as you give appropriate credit to the original author(s) and the source, provide a link to the Creative Commons license and indicate if changes were made.

The images or other third party material in this chapter are included in the chapter's Creative Commons license, unless indicated otherwise in a credit line to the material. If material is not included in the chapter's Creative Commons license and your intended use is not permitted by statutory regulation or exceeds the permitted use, you will need to obtain permission directly from the copyright holder.

An Analytic Lens Towards Electric Vehicle Policy of India

Mohit Dev[1(✉)], Barsha Paul[2], Apoorva Nawani[3], and Potupureddy Chola Priya[2]

[1] School of Planning and Architecture Bhopal, Bhopal, MP 462030, India
mohit.dev@spabhopal.ac.in
[2] pManifold Business Solutions Pvt. Ltd., Delhi 110074, India
[3] Independent Researcher, Lucknow, UP 226010, India

Abstract. Across the globe, governments are promoting electric vehicles (EV) because of their benefits such as improved air quality, reduced fuel consumption, and improved energy security. Government of India (GoI) has also launched a series of policies and schemes to help promote EV adoption. However, different states offer different incentives in the state landscape providing non-uniform economic benefits in the country level. The main findings magnify the huge discrepancies in the policies implemented by various countries, which do not aim towards a coherent strategy approach. This paper uses Qualitative Comparative Analysis (QCA) method to compare the EV policy's' qualitative parameters across 12 Indian states to propose a working approach for "best-practice" and "unified" EV policy in India, by reviewing individual state policies. This would help in better adoption, implementation, and monitoring of EV vehicles in India.

Keywords: Public Policy · EV Policy · Unified framework · QCA method · EV Landscape

1 Introduction

Globally, governments are promoting electric vehicles (EVs) to improve air quality, reduce fuel consumption, and to improve energy security. China accounts for 98% of the global e-bus market, whereas Norway has a 70% market penetration rate for electric cars [1]. Government of India (GOI) has also launched a series of policies and schemes to help promote the adoption of electric vehicles. The first national mission document, "National Electric Mobility Mission Plan 2020" was launched in 2013 to achieve 6–7 million sales of electric and hybrid vehicles by 2020 [2]. With a budget outlay of $120 million in 2015, the GOI introduced a policy framework, "FAME-I (Faster Adoption and Manufacturing of (Hybrid &) Electric Vehicles)" aimed at accelerating the development of EV infrastructure and encouraging private sector participation. FAME II was introduced in 2019 with a budget of $1.3 billion over three years [3]. However, with the presence of various state-wide EV policies in India, the economic benefits are also non-uniform at country level. Thus, there is a need of a unified EV policy of India. However,

---

A. Nawani—Independant Researcher.

© The Author(s) 2025
C. McNally et al. (Eds.): TRAconference 2024, LNMOB, pp. 695–701, 2025.
https://doi.org/10.1007/978-3-031-85578-8_94

it is important to note that urban transport is a state subject and therefore states have the power to form their own policy in line with the national policy. The aim of this is to produce a framework to form a unified EV policy in India.

2 Literature Review

The literature review focuses on three aspects: Public policy analysis, global EV scenario and Indian EV scenario. Public policy analysis is a problem-solving tool that helps in finding a practical solution for a problem by drawing from ideas, theories, and other findings from different disciplinaries. It is a multidisciplinary approach which aims in creating a critical assessment and communicate policy related knowledge [4, 5]. As there is a lack of unified EV policy in the Indian subcontinent, this study aims to review the EV policies of the different states of India with the help of Qualitative Comparative Analysis (QCA). Due to the situation of Covid-19, there was a limitation in collecting on-ground data. In such cases QCA is the most ideal analysis that can be done due to its ability to handle small datasets, capture complex relationships, provide a contextual understanding, focus on patterns, and incorporate qualitative information [6–8].

Global EV Scenario: The EV support policies of various countries all over the world have been studied. The eight EU countries, China, the USA which aim to implement EV adoption, from three perspectives: passenger, charging infrastructure, and e-mobility ecosystem. EU countries' EV policies primarily focus on financial and organizational instruments, falling into two categories. These policies originate from the national level of governance. At the country level, the focus is primarily on vehicles, rather than supporting charging infrastructure. The policies target the downstream of the vehicle value chain, including fiscal and non-fiscal incentives. Some countries also prioritize charging infrastructure and upstream financial incentives like research funding but do overlook the consideration of 'Services'. While these policies are effective for initial EV adoption, a different strategy and policy tools are necessary to support future scenarios and larger-scale implementation. China's market grew from 100 thousand units in 2014 to 650 thousand units in two years. Globally, there were 320 000 public charging stations in 2016, up 72% from 2015 [1, 9]. Governments and local authorities in these countries are implementing policies to increase EV usage. Currently, policymakers in these countries utilize measures such as supporting EV deployment, purchase subsidies, access restrictions, and Zero Emission Vehicle (ZEV) mandates. Mass production and research and development efforts have led to decreased costs and increased energy density. Comparative studies of EV policies globally reveal some drawbacks that hinder user adoption, including higher costs compared to conventional vehicles, slower charging, limited access to charging stations, and consumer concerns about range and charging times.

Indian EV Scenario: Many initiatives have been undertaken by the Indian government to upscale e-mobility over the previous decade, to encourage the acceptance of EVs in the country, including tax incentives for EV buyers/owners and the creation of public EV charging infrastructure. As a result, EV usage has started to expand over the last five years (industrial growth took place at a CAGR of 133 percent in the last five years) [10].

However, the set targets were not met. The following Table 1 summarizes the initiatives taken by the Indian government, their targets, and shortcomings.

Table 1. A brief on government initiatives.

Initiatives	Targets	Shortcomings
National Electric Mobility Mission Plan, 2013	6–7 million EVs on road by 2020	EV penetration in the last five years has been less than 1%. Policy measures were not sufficient
Faster Adoption and Manufacturing of Hybrid and Electric Vehicles (FAME I), 2015	NEMMP flagship programme to help EV market growth	Sanctioned funds inadequately used; yet it paved the way for electric vehicle adoption in India
FAME II (2019)	Aims to capitalize on the FAME I excitement to launch the EV sector	No mandate for EV adoption; no provision for fee-bate concept; additional riders for availing subsidy; no subsidy for private 4W; requirement of re-certification; requirement of indigenous component; there is no statutory organization who is responsible for the development if charging infra

3   Methodology

The aim of the paper is to produce a framework for a unified EV policy in India. The paper firstly includes understanding the concepts of public policy analysis and followed by literature review of the Global and Indian EV scenario. The 'within case' analysis is performed on the multiple state EV policies of India through 'qualitative comparative analysis' approach. Under the qualitative comparative analysis, each State EV policy is examined using 29 parameters, divided into seven segments. With a composite index method, these parameters, such as institutional mechanisms and policy mandates, are scored based on linear weightage. After evaluating each policy's advantages and disadvantages, an approach is proposed for a unified policy.

4   Analysis of State Policies and Findings

The chronology of the initiatives taken by GoI are shown in Fig. 1. The composite index method is used to rank 12 state policies (see and Fig. 2)[11]. In this method, a total of 29 parameters are categorized under seven segments, and each parameter is given a weight of one point. Based on these weights, each state's EV policy is scored, and the results are generated.

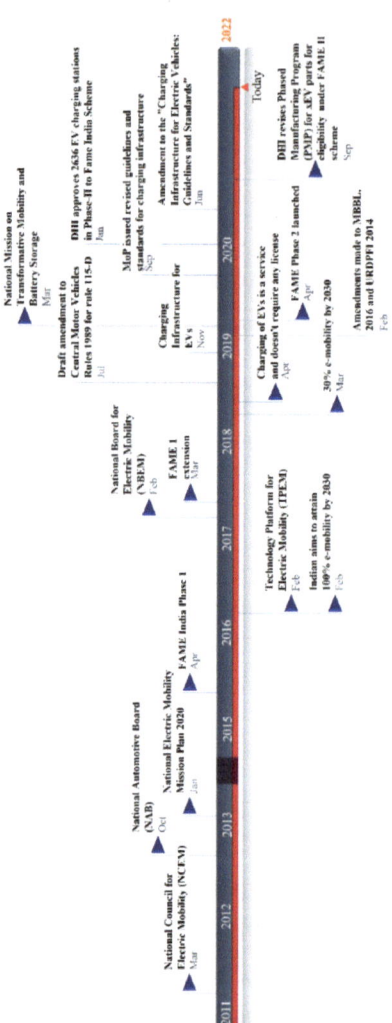

Fig. 1. Key national initiatives to boost EV usage; Source: Government Notifications.

Upon examination of various states' policies, it has been determined that Madhya Pradesh commands the superior score with respect to the parameter. Nonetheless, it is crucial to acknowledge that the mere adoption of specific/ higher number of policy levers does not validate a region's benchmark status. Equally important are the set electrification/ GHG mitigation targets, demand and supply networks, infrastructure availability, and appropriate monitoring mechanisms.

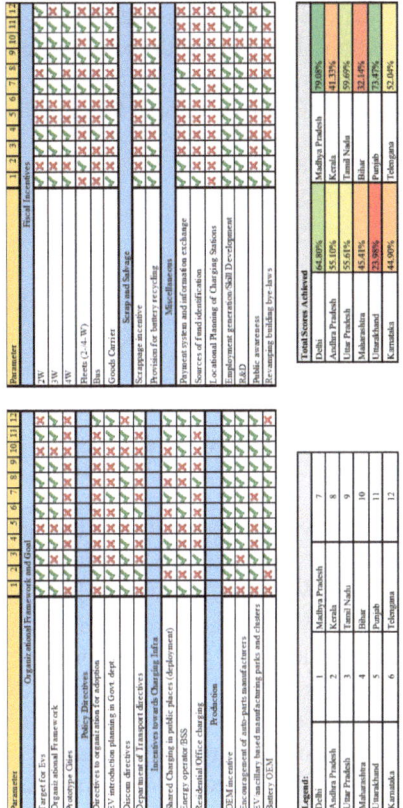

Fig. 2. Key national initiatives to boost EV usage; Source: Government Notifications

The landscape assessment along with the status quo review of the achieved targets of each state can be done in a further study to determine the causal loops between the ecosystem parameters.

5   Recommendations and Conclusion

In formulating a National Electric Vehicle (EV) Policy Framework (see Fig. 3), our first priority is to establish clear policy objectives aligned with India's National Determined Contribution (NDC), national energy, environmental, and transport goals. Key quantifiable benchmarks should be set in advance, encompassing desired shifts in the vehicular paradigm, such as the proportion of EVs in our total vehicle fleet, reduction of greenhouse gas emissions, and comprehensive goals for charging infrastructure development. Secondly, an exhaustive examination of India's current EV ecosystems is vital, providing an indispensable lens on market trends, consumer preferences, technological progress, and the efficacy of policy mechanisms across various states and EV segments. Subsequently, a holistic evaluation of the differential impacts of varied EV adoption scenarios is necessary to elucidate techno-commercial, socio-economic, environmental, and gender-oriented implications. Equipped with such knowledge, potential obstructions within current policy mechanisms can be identified and assessed, enabling a sharp focus on constructive regulations and incentives tailored for complete EV segments.

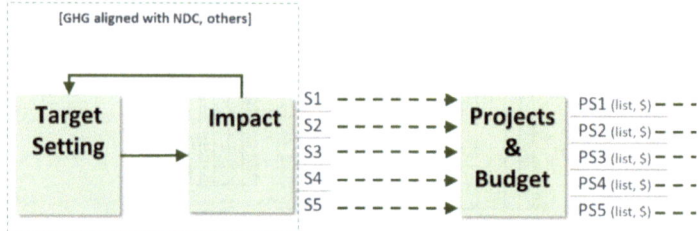

Fig. 3. A representation of process to be followed for Unified EV policy development for India

This strategic reorientation will address both the demand side, incorporating fiscal and non-fiscal incentives for EV acquisition and usage, along with disincentives for Internal Combustion Engine Vehicles (ICEVs), and the supply side with fiscal and non-fiscal incentives fostering EV manufacturing, retrofitting, and localization. Additionally, the establishment of standards and guidelines for EVs and ICEVs will be integral to this transition. An inclusive perspective accommodating cultural and social perception from all user and non-user cohorts is a fundamental element of this design. This comprehensive approach will result in a robust EV impact assessment, underpinning the influence of prospective policies on economic parameters (foreign exchange savings, revenue savings, EV budget), environmental impact (reducing GHG emissions, declining air pollution), and crucial social aspects, such as employment opportunities. Subsequent modifications through stakeholder consultations, coupled with the finalization of EV Policy Measures and Timelines, stakeholder roles and responsibilities, budget estimations, and resource management will culminate in a meticulously designed EV Policy of India. This policy will encompass (across all EV segments) National EV Targets, Infrastructure Requirement (Charger, Grid EVs), EV Policy Measures (Fiscal, non-fiscal), EV Budget & Cost to Government, EV Adoption Benefits, Resource allocation, Implementation Plan and Stakeholder roles & responsibilities, and Monitoring & Evaluation Mechanisms. The carefully crafted blueprint for a unified national EV policy in India encapsulates crucial elements such as market assessment, policy roadmap, feasibility analysis, and EV awareness and capacity-building. It ensures consistency across state policies, while simultaneously allowing adaptation based on distinctive local factors. In essence, the process requires strategic cohesion, regulatory uniformity, and context-driven flexibility. These components create a balanced framework that accommodates both uniformity and customization, thereby facilitating a performance-driven and sustainable national EV policy.

This study was conducted during Covid-19 situation, there it is based on the secondary data as limitation of the study. The study can enrich its findings more through the stakeholder interviews and consultation as way forward.

References

1. International Energy Agency, World Energy Outlook iea.org/weo. Int. Energy Agency, p. 763 (2017)
2. MoHIPE, NEMMP 2020 (2012). Accessed 20 May 2019. https://dhi.nic.in/writereaddata/Content/NEMMP2020.pdf
3. Ministry of Heavy Industries, FAME India Scheme (2019). https://pib.gov.in/PressReleasePage.aspx?PRID=1577880. Accessed 10 Jan 2021
4. Dunn, W.N.: Policy analysis structure (2019)
5. Cochran, C.L., Malone, E.F.: Public Policy: Perspectives and Choices, Third Edition. Lynne Reinner (2009)
6. Centre for Development Impact PRACTICE PAPER, Qualitative Comparative Analysis: A Valuable Approach to Add to the Evaluator's Toolbox? Lessons from Recent Applications (2016)
7. Ragin, C.C.: What is Qualitative Comparative Analysis (QCA)? www.u.arizona.edu/~cragin
8. Baptist, C., Befani, B.: Qualitative comparative analysis – a rigorous qualitative method for assessing impact (2015)
9. C40 Cities, Deadline 2020, p. 59 (2016). http://www.c40.org/researches/deadline-2020
10. Patnaik, S., et al.: Status quo analysis of various segments of electric mobility and low carbon passenger road transport in India, 39 (2021). https://niti.gov.in/sites/default/files/2021-04/FullReport_Status-quo_analysis_of_various_segments_of_electric_mobility-compressed.pdf
11. GNCTD, Delhi Electric Vehicles Policy. Dep. Transp. Gov. Natl. Cap. Territ. Delhi, 1–13 (2020). https://transport.delhi.gov.in/sites/default/files/All-PDF/Delhi_Electric_Vehicles_Policy_2020.pdf

Open Access This chapter is licensed under the terms of the Creative Commons Attribution 4.0 International License (http://creativecommons.org/licenses/by/4.0/), which permits use, sharing, adaptation, distribution and reproduction in any medium or format, as long as you give appropriate credit to the original author(s) and the source, provide a link to the Creative Commons license and indicate if changes were made.

The images or other third party material in this chapter are included in the chapter's Creative Commons license, unless indicated otherwise in a credit line to the material. If material is not included in the chapter's Creative Commons license and your intended use is not permitted by statutory regulation or exceeds the permitted use, you will need to obtain permission directly from the copyright holder.

Assessment of Parking Policy in Case of Tier II Cities in India: Case of Aurangabad, Maharashtra

Sanika Sahasrabudhe[1]([✉]) and Mohit Dev[2]

[1] M.Plan (TPLM), Department of Transport Planning, School of Planning and Architecture Bhopal, Bhopal 462030, India
sanika.s2801@gmail.com

[2] Department of TRAnsport Planning, School of Planning and Architecture Bhopal, Bhopal 462030, India

Abstract. Unprecedented development in Indian cities, including tier II and tier III driven by population growth and increased vehicle ownership has created urgent transport infrastructure and urban planning challenges. These challenges include traffic congestion, parking issues, pollution, and pedestrian safety concerns, all exacerbated by the growing number of private vehicles. This study focuses on evaluating parking policies in Chhatrapati Sambhaji Nagar (formerly Aurangabad), Maharashtra, India. It aims to develop a methodology for assessing on-street and off-street parking performance from user and management agency perspectives. The methodology includes a comprehensive review of parking policies across India and primary surveys involving license plate tracking, parking inventory, and interviews. The study analyzes various parking statistics and employs the Parking Performance Index (PPI) as a tool to assess parking issues. PPI measures user satisfaction, parking management, and service quality, considering factors like parking costs and search time. Clustering techniques define Parking Level of Service (PLOS) criteria, and a single index combines multiple parameters. The study underscores the importance of parking occupancy and highlights that areas with sufficient capacity still face congestion due to mismanagement and inadequate signage. Ultimately, the research aims to enhance parking policies and service quality in the studied city and other Indian urban areas.

Keywords: Parking Policy · Evaluation · Parking Performance Index

1 Introduction

Parking management is essential for achieving policy goals including reducing traffic congestion, improving the economy, or aiming to increase road safety. Parking is primarily a land use issue. As Parking management is a responsibility of Urban Local Bodies (ULBs) stated under twelfth schedule of 74th amendment act of 1992; list of items within the purview of the municipalities. These public conveniences including street lighting, parking areas, bus stops and public conveniences. In order to make cities more livable,

© The Author(s) 2025

C. McNally et al. (Eds.): TRAconference 2024, LNMOB, pp. 702–709, 2025.
https://doi.org/10.1007/978-3-031-85578-8_95

numerous city administrations are developing parking policies and regulations to ease parking pressure and congestion on roads and public places. It is important to evaluate the current parking system and understand the interaction of the criteria that regulate the level of service of parking facilities in order to optimize future planning and operation. The study aims to develop a framework for evaluating the parking regulations in tier II city in India by taking case of Chhatrapati Sambhaji Nagar (formerly Aurangabad), Maharashtra. The study identifies the locations within the boundary of Sambhaji Nagar, which have different land-use such as residential, business, commercial and core city area. Various parking statistics such as demand-capacity ratio, parking duration, parking load are determined to assess parking situations at the selected location. Moreover, based on existing literature, parking LOS indicators like D/C ratio, parking search time, walk time to destination, parking fee and safety and security are considered to estimate the Level of Service. The study employs clustering algorithms like K-means clustering, Fuzzy K-means clustering, Univariate and Agglomerative Hierarchical Clustering. Five LOS classifications; Excellent, Good, Fair, Poor and Very Poor are defined ranging from LOS A to LOS E. A single value index PPI is proposed which combines the thresholds of all five LOS indicators. The parking facilities can be evaluated thoroughly using this index.

2 Literature Review

A parking study was conducted in Delhi, India to evaluate the effectiveness of the parking system for different types of land use. However, the study lacked methods for assessing parking in developing nations like India where smart parking technologies and proper parking area markings are lacking. The study uses the clustering technique to calculate the Parking Performance Index (PPI) and assess the parking system quantitatively. Four parameters were considered: parking demand, D/C ratio, parking load, and parking duration. The study covered residential, commercial, and office areas. Data was collected through parking inventory survey, license plate survey, and questionnaire. Two clustering algorithms were used to classify the data into four PPI categories. The objectives of the study is to make the parking system more efficient and reduce reliance on private vehicles through policy implications (Pritikana Das, 2021). The lack of data and clear parking policies in Vadodara city has resulted in a situation where on-street parking in the central business district continues without any changes. To understand the parking demand, surveys were conducted using the license plate method, collecting information on factors such as gender, origin-destination, income, travel-time, and trip purpose. A fuzzy interference system-based model was developed using MATLAB to analyze the data and predict people's response to proposed parking policies. The study suggests that household income, trip frequency, and parking duration are factors that influence people's behavior towards on-street parking policies (Rutul M. Diyora, 2020). The term "level of service" refers to the quality of customer service provided by a facility currently in use. Parking has become a major concern in transportation planning due to increased use of private vehicles. It is important to assess parking lot conditions to determine how well they accommodate vehicles and how user-friendly they are. This study focuses on finding a method to measure the effectiveness of parking area operation in India, specifically the "Parking Area Level of Service (PALOS)". Factors such as parking costs, ease of use, walking distance, and demand/capacity ratio are considered as quantitative metrics for service level. Different clustering algorithms are used to define the PALOS criteria, with the K-means clustering approach being the most effective. A Parking Performance Index (PPI) is also developed to combine all the parameters. This study will be useful for planning professionals (Janak Parmar, 2019). The increasing number of cars in China's cities has made finding parking difficult. Evaluating parking services is important for transportation planning and management. This paper introduces the concept of level of service (LOS) based on a study in Beijing. A model is used to measure user satisfaction with parking quality. The model combines four variables to create a single index for assessing parking performance. Case studies demonstrate the usefulness of this technique (Yulong He, 2015). Das and Ahmed (2018) "conducted a study to define the level of service for on-street parking facilities. They defined three categories- PC (Parking Characteristics), DC (Design Characteristics), and SC (Safety Characteristics) to classify different elements of parking and developed LOS equation whose coefficients were estimated using AHP."

3   Methodology

The parking needs in a specific area are closely tied to its land utilization. The different types of areas having different land use have been selected as study area in Aurangabad, Maharashtra such as commercial, business area, core city market area.

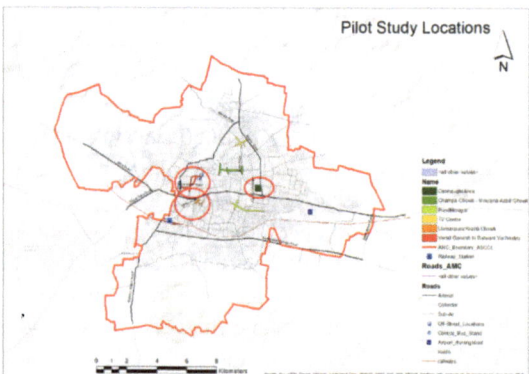

Fig. 1. Locations of Study Area in Aurangabad, generated using QGIS. QGIS is open-source software licensed under the GNU General Public License. Available at: https://qgis.org

The location of this areas has been shown in Fig. 1. These are the areas where there are issues with spillover, haphazard parking and illegal parking, parking on the carriageway, and other issues related to improper parking management, insufficient parking space at some locations, a lack of required parking signs, drivers' propensity to park as close to their destination as possible, and unkempt use of the available parking space. In order to analyze and assess the parking features and to propose parking regulations for the same, these areas have been chosen as a research area. To collect the data, various field surveys, including in-out surveys, license plate surveys, and parking inventory surveys, have been carried out. Different parking characteristics have been evaluated after an in-depth assessment. Finally, based on the observation and analysis, parking regulations have been proposed.

4   Estimation of Parking Performance Index (PPI)

D/C ratio, parking load, parking duration, and parking efficiency are four quantitative criteria considered. The first two parameters should be clear to everybody. The number of vehicles accumulated while parking is known as the parking load. It is expressed in terms of vehicle-hours and is the region beneath in the parking accumulation curve. The parking efficiency, sometimes referred to as the parking index, is a general indicator of how effectively a parking lot is used. It is a measure of how many bays are used in relation to their entire capacity within a specific period. Given that each unique car parks for a longer period, a high parking load predicts a lower efficiency. A methodology to derive a Parking Performance Index (PPI) in tier II city con-text is developed considering Level of Service (LOS) indicators- D/C ratio, search + park time, walk time to destination from parking spot, parking fees and safety and security in terms of presence of lighting and CCTV. As stated earlier, four unsupervised clustering algorithms; k-means clustering, fuzzy k-means clustering, univariate clustering, and agglomerative hierarchical clustering are used for the classification problem. These four are considered as partitional clustering approaches. The PPI is basically a measure that is used to evaluate the level of service provided by the parking lot. As the intent is to develop a single value index which combines all the five evaluation criteria, it is essential to place these criteria on an equal stage. To make these five indicators unitless so that they could be coalesced together, zero-dimension process is used using min-max normalization to make these indicators unitless as given in the research done for Delhi (Pritikana Das, 2021).

5   Data Analysis

In this study, two locations- Cannaught area and Gulmandi 2-wheeler parking are considered. PPI is calculated for these two locations and same methodology is followed for rest four locations in the city.

5.1   Cannaught Area

Cannaught area is one of the prominent mixed-use areas in the city of Aurangabad. The area is lined with mobile and gadget stores, eateries, and restaurants, Cannaught garden and business land-use like banks and offices. The average parking demand per hour for two-wheelers is 200 and for four-wheelers is 50 as seen in Fig. 2. To justify the hypothesis of relationship between the three variables selected amongst the four parameters a relationship graph between Demand-Capacity Ratio, Search Time in minutes taken by the user to find a parking spot after reaching to the destination and Walk Time in minutes taken by the user to reach their destination from the location of parking is plotted as shown in Figure 3 Occupancy vs search & walk time of Cannaught area Fig. 3. The Correlation Coefficient (R Value) is greater than 0 and less than 1 i.e. It indicates a positive relationship between the three selected variables. Which signifies the hypothesis that "as the Demand-Capacity Ratio will increase the Search Time of finding a free parking spot and Walk Time to the destination will increase" (Fig. 4).

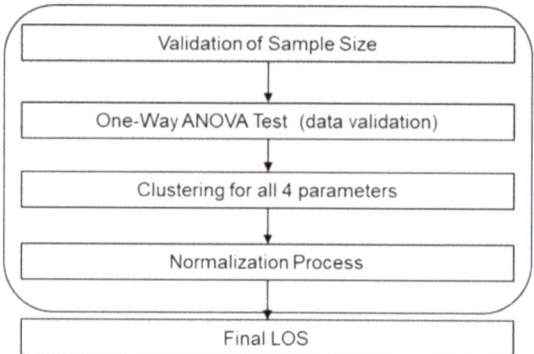

Fig. 2. Methodology for PPI

Fig. 3. Occupancy vs search & walk time of Cannaught area

Fig. 4. Vehicle Accumulation of Cannaught Area

5.2 Gulmandi Two-Wheeler Parking

It is off-street parking located in the core CBD area. It is only permitted for two-wheeler parking. The total area of the parking lot is 144 Sqm. In which maxi-mum capacity is

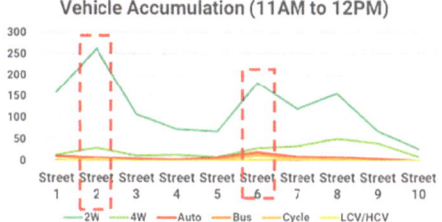

Fig. 5. Vehicle Accumulation of Gulmandi Area

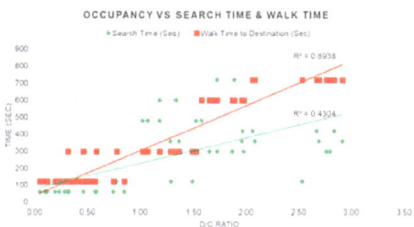

Fig. 6. Occupancy vs search & walk time of Gulmandi Area

of 216 vehicles. The peak hour as observed in the Fig. 5 peak time is from 05:30 PM to 07:00 PM in the evening because of the market. To support the hypothesis regarding the connection between the chosen three variables out of the four parameters, we have plotted a relationship graph depicting Demand-Capacity Ratio, User Search Time (in minutes), and User Walk Time (in minutes) as shown in the Fig. 6. Greater than 0 but less than 1 is the correlation coefficient, or R value. It suggests a favorable correlation between the three chosen variables. indicating the supposition that "as the Demand-Capacity Ratio will increase, the Search Time of finding a Free Parking Spot and Walk Time to the Destination will increase".

After validating the data, the next step involves clustering the five parameters into five performance ranges: Excellent, Good, Average, Poor, and Very Poor. These criteria are based on both quantitative (Demand-Capacity Ratio, Search Time, Walk Time, Parking Fee) and qualitative (Presence of Safety and Security) factors. Four clustering techniques—K-means, Fuzzy K-means, Univariate, and Agglomerative Hierarchical Clustering—were employed to classify the dataset into these categories, with the Silhouette index used to assess clustering quality. A higher Silhouette index indicates better clusters: -1 implies incorrect clustering, 0 signifies indifference, and 1 indicates well-separated clusters. K-means is an unsupervised learning algorithm, while Fuzzy K-means allows for cluster overlap. Univariate clustering groups data based on a single quantitative variable, and Agglomerative Hierarchical Clustering groups objects based on their dissimilarities. The cluster result with the highest silhouette value among the four algorithms was used to define the LOS criteria. That is Agglomerative Hierarchical Clustering (AHC) will be used for Clustering. After validating the clusters with Silhouette Score the final clustering for four Quantitate Parameters namely; Demand-Capacity

Ratio, Search Time, Walk Time and lastly Parking Fee are shown in Table 1 Final Clustering. For the one Quantitative Parameter Safety and Security no clustering will be required as its already on 1–5 Scale.

Table 1. Final Clustering

Criterion	Cluster Boundaries and Designation				
	Excellent	Good	Average	Poor	Very Poor
Parking Fee	<10	10 - 30	30 - 50	50 - 70	>70
D/C Ratio	<0.498	0.498 – 1.188	1.188 – 1.725	1.725 - 1.961	>1.961
Search Time (Sec)	<120	120 - 360	360 – 480	480 – 720	>720
Walk Time (Sec)	<120	120 - 300	300 - 600	600 – 720	> 720
Safety & Security	1	2	3	4	5

After clustering the parameters Min-Max Normalization Technique was employed to combine all five evaluation parameters in to single value of Parking Performance Index. Based on this approach, the estimated ranges for Parking Performance Index and respective LOS Category are defined in Table 2 PPI ranges. r^a represents the boundary value for Excellent condition that is a combination of values of the five parameters. Subsequently, r^b and r^c denote the boundary values for Good and average conditions respectively.

Table 2. PPI ranges

Boundary Condition	Range of PPI	Classification	LOS
$\leq (r^a)$	≤ 0.78	Excellent	LOS A
$(r^a) \sim (r^b)$	0.78 – 1.89	Good	LOS B
$(r^b) \sim (r^c)$	1.89 – 2.98	Average	LOS C
$(r^c) \sim (r^d)$	2.98 – 3.88	Poor	LOS D
$\geq (r^d)$	≥ 3.88	Very Poor	LOS E

The Parking Performance Index for Cannaught Area is Average i.e. Level of Service is B as the combined score is 1.70 which ranges between second range of 0.78 – 1.89. Because of the total score of 2.53, which falls within the range of 1.89 - 2.98. Parking Performance Index for Gulmandi Two-Wheeler Parking have Level of Service C which gives PPI as Average. Because of the total score of 3.15, which falls within the range of 2.98 – 3.88.

6 Conclusions

As previously justified in the hypothesis, there exists a significant relationship among the three parameters: Demand-Capacity Ratio, Search Time, and Walk Time to the Destination. This relationship becomes evident in the study application where we observe distinct patterns for On-Street and Off-Street Parking Locations. For On-Street Parking Locations, it is apparent that Search Time tends to be higher than Walk Time. This occurs because as the demand for parking increases, users spend more time searching for available spots, which ultimately lowers the Level of Service. In other words, a high demand for On-Street parking results in longer Search Times and, consequently, a decrease in the quality of service for users. In contrast, Off-Street Parking Locations exhibit a different behaviour. Even when the demand for parking is high, Search Time remains relatively low. However, Walk Time to the destination from these Off-Street locations tends to be much higher than for On-Street parking. This situation translates into a lower Level of Service for Off-Street parking due to the extended Walk Time users must endure to reach their destinations. In summary, the relationship between these parameters underscores the distinct challenges and impacts on the Level of Service for both On-Street and Off-Street parking scenarios. Understanding these dynamics is crucial for optimizing parking management strategies and improving the overall user experience.

References

Congress, I.R.: Guidelines for Parking Facilities in Urban Area, IRC:SP:12–2015. In IRC:SP:12–2015. New Delhi (2015)

ITDP. (n.d.). On-Street Parking Pricing A Guide To Management, Enforcement and Evaluation

Janak Parmar, P.D.: Development of level of service criteria for urban car parking system using clustering techniques. Research Gate (2019)

Pritikana Das, J.P.: A methodology for evaluating urban parking system: case study of Delhi (2021)

Rutul, M., Diyora, H.M.: Assessment of on-street parking for congested area of Vadodara City. Int. J. Eng. Res. Technol. (IJERT) ISSN: 2278–0181, 6 (2020)

Vehicle Registration & Population in the year 2019–2020. RTO Maharashtra (2020)

Yulong He, X.S.: Level of Service for Parking Facilities . J. Traffic Transp. Eng. (2015)

Open Access This chapter is licensed under the terms of the Creative Commons Attribution 4.0 International License (http://creativecommons.org/licenses/by/4.0/), which permits use, sharing, adaptation, distribution and reproduction in any medium or format, as long as you give appropriate credit to the original author(s) and the source, provide a link to the Creative Commons license and indicate if changes were made.

The images or other third party material in this chapter are included in the chapter's Creative Commons license, unless indicated otherwise in a credit line to the material. If material is not included in the chapter's Creative Commons license and your intended use is not permitted by statutory regulation or exceeds the permitted use, you will need to obtain permission directly from the copyright holder.

A Multicriteria Framework for the Evaluation of Qualitative and Quantitative Impacts of Sharing Mobility Services

Simona Muratori[1] , Magdalena Hykšová[2] , Vesna Janković-Milić[2] ,
Alessandro Luè[1(✉)] , André Maia Pereira[2] , Ivana Marjanović[3] ,
Valerio Mazzeschi[1] , Valerio Paruscio[1] , Jelena Stanković[3] , and Luca Studer[4]

[1] Poliedra-Politecnico di Milano, 20133 Milan, Italy
{simona.muratori,alessandro.lue}@polimi.it
[2] Faculty of Transportation Sciences, Czech Technical University in Prague,
110 00 Prague, Czech Republic
[3] Faculty of Economics, University of Niš, 18000 Niš, Serbia
[4] Dipartimento di Design, Politecnico di Milano, 20158 Milan, Italy

Abstract. In recent decades, there has been a continuous rise in interest surrounding shared mobility services. The deployment of these services has faced numerous obstacles during both the planning and operational stages. The objective of this paper is to define a comprehensive and general-purpose evaluation framework to support the impact assessment of sharing mobility policy/actions for both the monitoring of existing projects (ex-post) and the evaluation of simulated scenarios (ex-ante).

We propose a multi-criteria analysis able to consider both quantitative and qualitative impacts. Emphasis is given on structuring the problem by identifying the most significant economic, social, environmental, and mobility impacts of sharing services and giving examples of key performance indicators (KPIs) that can be employed to gauge them.

Finally, the paper presents the possibility of implementation in the city of Niš (Serbia) within the Horizon Europe project UR-DATA. Niš, the third largest city in Serbia, currently lacks a sharing mobility system.

Keywords: Multi-Criteria analysis · Policy Evaluation · Sharing Mobility · Decision Support Systems

1 Introduction

Over the recent decades, urban mobility has brought to the forefront an array of complex challenges, encompassing issues ranging from traffic congestion and air pollution to questions of equity, accessibility, safety, parking…

As a consequence, thanks to the interest at both European and local level, many actions, projects, researches and initiatives have been set up aiming at improving the urban mobility system (see for instance [1] and [2] for a glimpse on approaches to the

© The Author(s) 2025
C. McNally et al. (Eds.): TRAconference 2024, LNMOB, pp. 710–716, 2025.
https://doi.org/10.1007/978-3-031-85578-8_96

journey to work), and many among these focus on shared mobility [3]. Due to its novelty and dynamic nature, in many instances, decision-makers have found it challenging to effectively regulate the growth of these services.

In this context, it is essential to define a comprehensive and general-purpose evaluation framework to support the impact assessment of sharing mobility policy/actions (see [4] for a review) both the monitoring of existing projects (ex-post) and the evaluation of simulated scenarios (ex-ante). The ex-ante evaluation allows the comparison of alternative scenarios obtained by simulating the response of the mobility system as input parameters change, supporting the decision maker in his/her choices. The monitoring system shows the notable impacts resulting from the policy or actions' implementation, thereby facilitating the evaluation of their performance. This enables the early detection of unexpected consequences, allowing for timely corrective measures as needed. Moreover, employing a consistent evaluation framework for both monitoring and ex-ante assessments enables the possibility to adjust the decision-making structure, if necessary, in order to make more informed decisions in the future.

In the next Section, we structure the evaluation framework as a multi-criteria analysis, bringing particular attention to impact definition and key performance indicators. Finally, in Sect. 3 we discuss the application to the city of Niš in Serbia.

2   Structuring the Evaluation Framework

2.1   Multi-Criteria Analysis

The challenge in creating a sharing mobility service arises from the existence of multiple conflicting objectives that need to be addressed. Consider for instance the goal of expanding the service's geographical coverage, which aims to be maximized; this expansion leads to higher investment and operational expenses, which are objectives that should be minimized. Furthermore, various stakeholders have different objectives or perceive the same objective differently: using the same example, service users aim to maximize the service's geographical coverage, while service operators seek to optimize it to maximize their profit.

Lastly, the impacts of a shared mobility service encompass various aspects. Some of these impacts can be quantitatively measured, while others can only be qualitatively assessed. This variation could stem from constraints in measuring specific aspects using the resources at hand or from the inherent nature of these impacts.

Multi-criteria analysis (MCA) includes the methods seeking to explicitly take into account multiple criteria in holistically evaluating different decision alternatives having conflicting objectives and incommensurable impacts and to explore their values in decision making [5].

Various methodologies have been devised to help classify and rank options based on the criteria that decision-makers deem significant [5]. We can broadly categorize methods into two main groups, each characterized by a distinct approach:

- Definition of an overarching satisfaction function to aggregate the different criteria. The two most renowned approaches within this category include the Multi-Attribute Utility Theory [6] and the Analytic Hierarchy Process [7].

- Comparison of the various alternatives through pairwise assessments, utilizing concordance and discordance indices. An instance of this approach can be seen in outranking methods, such as Electre, developed by [8].

Each of these MCA methods possesses unique advantages and limitations, making it impossible to assert that one method is universally superior to the others, they simply work in different ways [9]. The selection of a method hinges on the specific characteristics of the problem at hand, such as whether most of the impacts are quantitative or qualitative, the intended purpose of the method, and the preferences of both the decision-maker and the analyst.

Hence, we do not intend to propose the adoption of a particular MCA method for evaluating all shared mobility services in every circumstance. Our emphasis, instead, is on structuring the problem by identifying the most significant impacts of such services and giving examples of key performance indicators (KPIs) that can be employed to gauge them.

2.2 The Impacts of Shared Mobility

The impacts of sharing mobility services are multifaceted [10], and can vary depending on the specific context and the stakeholders involved. Indeed, factors such as the population density of the area in question, existing congestion levels, the condition of the local transportation infrastructure, the current regulatory landscape, economic circumstances and income levels, and the diverse interests and preferences of stakeholders can lead to varying impacts.

We present an impact framework here, which can serve as a guideline for assessing all sharing mobility services. This framework begins with identifying the broad sectors influenced by these services, delves into pinpointing the primary impacts within each sector, and ultimately proposes a set of measurable KPIs that can be valuable in addressing the primary impacts in a majority of cases.

Impacts can be classified according to 4 main sectors: economic, social, environmental and mobility.

Shared mobility can imply various economic impacts, both positive and negative, on different stakeholders and sectors of the economy. Some of the key *economic impacts* are the following:

- Consumer savings
- Job creation
- Revenue generation for companies
- Impact on traditional taxi services
- Data economy
- Regulatory and compliance costs
- Impact on tourism and local economy

The social impacts of shared mobility can affect individuals, communities, and society as a whole. Some of the main *social impacts* are the following:

- Improved accessibility and inclusivity
- Reduced traffic accidents and road safety

- Impact on public transit
- Changes in travel behavior
- Privacy and data concerns
- Social interaction and community building

The environmental consequences of shared mobility can exhibit significant variability, contingent on factors such as the energy source powering shared vehicles, the rate at which vehicles are replaced, and the effectiveness of operational practices. While shared mobility can offer numerous environmental advantages compared to conventional modes of transportation, especially private car ownership, it can also have adverse effects. The most relevant *environmental impacts* are:

- Reduced greenhouse gas and air pollutant emissions/air quality improvement
- Reduced land use changes
- Resource efficiency
- Increased availability of data for monitoring environmental effects of mobility and optimizing it
- Promotion of sustainable transportation modes

Shared mobility undeniably exerts a significant influence on people's mobility patterns and access to transportation alternatives. It introduces new mobility options and reshapes the dynamics of travel for individuals and communities. Some of the key *impacts on mobility* are:

- Diversification of transportation options
- First/Last-Mile Connectivity
- Reduced Car Ownership
- Optimization of routes
- Increased use of public transit
- Easier access without the use of a private vehicle
- Transportation network integration and interoperability
- Increased availability and sharing of data for urban planning / traffic management

2.3   KPIs for an Electric Car Sharing Service

Measuring the impacts of sharing mobility services requires a set of KPIs that can take into account the relevant aspects of their effects. The selection of specific KPIs depends on the goals and priorities of the assessment, as well as the availability of data. Different stakeholders, including public decision makers, service providers, and researchers, may prioritize different KPIs to evaluate the impacts of sharing mobility effectively.

We take as example of KPIs selection, the case of designing an electric car sharing service in the city of Milano [11]. Starting form the formalization of a set of theoretical models, a system modelling of the service configurations was designed and implemented (see Fig. 1). The combination of configuration parameters (nodes on the left in Fig. 1) identifies different service options. The configuration parameters lead to some intermediate nodes that depend on the specific service configuration, and, in turn, determine the value of KPIs (nodes on the right in Fig. 1) that measure the extent to which the objectives are achieved. This approach enables the comparison of various service configurations by evaluating their outcomes and the resulting degree of objective fulfilment through KPIs.

The system modelling allowed to identify and quantify – through a stated preference model – the influence of a limited number of parameters, i.e. spatial localization, spatial flexibility, probability of finding a vehicle, and annual and hourly rates on the 5 KPIs identified as critical in this case:

- accessibility - variation of the level of accessibility to the urban mobility system;
- congestion - variation of the congestion level on the road network;
- local and global emissions - quantity of pollutants and greenhouse gas emissions;
- parking places - variation of the public space occupied by private car;
- net present value - economic performance of the car sharing service.

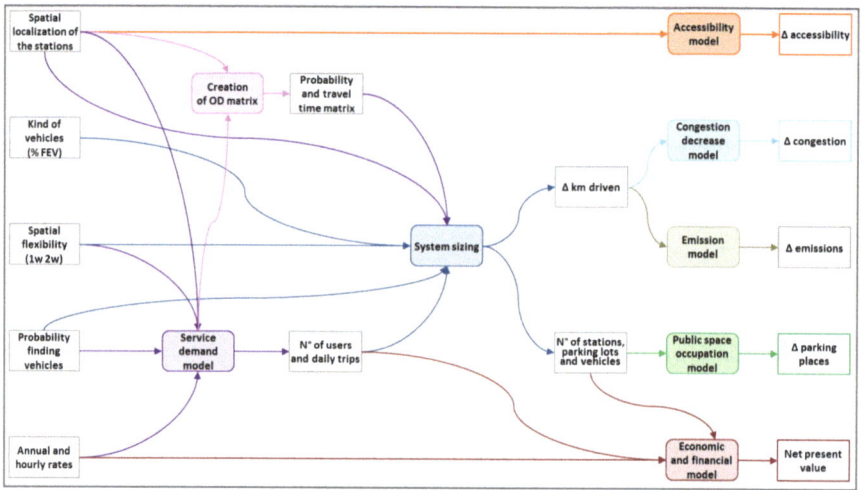

Fig. 1. System modelling of the service configuration [11].

3 Application to the City of Niš (Serbia)

Niš is the third-largest city in Serbia with a population of 250,000 inhabitants. Mobility and transport are of crucial importance for the City of Niš, especially as traffic volume has significantly increased in recent years, leading to substantial congestion, particularly during rush hours. Despite the evident challenges, no sharing mobility system has been implemented thus far, although it holds the potential to address some of the city's issues. Within the framework of the UR-DATA project—Twinning for excellence in Smart and Resilient Urban Development: Advanced Data Analytics Approach—the possibility of introducing sharing mobility services is currently under study.

Among the obstacles to implementation are an inadequate policy framework—only 2 out of 28 cities in Serbia have adopted a Sustainable Urban Mobility Plan—and a lack of citizens' awareness regarding the use of shared mobility beyond public transport. However, the main obstacle that UR-DATA is trying to overcome, in order to apply the framework described in this paper, is the lack of suitable data. There are data from public transport and taxi services, from public car parks in the city, and traffic accidents and violations, but those are not enough to build an O/D matrix, substantially needed for every impact evaluation. Therefore, an agreement has been signed with Telekom Srbija, the largest telecommunications company in Serbia and provider of a wide range of telecommunications services, including mobile telephone services. A database on urban mobility, including the O/D matrix for Niš, is being generated based on mobile phones data. That will enable to assess different shared mobility services and configurations, with the involvement of the city administration and the citizens.

Acknowledgements. The activities described in this paper received funding from the European Union's Horizon Europe Framework Programme (HORIZON) under grant agreement N. 101059994 (UR-DATA). The sole responsibility for the content of this publication lies with the authors and it does not necessarily reflect the opinion of the European Union. The European Commission is not responsible for any use that may be made of the information contained therein.

References

1. Shuttleworth, I., Feehan, C.: Active transport and the journey to work in northern Ireland: a longitudinal perspective 1991–2011. Econ. Themes **61**(1) 2, 19–39 (2023)
2. Teodorović, D., Nikolić, M., Work zone scheduling problem in the urban traffic networks, Econ. Themes **61**(1) 1, 1–18 (2023)
3. Machado, C.A.S., de Salles Hue, N.P.M., Berssaneti, F.T., Quintanilha, J.A.: An overview of shared mobility. Sustainability **10**(12), 4342 (2018)
4. Roukouni, A., Homem de Almeida Correia, G.: Evaluation methods for the impacts of shared mobility: classification and critical review. Sustainability **12**(24):10504 (2020)
5. Figueira, J., Greco, S., Ehrgott, M.: Multiple Criteria Decision Analysis: State of the Art Surveys, vol. 78. Springer, Boston (2005)
6. Keeney, R.L., Raiffa, H.: Decision analysis with multiple conflicting objectives. Wiley&Sons, New York (1976)
7. Saaty, T.L.: How to make a decision: the analytic hierarchy process. Eur. J. Oper. Res. **48**, 9–26 (1990)
8. Roy, B.: Multicriteria Methodology for Decision Aiding, vol. 12. Springer, Dordrecht (1996)
9. Løken, E.: Use of multicriteria decision analysis methods for energy planning problems. Renew. Sustain. Energy Rev. **11**(7), 1584–1595 (2007)
10. Roukouni, A., Homem de Almeida Correia, G.: Evaluation methods for the impacts of shared mobility: classification and critical review. Sustainability **12**(24), 10504 (2020)
11. Luè A., Colorni A., Nocerino R.: Cognitive mapping and multi-criteria assessment for the design of an electric car sharing service. In: M. André, Z. Samaras (eds.) Energy and Environment, John Wiley & Sons, pp 31–45 (2016)

Open Access This chapter is licensed under the terms of the Creative Commons Attribution 4.0 International License (http://creativecommons.org/licenses/by/4.0/), which permits use, sharing, adaptation, distribution and reproduction in any medium or format, as long as you give appropriate credit to the original author(s) and the source, provide a link to the Creative Commons license and indicate if changes were made.

The images or other third party material in this chapter are included in the chapter's Creative Commons license, unless indicated otherwise in a credit line to the material. If material is not included in the chapter's Creative Commons license and your intended use is not permitted by statutory regulation or exceeds the permitted use, you will need to obtain permission directly from the copyright holder.

National Roads 2040 and Active Mode Connectivity

Robbie English[1]([✉]), John Humphreys[1], Tahel Wexler[1], Keadi Adams[1], and Declan Keenan[2]

[1] AECOM Ireland, Dun Laoghaire, Ireland
Robbie.English@aecom.com
[2] Transport Infrastructure Ireland, Dublin, Ireland

Abstract. This research paper seeks to achieve three objectives: (1) assess the high-level impact of severance for Active Travel (AT) users in Ireland through a policy and literature review; (2) analyse a severance case study along the M50 in Dublin, exploring potential interventions; and (3) outline the implications of this study for transport planning practice and further research. Research objectives 1 and 3 are primarily literature and discussion based with research showing that a single definition for the term severance has not been agreed upon as multiple interpretations exist. To investigate M50 severance, residential address points from Geo Directory were mapped in GIS with Liffey Valley Shopping Centre as the destination identifying desire lines using the 'Near' tool and the points of severance.

Case study results are twofold. Firstly, there is North-South severance on either side of the River Liffey valley and the N4. This is due to a lack of AT accessible road connections over the river. Secondly, there is East-West severance between the Palmerstown community to the east of the M50 and the Liffey Valley shopping centre on the west side, due to the M50 not having a crossing point.

This research is focused on the M50 orbital motorway in Dublin, and associated areas like the Liffey Valley, to inform future research on how to identify severance issues and highlight its importance in project planning. Further research on this topic could develop robust standards to define severance in an Irish context, as a standard definition does not currently exist, and included across projects as part of improved project checks to ensure severance is avoided and mitigated.

Keywords: Severance · Ireland · Transport Planning · Active Modes · Motorway

1   Introduction

The Irish Design Manual for Urban Roads and Streets (DMURS) defines severance as "where the provision of road infrastructure bisects an area, making people movement within the area more difficult" [4]. Severance can lead to indirect routes which favour vehicular trips, discourages active travel (AT), and exacerbates transport poverty. In the early 2000s, Ireland rapidly improved its National Roads Network (NRN) - construction of motorways and dual carriageways. Due to myriad reasons, consideration of AT

© The Author(s) 2025
C. McNally et al. (Eds.): TRAconference 2024, LNMOB, pp. 717–722, 2025.
https://doi.org/10.1007/978-3-031-85578-8_97

infrastructure was not as prevalent at that time, including mobility culture, a disconnect between land use and transport planning, and a focus on value for money. As such, AT was often uncatered for in the design of infrastructure. The NRN is managed by Transport Infrastructure Ireland (TII) and the network facilitates the daily movement of people and goods across the state for various trip purposes. However, as AT has become increasingly significant, the inadvertent severance issues dividing communities and increasing trip distances for AT has become progressively apparent and needs to be addressed. As negative externalities of private motor vehicles become more accepted, and climate change a central focus of Government policy; transport planning approaches in Ireland shifted to prioritise sustainable travel modes. The Department of Transport's National Investment Framework for Transport in Ireland (NIFTI) prioritises investment in AT and public transport infrastructure. As part of this policy change, eliminating severance on National Roads is a key priority for TII, with their strategy National Roads 2040 (NR2040) seeking to improve conditions for AT. This paper explores the impact of severance linked to the National Roads network and the potential benefits from mitigating negative impacts.

This research paper seeks to achieve three objectives: (1) assess the high-level impact of severance for AT users in Ireland through a policy and literature review; (2) analyse a severance case study along the M50 in Dublin, exploring potential interventions and their benefits; and (3) outline the implications of this study for transport planning practice and further research.

2   Study Area and Methodology

The M50 is the most heavily trafficked road in Ireland, catering for up to 150,000+ vehicles per day [10]. The M50 is an orbital, segregated motorway, which connects the interurban radial routes on the outskirts of Dublin (Fig. 1), facilitating the movement of people, goods, and services. However, due to being a segregated motorway with AT restrictions as per legislation, it is also a linear barrier which can cause desire line severance due to the limited number of crossing points for AT users. This study uses a case study approach to investigate the impact of M50 sev-

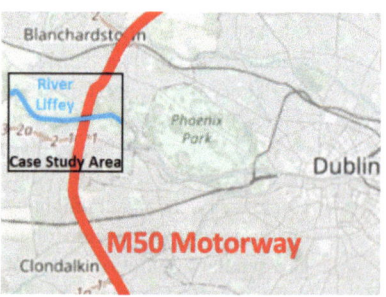

Fig. 1. Location of Case Study (image created using ESRI Mapping)

erance on AT. The case study for this paper is the locality around Liffey Valley Shopping Centre, which is a major trip destination. This is an interesting case study for severance as there is linear severance caused by the River Liffey to the north, and the M50 mainline to the east of the shopping centre.

To investigate M50 severance, residential address points from GeoDirectory were mapped in GIS with Liffey Valley Shopping Centre as the destination to identify desire lines using the 'Near' tool and the points of severance. Potential interventions were then mapped to consider the extent that severance could be mitigated. In addition to this, a path comparison was conducted to study severance caused by the River Liffey to the north with a separate origin destination pair using GIS and Google Maps to assess the change in the recommended path with and without proposed interventions across the river. Previous studies undertaken on severance on the N4 and M50 by Transport Infrastructure Ireland and the National Roads Authority were also consulted to inform the case study analysis.

3 Literature and Policy Review

3.1 Research Literature

As a concept associated with transport infrastructure, severance is frequently defined as a physical and/or psychological barrier that divided people, local services, or social connections within a community [7]. However, there is disagreement in the literature on its exact definition. Some authors used the term "severance" to describe the impact of traffic and use the term "barrier effect" for the impact of transport infrastructure [1]. It can be classified as being either primary by restricting movement from one side to another, or secondary by indirectly contributing to severance effects [7]. For the purposes of this paper, the primary classification for the term "severance" will be used.

Additionally, it has been revealed to influence the health and wellbeing of communities through their participation, or lack thereof, in AT [9]. Whilst some research states there is little evidence linking health to severance [6], AT is associated with positive health benefits [8]. Therefore, transport infrastructure which curtails AT may have negative health implications [7].

3.2 Policy Review

International policy is increasingly prioritising AT connectivity om transport infrastructure, such as roads, which may cause severance. The European Union is currently updating the Regulation of the European Parliament and of the Council Guidelines for the development of the trans-European transport network (TEN-T guidelines) to clearly recognise the need to consider active mobility when planning, implementing and operating infrastructure for long-distance transport. Proposed amendments to regulations emphasise the importance of severance mitigation for improved AT connectivity.

Recent Irish transport policy sets a clear intention to promote AT nationally, including specific actions for connectivity along National Roads. The National Development Plan (NDP) [5] encourages the removal of short, motorised trips from National Roads with an aim to ensure efficient network operation while providing a meaningful alternative to the private car. In line with the NDP, the National Sustainable Mobility Plan 2022–2035 [3] sets actions relating to National Roads, focused on improving road safety and provision of infrastructure to facilitate all road users, including sustainable mobility users, on the National Roads network.

In line with the EU and Irish legislation and policy, TII sets strong commitments in NR2040 [11], its National Roads Strategy, to delivering improved AT provision in all its projects and reducing the severance caused by some National Roads in urban areas. In the strategy, busy urban motorways, ring roads and town bypasses are recognised as potentially hostile environments for pedestrians and cyclists wishing to cross. Accordingly, TII commits to identify and remedy severance impacts from National Roads, through provision of safe crossing infrastructure for cyclists and pedestrians.

4   Results and Discussion

4.1   Objective 1: Literature and Policy Review Conclusions

A common theme between the literature and policy review is that a single definition for the term severance has not been agreed upon and the literature review has shown that there are multiple interpretations of severance. These multiple definitions can lead to uncertainty when trying to assess the impacts that severance can cause, as different definitions are associated with differing effects on communities and AT users. Yet, its broad definition of transport infrastructure acting as a barrier is widely accepted.

There has been a clear shift in focus within policy, to move away from prioritising cars and move toward prioritising AT users. New policy updates have made clear the intention that when planning new infrastructure, AT severance will be mitigated. However, interventions on existing transport infrastructure may require further guidance.

4.2   Objective 2: Liffey Valley Severance Case Study

The impact of severance in the case study area is twofold. Firstly, there is North-South severance on either side of the River Liffey valley and the N4. This is due to a lack of pedestrian/cyclist accessible road connections over the river, as the primary connection is the motorized-only M50. This river severance combined with a steep gradient, drastically in-creases AT trip distances compared to private vehicle travel times. Figure 2 demonstrates that the distance travelled by private vehicles is approx. 9 km for north-south trips, compared to 12 km for active modes, which could be reduced to 6/7 km with a new bridge. Secondly, there is East-West severance between the Palmerstown community to the east of the M50 and the Liffey Valley shop-

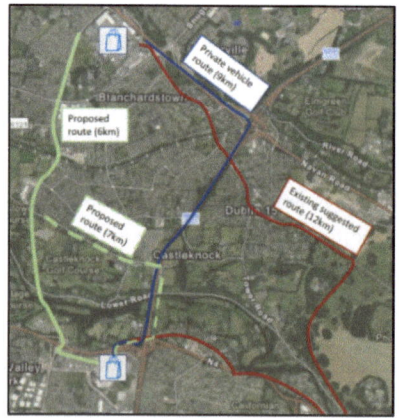

Fig. 2. North-South Distance Reduction (Image created using Google Maps)

ping center on the west side. At this location, there is a 1.4 km section of the M50 mainline which does not have an AT crossing point. Severance in both locations causes longer journey distances, making residents more likely to travel by cars than AT.

Severance can be alleviated by providing an additional AT crossing point. For this case study, Fig. 2 and Fig. 3 display two potential interventions to reduce severance in the case study area and their benefits. In the case of the North-South crossing (Fig. 2), this intervention would improve connectivity between Blanchardstown and Liffey Valley. For East-West crossing, an additional crossing would result in shorter trips to Liffey Valley for many households, indicated by yellow desire lines in Fig. 3.

4.3   Implications for Future Research

The case study undertaken as part of this study identified a single case of severance and proposed possible methods by which its impacts could be diminished. This case study highlighted how the consideration of active modes and severance at project level could greatly reduce the severance impacts of large National Roads. In the specific case of North South severance at the Liffey Valley, the lack of active mode infrastructure as part of the construction of the M50 bridge on the Liffey Valley did not decrease the distances to be travelled by active modes to cross the same obstacle. Under NR2040 this type of a severance issue could have been avoided as it would likely have been identified at the project level and mitigation measures implemented to cater for the needs of AT users.

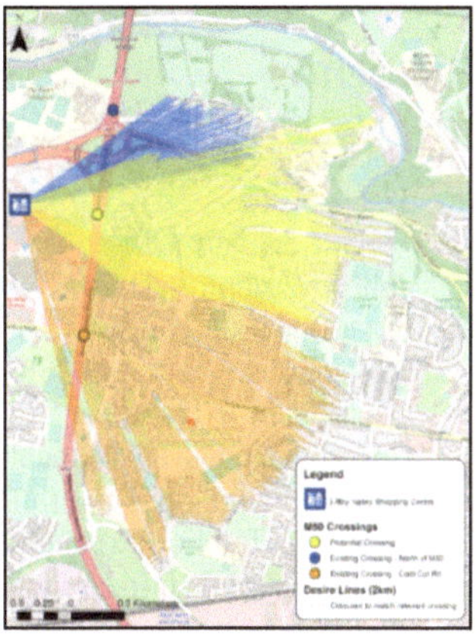

Fig. 3. East-West Desire Line Assessment (image created using ARCGIS Mapping)

The case study also highlighted how addressing existing and preventing future severance can support the transport sector to reach its climate targets by enabling mode shift to more sustainable modes, especially for short local trips such as the 44% of daily journeys which are less then 4km [2]. An extended alternative route for active modes or the lack of safe crossing points on National Roads (combined with gradients) make a private vehicle more convenient, quicker, and more attractive than active modes; in some cases, the private car could be the only reasonable mode option available.

Further research on this topic could develop robust standards to define severance in an Irish context, as a standard definition does not currently exist, and included across projects as part of improved project checks to ensure severance is avoided and mitigated. A method to identify severance across a wide area or measure existing severance at a location could be developed; this case study was a manual process, and it would be difficult to scale up to identify severance locations across a wider area. This type of new standard or method would allow practitioners/planners to identify severance quickly in a way that was comparable across sites.

5   Conclusion and Implications

Within the publication, NR2040 by TII, severance is recognised as occurring on some National Roads. NR2040 commits to identifying and addressing existing severance and to mitigating for possible future severance issues in future infrastructure projects. This is a step change in the approach to severance by Irish policy where active modes are now to be fully considered and the needs and safety of active mode users is to be incorporated into projects from the beginning.

NR2040 represents an important change in transport infrastructure policy for Ireland, especially as the transport sector has a large decarbonization target to achieve by 2030. It also compliments NIFTI, which prioritises active mode users and to enhance or optimise existing transport infrastructure instead of new infrastructure. To build upon this momentum shift, it will be useful for future research to set out severance definitions, especially in an Irish context and inform/establish a series of 'checks and balances' within project planning to ensure severance is avoided and mitigated.

References

1. Anciaes, P., Jones, P., Mindell, J.: Community severance: where is it found and at what cost. Transport Rev., 293–317 (2015)
2. Central Statistic Office, Travel Behavior Trends (2021). https://www.cso.ie/en/releasesandpublications/ep/p-ntstb/travelbehaviourtrends2021/distanceandduration/. Accessed Sept 2023
3. Department of Transport, National Sustainable Mobility Policy, p. 37 (2023). https://www.gov.ie/en/publication/848df-national-sustainable-mobility-policy/. Accessed Sept 2023
4. Department of Transport, Tourism and Sport, and Department of Housing, Planning and Local Government, Design Manual for Urban Roads & Streets (2013). https://assets.gov.ie/227051/cbe57ca9-b4c8-4aae-842f-79c805cfc639.pdf. Accessed Sept 2023
5. Dept. of Public Expenditure, NDP Delivery, Reform, National Development Plan, p. 72 (2021). https://www.gov.ie/en/publication/774e2-national-development-plan-2021-2030/. Accessed Sept 2023
6. Douglas, M., et al.: Health Impact Assessment of Transport Initiatives: A Guide. NHS Health Scotland, Edinburgh (2007)
7. Nimegeer, A., et al.: Experiences of connectivity and severance in the wake of a new motorway: implications for health and well-being. Soc. Sci. Med., 78–86 (2018)
8. Saunders, L., et al.: What are the health benefits of active travel? A systematic review of trials and cohort studies. PloS one **8**(8) (2013)
9. Smith, J., Gurney, A.: Community Effects of Traffic Congestion: a Review of the London Assessment Study Data, Berkshire: Transport Research Laboratory (1992)
10. Transport Infrastructure Ireland, National Roads Network Indicators (2022). https://www.tii.ie/tii-library/strategic-planning/tii-road-network-indicators/TII-National-Roads-Network-Indicators-2022.pdf. Accessed Sept 2023
11. Transport Infrastructure Ireland, National Roads 2040 (2023). https://www.tii.ie/tii-library/strategic-planning/national-roads-2040/TII-NR2040-Final-Report-EN-April-2023.pdf. Accessed Sept 2023

Open Access This chapter is licensed under the terms of the Creative Commons Attribution 4.0 International License (http://creativecommons.org/licenses/by/4.0/), which permits use, sharing, adaptation, distribution and reproduction in any medium or format, as long as you give appropriate credit to the original author(s) and the source, provide a link to the Creative Commons license and indicate if changes were made.

The images or other third party material in this chapter are included in the chapter's Creative Commons license, unless indicated otherwise in a credit line to the material. If material is not included in the chapter's Creative Commons license and your intended use is not permitted by statutory regulation or exceeds the permitted use, you will need to obtain permission directly from the copyright holder.

In the Loop: The Application of System Dynamics in Transport

Gillian Harrison[1]($\boxtimes$) , Astrid Gühnemann[2], Peraphan Jittrapirom[3],
Paul Pfaffenbichler[2], Simon Shepherd[1], and Jonatan J. Gómez Vilchez[4]

[1] Institute for Transport Studies, University of Leeds, Leeds LS2 9JT, UK
G.Harrison@leeds.ac.uk
[2] Institute of Transport Studies, BOKU, 1180 Vienna, Austria
[3] Nijmegen School of Management, Radboud University, 6525 AJ Nijmegen, Netherlands
[4] TE3 Modelling, 21023 Varese, Italy

Abstract. This paper is designed to present to the transport research community the wide existing body of work and potential future applications of the method of System Dynamics (SD) within transport research. We introduce the reader to the basic concepts of the SD method, and discuss previous reviews of academic SD papers, before setting out a systematic review on SD in transport. We find that over 400 papers have been published over the past 40 years, with over 80% in the past decade. The most popular topic is strategic and sustainable policy, but alternative fuel vehicle uptake and maritime applications have been increasingly popular over recent years. SD is also being adopted outside academia in both consultancy and governmental organisations, so we believe that this paper is a timely contribution to embedding SD as a leading emerging method within transport research, which will encourage all members of the community to consider adopting it within their toolkit.

Keywords: System Dynamics · Transport · Systematic Review · Insights

1 Introduction

In order to develop suitable solutions to deal with upcoming challenges for the transport system such as the climate crisis, demographic change or disruptive technological innovations such as digitalization and automation, there is a need for transport models that can map long-term and systemic effects. The method of System Dynamics (SD) is particularly well suited to capture temporal and sectoral interactions and to identify and account for rebound effects. It was developed over 70 years ago for the study of complex systems and has since been widely applied across many fields, including transport. Over the past 40 years, 447 academic papers on SD & Transport have been produced, with over 80% of those in the past decade. The method is being increasingly adopted outside academia by transport consultancies such as WSP, Steer and Goudappel, as well as being used to support policy making, including by national governments, e.g. both the UK and US Departments for Transport, and the European Commission Joint Research Centre.

© The Author(s) 2025
C. McNally et al. (Eds.): TRAconference 2024, LNMOB, pp. 723–729, 2025.
https://doi.org/10.1007/978-3-031-85578-8_98

Thus, the community of researchers working with SD in transport research is gradually growing. The International System Dynamics Society Special Interest Group in Transportation (SDS TSIG) was launched in 2016, and since has delivered 5 successful annual workshops (online during Covid), a seminar series, and a special issue of the Journal of Simulation, as well as hosting a mailing list of more than 200 interested parties. Further to this the SDS TSIG have engaged with the Association for European Transport to organise a thread on System Dynamics at the European Transport Conference for the past 4 years and a Transport thread at the International System Dynamics Conference for the past 2 years. A partial purpose of this paper is to grasp the opportunity to further widen the use of SD in the transport community, and to introduce newcomers to the method. This paper will introduce the concepts of SD and describe the applications of SD in transport, particularly over the last decade, through a systematic review of academic papers, as well as looking to identify future research directions.

2 An Overview of System Dynamics

As described by the System Dynamics Society (SDS), "System Dynamics (SD) is a (computer-aided) approach for strategy and policy design"[1]. The method comprises of both qualitative ("causal loop diagrams") and quantitative ("stock-flow") elements, based on complexity and systems thinking. The method can thus be used to aid communication of mental models between stakeholders, allowing the identification of conflict and consensus and the demonstration of potential policy impacts, as well as the development of both simple and complex simulation models for scenario analysis. The field of SD was first developed at MIT by Jay Forrester in the 1950's[2] in the field of management science to better understand business dynamics, with the first publication released in 1961[1]. Over the last 70 years, the method has been adopted across almost every field of research that requires system understanding, with special interest groups of the SDS in nearly 20 areas, including agriculture, education, energy, health, housing and resilience and structural racism. The complexities, uncertainties and trade-offs of the transport system mean that SD is seen as an ideal method for studying strategic transport policies [2, 3].

SD is based on an underlying theory that the structure of a system will lead to the observed behaviour, and it is by understanding this structure that will allow us to make changes or efficiencies within a system. Critical to SD are the concepts of closed feedback loops, time lags and stocks. An advantage of SD models are short runtimes, which allow comprehensive sensitivity analyses and alternative considerations and the simulation of a large number of explorative scenarios. This allows the identification of critical factors influencing the system behaviour, which can be used as starting points especially for more comprehensive and permanent transformations of systems.

---

[1] https://systemdynamics.org/what-is-system-dynamics/ (accessed 02/08/23).

[2] https://systemdynamics.org/wp-content/uploads/assets/infographic/SDS_infographic.png (accessed 02/08/23.

3   Previous Reviews of System Dynamics in Transport

Importantly, it is now an established modelling approach within the field of transport studies with many applications across freight & logistics, transport and land-use interaction, aviation, sustainable transitions and uptake of new mobility technologies and services. In particular, SD has been widely applied to the analysis and identification of sustainable and efficient transport policies and business models. Although having been utilised in academic transport research for over 40 years, more recently system dynamics models, and the related systems thinking approach is being increasingly adopted in consultancy and policy-making as part of a developing range of modelling tools used. There have been to date two general review studies of the use of SD in transport, one in 1994 [2] and the next in 2014 [3]. Alongside these, four subject-focused review studies have been published: vehicle fleet composition [4], sustainable transport policy [5], transport-health interactions [6] and freight decarbonisation [7].

4   Systematic Review of Transport and System Dynamics

A search for "transport*" and "system dynamics" within article title, abstract and keywords on Scopus (August 2023) revealed 2,240 results in total, of which 1,379 were journal articles[3] This however, is a crude test, as "transport' can be used in many other contexts, specifically within the life sciences. Further excluding all non-English language papers and any papers associated with life sciences[4], we have 1,028 articles. Transport may also be used in a different context in environmental and earth sciences and "system dynamics" may also have differing meanings in the material science/engineering context, but these subject areas were not removed as there is a chance of relevant content within them. These records were then individually assessed by the authors to ensure that they were using a system dynamics method and related to transport studies, and if this was not obvious then the records were excluded. This included studies which were related to technical/engineering papers (e.g. laerodynamics, materials). Some records (especially older ones) only had abstracts available, making it difficult to assess suitability. Following this process, we finally established 447 articles that fit our criteria. Shown in Fig. 1, the earliest article was published in 1981 and only 21 papers were published up until 2005, since when there has been a gradual growth in publications. There was a significant increase and peak of 64 papers in 2020, though with a small drop in subsequent years. We can only surmise on the reason for the 2020 peak, but can say that the profile of SD in transport has been raised by the SDS TSIG since their conception in 2016, hosting international workshops in 2018 and 2019 and produced a Special Issue of the Journal of Simulation on transport in 2020. The dip in subsequent years may well have been effected by researcher availability due to COVID.

We categorized these articles based on the categories of [3], though with some small amendments to better match the identified areas, as shown in Fig. 2. We note that the

---

[3] Document type limited to Article, Source type limited to Journal.

[4] Excluded Subject Areas: Agricultural and Biological Sciences, Chemistry, Medicine, Chemical Engineering, Biochemistry, Genetics & Molecular Biology, Pharmacology, Toxicology & Pharmaceutics, Immunology & Microbiology, Neuroscience.

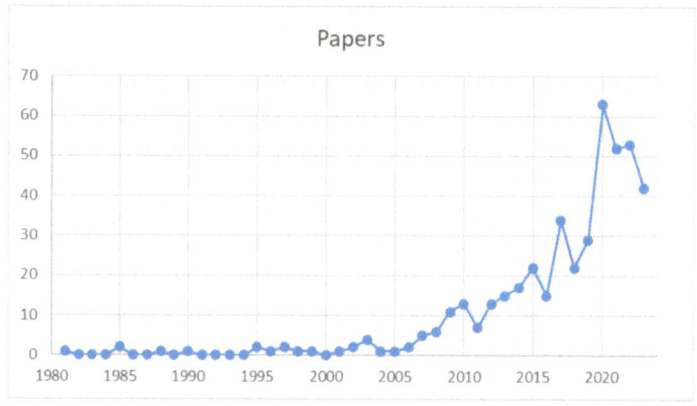

Fig. 1. SD & Transport papers up until 2023 (note 2023 is only 8 months)

largest category is what was fully termed 'Strategic and Sustainable Policy'. This contained a range of applications from specific urban policies related to congestion to wider energy-related emission policies that included transport. As such this category could be refined further in future iterations of this systematic review. The third largest category of Logistics, was similarly diverse, including many aspects related to general freight and specific supply chains so may be further categorized for more detail. The Other category contained a range of topics including, safety, health, emergency evacuations and new technologies (e.g. automated driving).

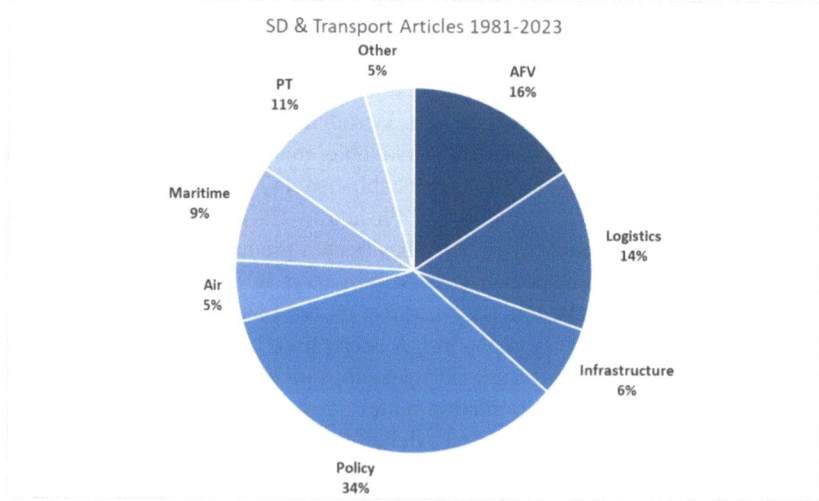

Fig. 2. Articles 1981–2023 (n = 447)(AFV: Alternative Fuel Vehicles; PT = Public Transport.)

5   New Insights Over the Last Ten Years

Given that a decade has passed since [3], and the increasingly wide-spread interest in the method, it would seem an opportune time for an updated review that not only provides insight for SD practitioners but also the wider interested community. In * MERGEFORMAT Fig. 3 we show the categorized articles over the past ten years (2023 excluded as not a full year). For the years 2013 to 2022 there are 322 results - 80% of all articles have been published up to 2022. Although this may seem like an impressive growth rate we recognise that it should be considered within the context of growth in the field of transport (and modelling) as well as general academic publications.

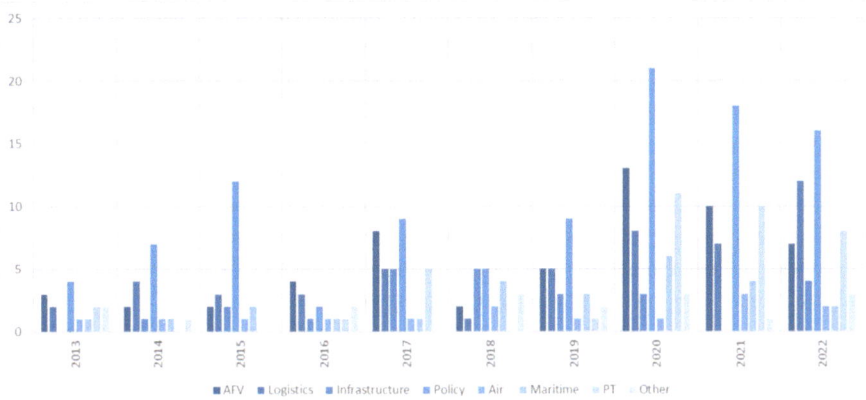

Fig. 3.   Articles 2013–2022 (n = 322)

Although there has been a growth in overall numbers over the past decade, the pattern across categories broadly follows. Policy based papers have dominated all but two years. This reflects the usefulness of SD in strategic policy analysis. Over this period, papers related to AFV uptake have surpassed Logistics as the second most popular topic. This is unsurprising for two reasons. Firstly, the general interest in AFVs has grown steadily since the early 2000s as (especially) electric vehicles are being strongly supported by policies across the world and introduced into our fleets. Secondly, a highly cited paper from 2008 was authored by John Sterman, seen as a leader in the SD community and thus drawing attention to AFV uptake as an SD application. Over more recent years, there has been an increasing interest in maritime, shipping and ports, though it is not clear why this is. There has not been a similar development in air transport, which may indicate an area with potential for growth. On the other hand, papers related to infrastructure and construction do not seem to have experienced the same level of growth in line with other topics, again an indication that researchers and policy agendas are now more focused on non-car options. What seems important is an increase in appreciation that the transport system is not only complex (multiple entities, several stakeholders, etc.) but is also a system within (or alongside) other systems (e.g. urban, energy, social, etc.), and SD is seen as an effective tool to deal with this complexity and explore transitions that include systems or responses not usually considered in other models.

6 Conclusions and Next Steps

In this paper we have given an overview of the use of SD in transport in order to not only show the variety of applications and future directions, but also promote the SD method to the transport research community. SD has been applied to complex, uncertain transport systems as there is a need to better understand problems, in order to design effective solutions. The benefits of structuring knowledge, identifying feedback loops and time lags, then developing short run-time simulations is ideal for this challenge. As such, SD complements (rather than replaces) alternative modelling approaches.

We recognize the limitations of our review, and intend to further refine the process in the future, including a consideration of geographical range of publications, further sub-division of topics and cross-categorisation. This will allow for the development of deeper insights such as where SD adds value compared to other approaches. In addition, the review will be extended to cover other databases and thus increase the robustness. Furthermore, we have not directly considered here how SD has been combined (or could be combined) with other modelling methodologies, such as Agent Based Modelling or Fuzzy Logic, and the use of new and emerging data. Although this paper has focused on academic papers, as mentioned in the introduction, SD is being adopted in consultancies and government organisations, who may not publish academic papers, so further exploration of their engagement and developing case studies of SD models making a "real life" impact on policy making would be beneficial to generate a broader view of the current state of SD in transport. The insights developed here will be built on further and can be used to support the further development of the use of SD in transport and embedding SD into the toolkit of all transport researchers.

References

1. Forrester, J.W.: Industrial Dynamics. MIT Press, Cambridge, USA (1961)
2. Abbas, K.A., Bell, M.G.H.: System dynamics applicability to transportation modeling. Transp. Res. Part A: Policy Pract. **28**(5), 373–390 (1994)
3. Shepherd, S.P.: A review of system dynamics models applied in transportation. Transp. B: Transport Dyn. **2**(2), 83–105 (2014)
4. Gómez Vilchez, J.J., Jochem, P.: Simulating vehicle fleet composition: a review of system dynamics models. Renew. Sustain. Energy Rev. **115**, 109367 (2019)
5. Fontoura, W.B., Ribeiro, G.M.: System dynamics for sustainable transportation policies: a systematic literature review. Urbe. Revista Brasileira de Gestão Urbana, 13 (2021)
6. Harrison, G., Grant-Muller, S.M., Hodgson, F.C.: A review of transport-health system dynamics models. J. Transp. Health **22**, 101138 (2021)
7. Ghisolfi, V., et al.: Freight transport decarbonization: a systematic literature review of system dynamics models. Sustainability **14**(6), 3625 (2022)

Open Access This chapter is licensed under the terms of the Creative Commons Attribution 4.0 International License (http://creativecommons.org/licenses/by/4.0/), which permits use, sharing, adaptation, distribution and reproduction in any medium or format, as long as you give appropriate credit to the original author(s) and the source, provide a link to the Creative Commons license and indicate if changes were made.

The images or other third party material in this chapter are included in the chapter's Creative Commons license, unless indicated otherwise in a credit line to the material. If material is not included in the chapter's Creative Commons license and your intended use is not permitted by statutory regulation or exceeds the permitted use, you will need to obtain permission directly from the copyright holder.

Structure of the Integration of Mobility as a Service (MaaS) with Urban Policy

Yurie Toyama[1]([✉]), Gen Hayauchi[1], Fumihiko Nakamura[2], and Yusuke Kanda[3]

[1] Nagoya University, Furocho, Chikusa-ku, Nagoya 464-8601, Aichi, Japan
mail@yurietoyama.com
[2] The University of Tokyo, 7-3-1 Hongo, Bunkyo 113-8656, Tokyo, Japan
[3] National Institute of Technology, Kure College, 2-2-11 Aga-Minami, Kure, Hiroshima 737-8506, Japan

Abstract. Mobility as a Service (MaaS) levels have been defined based on the degree of integration. However, many cases have not yet been integrated with the highest level, 'Integration of Societal Goal'. Therefore, this study focuses on the highest level, level 4, to guide the further development of MaaS. Based on a case studies, the relationship between the 'Societal Goal' and MaaS is a fundamental goal and objective to implement MaaS. In addition, the relationship between the integrations from levels 1 to 3 is not in linear steps. Nevertheless, they are choices of means to render MaaS successful under the societal goals. Based on this observation, we proposed a new diagram.

Keywords: Mobility as a Service (MaaS) · Urban Transport · Public Transport · Transportation Planning · Policy integration

1 Introduction and Literature Review

Currently, Mobility as a Service (MaaS) has been introduced in cities worldwide, and according to Arias-Molinares et al. (2020), the number of academic articles on MaaS has been increasing since 2014, but one research area in the various academic papers on MaaS is the definition of MaaS (Fig. 1).

Lyons et al. (2019) organized the degree of integration of MaaS around the user's level of cognition. Based on their highest level, level 5 represents complete integration under all conditions and modes.

Sochor et al. (2018) embodied MaaS levels based on a MaaS case study in which MaaS integrates multiple modes of transportation into a single service package and defined it as a level of 'integration'. The highest level, level 4, was defined as 'Integration of Societal Goals'. The contribution of Level 4 can be described as 'the added value is reduced private car ownership and use, a more accessible, livable city, etc. Incentives are implemented in the MaaS service (or implemented in individual services, as a Level 4 approach can be integrated at any level), i.e. reflected by how well local, regional, and/or national policies and goals are integrated into the service.'

© The Author(s) 2025
C. McNally et al. (Eds.): TRAconference 2024, LNMOB, pp. 730–736, 2025.
https://doi.org/10.1007/978-3-031-85578-8_99

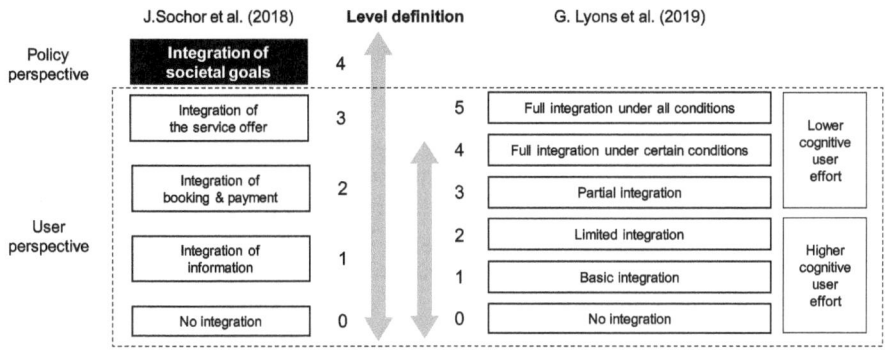

Fig. 1. MaaS levels on past research

Thus far, these proposed levels have influenced various reports and academic studies. For example, the OECD(2021), the Japanese Ministry (MLIT) (2022), and the Dutch government (2018) cited them. Their reports had a significant impact on MaaS implementation because the literature is referred to by industry and local policymakers. However, the 'Societal Goal' expected at level 4 is yet to be defined or discussed. Notably, many MaaS cases remain at levels 2 and 3 (Mulley et al., 2023), whereas MaaS is being expanded in various cities and regions worldwide. To achieve higher MaaS integration, common understanding and discussion on level 4 are required.

Therefore, this study focused on level 4 'Integration of Societal Goals', and conducted a specific interpretation of 'Societal Goals' in the context of MaaS.

Further, we elucidated the Societal Goals and proposed a new level structure considering the relationship between Societal Goals and MaaS.

2   Objective and Methodology

In Sect. 3, we summarized the goals and expectations of MaaS proposed by urban policymakers such as the government and the resulting important concept of 'Sustainability' to elucidate the Societal Goals in MaaS in Sect. (3.1). Section (3.2) summarized three aspects of sustainability, namely economic, environmental, and social, to elucidate 'Sustainability' as a policy goal in MaaS. Based on the understanding of 'Sustainability', we examined several MaaS projects presented by the Netherlands Institute for Transport Policy Analysis (2018) Sect. (3.3). The significance of MaaS in each case was investigated from the websites and literature of transportation operators and service providers that create and provide apps.

Finally, in Sect. 4, based on the result of Sect. 3, a framework of integration on MaaS is proposed.

3   Results: Societal Goals of MaaS

3.1   Goal in Urban Transportation Planning

Meyer and Miller (1984) describe goals and objectives in urban transport planning, which can be defined as a hierarchical structure in the order of value, goals, objectives, and indicators. Goals are generalised values and statements about the real environment, and objectives are specific and measurable statements. For example, in the context of MaaS, achieving additional seamless connections between public transport modes, reducing costs related to travel, and achieving sustainable mobility without reliance on private cars, are possible goals that could be realised by MaaS. Furthermore, if these goals are concretised into measurable statements and objectives are set, they include improving the ease of physical transfers, linking information, reducing payment costs, devising digital ticketing, and discouraging the use of private cars.

According to Hayauchi and Hidaka (2022), among the KPIs set by MaaS projects in Japan, a number of projects setting them were: the number of people transported by ticket/information provided by MaaS, the number of app downloads, and satisfaction with the service and intention to continue using it. Notably, the number of apps downloaded and satisfaction levels can be useful indicators when evaluating relatively short-term demonstrations. However, additional long-term evaluation indicators should be used if the objectives derived from the vision are set and corresponding evaluation indicators are designed.

3.2   Expectations from Governments for MaaS

To understand the Societal Goals, expectation from Japan, the U.S., and Europe are summarized regarding MaaS. In Japan, the Ministry of Land, Infrastructure, Transport and Tourism (MLIT) (2022) states that the motivation for MaaS is to improve the convenience of public transportation, secure and enhance mobility in tourist areas, and increase mobility that does not rely solely on private vehicles. In Europe, the major goals are compatibility, interoperability, and continuity of mobility among EU member countries under various environments (Sakai (2019)). Moreover, according to MaaS Alliance, headquartered in Europe, 'MaaS aims to be the best value proposition for users, societies, and the environment'. In the U.S., where the term 'MOD (Mobility on Demand)' is used instead of 'MaaS', ITS America has stated that 'MOD is a vision for an integrated multimodal network of safe, carefree, and reliable transportation options that are available to all, both personalized mobility and goods delivery'.

Given the policy in Japan to be sustainable in the sense of management independence and long-term service provision, in Europe to be sustainable in the sense of low environmental impact, and in the US to be socially sustainable, today's study and realisation of MaaS, being sustainable can be found as a common value. Therefore, the goals of MaaS should be based on values that aim in a sustainable direction.

3.3 'Sustainability' and MaaS

The three perspectives of sustainability, economical, environmental, and social, first introduced in the Brundtland Report(1987). Several studies have also evaluated sustainability from three perspectives (Purvis et al (2018)), and the impact of MaaS on sustainability has been summarized from economic, environmental, and social perspectives (Pritchard(2022)). Economic sustainability refers to reductions in vehicle costs and economic losses associated with accidents. Environmental sustainability refers to reductions in greenhouse gas emissions. Societal sustainability refers to factors such as improved well-being and access to education and employment opportunities.

Assuming that MaaS can contribute to sustainability from these perspectives, we assume the societal goals that MaaS should eventually integrate can also be categorized into these three perspectives if it is made concrete.

3.4 Goals and Initiatives for Sustainability in the MaaS Case Study

In organizing the case studies, we found that several cases have already been completed or have been taken over by other services. Table 1 presents the table organised in the Netherlands Institute for Transport Policy Analysis (2018), with additional columns confirming goal-setting leading to sustainability for this study. Sentences listed as goals on the websites and the reports of each service were examined. Those that have been updated to the latest information available at the time are shown in bold. Most of these cases are equivalent to level 2 or level 3 in Sochor's classification, but they have stated their goals in working on each MaaS. In some cases, the services had already been terminated as of when this study was conducted, and in such cases, if the data at that time were accessible, the goals would be described.

Interestingly, regardless of the level of Integration, the goals for achieving MaaS existed in six cities. Therefore, rather than having integration with social goals as a step beyond integration with services, which are in level 3, the goals existed for working on MaaS at any level. They were not necessarily achieved in the order of level 1 but could be realized simultaneously.

Furthermore, in recent years, digital integration and collaboration with mobility hubs are considered in some cases as important measures of MaaS together with apps (Vienna, Hamburg). Studies on the discussion of MaaS and mobility hubs (Aydin et al (2022)), the combination of these aspects with the real and digital world, and the relationship between mobility hubs and other real places cannot be ignored.

4 Discussion: Proposal for an Organizing System for the Relationship Between MaaS and Policy

To summarise the discussion in the previous chapters, a new diagram of MaaS and their societal goals is shown in Fig. 2. Table 1 indicates that goal-setting is not necessarily an extension of the order from levels 1 to 3. In other words, goal setting is independent of any of the levels. The combination of these means depends on a variety of factors, including each city's situation, goals, costs, and stakeholders. Levels 1 to 3 are not

Table 1. Case study on the Goals for Sustainable and Mean

Service*	Location*	Status*	Integration level*	Goals for Sustainable
moovel **Hamburg has own MaaS - hvv swith**	Hamburg and Stuttgart, Germany	Operational (2015-)	**Lv. 3**	Yes
myCicero, **Now as MooneyGo**	Italy	Operational (2015-)	Lv. 2	Yes
NaviGoGo	UK	Operational (2017-) **Project Discontinured**	Lv. 2	Yes
iDPASS	France	Operational (2017-)	Lv. 2	N/A
Tuup **Now as Foli**	Turku, Finland	Operational (2016-)	**Lv. 3**	Yes
Hannovermobil **Now GVH**	Hannover, Germany	Operational (2014-)	**Lv. 3**	Yes
EMMA (TaM) **NowM'Ticket and TaM**	Montpellier, France	Operational (2014-)	Lv. 2	N/A
Business passes	Netherlands	Operational	Lv.1 & 2	Yes
Smile	Vienna, Austria	**Discontinued**	Lv. 2	N/A
WienMobil Lab	Vienna, Austria	Operational (2017-)	Lv. 3	Yes
SHIFT	Las Vegas, US	**Discontinued**	Lv. 3	N/A
UbiGo	Gothenburg, Sweden	**Terminated in 2021**	Lv. 3	Yes
Whim	Helsinki, Finland	Operational (2016-)	Lv. 3	Yes

* Originally from Netherlands Institute for Transport Policy Analysis (2018), **Bold:** Modified by author

necessarily linear in this order, and level 2 or 1 can be realized simultaneously with level 3. Furthermore, we added integration with mobility hubs as a new integration element. This, like the integration of fares, was added in parallel because it should be addressed or not be dependent on the situation of the city. Finally, another position of policy is to define policy as a goal. In realizing levels 1 to 3, policies that assist countries and cities that have difficulty in realizing them with their current laws and systems will also be necessary. For example, in the Netherlands, in realizing MaaS, efforts to enable it to easily process data are being advocated in parallel.

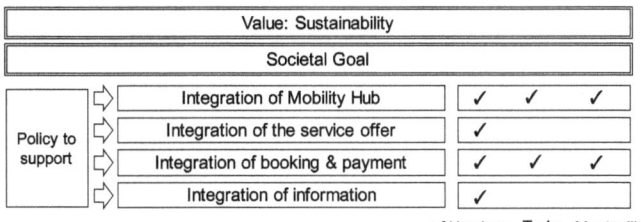

Fig. 2. Proposed diagram of MaaS integration with Societal Goals

5 Conclusion

In this study, we reinterpreted the relationship between the 'Societal Goals' and MaaS. The inadequate Interpretation of the societal goals, which is equivalent to Level 4, was considered an issue. Notably, the societal goals was considered a set of goals for the value of sustainability in modern urban transport planning. When the existing MaaS case studies were organised, goals were set at all levels. In recently, integration with mobility hubs have been discussed in MaaS as a means to realise sustainability.

Notably, a new diagram was proposed to highlight that the societal goals is set as a goal even in the early stages of MaaS. Moreover, it was emphasised that this will not necessarily be achieved in order from Level 1, but simultaneously. Furthermore, while newly defining integration with mobility hubs, it was defined that policies to accelerate specific measures for integration are available, and a new systematisation could be proposed based on this. By proposing this new diagram, we raise awareness of goals and the importance of evaluating these goals. In the future, we will verify this system by applying it to several cases and classify MaaS cases through a pattern of combinations, thereby contributing to the ease of reference to similar cases.

Acknowledgement. This study is based on the discussion in and partially supported by the Sub Research Committee of Comprehensive Research on MaaS with Practice, Evidence, and Theory at the Japan Society of Civil Engineers.

References

1. Arias-Molinares, D., García-Palomares, J.C.: The Ws of MaaS: Understanding mobility as a service from a literature review. IATSS Res. **44**, 253–263 (2020)
2. Lyons, G., Hammond, P., Mackay, K.: The importance of user perspective in the evolution of MaaS. Transp. Res. Part A: Policy Pract. **121**, 22–36 (2019)
3. Sochor, J., Arby, H., Karlsson, I.C.M., Sarasini, S.: A topological approach to mobility as a service: a proposed tool for understanding requirements and effects, and for aiding the integration of societal goals. Res. Transp. Bus. Manag. **27**, 3–14 (2018)
4. International Transport Forum, OECD, The Innovative Mobility Landscape – The Case of Mobility as a Service (2021)
5. MLIT, MaaS/AI demand service usecases, In Japanese (2022)
6. Netherlands Institute for Transport Policy Analysis, exploring mobility-as-a-service (2018)

7. Mulley, C., Nelson, J.D., Ho, C.Q., Hensher, D.A.: MaaS in a regional and rural setting: Recent experience. Transport Policy (2023)
8. Meyer, M. D., Miller, E.J.: Urban Transportation Planning: A Decision-oriented Approach, McGraw-Hill (1984)
9. Hayauchi, G., Hidaka, Y., Key Performance Indicators and future perspectives in Japanese MaaS projects: current situation in government-sponsored Projects from fiscal 2019 to 2021. In: 3rd International Conference on MaaS (ICoMaaS) (2022)
10. MLIT, Kotsu Seisaku Hakusyo, [Annual Report on Transport Policies] In Japanese (2023)
11. Sakai, K.: MaaS trends and policy-level initiatives in the EU. IATSS Res. **43**, 207–209 (2019)
12. MaaS Alliance, https://maas-alliance.eu/. Accessed 20 Sept 2023
13. ITS America, Peper on Mobility on Demand. Accessed 20 Sept 2023
14. Brundtland, G.: Report of the World Commission on Environment and Development: Our Common Future (1987)
15. Purvis, B., Mao, Y., Robinson, D.: Three pillars of sustainability: in search of conceptual origins. Sustain. Sci. **14**, 681–695 (2018)
16. Pritchard, D.J.: MaaS to pull us out of a car-centric orbit: Principles for sustainable mobility-as-a-service in the context of unsustainable car dependency. Case Studies on Transport Policy (2022)
17. Aydin, N., Seker, S., Özkan, B.: Planning location of mobility hub for sustainable urban mobility. Sustain. Cities Soc. (2022)
18. GVH, https://www.gvh.de/en/the-gvh/the-transport-association/. Accessed 10 Oct 2023
19. TaM, tam-voyages.com. Accessed 10 Oct 2023
20. MaaS Alliance. https://maas-alliance.eu/. Accessed 10 Oct 2023
21. Wienerlinien. https://www.wienerlinien.at/web/wl-en/wienmobil-app. Accessed 10 Oct 2023
22. Karlsson, I., Sochor, J., Strömberg, H.: Developing the 'service' in mobility as a service: experiences from a field trial of an innovative travel brokerage. Transp. Res. procedia **14**, 3265–3273 (2016)
23. Whim, https://whimapp.com/about-whim/. Accessed 10 Oct 2023

Open Access This chapter is licensed under the terms of the Creative Commons Attribution 4.0 International License (http://creativecommons.org/licenses/by/4.0/), which permits use, sharing, adaptation, distribution and reproduction in any medium or format, as long as you give appropriate credit to the original author(s) and the source, provide a link to the Creative Commons license and indicate if changes were made.

The images or other third party material in this chapter are included in the chapter's Creative Commons license, unless indicated otherwise in a credit line to the material. If material is not included in the chapter's Creative Commons license and your intended use is not permitted by statutory regulation or exceeds the permitted use, you will need to obtain permission directly from the copyright holder.

Asset Management Developments in Transport Infrastructure Ireland (TII)

Gerard O'Dea[1]([✉]), Kieran Feighan[2], Stephen Smyth[1], Rory Donnellan[3], and Pat Maher[1]

[1] Transport Infrastructure Ireland, Dublin, Ireland
gerard.odea@tii.ie
[2] PMS Pavement Management Services Ltd., Dublin, Ireland
[3] PMS Pavement Management Services Ltd., Athenry, Co. Galway, Ireland

Abstract. Transport Infrastructure Ireland's (TII) primary functions are to operate, maintain and extend the life of national roads, tunnels, and light rail infrastructure in Ireland. Among the priorities for investment include the use of asset management principles to manage assets safely, sustainably, efficiently, and effectively over their useful life. This paper describes TII's journey in asset management including the development of TII's recently published Asset Management Policy, Strategy and Framework, as well as the ongoing development of Strategic and Group-level Asset Management Plans for the various asset classes within TII to align with its overall strategic objectives and organisational goals. A best-fit approach has been developed which addresses TII's changing operational conditions and commitments in terms of sustainability, circular economy, and climate adaptation. A hierarchical structure based on the "line of sight" principal is presented in which TII proposes to achieve its overall asset management objectives in line with an ISO 55000 integrated asset management system approach.

Keywords: Transport Infrastructure · Asset Management · Policy · Strategy · Framework · Objectives · KPIs · Roads · Tunnels · Light Rail

1 Introduction and Background

Transport Infrastructure Ireland (TII) strives to provide sustainable transport infrastructure and services, thereby enhancing quality of life, fostering economic growth, and upholding environmental stewardship. TII is tasked with ensuring the provision of a safe and efficient transport network, which includes national roads, tunnels, and light rail transport networks. These networks are managed through a combination of direct management and concession contracts.

TII oversees c.5,300 km of national roads, which facilitates almost 50% of all road traffic and over 90% of freight traffic. With the replacement value of c. €31 billion, these national roads constitutes one of the most substantial asset classes under state ownership. Additionally, TII maintains the Luas light rail network, valued at c.€2 billion, in partnership with the National Transport Authority (NTA).

© The Author(s) 2025
C. McNally et al. (Eds.): TRAconference 2024, LNMOB, pp. 737–743, 2025.
https://doi.org/10.1007/978-3-031-85578-8_100

Given the considerable value of these assets, and the constant deterioration caused by traffic and environmental factors, a significant continuous reinvestment is required to preserve the integrity of initial investment. Thus, to ensure the efficiency and effectiveness of transport network, TII has developed the Asset Management policy, strategy, and framework. The Asset Management approach is tailored to serve the unique organisational needs, covering a multitude of light rail and road assets. It integrates the core concepts of Asset Management in activities while ensuring alignment with strategic objectives and goals outlined in TII's Statement of Strategy [1]. This integration yields the optimal outcomes from transport asset management practices and enables to plan for the most cost-effective and appropriate interventions.

The international standard (ISO 55001) specifies the foundational pillars for the establishment and implementation of asset management best practice [2], forming the basis for TII's asset management framework. A well-designed asset management system enhances accountability and governance for strategic national assets, while preserving asset value of the network and optimising performance. It necessitates coordinated planning throughout the asset lifecycle, from development to disposal or renewal.

In principle, effective asset management involves clear alignment between an organisation's statement of strategy and the asset management activities delivered by staff and strategic partners. This alignment, also known as 'Line of Sight', enables everybody to understand their role in achieving success. 'Line of Sight' translates organisational objectives into asset management policy, strategy, and objectives, which cascade down into more detailed asset management plans and activities (Figs. 1 and 2).

The following hierarchy sets out the structure in which TII proposes to achieve its overall asset management objectives in line with an ISO 55000 integrated asset management system approach [3].

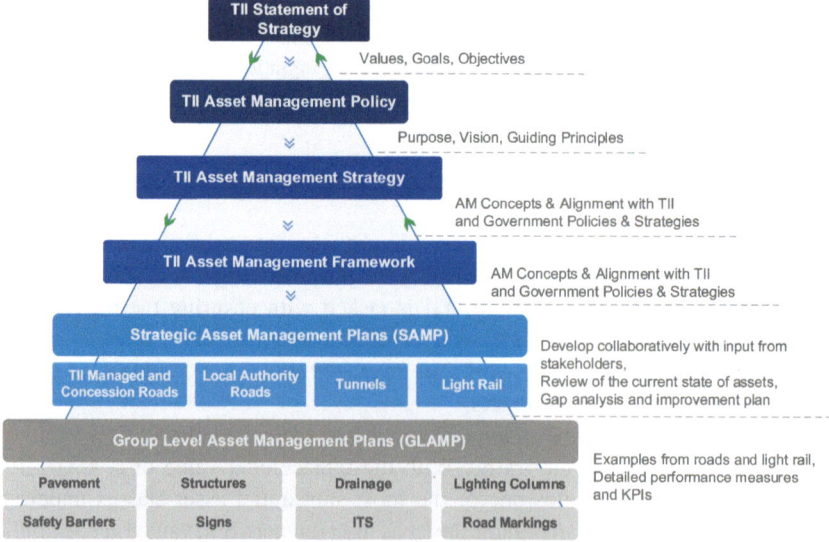

Fig. 1. TII Asset Management Hierarchy – Line of Sight

1.1 Asset Management Policy

TII's Statement of Strategy [1] establishes a commitment to protecting the significant investment that has been made in our national transport system. In line with that commitment, TII established the Asset Management Policy [4] that states: *"Assets will be managed in a sustainable manner through the development, implementation, and maintenance of an asset management approach that is risk-based and data-driven, enabling us to make informed decisions throughout the life of our assets."* The strategic outcomes from our Asset Management systems and processes are driven by six guiding principles:

- **Policy-driven** - Resource allocation is based on a well-defined set of organisational goals and objectives that inform all our asset management decisions.
- **Performance-based** - Asset management objectives are translated into system performance measures that are used for both operational and strategic management.
- **Founded on Quality Information** - Informed decision making and management over the full lifetime of the assets is based on relevant, accurate and reliable data.
- **Reliant on Analysis of Options and Trade-offs** - Decisions on how to allocate resources within and across various asset types are based on an analysis of how different options best align with relevant policies, risk mitigation strategies, and performance objectives over the life of the assets.
- **Providing Accountability and Feedback** - Performance results are monitored and evaluated for both efficiency and effectiveness.
- **Continual Improvement** - Performance feedback allows for continual assessment of results and improvement on past performance.

1.2 Asset Management Strategy

TII's Asset Management Strategy [5] builds on the Asset Management Policy. It describes the importance of asset management and its linkages to internal and external related TII, Department of Transport and Irish Government initiatives. It also introduces the concept of life cycle planning. All the documents are driven by the "line of sight" principle with consistency of approach throughout TII's structure. As Asset Management is continually evolving, TII will:

- Set strategic objectives and use performance measures to set targets and monitor progress.
- Make improvements to address changes to the data, processes, and systems that may be needed to advance asset management.
- Consider risks and uncertainties that can impact the level of service that is provided.
- Implement life-cycle strategies that consider needs over the whole life of an asset with an emphasis on preventative rather than reactive interventions.
- Generate optimal cost-effective decisions that:

  - Achieve strategic objectives.
  - Foster collaboration.
  - Promote systematic and sustainable approaches.
  - Satisfy relevant regulations and standards.
  - Reflect community values.

1.3 Asset Management Framework

The Asset Management Framework [6] builds on the Policy and Strategy by defining the key strategic objectives that will be achieved through asset management. TII have set eight strategic asset management objectives founded on its asset management policy:

1. **Safety** - To ensure transport networks are secure and safe for all.
2. **Reliability** - To ensure consistent and predictable journeys for the movement of people and goods across all modes.
3. **Condition** - To maintain, preserve, and extend the useful life of our transport assets.
4. **Maximising Value** - To plan, build, operate, and maintain the transport system, through collaboration and appropriately targeted investment choices.
5. **Customer** - To balance the diverse needs of our customers, achieving the best possible outcomes in terms of system quality and effectiveness.
6. **Sustainability** - To promote the sustainability of our transport systems while applying the principles of the circular economy.
7. **Environment** - To provide efficient and effective transport systems with positive environmental benefits.
8. **Resilience** - To proactively assess, plan, and invest in strategies that ensure the resilience of our network.

The AM Framework also sets out the data structures and processes that underpin the Asset Management approach, as well as laying out the new Governance Structure that will facilitate delivery of the Strategic objectives.

Strategic Asset Management Plans (SAMPs) and Group Level Asset Management Plans (GLAMPs) will provide more detailed plans on how the objectives will be achieved for light rail, local authority roads, and TII-managed and concession roads. These will include further detail around asset specific performance measures and KPIs.

There are significant areas of overlap with TII's Asset Management Framework, Sustainability Implementation Plan [7] and Climate Adaptation Strategy 2022 [8]. The strategic objectives of asset management are strongly aligned with the core guiding principles of sustainability and the strategic objectives of the Climate Adaptation Strategy. The Asset Management strategic objectives are also aligned with the aims of Ireland's Climate Action Plan 2023 [9] and will help to deliver on TII's Climate Action Road Map [10].

2 Performance Measures and KPIs

The increasing focus on accountability in transportation asset management is based in large part upon a growing need to demonstrate responsibility. Measuring progress over time against relevant and appropriate performance targets is viewed as evidence that the agency is responsibly using its limited resources to achieve performance that serves the public.

TII is approaching the issue of road and rail network management by recognising that the infrastructure networks are capital assets with measurable asset value. The aim of asset management under this approach is to conserve the asset value of the network. This concept is readily understood by a wide range of decision makers.

The current asset value of any road or rail network can be estimated with reasonable accuracy at a particular point in time, in the same manner as the balance sheet of a company. Lack of maintenance will result in the deterioration of the network by physical attrition because of climate and traffic, which implies a continuous decrease in its asset value. Investment in the rehabilitation of currently sub-standard roads and light rail sections or the addition of completely new roads and light rail sections leads to an increase in the asset value of the network.

The long life and large investments in assets necessitate the development of asset management plans that forecast future deterioration, maintenance, and enhancement needs. Monitoring asset value over time is used to demonstrate stewardship of assets. This information provides important input to a case for investing in the maintenance and upkeep of the transport infrastructure networks.

To ensure funding can go as far as possible, it is important for an organisation to track the performance of the asset at a network level. Key Performance Indicators (KPI) are commonly used to represent the health of the network, show progress towards goals, and measure the effectiveness of the organisation in managing its assets and link them back to the asset management objectives of the organisation.

The performance of the network is measured against multiple objectives. TII have defined eight asset management objectives in the AM framework document. The challenge in the Strategic Asset Management Plans (SAMPs) and Group Level Asset Management Plans (GLAMPS) is to define relevant KPIs and associated target levels that are directly related to each of the eight asset management objectives. Successful definition and implementation of these KPIs and targets provides the basis for demonstrating best practices in the management of our transport assets.

3   TII Asset Management Governance Structure

TII is establishing a structured system of management and governance to ensure consistency of approach while monitoring and reporting on asset management delivery across the organisation. The Asset Management Steering Committee (AMSC) is a subgroup of the Senior Management Team and reports to the Strategy Committee of the Board of TII. The AMSC is responsible for the development, update, and monitoring of asset management enhancements within TII. It is also responsible for ensuring that the purpose, vision, and guiding principles outlined in TII's asset management policy are embedded throughout the organisation, ensuring that the day-to-day asset management activities we deliver are aligned to our strategic objectives.

Technical Advisory Groups support the full implementation of asset management in TII. This includes developing performance measures and targets to be reviewed for approval by the steering committee; identifying and prioritising risks to TII's transport infrastructure; recommending changes to policies, procedures, and processes, and ensuring distinct groups and sections within TII work together to maximise the outputs from the resources employed (Fig. 2).

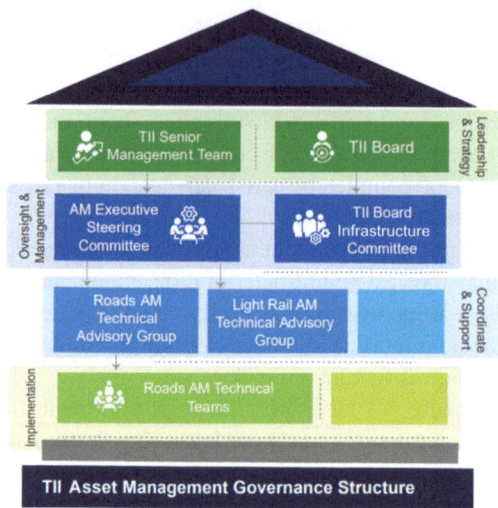

Fig. 2. TII Asset Management and Governance.

Asset Management Technical teams support the implementation and reporting of the activities that deliver effective asset management for TII through programmes carried out by key stakeholders working on behalf of TII. The activities include compiling and analysing data to support the development of performance targets aligned with TII's AM objectives and using asset management systems to perform gap and life cycle analysis that cover a range of funding scenarios. It also includes developing and managing programmes of protection and renewal for the various asset types and reporting on the outcomes of the programmes.

Adequate resources are vital to the successful implementation of an asset management improvement plan. There is likely to be a need for new skills and qualifications within TII to fully implement the asset management plans while ensuring line of sight from the Board and the Senior Management team all the way through to the day-to-day delivery of maintenance, renewal and protection programmes and activities on the ground. These resources include adequate training for agency staff to ensure sufficient skills and understanding to implement asset management throughout the organisation.

References

1. Transport Infrastructure Ireland, TII Statement of Strategy 2021 to 2025 (2020). Accessed 26 Sep 2023. https://www.tii.ie/tii-library/statements-of-strategy/statement-of-strategy-2021-2025/tii-statement-of-strategy-2021-2025-final.pdf
2. International Organization for Standardization, ISO 55000:2014 Asset management — Overview, principles and terminology (2014)
3. International Organization for Standardization, ISO 55001:2014 Asset management — Management systems — Requirements (2014)
4. Transport Infrastructure Ireland, TII Asset Management Policy (2021). Accessed 26 Sep 2023. https://www.tii.ie/tii-library/assetmanagement/documents/asset-management-policy.pdf

5. Transport Infrastructure Ireland, TII Asset Management Strategy (2021). Accessed 26 Sep 2023. https://www.tii.ie/tii-library/assetmanagement/documents/AM-Strategy(Dec2021)V1.3-08-09-2023.pdf
6. Transport Infrastructure Ireland, TII Asset Management Framework (2023). Accessed 26 Sep 2023. https://www.tii.ie/tii-library/assetmanagement/documents/AM-Framework-v4.2-08-09-2023.pdf
7. Transport Infrastructure Ireland, Sustainability Implementation Plan - Our Future (2021). Accessed 26 Sep 2023. https://www.tii.ie/tii-library/sustainability/TII-Sustainability-Implementation-Plan_Our-Future_EXTERNAL.pdf
8. Transport Infrastructure Ireland, Climate Adaptation Strategy (2022). Accessed 26 Sep 2023. https://www.tii.ie/technical-services/environment/changing-climate/Climate-Adaptation-Strategy-2022_v2.pdf
9. Government of Ireland, Climate Action Plan 2023, CAP23, Changing Ireland for the Better (2022). Accessed 10 Oct 2023. https://assets.gov.ie/270956/94a5673c-163c-476a-921f-7399cdf3c8f5.pdf#page=null
10. Transport Infrastructure Ireland, TII Climate Action Roadmap (2023). Accessed 10 Oct 2023. https://www.tii.ie/sustainability/TII%20Climate%20Action%20Roadmap/TII%20Climate%20Action%20Roadmap%202023.pdf

Open Access This chapter is licensed under the terms of the Creative Commons Attribution 4.0 International License (http://creativecommons.org/licenses/by/4.0/), which permits use, sharing, adaptation, distribution and reproduction in any medium or format, as long as you give appropriate credit to the original author(s) and the source, provide a link to the Creative Commons license and indicate if changes were made.

The images or other third party material in this chapter are included in the chapter's Creative Commons license, unless indicated otherwise in a credit line to the material. If material is not included in the chapter's Creative Commons license and your intended use is not permitted by statutory regulation or exceeds the permitted use, you will need to obtain permission directly from the copyright holder.

Analyzing Acceptance of Reduced Speed Limits on Greek Motorways: A Survey

Armira Kontaxi[✉], Christina Agourou, and George Yannis

National Technical University of Athens, 5 Heroon Polytechniou St., 15773 Athens, Greece
akontaxi@mail.ntua.gr

Abstract. This study explores the public acceptance of reduced speed limits on Greek motorways, investigating a decrease from 130 km/h to 120 km/h and 110 km/h. The inquiry is grounded on the urgent need to enhance road safety and achieve sustainable mobility, as high speeds significantly contribute to road accidents, air pollution, and fuel consumption, imposing economic burdens up to €8 billion annually on Greek society. Using a meticulously designed questionnaire, respondents were presented with scenarios involving different speed limits, evaluating their choices based on travel time, fuel consumption, and road accident risks. The resulting data was analyzed using binary and multinomial logistic regression models. Key findings indicate that involvement in property damage crashes, past traffic code violations, gender, and income are significant predictors in accepting speed limit reductions. Notably, individuals acknowledging the crucial role of speed in road accidents were more amenable to speed limit reductions. The study offers invaluable insights for policymakers and stakeholders, providing a nuanced understanding of public perception and acceptance of reduced speed limits, crucial for devising effective and widely accepted road safety interventions in Greece.

Keywords: Speed Limit · Motorways · Stated Preference · Logistic Models

1 Introduction

In recent decades, several studies have focused on the effects of speed and speed limit on road safety, showing that speed of travel is highly correlated with road accident frequency and severity (WHO, 2007; Aarts, 2011). In particular, driving on the high-speed motorway network can cause negative impacts such as air pollution, high fuel consumption and serious road accidents (Jin and Rafferty, 2021). In an effort to reduce the negative impacts of speed on road safety and promote safe and sustainable mobility for all, there is a global trend to implement lower speed limits (De Pauw et al., 2014). In addition to road safety, reducing driving speed can contribute to greener and more economical driving.

Reducing the speed limit on the Greek interurban road network, apart from its contribution to reducing road accidents and consequently injuries and mortality, is a measure that is estimated to bring significant economic benefits to the state. In particular, road accidents cost Greek society around €2.7 billion per year, while the actual cost could

© The Author(s) 2025

C. McNally et al. (Eds.): TRAconference 2024, LNMOB, pp. 744–750, 2025.
https://doi.org/10.1007/978-3-031-85578-8_101

potentially exceed €8 billion per year if the actual number of casualties and accidents with property damage alone are taken into account (ITF, 2020). In this context, the main objective of the present study is to investigate the acceptance of the reduction of the speed limit on Greek motorways from 130 km/h to 120 and 110 km/h.

2 Methodology

2.1 Data Collection

In order to achieve the study objective, a specially designed questionnaire was constructed. The questionnaire is divided into four sections covering a total of 43 questions. The questionnaires were collected in the form of an online survey via Google Forms. A total of 408 questionnaires were collected with the aim of gaining a representative sample in terms of gender, age and annual income. The first section of the questionnaire consists of questions regarding driving experience, main mode of transport, frequency of driving and possible involvement in road accidents. The second section examines the respondents' views on road accidents in Greece. In the third section, 10 different scenarios are introduced for a hypothetical three (3) hour out-of-town (long-distance) journey. Specifically, a choice between three alternatives is requested based on three parameters: travel time, fuel consumption, and the probability of being involved in a road accident with injury.

- Alternative 1: reduce the speed from 130 km/h to 110 km/h.
- Alternative 2: speed reduction from 130 km/h to 120 km/h.
- No change.

The following table shows a sample of one random scenario out of 10 (Table 1).

Table 1. Scenario example

Attributes	Alternative 1	Alternative 2	No change
Increase in Travel Time (minutes)	21	7	0
Reduction in Fuel Consumption (€)	9€	3€	0€
Reduction in the probability of road accidents with injury (%)	30%	10%	0%

2.2 Statistical Analysis

Logistic regression models are used when the dependent variable is distinct. Logistic regression is used to create models for predicting the influence of the presence or absence of certain characteristics in the selection of a particular alternative. In that context, utility function is used as a function of the attributes and other factors that affect the choice of the respondent.

More specifically, the utility function is defined as a mathematical model that describes the probability of the choice of each individual among alternatives based on the attributes. Based on the utility maximization context, as described by McFadden (1974) and Ben-Akiva and Lerman (1985), the utility of an alternative i (U_i) consists of a systematic part V_i and an error term ϵ, where the systematic part consists of (a) a vector of attributes α with attribute values Xiα for a given alternative i, and (b) their marginal effect on utility βiα and an Alternative Specific Constant ASC that captures systematic but non-explained variability in the data:

$$U_i = V_i + \varepsilon \tag{1}$$

where Vi is given by:

$$V_i = \beta' X_i + ASC_i \tag{2}$$

3 Results and Discussion

3.1 Binary Logistic Model

The binary logistic regression analysis presented in Table 2 illustrates how different variables influence the acceptance of speed limit reduction on Greek motorways.

Table 2. Binary logistic regression model

Parameters	β_i	p-value	Odds Ratio
(Intercept)	1.815	< 0.001	0,593
PROPERTY_CRASH	−0.883	< 0.001	0.414
VIOLATATIONS	−0.506	< 0.001	0.603
SPEED_ROLE	0.928	< 0.001	2.529
IF_TRAFFIC	−0.852	< 0.001	0.427
GENDER	−1.119	< 0.001	0.327
INCOME2	-0.541	< 0.001	0.582
INCOME3	−0.080	0.507	0.923
AIC	2960.5	-	-
X-squared	4.6897	0.196	-

The variables included in the model are defined below:

- PROPERTY_CRASH: Being involved in a crash with property damage
- VIOLATATIONS: Number of violations of the Road Traffic Code during the last 3 years.

- SPEED_ROLE: Importance of speeds for road crashes based on the respondents. 1 signifies low, 2 adequate and 3 high importance
- IF_TRAFFIC: Traffic existence as a factor of changing speed or the way of driving
- GENDER: the option "Woman" in the question "Choose your gender"
- INCOME2: 10.000 - 25.000€ as annual income
- INCOME3: Over 25,000 € as annual income

The odds ratios (OR) of the variables PROPERTY_CRASH, VIOLATIONS, IF_TRAFFIC, GENDER, and INCOME2 are less than 1 (0.414, 0.603, 0.427, 0.327, and 0.582 respectively), indicating a negative relationship with the dependent variable. Notably, individuals who have been involved in property damage crashes (PROPERTY_CRASH), those who have violated the Road Traffic Code in the past three years (VIOLATIONS), individuals considering traffic as a factor influencing their speed or driving style (IF_TRAFFIC), women (GENDER), and those with an annual income between 10,000 and 25,000€ (INCOME2) are less likely to accept the reduction in speed limits.

On the other hand, the variable SPEED_ROLE, which represents the perceived importance of speeds in road crashes by respondents, exhibits an odds ratio of 2.529. This odds ratio greater than 1 signifies a positive correlation with the dependent variable. Consequently, respondents attributing high importance to speeds in road crashes are more likely to accept the speed limit reduction on Greek motorways. The variable INCOME3, representing individuals with an annual income over 25,000€, has an odds ratio close to 1 (0.923) and a non-significant p-value (0.507), indicating that it does not significantly influence acceptance of speed limit reduction.

3.2 Multinomial Logistic Model

This section presents the findings of the statistical analysis on participants' choice of speed reduction alternative, based on the data collected in the third section of questionnaire survey (Table 3).

Table 3. Multinomial Logistic Model

Parameters	β_i	p-value	Odds Ratio
(Intercept):1	0.233	0.388	-
(Intercept):2	−0.477	0.023	-
Time	−0.026	0.005	0.450
Fuel	0.005	0.792	0.192
Accident	0.010	0.004	0.829
EXP:1	−0.413	0.001	1.997
EXP:2	−0.440	0.001	0.521
PROPERTY_CRASH:1	0.485	0.000	0.652

(*continued*)

Table 3. (*continued*)

Parameters	βi	p-value	Odds Ratio
PROPERTY_CRASH:2	0.757	0.000	1.534
STAT_DEATHS:1	0.826	0.000	0.690
STAT_DEATHS:2	1.023	0.000	-
SPEED_ROLE:1	−0.821	0.000	0.520
SPEED_ROLE:2	−0.402	0.036	-
MOTO_RISK:1	−1.119	0.000	1.398
MOTO_RISK:2	−0.785	0.000	456
GENDER:1	0.843	0.000	2.323
GENDER:2	0.508	0.000	1.662
INCOME:1	0.096	0.467	-
INCOME:2	0.551	0.000	1.734
Log-Likelihood	−2338.5	-	-
Likelihood ratio test: chisq	337.01	-	-

The variables included in the model are defined below:

- Time: increase of travel time
- Fuel: decrease of fuel consumption
- Accident: decrease of the probability of getting involved in a road accident with injury
- EXP: The experience of drivers in years
- PROPERTY_CRASH: Being involved in a crash with property damage
- STAT_DEATHS: Number of deaths from road crashes based on statistics
- SPEED_ROLE: Importance of speeds for road crashes based on the respondents. 1 signifies low, 2 adequate and 3 high importance
- MOTO_RISK: motorcycles risk level when the speed is within the limits of 130 km/h.
- GENDER: Gender of the participant
- INCOME: income of the participant

The parameter 'Time' has a negative coefficient (βi = -0.026), indicating that as travel time increases, the log-odds of the dependent variable decreases. Despite its non-significant p-value, the 'Fuel' variable's low odds ratio (0.192) suggests that changes in this predictor have minimal influence on the outcome. On the other hand, 'Accident', 'EXP:1', and 'PROPERTY_CRASH:2', with positive coefficients and significant p-values, positively influence the dependent variable. For instance, an increase in the 'Accident' predictor is associated with increased log-odds of the outcome. Variables 'EXP:1' and 'EXP:2' denote the experience of drivers, where 'EXP:1' positively influences the response while 'EXP:2' negatively correlates with it, signified by their odds ratios of 1.997 and 0.521 respectively.

Additionally, 'PROPERTY_CRASH:1' and 'PROPERTY_CRASH:2', representing involvement in property damage crashes, exhibit odds ratios of 0.652 and

1.534 respectively, indicating varying degrees of influence on the dependent variable. 'STAT_DEATHS:1' and 'STAT_DEATHS:2', accounting for the number of road crash deaths based on statistics, present positive coefficients and significant p-values, suggesting their substantial impact on the model. Further, 'SPEED_ROLE:1' and 'MOTO_RISK:1' variables display negative coefficients, highlighting their inverse relationship with the outcome variable. The coefficients for 'GENDER' and 'INCOME:2' are positive and statistically significant, demonstrating their positive association with the dependent variable, while 'INCOME:1' shows a non-significant p-value, implying negligible influence. These insights provide a comprehensive understanding of factors affecting the modeled outcome, crucial for developing informed strategies and interventions.

4   Conclusions

The research paper primarily aimed to explore the acceptance of reduced speed limits on Greek motorways, from 130 km/h to 120 km/h and 110 km/h. This exploration was grounded on existing literature emphasizing the adverse impacts of high speeds on road safety, air pollution, fuel consumption, and subsequent economic implications for the state.

Through the analysis, the binary logistic regression model pinpointed key variables, such as involvement in property crashes, traffic code violations, gender, and income, as significant determinants in accepting reduced speed limits. For instance, those involved in property damage accidents or having traffic code violations in the past three years were less inclined towards reduced speed limits. Conversely, respondents acknowledging the pivotal role of speed in accidents showed a higher acceptance rate for speed reductions. The multinomial logistic model further detailed how travel time, accident likelihood, driving experience, and the number of road crash deaths, among others, shaped the preferences of respondents. Factors like longer travel time, increased experience, and higher statistics of road deaths influenced resistance to speed limit reductions. These findings are instrumental for policy-makers and stakeholders in crafting effective and well-accepted interventions to promote road safety and sustainability in Greece.

Acknowledgement. This paper is based on work carried out in the framework of the project "Trendline - Technical Assistance for the development and collection of Road safety Key Performance Indicators", financed by the European Union under grant agreement No. MOVE/C2/SUB/2022–54/CEF/TA/SI2.892654.

References

1. Aarts, L., Van Schagen, I.: Driving speed and the risk of road crashes: a review. Accid. Anal. Prev. **38**(2), 215–224 (2006)
2. Ben-Akiva, M.E., Lerman, S.R., Lerman, S.R.: Discrete choice analysis: theory and application to travel demand, vol. 9. MIT Press (1985)
3. De Pauw, E., Daniels, S., Thierie, M., Brijs, T.: Safety effects of reducing the speed limit from 90 km/h to 70 km/h. Accid. Anal. Prev. **62**, 426–431 (2014)

4. International Transportation Forum (ITF), Speed and Crash Risk (2018)
5. ITF, Road Safety Report. Greece (2020)
6. Jin, J., Rafferty, P.: How the speed limit policy affects travel speed?: Quasi-experimental approach. Transp. Policy **103**, 2–10 (2021)
7. McFadden, D.: The measurement of urban travel demand. J. Public Econ. **3**(4), 303–328 (1974)
8. World Health Organization. Global Status Report on Road Safety (2007)

Open Access This chapter is licensed under the terms of the Creative Commons Attribution 4.0 International License (http://creativecommons.org/licenses/by/4.0/), which permits use, sharing, adaptation, distribution and reproduction in any medium or format, as long as you give appropriate credit to the original author(s) and the source, provide a link to the Creative Commons license and indicate if changes were made.

The images or other third party material in this chapter are included in the chapter's Creative Commons license, unless indicated otherwise in a credit line to the material. If material is not included in the chapter's Creative Commons license and your intended use is not permitted by statutory regulation or exceeds the permitted use, you will need to obtain permission directly from the copyright holder.

Lane Splitting Policy: A Second Experiment on Its Social Acceptability

Anita Bec-Gerion[1]([✉]) [ID], Chloé Eyssartier[1] [ID], and Marie-Axelle Granié[2] [ID]

[1] CY Cergy Paris Université, Cerema, MATRIS, Nantes, France
anita.bec-gerion@cerema.fr
[2] AME-MODIS, Université Gustave Eiffel, Univ Lyon, Lyon, France

Abstract. Lane splitting (LS) is the practice of riding a powered two or three-wheelers (PTW) between lanes when the traffic is slow or vehicles are stopped. Although lane splitting was forbidden in France before 2016, it was a common riding practice among PTW drivers. Two experiments (2016–2018 and 2021–2023) were conducted allowing lane splitting on highways and urban expressways in several counties. In those area, lane splitting was allowed and regulating by several specific conditions. The objective of this paper is to present the results of the second experiment by questioning the social acceptability of lane splitting within PTW drivers according to their driving experience. Drivers' experience will correspond in this study to previous LS practice and compliance with the LS rules before the beginning of the experiment. The sample is representative of the French population and composed of 817 PTW drivers. Analysis of PTW drivers profiles based on the Unified Theory of Acceptance and Use of Technology highlight differences in terms of social acceptability depending on whether or not LS is practiced, as well as the degree of compliance declared.

Keywords: Social acceptability · Transport policy · Lane splitting · Motorcyclist

1 Introduction

Lane splitting (LS) is the practice of riding a powered two or three-wheelers (PTW) between lanes when the traffic is slow or vehicles are stopped. The development of urban areas and the increasing traffic density has contributed to an increase in this practice as PTW drivers seek to travel between traffic lanes to avoid congestion [1].

Although this practice appears to be a common riding practice among French PTW drivers, from a legislative point of view this practice is not authorized in the traffic regulations rules. Indeed, the question of authorizing this practice raises social and public health issues. Despite the fact that the modal share of PTW drivers is less than 2%, the proportion of accidents involving this mode of transport in 2022 represents almost one fourth of people killed [2]. In addition, more than a third of accidents involving PTW drivers are the result of a collision with another vehicle [2]. In this context, it seems to be necessary to regulate this practice.

© The Author(s) 2025
C. McNally et al. (Eds.): TRAconference 2024, LNMOB, pp. 751–757, 2025.
https://doi.org/10.1007/978-3-031-85578-8_102

Since 2016, two experiments have been conducted in France (2016–2018 and 2021–2023) where the practice of LS was authorized and regulated in specific counties. These experiments aim to study the impact of LS on drivers' behavior and accident rates, but also to examine the social acceptability of this measure in order to authorize, or not, this practice on a national scale. The purpose of this article is to present the results of this second experiment by questioning how drivers' experience, through LS practice and compliance with the rules, could impact their social acceptability of the LS measure. The measure is composed of several rules regulating, for example, the maximum speed not to be exceeded when practicing LS, the type of behavior to be adopted (i.e. PTW drivers must indicate when entering and exiting lanes) or the type of roads on which it is permitted to practice LS.

1.1 Theoretical Background

Literature on LS is scarce and heterogeneous. On one hand, researches emphasize its ecological aspect, suggesting that it could reduce greenhouse gas emissions by making traffic flow more smoothly [3]. On the other hand, the causes of accidents arising from LS situations are often complex and multi-factorial [4]. The risk of accidents related to LS may be higher if the practice of LS is not regulated [1]. Although research highlights the need to regulate LS, regulation alone is not enough and it is essential that users accept and respect these rules.

Literature on social acceptability tends to define it as a concept of *attitude* [5]. It specifies that it would correspond to a prospective judgment of a measure that will be introduced in the future, where individuals will never have experienced these measures. However, if acceptability can be seen as the first stage in a process aiming at the appropriation and use of new devices, and more specifically, in the context of new public policies, compliance with new rules [6], then the use of new technology models such as the Unified Theory of Acceptance and Use of Technology (UTAUT) may be a relevant approach to study social acceptability [7]. This model is composed of four factors influencing behavioral intention and use of a new technology, or in the present case intention and stated compliance with the rules. *Performance expectancy* refers to people's beliefs about the consequences of the measure in terms of effectiveness in solving the problem. *Effort expectancy* refers to the degree of ease of use of the object or, in our case, the degree of ease to comply with the rule. *Social influence* refers to the impact of other people's behavior on one's own behavior. Finally, *facilitating conditions* correspond to individuals' beliefs about the existence of infrastructures to facilitate the use of a new system or compliance with a rule. In addition, the authors specify that *experience* acts as moderator variable in the model. Given that the experience of a new technology influences its acceptance [7], we could also argue that having experience of a measure would influence its acceptance. In the context of LS regulation, drivers' experience could refer to the habits of practicing LS or being used to comply with the rules. Hence, driver's experience that is drivers who used to practice LS before the experimentation and those who used to comply with the LS rules before the experiment should have a different level of social acceptability by comparison with drivers having different road experience.

Complementary factors linked to social acceptability can also be mentioned. Studies indicate that *affective attitude*, *perceived ineffectiveness* of the rule, *perceived benefits* of the measure, and *perceived danger* could also impact social acceptability [8].

1.2 Presentation of the Second Experiment

As transport policy is a key driver for promoting sustainable and safe transport, a first experiment (2016–2018) was conducted in eleven French counties, where LS was allowed on highways and urban expressways. Results highlight that both car and PTW drivers evaluate positively the experiment, with PTW drivers having a slightly higher positive attitude score than car drivers [8]. The French delegation for road safety and traffic (DSR) has decided to conduct a second experiment (2021–2023) in order to deepen its understanding of the potential benefits and risks associated with the practice of LS as well as its social acceptability of this measure in order to authorize, or not, this practice on a national level. Its aim is to clarify the results obtained in the *Gironde* department, where it was found an increase in accident rates on the LS network. This second experiment includes twenty-one counties where the practice of LS is allowed and regulated, while LS is forbidden in the rest of the country.

2 Aim

The objective of this paper is to present the results of the second experiment by questioning the social acceptability of this transport policy among French drivers. It aims to clarify how the different factors composing social acceptability can be expressed differently among PTW drivers, depending on their driving experience before the implementation of the measure. More specifically, social acceptability of LS regulations will be studied in relation to drivers' experience, namely whether or not PTW drivers already practice LS, and whether or not PTW drivers already comply with the LS rules.

As experience is a moderator of acceptance, we assume that there will be differences in PTW social acceptability for those who already practice LS and those who do not, as well as differences between those who state complying with the LS rules and those who do not.

3 Method

3.1 Population

Data collection took place between July 22 and August 2, 2021, before the beginning of the experiment. The sample is representative of the French population and composed of 817 PTW drivers, of which 543 claim to practice LS. The average age was 45.91 years and 264 (32.31%) participants were female.

3.2 Questionnaire

Participants answered anonymously an online survey distributed by a survey institute. The questionnaire was divided into different sections covering socio-demographic characteristics, driving habits, and factors relating to acceptability and compliance with LS rules.

The part of the questionnaire related to the acceptability of new technologies was based on the UTAUT factors previously described namely *performance expectancy, effort expectancy, social influence,* and *facilitating conditions.* For each question, respondents were asked to position themselves on a four-point scale ranging from 1 ("Strongly disagree") to 4 ("Strongly agree"). *Behavioral intention* to respect the rules and *stated compliance* with the rules were measured on 4-point scale such as 1-Never, 2-Sometimes, 3-Often, 4-Always.

Additionally, *affective attitude* were operationalized in terms of driving pleasure for PTW drivers. Finally, *perceived ineffectiveness* of the rule, *perceived benefits* of the measure and *perceived danger* of the LS practice were measured on four-point scale ranging from 1 ("Strongly disagree") to 4 ("Strongly agree").

For each of the factors described, an average score was calculated.

4 Results

Results firstly focus of social acceptability difference regarding PTW drivers who practice LS before presenting differences regarding PTW drivers who comply with the rules.

4.1 Social Acceptability for PTW Drivers Who Practice LS

Regarding the factors linked to social acceptability, results indicate that PTW drivers who practice LS (n = 543) have on average a higher scores than PTW drivers who do not practice LS (n = 274) on the factors *"Facilitating conditions"* (U = 57871, p < 0,001), *"Social influences"* (U = 61386, p < 0,001), *"Affective attitude"* (U = 59494, p < 0,001), *"Performance expectancy"* (U = 61777, p < 0,001) and *"Perceived ineffectiveness"* (U = 61833, p < 0,001). In other words, a greater proportion of PTW drivers who practice LS believe that the risk of being penalized for not following the rules is lower compared to PTW drivers who do not practice LS. Moreover, LS drivers perceive this practice like a common practice among the population of PTW drivers, and declare they enjoy practicing LS, despite the presence of the rules. Also, PTW drivers practicing LS consider it to be an effective way of improving safety and sharing the road compared to others drivers. However, they point out that regulating this practice will have little impact on the practices of PTW drivers.

For *"Perceived Danger"* (U = 85400, p < 0.001), *"Behavioral Intention"* (U = 98638, p < 0.001) and *"Stated compliance"* (U = 91268, p < 0.001), PTW drivers practicing LS have on average a lower scores than PTW drivers who do not practice LS. These results seem consistent with the practices of both groups. PTW riders practicing LS have a lower perception of the danger of LS than PTW riders not practicing LS and indicate that they also have less intention to comply with the rules of the LS. This point seems to be related to the perceived uselessness score of the rule which is more important for PTW drivers practicing LS. The non-respect of the rule seems to reside not in the difficulty of respecting the rules but rather in the perceived lack of usefulness of this framework.

No significant differences where found between groups for *"Effort expectancy"* and *"Perceived benefits"*.

4.2 Social Acceptability for PTW Drivers Who Respect the Rules

Regarding factors linked to social acceptability and compliance with the rules, results indicate that PTW drivers who "Always" (n = 218) comply with the rules have higher scores than drivers who "Often" (n = 242) or "Sometimes or Never" (n = 84) comply with the rules on the following factors: *"Behavioral intention"* (U = 317, p < 0,001), *"Effort expectancy"* (U = 152, p < 0,001), *"Social influences"* (U = 91, p < 0,001), *"Affective attitude"* (U = 90, p < 0,001), *"Performance expectancy"* (U = 97 p < 0,001) and *"Perceived benefits"* (U = 53, p < 0,001) and *"Facilitating conditions"* (U = 25, p < 0,001). In other words, PTW drivers who say they comply more frequently with the rules are on average more likely than the other drivers to declare that they intend to obey the rules and to believe that it is easy and pleasant to obey the LS rules. Also, they are more likely to believe that the practice of LS is common among the population of PTW drivers, and believe that this measure is effective in improving safety and sharing of the road. Participants who state complying with the rules believe that the risk of being sanctioned for not following the rules is lower compared to PTW drivers not always complying with the rules.

No significant differences where found between groups for others factors.

5   Conclusion

The objective of this paper was to present the results of the second experiment by questioning how the different factors composing social acceptability can be expressed differently among PTW drivers, depending on their driving experience before the implementation of the regulation. It aimed to clarify the importance of drivers' profiles when deploying a public policy measure in order to increase future acceptance. Indeed, since factors composing social acceptability appear to be associated with intention and compliance with the rules [7], focusing on how drivers' behavior prior to the implementation of the measure is related to social acceptability could help public authorities to improve their targeting and communication in order to increase future acceptance to the rules.

Results indicate that PTW drivers who already have an experiment with LS practice differs significantly from PTW drivers without any experiment of LS practice. Especially, PTW drivers practicing LS tend to have a lower perception of the sanction if they do not follow the rules as well as a lower intention to comply with the rules. It may be that these drivers, used to LS practice, have never been sanctioned before. Indeed, studies have shown that the subjective risk of detection is a one of the key element in the perceived level of sanction [6]. Moreover, it appears that although they found LS practice to be an effective way of improving safety and sharing the road, they do not feel the need for a regulation and find it unnecessary. A communication on the usefulness of the LS regulation and the development of actions to increase the subjective risk of sanction detection in particular targeted toward PTW drivers practicing LS could be of interest.

Regarding PTW drivers already complying with the rules, it is interesting to highlight that, in line with our previous findings, drivers who state not always complying with the rules have a higher perception of punishment than other drivers. This may be due to the fact that people who don't respect the rules have been sanctioned as a result of offending behavior, increasing their subjective risk of detection. Moreover, drivers complying with the rules indicate more frequently than the others that it would be easy to follow the rules and pleasant to drive despite the LS regulation. These elements could be used as a means of communication to improve compliance with the rules among other groups of drivers.

One of the main limitations of this work is its cross-sectional nature. It would have been interesting to follow up the participants after the start of the experiment to further investigate the association between drivers' experience, social acceptability, and acceptance of the rules.

In conclusion, this research highlights the relevance of a profile-based approach in the context of the social acceptability of a public road safety policy.

References

1. Guyot, R.: Circulation Inter-Files des Deux-Roues Motorisés. Rapport du ministère de l'intérieur, Paris (2012)
2. DSR: La circulation inter-files. Paris (2021)
3. Dorocki, S., Wantuch-Matla, D.: PowerTwo-wheelers as an element of sustainable urban mobility. Europe Land **10**, 618 (2021). https://doi.org/10.3390/land10060618
4. Clarke, D.D., Ward, P., Bartle, C., Truman, W.: The role of motorcyclist and other driver behaviour in two types of serious accident in the UK. Accid. Anal. Prev. **39**(5), 974–981 (2007)
5. Schlag, B., Schade, J.: Public acceptability of travel demand management. In: Rothengatter, T., Huguenin, R. (ed.) Traffic and Transport psychology, Elsevier (2003)
6. Cestac, J., Carnis, L., Assailly, J.-P., Eyssartier, C., Garcia, C.: Enquête sur le rapport à la règle chez les automobilistes français. Rapport final. Convention IFSTTAR-CEREMA-DSCR n°2200626575, p. 76 (2018)
7. Venkatesh, V., Morris, M.G., Davis, G.B., Davis, F.D.: User acceptance of information technology: toward a unified view. MIS Quart., 425–478 (2003)
8. Eyssartier, C., Lanfranchi, M., Ragot-Court, I.: Lane splitting experiment: comparative analysis of how this is accepted by motorcyclists and car drivers. In: TRA 2018, 7th Transport Research Arena: A digital era for transport, Vienne, Austria, p. 9 (2018)
9. Hamelin, F., Eyssartier, C: Les professionnels mobiles à l'épreuve des radars automatiques. Une acceptation entre stress contenu et identités au travail bousculées. Articulo-J. Urban Res., 7 (2011)

Open Access This chapter is licensed under the terms of the Creative Commons Attribution 4.0 International License (http://creativecommons.org/licenses/by/4.0/), which permits use, sharing, adaptation, distribution and reproduction in any medium or format, as long as you give appropriate credit to the original author(s) and the source, provide a link to the Creative Commons license and indicate if changes were made.

The images or other third party material in this chapter are included in the chapter's Creative Commons license, unless indicated otherwise in a credit line to the material. If material is not included in the chapter's Creative Commons license and your intended use is not permitted by statutory regulation or exceeds the permitted use, you will need to obtain permission directly from the copyright holder.

The Role of TRIMIS as a Policy Support Tool

Monica Grosso[1](✉), Ilias Cheimariotis[1], Marcin Stepniak[1], Chiara Lodi[2],
and Alessandro Marotta[1]

[1] European Commission -Joint Research Centre, Via E. Fermi 2479, 21027 Ispra, Italy
monica.grosso@ec.europa.eu
[2] Piksel S.r.l, 20126 Milan, Italy

Abstract. TRIMIS is a transport policy support tool that provides open-access information on transport research and innovation. It supports the vision of a clean, connected, and competitive European transport system and exemplifies the European Commission's commitment to open science. TRIMIS collects, curates, analyses, and disseminates European and non-European data on transport research and innovation. It also analyses technology trends and research and innovation capacities in the transport sector. TRIMIS maintains an open database of almost 9,000 transport research and innovation projects and programmes, including Horizon Europe, Horizon 2020, FP7, Interreg, Connecting Europe Facility, and projects funded by Member States. This paper aims to provide an overall understanding of TRIMIS, its objectives, ecosystem, and details on how it contributes to supporting EU transport policy.

Keywords: Transport policy · Research and Innovation · Policy tool · Open data · Open science

1 Introduction

The transport sector is at the heart of the European economy, representing about 5% of the total European Union (EU) gross value added (GVA) equal to EUR 555 billion in the EU-27 in 2020. It also provides employment to almost 10 million people in the EU-27, accounting for 5.2% of the total workforce (European Commission, 2022). However, it is well known that the transport sector is one of the main contributors to greenhouse gas (GHG) emissions, being responsible for 25% of the EU's total GHG emissions and showing a steady increase between 2013 and 2019 (European Commission, 2020a). The EU is strongly committed to achieving safer, smarter, more resilient, and inclusive mobility. In this framework, the decarbonization of the transport sector is key, aiming for a 90% emission reduction by 2050 as defined in the European Green Deal (European Commission, 2019) and further defined in the 2020 Sustainable and Smart Mobility Strategy (European Commission, 2020b). The 2021 "Fit for 55" legislative proposal, aiming to achieve climate neutrality in the EU by 2050, adds intermediate targets of at least a 55% net reduction in GHG emissions by 2030 (European Commission, 2021).

The role played by Research and Innovation (R&I) is key to achieving climate neutrality, and in this context, the EC adopted the Strategic Transport Research and

© The Author(s) 2025
C. McNally et al. (Eds.): TRAconference 2024, LNMOB, pp. 758–764, 2025.
https://doi.org/10.1007/978-3-031-85578-8_103

Innovation Agenda as part of the Europe on the Move package (European Commission, 2017a), which highlights the main transport R&I areas and priorities for clean, connected, and competitive mobility. European Partnerships between the EC and private and/or public partners are key to implementing Horizon Europe, the EU's key funding program for research and innovation. The EU Partnerships aim to address some of Europe's most pressing challenges through concerted R&I initiatives, helping to avoid the duplication of investments and contributing to reducing the fragmentation of the R&I landscape in the EU. Although other actions are present at the EU level (e.g., Horizon dashboard, CINEA Project Portfolio platform), the Transport Research and Innovation Monitoring and Information System (TRIMIS) represents the reference point to follow the EU R&I initiatives that need to be implemented and monitored, as well as supporting and boosting transport R&I needs. With these purposes, TRIMIS was developed in 2017 at the European Commission Joint Research Centre (JRC) (European Commission, 2017b). TRIMIS was advocated by the Directorate-General for Mobility and Transport (DG MOVE) and the Directorate-General for Research and Innovation (DG RTD) and established at the EC-JRC as an integrated transport policy support tool providing open-access information on EU and Member States (MSs) transport R&I. This paper presents the objectives, functionalities, and features of TRIMIS as a European platform and analytical policy support tool with an integrated approach to transport R&I monitoring and information. The focus of this paper is on the policy feedback objective, illustrating the overall framework that supports this activity, while earlier publications have presented the methodological background and applications of the TRIMIS project (Tsakalidis et al., 2018, 2020a, 2020b).

2   TRIMIS Ecosystem and Functionalities

The main objectives of TRIMIS are to collect, curate, analyse, and disseminate data on transport research and innovation (R&I) in Europe and beyond. Additionally, TRIMIS aims to analyse technology trends and R&I capacities in the transport sector and provide findings and recommendations, including policy feedback. The TRIMIS team at the JRC, supported by other relevant actors such as DG MOVE, DG RTD, and executive agencies like CINEA, translates these objectives into the TRIMIS outputs, which reach a broad spectrum of stakeholders including policy makers, the research community, the transport industry, and the general public (Fig. 1).

As part of its monitoring and information system, TRIMIS has developed and maintains a transport R&I database. This database includes almost 9,000 European, Member State, and international projects and programmes, making it a valuable asset of the TRIMIS project. TRIMIS is designed as an open-access information and knowledge management tool that provides a comprehensive overview of transport R&I. Its monitoring activities primarily focus on supporting R&I transport policy implementation by monitoring progress towards defined research targets, as detailed in this paper. Furthermore, TRIMIS defines and maps innovation capacity in the EU transport sector by examining financial and socio-economic indicators. It also identifies and maps relevant technologies, providing insights into future trends. Additionally, TRIMIS identifies useful technical, operational, and socio-economic key performance indicators (KPIs) and

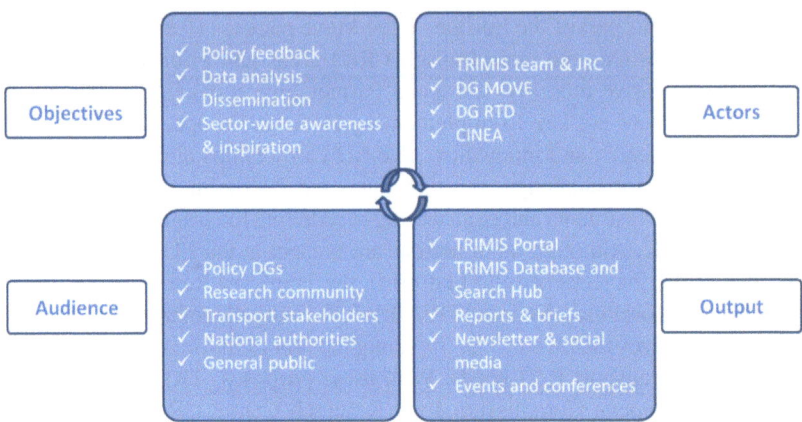

Fig. 1. TRIMIS Ecosystem

produces scientific reports that are of interest to the different TRIMIS stakeholders. TRIMIS's information system activities are supported by the dedicated portal (https://trimis.ec.europa.eu/), where TRIMIS scientific reports and briefs are available. The portal also promotes and disseminates relevant transport R&I news and events, which are highlighted in the TRIMIS Newsletter distributed to registered stakeholders. The TRIMIS team actively engages with different stakeholders through social media activities, as well as participation in conferences and events. Dedicated surveys and workshops are also organized with targeted groups of stakeholders on an ad hoc basis. Transport stakeholders may have various interests in TRIMIS, ranging from general information on the topic to disseminating their own results and gaining cross-sector awareness and inspiration for research or policy matters (Fig. 2).

3 TRIMIS as a Policy Support Tool

One of the main objectives of TRIMIS is to support the policy process throughout its entire cycle. TRIMIS does this by providing decision-making support information on the achievements of transport R&I activities, current trends, and research gaps that could be addressed in future projects and programmes. The workflow consists of sequential and interrelated steps, as illustrated in Fig. 3. The typical process begins with the definition of a research question or theme, which usually arises from scientific-policy interest and requires the contribution of the TRIMIS team. DG-MOVE and DG-RTD often initiate the query, but research questions and themes can also come from JRC scientific interests and EU transport-related institutions (e.g., EU-Rail). Once the research topic has been defined and agreed upon, the TRIMIS database is used to extract previous EU, MSs, and international projects that have addressed or are currently investigating the topic. Pre-selected keywords related to the topic are used to perform this activity, resulting in a selection of initial projects. The TRIMIS team then refines this selection to include only relevant projects. The list of selected projects is then analysed on two levels: an analytic review using transport, economic, technological, and policy tags, and a qualitative assessment of the research content. These activities lead to the development of analytics and infographics summarizing relevant indicators and the identification of main achievements and outcomes related to the research topic. The analysis is typically organized around sub-themes that include projects with similar content, such as technical or socio-economic aspects. The analyses are combined to create an overall assessment, which forms the basis for a scientific report, brief, or review that provides evidence of research achievements, as well as gaps and recommendations for future investigations. This output, usually presented as a Science for Policy report, serves as the foundation for policy reflections on the future transport R&I agenda and investments.

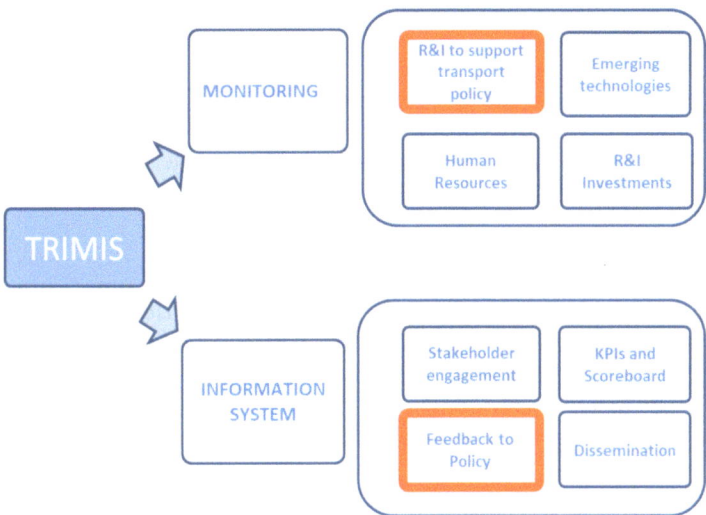

Fig. 2. TRIMIS main functionalities in the transport sector

Recent examples of TRIMIS' role as a policy support tool include its application to topics such as EU transport safety and resilience (Cheimariotis et al., 2023), urban mobility and logistics (Gkoumas et al., 2022), the role of R&I in public transport (Stepniak et al., 2022), and other key R&I-related transport topics. Publications on these topics are available on the TRIMIS portal.

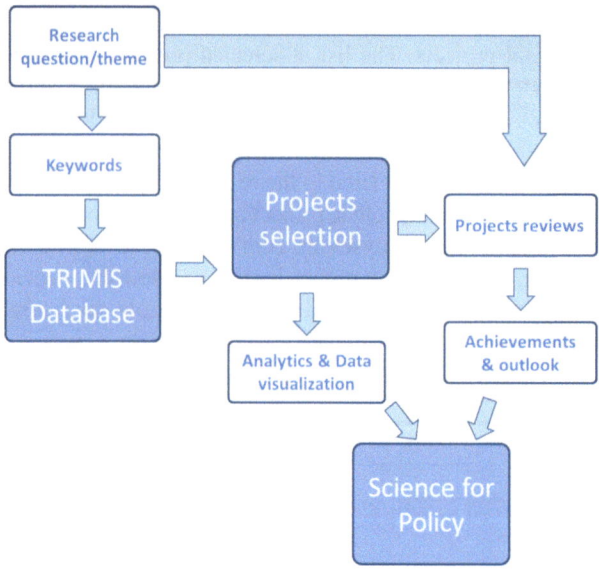

Fig. 3. TRIMIS Policy support workflow

4   Conclusions

This paper presents TRIMIS as a policy-supporting tool, illustrating the methodological approach used to produce scientific evidence that could be instrumental in future transport policy measures. TRIMIS has been acknowledged by the transport community, including policy makers and researchers, as a useful R&I instrument whose outcomes are used in policy reflections (e.g., EC feedback to policy exercises) and the definition of research state of the art. The increasing number of portal users is a clear indication of the general audience's appreciation. TRIMIS is continuously enhancing its features and functionalities, aiming for improvement in areas such as data gathering and quality, as well as increasing alignment between TRIMIS analysis and feedback and EU policy measures. The collection, monitoring, and quality of data within TRIMIS are key to producing reliable and useful analysis. The vast majority of transport data in TRIMIS comes from EU databases, such as CORDIS (https://cordis.europa.eu/it), which should guarantee quality data. However, there are projects that are not flagged as transport-related but actually have relevant topics within their scope. These projects typically fall under other EU programmes. TRIMIS is making efforts to establish optimized filters to capture such projects, and similar actions are taken on Member States and international projects that are not structurally collected by TRIMIS, relying on single initiatives. National contact points have been established in these cases, and constant engagement activities with stakeholders are put in place to expand the TRIMIS database and promote transport R&I projects. TRIMIS provides feedback to policy makers based on EU transport policy indications. However, monitoring the achievement of objectives and targets remains challenging as they are not always measurable or clear in the supporting documents of projects.

References

Cheimariotis, Stepniak, M., Gkoumas, K., Lodi, C., I., Marques dos Santos, F., Grosso, M., Marotta, A.: Research and Innovation in Transport Safety and Resilience — An assessment based on the Transport Research and Innovation Monitoring and Information System (TRIMIS), Publications Office of the European Union, Luxembourg (2023)

European Commission, 2017a. Europe on the move — An agenda for a socially fair transition towards clean, competitive and connected mobility for all, COM (2017) 0283 final

European Commission: 'TRIMIS Transport and Research and Innovation Monitoring and Information System' (2017b). (https://trimis.ec.europa.eu)

European Commission: The European Green Deal COM/2019/640 final; European Commission: Brussels (2019)

European Environmental Agency, 2020a. Transport and mobility, Publications Office of the European Union (2020). https://www.eea.europa.eu/en/topics/in-depth/transport-and-mobility. Accessed 18 Sept 2023

European Commission: Sustainable and Smart Mobility Strategy – Putting European transport on track for the future COM/2020/789, Brussels (2020b)

European Commission: Delivering the European Green Deal. Available online (accessed 18 September 2023) (2021). https://commission.europa.eu/strategy-and-policy/priorities-2019-2024/european-green-deal/delivering-european-green-deal/fit-55-delivering-proposals_en

European Commission: Statistical Pocket book 2022 – EU Transport in figures, Publications Office of the European Union, 2022 (2022) https://op.europa.eu/en/publication-detail/-/publication/f656ef8e-3e0e-11ed-92ed-01aa75ed71a1. Accessed 18 Sept 2023

Gkoumas, K., Stepniak, M., Cheimariotis, I., Marques Dos Santos, F., Grosso, M., Pekar, F.: Research and Innovation in Urban Mobility and Logistics in Europe, EUR 31197 EN. Publications Office of the European Union, Luxembourg (2022)

Stepniak, M., Gkoumas, K., Marques Dos Santos, F., Grosso, M., Pekar, F.: Public Transport Research and Innovation in Europe, EUR 31091 EN. Publications Office of the European Union, Luxembourg (2022)

Tsakalidis, A., Gkoumas, K., Pekar, F., Grosso, M., Haq, G., Marelli, L.: Towards an integrated European platform for monitoring and analysing transport research and innovation (TRIMIS). Proceedings of 7th Transport Research Arena TRA, pp. 16–19 (2018)

Tsakalidis, A., Gkoumas, K., Grosso, M., Pekár, F.: TRIMIS: Modular development of an integrated policy-support tool for forward-oriented transport research and innovation analysis. Sustainability 12(23), 10194 (2020a)

Tsakalidis, A., van Balen, M., Gkoumas, K., Pekar, F.: Catalyzing sustainable transport innovation through policy support and monitoring: the case of TRIMIS and the European green deal. Sustainability 12(8), 3171 (2020)

Open Access This chapter is licensed under the terms of the Creative Commons Attribution 4.0 International License (http://creativecommons.org/licenses/by/4.0/), which permits use, sharing, adaptation, distribution and reproduction in any medium or format, as long as you give appropriate credit to the original author(s) and the source, provide a link to the Creative Commons license and indicate if changes were made.

The images or other third party material in this chapter are included in the chapter's Creative Commons license, unless indicated otherwise in a credit line to the material. If material is not included in the chapter's Creative Commons license and your intended use is not permitted by statutory regulation or exceeds the permitted use, you will need to obtain permission directly from the copyright holder.

Future Workforce and Skills

Gender Imbalance in the Transport Sector – An International Review

Isobel Duxfield[1]([✉]) and Heather Allen[2]

[1] POLIS, Brussels, Belgium
i.duxfield@btinternet.com
[2] Median SRL, Brussels, Belgium

Abstract. According to the International Labour Organization's latest figures, women make up just 16.8% of those working in transport globally. This makes it one of the worst-performing industries worldwide- both across the public and private sectors, from boardroom to engine room. Improving gender diversity in the transport sector workforce is essential to achieve its innovation, sustainability and financial goals, while tackling the current, and profound, skills shortage. This paper provides an insight into the challenges and opportunities women face when they choose to work in transport. It is grounded in a review of published literature and primary data collection among a sample of men and women working in the sector. The findings offer an overview of women's global employment in transport, including all modes and sectors of transport. The study is presented as a toolkit highlighting the challenges women face when entering, working, and progressing in the transport sector, identifying five main areas or entry points where actions and measures could provide the most impact in accelerating gender balance and some 25 case studies. This paper presents the main findings from the study.

Keywords: Gender · Transport · Employment

1 Introduction

The transport sector currently faces many challenges including labour and skills shortages, harsh working conditions, industrial relations issues and an ageing workforce. Additionally, there are deep and widespread market changes in relation to automation and digitalisation, moves towards greener and cleaner transport options, reduced reliance on fossil fuels and sustainability. These issues present a threat, especially for low- and medium-skilled jobs, but also offer new opportunities[1]. Although statistics are difficult to obtain, the International Labor Organization (ILO) suggests only 16.8% of the global workforce in transport is female. In Europe, women represent less than 20 percent of

---

[1] Sustainable Mobility for All. 2023. Gender Imbalance in the Transport Sector: A Toolkit for Change. Washington DC, ISBN: 979-8- 9882420-1-7. License: Creative Commons Attribution CC BY 3.0 IGO.

© The Author(s) 2025
C. McNally et al. (Eds.): TRAconference 2024, LNMOB, pp. 767–780, 2025.
https://doi.org/10.1007/978-3-031-85578-8_104

the transport workforce[2].and a fragment of its leadership, both in the private[3] and public sectors[4]. These figures include storage and communication and a range of transport pipeline occupations[5] and do not reflect the deep differences and gendered job profiles within transport subsectors.[6]

The transport sector is now at a critical juncture, facing skills and labour shortages worldwide. There is a growing and widespread recognition[7] that recruiting and retaining women would improve the sector's financial and environmental resiliency. Indeed, greater diversity and gender equality are increasingly being seen as critical for increasing innovation, creativity, and addressing the human capital requirements for transport and helping it respond better to world demands for both passenger and freight service offers.

2 Background and Literature Review

2.1 The Global Picture of Gender-Balanced Employment

The United Nations' 2016 report "Leave no one behind: A call to action for gender equality and women's economic empowerment"[8] presented evidence and made a substantive case for actions to address the systemic constraints and persistent gaps in women's economic opportunities across all sectors. This remains at the heart of the International Agenda 2030 and the Sustainable Development Goals (SDGs)[9]. As global pressure to eradicate poverty and address climate change targets grow, enhancing women's economic participation must move from ambition to action at an accelerated pace. Closing the global gender employment gap plays a central role in economic prosperity, welfare, social cohesion, and environmental sustainability. Enabling women and girls to enter the formal job markets has been noted as one of the biggest opportunities[10] for

---

[2] The World Bank (2023) Making Way for Women in Transport and Logistics: Promising Practices in Europe and Central Asia (WWW), https://www.worldbank.org/en/news/feature/2023/01/26/making-way-for-women-in-transport-and-logistics-promising-practices-in-eur ope-and-central-asia.

[3] Lutte, R. (2021) Women in Aviation: A Workforce Report 2021 Edition.

[4] Inter-Parliamentary Union (2023) Women in Politics: 2023 (WWW) https://www.ipu.org/res ources/publications/infographics/2023-03/women-in-politics-2023.

[5] Additionally, the data includes storage and communication and a range of occupations such as transport via pipelines, supporting and auxiliary transport activities, and the activities of travel agencies, postal services, and telecommunications.

[6] Examples include aviation where there are deep differences between men and women occupying jobs such as pilots, crew and ground staff.

[7] Ng, W., and Bassan, D. (2022) Gender equality and the role of women in decarbonizing transport, OECD/ITF & FIA Foundation, Paris and London.

[8] UNESCO (2020) Global Education Monitoring Report – Gender Report: A new generation: 25 years of efforts for gender equality in education. Paris, UNESCO.

[9] UNHLP (2016) Leave No One Behind - A Call to Action for Gender Equality and Women's Economic Empowerment.

[10] World Bank. (2033) Women Business and the Law 2023. Washington, DC: World Bank. https://doi.org/10.1596/978-1-4648-1944-5. License: Creative Commons Attribution CC BY 3.0 IGO. https://wbl.worldbank.org/en/wbl.

human development and economic growth, contributing to achieving global development goals as outlined in the Sustainable Development Goals.

2.2 Transport in Focus

Transport is a sector which has a significant and persistent gender imbalance. It is also traditionally regarded as "*No place for women*"[11]. At the same time, many sub-sectors within transport, particularly logistics and public transport, are experiencing major shortages in the workforce[12]. More worryingly, traditional approaches to recruitment do not appear to be building a robust pipeline of talent. Occupational segregation is widely visible, and women remain heavily underrepresented in engineering and technical roles, and many jobs remain stereotyped as male occupations such as airline pilots, long haul maritime captains and crew, freight and maintenance jobs for aircraft, as well as decision-making, planning and operational positions in shipping, road, and rail. Although recognition for the need for action is growing, no significant improvement has been observed since 2010[13].

Women also remain a minority in political decision-making in transport, accounting for only 15.5% of ministers with transport portfolios across the 27 EU Member States[14]. As of February 2023, only 7 of the 64 member countries of the International Transport Forum[15] countries have female Ministers of Transport. According to the World Economic Forum Global Gender Gap report[16], 21% of the supply chain and transportation sector have women leaders; the fourth lowest score in the ranking of all industry sectors.

However, over the last decade, more attention toward issues surrounding gender balance in the workplace and women's employment experiences[17] has emerged and highlighted transport as a sector that would benefit from greater workforce diversity. A growing body of literature on this topic testifies to this. Examples include the publication of several key pieces of research into equality and discrimination[18], efforts by the

[11] International Road Union (2022) Driver Shortage Global Report 2022. https://www.iru.org/res ources/iru-library/driver-shortage-global-report-2022.

[12] International Labour Organization. (2013) Women In The Transport Sector, Geneva, https://www.ilo.org/wcmsp5/groups/public/---ed_dialogue/---sector/documents/briefingnote/ wcms_234882.pdf.

[13] International Labour Organization. (2013) Women In The Transport Sector, Geneva, https://www.ilo.org/wcmsp5/groups/public/---ed_dialogue/---sector/documents/briefingnote/ wcms_234882.pdf.

[14] IFALPA (2021) Why Diversity, Equity, and Inclusion Matter, 21POS09. https://www.ifalpa. org/media/3752/21pos09-women-in-aviation-why-diversity-equity-and-inclusion-matter.pdf.

[15] Sourced directly from ITF database.

[16] World Economic Forum (2022) Global Gender Gap Report 2022 Insight Report July 2022. https://www3.weforum.org/docs/WEF_GGGR_2022.pdf.

[17] Kronsell, A., Smidfelt, L., & Winslott Hiselius, L. (2016) Achieving climate objectives in transport policy by including women and challenging gender norms: The Swedish case, International Journal of Sustainable Transportation, 10:8, 703–711.

[18] Ng, WS and A. Acker (2020), "The Gender Dimension of the Transport Workforce", International Transport Forum Discussion Papers, No. 2020/11, OECD Publishing, Paris.

European Commission[19] and World Bank to collect data, and international representative bodies such as the International Transport Workers' Federation (ITF)[20] and European Transport Workers' Federation (ETF)[21] and networks including the Women's International Shipping & Trading Association, Women in Transport, and Women in Trucking which have exposed discriminatory practices and cultures, highlighting challenges, and started campaigning for action. The male domination of the sector and the repercussions for women's treatment in the workplace were confirmed in our study interviews and findings.

3   Methodology

This study was undertaken over a nine-month period from June 2022 to March 2023 and combined desk-based research, literature reviews, an internet-based survey, and 27 interviews with key players from the transport sector. The findings and recommendations were reinforced by over 25 case studies and examples highlighting what is being done to address the various issues women face when working in transport. The results were validated by an international steering group and the SuM4ALL Gender Working Group.

The online survey was completed by 300 individuals, from both the public and private sectors (regional, national, and local entities), international development agencies, academia, trade unions, professional and representative bodies and non-profit organisations from all the main modes of transport (civil aviation, maritime, road, rail, passenger, freight and logistics). The survey and interviews explored both women's and men's opinions and collected personal experiences of working in the transport sector from an organisational and individual perspective, and included views on the current state of international and national policy frameworks.

The survey was disseminated through the SuM4All and POLIS communications channels, social media, individual email outreach, at major international transport events and through international transport publications. Care was taken to ensure the survey and interviews covered all geographical regions, modes and subsectors as far as possible.

The interviews conducted enabled a deeper dive into many of the themes identified in the survey and the levers behind the positive action initiatives being deployed. The interviewees were sourced from the survey respondents and outreach to major players and key individuals. Their profiles included CEOs, human resource managers, academics, trade unions, leaders of women's transport networks, and others in leadership and managerial positions with insights into key barriers to gender equitable employment and positive actions to address this, as well as representatives from automotive and technology industries. Inputs from modes and regions with less representation in the survey were secured

---

[19] Ortega Hortelano, A., Grosso, M., Haq, A., Tsakalidis, A., Gkoumas, K., Van Balen, M. and Pekar, F. (2019) Women in European Transport with a focus on Research and Innovation, EUR 29833 EN, Publications Office of the European Union, Luxembourg.

[20] International Transport Workers' Federation (2020) Equality Testing in Twenty years on Transport, London.

[21] In March 2023, the ETF launched "Get ME home safely: A campaign for safe commuting to and from work for transport workers". Further information, https://www.etf-europe.org/activity/get-me-home-safely-safe-commuting-to-and-from-work-for-transport-workers/.

through further stakeholder outreach and peer reviews from the SuM4ALL International Gender Working Group.

The researchers were aware that the survey respondents only included those who had access to the internet (i.e. mainly managerial and other professional profiles), and they also observed a slight bias toward respondents coming from Europe and North America although 15% stated that they acted globally, and 36% of respondents described the scope of their work as international.

4   Results and Key Findings

4.1   The Need for More Diversity and the Direction of Policy

It is clear that transport has a gender balance issue. The survey respondents were specifically asked about their views on these aspects; and 97% agreed that the transport workforce needs to be more diverse. Furthermore, it was stated by many that the sector should consider wider diversity issues, not only gender.

Global consensus prevails that the changes in policies to support greater gender equality and women's empowerment are important and, for the most part, according to our respondents, the international policy environment was regarded as being relatively supportive, although this may vary depending on the geographic region and cultural norms. The survey exposed that many of the challenges women face go beyond national or international policy frameworks. Some 30% agreed—and a further 50% were inclined to agree—that the sustainable development goals (SDGs) and calls to action on gender equality and equity have helped increase the opportunities for women to enter the formal labour force, however, they repeatedly asserted that there were gaps in implementation and today they are not the key determinant for women to thrive in the transport sector. Additionally, 55% felt that policy frameworks provided support for more gender equitable employment in transport but only 15% asserting that this was "to a great extent".

4.2   Attracting and Retaining Women

Many survey respondents, of whom almost three-quarters were women, saw transport as an attractive sector to work in and had chosen transport as a conscious career choice. However, an equal number simply, in one respondent's words, "ended up" working there because they had the opportunity to work in the sector and took it (Table 1).

Table 1. Responses to a specific question posed in this survey.

Did you decide to join the transport sector as a specific career decision?	Percentage
Yes	46.5
No it happened by accident	47.2
other	6.3

Furthermore, nearly 50% of respondents felt that it was easy for women to join and work in transport if they had the right qualifications (Fig. 1). Some 44% felt that this depended on the type of job and for some job profiles it was possible, but not easy. More than 65% also stated that their organisations actively encouraged women to join while respecting equal opportunities requirements.

Attracting women to apply for positions was seen to be improving; 45.5% stated that they felt that it was easy for their organisations to attract women to work for them, however 30% were not sure that this was true and 20% disagreed. This suggests it may be easy for women to join some subsectors and/or apply for positions but that this is not yet widespread. It was clearly stated that technical positions were more difficult for women to obtain, and that junior positions were easier to fill than senior ones.

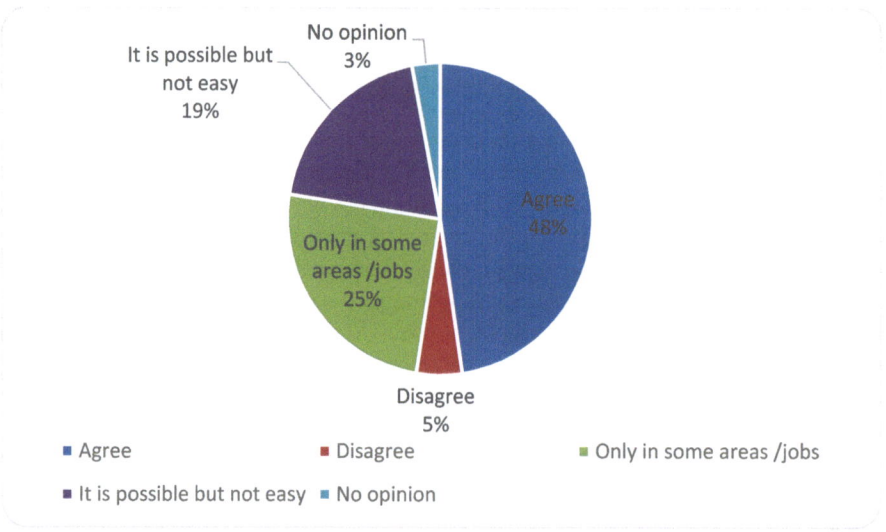

Fig. 1. On the ease for qualified women to enter and work in the transport sector. Answers to the question, *"Would you agree or disagree that women can easily join and work in the transport sector if they have the right qualifications?"* Question posed in the web-based survey.

Many organisations for which the respondents worked are focusing their attention on the retention of female employees and more than 55% of respondents' organisations had schemes to support women's job retention and career progression (Table 2).

Table 2. Percentage of respondents stating their organisations were implementing actions aimed at supporting gender-balanced or equitable employment.

Description of action	Percentage of respondents stating their organizations were implementing these actions
Flexible working	76%
Equal pay policies	58%
Gender equity policies	51%
Protocols and processes for addressing sexual harassment in the workplace	49%
Internal networks to support women or diversity	42%
Support for or assistance with childcare	26%
Mentoring schemes designed to fast-track or support women	19%
Blind CV selection	14%
Positive quotas	10%

Other examples include "return to work schemes", the availability of fridges for nursing mothers, increased options for flexible working, and sharing useful knowledge and broadening information on gender through internal communications channels.

5   Key Entry Points for Action

The survey findings revealed five key barriers which included recruitment and retention of women, as well as improved gender balance at leadership level; all of which would help make the working environment and lived experiences of women more positive. Measures and initiatives[22] taken to address them would increase opportunities for women to be able to add value to the sector, as well as, encourage other underrepresented groups to make their careers in the sector.

The five entry barriers are:

1. Gender stereotyping for certain jobs, especially technical ones
2. Discriminatory work cultures
3. Lack of flexible working and childcare provision
4. Invisible glass ceilings and restricted career options
5. Gender equity and its relationship with wider diversity, behaviour and culture change.

Some of these are not unique to the transport sector, but the survey findings show they are underpinning gender (in)equitable employment and identify how these are playing out in individual regions and modes within the sector. Case study examples of how to address these barriers are included in the full toolkit.

---

[22] Many of which can be found in the case studies of the complete toolkit.

5.1 Stereotyping Remains Strong for Certain Jobs, Locking Women Out of the Sector

Transport—among other related industries—continues to be recognised as being very masculine. Despite individual actions at institutional and organisational levels, progress is seen as being very slow. Simultaneously, transport itself is changing rapidly with new job functions emerging and workplace cultures across the various transport modes being challenged, which in turn are undeniably shaping how women engage with the sector. Stereotyping of certain jobs in the sector remains strong, particularly in engineering and technical domains, and organisations across the sector struggle to reach out to and recruit the growing body of qualified women in these areas.

Consequently, the current gender imbalances in the workforce stretches from junior roles to middle management and up to leadership positions. More targeted, innovative, and inclusive recruitment is needed. A large part of this problem stems from a lack of clarity around the skills and qualifications needed for specific jobs and functions and promoting this effectively using inclusive language. Many respondents cited difficulties with attracting and recruiting women and our extensive research reveals women's hesitancy in applying for jobs they do not consider themselves fully qualified for.

It is clear transport needs to change how it advertises and promotes vacancies, actively expanding its outreach to reach new audiences and counteracting the gendered stereotyping along with workplace culture (both were rated as the biggest barriers for women entering the sector in the survey). Simply featuring women in recruitment campaign material is not sufficient and must be combined with a range of wider measures to be fully effective, especially ensuring that the language used is inclusive and resonates with women.

Many organisations had gender-related internal policies, especially the larger ones and those with international activities. Several examples were found and can be read in more depth in the case studies section of the full toolkit. Over 65% of survey respondents asserted that their organisations had begun to take steps over the last few years to improve the recruitment and retention of women, including mentoring schemes, flexible working options, separate and dedicated training opportunities for women as well as training to address unconscious bias and more structured ways to communicate with potential and existing employees.

Interestingly, while staff shortages are often stated as a reason for the need to accelerate the recruitment of women, in some cases, it is proving detrimental to progress towards a more gender-balanced workforce by diminishing the focus on decent jobs, equal pay, and diversity and only looking at sheer numbers. This finding is essential for comprehending the importance of targeted gender outreach.

5.2 Discriminatory Work Cultures and Practices

The majority of respondents felt that the policy framework was "good enough" but that workplace cultures and behaviour change were seen as greater barriers for women to be able to enjoy the opportunities the sector had to offer.

Almost 50% of respondents stated that their organisations conduct training on inclusive language, discussions on stereotyping and unconscious bias; yet "male dominated surroundings", "uncomfortable" environments, "discrimination", and "prejudice" were routinely identified as leading challenges for women across the sector. Addressing practices with zero tolerance for sexual harassment in the workplace was a leading area where respondents wanted to see change.

The findings also indicate that women who enter at junior levels frequently leave the sector mid-career, especially after a second child. They leave due to a combination of circumstances but our findings suggest that the reasons are largely influenced by existing workplace cultures and practices. The survey and interview respondents reported that although equal pay was important, work/life balance and benefits such as flexible hours and schedules, health, safety and hygiene, and work family options—such as help or access to affordable childcare—were key determinants to retain women employees.

Social movements have provided an impetus and helped accelerate action. Individuals, particularly from the maritime sector, disclosed a significant change in female employees' readiness to speak out against inappropriate actions or encounters, as well as in the emergence of networks and platforms such as Safer Waves—a confidential support service for seafarers who have experienced sexual violence or gender discrimination. Violence and sexual harassment in the workplace is also a major issue to be addressed across all modes and job profiles.

5.3   Lack of Flexibility and Childcare Provision

Lack of flexible working opportunities and childcare facilities were also identified as key barriers to women's participation in transport impacting work–life balance and the retention of women especially as they start families. Difficulties coordinating professional employment and caring are a logistical and psychological challenge for women, who still fulfil most family-caring roles, and many respondents did not feel supported by their employer.

Survey and interview data revealed that regardless of maternity/paternity leave policies, balancing family and caring obligations remain major physical, and psychological, challenges for women, even in countries with some of the most generous provisions. Across the board, in transport and other sectors, pregnancy often forces women to abandon their careers and in this study, pregnancy was identified as a key cut-off point when women often leave the sector; or are obliged to move positions further entrenching occupational segregation and historical and inflexible practices.

The more progressive employers are increasingly making practical changes to schedules, rostering, and recruitment to make them more inclusive and gender sensitive. Alongside these changes, clarity around the rights, opportunities, and procedures for flexible work for both men and women are essential.

5.4   'Invisible' Glass Ceilings and Restricted Career Options

The transport sector remains resolutely male dominated at senior and decision making levels but some improvements were observed compared to a decade ago. Accelerating this pace is instrumental to diversify decision making and encourage more women into the sector.

Some organisations indicate that women are making headway in transport leadership. Indeed, almost 90% of respondents stated leadership and senior positions in their organisations are (or were in the recent past) increasingly being held by women. Over 60% of respondents indicated that the volume of leadership positions occupied by women has risen over the last decade, with many noting the last few years have seen particular progress. Interviewees also indicated that there is a strong business case for increasing diversity, citing how diverse teams can problem solve and find creative solutions more quickly, increasing productivity.

On the other hand, in some instances, higher levels of leadership within organisations were seen to be only 'walking the talk' when it suited them, and they were not yet leading by making the cultural changes needed further down the organisation. There was a dislocation between the equity vision of senior management and those in responsible positions further down the organisations. Indeed, it was noted that the success of policies depends largely on how they are implemented and adopted within an organisation.

The number of positive actions and measures the study was able to identify is encouraging; more than half of the organisations surveyed have schemes that support women's job retention and career progression and of this, over 50% hold equal pay policies, 20% have mentoring schemes designed to fast track or support women and just over 40% assist internal networks to champion women and wider diversity.

However, this research also discloses the insufficient scale of advance. In a third of instances, no progress has been made over the last decade. Such impasse was palpable across these interviews with women in leadership positions, many of whom were one of just a handful of women in such a position in their organisation. Racial diversity was also lacking, and women of colour were also not being given the same opportunities as their white counterparts.

The career pipeline for women to gain a foothold in leadership positions remains inadequate, some referred to this as 'leaky', and many good candidates resigned before reaching senior positions. Several employers reported that even when quotas for senior positions were imposed, not enough potential female candidates had reached senior levels within the organisation. This is often the result of a rigid system of promotion which appeared prevalent in the sector, which defends a historic culture of working one's way up through an organisation. If this is favoured over external recruitment for leadership roles, combined with the leaky pipeline syndrome, progress toward gender balance becomes almost impossible, although this is beginning to change.

It was also encouraging to see that as transport as a whole changes, new roles are being created and that this is reinforcing internal policies and actions. Nonetheless, so far this is largely a northern hemisphere phenomena and can not yet be considered a global trend.

5.5 Gender and Its Relationship with Wider Diversity, Behaviour and Culture Change

Although it is widely recognised that a more diverse workforce brings multiple benefits, there remains less understanding of the ways gender equity and wider diversity intersect, and the impacts of it on everyday workplace behaviours, especially in relation to transport. This research also revealed the need for clarity about the roles and responsibilities to make workplaces more inclusive for underrepresented groups from all backgrounds.

Respondents cited that even when diversity quotas are filled, and gender balance at the senior level is achieved, longstanding behaviours and cultures persist further down the organisation, and environments are still structured and dominated by masculine values and behaviours. This has implications not just for gender inequality but is detrimental for wider diversity too. Over half (53%) of survey respondents indicated that their organisation does not provide training to avoid stereotyping and unconscious bias within the workplace. Lack of collective dialogue about the impacts of individual action and the roles and responsibilities for creating more inclusive spaces has generated confusion and, in many cases, backlash, or hostility.

Discriminatory behaviours and cultures have tangible repercussions not just on women's wellbeing, but their physical ability to do their job, creating a workplace, which is not equipped for them. It was also noted that these types of behaviours persisted when employees, both men and women, did not have the voice or confidence to speak up or speak out against mistreatment for fear of repercussions.

The survey and interviews revealed that misconceptions remain about the changes in language required and the effects of small or unconscious actions. Few clear and safe forums exist to highlight concerns and hold exchange of experiences and guidance for colleagues to understand the impacts of their actions and adapt their behaviours.

The researchers heard how this limitation often unfolds in strong male dominated environments, such as in maintenance areas or bus depots, where female recruits felt male colleagues monopolised space or assumed responsibility for their tasks. Without clear avenues for meaningful conversations, people felt intimidated and were unable to voice grievances for fear of how it will be received and the potential repercussions of speaking out.

The transport sector must develop capacity for co-learning opportunities and clear dialogues on diversity and where male colleagues and those from majority ethnic groups can learn more about allyship and support in establishing more inclusive workplaces.

A more methodological and responsible approach is needed, starting with: (i) collection of gender-disaggregated data on employee experiences, (ii) systematically sharing these concerns with leadership, (iii) providing forums for discussion, and (iv) providing clear feedback channels.

6   Conclusions Preparing for a Gender-Balanced Future

The last several years have seen unprecedented social movements towards exposing women's lived experiences and pressing for more transformative change. The transport sector is at a critical juncture where there is renewed urgency for progress, accompanied by fresh opportunities for action. The transport workforce remains male-dominated, but it also faces critical shortages in key areas, examples include drivers for heavy duty vehicles (freight and logistics) and buses (for public transport). Furthermore, the full effects of the COVID-19 pandemic are yet to be ascertained, and initial reviews from across the globe suggest women's employment generally has been negatively affected[23].

Despite progress in global and national policy arenas, women working across the transport sector continue to face multiple impediments to thrive in the sector. Although international and national equal opportunities and antidiscrimination policy frameworks lay the foundations for change, decision makers in the transport sector cannot rely solely on these to generate the transformative shifts necessary at the rate required and behaviour changes are needed. Gendered stereotyping, discriminatory and unsafe workplace cultures, lack of flexible working, childcare provision and invisible glass ceilings were found to be the main constraints toward a more gender-balanced workforce.

Several respondents noted the transport sector's 'leaky pipeline'. This concept, which has also been identified in other sectors[24], refers to a talent drain in middle management, which for women often coincides with the life stage when family and caregiving responsibilities become a more conscious part of work-life balance. This has a serious knock-on effect for women's promotion to c-suite level, where there is then a lack of internal female candidates.

Despite the growing number of gender equality policies, initiatives and programmes established at organisational level and the generally positive sentiments shared in this research, there remain significant challenges to address the prevailing imbalance. Failure to do so will not only impact the workforce, but also the services delivered and the future sustainability of operations.

Targeted, sustained, and effective action is imperative, and change will demand comprehensive and sustained positive action from all those involved in the sector for women and other underrepresented groups to be able to contribute fully.

Simultaneously, we must recognise the spectrum of experiences of those who are underrepresented in the transport workforce and the degree to which a network or forum represents these experiences. In many parts of the world, especially in urban transport, women work under informal or casual contracts, which excludes them from social safeguards such as health insurance, pensions, maternity leave, or other benefits. The role of organised labour and trade unions in the protection of all workers, but especially women, is an area that has not been extensively researched yet for transport. Representation is necessary and serves to provide helpful support and opportunities for dialogue and exchanges, but does not alone provide the solution, which also requires industry and employer commitment and action. Nevertheless, fostering exchange between different networks and with leadership can further aid effective change.

The lack of data was also highlighted both across and within the sector. Thus, more regular gender-disaggregated quantitative data collection and qualitative information are essential, and should include lived experiences to ensure advances can be measured and persistent obstacles are identified. The volume and breadth of information and data on gender-equal employment strategies must also be shared more widely. Many do not recognise the value of their own good practices for others. There remains a lack

---

[23] UN (2022) Progress on The Sustainable Development Goals The Gender Snapshot 2022, (WWW) https://www.unwomen.org/sites/default/files/2022-09/Progress-on-the-sustainable-development-goals-the-gender-snapshot-2022-en_0.pdf.

[24] This has been identified in life sciences, academia, and technology sectors in particular.

of awareness about the positive action implemented by peers across the sector and how these can be replicated, and most do not look beyond national boundaries for inspiration, despite the common challenges faced regionally and globally. The practice of circulating Information on action, the achievements—and even failures—will help enable others to replicate and build on tried and tested methods.

Significantly, the common entry points for action revealed in this research demonstrate the potential for and value of cross-sector conversation and collaboration, learning from one another, and combining resources to enhance the scale and scope of positive action.

This research also revealed the importance of adapting the questions we ask and the language we use when seeking to understand women's experiences of employment. Combining questions on women's personal experiences and their opinions with quantitative questions on gender balance, identifies nuances, inconsistencies, and discrepancies between what women need and what is being provided. We recommend that future research builds on this approach to help direct more targeted action with a view to generating more impact and stimulating greater behaviour and cultures changes.

Acknowledgements. This paper is based on the findings of *"Gender Imbalance in the Transport Sector: A Toolkit for Change",* prepared by the Sustainable Mobility for All (SuM4All) Gender Working Group. This work was produced with the financial support of FIA Foundation Registered UK Charity No. 1088760 60. The report is part of the GRA in Action series funded by the World Bank, the German Federal Ministry for Economic Cooperation and Development (BMZ), the Foreign, Commonwealth, and Development Office (FCDO), and the Michelin Foundation.

The full toolkit which includes detailed case studies of 25 transport sector organizations, can be found here: https://www.sum4all.org/data/files/gender_imbalance_in_the_transport_sec tor_a_toolkit_for_change.pdf.

References

1. IFALPA: Why Diversity, Equity, and Inclusion Matter, 21POS09 (2021). https://www.ifa lpa.org/media/3752/21pos09-women-in-aviation-why-diversity-equity-and-inclusion-mat ter.pdf
2. International labour organization: Women In The Transport Sector, Geneva (2013). https://www.ilo.org/wcmsp5/groups/public/---ed_dialogue/---sector/documents/briefingn ote/wcms_234882.pdf
3. International Road Union: Driver Shortage Global Report 2022 (2022). https://www.iru.org/ resources/iru-library/driver-shortage-global-report-2022
4. International transport workers' federation: Equality Testing in Twenty years on Transport, London (2020)
5. Inter-Parliamentary Union, Women in Politics: 2023 (WWW) (2023). https://www.ipu.org/ resources/publications/infographics/2023-03/women-in-politics-2023
6. Kronsell, A., Smidfelt, L., Winslott Hiselius, L., Achieving climate objectives in transport policy by including women and challenging gender norms: the Swedish case. Int. J. Sustain. Transp. **10**(8) 703–711 (2016)
7. Lutte, R., Women in aviation: a workforce report 2021 edition (2021)
8. Ng, W., Bassan, D., Gender equality and the role of women in decarbonizing transport, OECD/ITF & FIA Foundation, Paris and London (2022)

9. Ng, W.S., Acker., A.: The gender dimension of the transport workforce. International Transport Forum Discussion Papers, No. 2020/11, OECD Publishing, Paris (2020)
10. Ortega Hortelano, A., et al.: Women in European Transport with a focus on Research and Innovation, EUR 29833 EN. Publications Office of the European Union, Luxembourg (2019)
11. UN: Progress on The Sustainable Development Goals The Gender Snapshot 2022, (WWW) (2022). https://www.unwomen.org/sites/default/files/2022-09/Progress-on-the-sus tainable-development-goals-the-gender-snapshot-2022-en_0.pdf
12. UNESCO: Global Education Monitoring Report – Gender Report: A new generation: 25 years of efforts for gender equality in education. Paris, UNESCO (2020)
13. UNHL: Leave No One Behind - A Call to Action for Gender Equality and Women's Economic Empowerment. (2016)
14. World Bank: Making Way for Women in Transport and Logistics: Promising Practices in Europe and Central Asia (WWW) (2023). https://www.worldbank.org/en/news/feature/2023/01/26/making-way-for-women-in-transport-and-logistics-promising-practices-in-eur ope-and-central-asia
15. World Bank: Women Business and the Law 2023. Washington, DC: World Bank. License: Creative Commons Attribution CC BY 3.0 IGO (2033).https://doi.org/10.1596/978-1-4648-1944-5, https://wbl.worldbank.org/en/wbl
16. World Economic Forum: Global Gender Gap Report 2022 Insight Report July 2022 (2022). https://www3.weforum.org/docs/WEF_GGGR_2022.pdf

Open Access This chapter is licensed under the terms of the Creative Commons Attribution 4.0 International License (http://creativecommons.org/licenses/by/4.0/), which permits use, sharing, adaptation, distribution and reproduction in any medium or format, as long as you give appropriate credit to the original author(s) and the source, provide a link to the Creative Commons license and indicate if changes were made.

The images or other third party material in this chapter are included in the chapter's Creative Commons license, unless indicated otherwise in a credit line to the material. If material is not included in the chapter's Creative Commons license and your intended use is not permitted by statutory regulation or exceeds the permitted use, you will need to obtain permission directly from the copyright holder.

Developing Digital Tools to Enhance Visual Inspection of Bridges During Extreme Climate Events

Kristopher Campbell[1,3]($\boxtimes$) , Myra Lydon[2] , Prof Su Taylor[1] ,
and Darragh Lydon[1]

[1] School of Natural and Built Environment, Queens University of Belfast, Belfast, Northern Ireland

[2] School of Engineering, University of Galway, Galway, Ireland

[3] Department for Infrastructure, Highways Structures Unit, Belfast, Northern Ireland

kristopher.campbell@infrastructure-ni.gov.uk

Abstract. Compared to the development of low carbon emission technologies such as EV, the adaptation of existing transport infrastructure to increase its ability to absorb and recover from the effects of climate change is relatively unexplored. Bridges form a critical link in our transport systems, and many have a design life exceeding 100 years. Therefore, they are one of the key climate-change relevant elements that needs to be considered when future proofing our transport networks. With substantial maintenance backlogs worldwide and limited budgets, the bridge management process is becoming increasingly critical to support connectivity across our regions. Existing bridge management systems have significant limitations. They often operate in isolation from other asset management tools and lack interoperability with geographical, societal, and climate-related data sets. These systems have limited capability to incorporate data from sensors, leaving engineering judgment as the primary decision-making factor within constrained time and financial resources. To address these challenges, it is essential to develop comprehensive bridge management systems that integrate with various data sources. This paper provides an overview of the current bridge management process in Northern Ireland and presents a number of interventions to improve the robustness of the decision-making process including enhanced interoperability with other data sources and improved inspector training using virtual reality.

Keywords: Bridge management · Inspections · Training · 3d · VR

1 Introduction

1.1 The Current Bridge Management Process

The effects of climate change on the performance of bridges within infrastructure systems are often overlooked in bridge management practice. Despite its direct relevance given the long design life of a bridge (+120 years), investment is currently set out under shorter time horizons. The future risk of bridge failure is further dramatically increased

© The Author(s) 2025

C. McNally et al. (Eds.): TRAconference 2024, LNMOB, pp. 781–786, 2025.

https://doi.org/10.1007/978-3-031-85578-8_105

due to the aggregation of several negative factors occurring simultaneously. Financial constraints across all asset owners precludes detailed risk management and the clarification of economic value remains the greatest barrier to future proofing our civil infrastructure. UK traffic has increased by ~ 30% in the last 20 years and in Northern Ireland structural maintenance funding for bridges has decreased by 50% in the same period. This consistent underfunding results in predominantly reactive management and limits capacity for strategic innovation in asset management.

Northern Ireland's (NI) road network is owned, operated, and maintained by the Department for Infrastructure (DfI). DfI maintain approximately 6000 bridges across the network with approx. 60%being more than 100 years old. This research provides a review of the current bridge management process and provides two digital solutions to support the bridge inspections process within the overall management lifecycle.

Visual inspections are the primary data source used by asset owners to determine the condition of the bridge stock. Given the importance of these inspections for allocating resources and planning maintenance, it is important that the individuals conducting the inspections are competent, experienced, and adequately trained. Inspectors need to identify defects correctly and understand the implications of them on the structure wholistically to ensure effective decisions can be made. It is not uncommon for one bridge engineer to be responsible for the inspection of several hundred bridges each year. On average bridges are inspected on a two-year cycle staggered to give annual inspection programmes. Limited resources, other workloads, weather and accessibility constraints puts additional pressure on inspectors in order to complete a full annual programme. Given the bridge safety risks associated with climate change, the planning and accurate completion of bridge inspections during extreme climate events is critical to future bridge safety. This research proposes two digital tools to support this process.

1. Formalised bridge inspector training which encompasses virtual reality simulation of inspection for bridge scour.
2. GIS based inspection planning dashboard with integrated flood risk data.

The following sections detail the development of these tool along with the implementation plan.

2   VR Bridge Inspector Training

2.1 Current Approach

The current process for bridge inspector training in NI has evolved through the following process. Individuals were posted to structures teams, these are not a dedicated bridge inspector role and consisted of other functions. They also did not require any specific level of tertiary education or experience specifically related to bridge inspections or engineering. Originally bridge inspectors were trained in-house with a mixture of slides and practical on-the-job training. New and less experienced staff would be partnered with more experienced staff to provide an extended period of on-the-job informal training reinforcing the classroom-based learning. Further to this, General Inspections could be conducted by any technical grade however Principal Inspections required at least one of the inspectors to be at a minimum management technical grade. These grades are arbitrary and did not take account of experience or professional/academic qualifications of the individuals. It is possible for inspectors to be in a management grade but have no formal academic engineering background or experience beyond the in-house training provided. In recent years, excessive staff rotations and retirements has accelerated the loss of experienced and in-house trained staff. The lack of recent dedicated training has resulted in the potential for informally trained individuals training new inspectors, resulting in inspections being conducted by individuals who have received no formalised training, only on-the-job type training. As a result, there is potential for the quality of inspection to reduce significantly and for unconscious bias or inconsistencies to creep into the way regional teams inspect assets. This can also lead to inspectors not being able to appreciate the significance of the defects that they might be recording on inspections and for timely interventions to be identified.

2.2 Proposed Training Course

To try and address the acknowledged concerns with inspections and to formalise the level of experience and training of bridge inspectors the National Highways Sector Scheme 31 (NHSS32) Bridge Inspector Certification Scheme (BICS) was introduced in 2016. BICS adheres to the requirements as provided in CS450 and certification is provided at both Inspector and Senior Inspector levels currently by Lantra.

Although the BICS scheme is an excellent marker for measuring the competency of bridge inspectors and widely accepted as the gold standard by Bridge Owners, the rigors of such a programme can be difficult to achieve by many local authorities. Staff movements through retirements, promotion and rotations often results in a limited ability to commit to the levels of training and experience to satisfy BICS certification. Given these challenges it has been necessary to consider a proportionate approach to training.

Bridge inspectors will complete a BICS aligned training course delivered over 4 days via a combination of classroom style presentations, workshop exercises and an on-site practical. The course covers essential topics and modules including an explanation of basic structural behaviour, structural types, identifying and classifying defects, confined space awareness and rope access arrangements. The trainer will progress through the slides pausing for class participation and questions to ensure understanding and shared knowledge. On completion of the training course, participants will complete a multiple-choice exam followed by an in-person interview and portfolio review to access competence. Only those who have fully attended the training and passed the formal assessments will be deemed competent by the Department for Infrastructure. It is important to note that this approach is for in-house inspectors only and is not extended to external bridge inspectors' providers. The course is designed to align with the requirements of the BICS certification, and individuals will be permitted to progress towards this qualification if they choose.

3   3D Training

Separate to the BICS aligned training course an additional training resource is being trailed as part of this review. A 3D scan of a bridge with known defects and scour issues has been undertaken (see Fig. 1). A 3D point cloud rendering of the structure will be used in a training room with a combination of 3D VR headsets to allow a virtual inspection in a classroom style environment. The environment for this testing will be developed in Unity, which is free for educational use. Alongside this will be the ability to manoeuvre around the model on a desktop environment using WebXR and undertake a training exercise simulating a bridge inspection while in a training room. The research on using VR as a means of instruction [1, 2], is generally supportive, particularly to replace dangerous or expensive laboratory trials [3] or as a means of increasing situational awareness of engineering environments [4]. With the availability of high-resolution devices having greatly increased in the last few years, the use of this type of training may become more widespread as the increased immersion and close synergy to learning outcomes is a significant benefit to this method of training. [5].

After the application has been tested by the user, it is important to get an idea of the user feedback based on their interactions throughout the experience. The feedback will be variable as it will be the individual's perception which proves to be difficult to measure. This questionnaire will be completed by the individual users after the VR experience. The questions will be implemented using a mixture of 5-point Likert scale from strongly agree to strongly disagree, to multiple-choice and text entry.

Fig. 1. 3D point cloud of a bridge site used for VR training. The image was rendered using Blender (www.blender.org).

4   Inspection Dashboard with Integrated Flood Risk

The goal of this aspect of the research is to develop a decision-support framework that is integrated with various datasets to measure how an organisations asset is performing currently and managed in the future. The creation of a GIS dashboard tool visualises the data available to asset managers and provides a valuable insight into bridge management data as part of a holistic decision-making process. The primary objective of the dashboard is to provide real time visualisation of bridge condition data and inspection status to the asset owner, (see Fig. 2).

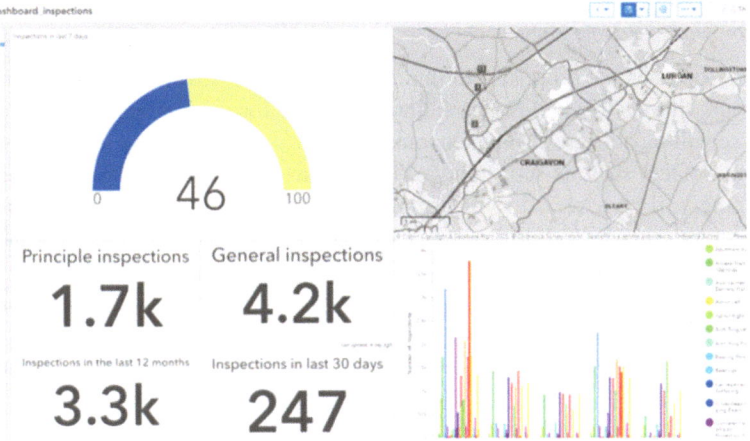

Fig. 2. Inspection status dashboard. Image created using ARCGIS (www.esri.com).

The dashboard consists of multiple tabs including bridge condition, traffic volume, collision data, rail bridge location and condition and rail strike. Of note are the data sets relevant to extreme climate events namely river flow (scour and gauges) and flood risk. The scour and gauges tab provides a unique link between bridge condition data and river flow data. The map highlights bridges that have been exposed to scour both historically and in current inspections. This data has been coupled with the river gauge information collected from 150 hydrometric stations across NI. This will allow bridge inspectors and asset managers an opportunity to identify bridges that could be at risk of damage from scour during a flooding event.

Related to the flow data, detailed flood models created in 2013/14 generated maps for areas at risk of flooding in NI. These are updated as changes are made to the infrastructure in their affected area, such as new flood defences and larger culverts. A real time link has been established that can indicate when a bridge could be at risk of inundation by overtopping that increases horizontal loading. The GIS based approach allows the identification of bridges that would be below the estimated flood level at any of the given flood events.

4.1 Planning Bridge Inspections in Extreme Climate Events

Bridge inspectors can interact with the dashboard to inform a broad scale risk management system that would allow asset managers to prioritise at risk bridges and plan inspections or inform relevant mitigation actions to prevent major damage or road closures during a flooding. During an event several stages can be considered such as, selecting bridges 'over water' (if scour or hydraulic loading a concern), select the bridge type (if considering certain construction types or condition), select bridges within a flood zone, select bridges with a history or existing issue with scour and selecting the size of the watercourse (Stralher rating, bigger rivers = larger flows).

5 Conclusions

This study outlines the proposals to improve the current state of consistency in bridge inspections within a local NI Roads Authority. A new formalised bridge inspector training course will be delivered which encompasses virtual reality to aid the demonstration of certain defects such as scour and various forms of cracking. The effectiveness of this training will be assessed to determine if it had any impact on the consistency of the inspectors. This will inform whether there may be geographic unconscious bias in inspection teams and how they might approach risk and certain defects.

To further improve asset owners' response to climatic events a GIS based inspection planning dashboard has been outlined that integrates various datasets including flood risk data. This can help generate targeted inspections based on real time river flow data and climatic changes. This may allow for bridges to be identified and proactive actions taken to address emerging risks before greater damage is caused.

References

1. Pridhvi Krishna, M.V., Mehta, S., Verma, S., Rane, S.: Mixed reality in smart computing education system. In: 2018 International Conference on Smart Systems and Inventive Technology (ICSSIT), IEEE, Dec. 2018, pp. 72–75 (2018). https://doi.org/10.1109/ICSSIT.2018.8748813
2. Hong, X., Lv, B.: Application of training simulation software and virtual reality technology in civil engineering. In: 2022 IEEE International Conference on Electrical Engineering, Big Data and Algorithms (EEBDA), IEEE, Feb. 2022, pp. 520–524 (2022). https://doi.org/10.1109/EEBDA53927.2022.9744801
3. Soliman, M., Pesyridis, A., Dalaymani-Zad, D., Gronfula, M., Kourmpe-tis, M.: The application of virtual reality in engineering education. Appl. Sci. 11(6), 2879 (2021). https://doi.org/10.3390/app11062879
4. David, J., Lobov, A., Lanz, M.: Learning experiences involving digital twins. In: Proceedings: IECON 2018 - 44th Annual Conference of the IEEE Industrial Electronics Society, pp. 3681–3686 (2018). https://doi.org/10.1109/IE-CON.2018.8591460
5. Try, S., Panuwatwanich, K., Tanapornraweekit, G., Kaewmoracharoen, M.: Virtual reality application to aid civil engineering laboratory course: a multicriteria comparative study. Comput. Appl. Eng. Educ.. Appl. Eng. Educ. 29(6), 1771–1792 (2021). https://doi.org/10.1002/cae.22422

Open Access This chapter is licensed under the terms of the Creative Commons Attribution 4.0 International License (http://creativecommons.org/licenses/by/4.0/), which permits use, sharing, adaptation, distribution and reproduction in any medium or format, as long as you give appropriate credit to the original author(s) and the source, provide a link to the Creative Commons license and indicate if changes were made.

The images or other third party material in this chapter are included in the chapter's Creative Commons license, unless indicated otherwise in a credit line to the material. If material is not included in the chapter's Creative Commons license and your intended use is not permitted by statutory regulation or exceeds the permitted use, you will need to obtain permission directly from the copyright holder.

The Effect of View, Latency, Speed and Performed Task on Remote Vehicle Operators

Christian Jernberg[✉] and Jan Andersson

VTI, Terra, 583 30 Linköping, Sweden
`christian.jernberg@vti.se`

Abstract. Autonomous vehicles (AVs) are being developed as an alternative to conventional transportation, which comes with great timing as there is a global shortage of skilled commercial drivers. However, the complexity of the driving task is extremely difficult for an AI system to solve, meaning that SAE level 5 vehicles are quite far away. During this interim period where we will have public roads with mixed traffic, remote operation could be a solution for facilitating the introduction of AVs.

As a remote operation station will be more similar to a simulator environment compared to a conventional cab, there might be a broader base for future recruitment of operators compared to drivers. This paper consists of subjective ratings from two simulator studies, comparing the performance of experienced drivers and experienced video gamers. Here we present the perceived performance and comfort of the simulator experiments, while comparing the groups in different conditions to see if one has an advantage over the other as potential employees.

Keywords: Remote operation · Autonomous vehicles · Simulator study · Subjective data

1 Introduction

As there is a global shortage of drivers for both buses and trucks that is expected to increase [1], it seems like deployment of autonomous vehicles (AVs) comes at a great time. However, due to the complex nature of the task of driving, combined with a large number of edge cases that an AI would have to be able to manage in order to reach SAE level 5, there will still be a need for a human in the loop for the foreseeable future [2].

As the technology level increases and the mobile networks are getting more stable, a remote operation station is a valid option to allow for one remote operator (RO) to either assist or drive a vehicles via link [3]. However, this changes the requirements the drivers will be physically removed from the environment in which they are operating.

This new work environment, combined with the lack of skilled drivers, could cause a need for companies to look elsewhere for future employees. As a RO could work in an office with a multi-screen setup instead of being inside the cab of a vehicle, there might be an advantage to having experience with similar screen setups. It has previously been

© The Author(s) 2025

C. McNally et al. (Eds.): TRAconference 2024, LNMOB, pp. 787–794, 2025.
https://doi.org/10.1007/978-3-031-85578-8_106

tested if people who are experienced in playing action video games could outperform other drivers in a simulator study, with some studies claiming that gamers show an increase in performance [4], while others show that gamers are showing an increase in risk-taking behavior [5]. As a remote operation station could very well look like a simulator, the current article aims to investigate how experienced gamers and experienced drivers perceived the complex task of driving remotely.

Apart from the cognitive difficulties that could emerge when an operator is physically detached from the vehicle being operated, such as lower situational understanding [6], there is also a range of technical difficulties that could arise, such as the issue of latency. Even though the technology is advancing, the need to send large amounts of data, such as video feeds and vehicle status, and to receive continuous control signals make the systems vulnerable to mobile network issues. There are still circumstances and locations where latency will be a problem, and the magnitude of the issues seems to be very dependent on the task being performed [7]. Previous studies seem to be made with participants that are aware of the premises of which they are operating, i.e., they were told that there would be added or variable latency. The current two studies performed for this article aimed at investigating how participants would perform if this was unknown to them. I.e., the change in latency during the studies were hidden from the participants in order to avoid intentionally adjusted behavior.

2 Method

This article is based on the combination of two sets of subjective data gathered in two separate simulator studies during 2021 and 2022. The objective data gathered from the simulator platform and the procedure of the simulator studies is presented in more detail in [8] and Jernberg et al. (under review).

2.1 Participants

The participants of the studies were recruited via declarations of interest sent out to a taxi company in Gothenburg and action gaming clubs at Chalmers. The main target was to divide the participants into two groups, one consisting of experienced drivers (EDR) and one consisting of experienced gamers (EGA). To do so, the results from the background questionnaire (see below, 2.2 Questionnaires) were used as a reference point to divide the participants a posteriori for the first experiment, and a priori for the second experiment. Table 1 shows the number of participants completing respective study and the threshold values for each group of participants in kilometers driven per year and hours of videogames played per week.

2.2 Questionnaires

Before the experiments, background information was gathered in order to establish the participants age and experience of driving as well as experience with action video gaming. The participants were also asked about how much they trusted autonomous vehicles or vehicles driven in other forms than conventional.

Table 1. Participant and group characteristics of the conducted experiments.

Variable	Experiment 1	Experiment 2
Total number of participants	30	30
Women/Men ratio	10/20	1/29
EDR / EGA ratio	13/17	15/15
Average age (EDR/EGA)	34.7 (42.9/28)	34.6 (39.9/29.4)
EDR threshold, km/year	1 500	2 500
EGA threshold, h/week	6	7

Participants drove three laps in experiment one or four laps in experiment two (see 2.3 Design and Procedure below) and between each lap, the participants were asked to rate, on five point Likert scales, how well they felt that they i) were in control of the vehicle, ii) how authentic the scenario felt and iii) how comfortable they were with driving the scenario. They were also asked an open question about what they felt affected them the most during the drive. Participants were also asked to grade their level of simulator sickness and how much that affected their driving on two seven-point Likert scales. Additionally, an abbreviated NASA TLX form [9] was filled out for each lap.

After the experiments, the test leader asked the participants if they had noticed anything different between the laps (apart from the change of view in experiment two). Then the test leader told the participants about the change of latency conditions (see 2.3 Design and Procedure below), explained the true purpose of the study (to test remote operation), and subsequently asked if they could rate in which order they believed that they had driven with the different latencies. As a last questionnaire, the participants were asked once more how much they would trust a vehicle controlled by any other means than conventional, as well as how they thought they would be able to perform the task.

2.3 Design and Procedure

When the participants arrived, they were shown to the simulator room to fill out paperwork for GDPR, informed consent and documents for the compensation (appx. 60 euro). Participants were also presented with a background questionnaire. Participants were informed that they were going to drive three laps in the first experiment, and four laps in the second, and that each lap would start with a short training session of four minutes before going into a route containing two environments. The environments were a high-speed rural road as well as a low-speed urban road with speed limits of 70 km/h and 50 km/h respectively. The participants were told to act as they would have done during normal driving, i.e., keep the speed limit and behave according to any situation that may occur. Table 2 shows the balanced parameters that changed for each lap.

During the first experiment, all participants drove three laps consisting of both environments, but with a different latency in each lap. During the second experiment, all participants drove four laps consisting of both environments, two in Normal View (one for each latency) and two for the elevated Roof View (one for each latency).

Table 2. Differences in driving conditions during each lap.

Variable	Experiment 1	Experiment 2
Environment conditions	Rural and Urban	Rural and Urban
Latency conditions	88 ms, 188 ms, 288 ms	88 ms, 238 ms
View conditions	Normal view	Normal and Roof

For each environment, five hazardous situations were designed in a way that was feasible both in an urban and a rural situation. These hazards were also paired with a proxy situation which were the same event, but performed in a way that the participants would not need to take any extra action (see Table 3). Each lap took approximately 12–16 min.

Table 3. Differences between hazards and proxy events.

Hazard	Proxy
H1: A car cuts in from a parking spot, right in front of the participant	P1: A car exits a parking spot far away from the participant
H2: A car in the opposing lane turns left right in front of the participant	P2: A car far away from the participant makes a safe left turn
H3: A vehicle drives into an intersection just in front of the participant	P3: A car drives through an intersection far away from the participant
H4: A child runs out from behind a bus, up to the lane of the participant	P4: A bus is parked at a bus stop, with no human present
H5: Two bicyclists are blocking the participants lane, and a vehicle is coming in the opposing lane	P5: Two bicyclists are blocking the participants lane with no traffic present in the opposing lane

The order of which environment, view and latency conditions the participants started with was orthogonally balanced and not told to the participants beforehand. The order of the hazardous events was also balanced.

After each lap there were two questionnaires to fill out. When the complete driving session was finalised, there were more questions (after the complete drive questionnaires) before participants were debriefed and thanked for their participation.

3 Results and Discussion

Table 4 presents the average rating for each group of participants after each driving condition. The questions asked were: Q1 – How in control of the vehicle are you; Q2 – How authentic is the scenario; Q3 – How comfortable is the drive; Q4 – How much simulator sickness do you feel; and Q5 – How much does the simulator sickness affect you. At the bottom of the table, the average rating of the NASA-TLX subcategories is presented. Due to the lack of space, not all results will be mentioned in detail.

While comparing the results for each study, a main effect of group (EDR vs EGA) did not appear with the exception of the questions regarding simulator sickness in experiment one. EDR consistently rated their feeling of simulator sickness higher than EGA, especially in the 288 ms latency condition, as well as on how much the simulator sickness affected their driving. For experiment two, the result was non-significant, and this result was no difference when age was considered as a covariate, indicating that there might be a threshold between the latencies of 238 and 288 ms where computer game experience could have a positive effect on how a participant feel in this setup.

There was also a main effect in experiment one for how the participants experienced the feeling of control of the vehicle, specifically that they felt less in control of the vehicle in the condition with the 288 ms latency. When plotting the Likert-scale ratings for Q1 - Control, the graph closely followed the plot for the NASA TLX subsection performance. This is true for experiment two as well, though no significant difference between the latency conditions were found. However, it shows a consistency with which the participants rate their perceived performance based on how well they feel in control of the vehicle they are currently operating. In Jernberg et al. (under review), one of the findings show that an increased latency also gives an increased reaction time that surpasses the manipulation of the system for certain tasks, showing that for the 288 ms condition there actually was a decrease in performance for some hazards.

It seems that there were more similarities than differences. For the NASA TLX, there were only the mentioned subcategory of Performance that showed a significant main effect of latency, group, or view. Some interaction effects were found between group and latency in experiment two, where EDR rated their mental and physical demand higher in the 238 ms latency conditions compared to the 88 ms, and their performance lower, showing that EGA might have a higher tolerance for latency changes even around the threshold of the simulator.

In both experiments, participants were asked how much they trusted a vehicle driven autonomously or by other means than conventional, and EGA consistently reported higher ratings when it came to trust of this technology, as well as for if they thought they could drive a vehicle remotely themselves.

Table 4. Average rating (standard error) for questionnaires by condition and group.

Latency/view	88 ms (exp 1)		188 ms (exp 1)		288 ms (exp 1)		88 ms normal view (exp 2)		138 ms normal view (exp 2)		88 ms upper view (exp 2)		138 ms upper view (exp 2)	
Group	EDR	EGA	EDR	EGA	EDR	EGA	EDR	EGA	EDR	EGA	EDR	EGA	EDR	EGA
Q1 – Control	3.8 (0.3)	3.6 (0.2)	3.3 (0.3)	3.6 (0.3)	2.5 (0.3)	3.2 (0.3)	3.2 (0.2)	3.9 (0.3)	3.1 (0.3)	3.8 (0.3)	3.6 (0.2)	3.7 (0.2)	3.3 (0.2)	3.7 (0.2)
Q2 – Authenticity	4.1 (0.2)	3.6 (0.2)	4.4 (0.2)	3.6 (0.2)	3.7 (0.4)	3.5 (0.2)	4.1 (0.2)	4.1 (0.2)	4.0 (0.2)	3.8 (0.3)	3.9 (0.3)	4.1 (0.3)	3.9 (0.2)	3.9 (0.2)
Q3 – Comfort	3.6 (0.3)	3.2 (0.2)	3.2 (0.3)	3.2 (0.3)	3.5 (0.2)	3.2 (0.3)	3.6 (0.3)	3.6 (0.3)	3.4 (0.3)	3.5 (0.3)	3.5 (0.2)	3.5 (0.2)	3.5 (0.2)	3.7 (0.2)
Q4 – Sickness	2.4 (0.5)	1.4 (0.1)	2.5 (0.4)	1.7 (0.2)	3.1 (0.5)	1.4 (0.1)	1.9 (0.4)	1.7 (0.3)	2.3 (0.4)	1.6 (0.3)	2.2 (0.4)	1.7 (0.3)	2.4 (0.4)	1.3 (0.2)
Q5 – Effect	2.1 (0.5)	1.2 (0.1)	1.9 (0.3)	1.3 (0.1)	2.0 (0.4)	1.1 (0.1)	1.5 (0.3)	1.3 (0.2)	1.8 (0.3)	1.2 (0.2)	1.8 (0.4)	1.5 (0.2)	1.9 (0.4)	1.1 (0.1)
Mental Dem.	10.8(1.1)	11.3(1.1)	11.1(1.4)	10.9(1.0)	10.2(1.1)	10.7(1.1)	7.7 (1.1)	9.4 (1.3)	9.6 (1.2)	8.4 (1.4)	8.1 (1.1)	11.1(1.4)	9.1 (1.2)	9.1 (1.4)
Physical Dem.	6.8 (0.2)	7.7 (1.0)	6.5 (1.0)	7.7 (1.1)	7.0 (1.0)	7.9 (1.2)	5.6 (1.1)	6.5 (1.1)	7.5 (0.9)	6.0 (1.2)	6.8 (1.0)	7.0 (1.2)	7.2 (1.0)	5.7 (0.8)
Temporal Dem.	5.9 (0.2)	7.4 (1.2)	6.5 (1.2)	7.2 (1.1)	7.1 (1.3)	8.1 (1.2)	5.1 (1.0)	6.9 (1.4)	5.7 (0.9)	7.6 (1.4)	6.5 (1.2)	8.3 (1.3)	5.8 (1.0)	7.6 (1.4)
Effort	8.4 (0.2)	9.4 (0.9)	7.4 (1.2)	9.2 (1.0)	9.1 (1.0)	9.2 (0.9)	6.1 (1.1)	7.8 (1.2)	7.7 (0.9)	8.1 (1.3)	7.2 (1.1)	8.8 (1.1)	7.1 (1.2)	7.5 (1.3)
Performance	13.5(0.2)	14.0(0.9)	13.0(1.2)	13.6(1.0)	11.5(1.4)	12.9(1.2)	14.0(0.8)	12.7(1.0)	11.7(1.0)	13.7(1.0)	13.3(0.5)	12.7(1.0)	12.7(1.0)	14.0(0.8)
Frustration	7.8 (0.2)	7.9 (1.0)	7.3 (1.3)	8.2 (1.2)	8.0 (1.3)	9.2 (1.2)	7.3 (1.3)	6.1 (1.2	8.7 (1.2)	6.1 (1.3)	8.1 (1.2)	6.5 (1.0)	8.1 (1.0)	5.7 (1.1)

4   Conclusion

One of the main focuses of the experiments were to investigate if there was a difference between the experienced drivers and the experienced gamers when it comes to performance in a simulator environment that could be a close proximity of how a remote operation station might look. These subjective ratings indicates that there might be a few difference in how they perceive their performance as well as how well they feel while driving a simulator, depending on the amount of latency the system would allow for. Even when taking age into account as a covariate, the same pattern occur, showing that the experience with a screen might be a factor. However, the effects are not big enough to declare one group as the "winner" to future prospective employers that are looking to increase their recruitment base. However, this also means that there is no clear loser and the laws and regulations that are in place today, where a remote operator needs a driving license for the type of vehicle that is being operated, could benefit from a look-over.

The next step that needs to be investigated is how these results transfers into a setting with actual live video feeds instead of a simulated environment.

References

1. International Road Transport Union: IRU Driver Shortage Global Report (2022)
2. Beiker, S.: Unsettled issues in remote operation for on-road driving automation. SAE Edge Research Report (2022)
3. Andersson, J., Rizgary, D., Söderman, M., Vännström, J.: Public report HAVOC - Heavy Vehicle Operation Centre. https://www.vinnova.se/globalassets/mikrosajter/ffi/dokument/slu trapporter-ffi/trafiksakerhet-och-automatiserade-fordon-rapporter/2020-02953eng.pdf?cb= 20220705105235. Accessed 25 Sept 2023
4. Howard, J., Bowden, V.K., Visser, T.: Do action video games make safer drivers? The effects of video game experience on simulated driving performance. Transport. Res. F: Traffic Psychol. Behav. **97**, 170–180 (2023)
5. Stinchcombe, A., Kadulina, Y., Lemieux, C., Aljied, R., Gagnon, S.: Driving is not a game: Video game experience is associated with risk-taking behaviours in the driving simulator. Comput. Hum. Behav. **69**, 415–420 (2017)
6. Zhao, L., Nybacka, M., Rothhämel, M.: A survey of teleoperation: driving feedback. In: 2023 IEEE Intelligent Vehicles Symposium (IV) (2023)
7. Neumeier, S., Wintersberger, P., Frison, A-K., Becher, A., Facchi, C., Riener, A.: Teleoperation: the holy grail to solve problems of automated driving? sure, but latency matters. In: Proceedings of the 11th International Conference on Automotive User Interfaces and Interactive Vehicular Applications, pp. 186–197. Association for Computing Machinery, New York (2019)
8. Jernberg, C., Sandin, J., Ziemke, T., Andersson, J.: The effect of latency, speed and task on remote operation of vehicles. Transp. Res. Interdiscip. Perspect. **26** (2024). https://doi.org/10.1016/j.trip.2024.101152
9. Hart, S.: Nasa-task load index (NASA-TLX); 20 years later. In: Proceedings of the Huvman Factors and Ergonomics Society 50th annual meeting, pp. 904–908 (2006)

Open Access This chapter is licensed under the terms of the Creative Commons Attribution 4.0 International License (http://creativecommons.org/licenses/by/4.0/), which permits use, sharing, adaptation, distribution and reproduction in any medium or format, as long as you give appropriate credit to the original author(s) and the source, provide a link to the Creative Commons license and indicate if changes were made.

The images or other third party material in this chapter are included in the chapter's Creative Commons license, unless indicated otherwise in a credit line to the material. If material is not included in the chapter's Creative Commons license and your intended use is not permitted by statutory regulation or exceeds the permitted use, you will need to obtain permission directly from the copyright holder.

On Tackling Skill Shortage in the Rail Sector: The STAFFER ERASMUS+ project

Alice Consilvio⓿, Angela Di Febbraro⓿, and Nicola Sacco⁽✉⁾⓿

University of Genoa, 16145 Genoa, Italy
{Staffer.coordination,Nicola.sacco}@unige.it

Abstract. The rail sector is experiencing an aging workforce that is approaching a well-deserved retirement, while rail enterprises, in many European Member States, are reporting difficulties to find railway engineers. Educational and training pathways are essential to prepare students, transitioning employees and current rail professionals for fulfilling and impactful careers in the sector for the coming decades. The EU project STAFFER, "Skill Training Alliance For the Future European Rail system" is the Erasmus+ Sector Skills Alliance for rail consisting of 31 full partners and 17 associated partners from 12 EU countries.

STAFFER is aimed at developing a holistic Blueprint strategy to recognize present and new skill needs suitably and timely.

In synergy with the European Year of Skills, a common methodology for the identification of current and future skills and competence needs for the whole complex rail system is developed.

Keywords: Future Rail workforce · Skills Gaps · Rail Digitalization · Training and Mobility Programmes · Rail Employability and Career Opportunity

1 Introduction

Europe's railways represent the least carbon intensive mode of mass transit, contributing only 0,4% of the transportation sector's considerable greenhouse gas emission [1]. This attribute, combined with the fact that rail has managed to be the only mobility service to become more energy efficient while increasing capacities, has led the European Commission (EC) to seek to inspire a doubling of passenger and freight rail traffic by 2030 and 2050, respectively [2]. In addition, rail is expected to serve as the backbone of the future multimodal European transport system according to the 2050 vision of the European Rail Research Advisory Council [3], while, in parallel, a digital transformation of the sector is ongoing and rail industries are attempting to deploy rapidly evolving digital technologies.

Another important challenge, currently faced by the sector, is the overcoming of existing obstacle to long-distance cross-border rail services [4].

In this context, tackling the imminent skill shortage in the rail sector is crucial. The sector not only needs to address the ageing workforce, but also the fact that highly skilled people, with IT and engineering profiles, tend not to choose a career in rail. Moreover, rail

© The Author(s) 2025
C. McNally et al. (Eds.): TRAconference 2024, LNMOB, pp. 795–801, 2025.
https://doi.org/10.1007/978-3-031-85578-8_107

companies struggle even to fill lower-skilled job profiles, as the demand for rail services increases with an increasingly climate conscious customer base. Finally, the rail sector needs to diversify the current workforce by attracting young people and women, which only account for around 21% of the railway sector workforce [5].

The EU project STAFFER, "Skill Training Alliance For the Future European Rail system", is the Erasmus+ Sector Skills Alliance for rail consisting of 31 full partners and 17 associated partners from 12 EU countries. The project gathers the different parts of the complex rail world: infrastructure managers and operators, industry suppliers, educational institutions, and other different associations and organisations. It started on 1 November, 2020, upon approval by the EU Education, Audiovisual and Culture Executive Agency (EACEA) and runs over four years. STAFFER proposes to create an innovative broad European framework where the different stakeholders can deal together with skill gaps and employment risks, collaborating in defining a common perspective on the challenges to face. No similar framework has ever existed, and it will lead to consider rail to be a unique system of systems, which in turn will allow to improve supply, demand, use and development of skills in the sector.

Its objective is to help identify the main existing skill gaps and assess the future skill needs in the rail sector. Vocational & Education Training (VET) institutions and technical universities will propose adaptations to curricula, training and educational programmes, so as to address new technological developments and trends [6].

2   Methodology for Skills Needs Identification

A common methodology for the identification of current and future skills and competence needs for the whole complex rail system is developed, considering the future vision of the rail sector from the point of view of rail operators and infrastructure managers and rail suppliers.

The common methodology for skills needs identification is based on the following steps [7]:

1. Trends identification, which consists of selecting the relevant trends that are affecting the sector in terms of skills needs.
2. Occupational profiles identification, which regards the selection of rail occupational profiles affected by skill mismatch.
3. Skills gaps identification which deals with the skills needs related to the identified profiles.

The "trends identification" was based on the analysis of existing studies and in-depth interviews.

The results and findings of the literature review (projects/studies/initiatives) provided an overall view of the development of the railway sector and showed close interaction of several trends. The "trends identification" was then completing by conducting a survey among railway operators and suppliers. Through this, the level of impact of each macro and micro trend in the working field of railway suppliers and operators was determined. The survey provided a general insight into perceptions of trends, professional profiles and assessment methods of skills needs in the sector.

The "occupation profiles identification" was carried out, starting with the analysis of the ESCO, the (multilingual) classification of European Skills/Competences, Qualifications and Occupations (https://esco.ec.europa.eu/en/classification). Compiling the STAFFER list of rail profiles, together with the information from ESCO, helped to gain further insights about the validity of these job profiles regarding their descriptions, skills and competences. It also proved that some of the profiles identified within STAFFER are new job profiles, not already defined in ESCO.

Regarding the proposed methodology for "skill gaps identification", the following procedure has been followed:

- Analyze micro and macro trends affecting the market and the industry.
- Project and evaluate the impact of the trends on respective business products/ services and processes.
- Break down the evaluated products/ services and processes in corresponding tasks and evaluate the changes on a task level.
- Assign relevant occupational profiles to the task level and discuss the skill gap.
- Decide on workforce adaptations: re-skilling/upskilling or separation/ new hiring.

Two parallel activities were carried out, to collect the perspectives of both operators/infrastructure manager and suppliers.

In order to learn more about specific challenges and needs as regards future skills and competences that are specific to railway operation, a number of research activities were carried out, including desk research and focus groups with company practitioners, cross-border railways, rail freight corridors and transnational cooperation. The research looked at current and foreseeable future challenges in three areas:

- Cross-border rail traffic, with particular reference to EU rail freight corridors.
- Language and communication issues in cross-border traffic.
- Skills requirements of railway personnel, including trainers and managers, as well as skills and competences of professional profiles affected by digitalisation, big data, cybersecurity as well as environmental policies.

Specific challenges and needs related to the suppliers were, instead, identified with an approach based on best practice analysis, experts' interviews and data scrolling.

The scope of the data scrolling was to explore the actual talent demand on the job market for rail suppliers, as well as an analysis about skills and their request development over the past four years. These big data analyses are especially interesting to get insight of big trends in skills that are reflected in job postings.

3   Results

The results of the first phase about trend identification, conducted through literature research and surveys, revealed as main macro trends "technical changes", "society", "environment and politics" and "economics and market". For each trend, specific micro trends were assigned and their relative importance was estimated. The surveys results show that the most important micro trends came from the "environment and politics" and "economics and market" macro trends. The importance of the "environment and politics" trend can be explained by a growing environmental mindset to reduce emissions and the fact that the railway sector is seen generally as an environmentally friendly mode of transport. Within the "economics and market" macro trends, it is worth highlighting the micro trend related to new mobility solutions, such as mobility as a service or mobility on demand, which play an important role for the integration of rail services into multimodal transport solutions.

The results about the analysis of the occupational profiles show that the profiles with the highest need for adaptation include engineers and digital and IT technicians, as a result of new digital technologies and the increasing complexity of the rail system, with a consequent increased necessity for data security and an increased demand for data analysis, as shown in Fig. 1.

From the operators and infrastructure manager point of view, 30 occupational profiles were identified in different domains of railway operation and infrastructure management on the basis of ESCO, the (multilingual) classification of European Skills/Competences, Qualifications and Occupations.

Three occupational profiles /groups were selected for an in-depth analysis as they are particularly relevant in terms of the impact of main trends, such as digitalization, and expected changes due to emerging technologies, such as automated operation, digital traffic control, disruptive changes in infrastructure and maintenance of rolling stock as well as construction (BIM, digital twins, predictive maintenance, etc.). These three groups are:

- Train drivers (including train driver instructors).
- Occupational profiles in traffic control centers (dispatchers, signal box staff, etc.).
- Occupational profiles in infrastructure and maintenance, including engineering profiles.

Regarding the third phase of skills gaps identification, a survey was conducted during May and June 2021, promoted by the European rail associations for operators and infrastructure managers CER and rail suppliers UNIFE, which gathered more than 80 responses from 19 European countries.

When asked which skills are missing and are necessary in the future, participants indicated "applying and handling digital technologies", "skills and overall knowledge about the railway systems" and "new collaboration forms" as the most important trend, as depicted in Fig. 2. Moreover, hard and soft skills, such as "language skills" or "innovative and open learning attitudes" and "cultural skills" were also deemed important.

From the rail operators and infrastructure managers point of view, responses showed that there is a general trend from medium to more advanced skill levels across all occupational profiles.

The main drivers of an upward skill shift are new technologies, digitalisation and automation. A detailed about the skill needs for train driver is shown in Fig. 3.

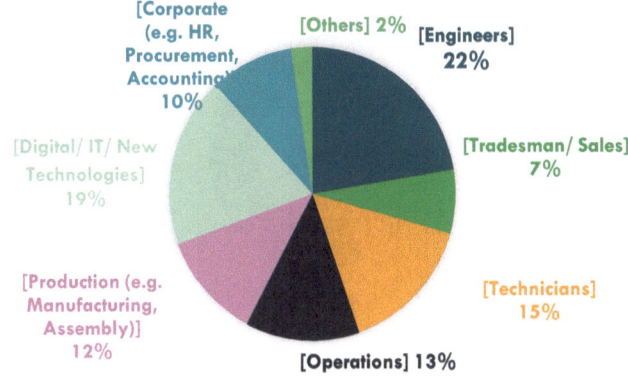

Fig. 1. Occupation groups with the highest need for skill adaptation

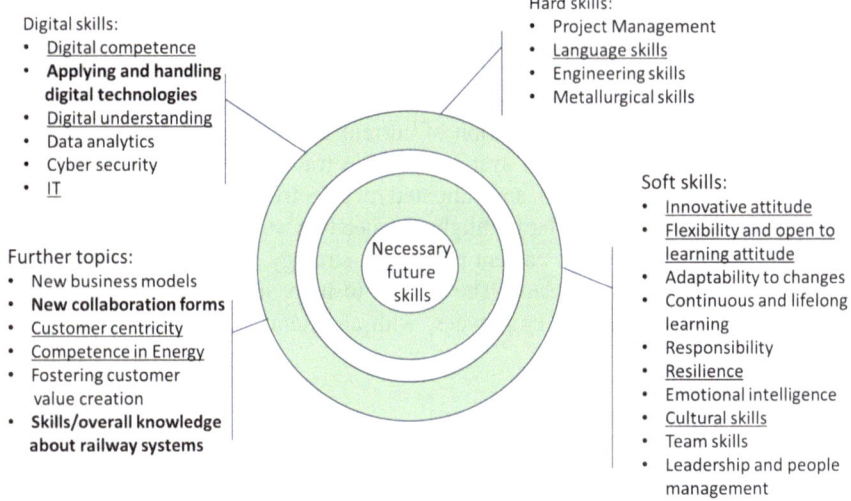

Fig. 2. Necessary future skills of the workforce

Another finding is that greater efficiency and employability in cross-border rail could be achieved with shifting mindsets from national-focused to more European-focused. This would require more European railway knowledge and intercultural competences, including soft skills like communication, problem-solving, autonomy and resilience, learning motivation and openness. From the suppliers' point of view, five most relevant future skills needs were identified, which are highly related with each other: "collaboration", "problem solving", "holistic understanding & system thinking", "bridging of traditional and digital approaches" and "life-Cycle Management".

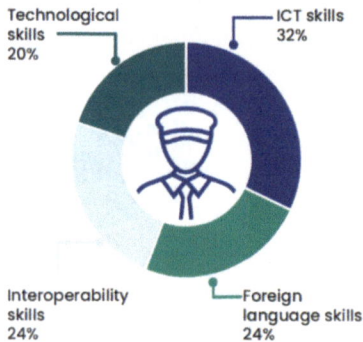

Fig. 3. Train driver: fundamental future skill needs as highlighted by survey participants.

4   Conclusions

STAFFER is defining a long-term and sustainable strategy to fill the gap between the supply of and the demand for a suitably skilled workforce. In addition, a long-term action plan is going to be developed and endorsed to roll-out the project results at the different territorial levels. Upon the identification of current and future skills and competence needs for the whole complex rail system, suitable training and education paths and curricula are designed, developed, and validated for effectively increasing employability and career opportunities. Results highlight the need of a better understanding of the existing mismatch between the current promotion strategy and what the target groups find attractive about the rail sector. The goal is to help making the rail sector more attractive to current and future employees, with an attention to diversity and gender balance issues.

References

1. European Environment Agency, Transport and environment report 2022 Digitalisation in the mobility system: challenges and opportunities (2022)
2. European Commission, COM (2019) 640, The European Green Deal, 11.12.2019 (2019)
3. ERRAC, Rail 2050 Vision (2017)
4. European Commission, COM (2020) 789, Sustainable and Smart Mobility Strategy – putting European transport on track for the future, 9.12.2020 (2020)
5. European Commission, Directorate-General for Mobility and Transport, Long-distance cross-border passenger rail services: final report, Publications Office of the European Union (2021). https://data.europa.eu/doi/10.2832/019365
6. STAFFER Mid-Term Report (2022). https://www.railstaffer.eu/wp-content/uploads/2022/10/STAFFER-Mid-Term-Report-Web.pdf
7. STAFFER D1.1 Identification of current and future skills and competence needs (2021). https://www.railstaffer.eu/publications-resources/deliverables/

Open Access This chapter is licensed under the terms of the Creative Commons Attribution 4.0 International License (http://creativecommons.org/licenses/by/4.0/), which permits use, sharing, adaptation, distribution and reproduction in any medium or format, as long as you give appropriate credit to the original author(s) and the source, provide a link to the Creative Commons license and indicate if changes were made.

The images or other third party material in this chapter are included in the chapter's Creative Commons license, unless indicated otherwise in a credit line to the material. If material is not included in the chapter's Creative Commons license and your intended use is not permitted by statutory regulation or exceeds the permitted use, you will need to obtain permission directly from the copyright holder.

Developing a Qualified Workforce for Greener Ports

Theocharis Tsoutsos$^{(\boxtimes)}$ ⓘ, Nikolaos Skarakis ⓘ, and Stavroula Tournaki ⓘ

Renewable and Sustainable Energy Systems Lab, School of Chemical and Environmental Engineering, Technical University of Crete, University Campus, 73100 Akrotiri, Chania, Greece
ttsoutsos@tuc.gr

Abstract. Ports are key components of the supply chain management and overall economy, by granting access to resources via transportation and services, while pursuing safety and security. Europe accommodates more than 1,200 ports, associated with adverse environmental impact, due to greenhouse gas emissions and air pollutants. Although the EU prioritises the greening of ports, its Blue Economy workforce, which represents about 5.4 million jobs, lacks knowledge, skills and competences required to promote green pathways and to employ efficiently recent technologies towards the European Green Deal. Against this background, the BLUE PORTS project enhances the capacity of the European ports' human capital in small and medium ports by introducing a tailored training and certification scheme that covers green technology solutions, integration of green products/services and business models, that help minimize the environmental footprint of ports' activities and drive their green transition. This new operational certification scheme, in accordance with ISO/IEC 17024, will be tested and validated in selected Mediterranean and Black Sea ports. The mutual recognition across Europe of the BLUE PORTS scheme, based on the European Accreditation Mutual Agreement will offer professional advantage, enhanced career and mobility opportunities, ultimately creating and sustaining new, greener jobs in the port industry.

Keywords: green ports · personnel certification · upskilling ports workforce · green skills · Mediterranean ports

1 Introduction

Ports are key components of the supply chain management and overall economy on a regional, national, and international level by granting access to resources via transportation and services. They also provide safety and security with respect to passenger mobility, accessibility, labour and health aspects. The European Blue Economy currently represents around 5.4 million jobs and is set to double its employment by 2030 [1]. Sea transport accounted for 46% of goods traded between the EU and the rest of the world, or 74% measured in volume (Fig. 1) [2].

However, maritime transport is a growing source of greenhouse gases (GHGs). In 2018, global shipping emissions represented 2.9% of the ones caused by human activities. Maritime transport represents 3–4% of the EU's total CO_2 emissions [3], of which

© The Author(s) 2025
C. McNally et al. (Eds.): TRAconference 2024, LNMOB, pp. 802–808, 2025.
https://doi.org/10.1007/978-3-031-85578-8_108

6–7% are generated at berth [4]. Meanwhile, small and medium-sized ports (SMSPs) reportedly lack environmental credentials, security and policy compliance [5, 6].

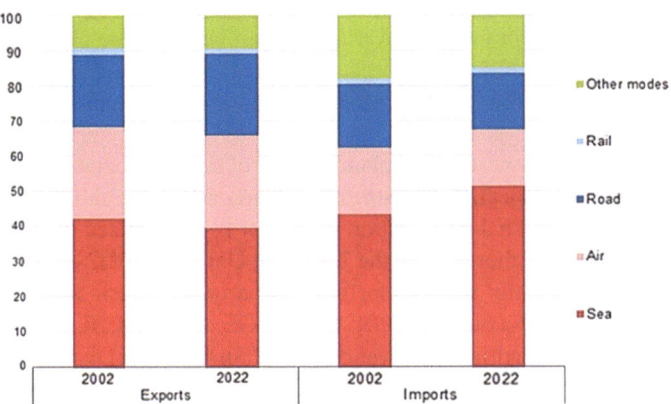

Fig. 1. Value of extra-EU trade in goods, by transportation mode (%, 2002 and 2022) [2].

To address the pressing need for marine conservation and sustainable development, various initiatives, such as the EGD or the Sustainable and Smart Mobility Strategy of the European Commission, have established targets for 2030 and 2050. Furthermore, some of the UN's 17 Sustainable Development Goals (SDGs) are already in port-related programs, such as the "World Port Sustainability Program" (WPSP) or the "EcoPorts Initiative" by the European Sea Ports Organization (ESPO) [7].

Still, according to the existing scientific literature, there is a lack of tools that could facilitate the shift of SMSPs towards sustainable development. Today, the Blue Economy workforce lacks the knowledge, skills and competences required to promote green pathways and to efficiently employ recent technologies/techniques in alignment with the European Green Deal policy, as in most countries, there is no relevant training and recognized accreditation scheme for ports and nearby facilities. Such a transition requires the training of professionals who work in ports on environmental, climate and energy issues. This can be a step closer to unlocking the full potential of SMSPs.

2 State-of-the-Art

Over half a decade, the EU has funded several maritime projects concerning education and upskilling in blue careers, supporting the Atlantic, Black Sea and West Mediterranean Sea basins strategies. Most training programs have been provided through graduate courses, regarding safety, security, logistics, management, innovation and technology for blue careers, targeting young people, often graduate students, who sought to gain the necessary skills to enter the maritime industry. The gap between academia and the private sector has been merely covered by apprenticeships. Local authorities have also been engaged in the schemes developed, expanding the training networks. The rise of digitalization has introduced new toolkits (e-platforms, webinars, contests etc.) [8].

Going beyond the impact of other initiatives, the project BLUE CAREERS IN NET ZERO ENERGY PORTS (acronym: BLUE PORTS), funded by the EMFAF programme 2021–2027, supports the development of the blue skills for the human capital of the ports' business-involved personnel, i.e., administration, operators and related authorities, for SMSPs. Emphasis is drawn on covering the gap between the current services for passengers and goods and the EGD targets, including the Sustainable Blue Economy Strategy, the EU Offshore Renewable Energy strategy, and the Circular Economy Action Plan.

The consortium is composed of a balanced team of complementary expertise including public bodies and respective ministries, academia and research institutions, small-medium ports and accredited certification bodies. It is coordinated by the Renewable and Sustainable Energy Laboratory of the Technical University of Crete ReSEL@TUC, and covers a geographically wide area (from Iberia to the Black Sea), comprising a total of 8 partners from 5 countries (GR, ES, IT, GE, TN) and 26 supporting bodies representing the entire quadruple innovation helix framework: (i) marine and maritime educational/academic institutions and/or vocational training providers; (ii) Blue Economy industry-economy; (iii) civil society-environmental organizations; (iv) governance of all local, regional, national, transnational level (see Fig. 2).

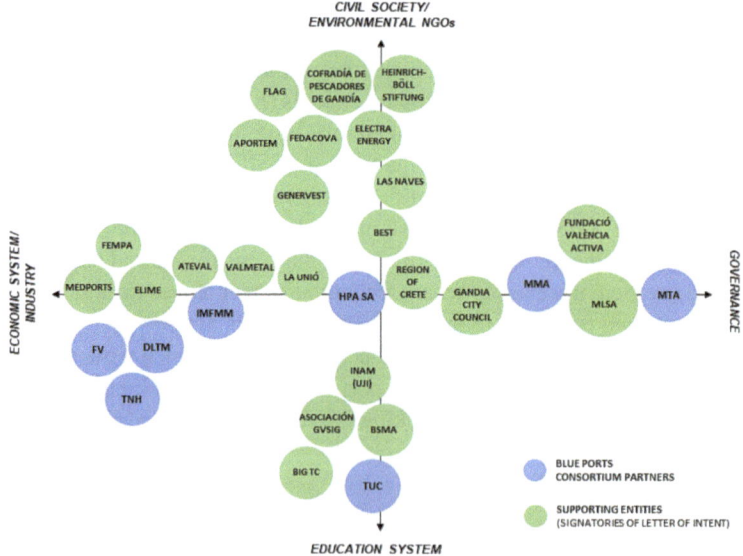

Fig. 2. BLUE PORTS consortium engages the quadruple innovation helix model.

The BLUE PORTS certification scheme's definition and requirements are the result of direct consultation amongst competent National Technical Committees, bringing together representatives of the full value chain and building trust and confidence between the consortium, the ports sector and national policy-makers. NTCs transfer the industry's knowledge/experience and assist to in the identification of skills gaps, requirements, priorities and constraints of the actual market operations, and monitor the scheme implementation and efficiency.

A fully digitised training material for trainers and trainees explores the following themes:

- Zero emission port / Green energy dynamic autonomy for ports.
- Integrated waste management (zero waste, circular economy).
- Offshore green power generation (wind, wave, tidal, solar).
- Sustainability assessment analysis tools.
- Performance Analysis, Economics and Environmental issues.

In parallel, special courses are to be developed for young engineers and managers on the Blue Economy in cooperation with youth's EU associations. Career development-oriented events will further motivate the young and new workforce, as a critical pillar to cover the continuously increasing gap in the skills required to enable the application of the Blue Economy Strategy in the Mediterranean and Black Sea.

To overcome distance barriers and save GHGs, a synchronous and asynchronous platform, as a fully digital examination system are developed, to be tested in four countries, Greece, Spain, Italy and Georgia in accordance with ISO/IEC 17024. The BLUE PORTS scheme is expected to gain mutual recognition through national governmental and accreditation authorities, on the basis of the European Accreditation Multilateral Agreement (EA MLA), which is internationally recognised by IAF (International Accreditation Forum) and ILAC (International Laboratory Accreditation Cooperation), thus inspiring market confidence by ensuring consistency, transparency and acceptability.

3 Methodology

BLUE PORTS seeks to cover the needs of European, North African and Black Sea ports, with an overarching ambition to develop a skilled workforce according to multinational criteria and standards. The project delivers a pool of 200 skilled/certified Blue Economy staff. Additionally, 100 + specialized young students and graduates will be involved, leveraging existing academic networks. It also envisions hosting informational events and a case study competition in collaboration with students' associations. This multi-faceted approach will inspire current and future Blue Economy labour and young talents to devise efficient solutions for ports and accelerate the green shift of shipping. The methodology for implementing BLUE PORTS is organized into distinct stages:

1. *Identifying existing experience-Baseline;* analysing the tools and experience available at EU, national and regional levels.
2. *Establishing consultation structures (4-helix involved);* bringing together organizations of the public and private sector.
3. *Training and certification scheme's definition;* with respect to scope, expected performance, employability and career development-professional rights.
4. *Developing digital training tools;* following the Industry 4.0 requirements.
5. *Training testing and evaluation* in 4 of the participating countries.

6. *Certification scheme validation* on the real ground.
7. *Mainstreaming the certification scheme;* expanding the impact of the project in regions within and outside the consortium.

4 Results

BLUE PORTS is expected to deliver the following impactful outcomes:

- *Skills enhancement and training innovations* through a comprehensive training program, accessible to port personnel, authorities, universities, and associations, as well as new talents entering the field.
- *Reskilling and upskilling opportunities* to facilitate career advancement, job mobility, and talent retention within the Blue Economy sectors.
- *Promoting employability and adaptability* by equipping students with digital, green, soft, transversal, inter-/multi-disciplinary, managerial, and entrepreneurial skills.
- *Hands-on learning* as almost 30% of the total training course will be dedicated to practical examples drawn from selected cases.
- *Increasing awareness and attractiveness* of blue career opportunities, informing target groups about the certification mark and creating added value for it.

The medium-long term benefits of BLUE PORTS efforts include:

- reduction in GHGs emissions,
- common quality standards on Blue Economy staff training throughout EU Member States,
- competent staff in an ever-growing market,
- more confidence in the technology of opinion leaders, planners and consumers and
- improved and cost-efficient maintenance procedures.

Built on assumptions based on real data and experience of TUC, the energy, environmental and social impact are expressed as the following quantified indices (Table 1).

By enhancing the capacity of their workforce, ports can achieve streamlined operations, regulatory compliance, and increased competitiveness. This ultimately ensures innovation, fostering sustainable practices and enhancing the port's reputation. As ports with certified personnel gain recognition, they attract more partnerships, investments, and collaborations, that bring them closer to channelling the Green Deal in the maritime industry and achieving the green transition.

5 Conclusions

All in all, there is still a lot of ground to cover when targeting the greening of ports, in terms of energy efficiency and climate change. In order to enable the Blue Economy efforts to align with the EGD's objectives, by the end of 2021, 874 million euros (18% of the total EMFF funding committed to date) had been allocated to climate change projects [8].

Apart from the cooperation of Academia, Governance and Industry, the involvement of the community (civil society) appears limited in most projects. Moreover, the greening

Table 1. Environmental and social impact of BLUE PORTS, based on performance indicators.

Performance Indicators	By the end of the action (short term)	By 2030 (medium term)	By 2050 (long term)
Energy saving & renewable energy generation (MWh/y)	15,000	165,000	1,650,000
Primary energy savings (MWh/y)	34,800	382,800	3,828,000
GHGs reduction (Mts CO_2/y)	14,835	163,185	0
National financial measures triggered to encourage participation in the training courses	4 (at least 1/country)	20	

of future maritime jobs has been heavily stressed mostly as young people's concern [9], but to realize this transition, current professionals also need to be upskilled. Broadening the target group of this process to include ashore port employees is yet to be done. Besides, these skills are rarely certified, so forming a certification scheme of international validity would evoke new credentials. This, in turn, may potentially lead to the creation and maintenance of new working positions.

The BLUE PORTS project aspires to cover these gaps. By bringing together a quadruple-helix-based consortium of EU and non-EU countries, it expands its activities to an international level, which could benefit not only the Mediterranean Basin, but the Black and Caspian Seas, too, through a smart specialization approach. Meanwhile, it manages to both support and promote competence and EU's digitalization-oriented frameworks, via the upskilling certification scheme. Its dissemination/exploitation strategies, alongside synergies with relevant projects, will initiate a snowball effect, potentially introducing and sustaining a digital qualification in the ports industry, thus forming new, greener jobs.

References

1. European Commission. https://s3platform.jrc.ec.europa.eu/blue-growth, Accessed 20 Sep 2023
2. Eurostat, https://ec.europa.eu/eurostat/statistics-explained/index.php?title=International_trade_in_goods_by_mode_of_transport, Accessed 20 Sep 2023
3. European Commission. https://climate.ec.europa.eu/eu-action/transport/reducing-emissions-shipping-sector_en, Accessed 20 Sep 2023
4. European Parliament Think Tank. https://www.europarl.europa.eu/thinktank/en/document/EPRS_ATA(2022)729395, Accessed 20 Sep 2023
5. Sifakis, N., Tsoutsos, T.: Planning zero-emissions ports through the nearly zero energy port concept. J. Clean. Prod. **286**, 125448 (2021)
6. Argyriou, I., Daras, T., Tsoutsos, T.: Challenging a sustainable port. a case study of Souda port, Chania, Crete. Case Stud. Transport Policy **10**(4), 2125–2137 (2022)

7. Gerlitz, L., Meyer, C.: Small and medium-sized ports in the ten-t network and nexus of Europe's twin transition: The way towards sustainable and digital port service ecosystems. Sustainability **13**(8), 4386 (2021)
8. EMFF. https://emff.easme-web.eu/, Accessed 20 Sep 2023
9. European Commission. https://oceans-and-fisheries.ec.europa.eu/publications/emff-implementation-report-2021_en, Accessed 20 Sep 2023

Open Access This chapter is licensed under the terms of the Creative Commons Attribution 4.0 International License (http://creativecommons.org/licenses/by/4.0/), which permits use, sharing, adaptation, distribution and reproduction in any medium or format, as long as you give appropriate credit to the original author(s) and the source, provide a link to the Creative Commons license and indicate if changes were made.

The images or other third party material in this chapter are included in the chapter's Creative Commons license, unless indicated otherwise in a credit line to the material. If material is not included in the chapter's Creative Commons license and your intended use is not permitted by statutory regulation or exceeds the permitted use, you will need to obtain permission directly from the copyright holder.

Traffic Simulation and Safety Assessment Requirements for Enhancing Road Safety Prediction Tools

Maria G. Oikonomou[1](✉) ⓘ, Apostolos Ziakopoulos[1] ⓘ, Shanna Lucchesi[2,3] ⓘ, Monica Olyslagers[3], and George Yannis[1] ⓘ

[1] Department of Transportation Planning and Engineering, National Technical University of Athens, 5 Heroon Polytechniou Street, 15773 Athens, Greece
moikonomou@mail.ntua.gr
[2] Transport System Laboratory, Universidade Federal do Rio Grande do Sul, Porto Alegre, Brazil
[3] International Road Assessment Programme, The Ring, Bracknell RG12 1AX, UK

Abstract. Improving road safety prediction tools requires assessing established traffic simulation tools and safety assessment methods. Enhancing these tools with innovative data sources and methods can significantly reduce urban crashes and their impact. To achieve this, it is imperative to identify the requirements and gaps of relevant stakeholders in terms of professional road safety analysis tools. The present study aims to utilize association rule mining to determine underlying profiles of local stakeholders who are identified as hands-on practitioners. To accomplish this objective, a dedicated survey was conducted, and the data were analyzed to discover meaningful links among stakeholder characteristics through the characteristics mined using the Apriori algorithm. The results provide a quantification of the frequency and relationships between stakeholder responses, indicating connections between education levels, work regions, experience levels, and stakeholder needs related to road safety prediction tools. The study insights offer a quantitative perspective on the interconnections and dependencies among different stakeholder attributes, shedding light on potential patterns and preferences that can guide decision-making in the context of road safety improvements.

Keywords: Traffic simulation · Road safety assessment · Proactive Road safety tools · Apriori algorithm · Association rule mining

1 Introduction

There is an essential need for the improvement of road safety prediction tools through the assessment of established and proven traffic simulation tools and road safety assessment methods, with the aim of subsequently enhancing these tools using innovative data sources and methods.

Incorporating behavior into road safety assessment poses a significant data challenge. A survey of National Road Administrations revealed that 60% reported limited

© The Author(s) 2025
C. McNally et al. (Eds.): TRAconference 2024, LNMOB, pp. 809–816, 2025.
https://doi.org/10.1007/978-3-031-85578-8_109

data availability to assess factors related to user behavior [1]. Advanced crash count prediction models have been developed that distinguish between engineering-related and behavioral-related crashes [2, 3]. On the topic of road design, data scarcity often hampers the reliable implementation of road geometric designs, leading to limited safety performance. Traffic microsimulation proves ideal for such investigations, being widely utilized and enabling the testing of various configurations in safe environments [4, 5]. On the other hand, it is argued [6] that there are still no suitable methodologies and reliable surrogate indicators for simulation-based safety studies due to the absence of complete models for simulating potential crashes. To date, few studies have proposed alternative methodologies [7].

It is imperative to enhance these tools with innovative data sources and novel methods to maximize road safety gains. Infrastructure improvements should carefully account for induced demand, which arises from the intricate interconnectivity of transportation systems [8]. While Machine Learning (ML) methods are gaining prominence in travel behavior research, it is essential to foster greater cohesion in parallel research methodologies [9]. Additionally, alternative approaches, such as the Star Rating protocol developed by the International Road Assessment Programme, offer a proactive means to assess road safety even in the absence of crash data [10].

This research aims to support the improvement of road safety prediction tools that will result in valuable outcomes for stakeholders, leading to a real and observable impact in reducing crashes and mitigating their consequences. Specifically, the objectives are to utilize association rule mining to determine underlying profiles of local stakeholders that are identified as hands-on practitioners.

2 Methodology

2.1 Stakeholder Questionnaire

In order to capture the current practices, needs and gaps from the perspective of transport managers and municipalities, an online stakeholder survey was designed. The survey included questions related to several new metrics, models (e.g., behavioral) and techniques integration (e.g., road safety assessment, traffic microsimulation) considering factors such as human behavior, modal shift, and improved data exploitation through machine learning methodologies.

The total number of complete stakeholder responses received was 50, while the distribution of respondents covered 36 different cities. Specifically, the majority of respondents (41 in total) work in 29 European cities. In addition, most respondents (22%) are private sector employees involved in road safety and 17% of respondents use road safety assessment methodologies in their daily activities, while 10% use macroscopic traffic simulation. It is important to note that expert respondent samples are considerably more difficult to gather compared to layman samples, and as such the present sample size is considered decent.

The answers analyzed within this study, which are described in Table 1, concerned the area where stakeholders work, their education level, their experience on their role, their needs related to traffic microsimulation as well as their needs on road safety assessment methods, their expected frequency of using an integrated analytic tool and their evaluation of how much an integrated analytic tool could impact real-crash numbers. In Table 1, the variables analyzed and their short description are included.

Table 1. List of examined variables.

Variable	Description	Values
AREA	Work area	1 (Central or Eastern Europe), 2 (Southern Europe), 3 (Western Europe), 4 (Northern Europe), 5 (Rest of the world)
EDUCATION	Education level	1 (Secondary education diploma), 2 (University/College-level degree/diploma), 3 (Master's degree), 4 (Doctoral degree/Postdoctoral studies)
EXPERIENCE	Work experience	1 (<5 years), 2 (5–10 years), 3 (10–15 years), 4 (15–20 years), 5 (>20 years)
MICROSIM_ALL_NEEDS	Importance of incorporating in traffic simulation: (i) infrastructure safety information; (ii) modal shift information; (iii) induced demand models; (iv) human behaviour models and (v) accuracy improvement	1 (Not important) - 10 (Very important)
ROAD_AS_ALL_NEEDS	Importance of incorporating in road safety assessment methods: (i) traffic microsimulation information; (ii) modal shift information; (iii) induced demand models and (iv) AI/ML models	1 (Not important) - 10 (Very important)
INTEG_TOOL_USE	Expected usage of an integrated analytic tool	1 (Never) - 10 (Very Frequently)
INTEG_TOOL_IMPACT	Expected impact of an integrated analytic tool on real crash numbers	1 (Not at all) - 10 (Extremely)

2.2 Apriori Algorithm

In order to analyze the survey data and discover meaningful links of stakeholder characteristics, association rule mining was implemented with the Apriori algorithm. This algorithm, a seminal contribution to data mining, was introduced by Agrawal & Srikant [11]. Its primary purpose is to unearth frequent item-sets within transactional datasets, paving the way for the discovery of association rules. Association rules, characterized by an antecedent part (e.g., "if X occurs") and a consequent part (e.g., "…then Y occurs as well"). The rules generated by the Apriori algorithm are assessed based on three key parameters:

- Support: This parameter measures the probability that both X and Y occur together.
- Confidence: Confidence expresses the conditional probability that Y occurs given that X has occurred. It quantifies how often Y follows X.
- Lift: Lift measures the ratio between support and confidence. A lift value of 2, for example, indicates that the likelihood of encountering X and Y together is twice the likelihood of encountering Y alone.

3   Results and Discussion

3.1 Traffic Simulation User Profiles

In order to discover meaningful associations of stakeholder characteristics related to their opinions on the requirements of traffic simulation, at least a minimum support and confidence value was required, to filter out rules that either concern less frequent characteristics or are less informative. After several trials during the modelling process, the selected values were 0.05 and 0.40 for support and confidence respectively. The resulting association rules with the top five scoring lift values are presented in Table 2. Lift values indicate the strength of association between the antecedent and consequent of each rule, and higher lift values suggest stronger associations.

Table 2. Top five association rules for stakeholder needs related to traffic simulation.

Rules	Antecedent		Consequent	Support	Confidence	Coverage	Lift	Count
[1]	{MICROSIM_ALL_NEEDS = 5}	=>	{EDUCATION = 3}	0.06	0.750	0.08	1.293	3
[2]	{MICROSIM_ALL_NEEDS = 8}	=>	{EDUCATION = 3}	0.10	0.833	0.12	1.437	5
[3]	{AREA = 3}	=>	{MICROSIM_ALL_NEEDS = 10}	0.08	0.667	0.12	1.961	4
[4]	{EXPERIENCE = 1}	=>	{MICROSIM_ALL_NEEDS = 10}	0.08	0.500	0.16	1.471	4
[5]	{MICROSIM_ALL_NEEDS = 7}	=>	{EDUCATION = 2}	0.08	0.500	0.16	1.471	4

Based on the provided results, the first rule reveals that, stakeholders who state that there is a moderate need (5 out of 10) of traffic simulation tool enhancement hold a master degree title with a 75% confidence. Similarly, those who state that there is an adequate need (8 out of 10), also hold a master degree title with an 83% confidence, based on the second rule. The third rule indicates that, there is a 67% confidence that a stakeholder who works in Western Europe believes that there is a highly need of simulation tool enhancement. Based on the fourth rule, for stakeholders who have experience under 5 years in their role, there is a 50% confidence that highly need a traffic simulation enhancement. Finally, it seems that stakeholders stated that there is a 7 out of 10 need of the enhancement hold a University or College-level degree or diploma title with a 50% confidence.

3.2 Safety Assessment User Profiles

Similarly, to uncover significant associations among stakeholder characteristics regarding safety impact assessment methodologies, the chosen parameter values were set at 0.05 and 0.40 for support and confidence, respectively. The resulting association rules, featuring the top five lift values, are detailed in Table 3.

Table 3. Top five association rules for stakeholder needs related to road safety assessment.

Rules	Antecedent		Consequent	Support	Confidence	Coverage	Lift	Count
[1]	{ROAD_AS_ALL_NEEDS = 7}	=>	{AREA = 2}	0.06	0.500	0.12	1.471	3
[2]	{ROAD_AS_ALL_NEEDS = 7}	=>	{EDUCATION = 3}	0.08	0.667	0.12	1.149	4
[3]	{AREA = 3}	=>	{ROAD_AS_ALL_NEEDS = 10}	0.10	0.833	0.12	2.604	5
[4]	{ROAD_AS_ALL_NEEDS = 5}	=>	{EXPERIENCE = 2}	0.06	0.500	0.12	2.778	3
[5]	{ROAD_AS_ALL_NEEDS = 5}	=>	{EDUCATION = 3}	0.08	0.667	0.12	1.149	4

The first rule reveals that, for stakeholders who state that there is a 7 out of 10 need of road safety assessment methodologies enhancement there is a 50% confidence that they work in Southern Europe. Similarly, those who identify the same need, hold a master degree title at 67% confidence level, based on the second rule. The third rule indicates that, there is an 83% confidence that a stakeholder who works in Western Europe believes that there is a highly need of road safety assessment enhancement. For stakeholders who believe that is a moderate need of this enhancement, have an experience between 5 to 10 years in their role with a 50% confidence level as well as hold a master degree title at 67% confidence level, based on the fourth and fifth rule respectively.

3.3 Integrated Analytic Tool Expected Usage

In a similar vein as the previous analyses, in order to reveal noteworthy links among stakeholder attributes concerning the expected usage of an integrated analytic tool, support and confidence parameters were established at values of 0.05 and 0.30, respectively. The ensuing association rules, which encompass the top five lift values, can be found in Table 4.

Table 4. Top five association rules for stakeholder expectation to use an integrated analytic tool.

Rules	Antecedent		Consequent	Support	Confidence	Coverage	Lift	Count
[1]	{INTEG_TOOL_USE = 6}	= >	{EDUCATION = 3}	0.06	0.750	0.08	1.293	3
[2]	{INTEG_TOOL_USE = 9}	= >	{EXPERIENCE = 4}	0.06	0.100	0.20	2.308	3
[3]	{INTEG_TOOL_USE = 9}	= >	{EDUCATION = 3}	0.06	0.600	0.10	1.034	3
[4]	{INTEG_TOOL_USE = 10}	= >	{AREA = 3}	0.06	0.500	0.12	4.167	3
[5]	{AREA = 3}	= >	{INTEG_TOOL_USE = 10}	0.06	0.500	0.12	4.167	3

It seems that stakeholders who claim there is a moderate possibility (6 out of 10) of using an integrated analytic tool hold a master's degree title with a 75% confidence level. On the other hand, those who state that it is very likely to use this kind of tool (9 out of 10) have between 15 to 20 years of experience in their roles and also hold a master's degree title with 10% and 60% confidence, respectively. Finally, there is a 50% confidence that a stakeholder who is certain they will use a tool like this works in Western Europe, and conversely, a stakeholder working in Western Europe will use the tool.

3.4 Integrated Analytic Tool Expected Impact on Safety

In the pursuit of uncovering significant associations among stakeholder characteristics related to the expected impact of an integrated analytic tool on safety, parameter values of 0.05 and 0.40 were opted for support and confidence, respectively. The resulting association rules, highlighting the top 5 lift values, have been elaborated upon in Table 5.

Based on the first rule, stakeholders who believe that an integrated analytic tool will possibly affect real-crash numbers, they hold a master degree title with a 50% confidence. The second rule indicates that, there is a 50% confidence that a stakeholder who works in Western Europe believes that there is a moderate possibility of this kind of tool to impact crashes. In addition, stakeholders who state that is very likely for an integrated analytic tool to impact crash numbers, they have experience between 10 to 15 years at a 67% confidence level and they work in Central or Eastern Europe at a 50% level, based on the third and fifth rule, respectively. Based on the fourth rule, for stakeholders who have experience between 10 to 15 years in their role, that an integrated tool is very likely to affect road safety with a 40% confidence.

Table 5. Top five association rules for stakeholder expectation of an integrated tool impact.

Rules	Antecedent		Consequent	Support	Confidence	Coverage	Lift	Count
[1]	{INTEG_TOOL_IMPACT = 8}	=>	{EDUCATION = 3}	0.06	0.500	0.12	0.862	3
[2]	{AREA = 3}	=>	{INTEG_TOOL_IMPACT = 7}	0.06	0.500	0.12	3.125	3
[3]	{INTEG_TOOL_IMPACT = 9}	=>	{EXPERIENCE = 3}	0.08	0.667	0.12	3.333	4
[4]	{EXPERIENCE = 3}	=>	{INTEG_TOOL_IMPACT = 9}	0.08	0.400	0.20	3.333	4
[5]	{INTEG_TOOL_IMPACT = 9}	=>	{AREA = 1}	0.06	0.500	0.12	1.563	3

4   Conclusions

Overall, the association rules indicate a relationship between education levels, work regions, experience levels, and stakeholder needs related to road safety prediction tools. These findings transcend mere data correlations; they offer a profound understanding of the interplay among these diverse stakeholder attributes. Consequently, these findings are able to offer a quantitative perspective on the interconnections and dependencies among different stakeholder attributes, shedding light on potential patterns and preferences that can guide decision-making in the context of road safety improvement. In essence, this study not only provides valuable insights into the associations between stakeholder attributes but also serves as a crucial resource for constructing more effective, data-driven road safety strategies that can ultimately improve safety and make roadways safer for all.

It should be highlighted that this profiling is not a mandate to exclude stakeholders with different characteristics from using analytic tools. Rather, it is a way to: (i) gauge the most likely stakeholder audiences, (ii) tailor the developed tools to their needs but also their capacities and (iii) develop additional venues, resources and tools to reach different stakeholder profiles.

Despite the valuable insights gained from the analysis, several limitations must be acknowledged. Firstly, this study is confined to the scope of available data, potentially overlooking other influential factors in road safety. Additionally, the integrated tools will be developed within PHOEBE in the future, and as such, their characteristics are not readily available for the stakeholders to interact with. While association rules unveil correlations, they do not establish causality, prompting the need for further research into causal relationships among the identified associations. To address these limitations and contribute to the advancement of road safety research, future studies can explore causal modeling techniques, longitudinal analyses, cross-cultural comparisons, and the integration of real-time data through advanced analytics. By embracing these research directions, we can enhance our understanding of road safety dynamics and pave the way for more effective, data-driven interventions, ultimately fostering safer roadways for all.

Acknowledgement. The present research was carried out within the research project "PHOEBE - Predictive Approaches for Safer Urban Environment", which has received funding from the

European Union's Horizon Europe research and innovation programme under grant agreement No 101076963.

References

1. Yannis, G., et al.: Use of Accident prediction models in road safety management - an international inquiry. Trans. Res. Proc. **14**, 4257–4266 (2016)
2. Afghari, A.P., Washington, S., Haque, M.M., Li, Z.: A comprehensive joint econometric model of motor vehicle crashes arising from multiple sources of risk. Anal. Methods Accident Res. **18**, 1–14 (2018)
3. Shaon, M.R.R., Qin, X., Afghari, A.P., Washington, S., Haque, M.M.: Incorporating behavioural variables into crash count prediction by severity: a multivariate multiple risk source approach. Accid. Anal. Prev. **129**, 277–288 (2019)
4. Ghanim, M., Kharbeche, M., Hannun, J., Hannun, J., Shamiyeh, K.: Safety and operational performance of signalized roundabouts: a case study in Doha. Proc. Comput. Sci. **170**, 427–433 (2010)
5. Bahmankhah, B., Macedo, E., Fernandes, P., Coelho, M.C.: Micro driving behaviour in different roundabout layouts: pollutant emissions, vehicular jerk, and traffic conflicts analysis. Trans. Res. Proc. **62**, 501–508 (2022)
6. Astarita, V., Festa, D.C., Giofrè, V.P., Guido, G.: Surrogate safety measures from traffic simulation models a comparison of different models for intersection safety evaluation. Trans. Res. Proc. **37**, 219–226 (2019)
7. Oikonomou, M.G., Ziakopoulos, A., Chaudhry, A., Thomas, P., Yannis, G.: From conflicts to crashes: simulating macroscopic connected and automated driving vehicle safety. Accid. Anal. Prev. **187**, 107087 (2023)
8. Cascetta, E.: Transportation Systems Analysis, vol. 29. Springer US (2009)
9. Hillel, T., Bierlaire, M., Elshafie, M.Z.E.B., Jin, Y.: A systematic review of machine learning classification methodologies for modelling passenger mode choice. J. Choice Model. **38**, 100221 (2021)
10. Lawson, S.: Crash rate–Star Rating comparisons. Review of available evidence. Working Paper 504.2, EuroRAP, Brussels. (2011)
11. Agrawal, R., Srikant, R.: Fast algorithms for mining association rules. In Proceedings of 20th International Conference on Very Large Data Bases, vol. 1215, pp. 487–499 (1994)

Open Access This chapter is licensed under the terms of the Creative Commons Attribution 4.0 International License (http://creativecommons.org/licenses/by/4.0/), which permits use, sharing, adaptation, distribution and reproduction in any medium or format, as long as you give appropriate credit to the original author(s) and the source, provide a link to the Creative Commons license and indicate if changes were made.

The images or other third party material in this chapter are included in the chapter's Creative Commons license, unless indicated otherwise in a credit line to the material. If material is not included in the chapter's Creative Commons license and your intended use is not permitted by statutory regulation or exceeds the permitted use, you will need to obtain permission directly from the copyright holder.

MUNER – Italian Motor Valley Excellence for Education and Innovation in Automotive and Sustainable Mobility

Francesco Leali[1]([✉]), Leonardo Guglielmetti[2], Giulia Tagliazucchi[3], Soundous Drissi[2], and Teresa Serra[2]

[1] Enzo Ferrari Department of Engineering, University of Modena and Reggio Emilia, 41125 Modena, Italy
francesco.leali@unimore.it
[2] Motorvehicle University of Emilia-Romagna, 41125 Modena, Italy
{info,muner,communication}@motorvehicleuniversity.com
[3] Marco Biagi Department of Economics, University of Modena and Reggio Emilia, 41121 Modena, Italy
giulia.tagliazucchi@unimore.it

Abstract. The MUNER Association is finalized to carry out innovative teaching-by-learning methodologies in a multidisciplinary approach. Based on this framework, it offers three inter-universities Master's Degree programs on the Automotive theme, with the aim of attracting international talent and training them on extremely specialized topics. These programs were pioneers in the adoption of teaching methods characterized by a strong contamination of the academic world with the industry realities, by a strong interdisciplinary approach, especially on the themes that characterized the technological innovation in the automotive sector. The present contribution discusses how innovative educational programs can be designed and implemented in collaboration between universities and industry.

Keyword: Education · Motorvehicle Industry · Competence gap

1 Introduction

Among the possible, the human capital that is part of a company is certainly one of its main resources, and it is proven to assume a central role in affecting performance and innovation (e.g., Felicio et al., 2014; Wang et al. 2019). In general terms, "human capital is basically a way that is attached to education, training, and other professional initiatives to enhance the levels of knowledge, skill, abilities and social assets of an employee" (Alnachef and Alhajjar 2017, p. 1154), and therefore refers in a broad sense to the know-how of a company. The link between the human capital and performance, both financial and non-financial, is proven by the literature, as is its role in boosting the innovative performance. This is not limited to those who are already in the company, the employees, but ultimately extends to new arrivals and hires. More specifically, "by combining complementary knowledge and skills from new and incumbent employees

© The Author(s) 2025
C. McNally et al. (Eds.): TRAconference 2024, LNMOB, pp. 817–822, 2025.
https://doi.org/10.1007/978-3-031-85578-8_110

(March 1991), an organization can increase its human capital resources, which in turn enable it to achieve competitive advantages" (Wang et al., 2019, p. 102). New hires therefore become fundamental to bring in new knowledge, perspectives and abilities, which are essential to enrich the human capital of the firm, and which become essential to face the innovative changes taking place in an industry. In this vein, scholars agree that newcomers' fresh ideas and up-to-date competencies enable the organization to adapt more adequately to changing environments (Dalton and Todor 1979). Greater emphasis is then placed on the education and training of the individuals (Nafukho et al. 2007) and of the young about to enter the firms.

Attention must therefore shift to the actors responsible for training and education, and in particular for second-level higher education: the universities. Universities could indeed play an important role in targeting and managing the technologically innovative processes occurring in a specific industry, not only with technology transfer activities, but also by rethinking their teaching and education models. Universities have been called to operate "for the sake of solving specific and compelling problems and challenges confronting society" (Audretsch 2014, p 317), implying for local universities to take on different roles and to collaborate at local and national levels with firms and policy makers (Tagliazucchi et al. 2021). Such collaborations can also materialize in innovative educational models, created to meet the needs of the local industry – such as the automotive as in the case of the Italian motor valley – to face waves of change that evolve globally. To date the automotive industry and the mobility sector are indeed currently experiencing strong and important changes. The acronym ACES - which stands for Automation, Connectivity, Electrification, Sharing - summarizes these changes, which affect both the technological sphere of the vehicle and its use, and the rethinking of transport modes by the user. If, on the one hand, automation, connectivity and the search for alternative powertrains to those of fossil origin are changing the conception and design of the vehicle itself; on the other hand, the issues related to a more sustainable mobility are leading to a rethinking of the ways in which goods and people are moved.

Structured examples of university-industry collaborations that lead to the creation of specific degree paths to target specific innovation and competencies needs are rare, and empirical contributions that analyze them are scarce. The present contribution aims then to discuss how innovative educational programs can be designed and implemented in collaboration between universities and industry, in giving a concrete response to the needs in terms of competencies – soft and hard skills - to a constantly evolving territory and industry.

2   The Case: MUNER

In 2017, under the impulse of the Emilia Romagna Region, the main public and private actors of the Motor Valley have put to system their skills and experiences, leading to the foundation of the Motorvehicle University of Emilia-Romagna (MUNER). The Motorvehicle University of Emilia-Romagna was born as an Association in order to create a synergic link between the universities based in Emilia-Romagna (University of Bologna, University of Ferrara, University of Modena and Reggio Emilia, University of Parma) and the local automotive companies that represent the excellence of Made in Italy in the world by designing and manufacturing high-performance vehicles (Automobili Lamborghini, Dallara, Ducati, Ferrari, HaasF1Team, Maserati, Pagani and Scuderia AlphaTauri), and their most qualified suppliers of technology (Marelli, HPE Group). Later on the group of companies expanded to include: ART, AVL, Bosch, CNH, Danisi, Pirelli, STMicroelectronics.

The Association is finalized to carry out innovative teaching-by-learning methodologies in a multidisciplinary approach. Based on this framework, it has launched three inter-universities Master's Degree Programs on the Automotive theme, with the aim of attracting international talent and training them on extremely specialized topics. To date, there are eight curricula activated, which refer to three different Master's Degree Programs. More specifically, the curricula activated are the following:

- The curriculum in Electronic and Communication Systems (ECS) is part of the Master's Degree Program in Electronic Engineering for Intelligent Vehicles (EEIV) and it is focused on topics related to advanced automotive electronic engineering.
- The curriculum in Autonomous Driving Engineering (ADE) is part of the Master's Degree Program in Electronic Engineering for Intelligent Vehicles (EEIV). It integrates topics related to advanced automotive electronic engineering with those specific to autonomous driving.
- The curriculum in Advanced Motorcycle Engineering is part of the Master's Degree Program in Advanced Automotive Engineering (AAE). It aims to train real experts in the design and development of high-tech motorcycles, with a focus on electronic engineering, endothermic powertrains and industrial design.
- The curriculum in Advanced Powertrain (AP) is part of the Master's Degree Program in Advanced Automotive Engineering (AAE). This specific Program features two different specializations that provide an in-depth knowledge of powertrain design, production, planning and control of advanced powertrains both electric and endothermic.
- The curriculum in Racing Car Design (RCD) is part of the Master's Degree Program in Advanced Automotive Engineering (AAE) and it provides a high level of training on the dynamics of the vehicle and on the most innovative materials related to the design of racing cars.
- The curriculum in Advanced Sportscar Manufacturing (ASM) is part of the Master's Degree Program in Advanced Automotive Engineering (AAE). The Program aims to train professionals in the development of production systems in the automotive field, from the design of new industrial plants to the new technologies of the industry 4.0.
- The curriculum in High Performance Car Design (HPCD) is part of the Master's Degree Program in Advanced Automotive Engineering (AAE) and it aims to train professionals able to design high performance road vehicles, with strong skills in vehicle design, dynamics and NVH, aerodynamics and electronics.
- The Master's Degree Programs in Electric Vehicle Engineering (EVE) aims to train real experts in the design and development of high-tech electric vehicle, with particular emphasis on the electric subsystems that are embedded in future vehicles.

These Programs were pioneers in the adoption of teaching methods characterized by a strong contamination of the academic world with the industry realities, by a strong interdisciplinary approach, especially on the themes that characterized the technological innovation in the automotive sector, and by the constant "hands-on" experimentation of the theoretical contents. The Programs hence put in practice a "Learning by doing" methodology of teaching by providing practical experiences within university laboratories and activities within the Motor Valley companies in addition to the fundamental theoretical lectures. The programs also include an alternation of teachers coming both from the university and research world and from the companies themselves, bringing the excellence of both theoretical and basic knowledge as well as practical and hands-on.

By expanding up its role, the Association has also recently engaged in summer and winter schools: dedicated to high school students (advanced engineering summer program Italian Motor Valley Experience), to female bachelor students from the European community (Women in Transport), to master students (International Summer School in Industrial Engineering for Advanced Automotive) and PhD (Future of Automotive for Intelligent Mobility). These programs have the aim of expanding the consolidated educational model on master's degree programs to specific users, identified as of interest and key importance, at different levels of educational order.

3   Results: A Growing Number of Students

From its inception to date, MUNER has experienced strong growth, attracting the best national and international students within its academic programs. The results in terms of enrolled students are detailed below.

3.1   Master's Degree Programs

Over time, the number of students enrolled in MUNER programs has grown, with positive results both in terms of academic results achieved upon graduation and employment immediately after graduation.

The first academic year of MUNER, 2017–2018, started with enrolling around 60 students in two programs. The number of enrolled students grew over the last six year to reach over 160 students enrolled in the three programs, in 2023–2024, marking the growth rate of 166%.

Up to date, over 900 students have studied at MUNER Master's Degree Program. About 400 of these are now alumni. Upon graduation, around 90% of graduates gets immediately hired or involved in professional or academic activities. This latter included University research and PhD programs. A significant portion of hired graduates work MUNER's prestigious companies.

3.2   Summer and Winter Programs

Orientation and integrative courses in the form of summer and winter schools have also seen strong development and interest.

The advanced engineering summer program Italian Motor Valley Experience is designed to offer international high-school students the possibility of a residential and immersive experience within the Italian motor valley. It is held at the Enzo Ferrari engineering department of the University of Modena and Reggio Emilia, open to a limited number of students with a strong STEM background and interest in the world of two and four wheels, it integrates cultural and educational activities. The pilot edition, held in 2019, saw the participation of a group of seven students from California (US). In the following years it saw significant growth: in 2021 there were 25 enrolled students in the on-line edition – due to COVID-19 ban of travelling, coming from Europe and the United States; in 2022 there were 22 enrolled students in the on-site edition, coming from Europe, the United States and Canada; in 2023 there were 21 enrolled students in the on-site edition, coming from Europe, the United States, Canada, Latin America. The 2021 and 2022 editions are also part of the activities of the "MUNER International School of Higher Education in Automotive for Intelligent Mobility", a project financed by the Emilia Romagna Region and approved with Resolution no. 1251 of 07/22/2019.

The summer program Women In Transport offers motivated and passionate female students an immersive experience in the automotive world, with the ultimate aim of overcoming gender stereotypes. It is promoted and accredited within the "Women in Transport - EU platform for change" platform. In its first edition, in 2022, 18 students were enrolled from 9 different European countries, hosted at the Enzo Ferrari engineering department of the University of Modena. The following edition, in 2023, saw the participation of 17 students, hosted at the University of Bologna. Both editions are part of the activities of the "MUNER International School of Higher Education in Automotive for Intelligent Mobility", a project financed by the Emilia Romagna Region and approved with Resolution no. 1251 of 07/22/2019.

The International Summer School in Industrial Engineering for Advanced Automotive aims to improve the training of future engineers, providing specialized knowledge in the mechanical, automotive and production fields, tailored to the needs of the thriving automotive industry, which is particularly present in the Emilia-Romagna region. Since 2021, the summer program host yearly between 30 and 40 students. The winter program Future of Automotive for Intelligent Mobility aims at introducing young professionals to some of the most advanced trends in the automotive sector and focuses on fostering the skills and attitudes needed for young engineers to contribute to an unprecedented paradigm shift in the industry. In its 2022 edition the program hosted 23 participants.

4   Discussion and Conclusion

Among the ways of integrating fresh technological competencies within the firm, the role of human capital is prominent (March 1991), and in particular the newly hired (Wang et al., 2019). To face concurring waves of technological innovation changes, automotive companies need to focus on the newly and fresh competences to maintain their competitive position (Teece 2018). The education background of newly-graduates employed into the firms is then a key issue, and it becomes prominent not only to involve companies in current educational paths, but to implement new educational models based on hands-on experience, learning-by-doing approaches, and a mix of theoretical knowledge from university professors and researchers and practical knowledge from industry professionals. The main objective of MUNER concerns precisely the implementation of a new educational model that hinges on these pillars. By postulating that the individual education background of newly-graduates can be transformed into competencies in support of innovative processes in firms, specific actions can be designed to rethink the second-level educational in collaboration between universities and the local industry. Ultimately, these actions in support of new model of second-level educational offer may have a positive effect in addressing technologically innovative challenges for the benefit of the local innovative system, and in promoting new skills and knowledge development and diffusion in key areas, such as carbon reduction and sustainable transportation.

References

Alnachef, T.H., Alhajjar, A.A.: Effect of human capital on organizational performance: a literature review. Inter. J. Sci. Res. **6**(8), 1154–1158 (2017)

Audretsch, D.B.: From the entrepreneurial university to the university for the entrepreneurial society. J. Technol. Transf. **39**(3), 313–321 (2014)

Dalton, D.R., Todor, W.D.: Turnover turned over: an expanded and positive perspective. Acad. Manag. Rev. **4**, 225–235 (1979)

Felício, J.A., Couto, E., Caiado, J.: X: Human capital, social capital and organizational performance. Manag. Decis. **52**(2), 350–364 (2014)

March, J.G.: Exploration and exploitation in organizational learning. Organ. Sci. **2**, 71–87 (1991)

Nafukho, F.M., Hairston, N., Brooks, K.: Human capital theory: Implications for human resource development. Hum. Resour. Dev. Int. **7**(4), 545–551 (2004)

Teece, D.J.: Tesla and the reshaping of the auto industry. Manag. Organ. Rev. **14**(3), 501–512 (2018)

Tagliazucchi, G., Marchi, G., Gherardini, F., Leali, F.: The multiple roles of universities in U-I collaborations: The case of Emilia-Romagna motor vehicle industry. J. Eng. Tech. Manage. **62**, 1–16 (2021)

Wang, T., Zatzick, C.D.: Human capital acquisition and organizational innovation: atemporal perspective. Acad. Manag. J. **62**(1), 99–116 (2019)

Open Access This chapter is licensed under the terms of the Creative Commons Attribution 4.0 International License (http://creativecommons.org/licenses/by/4.0/), which permits use, sharing, adaptation, distribution and reproduction in any medium or format, as long as you give appropriate credit to the original author(s) and the source, provide a link to the Creative Commons license and indicate if changes were made.

The images or other third party material in this chapter are included in the chapter's Creative Commons license, unless indicated otherwise in a credit line to the material. If material is not included in the chapter's Creative Commons license and your intended use is not permitted by statutory regulation or exceeds the permitted use, you will need to obtain permission directly from the copyright holder.

Author Index

© The Editor(s) (if applicable) and The Author(s) 2025
C. McNally et al. (Eds.): TRAconference 2024, LNMOB, pp. 823–827, 2025.
https://doi.org/10.1007/978-3-031-85578-8